Cover painting: On the Outskirts of Saturn's Rings. The globe of Saturn (left of center) is lit in crescent phase by the distant sun (right). The edge-on ring system, composed of ice particles, forms a white band across the center of the picture. It stretches thousands of kilometers into the distance, forming a remote "ring-horizon." Scientists have discovered that fine structures in the rings are controlled by small, embedded moonlets, two of which are seen in the foreground. Distant parts of the rings cast shadows on the lower part of Saturn's crescent. Faint reddish aurora displays are visible near Saturn's poles. (Painting by author.)

The painting on the inside of the front cover shows the red giant star Antares and its blue companion star as seen from an imaginary planet. (Painting by author.)

The photo on the opposite page shows a cluster of blue-white stars, clouds of red-glowing hydrogen gas, and silhouetted clouds of dark dust, marking a region where stars are forming today in our galaxy. This formation is just one of many clouds and star clusters dotted along the spiral arms of our galaxy. It is thus intermediate in scale between the smaller systems discussed in this book (such as our solar system and other stars) and the larger systems (such as galaxies and clusters of galaxies). For a different view of this and surrounding regions, see Figure 16-12. (Photo copyright Anglo-Australian Telescope Board.)

The painting inside the back cover: Children of Earth. As humanity expands from the earth into space, we may develop an interplanetary culture. Here, perhaps a century hence, space cities ring the earth's equatorial plane. They may house research centers, refineries for asteroid metals, solar power collecters, zero-gravity manufacturing centers, space telescopes, and way stations on the road to other planets. Transferral of such facilities from Earth to space may allow the earth's environment to relax back to a less polluted state. Milky Way crosses the background behind the earth. (Painting by author.)

The photos immediately preceding that painting: From the earth to the universe. Top: *Space: the endless frontier. A telescopic photo of part of the Milky Way shows massed stars including open cluster NGC 6520 (bluish stars, left center). An opaque dust cloud, Barnard 86, is silhouetted against the background stars. New stars may one day form inside it. (Photo by David F. Malin, copyright Anglo-Australian Telescope Board.)* Bottom: *Wide-angle view of the sky, an hour before dawn, from the 14 000-ft summit of Mauna Kea in Hawaii. Stretching upward from the red-glowing horizon is the band of zodiacal light—sunlight reflecting from interplanetary dust. Sloping up from the lower right, crossing the zodiacal light, is the Milky Way—the band of distant stars defined by our galaxy's disk. Fisheye lens distorts the horizon (at bottom) into a curve; an observatory dome is silhouetted against the sky. (Photo by author, 35-mm camera, 15-mm lens, f2.8, 20-min exposure, 3 M ASA1000 slide film.)*

ASTRONOMY THE COSMIC JOURNEY

1987 EDITION

William K. Hartmann

Wadsworth Publishing Company
Belmont, California
A Division of Wadsworth, Inc.

Astronomy Editor: Jack C. Carey
Development Editor: Mary Arbogast
Designer: MaryEllen Podgorski
Print Buyer: Karen Hunt
Copy Editors: Jonas Weisel, Don Yoder
Art Editors: Toni Haskell, Marty Kongsle
Technical Illustrators: Catherine Brandel, Victor Royer, Darwen and Valley Hennings, Lisa Sliter, Salinda Tyson
Cover Painting: William K. Hartmann

Printed in the United States of America 48

2 3 4 5 6 7 8 9 10---91 90 89 88 87

Library of Congress Cataloging-in-Publication Data

Hartmann, William K.
Astronomy, the cosmic journey.

Bibliography: p.
Includes index.
1. Astronomy. I. Title.
QB45.H32 1987 520 86-28131
ISBN 0-534-07938-5

For my parents
with thanks for encouragement
and for
Carl Bauer, Walter Haas, and Nicholas Wagman
with thanks for early lessons
in astronomy.

CONTENTS IN BRIEF

Astronomical distance and scale are illustrated in this brief overview of the Contents

scale: 100 km (10^5m)

PART A

The Early Discoveries

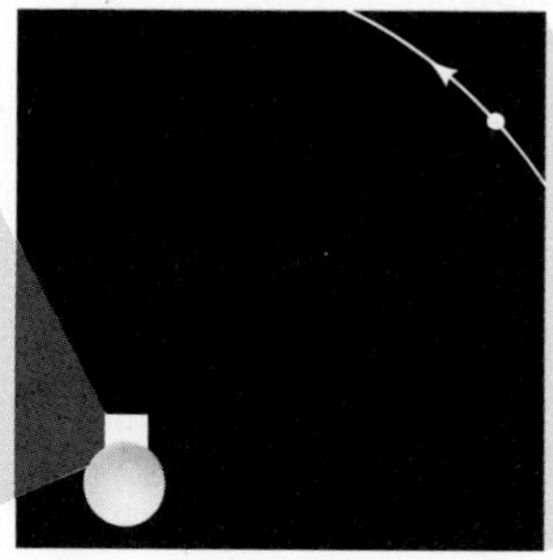

scale: 100 000 km (10^8m)

PART B

Exploring the Earth–Moon System

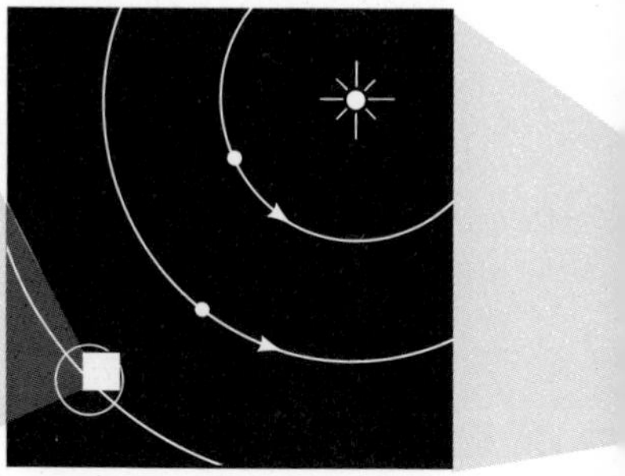

scale: 1 AU (10^{11}m)

PART C

The Solar System

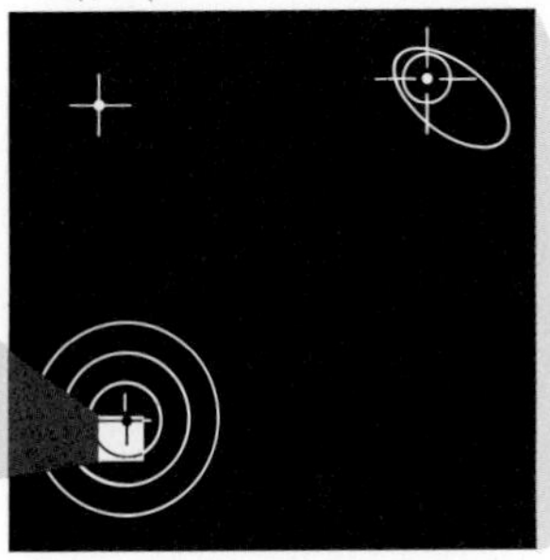

scale: 1 parsec (10^{16}m)

PART D

Stars and Their Evolution

scale: 100 parsecs (10^{18}m)

PART E

Environment and Grouping of Stars

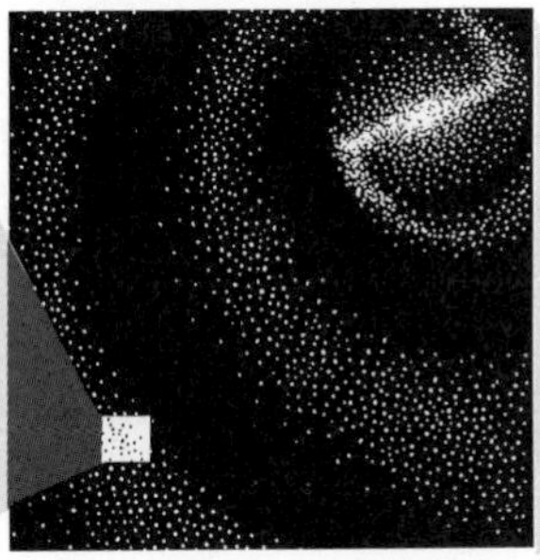

scale: 10 kiloparsecs (10^{20}m)

PART F

Galaxies

scale: 1 megaparsec (10^{22}m)

PART G

Frontiers

DETAILED CONTENTS

PART B

Exploring the Earth–Moon System

49

FOR TEACHERS AND STUDENTS

With human footprints on the moon, radio telescopes listening for messages from alien creatures (who may or may not exist), technicians looking for celestial and planetary sources of energy to support our civilization, orbiting telescopes' data hinting at planetary systems around other stars, and political groups trying to figure out how to save humanity from nuclear warfare that would damage life and climate on a planet-wide scale, an astronomy book published today enters a world different from the one that greeted books a generation ago. Astronomy has broadened to involve our basic circumstances and our enigmatic future in the universe. With eclipses and space missions broadcast live, and with NASA, Europe, and the USSR planning and building permanent space stations, astronomy offers adventure for all people, an outward exploratory thrust that may one day be seen as an alternative to mindless consumerism, ideological bickering, and wars to control dwindling resources on a closed, finite Earth.

Today's astronomy students not only seek an up-to-date summary of astronomical facts; they ask, as people have asked for ages, about our basic relationships to the rest of the universe. They may study astronomy partly to seek points of contact between science and other human endeavors: philosophy, history, politics, environmental action, even the arts and religion.

Science fiction writers and the special effects artists on recent films help today's students realize that the unseen worlds of space are real places—not abstract concepts. Today's students are citizens of a more real, more vast cosmos than conceptualized by students of a decade ago.

In designing this edition, the Wadsworth editors and I have tried to respond to these developments. Rather than jumping at the start into the murky waters of cosmology, I have begun with the viewpoint of ancient people on earth and worked outward across the universe. This method of organization automatically (if loosely) reflects the order of humanity's discoveries about astronomy and provides a unifying theme of increasing distance and scale.

The arrangement of this book, then, aims to give an unfolding, ever-expanding panorama of our cosmic environment. We hope it unfolds like a story in which each chapter provides not only a new facet, but also a growing understanding of the relationships among the elements of the whole.

The subtitle refers to three separate cosmic journeys that we undertake simultaneously. First, we travel through historical time, where we see how humans slowly and sometimes painfully evolved our present picture of the universe. Second, we journey through space, where we see how our expanding frontiers have revealed the geography of the universe. Beginning with an earth-centered view, we study the earth–moon system, the surrounding system of planets, the more distant surrounding stars, our own vast galaxy, and the encompassing universe of other galaxies. Finally, we travel back through cosmic time. Familiar features of the earth are typically only a few hundred million years old. The solar system is about 4.6 billion years old. Our galaxy may be 10 to 13 billion years old. The universe itself began (or began to reach its present form) perhaps 10 to 18 billion years ago.

Because astronomy touches many areas of life and philosophy, I have allowed the text to encompass a wide range of relevant topics, including space exploration, financing of science, cosmic sources of energy, the checkout-counter's barrage of astrology and other pseudo-science, and the possibility of life on other worlds, as well as the conventional "hard science" of astronomy. This variety of topics shows how basic scientific research touches all areas of life—I hope in a way that lets readers ponder the relation between science and priorities in our society.

The arrangement of text material into eight parts and twenty-six chapters should give instructors some flexibility in tailoring a course according to their interest. For example, those who are not much interested in historical development could use *Part A* only as assigned outside reading.

Each part gives some historical background, describes recent discoveries and theories, and then discusses advances that might occur if society continues to support research. This more or less chronological approach has several purposes. Since there is often a certain logic to the order in which discoveries were made, historical emphasis may help readers remember the facts. Second, historical discussion allows us to introduce basic concepts in a more interesting way than by reciting definitions. It makes life richer to realize that some seemingly modern concepts descend from knowledge of ancient millenia, and have thousand-year-old names. Third, there is a widespread fallacy that the only progress worth mentioning is that of the last few decades. Astronomy, of all subjects, shows clearly that, to paraphrase Newton, we see as far as we do because we stand on the shoulders of past generations. Exploration of the universe is a continuing human enterprise. As we try to maintain and improve our civilization, that is an important lesson for a science course to teach.

Another philosophy we have followed is to treat astronomical objects in an *evolutionary* way, to show the sequence of development of matter in the universe. Stars, pulsars, black holes, and other celestial bodies are linked in evolutionary discussion, rather than listed as different types of objects detected by different observational techniques. I have also not hesitated to mention nonscientific approaches to cosmology and evolution, such as "creationist" concepts recently encouraged in two states' school systems by their state legislatures, but later thrown out after courtroom battles.

As a result of Wadsworth's extensive market surveys during preparation of this edition, we have adopted several noteworthy design concepts. The level of physics presentation is upgraded from the second edition, and the number of optional basic equations expanded from five to nine. (A shorter, nonmathematical, descriptive edition of this book is also in preparation.) Mathematics is reduced to a minimum in the text and discussed primarily as applications of the nine optional basic equations. These are described *qualitatively* in the main text, so that the book can be used in an entirely nonmathematical way. The basic equations are also described more *quantitatively* in optional boxes. Applications are illustrated there. For a more mathematical approach, instructors can emphasize these (and discuss the *Advanced Problems* at ends of chapters, which are mostly applications of the basic equations). At the request of numerous users, introductory material on light, atoms, and spectra has been integrated into the text as a new Chapter 7 (including Essay A in the previous editions). Thus, the student gets an overview of spectroscopy and atomic structure just before encountering spectroscopy in the study of planets. Then, in Chapters 13 and 14, as the sun and stars are introduced, this basic physics primer is expanded to cover topics more applicable to stars. Thus, the student gets a stepwise, perhaps gentle, introduction to some of the basic physics underlying astronomy. We have also included supplemental *Enrichment Essays* that can be used or omitted at will. We have cross-referenced the color plates thoroughly in the text. To reduce costs, we have limited our color photographs primarily to examples that illustrate the role of color in understanding astronomical phenomena.

More specifically, teaching aids incorporated in the book include the following.

TEXT

1. In addition to the classic large telescope photos and recent NASA photos, I have included three other categories of illustrations:

a. *Photos from recently published research papers, kindly provided by various authors and institutions.*

b. *Photos by amateurs with small and intermediate instruments, often used to show sky locations of well-known objects in the large-telescope photos. These can help readers to visualize and locate these objects in the sky, a difficult task if based on classic large-telescope photos alone. Photographic data provided with many of these pictures may be used in setting up student projects in sky photography.*

c. *Scientifically realistic paintings show how various objects might look first-hand to observers in space. Discussion of features shown in the paintings illustrates a synthesis of scientific data from various sources.*

2. Key concepts are shown in **boldface** type. These are repeated in *Concepts* lists at the ends of chapters as aids to review. Definitions of key concepts are included in an expanded *Glossary* at the end of the book.

3. The nine optional basic equations are introduced in the text as needed, but are set off in boxes by a light gray screen for optional use. The nine boxes discuss:

I. *The Small Angle Equation,* useful for calculating apparent sizes of objects at known distances.

II. *Newton's Universal Law of Gravitation,* illustrating the simplicity of gravitational attraction between bodies throughout the universe.

III. *Circular and Escape Velocities,* useful for deriving speeds or masses in co-orbiting systems (planetary, binary star, galactic).

IV. *The Definition of Mean Density,* a simple concept for gaining information about the nature of material inside planets and stars.

V. *Measuring Temperatures of Astronomical Bodies: Wien's Law,* which shows how radiation measurements can reveal the temperatures of distant objects.

VI. *Typical Velocities of Atoms and Molecules in a Gas,* by which we characterize temperature, as well as collision energies when the atoms or molecules smash into each other. These energies, in turn, control the types of chemical or nuclear reactions that can occur.

VII. *The Doppler Shift: Approach and Recession Velocities,* which shows how spectral measures can reveal radial velocities of distant objects.

VIII. *The Stefan–Boltzmann Law: Rate of Energy Radiation,* which shows how temperature and luminosity measurements can reveal sizes of radiating sources.

IX. *The Relativistic Doppler Shift,* a modification of item VII, to explain phenomena that occur at high speeds, approaching the speed of light.

4. Limited numbers of references to technical and nontechnical sources appear in the text. They are there partly to help students and teachers find more material for projects, and partly to help instructors emphasize that statements should be verifiable. These sources are included in an expanded *References* section.

END OF CHAPTER MATERIALS

1. *Chapter Summaries* review basic ideas of the chapter and sometimes synthesize material from several preceding chapters.

2. *Concepts* lists include the important concepts appearing in **boldface** in the text. Reviewing the *Concepts* is a good way for the student to review the content of each chapter.

3. *Problems* are aimed at students with nonmathematical backgrounds.

4. *Advanced Problems* usually involve simple arithmetic or algebra and are usually applications of the nine basic equations. These can be omitted in nonmathematical courses.

5. *Projects* are intended for class use where modest observatory or planetarium facilities are available. The intent is to get students to do astronomical observing or experimenting.

SUPPLEMENTARY MATERIAL

1. *Enrichment Essays* can be used or not as instructors wish. These include essays on:

A. *Pseudoscience and Nonscience*
B. *Astronomical Coordinates and Timekeeping Systems*

2. Appendixes are included on:

1. *Powers of Ten*
2. *Units of Measurement*
3. *Supplemental Aids in Studying Astronomy*

3. The *Glossary* defines all terms included in the *Concepts* lists, as well as other key terms.

4. The *References* section includes all sources mentioned in the text and others used in compiling the text. Nontechnical references useful for student papers are starred (*); widely available journals and magazines are emphasized in this group, including most astronomy articles appearing in *Scientific American* in recent years.

5. Two indexes are included, an *Index of Names* and an *Index of Terms.*

6. *Star Maps* for the seasons are found after the indexes. Since more detailed, larger maps are usually available in classrooms or laboratories, these have been simplified, emphasizing the plane of the solar system and the plane of the galaxy, and major constellations mentioned or illustrated in the text.

ACKNOWLEDGMENTS

My thanks go to many people who helped produce this book. I have tried to incorporate as many of their suggested corrections and improvements as possible, although final responsibility for weaknesses remains mine.

Included among two groups of reviewers who criticized parts of the text for research comprehensiveness and teaching potential were: Helmut A. Abt, Bill T. Adams, Jr., Richard A. Bartels, Henry E. Bass, Bart J. Bok, Lee Bonneau, David R. Brown, Warren Campbell, Robert J. Chambers, Clark R. Chapman, Kwan-Yu Chen, Clark G. Christensen, Dale P. Cruikshank, Donald R. Davis, Robert J. Doyle, John Fix, Benjamin C. Friedrich, F. Trevor Gamble, Owen Gingerich, Edward Ginsberg, Jeff Goldstein, Richard Greenberg, Hubert E. Harber, Alan Harris, George H. Herbig, Ulrich O. Herrmann, J. R. Houck, William W. Hunt, Nathan Krumm, Gerald H. Newsom, Charles Peterson, Terry P. Roark, John M. Samaras, John Schopp, Richard Schwartz, Myron A. Smith, Richard C. Smith, Ronald Smith, Michael L. Stewart, Kjersti Swanson, Donald Taylor, James E. Thomas, Scott D. Tremaine, Peter D. Usher, Stuart J. Weidenschilling, Walter G. Wesley, Ray J. Weymann, Richard M. Williamon, R. E. Williams, and Warren Young.

I also thank Linda Dietrich, Gayle Hartmann, Floyd Herbert, and Roger Howlett for discussions and criticisms that affected the presentation of material; and finally, a wide circle of colleagues, especially in Tucson and at the University of Hawaii, for helpful discussions.

In locating and providing photographs, Dale Cruikshank, Walter Feibelman, Steven Larson, Alfred McEwen, Alan Stockton, Don Strittmatter, and Dave Webb were helpful. Don Bane and Jurrie Van der Woude (Jet Propulsion Laboratory), David Moore (Kitt Peak National Observatory), and Holly Chaffee (Arizona State Museum) gave kind assistance with their institutional files. I especially thank Chesley Bonestell, Don Dixon, David Egge, James Hervat, Paul Hudson, Pamela Lee, Ron Miller, Jim Nichols, Kim Poor, Adolf Schaller, Andrei Sokolov, Richard Sternbach, and Brian Sullivan for stimulating discussion about paintings of astronomical subjects. Floyd Herbert and Mike Morrow helped me in making several wide-angle sky photos with clock-driven mounts.

I thank Karla Hankey and Paula McBride for assistance in coordinating this and other research projects during preparation of the final manuscript of the third edition.

The staff at Wadsworth Publishing Company has worked hard to create a useful and beautiful product, and made the job a pleasure at the same time. I thank Tom Nerney, Steve Rutter, Gary Mcdonald, MaryEllen Podgorski, and MaryLee Mackichan for their friendship and professional help in shepherding the third edition through various editorial stages. Thanks to all of them as well for exceptional work on production and design, cheerful dispositions, hospitality, and care for the subject, not to mention the harried author.

William K. Hartmann
Tucson, January 1985

A NOTE FROM THE PUBLISHER

Since 1984, when the third edition of *Astronomy: The Cosmic Journey* was written, astronomers have increasingly come to accept black holes as the power sources at the center of active galactic nuclei, they are analyzing exciting new information about the structure of Uranus' rings, and they've photographed the nucleus of a comet. New telescopes, detectors, and computer processing techniques are allowing them to map the largest known structures of the universe in ever more revealing detail.

The rapid pace of such discoveries makes it difficult for people to keep up with the most recent advances. It's particularly challenging for textbooks to remain current because of the time needed to plan and publish them. This 1987 Edition of *Cosmic Journey* is our contribution toward providing instructors and students with the most accurate and up-to-date information available in this continually changing field. Incorporating the latest astronomical discoveries and research, William Hartmann has solidly updated sections of the text and included many new photographs and illustrations. He has also updated the bibliography, to direct students to the latest review articles. Yet the compelling writing, stunning visuals, and scientific accuracy of the content remain the same.

Among the recent events chronicled in *Astronomy: The Cosmic Journey,* 1987 Edition are new data and photographs from the various space probes sent to investigate Comet Halley (Chapter 11), new information and pictures from the Voyager encounter with Uranus (Chapter 10), and new data from Soviet radar mapping and balloon probes of Venus (Chapter 8). Some proposed missions are described, and the impact of the Challenger disaster is assessed (Chapter 4). New results on star formation and circumstellar material from IRAS and from the Epsilon Aurigae eclipse are discussed (Chapter 16). Hartmann also updates theoretical work on the moon's origin (Chapter 5), supernovae models (Chapter 17), our understanding of the galactic center (Chapter 21), models of the evolution of different galaxy shapes and galactic cannibalism (Chapter 22), the latest Hubble constant measurements (Chapter 23), the missing mass problem, and the filamentary arrangement of galaxies and galaxy clusters (Chapter 24).

We view this revision primarily as a necessary update, rather than as a major change in the text itself (which could require substantial revision of an instructor's course notes). Therefore, we have been careful to keep the original section-by-section sequence, chapter length, and book length intact, and have integrated the new material smoothly into the book's original structure. We sincerely believe these changes will assist instructors in their efforts to teach a current, accurate, and effective course in astronomy. And we hope that students will continue to be fascinated by the *Cosmic Journey* of learning about the scientific wonders of the universe.

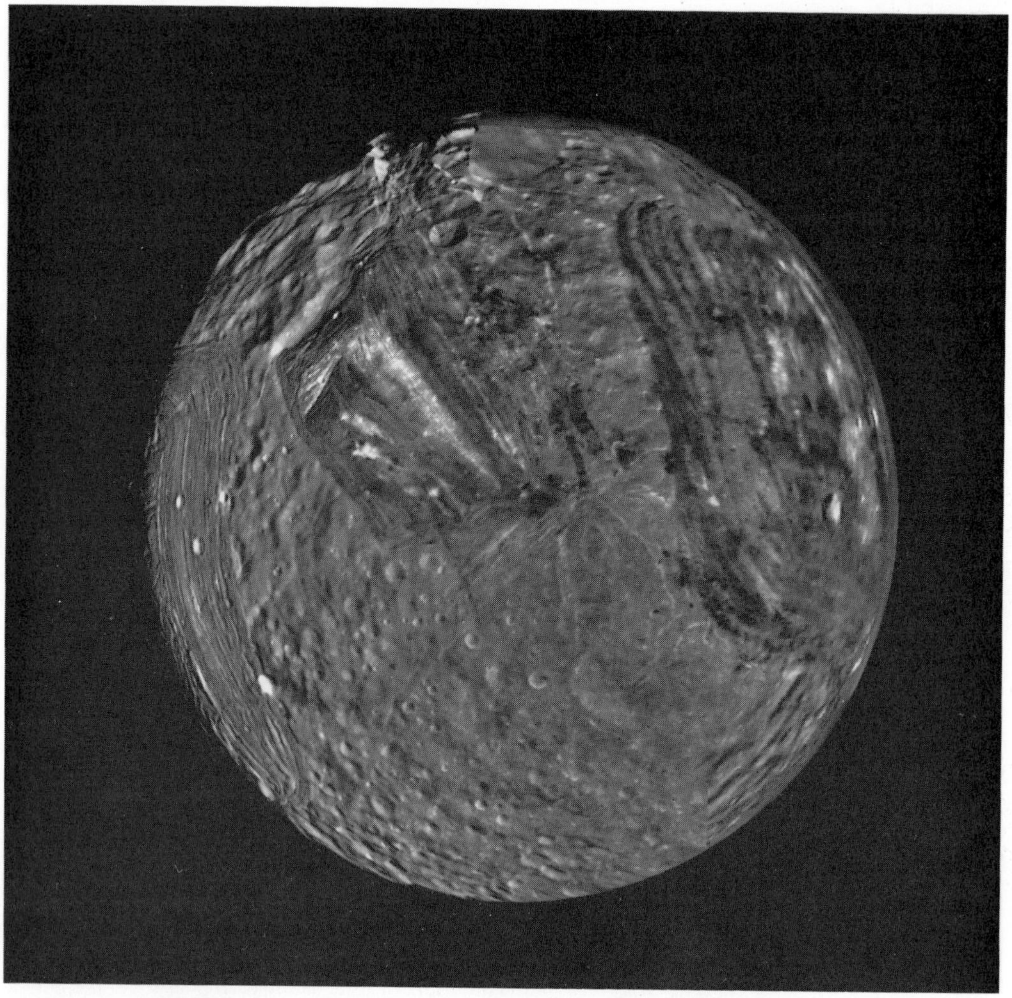

On January 24, 1986, Voyager 2 flew by the strange moon Miranda of the planet Uranus. The spacecraft was more than 18 astronomical units (1.7 billion miles) from Earth, and only about 35 000 mi from Miranda, which is about 480 km (300 mi) across. Smallest details visible on this print are about 800 m (½ mi) across. Miranda's surface is composed of ice and dark dust. Ancient regions (bottom center) have been cratered by meteorite impacts for several billion years, but other regions (left, top, right) have been resurfaced by fractures, grooves, and canyons marked by dark stripes. The cause of these fractures and stripes is not well understood. See Chapter 10 for more detail on this and other icy moons of the outer solar system. (NASA photo, processed by U.S. Geological Survey; courtesy Jet Propulsion Laboratory of Caltech.)

INVITATION TO THE COSMIC JOURNEY

If you awoke one day to discover that you had been put on a strange island, your first project after getting food and water would be to try to find out where you were. This is just what has happened to all of us. We have all been born on an island in space: the earth. We are all passengers on a cosmic journey. Astronomy is the process of finding out where we are—our place in space and our position in the unfolding history of the universe.

This definition of astronomy would not always have satisfied people, because at one time we thought we knew where we were: at the center of the universe, with the sun, stars, planets, and satellites all revolving around us. Modern astronomy contradicts that old theory. The definition is not fully satisfactory now, either, because it doesn't really explain what astronomers do.

Our Definition of Astronomy

Paradoxically, the more we learn, the harder astronomy is to define. When astronomy was restricted to observations by earthbound viewers, it was easily defined as "the science of objects in the sky." But now that we have actually begun to explore space, other disciplines have become involved. Is the astronaut who chips a rock sample off the moon practicing astronomy? What about the researcher who studies the geology of the sample once it reaches the earth? What of the nuclear physicist who measures properties of nuclear reactions going on in the center of stars? What about the biologist interested in life on Mars? Their work should be included in our definition of astronomy.

There is a popular misconception that as various scientists pursue new research, knowledge proliferates into an endless complex of specialized disciplines. Indeed, students are usually taught this way, with advanced courses probing into narrower and narrower specialties. According to this idea, research is like climbing a tree, starting near the trunk and proceeding to ever more specialized branches of knowledge.

But science is also like the process of working our way *down* the tree of knowledge. At first we see a bewildering variety of seemingly unrelated phenomena—twigs of knowledge. After analyzing these twigs, we see that they meet in a branch. As workers map details on the branch, someone with vision may discover that it is joined to another branch. By grasping the relations between the major branches closer to the trunk, we can better understand the higher branches and twigs. As we work our way toward the trunk, more and more branches join. This is why different specialties are connecting with astronomical research; astronomy is becoming more general.

A historical example illustrates this process. Around 1600, Galileo and others conducted many experiments to understand seemingly unrelated types of motion, such as the motion of falling objects and the motion of the moon. The wide variety of motions and accelerations seemed impossible to explain in a simple way. However, by about 1680, Isaac Newton realized that they were connected by simple relationships between mass, acceleration, and gravitational attraction. Suddenly, as a result of Newton's insight, ordinary students could understand motions better than the Galileos of the past.

In the same way, today's students can grasp the relationship between the earth and the cosmos better than the explorers who mapped astronomical frontiers in the past. Better than any generation in human history, we can begin to sense where we are, at least in relation to other physical objects in the universe.

This book treats astronomy not as a narrow set of academic observational results or abstruse physical laws, but as a voyage of exploration with practical effects on humanity. A broad view of the major discoveries of the last years, the last decades, and even the last centuries is at least as important as the latest factual detail, because the broad view helps us understand how scientific research will continue to affect us and perhaps offer us new options for living. In an era of energy crises, food shortages, and environmental threats, when badly applied technology has created such horrors as biological weapons and hydrogen bombs, educated people are called on to make judgments about their own actions—the kinds of jobs they have, the materials they consume, and the actions of organizations they work for. In short, we are called on to judge what kind of civilization should continue. If so, we all need to know how our world is affected by its cosmic surroundings and to understand the scientific and social procedures available for obtaining and extending that knowledge.

Therefore, this book is organized according to the following definition of **astronomy**:

Astronomy is the study of all matter and energy in the universe, emphasizing the concentration of this matter and energy in evolving bodies such as planets, stars, and galaxies, and fully recognizing that we observers—humanity—are part of the universe, and that our home, the earth, is only one of many places in the universe, but also the special point from which our voyage of exploration has started.

The reason for emphasizing humanity is that astronomy influences how we think about ourselves and our role in the universe. The growing perception of environmental problems in the 1970s and 1980s has shown that we *must* look at the earth as a single astronomical body in order to survive. As the poet Archibald MacLeish said after one of the Apollo flights, we have now seen the earth as "small and blue and beautiful" and ourselves as "riders on the earth together." Thus this book regards astronomy as including space flights and studies of rocks from our neighbor planets, just as much as telescopic observations of remote galaxies.

A Survey of the Universe

The ensuing chapters start with subjects on earth and then move outward through space. However, before beginning, we give a brief preview of the universe, starting at the largest scale.

The **universe** is everything that exists. To the best of our knowledge, the universe consists of untold thousands of clusters of galaxies. **Galaxies** are swarms of billions of stars. Galaxies differ in form. Some are football shaped, some irregular. Our galaxy, like many others, is a disk with about a hundred billion stars arrayed in curving spiral arms. Most galaxies, including ours, are surrounded by a halo of **globular star clusters**, each a spheroidal mass of hundreds of thousands of stars. If we made a model of our galaxy big enough to cover North America, the earth would not be as big as a basketball, or even a BB. It would be about the size of a large molecule.

Scattered inside our galaxy, mostly in the spiral arms, are groupings of stars called **open star clusters** and clouds of dust and gas called **nebulae**. Scattered at random are individual stars. Each **star** is an enormous ball of gas, mostly hydrogen. In the center of a typical star, atoms are jammed together so closely that their nuclei interact to create the heat and light of the star.

Three hundred million billion kilometers (200 000 000 000 000 000 mi) from the center of our galaxy and 40 thousand billion kilometers from the nearest star is the **sun**. A hydrogen sphere more than a million kilometers in diameter, the sun is the center of the **solar system**—the small system made up of the sun and its family of orbiting rocky and icy **planets**. Circling around most planets are smaller bodies, called **satellites**. Between planets are various other small bodies, called **asteroids** and **comets**.

About 150 million kilometers from the sun is the **earth**. The closest two planets approach the earth within about 50 million kilometers. The **moon** is the earth's satellite, only 384 000 km away. The earth itself is nearly 13 000 km across, a tiny speck among myriads of larger and smaller specks in the universe.

A Word About Mathematics

The preceding survey of the universe shows how hard it is to express astronomical quantities in ordinary units. For this reason the system of expressing numbers as **powers of ten** is useful. For example, the distance from the sun to the galactic center, instead of being written as the unwieldy number 300 000 000 000 000 000 km, becomes simply 3×10^{17} km; the distance from the earth to the sun is 1.5×10^{8} km. A reader uncertain about this system should consult Appendix 1 at the end of the book. We will use this system occasionally, as convenience dictates.[1]

When using any equation to calculate physical quantities, you must remember to use consistent sets of units. For example, the English system of inches, ounces, and so on, must not mix in the same equation with the metric system. The metric system, long used by scientists, is becoming universal and we will use it here, often giving English equivalents. The metric system is *much* easier to use than the English system because its units come in multiples of ten. A centimeter = 10 millimeters, a meter = 100 centimeters, and so on. Gone are complicated conversions like 1 mile = 5280 feet. Specifically, in this book we will give preference to the **"SI" metric system** of units—those adopted in the Système International (International System), such as meters, kilograms, and so on. In some cases, astronomical units such as "parsecs" will be explained and adopted.

A Note About Names of People

One of the things that distinguishes science from the pseudoscience now so popular is that scientific assertions are backed by evidence. Because much of this evidence has already been published elsewhere, it is not always discussed in detail, but is merely referenced. This is

[1]We also use the term *billion* here to mean 1 000 000 000, or 10^{9}, which is the usual American meaning. The British use *billion* to mean 1 000 000 000 000, or 10^{12}.

usually done by listing the name of the author and the date of publication (for example, "Smith, 1982"); the reference is then listed in the back of the book so that a reader who wants more information can locate the source. With so much wildly distorted pseudoscience being published today under the guise of nonfiction, this referencing of evidence is important in helping you recognize whether you have reliable material in your hands. We have thus included many references here for the use of both students and teachers.

Students are sometimes distressed to find these names, thinking they are supposed to learn names and dates. The names are only to document assertions, not to be memorized. The emphasis in this book is on the exciting nature of the universe that astronomy is revealing, not on the astronomers themselves. However, many of the Greek and Renaissance astronomers mentioned in Chapters 2 and 6 are major figures in our Western intellectual heritage, and students should learn about them.

A Hint on Using This Book

When reading a text, many students underline everything important in sight. My publishers and I have taken care of part of that job by putting terms that represent important concepts in boldface and then listing them in order at the end of the chapter. A useful study procedure is to read a chapter through and then go over the concept list. Try to define each term out loud to yourself and use it in a sentence. If you have trouble with a term, go back and check it in the chapter or look it up in the glossary. If you are comfortable with all the terms in each concept list, you will have gone a long way toward mastering the material and will be able to enjoy many articles about astronomy in newspapers and magazines.

Face to Face with the Universe

The universe is immense; we are perhaps audacious even to attempt studying it. Yet something noble in the human spirit urges us to seek knowledge of the universe and our place in it. Throughout history, scientific discoveries in the heavens have sometimes illuminated and sometimes conflicted with other religious, metaphysical, and philosophical conceptions. Persecutions, legal action, even wars have resulted from the challenge to old beliefs by new knowledge. In this book we present scientific theories and the *evidence* for them with the understanding that our knowledge is limited and slowly improving. New evidence is what usually distinguishes new knowledge from old superstition. From astronomy we get a new cosmic perspective from which many conflicts of the past, often founded on amazingly few facts, now seem inconsequential. For example, we have given up the medieval arguments over how many angels can dance on the head of a pin, and stopped persecuting those who advocate the theory that the earth moves around the sun. Unfortunately, world news shows that we seem still unable to give up going to war over ideologies, whether Christian vs. Moslem vs. Jew or capitalist vs. communist. Indeed, astronomy unifies humanity by showing us that we are all in this together, facing the unknown—a word that aptly describes much of the universe, its past, and its future.

The universe is no abstract concept; it consists of real, physical places. This book will offer descriptions of those places, often with photos or scientifically realistic paintings to show how we think they look.

Suppose that we were able to explore these places and that in all these vast reaches, among all the stars and galaxies, we found no one else—no living creatures anywhere except ourselves. Or suppose that we discovered nonhuman intelligent life—alien civilizations, perhaps advanced and incomprehensible or perhaps very simple. Our contacts with them could alter our history far more profoundly than the first contacts between the American Indians and Europeans in 1492.

These are two wildly different possibilities. One is likely to be true. Either boggles the mind. We are just barely entering the century in which we may be able, if research continues, to decide which scenario is more realistic.

Concepts

astronomy	star	comet
universe	sun	earth
galaxy	solar system	moon
globular star cluster	planet	powers of ten
open star cluster	satellite	SI metric system
nebula	asteroid	

Problem

1. Scientific discoveries are sometimes described as conflicting with religious or philosophical beliefs. Before continuing this book, examine your present beliefs or expectations and comment on which of the following views best match your own:

a. There are concepts in astronomy that conflict with my own religious or philosophical views.

b. Astronomical concepts are consistent with my religious or philosophical views; there is no problem.

c. If there is a conflict, scientific hypotheses will evolve and eventually become consistent with my religious or philosophical views.

d. Science measures the reality of the physical world only; there is no need for this to be consistent with an inner psychological or spiritual reality.

e. Science tries to reflect how all nature behaves; thus science and religious philosophy of daily life should be consistent with each other to contribute to a well-integrated personality.

P A R T A

The Early Discoveries

To paraphrase Brahms
in his reference to Beethoven,
we hear the tramp of the giant
behind us.

Sir Fred Hoyle,
commenting on the
builders of Stonehenge

C H A P T E R 1

PREHISTORIC ASTRONOMY: ORIGINS OF SCIENCE AND SUPERSTITION

In 1906 the American astronomer Percival Lowell wrote:

Smoke from multiplying factories . . . has joined with electric lighting to help put out the stars. These concomitants of an advancing civilization have succeeded above the dreams of the most earth-centered in shutting off sight of the beyond, so that today few city-bred children have any conception of the glories of the heavens which made of the Chaldean shepherds astronomers in spite of themselves.

This observation seems even more apt today. It is difficult to realize how important the sky once was in daily affairs. In addition, we have forgotten how much we depend on ideas that may seem obvious today, but were developed only through centuries of struggle to understand nature.

There are several reasons to review the ancient discoveries. First, they were often very basic. We forget how hard they were to recognize. For example, we think it obvious that the earth moves round the sun, but early scholars were severely criticized for saying so.

Second, discoveries were often made just when society reached a point where they *could* be made, because the needed technical devices or theoretical concepts had just been invented. Therefore, a historical approach helps place important concepts in a logical order.

Third, many common ideas today come from astronomical traditions of the past. For example, although we favor a decimal numerical system, we have 24 hours in a day instead of 10, and 360 degrees in a circle instead of 100. The reason comes from ancient astronomy and mathematics.

The Earliest Astronomy: Motives and Artifacts (c. 30 000 B.C.)

Archaeology and anthropological studies of present-day primitive tribes[1] shed light on the earliest astronomical practices. Imagine yourself an intelligent hunter of 30 000 years ago. Agriculture is not yet practiced. You have to depend on hunting and gathering for food. Celestial phenomena are not your most immediate concern, but you know that certain celestial cycles are important. You want to know when berries on the mountain will ripen and when migratory birds will arrive at a nearby lake. Such knowledge requires cumulative day counts and knowledge of the seasons, best derived from sky phenomena. When you travel far, you want to be able to find your way home. Celestial objects are your only reliable guides in unfamiliar landscapes.

If you are a woman, you also want to know in what season the child you may carry will be born and when your next menstrual period will occur. Living mostly outdoors, women undoubtedly used the $29\frac{1}{2}$-d cycle of the moon's phases to keep track of their 27- to 30-d menstrual cycles. Some experiments indicate that menstrual cycles of women sleeping in the light of an artificial moon become synchronized to its phases, and some scientists have even speculated that the human menstrual cycle evolved to match the period of the moon's phases (Luce, 1975).

For such reasons, early people began to keep records of events in the sky. There may be actual physical evidence of the earliest such records. Curious notations, found on thousands of artifacts dating as

[1]Time grows short for these studies. The last few pockets of primitive culture are being discovered and integrated into their surrounding cultures by scientific expeditions and even guided tours.

far back as 30 000 B.C., scattered over Europe, Africa, and the Soviet Union, may be the beginnings of calendar systems—perhaps counts of the number of days between phases of the moon, such as new moon (the thinnest crescent, Figure 1-1) and full moon (Marshack, 1972). The makers of these tools were not the brutish cave dwellers of some stereotypes; their craftsmanship is demonstrated by the magnificent paintings of their prey—mammoths, boars, reindeer—that they left on European cave walls.

This interpretation is supported by contemporary calendar sticks made by modern aborigines. For example, a lunar calendar stick carved by Indian Ocean islanders shows a series of grooves almost identical to the Paleolithic examples of 30 000 years earlier. And North American Indian calendar sticks similarly record historical events and other natural events by sequences of carvings.

The calendar stick in Figure 1-2, for example, records an earthquake and other events. Since the Indians who made it had no written language, it can be "read" only by the carver. It is dramatic evidence of how records of astronomical events may have first accumulated, thus allowing the discovery of astronomical cycles.

Calendar Refinements (10 000–3000 B.C.)

Around 10 000 B.C., the domestication of animals and the cultivation of crops began. This agricultural revolution brought two new incentives for understanding the sky. First, to know when to plant, people had to be able to predict seasonal changes days in advance. Second, in the stable villages made possible by agriculture, people kept records; the mere existence of records gave thinkers an incentive and opportunity to discover seasonal and annual cycles of celestial events.

This stage of development is shown in historical records of Indians of the American Southwest. Papago Indians, for example, named the 12 lunar cycles (months) that made up the year—among them a "Green Moon" (March, plants leafing), and a "Hungry Moon" (May, no wild foods). This was obviously the beginning of a calendar system. But since each lunar cycle is about $29\frac{1}{2}$ days, 12 such cycles would be only 354 d, not a full year. Thus, the lunar-based calendar must be readjusted each year.

People began to get a better sense of the yearly cycle by counting days or moon cycles between certain dramatic annual events. In Egypt, for instance, the Nile flooded at a certain time each year. At high northern latitudes people looked eagerly for the sun to return to the northern sky after the cold, short days of winter. In some

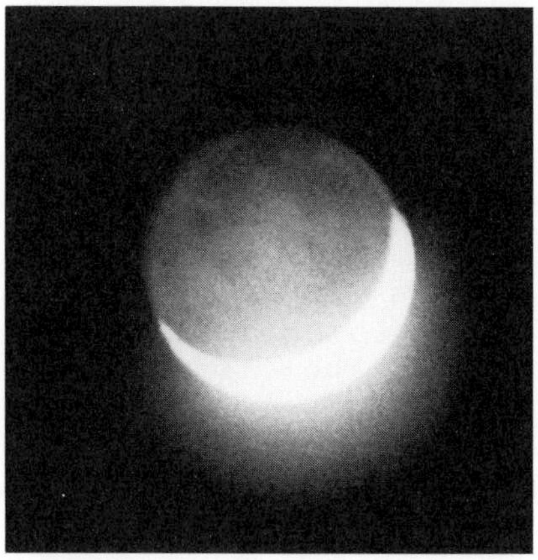

***Figure 1-1.** Paleolithic evidence suggests that ancient people tabulated the phases of the moon. The configuration shown here, called "the old moon in the new moon's arms," is seen in the evening sky a few days after a new moon, when the portion not lit by the sun is dimly lit by light reflected off the earth. (Photo by author.)*

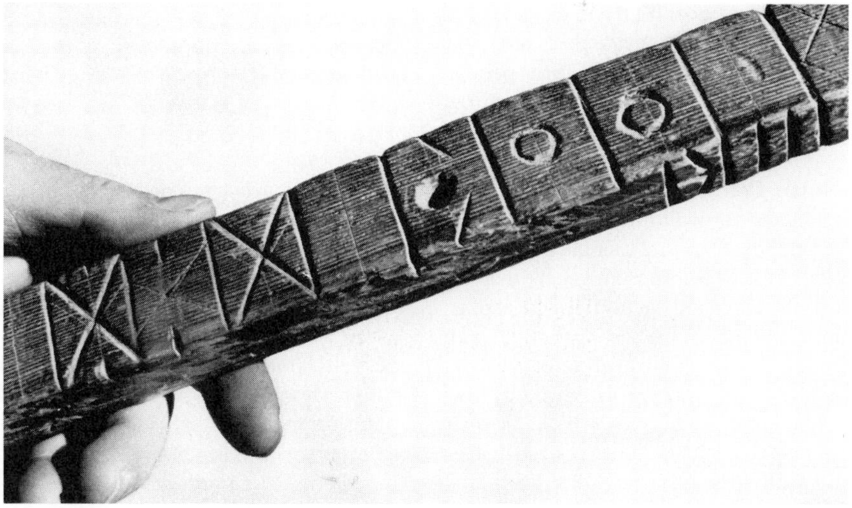

***Figure 1-2.** Portion of a calendar stick carved by Papago Indians of Arizona. The stick, begun in 1841 and passed on to the carver's son, carries a record of social and natural events, including an earthquake, for the years 1841 to 1939. (Calendar stick from collection of Arizona State Museum, University of Arizona. Photo by author.)*

early societies, day counts of this cycle gave estimates of about 360 d, later refined to 365 and then $365\frac{1}{4}$.

Accumulated observational experience revealed the correlation between this annual cycle of seasons and the apparent movements of the sun. The sun does not follow the same path across the sky every day. In the summer at northern latitudes (for example, Europe and the United States), the sun is high at midday; in the winter, it is low in the southern sky. Thus, careful observations of the sun's motion relative to other features of the sky would have helped determine the exact number of days in the year and the exact dates of such observations.

Organized observations of this kind were made throughout the world thousands of years ago. Mesopotamian calendars were being designed by 4000 to 3000 B.C. In recent decades, anthropologists have found modern pretechnological people using similar means to create sophisticated systems of time measurement.

Figure 1-3. *Two notable constellations of the northern sky were originally labeled the greater and lesser bears. Today they are more commonly known as the Big Dipper and the Little Dipper. They are important to outdoor people, because the star at the end of the Little Dipper's handle is Polaris, "the North Star." It always hangs directly above the north point on the horizon, as seen from the Northern Hemisphere. It can be found by using "the pointers"—the two end stars in the Big Dipper's bowl. (Photo by author, 42-min exposure with wide-angle 24-mm lens, f2.8, on commercially available 2475 recording film.)*

Other Early Discoveries

Before describing ways in which these principles were utilized and extended, we pause to describe other aspects of the sky discovered or defined by early people. These aspects are important aids in describing many astronomical phenomena.

North Celestial Pole. Camp out for a night and you will discover that the stars slowly wheel around a single point in the sky. In the Northern Hemisphere that single point is called the north celestial pole; in the Southern Hemisphere, the south celestial pole. The wheeling of the stars is due to the earth's rotation on its axis, and therefore one complete circuit takes about 24 h (hours). The two **celestial poles** can be thought of as the projections of the earth's axis onto the sky, or the spots targeted by vertical searchlights at the North and South Poles. You can visualize the sky as a giant dome overhead, as seen in Figure 1-3. Recognizing the celestial pole helped in navigation, since it indicates the north or south direction.

North Star. A bright star, **Polaris** (shown in Figure 1-3) happens to lie near the north celestial pole. Thus, it stands almost still all night above the northern horizon as other stars wheel around it. It thus serves as an excellent reference beacon. For nearly 1000 years, Polaris has been located within a few degrees of the north celestial pole, but in other times other stars have been nearer the pole. If a bright star is located near the north celestial pole, it is called the **North Star.**

Zenith and Meridian. Wherever an observer stands, the point directly overhead is called the **zenith.** The **meridian** is an imaginary north–south arc from horizon to horizon through the celestial pole and the zenith. The zenith and meridian are shown in Figure 1-4. This line marks the highest point that each star reaches above the horizon on any night; it is the dividing line between the rising stars and setting stars. For this reason it is especially useful in timekeeping. In the daytime, the moment when the sun crosses the meridian is called noon; in very early times, therefore, people used this astronomical concept to divide the day.

Celestial Equator. As shown in Figure 1-4, the imaginary circle lying 90° from both the north and the south celestial poles is the **celestial equator.** It is the projection of the earth's equator onto the sky and is parallel to the daily east-to-west paths of

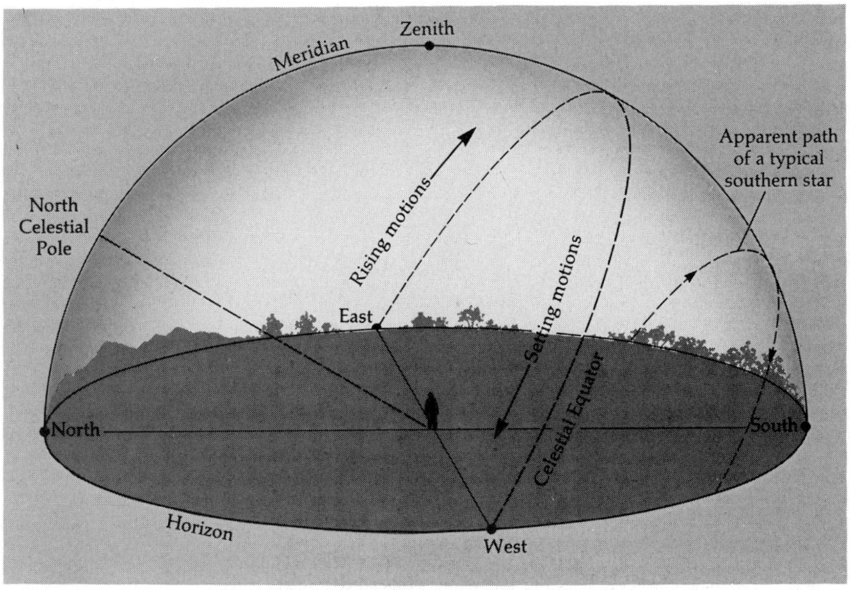

Figure 1-4. *General properties of the sky as defined by the earth's daily rotation.*

the stars as they wheel around the celestial poles. The celestial equator helps divide the sky into easily recognizable portions.

The 360° Circle. Our division of the circle into 360 units, or degrees, follows a Babylonian practice of about 3000 years ago. It probably comes from the ancient estimate of 360 d for one year, and from the fact that 360 is easily divisible into 2, 3, 4, 5, or 6 equal parts. The decimal (10-based) system of counting and writing numbers later replaced the sexagesimal (60-based) system in most uses *except* the expression of angles and time.

The Ecliptic. By observing the relative positions of sun and stars at dawn or dusk, one can establish that the sun appears to shift nearly a degree to the east each day relative to the stars, as shown in Figure 1-5. (This can be confirmed by noting that the sun "moves" 360° in 365 d.) The cause of this shift is the earth's motion around the sun, as shown in Figure 1-6. Ancient observers found that the sun traces the *same path* through the stars year after year. The path differs from the celestial equator (being tipped to it by $23\frac{1}{2}°$ and crossing it at two points). This path of the sun among the stars is a circle extending all the way around the sky and is called the **ecliptic.** We now know that the ecliptic is the same as the plane of the earth's orbit around the sun, as can be visualized in Figure 1-6.

The Planets. Ancient observers—priest-astronomers and wakeful shepherds alike—found that five of the brighter star-like objects were not fixed like the rest. They came to be known in the Western world as planets, from the Greek word for "wanderers."[2] Not until after the telescope was invented, in about 1610, was there any proof that these **planets** were globes like our earth or that three faint ones had gone undiscovered. For this reason the ancients spoke of five planets, whereas we speak of nine (counting our own).

[2]Many of our technical terms have Greek roots because the modern Western tradition is directly linked to Greek civilization through the Roman and medieval worlds. However, many of the concepts discussed here were discovered in pre-Greek times and later renamed by the Greeks. Roman and medieval writers, often unaware of the earlier work, revered the Greek thinkers and kept their terms.

The Zodiac. Ancient observers tracking the paths of the planets from night to night among the stars discovered that the planets never stray out of a zone about 18° wide centered on the ecliptic. Many star patterns, or constellations, along this zone were said to resemble animals, so the zone came to be called the **zodiac** (after Greek for "animals"—the same root as *zoo.*) The constellations located within the 18° zone came to be known as the signs of the zodiac. They include such familiar examples as Scorpius (the scorpion) and Leo (the lion—see star maps on the last pages of the book).

Heliacal Risings and Settings. Any given star rises and sets slightly earlier each night. The **heliacal** (he-LIE-ah-cal) **rising** of a star *occurs on the first day each year when the star can be seen just before dawn.* (*Heliacal* means "near the sun," from the Greek *helios,* "sun.") One day earlier, the star rises a few minutes *after* the sky gets too bright. On the day of heliacal rising, the star glimmers near the horizon just a minute or so before the sky gets too bright. On the day after, the star rises slightly earlier and can be seen for several minutes before the sky lightens. Similarly, **heliacal setting** *occurs on the last day of the year when the star can be seen at dusk.* On the next day, by the time darkness falls, the star has already set.

Why would anyone living in, say, 4000 B.C. care about the heliacal rising or setting of a certain star? The answer is that each heliacal rising or setting occurs on the same date every year. Depending on the star, that date might be the first day of spring or the day when planting should start. Without clocks, calendars, and morning newspapers on the doorstep, heliacal risings and settings would tell the date with an accuracy of a day or two. It is scarcely surprising that the ancient Egyptian calendar began with the heliacal rising of Sirius, which marked the beginning of the Nile's annual flooding.

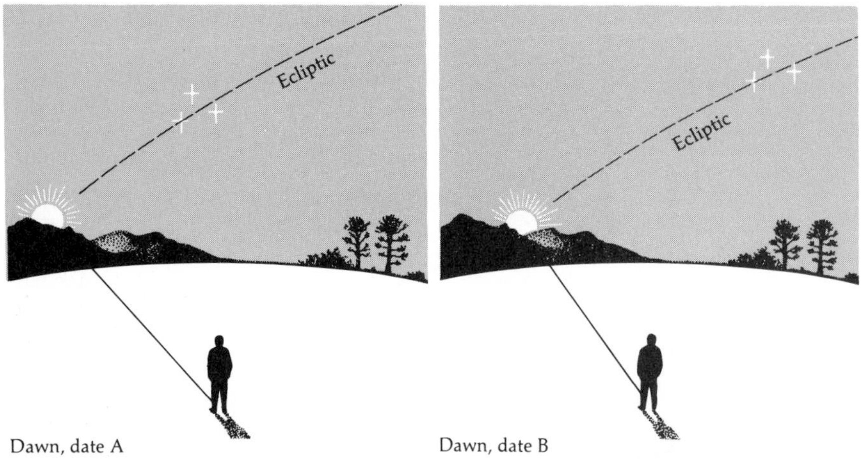

Figure 1-5. *At dawn on date A, the sun is near certain bright stars. A few days later (date B) the sun has shifted eastward along the ecliptic, and the stars appear farther from the sun at dawn.*

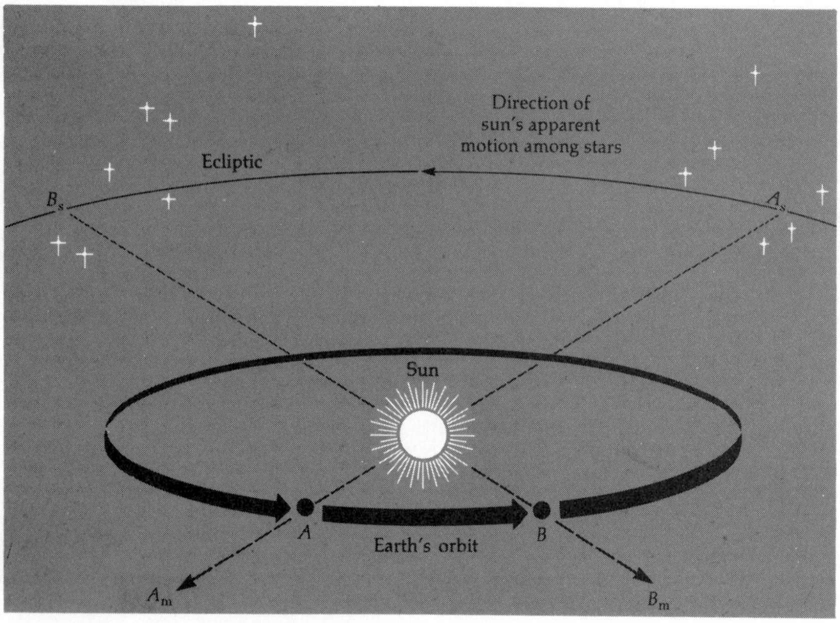

Figure 1-6. *Explanation of observations shown in Figure 1-5. As the earth moves in a few weeks from point A to point B in its orbit, the sun first appears among background stars A_s then against stars B_s. Similarly, the midnight sky, opposite the sun, first contains stars in direction A_m, and later those in direction B_m.*

Similarly, agriculture in Java was once scheduled according to the heliacal rising of Orion's belt. Australian aborigines begin their spring when the Pleiades, a prominent star cluster, rise in the evening sky. Papago Indians of Arizona also divided their agricultural year according to the Pleiades (Castetter and Bell, 1942):

Pleiades rising in summer, start planting;
At the zenith at dawn, too late to plant more;
Past the zenith, time for corn harvest;
One quarter down from the zenith, time for deer hunting;
Setting, time for harvest feast.

Origin of the Constellations

As memory aids for learning and locating stars, early observers named groups of stars for their resemblances to familiar animals, objects, and mythical characters—such as Scorpius (which really resembles a scorpion; see Figure 1-7), Leo (the lion), and Orion (the hunter). These groups are **constellations.** People in different cultures saw different patterns, often derived from their mythologies; for example, the ancient Chinese constellations differ from Western constellations. The star group called Ursa Major was seen as a bear in Europe, Asia, North America, and even ancient Egypt, where there are no bears; for this reason, Gingerich (1984) suggests the bear identification may go all the way back to ice-age Euro-Asia, from where it spread.

Most of our own familiar constellations came out of the Near East. As early as 3300 B.C., Leo and Taurus often appear in struggle on Mesopotamian artifacts. Star symbols sometimes shown on Leo's shoulders indicate that the artisans had the constellation in mind rather than a real lion. The struggle motif may refer to Leo chasing Taurus out of the sky: As Leo rises, Taurus sets. These images originated when the risings and settings of Leo and Taurus coincided with equinox and solstice dates. Later, they gained a mythical importance of their own and persisted in art up till A.D. 1200 (Hartner, 1965).

Many constellations are known from *historic* records at least as early as 420 B.C., but they evidently date from much earlier (Gingerich, 1984). The English astronomer Michael Ovenden (1966) did some astronomical sleuthing to find out who originally formalized most of our present constellations, when, and why. The most ancient constellations fill only the northern sky, leaving an empty zone around the south celestial pole.[3] This suggests that the ancient constellation makers couldn't see this southern zone from their northern latitude. As shown in Figure 1-8, for an observer at any given northern latitude $L°$, the sky is divided into three zones: the *north* **circumpolar zone,** $L°$ from the north celestial pole, containing stars that never set; the **equatorial zone**, containing stars that rise and set; and the *south* circumpolar zone, $L°$ from the south celestial pole, containing stars that never rise. The size of the south circumpolar zone is the clue to the northern latitude $L°$ of the ancient observers. (To see this, study Figures 1-8 and 1-9 together. Figure 1-10 shows the north circumpolar zone from a latitude of 20°.)[4] Using this and other clues, Ovenden obtained a latitude of $36\pm\frac{1}{2}°$ N for the ancient constellation makers.

He estimated the date of their activity by using a more complicated fact about the sky. Because of lunar and solar forces acting on the earth, the spinning earth slowly wobbles in space like a spinning top, and the axis thus points in different directions during different millennia. This wobble is called **precession.** Due to precession, the positions of the north and south celestial poles and celestial equator slowly drift among the stars, as seen in Figure 1-11. During the 26 000 years required for one such circuit, different stars become the North Star.[5] The ancient constellations show a rough symmetry around the star Alpha Draconis, about 25° from Polaris. Since this star was the north star around 2600 B.C., Ovenden concluded that most constellations were designed about 2600 ±800 B.C.

These findings clarify the purpose of the constellations. Many people have supposed the constellations were dreamed up by uneducated shepherds or sailors who amused themselves by finding images in random star groupings. Now it appears that the original constellations were more carefully designed to help people learn the positions of the celestial pole, equator, and coordinates of about 2600±800 B.C. The purpose may have been to help teach navigation. This idea accounts for the seemingly forced representations of some constellations. For example, Hydra, a peculiar, elongated constellation of 25 faint stars, lies accurately along the celestial equator of about 3000 B.C. Certain additional evidence suggests the designers were seafarers.

What seafaring civilizations lay near 36° N in the period 2600±800 B.C.? The Minoan sailors of Crete (35° N) are the most likely suspects. Vestiges of Minoan culture were transmitted to the Greeks and thence to us, consistent with this theory. In summary, many constellations may be Minoan creations handed down to us from around 2600 B.C., with still earlier elements incorporated into them. We should not assume that "it all started with the Greeks." We have received knowledge and traditions from even earlier societies!

As shown in the four star maps at the end of this book, different constellations are visible in the evening sky during different seasons. Today they no longer have mythical significance, but it is pleasant to greet them each year as old friends who signal the arrival of each new season.

Figure 1-7. *The ancient, sinister constellation of Scorpius gives a clue that the constellations originated in warm, low-latitude lands where the scorpion was known. The stars nicely outline the scorpion's claws (top) and its curved tail tipped by a double star as the stinger (bottom). The brightest object is the planet Jupiter, which happened to be passing through Scorpius when this picture was made in 1983. Its brilliance shows why it was named for the king of the gods. Glowing clouds (bottom) are masses of distant stars in the Milky Way. (Photo by author, 15-min exposure on commercially available 2475 recording film, 55-mm lens, f2.8, 35 mm camera.)*

The Seasons: Solstices, Equinoxes, and Their Applications

Most people think the sun rises in the east and sets in the west. This is roughly true, but it is only an approximation. As the

[3]Constellations now filling this southern zone, such as the Southern Cross and the Telescope, were not charted until European navigators sailed these waters in the 1600s and 1700s. In 1930 the International Astronomical Union revised the boundaries of all constellations for easier record keeping. Ancient names were kept, but the old, irregular boundaries were replaced by neater, geometric ones.

[4]To make such a photo yourself, see Project 11 at the end of this chapter.

[5]This effect of *precession* is detailed in the next chapter, where Figure 2-10 shows Alpha Draconis as the North Star in 2600 B.C.

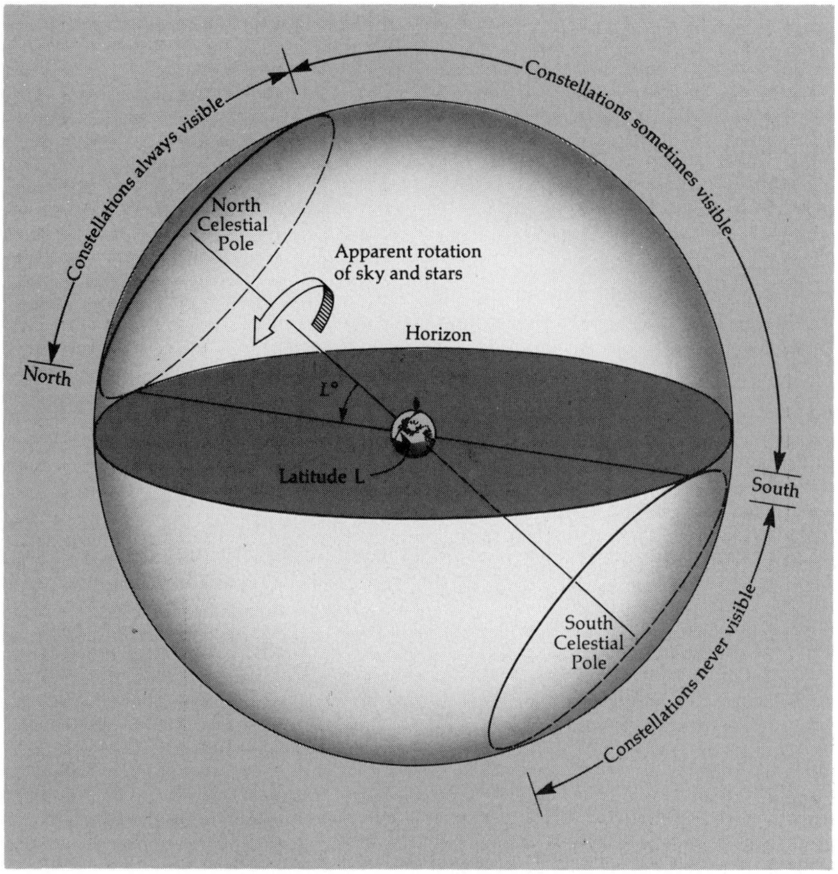

Figure 1-8. *For an observer at latitude L in the Northern Hemisphere, no stars and constellations within L° of the north celestial pole ever set. Stars in a middle zone rise and set, and a southern zone within L° of the south celestial pole is never seen.*

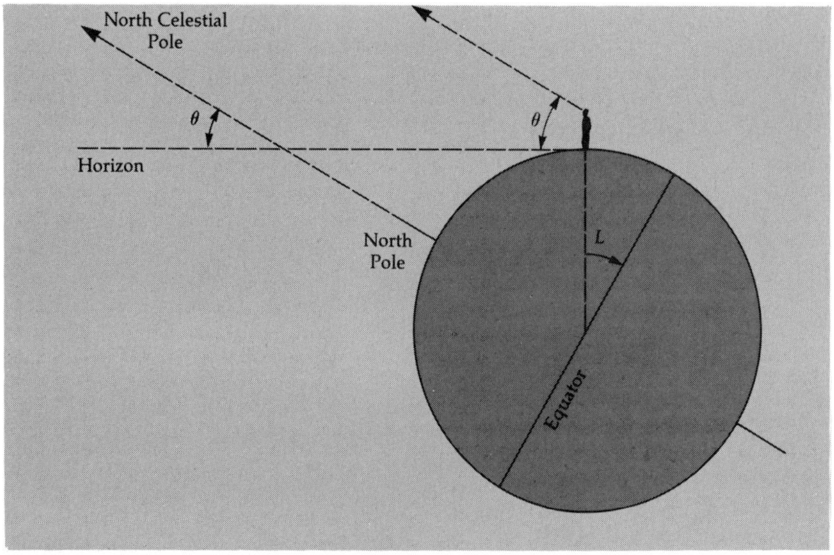

Figure 1-9. *How to determine your latitude astronomically. Angle L is the latitude, by definition. The elevation θ of the north celestial pole above the horizon equals the latitude L. To understand the principle better, try drawing the diagram for an observer at the equator (θ = L = 0°) and at the pole (θ = L = 90°). For convenience, keep the horizon line horizontal and the observer at the top of the globe.*

Figure 1-10. *Time exposure showing the circumpolar stars over Mauna Kea observatory in Hawaii, latitude 20°. Polaris is the bright star just left of the north celestial pole (center of star trails), which is 20° above the horizon. Stars within 20° of the pole never set. Bright stars in the right half of the picture form the Big Dipper; they are members of a cluster of stars relatively close to the solar system. (3.8-h exposure with wide-angle lens, f2.8, Tri-X film. Photo by author. Note wide-angle lens slightly distorts concentric paths of stars.)*

seasons progress, the sun rises and sets at different points on the horizon. Consider an observer in the United States or elsewhere in the Northern Hemisphere. In summer, the sun rises in the *north*east, passes high overhead at noon, and sets in the *north*west. In winter, the sun rises in the *south*east, passes low in the southern sky at noon, and sets in the *south*west. Only on two dates, the beginning of spring and the beginning of autumn, does the sun rise exactly due east and set exactly due west (see Figure 1-4, where such a path is shown).

The reasons for this have to do with the **cause of the seasons**: the tip of the earth's rotation axis by $23\frac{1}{2}°$ to the plane of the ecliptic, as seen in Figure 1-12. As the earth moves around the sun, the axis always stays pointed in the same direction—always toward the North Star (or, to be more precise, toward the north celestial pole). Thus, as seen in the figure, the Northern Hemisphere is tipped toward the sun for part of the year, and away from the sun during the other part of the year. When the Northern Hemisphere is tipped toward the sun, the sun rises and sets further around toward the northern part of the horizon; the sun also rises higher in the sky and shines more directly down on us; the weather grows warmer and we call the seasons spring and summer.

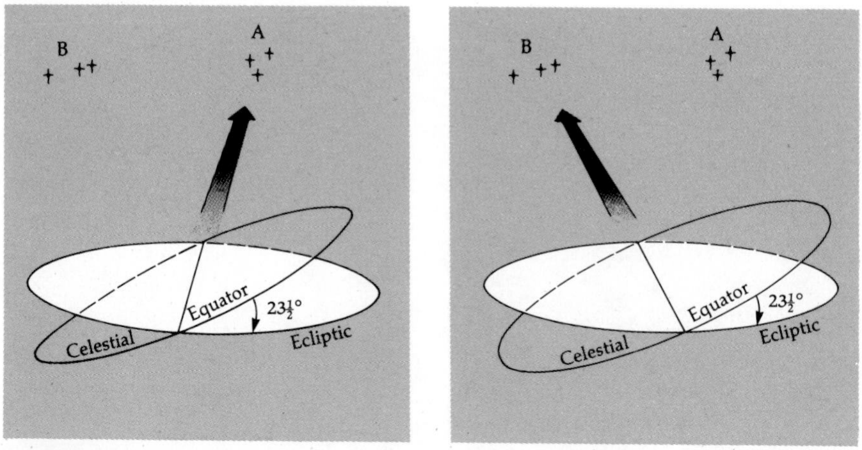

Figure 1-11. *The effect of precession, or rotation of the plane of the celestial equator. At different dates, A and B, turning the 26 000-y precessional cycle, the intersection of the celestial equator and ecliptic is seen in different positions, A and B, among the stars.*

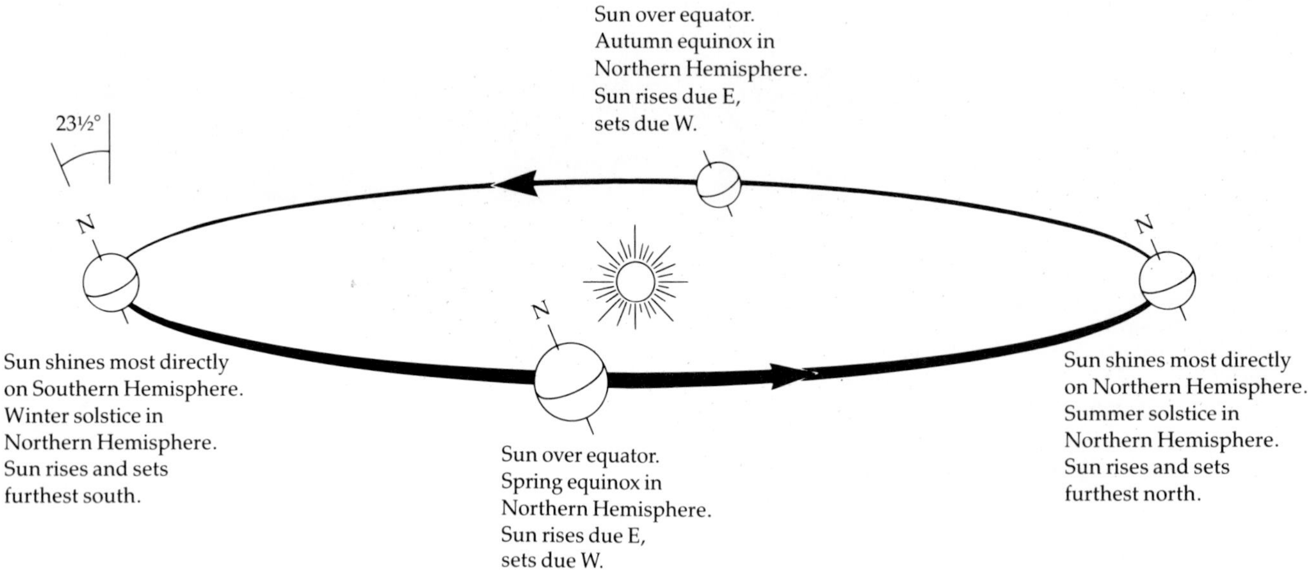

Figure 1-12. *The cause of the seasons. The earth's axis is tipped at $23\frac{1}{2}°$ to its orbit plane in a constant direction as it moves around the sun. Therefore the sun shines more on the Northern Hemisphere during one season (northern summer) and more on the Southern Hemisphere six months later (northern winter). See text for further discussion.*

Thus, there are four special days each year, called **equinoxes** and **solstices**:

Spring Equinox (First Day of Spring, About March 21). Sun crosses the celestial equator moving north.[6] Rises due east, sets due west. An important day to ancients in the Northern Hemisphere because it marked the return of the sun to "their" sky, bringing warmth. Also called the vernal equinox.

Summer Solstice (First Day of Summer, About June 22). Sun reaches point farthest north of celestial equator. Rises and sets farthest north. Most hours of daylight, ensuring warm weather. Most important day of year in many ancient northern calendars.

Autumn Equinox (First Day of Autumn, About September 23). Sun again crosses celestial equator, moving south. Colder weather on the way.

Winter Solstice (First Day of Winter, About December 22). Sun farthest south of celestial equator. Rises and sets farthest south. Fewest hours of daylight.

Part of this sequence can be seen in Figure 1-13, which is a series of photos looking west at sunset, and taken one month apart. In the first photos, around winter solstice, the sun sets in the southwest (left), but as we move toward summer, the sunset position slides to the north (right). We might call this cyclical shift in the sunset and sunrise positions the **solstice principle**.

ANCIENT APPLICATIONS OF THE SOLSTICE PRINCIPLE

These concepts were very important to primitive people because they had no dated newspapers arriving on their doorsteps and no calendars to hang on their walls. Observations of solstices and equinoxes offered a way to reset the calendar each year. For example, if you lived in an agricultural community, and tradition dictated that the best time for planting was mid-April, how would you know when mid-April had arrived? Priest-astronomers might have begun counting days at the time of spring equinox and might know that the optimum planting time was 24 days after equinox.

Several lines of evidence indicate that such ideas were practiced. For example, Spanish explorers recorded that Peruvian Inca Indians determined the date of solstice in this way and conducted ceremonies at that time, sending out runners to spread the word through the Inca empire of the beginning of a new year. Similarly, Pueblo Indians in the Southwest observed sunrise and sunset positions to determine the solstice or equinox dates.

But one of the most extraordinary lines of evidence reaches back much earlier in time. Around 2800–2200 B.C. at **Stonehenge**, in England, prehistoric builders constructed a large circular embankment with a broad avenue leading outward about $\frac{1}{2}$ km ($\frac{1}{4}$ mi) directly toward the horizon position of summer solstice sunrise, as mapped in Figure 1-14. Subsequently, as late as 2100 to 1500 B.C., additional huge stones were brought in to construct a monument in the center of the ring. These included 30- to 50-ton stone blocks brought as far as 30 km (20 mi) and some 5-ton stones hauled about 380 km (240 mi). These stones and the outer ring can be seen in Figure 1-15. A large stone was placed in the avenue in such a way that an observer at the center would see the sun rise over the stone *on the day of summer solstice* as shown in Figure 1-16 and Color Plate 1. Archaeologists have found additional post holes near this stone, suggesting that a grid of wooden posts might have allowed "fine-tuning" of the observations to determine on which day the sun rose furthest north. As shown in the map in Figure 1-14, other astronomical alignments have also been found in Stonehenge, suggesting that the builders may have been much more sophisticated, keeping track of motions of the moon as well as the sun, and possibly predicting eclipses!

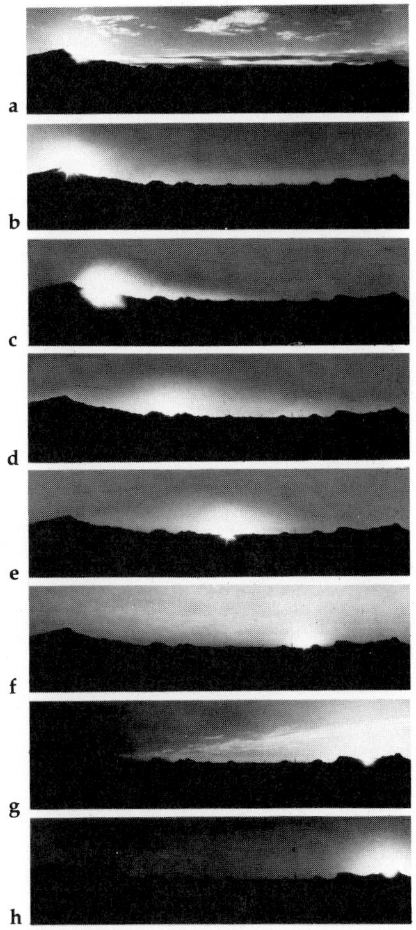

Figure 1-13. *Movement of the sunset position along the horizon during an 8-mo period. Moving south in (a), the sun reaches winter solstice in (b). Here the setting position reverses direction and proceeds north again in (c), reaching the vernal (spring) equinox in (e). Summer solstice is reached in (h). The author's former home, from which these pictures were made, happened to occupy the focus of a "natural Stonehenge," where the two solstice positions were marked by prominent mountain peaks. Such situations may have triggered the prehistoric idea of building solstice markers to reset calendars twice a year. (Photos by author.)*

[6] As a memory aid, recall that the sun crosses the *equ*ator on the *equ*inox.

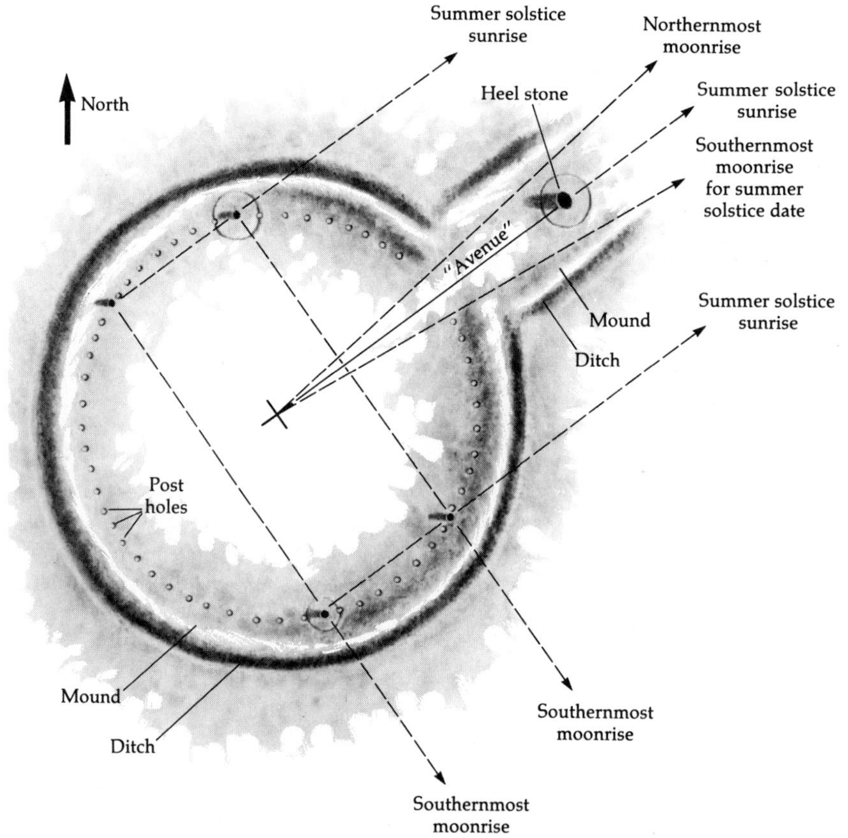

Figure 1-14. *Aerial plan of the original Stonehenge construction around 2500 B.C., consisting of an outer ditch and mound, an avenue toward summer solstice sunrise, and a ring of posts or stones. Selected moonrises could also be observed.*

Of course, we don't know whether Stonehenge was used like a practical observatory to *measure* dates of astronomical events like summer solstice, or more as a ceremonial temple where, perhaps, solstice-related rituals were held. There is some evidence that the earliest builders had precise observing in mind, and that builders in later centuries were more interested in elaborate ritual, because the huge stones that were among the latest additions seem to show less clear astronomical alignment. Perhaps they were added just to enhance a sense of mystic magnificence to what had become a quasi-religious temple. In any case, more than 4000 years ago, the builders of Stonehenge had a strong awareness of the solstices.

The fact that Stonehenge is designed around a solstice alignment was pointed out as long ago as 1740 by English scholar Dr. W. Stukely. Yet in the 1880s and 1890s, when American astronomer Samuel Langley and English astronomer J. N. Lockyer spoke of Stonehenge as an astronomical observatory, most scholars thought ancient people could not have been sophisticated enough to orient large structures this way. Today we have much more evidence that early civilizations did make sophisticated use of the solstice principle and other astronomical knowledge.

For example, one of the largest temples ever built, the Temple of Amon-Re in Karnak, Egypt, built around 1400 B.C., covered twice the area of St. Peter's in Rome. Its central hallway, about 370 m (1200 ft) long, was aligned within $\frac{1}{2}°$ of the summer solstice sunset position (Figure 1-17). Other Egyptian temples had similar orientations, whereas still others were oriented toward heliacal risings and settings of certain stars. Because of precession, the rising and setting points of stars shift slowly over the centuries. Some of these temples show evidence of rebuilding as if to correct for such shifts, even though the rebuilding often ruined the symmetry of the temple by changing the angle of the central hall. The Egyptians are known to have made astronomical observations: Chicago's Oriental Institute has part of a device for charting star positions that was made by the young Pharaoh Tutankhamen (the famous "King Tut") around 1350 B.C.

Archaeologists have recently published a flood of additional analyses of temples throughout the world with suspected astronomical alignments. Among these are temples of Angkor Wat, Cambodia (c. A.D. 1150); Tihuanaco, Bolivia (A.D. 375–725); the Caracol, in Yucatan (c. A.D. 1000); Casa Grande, Arizona (c. 1350); Cuzco, Peru (c. 1540); and Medicine Wheel,

Figure 1-15. *A wide-angle view of Stonehenge. The largest stone structures (right) are among the latest (2000 B.C. or later). The fence at left marks an earlier earth bank surrounding the structure, and the white spots mark post or stone holes; these features and an associated avenue leading toward summer solstice sunrise (far side of monument) are believed to date from about 2500 B.C. (Photo by author.)*

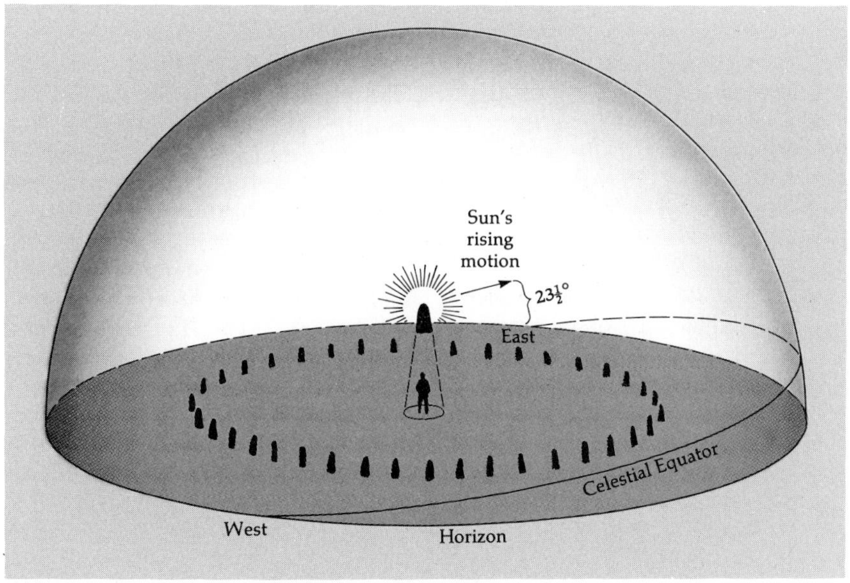

Figure 1-16. *The Stonehenge principle. The sun's position at solstice is $23\frac{1}{2}°$ from the celestial equator. The rising (or setting) position on this date can be marked by a prominent pillar. Other markers in a ring could allow observation of other selected risings and settings during the year.*

Figure 1-17. ***The ruins of the long hall in the Egyptian temple of Amon-Re, built around 1400 B.C. Nineteenth-century astronomer-archaeologist Norman Lockyer used this illustration when he pointed out in 1894 that this main hall is aligned toward the summer solstice sunset position.***

Wyoming (1500–1760). Evidence includes special alignments or window designs that mark solstices or other events and legends or historical records of astronomical observations or ceremonies being carried out at the sites. Color Plate 1 shows a view through a peculiar cylindrical window at the Casa Grande ruin, apparently carefully designed so that the setting sun was framed in a diagonal view only at summer solstice. Pueblo Indians of the Southwest still used a similar system in the 1800s. Evidence of this sort indicates how numerous astronomical traditions sprang up independently around the world.

Even Christian cathedrals may have inherited some of these traditions from the ancient Middle East. The old (330–1506) and new (1506–) cathedrals of St. Peter in Rome were reportedly oriented so accurately to the east that on the morning of the vernal equinox "the great doors were thrown open and as the sun rose, its rays passed through the outer door, then through the inner door, and penetrating straight through the nave, illuminated the High Altar" (Lockyer, 1894). Certain other Christian cathedrals and missions were oriented toward sunrise on the day of the saint to whom they were dedicated. Ghostly fingers of ancient practices reach into the present to touch us!

MODERN APPLICATIONS OF THE SOLSTICE PRINCIPLE

A totally different application of the solstice principle is appearing in modern times. As fossil energy supplies dwindle, we are more concerned about utilizing solar energy in our buildings. The cheapest way is by clever building orientation and roof overhang, to let sunlight in during winter and keep it out during summer.[7] In a sunny climate, most of the heat input through windows and walls comes in afternoon and morning when the sun is low and strikes these surfaces full on. (Midday heat input through roofs must be minimized by thick insulation.) As shown in Figure 1-18, house design can utilize the solstice principle. Southeast- and southwest-facing windows, for example, can be unshaded to provide "free" warmth around winter solstice, and northwest-facing windows can be covered with external shields in summer to prevent late afternoon sunlight from pouring unwanted heat into the house.

Astrology: Ancient Origins of a Superstition

If you lived in a society where the calendar was determined by observations of the stars, where priest-astronomers counted days from the summer solstice and studied heliacal risings in order to announce when to plant crops, and where the death of a king was remembered as happening in "the year when Mars passed through the constellation of Gemini," you could see how easy it might be to make a serious philosophical error. Instead of realizing that the stars offered a practical way to date and coordinate human affairs, you might come to believe that the stars *control* or at least influence human affairs. This belief is called **astrology.** We call this belief a superstition for the same reason that we say it is superstitious to believe in invisible fairies in your garden: No experimental evidence supports the belief, and predictions based on the belief have no better than random accuracy.

An occupational hazard of being an astronomer is to be introduced at parties as an astrologer. This mistake is annoying because astrology is not part of modern astronomy, but a pseudoscience associated with astronomy as it was practiced 3000 years ago.

The pseudoscientific nature of astrology can be understood best by exploring astrology's roots. Astrology can be traced back at least 3000 years, when it flourished with other ancient magical beliefs. A common form of ancient magic was to associate patterns in nature with patterns of human events. Most people today would scoff at having their futures read from flight patterns of migrating birds or patterns in the bloody intestines of freshly sacrificed animals. Yet these were popular forms of divination in Babylon and Rome when astrology was flourishing, and they have the *same basic logic.* In astrology, the pattern in nature is that of the planets and the imaginary figures of the constellations.[8]

Archaeological research suggests that early people correctly used patterns in the sky in a practical way to foretell *natural*

[7]This is called passive solar utilization because it requires no further machinery or energy consumption. Active solar utilization implies further machines, such as fans to distribute warm air once it is heated by the sun.

[8]An interesting example of this all-too-easy mental connection—between changing patterns in the sky and mystic patterns of human relationships—was given by French diarist Anaïs Nin during a 1935 visit to a planetarium. A practicing psychoanalyst, she had been thinking of the "human tangles" of her patients. Her diary records: "The planetarium . . . restored to me

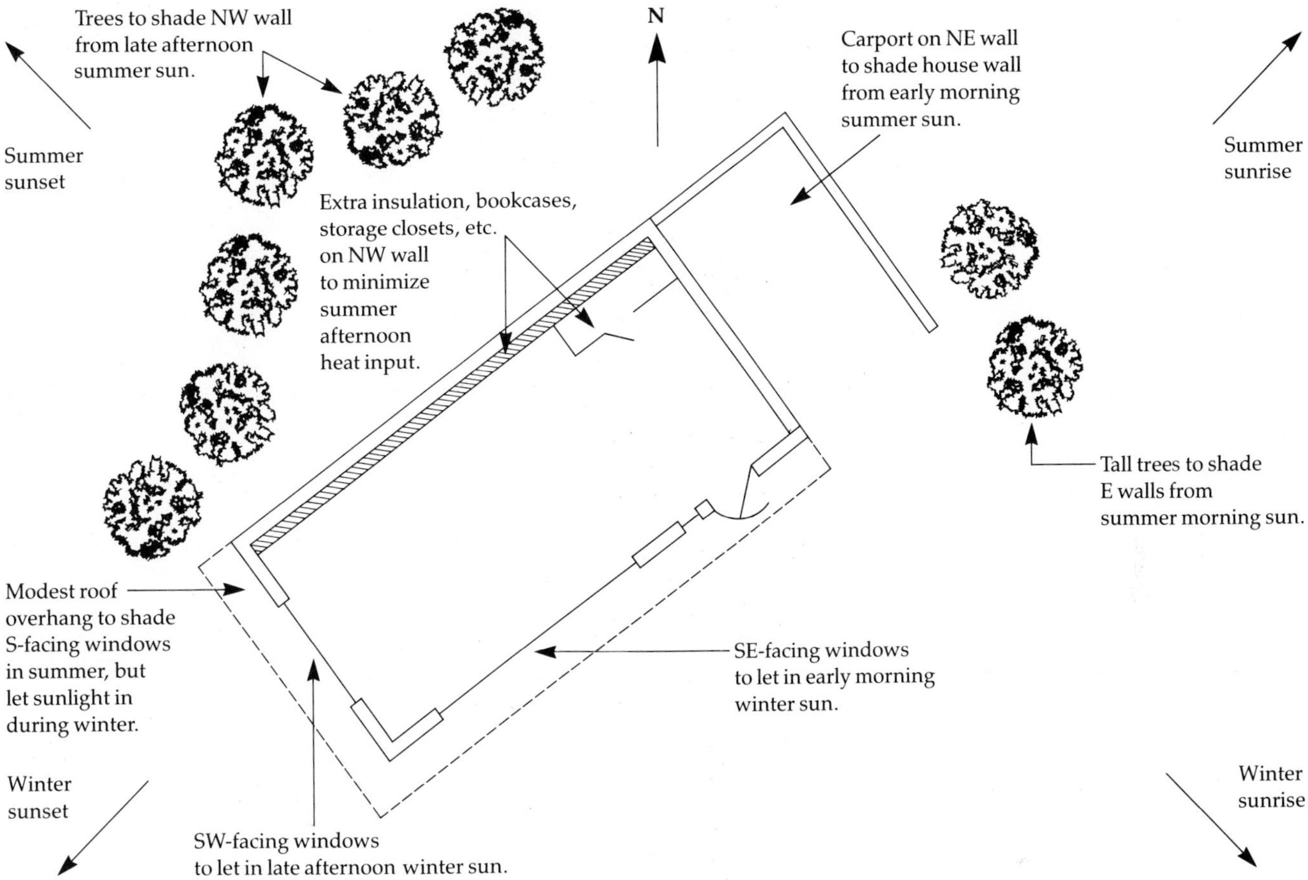

Figure 1-18. *The solstice principle as applied to environmental architecture and landscaping. Northeast and northwest walls are shaded to reduce heat from the summer morning and afternoon sun. Southern windows allow sunshine to enter and heat the house in winter.*

events, such as the change in seasons. Egyptian priest-astronomers, as we saw, designed their calendar to begin with the heliacal rising of Sirius, which was also the date of the Nile's flooding. Observation of the heliacal rising would thus help predict the time of the flood, an important prediction for agricultural planning.

When priest-astronomers were observing signs in the sky to calibrate the calendar or determine when to plant crops, it must have been easy for the "man in the street" to conclude mistakenly that celestial signs forecast *human* events.

Human minds are quick to assume that if one event follows another, the second is caused by the first. Sometimes this is true and sometimes not. Centuries ago, logicians identified and named this error with the Latin phrase *"post hoc, ergo propter hoc"* ("following this, therefore *because* of this"). Yet this type of error is still common today. Historical evidence suggests that astrology grew out of this error. For example, if a king died a few days after an eclipse, some early observers reasoned that an eclipse causes or foretells the death of a king.

There is some evidence to suggest that early astronomical knowledge later degenerated into myth and pseudoscience. For example, the best astronomical alignments at Stonehenge were probably those constructed first, around 2500 B.C. Later construction, around 1800 B.C., muddled the design. In the nineteenth and twentieth centuries we see the monument used for mystical rites by costumed Druids and self-proclaimed witches. Here is an example of the loss of knowledge of original astronomical purpose.

In the same way, records of astronomical events, originally gathered by accurate observation, may have degenerated as later interpreters added mystical interpretations of astronomical events to accounts of historical incidents, describing them as if one controlled the other.

a sense of space and I could detach myself from the haunting patients. . . . I saw, instead of stars, relationships moving like constellations, moving away and towards each other . . . turning, moving, according to an invisible design, according to influences we have not yet been able to measure, analyze, contain."

HISTORICAL DEVELOPMENT OF ASTROLOGY

Striking examples of this error can be found in ancient historical records. A Babylonian text of about 1600 B.C. describes motions of Venus and historical events together, assuming that the same events will occur again in the future whenever Venus presents the same patterns (paraphrased from Pannekoek, 1961):

If in the month of Abu, on the sixth day, Venus appears in the east, rains will be in the heavens and there will be devastation. Venus remains in the east until the tenth day of Nisannu and disappears on the eleventh. Three months she is not seen. On the eleventh of Duzu, Venus flares up in the west. Hostility will be in the land; the crops will prosper.

Similarly, a letter of an Assyrian astrologer to his king in 668 B.C. shows the same misconception still current a thousand years later (Pannekoek, 1961):

When Jupiter appears in the third month the land will be devastated and corn will be expensive. . . . When Jupiter enters Orion the gods will devour the land.

Mesopotamian astrology became a popular fad in the Greek world. Around 350 B.C. a Greek philosophical school, called *catasterism,* identified individual stars with certain of the humanlike Greek gods. Of course, implicit in the concept of a god is the idea that divine will prevails over human will. So it was a short step to the idea that the stars determine the fate of humans. The practice of forecasting personal fortunes spread from heads of state to sophisticated Greek households seeking diversion and finally to commoners.

Astrology also spread rapidly eastward. In a Chinese example as recent as 1882, an unsuccessful astrological attempt to stave off the dire consequence of a comet is blamed on government bungling (Stephenson and Clark, 1977):

When a [comet] was seen last year, an imperial decree was written to the palace and court officers, ordering them to perform their respective duties conscientiously. In . . . this month, the comet was seen again in the southeast. This must be due to the frequent mistakes committed by those employed in the administration.

In medieval times, astrology was banned by the Church as magic, yet it persisted, deeply ingrained in people's thinking. For example, historian Owen Gingerich (1967) documents how astrology was apparently used to pick dates for laying cornerstones of palaces and churches, and how a palace built in 1563 was oriented toward sunset on the date of the battle it commemorated. These traditions suggest a holdover from the ancient patterns of astronomical temple orientation at Stonehenge and in Egypt, but with the original purpose forgotten and replaced by ritual.

Since the Renaissance, astrology has continued to flourish alongside scientific thought. Today the ancient magic of 3000 years ago arrives on our doorsteps in the astrology columns of our daily papers!

PROBLEMS WITH ASTROLOGY

A test of astrology is whether it is consistent with observations. The basic practice is to forecast the influences for a particular day by casting a *horoscope*. The horoscope is usually presented as a circular chart, showing the positions of the planets, sun, and moon with respect to the constellation signs at the moment being studied. For example, since Venus was the goddess of love, prominence of Venus might indicate a loving influence. The prominence of the war god Mars might suggest an aggressive influence. This technique follows the rules laid down by astrologers centuries ago.

But application of the ancient rules creates an embarrassing problem for astrologers today. Astrologers nearly 4000 years ago designated 12 signs, or portions of the zodiac, corresponding to the 12 zodiacal constellations. Like the planetary gods and goddesses, each constellation and its sign were said to have a certain personality trait. In 1867 B.C., the point where the sun crossed the celestial equator on March 21 (spring equinox) and entered the northern hemisphere was in the constellation Aries and was called the "first point of Aries." The twelve signs were measured from this point. Thus a person born between March 21 and April 19 was said to be an "Aries," since the sun lay in this constellation and thus in this sign. The original astrologers of that era thought this pattern was fixed. But around 130 B.C., Hipparchus discovered precession, which shifts the first point of Aries relative to the constellations. Thus the signs no longer correspond to their respective constellations!

Astrologers during Ptolemy's time tried to patch up the scheme by claiming that the signs were more important than the actual constellations. They defined the signs as 30° intervals measured from the March equinox point. Astrologers still claim that a person born between March 21 and April 19 is an Aries, but the sun during most of this period is really in Pisces! Thus astrology has become removed from the realities of the sky.

Another embarrassment is that we know know that the stars move independently; thousands of years from now the constellation patterns themselves will have changed. Although one could patch up astrology further by inventing new constellations and associated psychological characteristics to cover this problem, the discrepancy shows that astrology as it is now practiced is not consistent with observed reality.

Another example will arise in the near future. When astronauts land on Mars, they will see no war god Mars in the sky. They will be standing on him! Instead, they will see a new planet, Earth, crossing the zodiac. Will astrologers invent a new personality influence associated with Earth's position in the Martian sky when the first baby is born on Mars? The old astrological rules will hardly apply!

DOES ASTROLOGY WORK?

A more scientific test of astrology is to examine the success of its predictions. It has failed in many spectacular cases. For example, a grouping of all known planets in Libra in 1186 led astrologers to predict a disastrous hurricane, because Libra was associated astrologically with the wind. People dug storm cellars in Germany; the archbishop of Canterbury ordered fasting; the palace in Byzantium was walled up; people fled to caves in the Near East. But the conjunction of planets passed without incident. Similarly, in 1524, all known planets clustered in Aquarius, the Water Bearer. This time astrologers predicted a second Deluge, but the month of the conjunction passed without disaster and was reportedly drier than usual (Ashbrook, 1973).

Some statistical studies of astrological predictions have been reported to see if these predictions are more successful than random predictions. Jerome (1975) reviewed several of these studies and concluded "legitimate statistical studies of astrology have found absolutely no correlation between the positions and motions of the celestial bodies and the lives of men."

Still another test would be to search for forces exerted by planets or stars by which astrological influences could be manifested. Some modern apologists for astrology argue that these forces might be like gravity. So far, no such mysterious forces have been detected. Furthermore, we can calculate that the gravitational influences of nearby objects on a newborn baby, such as that of the mother during birth, are greater than the gravitational influences of the planets!

WHAT SHOULD WE DO?

Because its logic is identical to that of ancient magic, because it is inconsistent with the twentieth-century sky, and because it fails to predict accurately, we conclude that astrology does not warrant our belief or our attention in planning our lives.

We might just smile at astrology as another amusing human foible if it weren't for the nagging suspicion that it threatens healthy civilization. For example, a 1975 Gallup poll indicated that about 12% of all Americans take astrology quite seriously. In an age of space exploration and serious nuclear threats, do we really want ancient magic to be an influence in human affairs? It is sad to realize that millions of people seeking advice about real problems in their lives spend their (sometimes meager) money on the products of astrologers and pseudopious con artists. As astronomer Bart Bok (1975) pointed out, "The astrologer can refine his interpretations to any desired extent—the end product becoming increasingly more expensive as further items are added." Nowadays, computers are used to calculate planetary positions while casting horoscopes, giving a scientific gloss that misleads those who confuse scientific tools with scientific method. Thus, naive members of the public are fooled into thinking that astrology has some scientific basis.

Philosophically, the continuing practice of astrology helps remind us that the human mind has an amazing capacity to come up with ideas—many of which are wrong—and then to develop intricate systems of thought based on those ideas without ever establishing that the ideas are correct.[9]

Essay A, in the back of this book, describes further examples of pseudoscience that are merchandised as scientific truth in books and tabloids at supermarket checkout stands. Essay A also discusses in more detail how to distinguish science from pseudoscience, not through the hypotheses themselves, but rather through the way *evidence* is used to confirm or refute the hypothesis.

If we agree that astrology is little more than ancient superstition that wastes people's money and mental energy, should we try to suppress it? Newspapers routinely reject suggestions that they drop astrology columns, which they claim are harmless entertainment that promotes sales. Of course, most advice in astrology columns is innocuous. It might be said of astrology, like fairy stories, that if you believe it hard enough, it will work. For example, if you read morning advice to avoid frivilous expenses today, you may indeed be more prudent than usual. But one would hope we could find better sources of inspiration than ancient magic that is pseudoscience at best and fraudulent waste at worst. In a free country the answer is not suppression. In a free country with a free press, we hope that educated, pragmatic citizens will pick and enjoy the best from among competing published materials, and look for evidence of accuracy before committing themselves to particular systems of belief.

[9] I once met an industrious young man of no great income who had amassed a library of books on arcane details of various authors' astrological systems. His goal was to make more accurate astrological forecasts to guide his own life and to perfect a gambling system he could apply in Las Vegas. He had once had a string of successful bets in a football pool and was convinced that a fatal bet in which he lost all his winnings was merely the result of an erroneous reading of the horoscope. As a student I rented a room from an ill, elderly woman of rather low income. I was aghast to discover that she had accumulated a substantial library of books and tracts purchased from a radio and TV evangelist. The books included drawings of various horned and tailed "demons," which, according to the books, were responsible for various illnesses. It was depressing to see the energy and money she needed for life being squandered in this way.

Eclipses: Occasions for Awe

So far we have treated the moon and the sun as independent bodies that could be tracked for practical purposes. Cycles of the moon divide the year into about 12 equal months; solstices and equinoxes, defined by sunrise positions, divide the year into four seasons. Ancient people who discovered these facts may have gained a sense of well-being: At least *some* things in their environment could be counted on from year to year. Calendars could be made and agriculture regulated. The predictable, friendly sun could be worshiped as a deity, always providing light and warmth.

What, then, could be made of **eclipses**, the sporadic occasions when the sun or more commonly the moon disappears while above the horizon? In the few minutes of a total solar eclipse the sky turns dark, stars can be seen in daytime, and an uncanny chill and gloom settle over the land, as indicated in Figures 1-19 through 1-22. During the hours of a total lunar eclipse, the moon turns blood-red. Small wonder that the Greek root of eclipse, *ekleipsis,* means abandonment, or that the ancient Chinese pictured a solar eclipse as a dragon trying to devour the sun. Eclipses must have seemed an unpredictable menace to the scheme of things.

We know eclipses awed early people. In Greece, the poet Archilochus, observing the solar eclipse of April 6, 648 B.C., was moved to write: "Nothing can be sworn impossible . . . since Zeus, father of the Olympians, made night from mid-day, hiding the light of the shining sun, and sore fear came upon men." Herodotus reports that a war between the Lydians and the Medes stopped when an eclipse surprised the competing armies during a battle in the 580s B.C. A hasty peace was cemented by a double marriage of couples from the opposing camps.

Knowledge of what causes eclipses allayed people's fears of them. And if people knew how to predict eclipses they could become very powerful. For instance, in February 1504, Christopher Columbus was having trouble convincing Jamaicans that they were obliged to feed him and his crew. Seeing that his almanac foretold an eclipse of the moon on February 24, Columbus warned the Indians that the Christian god would punish them by turning the moon to blood. The scoffing natives changed their attitude when the eclipse began on schedule.

Figure 1-19. *The March 1970 total solar eclipse as seen from the village of Atatlan, Mexico. (Photo by author.)*

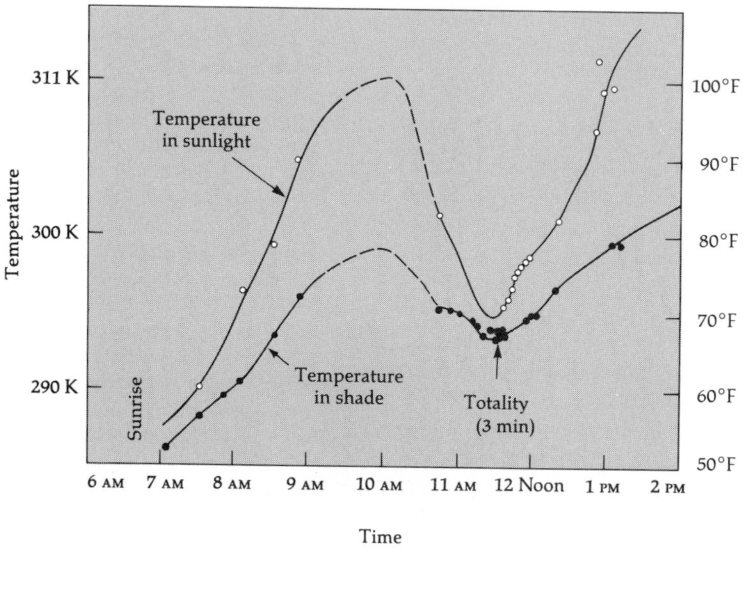

Figure 1-20. *Temperatures recorded during the total eclipse of the sun in Mexico, March 1970. The peculiar cooling would have been just one of many awe-inspiring effects of eclipses experienced by ancient witnesses.*

***Figure 1-21.** The totally eclipsed sun photographed with an 11-cm (4.4-in.) aperture telescope–filter assembly specially designed to enhance the glowing hot gas of the sun's outer atmosphere, or corona. The corona is visible to the naked eye only during eclipses. The planet Venus appears nearby, being on the far side of the sun during this particular eclipse. This eclipse was photographed at 4 km (13 000 ft) altitude in the Bolivian Andes in November 1966. (National Center for Atmospheric Research.)*

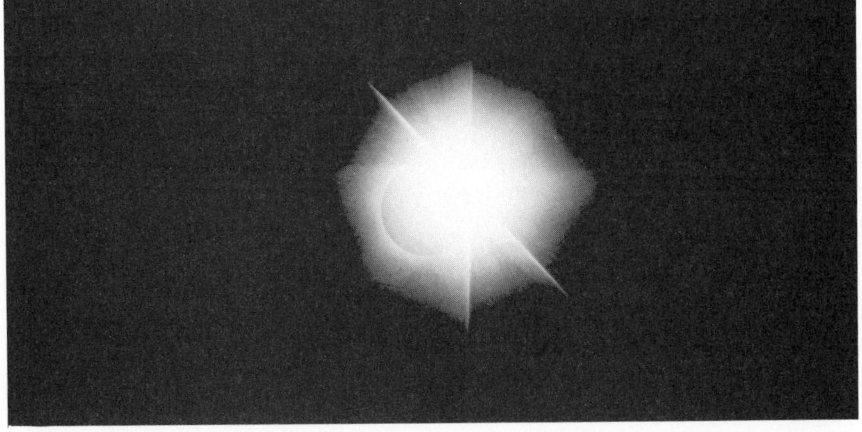

***Figure 1-22.** The "diamond-ring effect" occurs when the first or last brilliant solar rays pass through mountain valleys at the edge of the moon as the moon moves across the sun's face. Rays emanating from the diamond are due to optical effects in the 35-mm camera used to record this view in March 1970. (Photo by author.)*

This event may have inspired Mark Twain, whose Connecticut Yankee established his authority in King Arthur's court by predicting a total solar eclipse and then "commanding" the eclipse to end. This imaginative scene suggests the political prestige gained by some of the first priest-astronomers who discovered how to predict eclipses.[10]

In addition to providing us with interesting historical reading, the ancient astronomers who recorded eclipses centuries ago provided another valuable service: their records are used today to help analyze a slight increase in the length of the day, at a variable rate of about 0.002 sec per century (Stephenson, 1982).

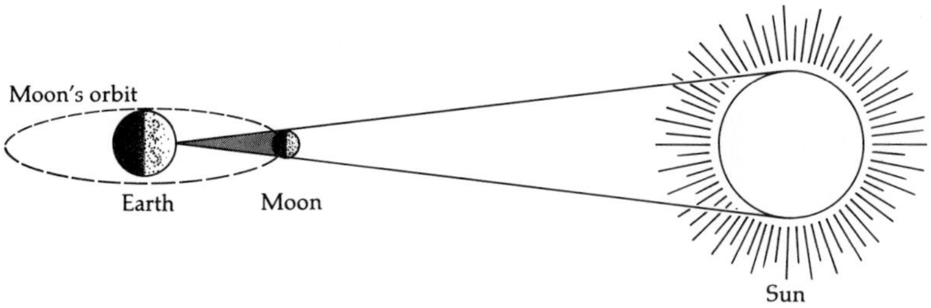

Figure 1-23. *Geometry of an eclipse of the sun (not to scale). The shadow of the moon falls on the earth.*

CAUSE AND PREDICTION OF ECLIPSES

Whatever their emotions and motives, some early people learned how to predict eclipses—and this was an early step toward recognizing that we live on a world among worlds orbiting in space.

To understand how eclipses can be predicted, we should first understand the **causes of eclipses.**

Two important things to remember about eclipses are: (1) They happen when the shadow of one celestial body falls on another, and (2) what you see during an eclipse depends on your position with respect to the shadow. If you see an eclipse at all, you are either in the shadow (a celestial body has come between you and the source of light) or you are looking at the shadow from a distance as it falls on some surface.

[10]Even animals are affected by eclipses. I have seen roosters crowing and birds flocking to trees at midday because of the gloom during a solar eclipse, causing the creatures to believe evening has fallen. There is a story about an eclipse in Colorado during which an astronomer arrived too late to get a prime observing site. He hurriedly set up his equipment in an empty chicken coop to protect his instruments from the wind, and then spent most of the eclipse trying to shoo away the chickens, who dutifully reported to the roost when darkness fell.

Earthbound observers see two types of eclipses: solar and lunar. **Solar eclipses** occur when the moon, on its $29\frac{1}{2}$-d journey around the earth, happens to pass between the earth and the sun as shown in Figure 1-23. A **total solar eclipse** occurs if the moon completely covers the sun as seen by an earthbound observer. A **partial solar eclipse**, illustrated in Figure 1-24, occurs if the moon is "off center" and covers only part of the sun. By coincidence, the moon happens to have the same angular size as the sun—about $\frac{1}{2}°$, or a little less than the angular size of a little fingernail at arm's length.[11] Therefore, the moon usually just covers the sun during a solar eclipse. But if the moon happens to be at the farthest point in its orbit, it has a smaller angular size than usual and doesn't quite cover the sun. This causes an **annular solar eclipse**, in which the observer sees a ring (Latin *annulus*) of light, which is the rim of the sun, surrounding the moon.

Lunar eclipses occur when the moon, on its journey around the earth, passes through a point exactly on the opposite side of the earth from the sun. This point lies in the shadow cast by the earth (Figure 1-25). It takes the moon a few hours to pass through the earth's shadow, during which time the moon turns an astonishing copper-red from sunlight passing through the earth's atmosphere.

[11]If this concept is unclear, see the discussion of the small-angle equation in Optional Basic Equation I in Chapter 2.

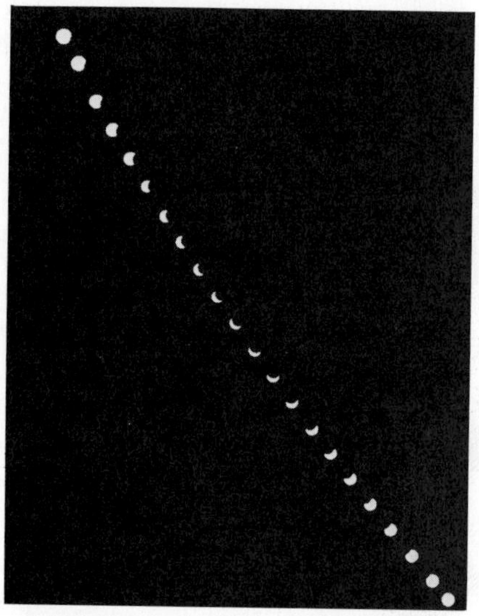

Figure 1-24. *Multiple exposure showing the sun's movement across the sky during a partial solar eclipse on July 10, 1972. At maximum, 72% of the sun was covered by the moon. Photos were made 6 min apart with a fixed 35-mm camera and a dark filter. (NASA photo by A. K. Stober.)*

Eclipses can happen in other parts of the solar system. For example, Figure 1-26 is a photo of a solar eclipse as seen from the moon's surface, and Figure 1-27 shows an eclipse on Mars.

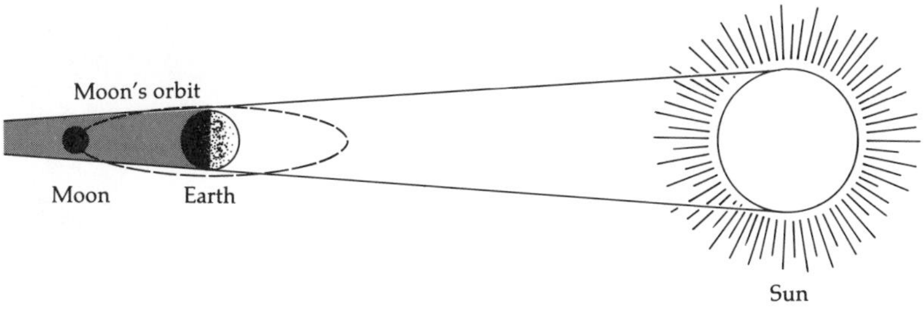

Figure 1-25. *Geometry of an eclipse of the moon (not to scale). The moon passes through the earth's shadow.*

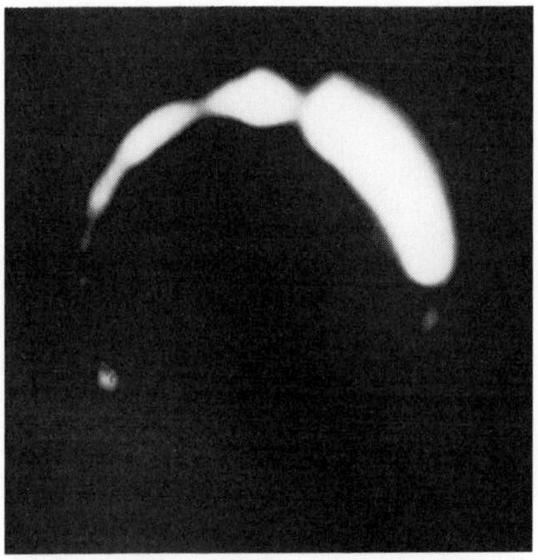

Figure 1-26. *An eclipse of the sun by the earth, photographed from the surface of the moon. This television image, made by Surveyor III in 1967, shows a ring of light that is the earth's atmosphere, back-lit by the sun, which lies behind the black disk of the earth. The brightest light is coming through the atmosphere over eastern Asia; local clouds affect the amount of light transmitted. (NASA.)*

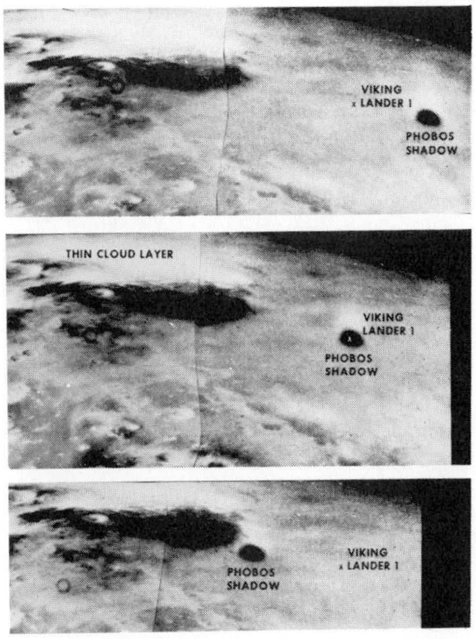

Figure 1-27. *Oblique orbital views of an eclipse happening on the surface of Mars. The penumbral shadow of the small Martian moon Phobos moves from right to left. The shadow, about 90 km wide, passed over the site of the Viking 1 lander, helping scientists locate the lander's position. (NASA Viking 1 orbiter photo.)*

UMBRAL AND PENUMBRAL SHADOWS

In solar and lunar eclipses, the shadows of the earth and the moon are not sharply defined. Because the sun as seen from the earth has an angular size of $\frac{1}{2}°$, sunlight is slightly diffuse; its rays come from slightly different directions.

Thus, shadows—including eclipse shadows—cast by sunlit objects near the earth are not sharply defined. The inner, "core" area of a shadow—the part that receives no light at all—is named from the Latin word for shadow, **umbra**. The outer, fuzzy boundary is called the **penumbra.** An observer in the umbra sees the sun entirely obscured; an observer in the penumbra sees the sun only partly obscured. Figure 1-28 shows some varieties of solar eclipses, illustrating these principles. If you hold your hand at eye level above smooth ground in sunlight, its shadow will have a central, dark umbra and a penumbra about a centimeter wide. If you hold your hand high and spread your fingers, their shadows will be indistinct. An ant in the umbra would see a total solar eclipse; an ant in the penumbra would see a partial solar eclipse.

The relative sizes of umbra and penumbra depend on the distance between the shadow-casting body and the surface on which the shadow appears. During a lunar eclipse, shown in Figure 1-29, the earth's shadow on the moon has an umbra several times the moon's diameter, and a considerably larger penumbra. Total lunar eclipses, with the moon in the umbra, can last up to $1\frac{3}{4}$ h.

On the other hand, the moon's umbral shadow on the earth does not exceed a 267-km (167-mi) diameter. Due to the motion of this shadow across the earth, total solar eclipses cannot last more than $7\frac{1}{2}$ min. In an annular eclipse, the moon is too far away to produce any umbral shadow, and the eclipse has no true total phase, as seen in Figure 1-28.

FREQUENCY OF ECLIPSES

To be able to predict eclipses, we must understand the intervals between them. Discovery of these intervals several thousand years ago marks the beginning of our ability to predict seemingly mysterious celestial events. To understand the technique, imagine a viewpoint in space. If the moon's orbit around the earth lay exactly on the plane of the earth's orbit around the sun, the moon would pass exactly between the earth and the sun every time around, as can be visualized from Figures 1-23 and 1-25. It would also enter the earth's shadow on every pass, and a solar and a lunar eclipse would occur every month.

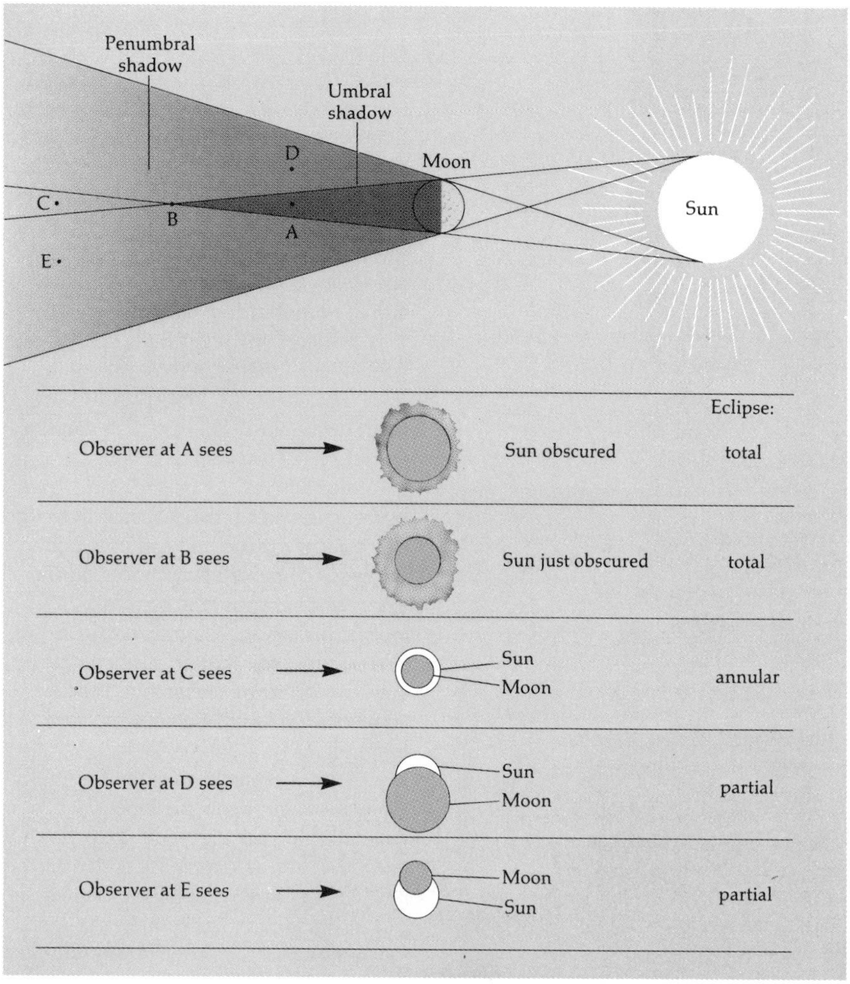

Figure 1-28. *Geometry and configurations of solar eclipses by the moon. Top diagram shows umbral and penumbral shadows of the moon. Observers at different points see different kinds of eclipses, as shown here. By chance, the tip of the umbra* (B) *lies very close to the earth's surface.*

a

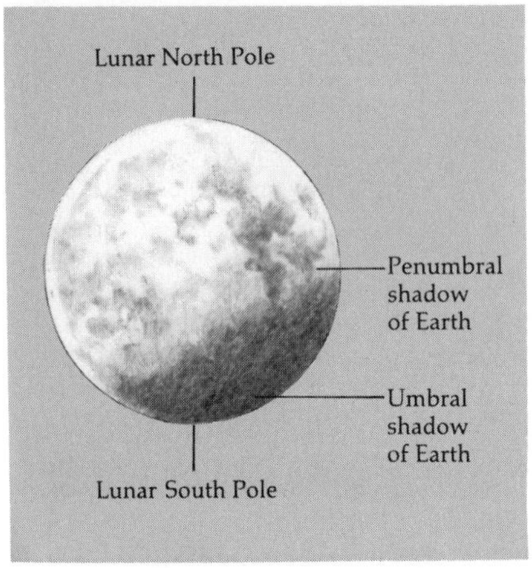

b

Figure 1-29. (a) *A photograph of a partial lunar eclipse.* (b) *A diagram of same, showing umbral and penumbral shadows.*

But the moon's orbit is tipped by an angle of 5° out of the earth's orbital plane, as sketched in Figure 1-30. Therefore the moon is likely to pass "above" or "below" the earth–sun line and the earth's shadow. Since the two orbital planes must, by simple geometry, intersect in a line, there is a line (NN′ in Figure 1-30) that contains the **nodes**, the only two points where the moon passes through the earth's orbital plane. This line between the two points is called the **line of nodes.** To produce an eclipse, the moon must be near one of the nodes, and even then *an eclipse will occur only if the line of nodes is pointing at the sun.*

When is the line of nodes pointed at the sun? If no external forces acted on the moon's orbit, the orbital plane would stay fixed with respect to the stars, and the line of nodes would line up with the sun twice a year, as can be seen in Figure 1-31. Eclipses *could* occur at these two possible times each year, but *only* if the moon moved through points N and N′ at these moments. Predicting eclipses would require alertness during only two intervals per year.

But there is one last complication. The moon's orbit is not fixed with respect to the stars. It is disturbed by various forces. As a result of these forces, the moon's orbit shows what is called a **regression of nodes:** The orbital plane swings slowly around, always keeping its 5° tilt to the earth's orbital plane. This regression of nodes is analogous to the precession of the earth's axis. The line of nodes NN′ rotates slowly with respect to the stars, taking 18.61 y

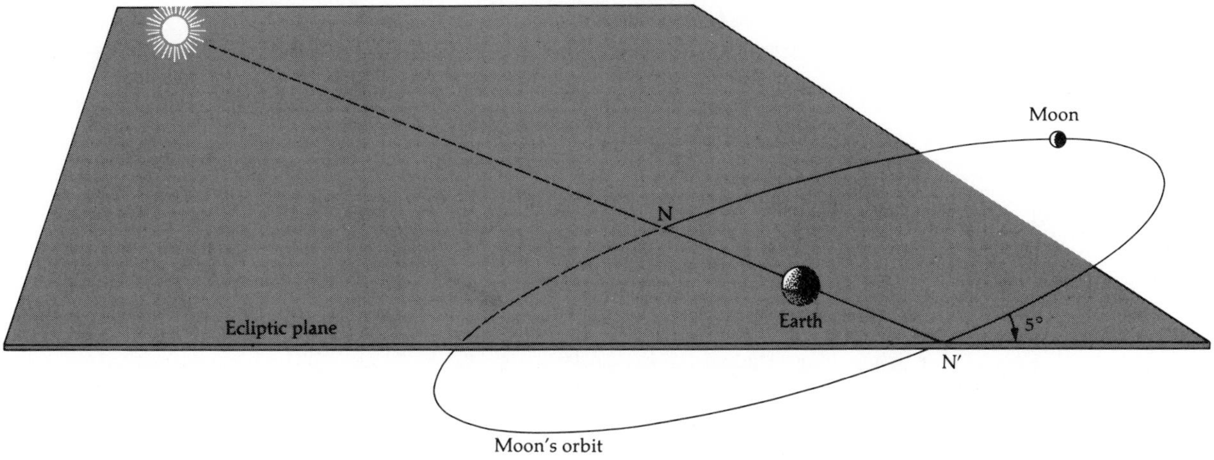

Figure 1-30. *The moon's orbit lies out of the ecliptic plane. Thus, only at points N and N′ can the earth, moon, and sun be aligned to produce an eclipse.*

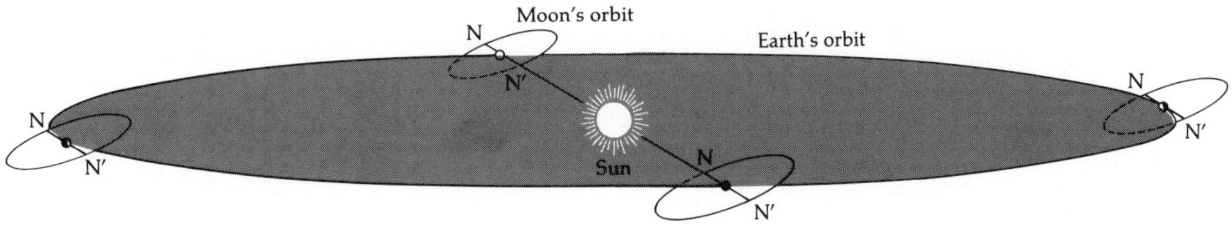

Figure 1-31. *As the earth moves around the sun, there are only two periods each year when the nodal line NN′ of the moon's orbit aligns with the sun so that eclipses can occur.*

(years) to complete one rotation. Therefore, the line of nodes aligns itself with the sun twice in a period somewhat less than a year. This period, called the *eclipse year,* turns out to be 346.6 d.

Thus several simultaneous cycles must be in phase to produce an eclipse: the 29.5-d cycle of lunar revolution with respect to the sun; the 1-y cycle of the earth's revolution; and the 18.6-y lunar regressional cycle. These overlapping cycles cause subtle periodicities in the occurrence of eclipses. These periodicities helped ancient astronomers discover how to predict slopes, as we will see in the next sections.

DISCOVERY OF THE SAROS CYCLE (c. 1000 B.C. ?)

From the information just given, we can understand one periodicity among eclipses. The moon passes nearest the sun's direction every 29.5 d (month), and the node N lines up with the sun every 346.6 d (eclipse year). A lunar or solar eclipse occurs if the two cycles coincide. How often do they coincide? Using figures more exact than those just given, we find that 223 lunar months equal 6585.321 d, while 19 eclipse years equal 6585.781 d. Thus, the two cycles come almost exactly into phase (to within only 0.46 d) every 6585 d, or 18 y, 11 d.

This interval is called the **saros cycle** (a Greek name from an earlier Assyrian–Babylonian word). Early astronomers discovered that if, in a given year, a particular sequence of eclipses occurred, a similar sequence would probably occur after one saros. (Shorter periodicities, such as 41 and 47 mo [months], also exist but are less accurate. Various periodicities were discovered and used by early astronomers to predict eclipses.)

Because the moon's umbral shadow on the earth is so small, a fixed observer has only a small chance of seeing any given solar eclipse. But because the earth's umbra is large and because the moon can be seen from the whole night hemisphere of the earth, half the earth will see every lunar eclipse (barring cloud cover). Total lunar eclipses are therefore relatively common for any observer, and they were thus easier than solar eclipses for ancient astronomers to predict using the saros and other cycles. Modern astronomers use computers and orbital theory, not cycles, to predict all eclipses accurately.

On the average a given location may witness a lunar eclipse nearly every year and a partial solar eclipse nearly every other year, but a total solar eclipse only about once every four centuries. Many of the dates are related by cycles.

Let us now return to the viewpoint of ancient earthbound peoples. Once records were made and kept, astronomers could benefit from their community's past observations of eclipses. If they discovered periodicities, they could make rough predictions. We know this happened by 700 B.C. in the Mediterranean area and probably by about A.D. 500 in Central America, because records found in Assyrian libraries and Mayan cities discuss the prediction of eclipses.

Table 1-1. *Total Solar Eclipses, A.D. 1984–2010*

Date	Duration of Totality (min)	Selected Regions Where Totality Visible
1984 Nov. 22	2.1	New Guinea, Chile
1987 Mar. 29	0.3	Atlantic, Africa
1988 Mar. 18	4.0	Philippines
1990 Jul. 22	2.6	Finland, N. Asia
1991 Jul. 11	7.1	Hawaii, Central America, Brazil
1992 Jun. 30	5.4	S. Atlantic
1994 Nov. 3	4.6	Chile, Brazil
1995 Oct. 24	2.4	Iran, India, Vietnam
1997 Mar. 9	2.8	NE Asia
1998 Feb. 26	4.4	Central America
1999 Aug. 11	2.6	Europe, India
2001 Jun. 21	4.9	S. Africa
2002 Dec. 4	2.1	S. Africa, Australia
2003 Nov. 23	2.0	Antarctica
2005 Apr. 8	0.7	S. Pacific
2006 Mar. 29	4.1	Africa, USSR
2008 Aug. 1	2.4	Siberia, China
2009 Jul. 22	6.6	India, China, S. Pacific
2010 Jul. 11	5.3	S. Pacific

Note: Brackets show pairs of eclipses separated by one saros cycle.

THREE EXAMPLES OF ECLIPSES IN HISTORY

Thales (c. 580 B.C.). Historical records indicate that one of the first known astronomers, Thales of Miletus, predicted the eclipse that stopped the battle between warring Greek factions in the 580s B.C. Thales may have made this prediction by knowing of the saros cycle from Mesopotamian records, or he may have discovered a useful 3-saros periodicity of 669 lunar months, which would have been prominent in the eclipse records of his region for the preceding 125 years.

You can repeat his discovery of this cycle by studying Table 1-1, which lists total solar eclipses for the last quarter of this century. Table 1-2, incidentally, lists some current lunar eclipses.

Mayans (c. A.D. 400). We know that Mayan astronomers of Mexico and Guatemala recorded and predicted eclipses long before European contact. One of three priceless Mayan manuscripts left after the Spanish conquest is a record of observed and predicted solar eclipses as well as other astronomical information, such as the motions of Venus. Tragically, because the Spaniards burned 27 other Mayan manuscripts in 1562, we may never know the extent of Mayan knowledge.

While the Mayan calendrical system had a 365-d yearly count and a so-called long count, or cumulative tally of days since a starting date about 3000 years previously, it also had a cyclic count useful for predicting eclipses. A combination of word names and number names for days (such as our usage *July 4*) produced a 260-d cycle called the Mayan Sacred Round. In other words, the day names repeated every 260 d. As we have seen, the nodes of the moon's orbit are aligned with the sun (in position for potential eclipses) once every 173.31 d. Eclipses could thus occur 3 times in 519.93 d, less than 2 h short of two Mayan Sacred Rounds. The Sacred Rounds

Table 1-2. ***Selected Lunar Eclipses over the United States***

1986, April 24 (Total)
1987, October 7 (Partial)
1988, August 27 (Partial)
1989, August 17 (Total)
1991, December 21 (Partial)

thus let Mayan priests be aware of possible eclipses.

Why did the Mayans (and perhaps others) develop an eclipse-based calendar? One motivation may have been the fact that *five* solar eclipses were visible in the Mayan area during the 14 y from A.D. 331 to 344: a total eclipse in 331, a partial eclipse in 335, near total eclipses in 338 and 344, and an annular eclipse in 342 (Harber, 1969).

Such a cluster of eclipses may have impelled the Mayan priest-astronomers to keep records so that they could predict the sun's next disappearance. In fact, the first astronomical observations in the Mayan manuscripts date from about this time. Interest in eclipses may have been further spurred in A.D. 495, when two partial solar eclipses were visible in Central America only 30 d apart, with a lunar eclipse between them (Owen, 1975). Mayan astronomy was supported by the state and highly organized, as shown by a convention of prehistoric Central American astronomers at Copán, Honduras (probably on May 12, 485) to discuss the calendrical system.

Although the Mayan priests probably could not predict every eclipse, once they got their calendrical system working they might have had the best of both worlds. When they correctly predicted an eclipse, their power and knowledge were proved; when predicted eclipses did not occur, they could attribute this happy omen to the power of their rituals. The successes and failures of Mayan astronomy may thus have worked together to nourish the cultural traditions and help sustain the society for centuries.

We should also note that some scholars attribute the moon alignments at Stonehenge (Figure 1-14) to eclipse-predicting efforts dating as early as 2000 B.C., but this conclusion is quite controversial.

The Crucifixion. An interesting example of the use of eclipses to date historical events is given by Humphreys and Waddington (1983), who attempt to find the date of the crucifixion of Jesus. References in the books of Matthew, Mark, Luke, and John suggest the date of Nisan 14 or 15 in the Hebrew calendar, on or just before Passover. Calculating back through calendar revisions to find years when this fell on a Friday, the two scholars restrict the number of possible dates to four Fridays between April 11, A.D. 27, and April 23, A.D. 34. Additional historical references suggest restricting the date further to April 7, 30, or April 3, 33. At this point the scholars invoke a speech of Peter, reported in Acts, which refers to a blood-red moon, possibly referring to an event at about the time of the crucifixion. Calculating the dates of lunar eclipses, they find that a partially eclipsed moon rose over Jerusalem on the night of Friday, April 3, A.D. 33, which they conclude was the date of the crucifixion.

NEW YEAR'S DAY AND AN "ALIEN" CALENDAR

Why does the New Year begin on January 1? Is there anything special that happens on that date in terms of Earth's motion around the sun? No. The date seems associated with ancient European celebrations of the beginning of the sun's motion back toward the north at the time of winter solstice, about December 22.

There are remnants of older traditions. In Europe, some Christians believed the year should start on March 25, a feast day commemorating the conception of Jesus. (Was this a date fixed by early priests to co-opt still earlier pagan celebrations of spring equinox, around March 21?) In *Tess of the D'Urbervilles* novelist Thomas Hardy records that farm laborers' contracts in nineteenth-century England were still calculated from that date, called "Lady's Day."

Closer to the equator, the Mayans celebrated the new year, and the beginning of the modern creation cycle, on July 26! What could have led them to this choice? Once again, it was astronomical observation, but in a non-European tradition. In the tropics ($23\frac{1}{2}°$ north latitude to $23\frac{1}{2}°$ south), but not in Europe, the sun can pass directly overhead at noon. Perhaps because their horizons were obscured by dense jungle, the Mayans paid as much attention to zenith observations as to observations of the sun's position on the horizon. At the latitude of a major Mayan observatory, Izapa, in Yucatan, the sun passes through the zenith at noon on July 26. A cleverly designed stone pedestal in front of the observatory pyramid reveals this configuration by its shadows (Thomsen, 1984).

Summary

Unaided by telescopes, unknown geniuses of prehistoric times made many of the most basic astronomical discoveries. They (1) recognized and tracked five planets; (2) discovered the celestial poles and the four directional coordinates defined by daily star motions; (3) learned to use heliacal and solstitial risings and settings to formulate calendars; (4) designed constellations as memory aids for learning the sky; (5) recognized the ecliptic and the zodiac as the paths of the sun and planets, respectively;

(6) may have recognized some effects of precession, which causes stars to shift their positions relative to the celestial poles and the celestial equator; and (7) discovered eclipse-related cycles.

Stable societies encouraged astronomy, and vice versa. Astronomy probably contributed to civilization by creating calendars and encouraging record keeping. The discovery of eclipse cycles in particular required record keeping over many years.

Around 2600 B.C. there may have been a golden age in the Mediterranean world, when modern constellations were sighted and named on or near Crete, the Pyramid of Khufu was carefully oriented in Egypt, and the Stonehenge solstitial observatory was built in England. By around 1400 B.C. building temples with astronomical alignments was a well-developed art in Egypt.

A similar development under way in America since the first millennium of the Christian era declined due to uncertain causes (climatic change?) shortly before the European invasion.

While the recognition of solstices, eclipse cycles, and so on, did have applications in ancient societies, it did not lead to visualization of the earth's movement through space. As we will see in the next chapter, the first glimmers of a sun-centered system came in Greek times, around 200 B.C. These concepts were not fully clarified until the most recent "moments" in the long story of human civilization. Of the 150 generations that have lived since astronomy emerged around 3000 B.C., only the last dozen generations (since about A.D. 1600) have well understood the concept of the earth as a ball moving around the sun.

Concepts

celestial poles	solstice
Polaris	solstice principle
North Star	Stonehenge
zenith	astrology
meridian	eclipse
celestial equator	causes of eclipses
ecliptic	solar eclipse
planet	total solar eclipse
zodiac	partial solar eclipse
heliacal rising	annular solar eclipse
heliacal setting	lunar eclipse
constellation	umbra
circumpolar zone	penumbra
equatorial zone	node
precession	line of nodes of moon's orbit
cause of the seasons	regression of nodes of moon's orbit
equinox	saros cycle

Problems

1. Suppose you are standing facing north at night. Describe, as a consequence of the earth's rotation, the apparent direction of motion of each of the following:

a. A star just above the north celestial pole
b. A star just below the north celestial pole
c. A star to the left of the north celestial pole
d. A star on the northern horizon

2. Answer the parts of Problem 1 but for an observer in the Southern Hemisphere facing south and looking near the south celestial pole.

3. What is your latitude? What is the approximate elevation of Polaris above your horizon?

4. What is the radius of the north circumpolar zone of constellations as seen from your latitude?

5. How many degrees from the south celestial pole is the southernmost constellation that you can see from your latitude?

6. Solar eclipses are slightly more common than lunar eclipses, but many more people have observed lunar eclipses than solar eclipses. Why?

7. Does the moon cast an umbral shadow on earth during an annular solar eclipse? Why or why not?

8. How would a lunar eclipse look if the earth had no atmosphere? Compare the earth's appearance from the moon during such an eclipse with its appearance from the moon in reality with its actual atmosphere.

9. Why would lunar and solar eclipses each occur once per month if the moon's orbit lay exactly in the ecliptic plane?

10. Compare ancient European and American cultures of around A.D. 200 to 1000.

a. How much did each know about eclipse phenomena?
b. How many years apart did they achieve a comparable ability to predict eclipses?
c. To understand the causes of eclipses?
d. During this period, were the two cultures' advances in technology similar to their astronomical advances?
e. Do you think that either of these two types of advances (or the two combined) offer a valid measure of cultural achievement?

ADVANCED PROBLEM

11. Using the following diagram, prove that the angle of elevation λ of the north celestial pole above the horizon equals the latitude of the observer. Why does the angle from the celestial equator to the zenith equal the latitude?

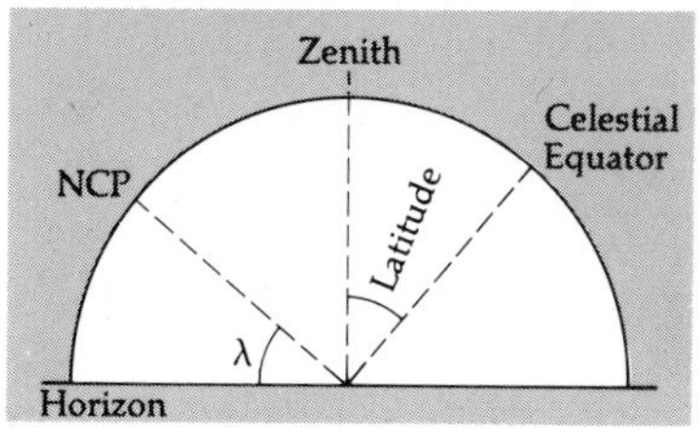

Projects

1. Starting on the date of the new moon (shown on many calendars), observe the sky at dusk and record whether the moon is visible. Repeat each evening for several weeks and record the moon's appearance. Repeat at the next new moon. How many days is the moon visible between the new moon and the first quarter? Between the first quarter and the full moon? How might these counts relate to clusters of grooves, such as 3, 6, 4, 8, . . . , found on prehistoric tools? Can you prove or only speculate that the prehistoric records are lunar calendars?

2. From the preceding project, determine how much later the moon sets or rises each night.

3. If a planetarium is available, arrange for a demonstration showing the following:

- **a.** The position of the north celestial pole
- **b.** The position of the celestial equator
- **c.** The daily motion of the stars
- **d.** The prominent constellations
- **e.** The daily motion of the sun with respect to the stars
- **f.** The position of the ecliptic
- **g.** Planetary motions

4. Measure the angle of elevation λ of the north star above the horizon. A protractor can be used as shown to make the measurement. Compare the result with your latitude.

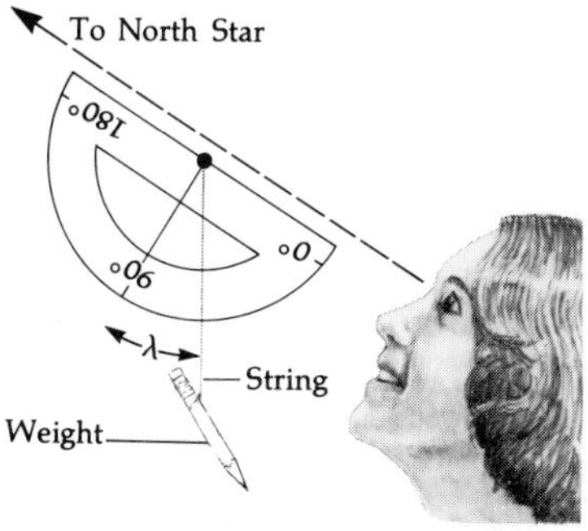

5. Identify a bright planet and draw its position among nearby stars each night for several weeks. (A sketch covering about 10° × 10° should suffice.) Does the planet move with respect to the stars? How many degrees per day? (The latter result will differ from one planet to another and from one week to another.)

6. Find a viewing area with a clear western horizon and determine the date of heliacal setting for some bright star or star group. If several students work independently on the same star, compare results. With how many days' uncertainty can the heliacal setting date be identified?

7. From a viewing area with a clear western horizon, chart sunset positions with respect to distant hills, trees, or buildings for several days and demonstrate the motion of the sunset point from day to day. Do the same for a few days around winter or summer solstice and demonstrate the reversed drift of the sunset position. How accurately can you measure the solstice date in this way?

8. Using a light bulb across the room for the sun, a small ball for the moon, and yourself as a terrestial observer, simulate total, partial, and annular solar eclipses.

9. Using the same props, demonstrate a total eclipse of the moon. Show why a lunar eclipse occurs only during a full moon.

10. If an eclipse of the moon occurs while you are taking this course (see Table 1-2), observe it. Compare the visibility of surface features on the moon before the eclipse, in the penumbra, and in the umbra. Confirm that the curved shadow of the earth defines a disk bigger than the moon. What color is the moon and why? Would the moon be lighter or darker during eclipse if there were an unusually large number of storm clouds around the "rim" of the earth, as seen from the moon?

11. Make a star trail photo like that of Figure 1-10 by setting your camera at night on some interesting foreground, opening the lens wide open, focussing on infinity, and exposing on a "fast" film, like Tri-X, for an hour or more.

C H A P T E R 2

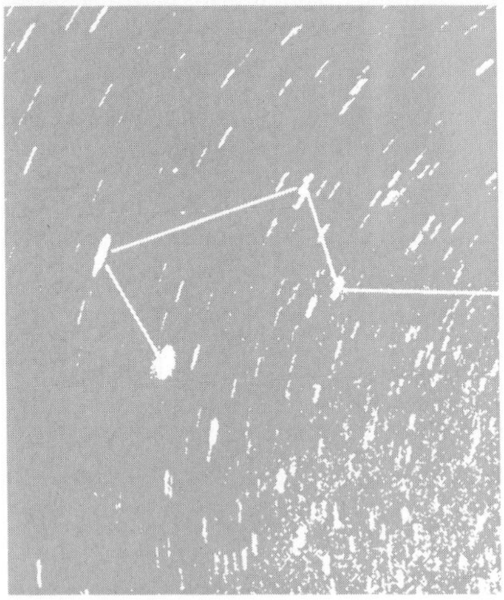

HISTORIC ADVANCES: WORLDS IN THE SKY

In their epic *Gilgamesh,* the Mesopotamians forecast that their own works would vanish as the wind. As they predicted, their names were lost and many of their discoveries degenerated into myth. But some of their knowledge of nature reached the Greeks.

The Greeks, with their blend of practicality and imagination, expanded this knowledge rapidly. Historical records give the names and sometimes the biographies of many Greek astronomers. Partly because of their work, we now accept without question the earth's roundness, the nearness of the spherical moon, and other commonplace but subtle ideas.

If you doubt the sophistication of these early scientists, try to measure the diameter of the earth and the relative distances between the earth, the moon, and the sun without any telescopes or electronic devices. *They* did.

Because discoveries often depend on an underlying conceptual or philosophical framework, in this chapter we look at the attitudes toward nature that the Greeks inherited and developed. We finish with some notes on astronomy in non-Western cultures.

Early Cosmologies and Abstract Thinking (2500–100 B.C.)

Neither Greek scholars nor their systematic observations burst upon the Mediterranean scene from nowhere. The *idea* of thinking about abstract physical concepts can be traced back to early **cosmologies**, or theories about the origin and nature of the universe. For example, certain Egyptian cosmologies assigned different roles to different godlike personages who interacted with the real world. The Memphite theology (c. 2500 B.C.) spoke of an intelligence that organized the "divine order" of the universe (see Figure 2-1). According to Wilson (1951), this theory's "insistence that there was a creative and controlling intelligence, which fashioned the phenomena of nature and which provided, from the beginning, rule and rationale, was a high peak of pre-Greek thinking." This kind of thinking represented a step toward astronomical science, because it assigned different gods, intelligences, or forces to different kinds of natural events and sought relations between them. Early naturalists expressed these relations in myths; later naturalists expressed them in "laws," or generalizations derived from repeated observations.

It is interesting to trace how cosmological thought evolved. Many ideas seem to have carried over directly from the world of 3000 years ago to us today. For instance, the dominant god among all Egyptian gods came to be Amon-Re, to whom the solstitial temple at Karnak was dedicated (see Chapter 1). Around 1350 B.C. the revolutionary Akhenaten, husband of the famous Queen Nefertiti, created a new religion based on the idea of a single god, Aton, the solar disk. Akhenaten's heresy was short-lived because he was overthrown by priests of the old religion. They installed a new young pharaoh, Tutankhamen, whose treasure-filled tomb astonished the world when it was discovered in 1922.

Out of this world, around 1300 B.C., escaped a tribe of nomads whose book of monotheistic religious thought is the core of Western religious principles today. Some scholars believe their Psalm 104 may be a direct transcription of a hymn to the sun written by Pharaoh Akhenaten himself and still preserved (Pritchard, 1955). The pharaoh praised the sun in almost Biblical terms: "How manifold it is, what you have made! . . . You created the world according to your desire. . . ."

The Hebrews recorded a similar cosmology: "In the beginning God created the heavens and the earth. . . ." This cosmological theory asserted that the earth was created in six stages, or days: (1) light, day, night; (2) sky; (3) dry land, ocean, plants; (4) sun, moon, stars; (5) sea creatures, birds; and (6) humans. This cosmology dominated much Western art and science (Figure 2-2).

Cosmological theories of this type, which were especially common in the Middle East, stimulated new questions about relations between phenomena in the universe. Out of this thinking came an important new idea. Regardless of the question of gods, facts could be learned about nature by systematic observations and experiments from which repeatable results could be obtained. A naturalist in Alexandria could get the same results as a naturalist in Athens. **Science** (from Latin "to know") is simply the process of learning about nature by applying this technique: Questions are formulated that can be answered by observations or experiments, which are then carried out.

This system of scientific observation and recording of nature was developed to the greatest extent in Greece. In addition to richly provocative Egyptian and Hebrew philosophies, the Greek world received a legacy of astronomical concepts, such as the ecliptic, the zodiac, solstices, the saros cycle, as well as generations of astronomical observations of eclipses and planet motions from Mesopotamian (and perhaps European) sources. This inheritance, plus the Greeks' inclination to philosophize about natural phenomena, led to a Greek Renaissance.

Figure 2-1. *An Egyptian conception of the universe. Stars are distributed in different shells around the universal center. The shells are separated by boundaries depicted as goddesses. This conception may have carried over to Greek and medieval times, when the planets and stars were also pictured as distributed in concentric shells.*

The System of Angular Measurement

One of the most important inheritances which the Greeks received from the earlier world was the sexigesimal (60-based) system of measuring angles and time, which was mentioned in Chapter 1. Just as there are 60 minutes of time in one hour and 60 seconds in a minute, the system of angular measurement uses the following definitions and symbols:

1 **degree** $= 1° = \frac{1}{360}$th of a circle
60 **minutes of arc** $= 60' = 1°$
60 **seconds of arc** $= 60'' = 1'$

To give a better idea of 1 second of arc: It is the angular size of a tennis ball seen at a distance of about 8 miles.

With this numerical system for measuring angles, the Greeks developed not only rules of geometry (such as Euclid's geometry and Pythagoras' famous theorem about the hypotenuse of a right triangle), but also ways of measuring phenomena in the sky. With sighting devices, Greeks and other early observers measured the positions of planets relative to the fixed pattern of background stars, and the number of degrees of the sun above the southern horizon as it crossed the meridian at noon in different seasons. These measures first revealed the *detailed* systematics of the movements of celestial bodies.

An important concept is the difference between linear measure and angular measure. **Linear measure** gives the actual length of something in linear units such as inches, meters, or miles. **Angular measure** gives the apparent angle covered by an object at a given distance from the observer, in angular units such as degrees. The verb **subtend** refers to the angle covered by such an object: We say a distant object subtends 1°. A useful rule of thumb is that your thumbnail at arm's length subtends about 1°. The disks of the sun and

Figure 2-2. *English painter William Blake's version of the Western creation myth reflects the idea that the universe was created according to rational principles. This idea contributed to the birth of modern science, for it allowed philosophers to study the evolution of the universe without considering sudden changes in natural laws that might be caused by a pantheon of capricious gods. (Photograph courtesy of the Bettmann Archive, Inc.)*

moon always subtend $\frac{1}{2}°$. The pointers in the Big Dipper (see Figure 1-3) subtend about 5°. As the Greeks knew, when you see a distant object (such as a ship at sea or the moon), you can directly measure not its linear size or distance, but only its angular size. We unconsciously *infer* linear distance of many objects by recognizing the object (such as a ship) and knowing roughly how big it is; we similarly *infer* the linear size by estimating an object's distance and judging it must be as big as a house, a dog, and so on. The treachery of such inferences shows up, for example, when people report unfamiliar aerial objects such as a bright meteor, called a fireball. People commonly report that a fireball "looked as big as a dinner plate," but this statement is literally meaningless; they really perceive only angular size, not linear size. To specify angular size correctly, they would have to say, "It looked as big as a dinner plate at a distance of 50 feet;" or "It looked twice the angular size of the moon." Similarly, the common report that a fireball "must have landed just over the hill" is almost always wrong. The speaker misjudges the distance because he *assumes* a certain linear size after he *observes* only the angular size. Fireballs are typically in the upper atmosphere 60 mi from observers!

As we will see in a moment, Greek thinkers carefully separated angular measures from linear measures, and used angular measures together with clever logic to estimate linear sizes and distances of the sun, moon, and other objects.

The word **resolution** refers to the smallest angular sizes that can be discriminated with given optical systems. The human eye, for example, can resolve an angle of about 2′; thus, we can see details covering about $\frac{1}{15}$ of the lunar disk. The largest planetary disks subtend only about 1′ and are thus too small for the eye to resolve.

These principles help us analyze photographs and other images. For example, as shown in Figure 2-3, a typical snapshot subtends an angle of about 40°, about the portion of a scene that the eye concentrates on. A 35-mm camera is so named because it uses film 35 mm wide and makes

Figure 2-3. These views toward the northeast from the center of Stonehenge (compare map, Figure 1-14) illustrate angular measurements applied to photography with different lenses. (a) *This view using an ultra-wide-angle "fisheye" lens subtends a horizontal angle of 120°.* (b) *This view with a standard wide-angle lens subtends 65°.* (c) *This view with a normal lens subtends 40°; this is about the field of view of most snapshots, postcard views, paintings, and so on.* (d) *This telephoto view subtends only 15°. (Photos by author with a 35-mm camera. Lens focal lengths of* a–d: *15 mm, 24 mm, 50 mm, and 135 mm, respectively. Curvature of pillars at edges of* (a) *is distortion common with ultra-wide-angle lenses.)*

a negative about 35 mm in width. With modern films, the typical 40°-wide snapshot can resolve 2′ details and thus presents a view comparable to what the eye sees. However, 40° views made with smaller cameras (such as instamatic and disk cameras) are limited by the inherent graininess of film and often do not resolve details as small as 2′. Thus they look grainy and do not resolve as much as the eye can see from the same spot. TV images suffer a similar limitation, resolving less detail than the eye can see in the typical 40° view. Conversely, modern motion pictures photographed and projected with 70-mm film have a dramatically realistic presence because, like life, they can present the eye with a 40° view containing more detail than the eye can resolve.

The ancient system of angular measure, using degrees, minutes of arc, and seconds of arc, is still used today by surveyors, engineers, navigators, astronomers, and others. Modern large telescopes can often resolve details as small as 0.5″, but rarely can do better than this because of the shimmering quality of heat waves in the atmosphere. The planned Space Telescope to be launched into orbit in the late 1980s should routinely be able to resolve 0.05″ to 0.1″. Later in this book we will encounter more applications of angular measurement.

Early Greek Astronomy (c. 600 B.C. to A.D. 150)

Around 600 B.C. the Greeks began vigorously applying logic and observation to learn about the universe. They talked more of tangible physical "elements" and less of metaphysical relations. They used geometric principles to measure cosmic distances as well as farmyards.

One of the first known Greek thinkers was Thales of Miletus (a Greek-dominated town in present-day Turkey). Living about 636–546 B.C., Thales was a noted statesman, geometer, and astronomer. He is best known for predicting the peacemaking solar eclipse mentioned in Chapter 1. He probably knew some Mesopotamian astronomical concepts—perhaps the saros cycle, the lengths of seasons, and the daily changing position of the sun among the constellations of the ecliptic. Thales also reportedly speculated that the sun and stars were not gods, as was then usually thought, but balls of fire. Of course, Thales could not prove his idea, but he got other Greeks thinking in terms of tractable, physical ideas.

Thales' school produced several notable thinkers. Anaximander (611–547 B.C.) made astronomical and geographical maps; speculated on the relative distances of the sun, moon, and planets from the earth; and argued that the matter from which things are made is an eternal substance. Heraclitus (535?–475? B.C.) made this remarkable comment:

This ordered cosmos, which is the same for all, was not created by any one of the gods or by mankind, but it was ever and shall be ever-living Fire, kindled in measure and quenched in measure. . . . The fairest universe is but a dust-heap piled up at random.

THE PYTHAGOREANS: A SPHERICAL, MOVING EARTH (c. 500 B.C.)

Pythagoras (flourished 540–510 B.C.), famous for his theorem on right triangles, was also one of the first experimental scientists. Pythagoras proposed the unusual idea that the earth is spherical. He may have gotten this idea by studying the phases of the moon. The line separating the lit side of the moon (or any planetary body) from the unlit side (the **terminator**) changes its curvature as the moon's phases progress, thus revealing that the moon is spherical rather than flat, as shown in Figure 2-4. By analogy, then, the earth and other bodies would also be spherical.

In southern Italy, Pythagoras founded a school that had wide influence around 450 B.C. It is unclear, however, which thinkers should be credited with which ideas in this school. Pythagoras himself put the earth at the center of the universe, but later Pythagoreans proposed that it moves, like the moon and the planets, around a distant center. The universe is spherical with a central "fire" containing a force that controls all motion. Around it, in order outward from the center, move the earth, the moon, the sun, the five planets, and the stars. This system predates by more than 2000 years Copernicus' revolutionary, correct model of the planets moving around the sun (see Chapter 6). The idea of a spherical earth persisted among some Greeks, though it was not universally accepted.

Anaxagoras (500?–428 B.C.) is credited with deducing the true cause of eclipses. Thereafter, the observed roundness of the earth's shadow on the moon (Figure 2-4) undoubtedly helped to establish the theory that the earth itself is a spherical body. After residing in Athens for 30 years, Anaxagoras was charged with impiety and banished for saying that the sun was an incandescent "stone" even larger than Greece.

PLATO (c. 400 B.C.)

Though known primarily as a philosopher, not a scientist, Plato (427?–347 B.C.) reasoned that astronomy contributed to the civilization of humanity. In *Timaeus,* he said that philosophy came from astronomy:

Had we never seen the stars, the sun, and the heavens, none of the words we have spoken about the universe would have been uttered. But now the sight of day and night, and the revolutions of the years, have created number and given us a concept of time as well as the power of enquiring about the nature of the universe: and from this source we have derived philosophy. No greater good ever was or will be given by the gods to mortal man.[1]

ARISTOTLE: THE EARTH AGAIN AT THE CENTER (c. 350 B.C.)

The most influential Greek scientist-philosopher was Aristotle (384–322 B.C.). His views were built on earlier knowledge but were biased in favor of absolute symmetry, simplicity, and an abstract idea of perfection. Aristotle's universe was spherical and finite, with the earth at the center. Planets and other bodies moved in a multitude of spherical shells centered on the earth. The shells were supposed to turn with varying rates, which explained the observed changeable motions of the planets.

Aristotle is credited with founding modern scientific investigation. His school at Lyceum (a grove near Athens) contained a library, a zoo, and lavish physical and biological research equipment paid for by his onetime pupil Alexander the Great, then ruler of Greece. In the Middle Ages, when research lapsed, Aristotle came to be regarded as the final authority, and so his rejection of the Pythagorean idea and his placement of the earth at the center of the solar system turned out to delay progress in astronomy. However, there was little reason for him to choose a sun-cen-

[1]In the 1800s Ralph Waldo Emerson extended this thought with the following lines: "If the stars should appear one night in a thousand years, how would men believe and adore, and preserve for many generations the remembrance of the city of God?" Science fiction writer Isaac Asimov ironically reversed this thought in a 1941 story called "Nightfall." He imagined a planet in a multistar system with six suns. Only once in 2050 y did the orbits of all the suns bring them all to one side of the planet, plunging the other side into darkness so that the stars came out. Instead of inspiring worship, this event so frightened the inhabitants that they burned everything around them to give them light. As a result, civilization on this planet was burned to the ground every 2050 y and had to start over again.

a b

Figure 2-4. (a) *The phases of the moon correctly suggested to some Greeks that the moon is not a disk but a sphere illuminated by the sun.* (b) *The earth's curved shadow on the moon during every lunar eclipse also suggested that the earth is spherical.*

tered over an earth-centered system, since either view was consistent with observations known in his time.

Aristotle was right about several important astronomical ideas, however.

1. He thought the moon is spherical.

2. He argued that the sun is farther away than the moon because:

- **a.** the moon's crescent phase shows that it passes between the earth and the sun, and
- **b.** the sun appears to move more slowly in the sky than the moon (this second argument is not rigorous, but the first is).

3. He thought the earth is spherical because:

- **a.** the curvature of the moon's terminator rules out it being a disk, and the earth is likely to be like the moon in this respect, and
- **b.** as a traveler goes north, more of the northern sky is exposed while the southern stars sink below the horizon—a circumstance that would not arise on a flat earth.

The apparent motions of the sun, the moon, and the stars around the earth could be explained, said Aristotle, either by their actually moving around us or by the earth moving. But Aristotle concluded that the earth is stationary and gave a very powerful argument. If the earth were moving, we ought to be able to see changes in the relative configurations of the various stars, just as, if you walk down a path, you see changes in the relative positions of nearby and distant trees. If you line up a tree in the middle distance with a very distant tree and then step to one side, the nearby tree will seem to shift to the side of the distant one. Such a shift in position due to motion is called **parallax**, or a *parallactic shift.* If the earth were moving in a straight line, we would see a continuous parallactic shift of the nearer stars with respect to more distant stars; and if the earth were moving around some distant center, we would see a periodic parallactic shift back and forth among the stars. But a visual survey of the stars and the constellations over time showed no evidence of such a shift. So, reasoned Aristotle, the earth must not move.

The reasoning was sound, but the stars are too far away to produce noticeable parallactic shifts for the unaided eye during a human lifetime. In the same way, distant mountains show little parallactic shift from a car speeding down an interstate highway, even though nearby trees whiz by in seconds. Stellar parallaxes were sought for years and not discovered until 1838.

Aristotle died shortly after being forced to leave Athens for allegedly teaching that prayer and sacrifices to the Greek gods were useless.

ARISTARCHUS: RELATIVE DISTANCES AND SIZES OF THE MOON AND THE SUN (250 B.C.)

Aristarchus (310?–230? B.C.) of Samos (an island off present-day Turkey) cleverly extended the Greek methods of seeking quantitative data. His only surviving work is "On the Sizes and Distances of the Sun and Moon," although his other astronomical works are quoted by other Greek authors.

He devised a way to measure the relative distances of the sun and moon from earth, based on the geometry of the moon's orbit and phases. From this, he correctly inferred that the sun is much farther away than the moon.

Aristarchus also formulated a way to measure the relative sizes of the earth and the moon. We can reconstruct his calculations, which used what he knew about eclipses. Assuming that a total eclipse of the moon was in progress (Figure 2-5), he drew a circle to represent the earth. Because he knew the sun's angular diameter is $\frac{1}{2}°$, he could represent the earth's umbral shadow by lines leading away from the earth and converging at an angle of $\frac{1}{2}°$. Aristarchus also knew that the moon looks $\frac{3}{8}$ as big as the earth's shadow, so that wherever he placed the moon in the earth's shadow, its diameter would have to be $\frac{3}{8}$ the distance across the shadow at

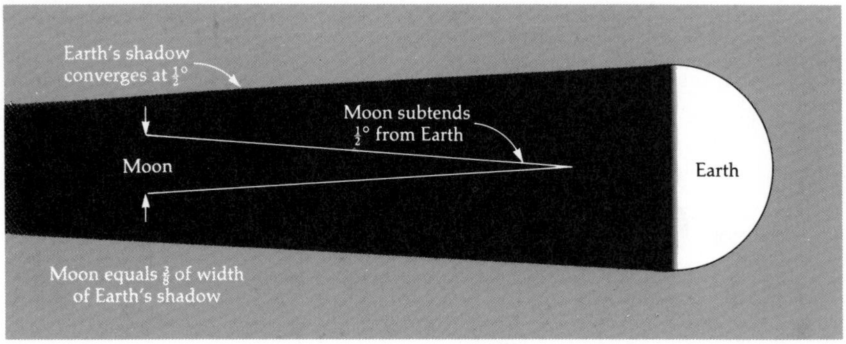

Figure 2-5. *The geometry by which Aristarchus estimated the relative sizes of the moon and the earth (not drawn to true scale).*

that point. But where to place the moon's circle? Aristarchus knew that the moon, like the sun, has an angular diameter of $\frac{1}{2}°$, as seen from earth. These criteria specified the position and size of the moon, as trial and error with a diagram like Figure 2-5 will show.

What about the sun? By estimating that the sun is 18 to 20 times as far away as the moon, Aristarchus applied the observation that the sun's angular diameter is $\frac{1}{2}°$, like the moon's, thus fixing its relative size. Aristarchus concluded that the moon is one-third as big as the earth, and that the sun is about 7 times as big as the earth. The correct figures are closer to $\frac{1}{4}$ and 100, but Aristarchus was on the right track.

Aristarchus made still another contribution. Because he thought the sun is much bigger than the earth, he guessed (without many supporting observations) that the sun, not the earth, must be the central body in the system. For this an outraged critic declared he should be indicted for impiety.

Although Aristarchus made some quantitative errors, he was nonetheless far ahead of later scholars who thought the earth is flat. Aristarchus correctly visualized the moon in orbit around a spherical earth and the earth in orbit around the sun, and developed a method of measur-

OPTIONAL BASIC EQUATION I

The Small-Angle Equation

Angles and linear measures can be combined in an extremely useful and simple equation called the **small-angle equation,** which involves the angular size of an object, its linear size, and its distance. If any two of these quantities are known, the third can be calculated. Let us call the angular size α, expressed in seconds of arc. Let the diameter of the object be d and its distance D. Then the small-angle equation is:

$$\frac{\alpha}{206\,265} = \frac{d}{D}$$

The number 206 265 is called a *constant of proportionality;* it stays the same in all applications of the equation.*

Consider an example. Suppose a friend who is 2.0 m tall is standing across a field, where he subtends an angle of $\frac{1}{2}°$, or 1800″, as shown in Figure 2-6. How far away is he? We want to solve the equation for D. Rearranging the equation, we have

$$D = 206\,265\; d/\alpha.$$

Using SI metric units, we would write d = 2.0 m. Thus, expressing the equation in SI metric units and powers of 10, we would have

$$D = \frac{206\,265\, d}{\alpha}$$

$$= \frac{2.06(10^5)2}{1.8(10^3)} = 2.3 \times 10^2\,\text{m} = 230\,\text{m}.$$

*The number 206 265 is actually the number of seconds of arc in 1 radian (rad). A radian is an angle of about 57°.3, defined as the angle subtended at the center of a circle by one radius of the circle laid along the circumference. The radian has many applications in geometry.

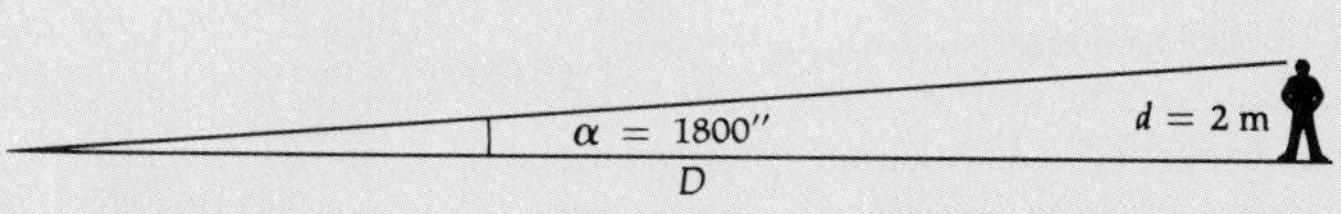

Figure 2-6. *An application of the small-angle equation. If your friend is 2 m tall and subtends an angle of $\frac{1}{2}°$ (or 1800″), his distance D is 230 m.*

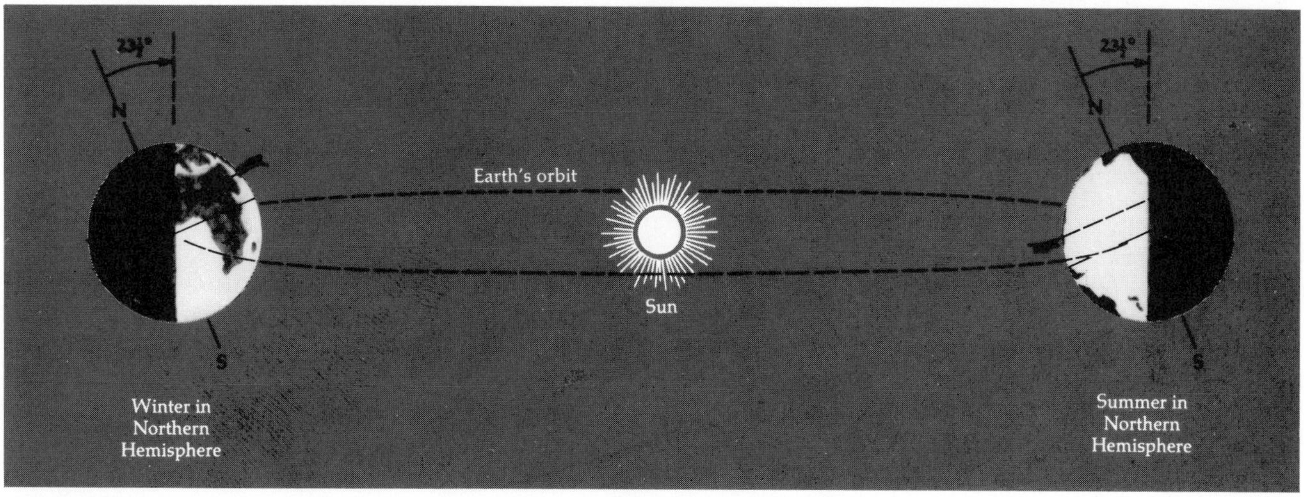

Figure 2-8. *The seasons are caused by the earth's inclination of $23\frac{1}{2}°$ to the plane of its orbit around the sun. Eratosthenes measured the difference between the sun's noontime elevation in summer and winter at a fixed northern latitude (dashed straight line). The observer finds the sun more nearly overhead in summer. The angular difference between these two elevations is twice the earth's inclination.*

Your friend is about $\frac{1}{6}$ of a mile away.

As the Greeks realized, exactly the same geometry can be used to investigate astronomical distances. The Greeks could not get good measurements of the moon's diameter, but only its angular size α, which is roughly $\frac{1}{2}°$, or 1800″. If we use the modern knowledge that the moon is about 3500 km in diameter, we can estimate its distance just as we did for the friend's distance above (Figure 2-7). In SI metric units, d would be 3.5×10^6 m. The equation would read:

$$D = \frac{206\,265\,d}{\alpha} \simeq \frac{2.06(10^5)3.5(10^6)}{1.8(10^3)}$$

$$\simeq 4 \times 10^8\text{ m} \simeq 4 \times 10^5\text{ km},$$

or about 400 000 km.

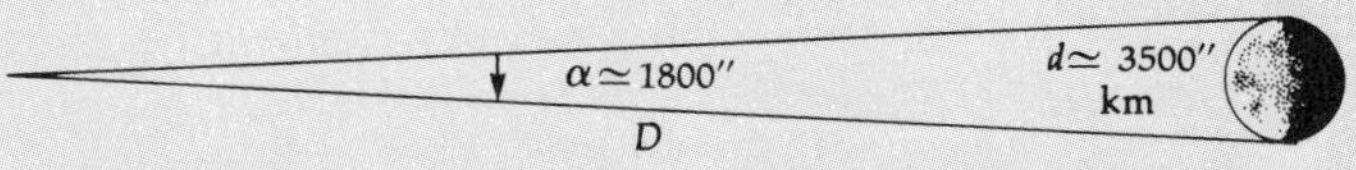

Figure 2-7. *The same geometry as in Figure 2-6 can be applied to measure interplanetary distances, such as the distance to the moon.*

Several mathematical notes should be observed. First, the symbol ≃ means "approximately equal to"; it is useful whenever approximate values (such as $\frac{1}{2}°$) are involved.

Second, this calculation shows the economy of writing the numbers as powers of ten, for example, 3.5×10^6 m instead of 3 500 000 m.

Third, the answer is given only to an accuracy of one significant figure. **Significant figures** are the number of digits known for certain in a quantity. For example, π to one significant figure is 3; to three significant figures, 3.14. Generally, an answer to a calculation should have no more significant figures than does the least accurate number in the equation.* In this case that number was the angular size of the moon, $\frac{1}{2}°$, or 0.5°, with only one significant figure.

A fourth note is that, following our rule given in the prologue, we convert data to SI metric units. (In our small-angle equation this is unnecessary, since the dimensions of d and D are the same and cancel out; but it is a useful habit.)

*This statement is especially important to users of electronic hand calculators, which blindly print answers with seven or eight significant figures, even when the accuracy of the input is only two significant figures.

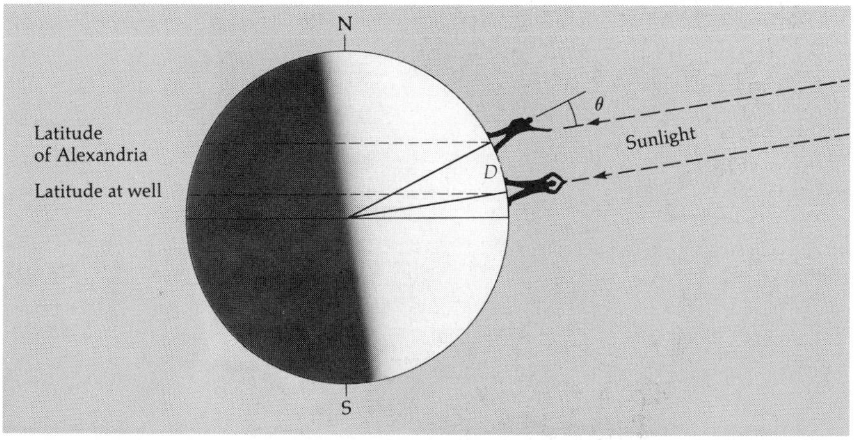

Figure 2-9. *The geometry of Eratosthenes' measurement of the size of the earth. When the sun was directly over a certain well, Eratosthenes measured the angle at Alexandria, a known distance D away, and found D was $\frac{1}{50}$ of the way around the earth.*

ing interplanetary distances. These ideas were not confirmed for another 2000 y!

ERATOSTHENES: EARTH'S SIZE (200 B.C.)

As Greece declined and Rome prospered, Greek scholars became resident intellectuals in many parts of the Mediterranean world. Eratosthenes (276?–192? B.C.) was a researcher and librarian at the great **Alexandrian library** in Egypt. He reportedly completed a catalog of the 675 brightest stars and measured the $23\frac{1}{2}°$ inclination of the earth's polar axis to the ecliptic plane, as shown in Figure 2-8 on page 39. As described in Figure 1-12, this is the tilt that causes our seasons.

Eratosthenes is most famous for using angular geometric relations to measure the earth's size. Told that at summer solstice the sun shone directly down a well near Aswan, he noted that the sun's direction was off vertical by $\frac{1}{50}$ of a circle on the same date at Alexandria (Figure 2-9). He realized this difference had to be due to the curvature of the earth and concluded that the earth's circumference was 50 times the distance from Alexandria to the site of the well. His estimate was probably within 20% of the right answer. This Greek master clearly understood the shape and approximate size of the earth 1700 years before Columbus.

HIPPARCHUS: STAR MAPS AND PRECESSION (c. 130 B.C.)

From his observatory on the island of Rhodes, Hipparchus (160?–125? B.C.) observed the positions of astronomical bodies as accurately as possible and compiled a catalog of some 850 stars. His

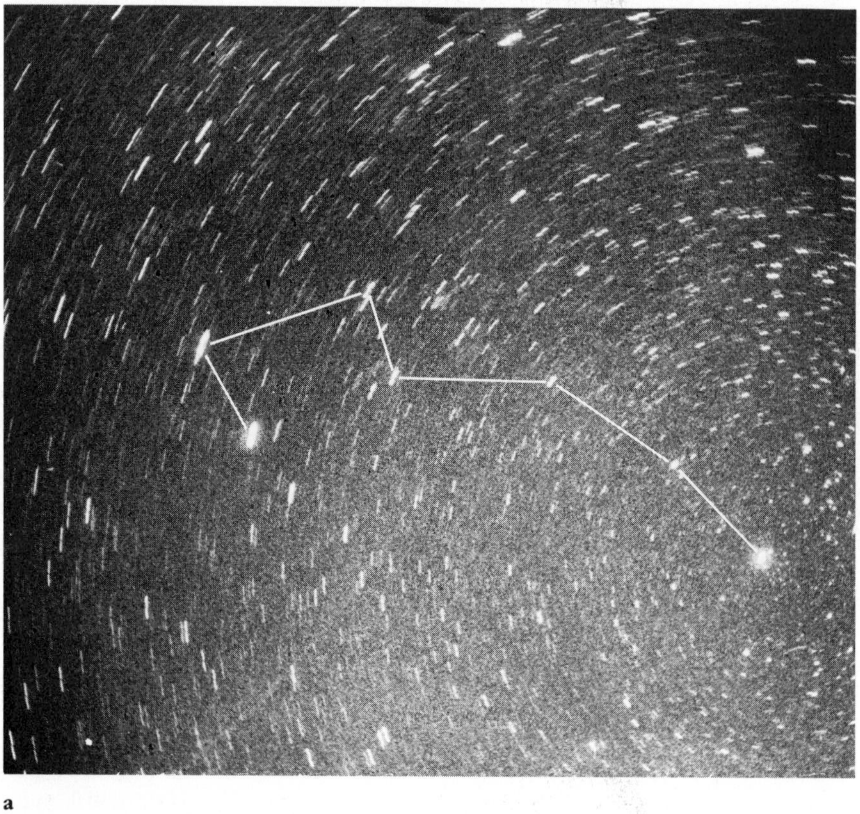

a

Figure 2-10. (a) *This 10-min exposure shows stars in the north polar region, trailed by their daily motion around the north celestial pole near Polaris, the bright star in the lower right. (Photo by author.)* (b) *Map of stars in the same region shows the Little Dipper (prominent in* [a]*) and the changing path of the north celestial poles in previous centuries. Because of precession the north celestial pole 2000 y ago did not lie near a prominent star. About 4600 y ago Alpha Draconis (lower left) was the North Star. The north celestial poles move in a circle around the ecliptic pole (top).*

exhaustive observations—all done, of course, without a telescope—along with material he inherited from Babylon, enabled him to predict with reasonable accuracy the position of the sun and moon for any date. Hipparchus has been called antiquity's greatest astronomer.

The most important discovery attributed to Hipparchus is *precession* (though astronomers of earlier centuries may have been aware of its effects; see Chapter 1, p. 11). Comparing his own measurements of star positions with materials handed down to him from centuries before, Hipparchus found that, with respect to the background stars, there had been curious shifts in the positions of the north celestial pole, the vernal and autumnal equinoxes, and other coordinates. The whole celestial equator was oriented somewhat differently with respect to the stars! Could the old maps be wrong? Hipparchus concluded instead that the whole coordinate system of the celestial equator and the poles was drifting slowly with respect to the distant stars. This drift came to be known as **precession** or *precession of the equinoxes.*

In modern terms, *precession is the result of a wobble of the spinning earth due to forces produced by the sun and the moon.* Just as a spinning top describes a conical wobble when it is pulled downward by the force of the earth's gravity, the spinning earth's polar axes describe a conical wobble with respect to the fixed stars; hence each celestial pole describes a circle among the stars, shown in Figure 2-10(b). The circle, $23\frac{1}{2}°$ in angular radius,

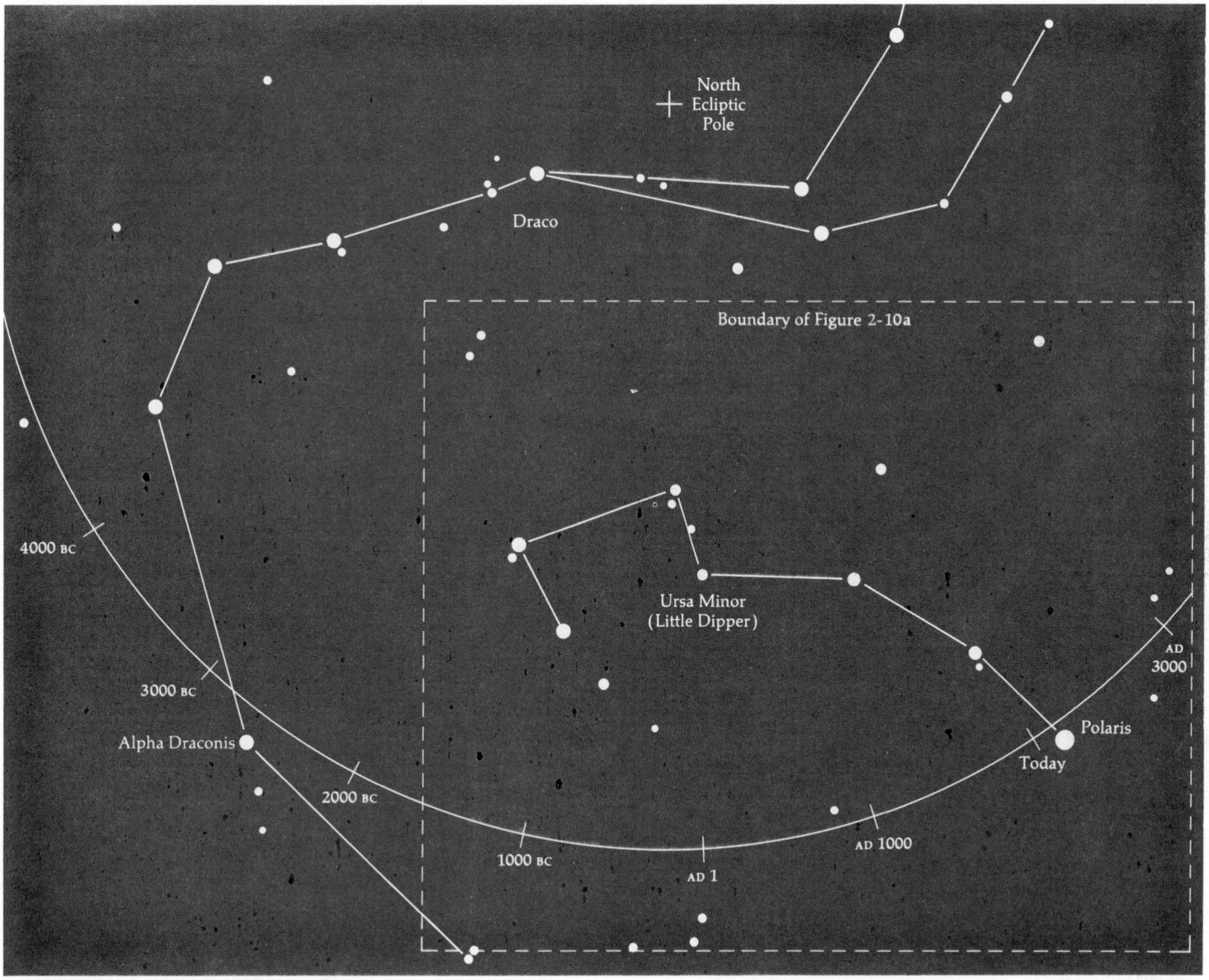

b

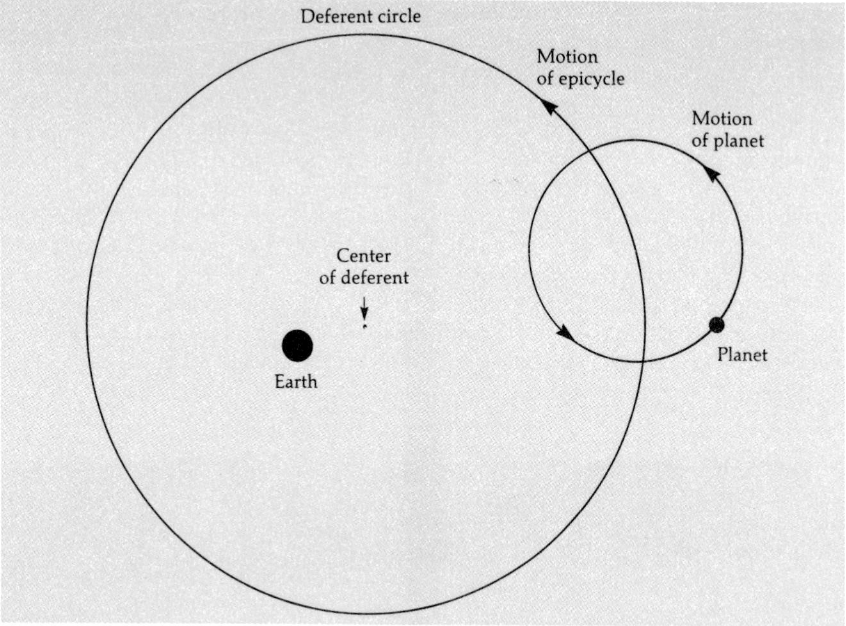

Figure 2-11. ***Ptolemy's system for explaining planetary motions. Ptolemy thought each planet moved in a circular epicycle, whose center moved in a circular orbit (deferent) around a point near the earth.***

is centered on the ecliptic poles. Thus, as mentioned in the last chapter, in different millennia different stars become the North Star; a complete cycle takes about 26 000 y. As seen in Figure 2-10(b) the star Alpha Draconis was the North Star around 2500 B.C., when Stonehenge was being built. All star coordinates, which are measured with respect to the celestial equator, therefore change slightly each year.

Hipparchus also contributed to the description of solar and planetary motions. Although a few of Hipparchus' contemporaries accepted Aristarchus' idea that the earth moves around the sun, Hipparchus and most other astronomers thought this model unnecessarily complex. His own observations showed that the sun's apparent motion with respect to the stars is not uniform from day to day; he therefore concluded that the sun moves around the earth in a circular orbit whose center is slightly offset from the earth.

Like so many ancient works, most of Hipparchus' writings have disappeared except in others' reports—according to which Hipparchus also studied the relative distance from the earth to the moon and the sun. He calculated that the moon is $29\frac{1}{2}$ earth diameters away, close to the correct value of about 30. Hipparchus apparently realized that the sun must be much farther away than Aristarchus' estimate of 18 to 20 times the lunar distance.

PTOLEMY: PLANETARY MOTIONS (A.D. 150)

Claudius Ptolemy (flourished c. A.D. 140) was another scholar associated with the Alexandrian library. His fame as an astronomer is based on a 13-volume work *The Mathematical Collection.* Passed on to the Arabs after the destruction of the library, the work became known as *al-Megiste* (The Greatest). European translations were called the *Almagest,* and the book was famous for more than a thousand years.

Ptolemy extended the star catalog to 1022 entries, correcting older reported star positions to compensate for precession.[2] But his best-known contribution was a method for predicting the positions of the sun, moon, and planets, called the **epicycle theory**, or *Ptolemaic theory*. Following the tradition of his day, Ptolemy incorrectly assumed that the earth is near the center of the planetary system, as shown in Figure 2-11. In order outward from the earth, he placed in *circular* orbits the moon, Mercury, Venus, the sun, Mars, Jupiter, and Saturn. To explain why planets apparently do not move at uniform rates, Ptolemy devised combinations of circular motions. Each planet, he said, moves in a circle called the deferent, whose center is offset from the earth. The planet does not move strictly on this circle but in a smaller circle called an epicycle, whose center moves along the deferent just as the moon moves around the earth while the earth moves around the sun. The epicycle theory was simply a case of adding "wheels within wheels" until there were enough wheels to explain the observed irregular movements of the planets. Astronomers who applied this theory and its later changes during the Middle Ages could predict the actual positions of planets within a few degrees. At that time, such predictions were used more often for casting astrological horoscopes than for astronomical studies of the universe!

In spite of its usefulness in ancient times, Ptolemy's theory was incorrect. Today we know that planets move in elliptical orbits around the sun, not in circles around the earth. Ptolemy has been criticized for abandoning Aristarchus' earth-centered system and for biasing his theory toward a supposed "perfection" of the circle, thus delaying the introduction of the true elliptical orbits. But the system fitted the observations available in Ptole-

[2]Correction for precession continues today. For precise setting of a large, modern telescope, a star's coordinates published for "epoch 1950" or any other year must be corrected for the current date, or the star could be missed. The correction is generally done by a computer operating as part of the telescope.

my's time. Aristotle's old argument still seemed true: The earth could not move around a distant center, because no one had observed parallax. Certainly no one had yet *observed* elliptical motions. And Ptolemy's system worked. The trouble with Ptolemy's choice of Aristotle over Aristarchus lies not in its being a mistake, but in its historical effects: The *Almagest* became the Bible of ancient astronomy, and the erroneous earth-centered system held sway for 1400 y.

The Loss of Greek Thought (c. A.D. 500)

Alexandria, where Cleopatra first fascinated Julius Caesar and Mark Antony around 47 B.C., was the world's intellectual center by A.D. 250. With Rome's fall and the world in disorder in 410, maintaining the great library of ancient discoveries became more and more difficult.

Among the last guardians of the old knowledge in Alexandria was one of the first known woman astronomers, Hypatia (c. A.D. 375–415). Widely admired for her intelligence and beauty, she wrote a commentary on Ptolemy's work and invented astronomical navigation devices, but was murdered by a mob during one of the riots that plagued Alexandria during its decline. In A.D. 640, after a 14-mo siege by the Arabs, Alexandria fell. The library buildings were burned and the best collection of Greek knowledge was lost. Because there was no printing, there were few other reference books. Much of the Alexandrian knowledge passed into the hands of the Arabs, who further developed mathematics and astronomical instruments (Gingerich, 1986). But in the West, the rate of new discoveries declined, and Europe slipped into the Dark Ages.

Ancient Astronomy Beyond the Mediterranean

Exemplifying Arab advancement, the next recorded measurement of the earth's circumference was made near Baghdad in 820. The result was only 4% too large. Similarly, the Arab astronomer Muhammad al-Battani (c. 850–929; known in Europe as Albategnius) made only a 4% error in his measurement of the eccentricity of the earth's orbit (he would have called it the sun's orbit around the earth). Around A.D. 1000 the Islamic empire had spread to Spain, and astronomical tables were published with the reference longitude in Córdoba.

Although it is the Mesopotamian–Greek–Arab development of astronomy that most clearly contributed to modern Western conceptions, other astronomical centers, initially independent in their development, eventually influenced the West.

Figure 2-12. ***An imaginary scene showing ancient Oriental observations of a comet. A teacher is discoursing on the significance of the celestial visitor. (Painting by Chesley Bonestell.)***

ASTRONOMY IN INDIA: A HIDDEN INFLUENCE

Astronomical practices in India date back to about 1500 B.C. The first known astronomy text, describing planetary motions and eclipses and dividing the ecliptic into 27 or 28 sections, appeared around 600 B.C. By this time India had contact with the Mesopotamian and Greek worlds, and influences probably traveled both ways.

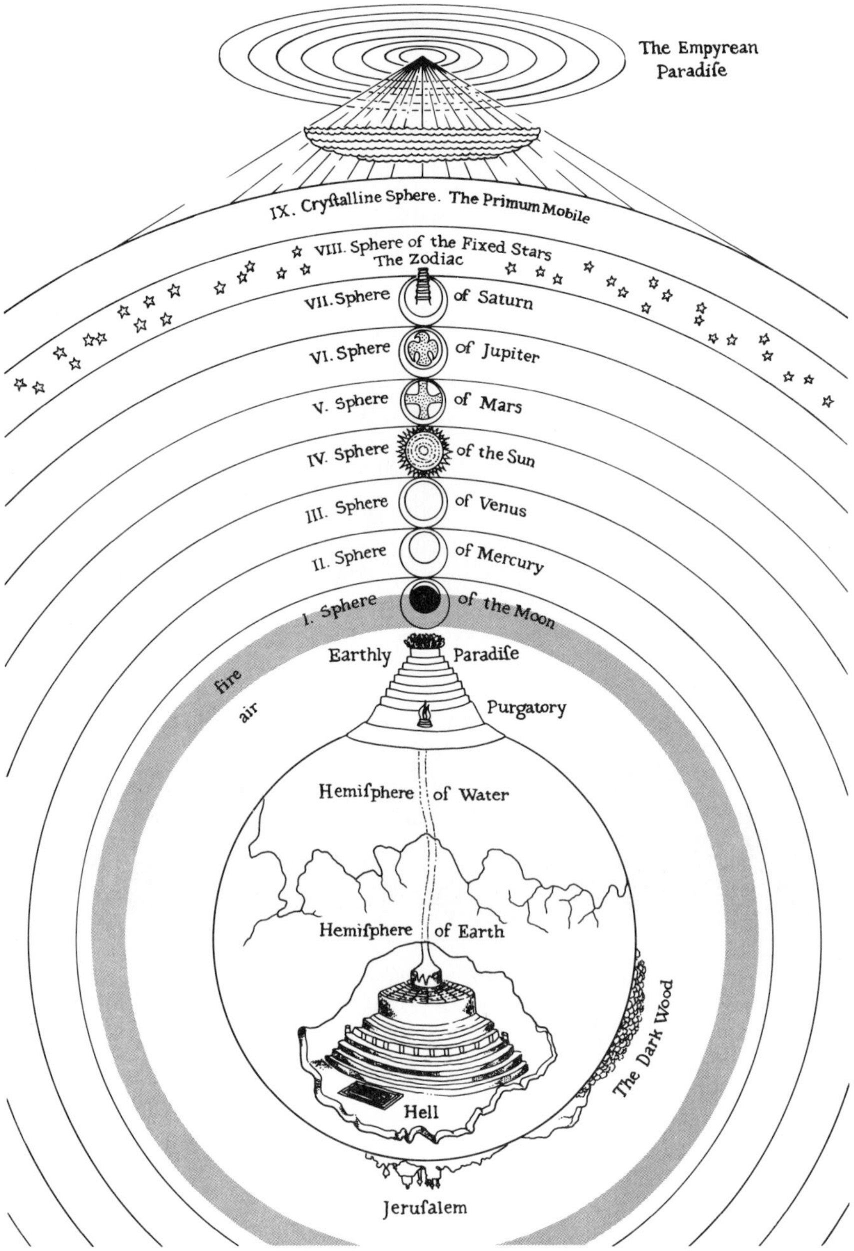

Figure 2-13. *A diagram of the universe, after a medieval sketch based on a description by Dante in the* Divine Comedy, *c. 1310. This diagram shows the persistence of Ptolemy's conception 1160 y after he did his work. The heavens are arranged in seven concentric planetary spheres with the earth at the center. Beyond lies the starry zodiac and an outer "crystalline sphere" marking the boundary of the universe.*

Texts dating from around A.D. 450 use Greek computational methods and refer to the longitudes of both Alexandria and Benares, a major Indian astronomical center. Arabs who later visited India wrote of Brahmagupta (588–660?) as one of the greatest Indian astronomers, who reportedly helped introduce Greco-Indian astronomy to the Arabs.

Unfortunately, most records of this fertile early period of Indian astronomy were destroyed during invasions in the 1100s. The great center at Benares was destroyed in 1194, and various university libraries of Buddhist and other ancient literature were burned in religious wars. A massive observatory—one of the world's five major observatories by the 1700s—was reestablished at Benares in later centuries (and damaged again by invading religious fanatics).

ASTRONOMY IN CHINA: AN INDEPENDENT WORLD VIEW

In legend, Chinese astronomers were predicting eclipses before 2000 B.C.; scholars estimate the time as being closer to 1000 B.C. Thus, Chinese astronomy (symbolized in Figure 2-12, page 43) flourished about the same time as Greek astronomy. Both were probably influenced by the same Middle Eastern cultures. Excellent Chinese observations include the world's best lists of the mysterious "guest stars," mostly novae and supernovae, recorded from 100 B.C. to the present and still consulted by modern astronomers.

Chinese conceptions of the universe were strikingly modern. As early as 120 B.C. a Chinese text stated (quoted by Needham, 1959): "All time that has passed from antiquity until now is called *chou;* all space in every direction . . . is called *yü.*" The Chinese term for the universe, *yü-chou,* was thus similar to the modern concept of space–time, a four-dimensional continuum encompassing all that has existed.

Another Chinese statement of this period is apt (quoted by Needham, 1959): "The Earth is constantly in motion, never stopping, but men do not know it; they are like people sitting in a huge boat with the windows closed; the boat moves but those inside feel nothing." Contrast this with Aristotle's view of the same era, which, for most Westerners, put a stationary earth at the universe's center. Unfortunately, these advanced ideas had little influence on Western astronomy until after the Renaissance.

The Chinese astronomer Yü Hsi reportedly discovered precession independently, stating in A.D. 336, "I think that the heavens are infinitely high." At that time in the West, Ptolemaic groundwork was being laid for the medieval conception that planets moved in as many as 10 separate physical shells that surrounded the earth and held up the stars, as shown in Figure 2-13.

Of course, some of these Chinese ideas were only speculations and not data obtained from experiments or observations. Experimental data were not available anywhere at that time.

ASTRONOMY IN POLYNESIA: STARS AS GUIDEPOSTS

Though they had cosmological myths, the Polynesians were not so much interested in understanding the relations of worlds in space as in plotting the relations of islands in the trackless Pacific. Driven by this need, Polynesian navigators developed amazing skills. Lacking the magnetic compass, they steered toward chosen directions by learning where certain bright stars rise and set on the ocean horizon. Those directions remain constant for each star (except for the very slow changes of precession occurring over several generations). Modern Polynesians can sail such a constant setting that they can accurately measure subtle nighttime shifts in wind direction on the open ocean even though they themselves are moved by the wind (Lewis, 1973).

Polynesian navigators found their latitude by noting which stars passed directly over their masts. For example, in A.D. 1000, if Aldebaran passed overhead, the boat was about 5° south of Hawaii's latitude; if Arcturus, about 100 miles north. This method was accurate to about 0.2° to 0.5°, or about 12 to 30 miles. To reach, say, Hawaii from Samoa, one could simply sail north to the correct latitude, then east until the island was sighted.

AMERICAN INDIAN ASTRONOMY: NATIVE SCIENCE CUT SHORT

Many people still underestimate the sophistication of American Indian civilization. American natives, as noted in Chapter 1, built astronomically aligned observatories, tracked and recorded planetary positions, and devised calendars based on eclipse cycles. Much of this "protoscience" developed between A.D. 100 and 1200, especially in Central America. The influence of Central American civilization reached as far north as the "frontiers" of the Mississippi Valley and the New Mexico–Arizona deserts. Cultural decline occurred around 1300 throughout Central America and the northern frontiers, perhaps due to climate changes. Many astronomical traditions were preserved, however, and European explorers found American natives in various regions conducting astronomical observations of solstice dates, planetary positions, and so on.

Particularly interesting is the astronomy flourishing in the American Southwest when the Spanish arrived, because it has been preserved and practiced until modern times in the New Mexico pueblos and Arizona Hopi villages. Ethnographers of the 1800s and early 1900s recorded many of the ancient traditions passed down by Indian priests. With interest in archaeoastronomy surging since the 1970s, many of these records are being restudied in order to interpret possible astronomical functions of ancient Indian ruins. Zeilik (1985) gives a good survey of Pueblo astronomy. Priests were charged with studying sunrise and sunset positions in order to *predict* dates such as solstices for religious ceremonies and to set times for planting crops. Using the principles of Stonehenge, they determined dates either by observing the sun's position on the horizon from a cetain spot relative to distant mountains or by measuring the sun's rising or setting position by observing the pattern of light and shadow cast through windows onto special markers placed in opposing walls. The latter method supports the theory that specially shaped windows found in the ruins of Casa Grande, built in central Arizona around A.D. 1350, were solstice observing sites, as shown dramatically in Color Plate 1.

Summary

We have seen how various cultures, preliterate and technological alike, moved toward certain concepts about the earth as a world among other worlds in space. Some of these concepts were purely practical; some, abstract. These movements came in fits and starts, and were scattered throughout the world. Progress toward knowledge has not been continuous. Cultures have advanced and regressed, depending on their stability and vigor.

Many of the discoveries reviewed here dealt with the relation of the earth, the moon, and the planets. Table 2-1 furnishes a good review of key observations by the Greeks. Their advances, among all those of antiquity, were most important in influencing Western scientific thought. Although conceptual models of the universe differed from culture to culture, all cultures moved toward discovering astronomical relationships.

Table 2-1. *Astronomical Discoveries of the Greeks*

Observation	Inference	Observer Commonly Quoted
Curved lunar terminator	Moon round	Pythagoreans
Round shadows during lunar eclipses	Earth round	Pythagoreans
Crescent phases of moon	Moon between earth and sun	Aristotle
Different stars at zenith at different latitudes	Earth round	Aristotle
No evident stellar parallax observed by naked eye	Distance earth moves is small compared with distances to stars	Aristotle
Relative sizes and angles of moon and earth's shadow	Moon smaller than earth; sun bigger than earth	Aristarchus
Angle from first quarter to last quarter moon slightly less than 180°	Sun tens of times farther away from earth than moon	Aristarchus
Relation between angular shift of zenith position among stars and linear distance traveled on earth	Calculable circumference of earth	Eratosthenes
North celestial pole's shift with respect to constellations	Precession	Hipparchus

Concepts

cosmology
science
degree
minute of arc
second of arc
linear measure
angular measure
subtend
resolution
terminator
parallax
small-angle equation (from Optional Basic Equation)
significant figures (from Optional Basic Equation)
Alexandrian library
precession
epicycle theory

Problems

1. How critical can we be of early theorists who believed the earth is at the center of the universe? Explain, considering the following questions.

a. Did they have any basis for not putting the earth at the center?
b. Did either possibility fit the available observations?
c. Did any of the Greeks *prove* that any celestial bodies revolve around a central earth or do not do so?

2. As the moon goes through its phases,

a. why is its terminator usually curved?
b. at what lunar phase is it straight?
c. at what lunar phase is it not seen?

3. Contrast the types of astronomical observations available to the Greeks with those available today.

4. Do you agree with the quotation from Plato implying that astronomical events may have influenced the beginnings of philosophizing about the universe? Could astronomical phenomena have influenced religious concepts?

5. Do you think Aristotle's faith in symmetry and "perfection" helped or hindered his investigations of the universe?

6. Why is the moon's umbral shadow on the earth much smaller than the moon itself? (*Hint:* See Figure 2-5.)

7. How does Hipparchus' discovery of precession prove the existence of earlier, careful astronomical records of stars' positions?

8. Do you think destructive events such as the pillaging of the Alexandrian library or the sacking of the observatory and library at Benares are significant or insignificant in world history? (The answer, of course, requires that you define *significant.*)

9. Does the sun pass through the zenith every day on the equator? If not, on what dates does it do so?

ADVANCED PROBLEMS

10. At latitude 40° N will the sun ever pass through the zenith? At latitude $23\frac{1}{2}$° N? If so, on what date(s)?

11. The angular diameter of the sun is roughly $\frac{1}{2}$° and its distance from earth is about 150 million kilometers.

a. Use the small-angle equation to estimate the sun's diameter.
b. Could Aristarchus determine the ratio of the sun's diameter to its distance?
c. Why could the Greeks, such as Aristarchus, not determine the sun's linear diameter or distance?

12. American TV pictures have about 435 resolution elements (sometimes called picture elements, or "pixels") along the width of the picture.

a. Assuming a TV picture is photographed with a lens giving the standard "snapshot" field of view of 40°, calculate the angular size of the smallest details resolved, and compare this with what the eye could see from the same viewpoint.

b. European television systems mostly have 20 to 50% more scanning lines across the screen and more resolution elements than American TVs (except in England, where the number is about 23% less). Comment on the resolution and sharpness of these TV images.

13. A backyard telescope with aperture greater than about 15 cm (6 in.) can reveal features with an angular diameter 1″. The moon is about 400 000 km away. What size lunar crater could you see with such a telescope?

14. If you landed on a small satellite and found that walking in a 1-km straight line caused the stars in front of you to rise 1° farther above the horizon (while stars overhead also shifted by 1°), what would be the satellite's circumference? What would be its diameter?

Projects

1. Using a distant, strong light source such as the sun or a light bulb, a small ball to represent the moon, and your eye to represent a terrestrial observer, show that crescent phases of the moon prove that it passes between the earth and the sun.

2. If you travel far enough during vacation to change your latitude significantly, compare measurements of the elevation of the north star (or the sun during daytime, taking care not to stare directly at it) made from various points in your trip. A sighting device like that described for the problems in Chapter 1 can be used. (Point the device at the sun by watching its shadow; don't look directly at the sun.) How accurately can latitude be determined in this way? Measure the number of kilometers or miles corresponding to your change of latitude, and estimate the circumference of the earth (this method is similar to Eratosthenes'). The project can be done as a class effort, with different people's reports of elevation angles plotted against their latitude to give a curve showing how elevation angle changes with latitude. Coordinate with your instructor.

3. With a camera and fairly fast black-and-white film, such as Tri-X, make time exposures of the night sky. Try different exposures, such as 1 min, 5 min, and 1 h. Make one series including the North Star, one toward the eastern or western horizon, and one toward the southern horizon. Explain the patterns made by the trails.

4. During a camping trip or late-evening outing, pick an equatorial constellation in the sky and follow its motion. Make a series of sketches at different hours, showing its position with respect to the horizon. Do the same for a circumpolar constellation and contrast the results.

5. Using star maps, trace out the position of the celestial equator in the sky. Compare this with the position of the sun's path, the ecliptic.

PART B

Exploring the Earth-Moon System

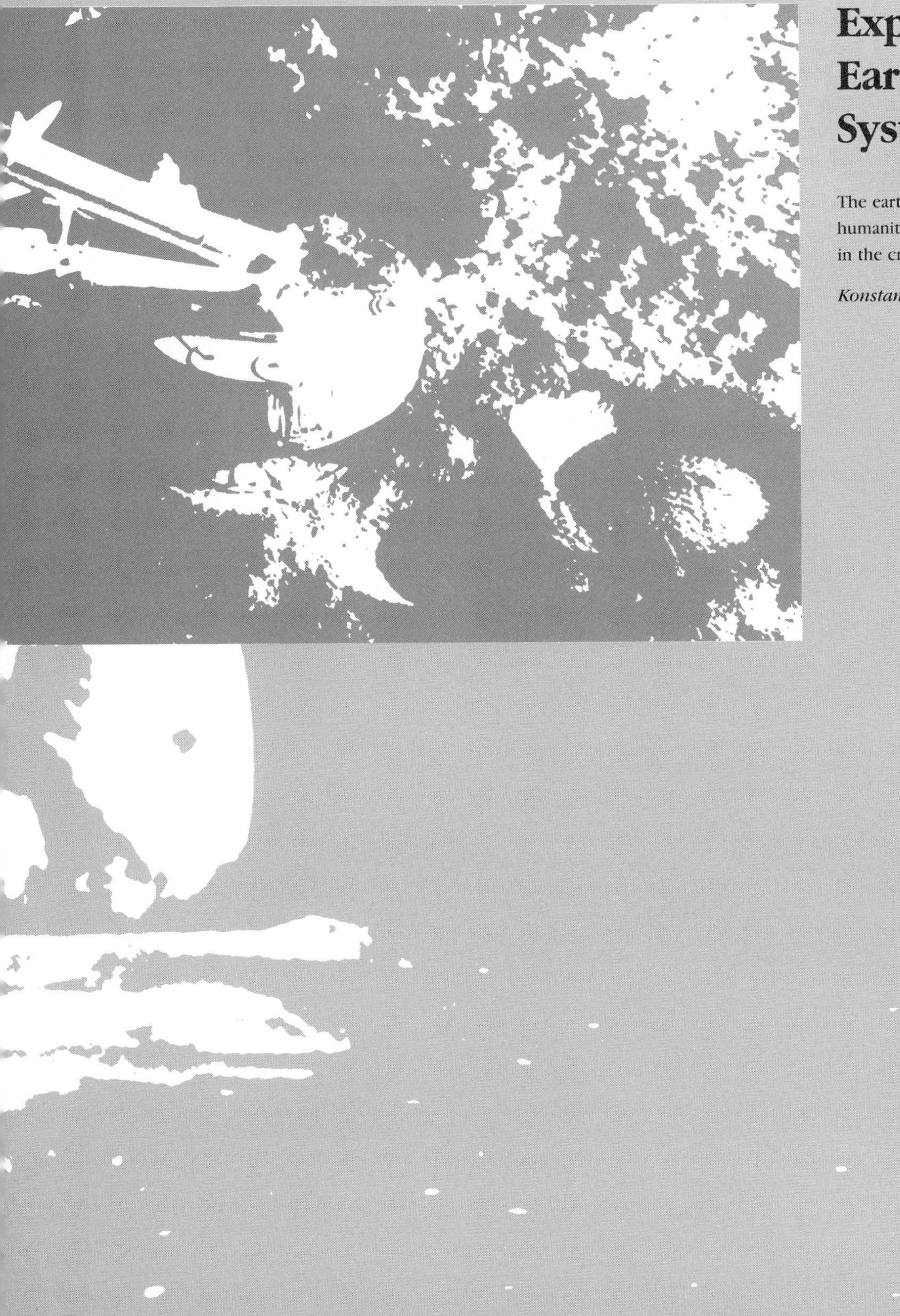

The earth is the cradle of humanity, but one cannot live in the cradle forever.

Konstantin Tsiolkovsky, 1899

CHAPTER 3

THE EARTH AS A PLANET

Some readers might wonder, "Why discuss the earth in an astronomy book? Why not relegate this material to a course in geology?" The answer is dramatized by Figure 3-1. The soil in your backyard is just as much a sample of planetary material as a piece of the moon. People have taken a long time to realize this, and many people still don't recognize the implications of the astronomical discovery that we are spinning through space on a small, finite ball, as seen in Color Plate 2.

When Renaissance astronomers realized that other planets were worlds, they assumed that they closely resembled the earth. This led to the idea of "the plurality of worlds": the assumption that other planets had climates and life forms like the earth's. Now we realize this is only partially true. Our spaceships have brought us photos of other worlds, showing lonely lava flows, rugged craters, dusty winds, and scudding clouds. We realize now that the earth is just one planet out of many, one natural laboratory in which one set of "input" conditions has acted to produce one set of rock types, one kind of climate, and a certain set of life forms. If we can understand how the earth's conditions have evolved, we will be better equipped to understand how other input conditions have produced different environments on other planets.

Conversely, astronomical exploration of the planets has helped us evolve a new picture of the earth. We now know that the earth's crust is not a static platform for our activities but is constantly crumpling due to forces arising in the earth's interior. These processes have much to do with exposing the earth's mineral resources, upon which we have built much of our technological civilization. These same processes, which are rare in the sola system, also help us understand differences between the earth and other planets (Color Plates 11 and 12).

Furthermore, modern astronomy has revealed that the earth's landscape and climate are also influenced by astronomical forces from outside the earth. The purpose of this chapter, then, is to describe some astronomical insights and evidence about planetary evolution that have been discovered primarily by studying the earth as a planet.

The Earth's Age and Origin

The earth has erupting volcanoes, surging oceans, shattering earthquakes, sliding glaciers, and a tempestuous atmosphere. Why isn't the earth just static? What makes the earth evolve? How did it come into being? How old is it?

A variety of astronomical and geologic evidence indicates that the earth was formed about 4.6 billion (4.6×10^9) years ago from particles orbiting the sun and that the earth displays many ongoing evolutionary processes caused by both internal heat and external forces. Chapter 12 will detail the process that formed the earth and all the other planets; here we will emphasize evolutionary processes that will help us understand the conditions we find not just on the earth but also on other planets.

DEBATES ABOUT THE EARTH'S AGE

Our view of the earth as an evolving planet of great age has not come easily. Until a few centuries ago, the earth's age was hardly even considered, except for vague ideas that it was very old. In 1646, English scholar Sir Thomas Brown wrote that determining the earth's age "without inspiration . . . is impossible and beyond the Arithmetick of God himself." Around 1650, several other scholars hypothesized that the earth's age and history could be deciphered from references in ancient scriptures. Archbishop James Ussher used this method and calculated that the whole cosmos was formed on Sunday, October 23, 4004 B.C., and that humanity was created on Friday, October 28.

However, also in the seventeenth century, naturalists realized that if sediments accumulated at the rate measured at the mouths of European rivers, many more millennia would have been required to accumulate the sediment deposits actually observed.

Even as evidence for the earth's great age accumulated, the idea of geologic evolution, or change, was as controversial as the later question of biological evolution. Many naturalists argued that all landforms were created by sudden catastrophes, like earthquakes, rather than by very long term processes. As late as the 1700s other people argued that fossils and strata deposits interpreted by geologists as signs of evolving climates and landforms were merely "devices of the Devil" put in rocks to mislead us. Even today some "creationists" argue that fossils and other geologic features were created and put in the ground all at once. By the same logic one might imagine that all "ancient" artifacts and history books were fabricated and placed in museums in, say, 1836. Scientists reject such views on the grounds that they have little productive consequence; they don't lead us toward new observations or possibilities for new discoveries.

Scientists who studied our planet from the 1500s to the 1800s were influenced by medieval conceptions that the earth started in an Eden-like state (Davies, 1969): "a magnificent, lush estate which God had stocked with everything necessary for human well-being before admitting man as a freehold tenant." Those who accepted the earth's evolution felt that it had been a process of shriveling, as an apple dries out. Mountains were evidence of the earth's decline in old age, "even as warts, tumours, wenns, and excrescencies are engendered in the superficies of men's bodies." This view even affected perceptions of the landscape. When John Dennis crossed the Alps in 1693, he reported not the majestic panoramas that impress us, but "only a vast, but horrid, hideous, and ghastly ruins."

The new concept of changing geologic features fascinated Victorian thinkers. Tennyson wrote, "Where the long street roars hath been the stillness of the central sea . . . the solid lands, like clouds they shape themselves and go."

Study of landscapes and rocks during the 1800s and 1900s contradicted theories of sudden catastrophes or mere passive shriveling as mechanisms forming most geologic features. For example, study Figure 3-2. Several thoughts emerge. First, such a smooth fold would not result from a sudden catastrophic compression. Rocks shatter in such an event. This fold suggests a slow evolution of the earth's landscapes. Second, at measured deformation and erosion rates, the earth must be much more than a few thousand years old to allow time to produce such folds and then to allow erosion to expose them. Third, enormous forces must be at work to compress and crumple regions the size of mountain chains. Photographs of Earth taken from space enable us to sense this more clearly, as seen in Figure 3-3. At the

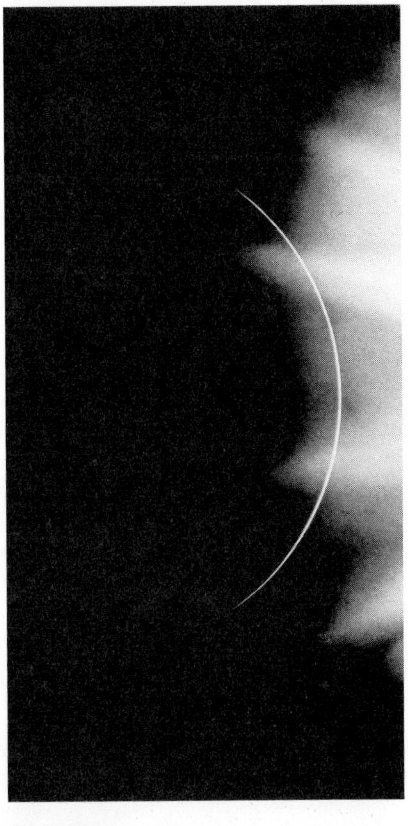

Figure 3-1. *The earth as a planet. The crescent earth was photographed by Apollo astronauts. It is back-lit by the sun nearby just out of the frame to the right. (NASA.)*

Figure 3-2. *Folded strata show that rock layers can deform plastically when subjected to stresses over long periods of time, rather than shatter as they would do if subjected to comparable sudden stresses. Note figures (top) for scale. (Lulworth Cove, southern coast of England. Photo by author.)*

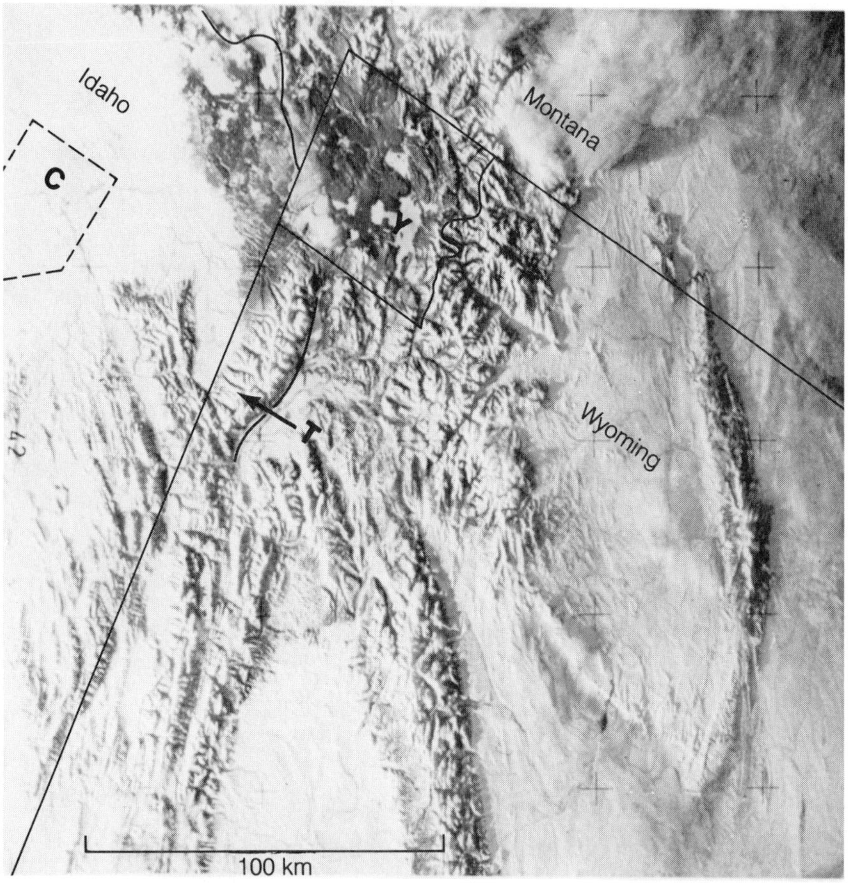

a

b

Figure 3-3. *Orbital and ground views of the Rocky Mountains.* (a, *left) View with a hand-held camera by a Skylab astronaut shows crumpled, eroded mountains unique to Earth.* Y = *Yellowstone Park, a crustal hot spot with active geysers;* C = *area of recent volcanism near Craters of the Moon National Monument;* T = *large fault forming east side of the Grand Teton Mountains. The arrow marks the direction of view of these peaks in part* (b). *(NASA.)* (b, *right) Eastern face of the Grand Teton Mountains, exposed by faulting of the earth's crust. The fault has exposed a vertical fissure through which basaltic lava ascended toward the original surface (far above the present surface, which has been lowered by erosion of the mountain). (Photo by author.)*

same time, observations of volcanoes, earthquakes, movements along fractures, and other features show that mountain formation continues even today. Earth is still actively evolving.

MEASURING ROCK AGES BY RADIOACTIVITY

All questions of the earth's age and evolution rates were clarified by the discovery of a reliable technique to determine the ages of rocks. This subject might seem unrelated to astronomy, but remember that Earth's rocks, like lunar rocks and meteorites, are cosmic materials. We live on a large space-rock.

In 1896 the French physicist Antoine Becquerel accidentally left some photographic plates in a drawer with some uranium-bearing minerals. Later he opened the drawer and found the plates fogged. Being a good physicist, he did not dismiss the event but investigated. He found that the uranium emitted "rays," which, like Roentgen's X rays of 1895, could pass through cardboard. The rays turned out to be energetic particles emitted by unstable atoms. Radioactivity had been discovered.

A **radioactive atom** is an unstable atom that spontaneously changes into a stable form by emitting a particle from its nucleus. The original atom thus becomes either a new *element* (change in the number of protons in the nucleus) or a new *isotope* of the same element (no change in the number of protons but a change in the number of neutrons). The original atom is called the **parent isotope** and the new element is called the **daughter isotope**.

The time required for half of the original parent isotopes to disintegrate into daughter isotopes is called the **half-life** of the radioactive element. If a billion atoms of a parent isotope were present in a certain mineral grain in a rock, a half billion would be left after one half-life, a quarter billion after the second half-life, and so on. As examples, half the rubidium-87 atoms in any given sample change into strontium-87 in 50 billion years; uranium-238 changes into lead-206 with a half-life of 4.51 billion years; potassium-40 changes into argon-40 with a half-life of 1.30 billion years; and carbon-14 decays into nitrogen-14 with a half-life of only 5570 y.

Early in the twentieth century, physicists realized that here was a way to determine the ages of rocks. Suppose we could determine the *original* number of parent and daughter isotope atoms in a rock. In part, this can be done by measuring numbers of stable isotope atoms, which usually occur in certain proportions to the unstable isotope atoms in a given fresh mineral. Then, if we simply count the *present* numbers of parent and daughter isotope atoms in the rock, we can tell how many parent atoms have decayed into daughter atoms and hence tell how old the rock is. If half the parent atoms have decayed, the age of the rock equals one half-life of the radioactive parent element being studied. This technique of dating rocks is called **radioisotopic dating**, since radioactive atoms that decay are called radioisotopes.

It is important to note that the quantity being determined in radioisotopic dating of rocks is time—the time since the rock began to retain the daughter element. In most cases being discussed here, this is the time since the rock solidified from an earlier molten material, such as lava. Once the rock solidifies, any daughter isotope atoms are trapped in the solid mineral structure.

MEASURING THE EARTH'S AGE

To measure the earth's age, you might expect that we would merely search for the oldest known rock on the earth and assume that it dates from the earth's creation. In reality it is not so simple. The earth is geologically so active that rocks dating from the earth's formation have long since eroded away. For instance, many rocks in the Rocky Mountains of the United States formed about 60 million years ago or less; many rocks in the eastern United States formed a few hundred million years ago. Older, more stable parts of North America exist around Hudson's Bay, where rocks a few billion years old have been found. Such stable, ancient regions yield the oldest known rocks. These regions, which have been found in several continents, are called **continental shields**, after their flat, circular shapes. The oldest known earth rocks, formed about 3.9 billion years ago,[1] were found in Greenland, the eastern extension of the Canadian Shield. Rocks that have survived so long are extremely rare.

These results mean that the earth must have formed more than 3.9 billion years ago. Without additional information, we would not be able to fix the age exactly. However, as we will see in the next few chapters, more information comes from outside the earth. Lunar rocks and meteorites show that *all the planets formed within a relatively short interval (about 50 to 90 million years) about 4.6 billion years ago* (Pepin, 1976). The **age of the earth** is therefore believed to be 4.6 billion years.

The Earth's Internal Structure

The world's deepest drill hole as of 1984 was a 12-km shaft in the Soviet Union (Kozlovsky, 1984). It has yielded interesting data on deep rocks, but no drill hole is deep enough to reveal the earth's deep internal structure. Our best clues about the interior come, instead, from waves that pass through the earth's material. If you snap your finger on the surface of a calm swimming pool, waves radiate across the surface from the disturbance. Observers at other edges of the pool can gain information about the disturbance by observing the waves; for example, they can locate the disturbance by comparing the directions from which the waves come.

In the same way, waves are generated in the earth by earthquakes, and these carry information not only about the earthquake but also about the earth's material. Waves traveling through the earth (or other planets) are called **seismic waves**. Some travel along the surface and others penetrate through the earth. The velocity and characteristics of the waves depend on the type of rock or molten material that they traverse. For example, as shown in Figure 3-4, waves that travel through the earth from an earthquake on the far side traverse deep interior regions. Such waves differ from waves that travel a shorter distance. Waves from adjacent earthquakes reveal distinct differences in rock layers at different depths.

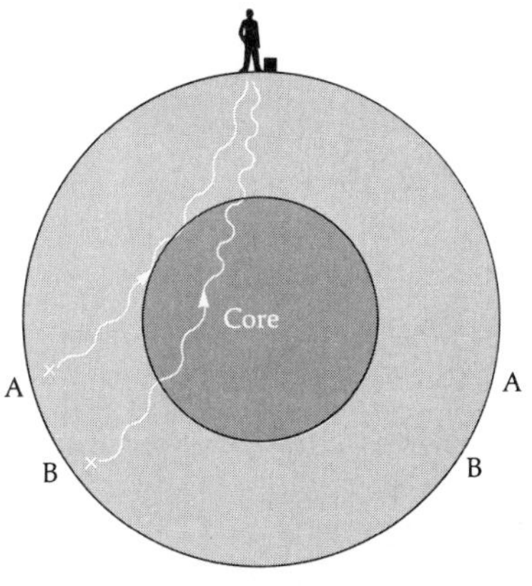

Figure 3-4. *The earth's core was discovered in 1906 when observers discovered that earthquakes on the far side of the earth (beyond* B *from the observer) produce seismic waves distinctly different from those of nearer earthquakes (closer than* A*). Waves from* B *and beyond are altered by passing through the core. The wave paths are curved because rock layers at different depths have different properties.*

[1]Expressing geological ages is a continuing editorial problem. This book will use the most common American convention: 1 billion years = 10^9 years = 1 000 000 000 years (abbreviated "1 b.y."). This would be fine except that the English use 1 billion to mean 10^{12}! Therefore some authors have adopted 10^9 years = 1 aeon (or eon). Other authors have adopted the international prefix "giga-," which stands for 10^9, and they use 10^9 years = 1 Gy (see Appendix 1).

Such studies have revealed two important different types of layering in the earth: chemical and physical. Chemical layering refers to layers of different composition. Physical layering refers to layers of different mechanical properties, such as rigid layers versus plastic or fluid layers.

CHEMICAL LAYERING OF THE EARTH: CORE, MANTLE, CRUST

Chemical layering was the first type of layering recognized. Seismic and other data indicate that the earth contains a central **core** of nickel-iron metal. The core's radius is about 3500 km, just over half the earth's radius, as shown by the dark material sketched in Figure 3-5.

The core is surrounded most of the way out to the surface by a layer of dense rock, called the **mantle**. Near the surface, the densities of the rocks are typically lower. The **crust** is a thin outer layer of lower-density rock about 5 km thick under the oceans and about 30 km thick under the continents.

The core-mantle-crust structure gives us important clues about the history of the earth and other planets. First, it shows the importance of **differentiation processes**—processes that separate materials of different composition from one another. Most geologists believe that the key differentiation process in the earth was melting of much of the inner rock material after the earth formed. The source of the heat was radioactive minerals trapped in the earth as it formed. Gradually those minerals released heat as radioactive atoms decayed. The interior of the earth was so well insulated by overlying rock that the heat could not escape. The temperature rose until the rock melted. When the rock melted, heavy portions like metals were able to flow downward toward the center, while lighter, low density minerals could float toward the surface, where they eventually solidified into a crust of low density rock.

As a more detailed example of this theory, we note that one of the lowest density and most common minerals to form in cooling, molten rock is called feldspar. If the theory is right, feldspar should have formed and floated toward the surface. Indeed, we find that surface rocks of both the earth and moon are extremely rich in feldspar. One of the rock types formed from mixtures of minerals rich in feldspar is called **basalt**. Basalt is the common lava that erupts from volcanoes that tap the crust and upper mantle, and basalt is also common on the moon's surface.

A still lower density type of rock is **granite**, the light-colored rock commonly formed from molten materials in continents. In some ways, continents seem to be a low density granitic "scum" floating on the denser rocks of the basaltic lower crust and upper mantle.

A simple example of differentiation that can produce such layering occurs during the smelting of metal ores. When the ore is melted, the metal sinks to the bottom of the vat (core) while the bulk of the rock fills the upper part of the vat (mantle). On the surface floats a thin scum of slag, or low density rock (crust).

Support for these ideas comes from meteorites, the stone and metal fragments that fall onto the surface of Earth from space. These will be discussed further in Chapter 11, but we note here that they are believed to be fragments from the surfaces and interiors of small planetary bodies broken during collisions. Interestingly, they show many of the rock types we have discussed: Nickel-iron meteorites are probably metal fragments from cores of these small worlds, similar to Earth's core. Basaltlike meteorites are probably fragments of mantles and crusts. Thus we see that many worlds may have gone through a differentiation process that led to structures like the earth's.

PHYSICAL LAYERING OF THE EARTH: LITHOSPHERE AND ASTHENOSPHERE

A second type of layered structure has been discovered involving layers of different rigidity. This type of layering is extremely important in determining what types of features we see in the landscapes around us on Earth, and on other planets, as we will see in a moment.

Layering by rigidity was created during the cooling of the earth. As you can imagine from the example of the smelter vat, the interior stays molten for a long time because it is hard for the heat to get out, but the surface cools fast because it is exposed to the atmosphere and surrounding space, and the heat can easily radiate away. Thus, if a planet were melted and left to cool, a solid layer would form on the surface, underlain by a still-molten liquid layer. This picture is complicated by the fact that the layers at different depth have different compositions (due to the differentiation discussed above) and are also at different pressures. They may thus have different solidifying temperatures, and they may actually form alternating layers of solid material, molten material, or partly molten slushy material. A partly molten layer can by visualized as a layer in which some low-melting-point minerals are melted, but high-melting-point minerals are not. (For example, if you heated a mixture of ice and sand, you would get a mixture of solid sand grains with moisture between them.)

The solid layer at the surface of the earth or any other such planet is called the **lithosphere** (from Greek roots meaning "rock shell"), and it is underlain by a a partly melted layer that is much less rigid and less strong. The underlying partly melted layer is called the **asthenosphere** (from Greek roots for "weak shell").

The reason the lithosphere and asthenosphere are important is that their interplay determines the surface features of Earth, such as mountains, seafloors, and continents. To understand this, we need to review some principles of heat flow, to describe how heat gets out of the hot interior of the earth or any other planet, once the planet is heated by radioactivity. Heat always flows from hot to cooler regions, and can flow by three methods. The first is radiation; an example is the radiant heat that reaches us through space from the sun. The second is conduction; an example is the flow of heat through a metal cooking pot, whose handle might grow too hot to touch. The third is convection, which is heat flow by movement of the heat-carrying medium; an example is the

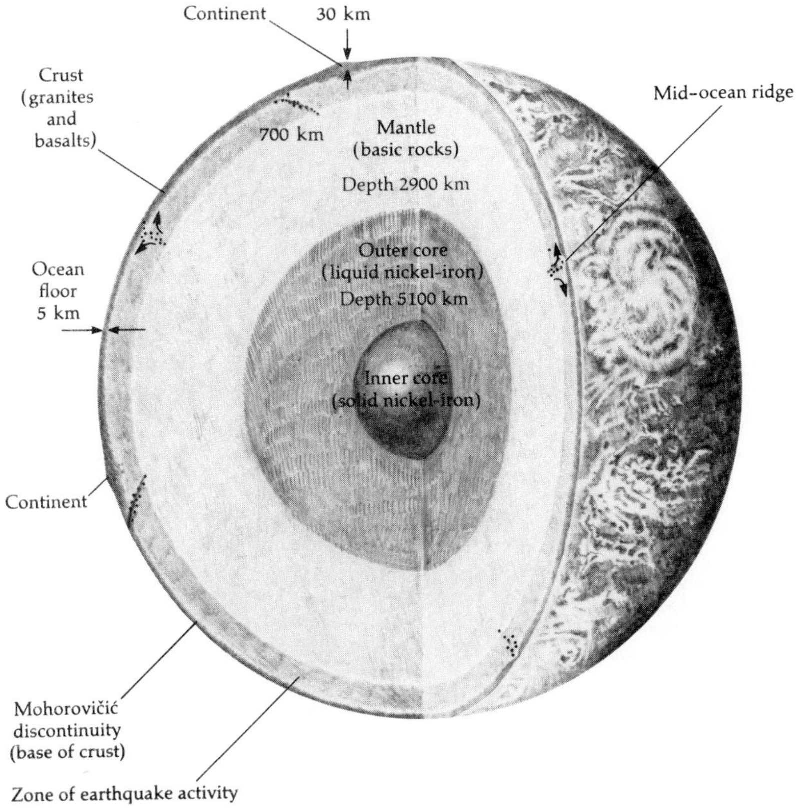

Figure 3-5. *A simplified diagram of the earth's interior structure as revealed by modern seismic studies. Dots indicate typical earthquake positions.*

buoyant ascent of warm air in a thundercloud, or of a hot-air balloon. If you put an inch-deep layer of cooking oil in a pan and heat it from below on a stove, you may be able to see patterns of currents, called convection cells, set up in the oil as it convects.

The relation of these ideas to a planet is as follows. Once a planet heats up, its insides cool by all three processes. Heat radiates from the surface into space; heat is conducted outward from the hot center through rock; and convection may occur in some of the more fluid layers where currents can flow, like the convection cells in the cooking oil. As the planet cools, the surface layers solidify, forming a relatively rigid surface layer of rock—the lithosphere. In the earth, the relatively rigid lithosphere is a layer about 100 km deep, and the relatively fluid asthenosphere is the underlying layer from about 100 to 350 km deep. Even though the asthenosphere is not totally molten, it seems to be plastic enough for sluggish convection currents to flow. These currents drag on the underside of the lithosphere, setting up stresses in it and tending to crack it into separate, large-scale pieces.

Just like arctic explorers crossing ice floes that are floating on the ocean, we are living on a rigid layer that "floats" on a more fluid base. Of course, we are not in quite so much danger of a cracking floor as the arctic explorer, but cracks do occur. Figure 3-6 shows a result. As the asthenosphere shifts, it can stretch the lithosphere only so far before the brittle lithosphere cracks. This is what we perceive as an earthquake.

Lithospheres and Plate Tectonics: An Explanation of Planetary Landscapes

Tectonics is the study of movements in a planetary lithosphere, such as the movements that cause earthquakes, mountain-building, and so on. For more than a century, geologists studied the earth's tectonics without recognizing the underlying nature of the forces that cause these movements, as just outlined. Only since the 1960s have geologists pieced together a real understanding of how these principles affect the earth, and how they apply to other planets. This new understanding is called the theory of **plate tectonics**.

As the asthenosphere drags on the lithosphere, it cracks the lithosphere into large, continent-scale pieces called *plates*. Further asthenosphere movements tend to drag and jostle the floating plates, sometimes pulling them apart from each other and sometimes pushing them into each other. Cracks along the margins of plates are usually the sites of volcanoes and earthquakes, since molten magma from below can squeeze up to the surface through the cracks, and since plate collisions cause stresses as plates rub together, eventually leading to rock fracturing in the form of earthquakes. This can be seen in a map of volcanoes and earthquakes, such as Figure 3-7. The string of shallow earthquakes down the mid-Atlantic (marked also by volcanoes of Iceland and the Azores) marks a plate boundary where new lava is rising from below. On the ocean floor it erupts and piles into a feature mapped by oceanographers in the 1960s—the mid-Atlantic ridge. Eruption of new lavas at these sites pushes apart on the neighboring plates, causing the Americas to drift westward and Europe, eastward, relative to the plate boundary.

On the west coast of the Americas, the American continental plate collides with the Pacific plates, crumpling and riding up over them as the margins of the Pacific plates slide under the American plates. This is the explanation of the major western American mountain chains like the Rockies, Sierras, and Andes: They mark the crumpling of colliding plates. Fractures caused by earthquakes are called **faults**, and colliding plate margin regions such as California are thus laced with faults like the famous San Andreas fault (shown in Color Plate 3). One of the many offsets along this fault caused the devastation in Figure 3-6.

Other scenes of plate tectonic activity can be glimpsed in Figure 3-7. For instance, the plate containing the subcontinent of India years ago drifted northward, as shown in Figure 3-8. It collided with the Asian plate and caused the massive crumpling that created the Himalayas, marked by shallow earthquakes.

Much geological evidence supports the theory of plate tectonics. This includes matches of geologic provinces of equal date and rock type in eastern South America and western Africa, showing that they were once part of the same land mass, as seen in Figure 3-8. Traces of glaciation and other evidence in southern Africa indicates that this land mass drifted north from a position once much closer to the icebound south pole. The theory of plate tectonics also explains why rock units on Earth older than one or two billion years are so rare. Most older surfaces have been crumpled beyond recognition or driven downward under other plates, only to be remelted, mixed with mantle material, and perhaps reerupted as new lavas. Some of these complex processes of the earth's active geology can be seen in the cross section seen in Figure 3-9.

Figure 3-6. *Occasionally nature reminds us that we are living on a thin, brittle crustal planetary lithosphere that is often broken by motions of underlying semifluid layers. This devastation was caused in 1906 in San Francisco by an earthquake and resultant fire. The quake involved movement on the San Andreas Fault, which passes under the downtown area. (Photo courtesy S. M. Larson.)*

Now we can see a connection with the landscapes on other planets, which will be discussed further in later chapters. Smaller worlds, like Mars and the moon, do not have well-developed crumpled mountain ranges or plate boundaries shown in Figure 3-9. The reason goes back to the basic principles of heat flow and lithosphere formation. Smaller worlds cool faster than big worlds. Therefore, their lithospheres get thicker in the same amount of time. Thus, their surfaces are more stable and more protected against asthenosphere currents far below. Lava does not so frequently gain access to the surface. Convection can't so easily drive plates apart or cause them to drift into each other. Therefore, the surfaces of Mars and the moon have much more ancient structures (such as ancient impact craters) than the earth does.

Other Important Processes in the Earth's Evolution

Volcanism is the eruption of molten materials from a planet's interior onto its surface. On Earth, the asthenosphere contains pockets of partly melted materials, as indicated by seismic wave analysis. This underground molten rock, called **magma**, is under pressure, often charged with gas such as steam, less dense than surrounding rock, and highly corrosive. Therefore it tends to work its way to the surface, especially in regions where fractures provide access. When it reaches the surface, it erupts and is then called **lava**. It may shoot in a foamlike form into the air (due to pressure from dissolved gas, like the spray from an agitated can of pop when it is opened) or ooze out and flow for many kilometers across the ground. If enough lava is erupted, it may accumulate into volcanic mountains. During intervals ranging from years to millions of years, volcanism thus creates landforms ranging

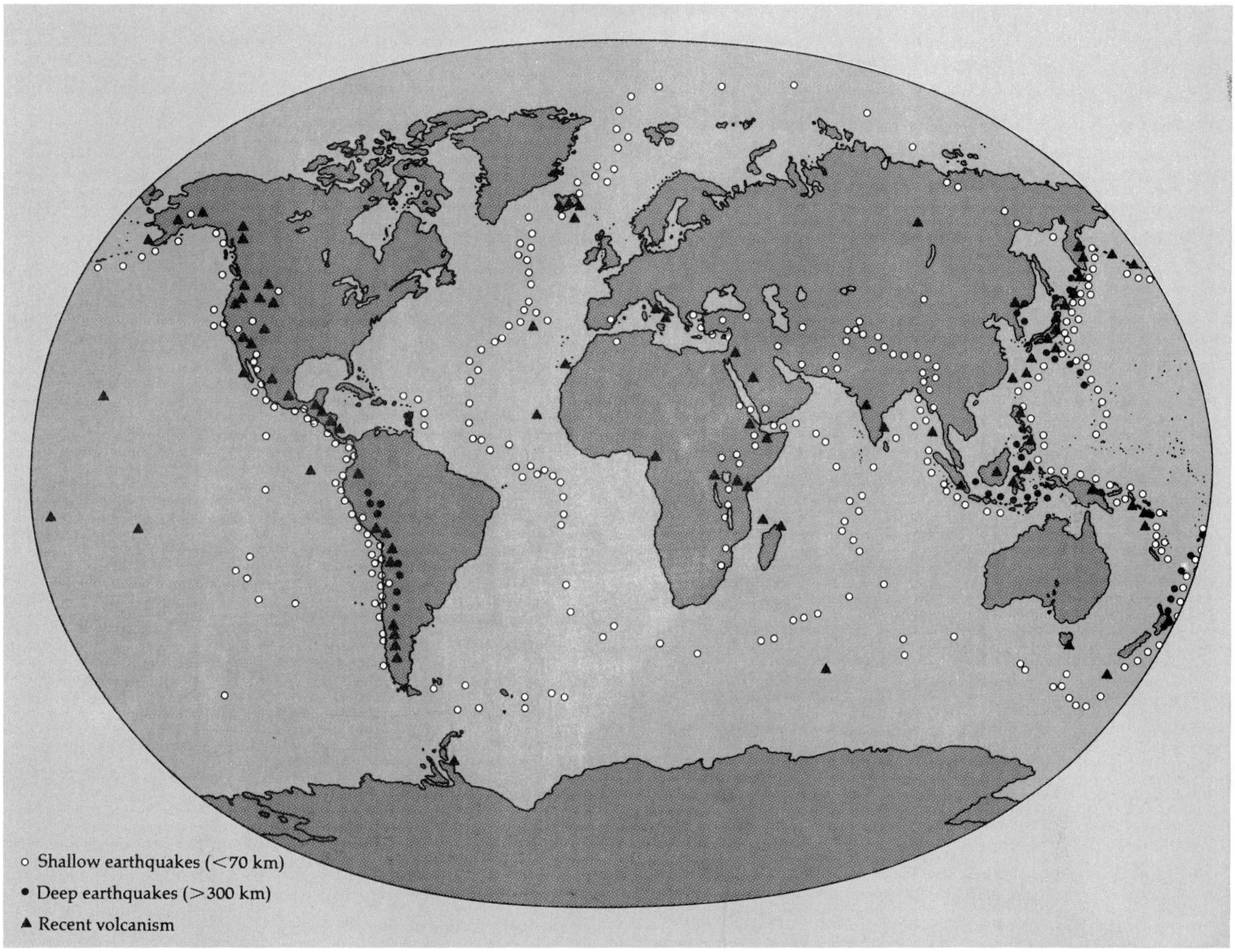

Figure 3-7. *A map of the distribution of earthquakes and volcanoes reveals that lines of features define active margins of tectonic plates. For example, the line of shallow earthquakes down the middle Atlantic lies along the Mid-Atlantic Ridge, the fracture that allowed the Americas to break off from Europe and Africa.*

from flat lava flows to craters to clusters of volcanic peaks, as indicated by Figure 3-10.

Space exploration has shown that volcanism is one of the most important processes forming landscapes on other planets. Some planets have huge lava flows and volcanoes. Study of the earth's volcanoes helps us understand these alien landscapes. Conversely, study of the other planets' volcanic features helps us understand relations we see among terrestrial volcanoes.

Among all known planets, the earth undergoes the most active processes of landform destruction. Largely, this activity is due to the earth's thick atmosphere and flowing water, which other planets lack. **Erosion** includes all processes by which rock materials are broken down and transported across a planet's surface; such processes include water flow, chemical weathering, and windblown transport of dust. **Deposition** includes all processes by which the materials are deposited and accumulated; such processes include deposition of sediments in lake and ocean bottoms and dropping of windblown dust in dune deposits.

On the earth, erosion and deposition, especially by flowing water, are the most important landscape-forming processes, but this is not true on all other planets.

The extreme activity of erosion and deposition on the earth is responsible for wearing away or covering up the most ancient rocks and for wearing away ancient impact craters so rapidly that they were not even recognized on the earth until the last few decades. Similarly, most familiar mountain ranges do not display their initial forms, which may have been caused by volcanism, fracturing, and folding; instead, most mountain ranges are merely resistant rock "cores" left after uplift and erosion of more massive units of crustal strata.

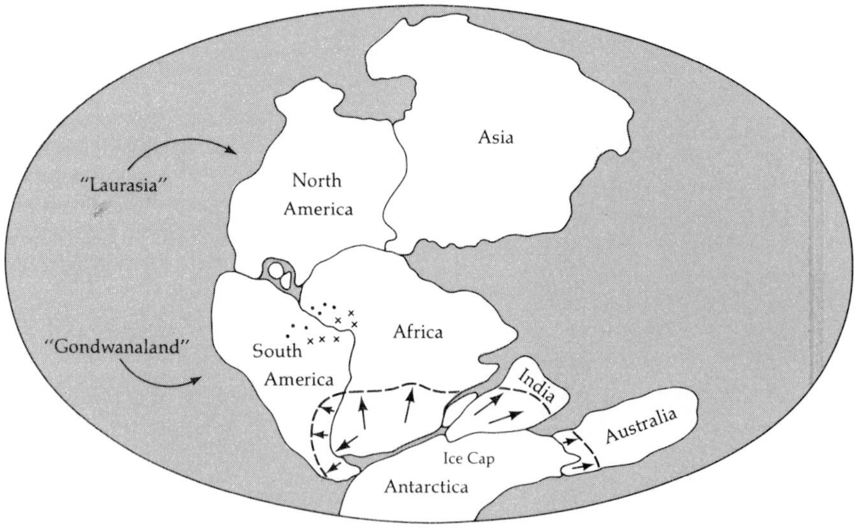

a 300 million years ago

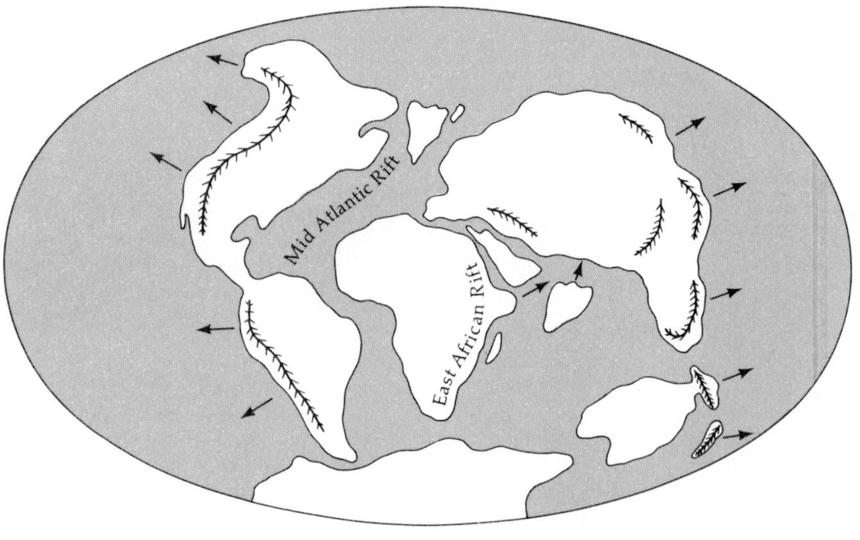

b 50 million years ago

Figure 3-8. *(a) Configuration of the continental masses 300 million years ago at the beginning of the breakup of Pangaea. Dots and x's show fit of geologic rock types once continuous but now on opposite sides of the Atlantic. Dashed line bounds region with geologic evidence of glacial movement along arrows, indicating that these regions were once covered by the south polar ice cap. (b) Configuration roughly 50 million years ago as continental masses were drifting toward their present positions. This shows mountain masses along the leading continental edges, which are collision zones between plates.*

The Earth's Magnetic Field

Physicists use the term **field** to describe any property that can be measured throughout a volume of space. For instance, we know the earth's gravity field extends far out beyond the moon because we can measure its effects on the motions of the moon, spaceships, and other planets. Similarly, if we move a compass throughout the region around the earth, we discover that there is a **magnetic field** defined by the fact that the compass is deflected in a direction roughly toward the north pole. The compass does not point exactly at the true north pole (defined by the earth's spin), but at a point called the magnetic north pole, which moves slightly from year to year and is presently located among Arctic islands off northern Canada.

What causes the earth's magnetism? The interior of the earth cannot be magnetized like a giant bar magnet, because the interior regions are too hot to maintain magnetism. Rather, geophysicists believe that liquid iron in the outer core is slowly convecting, and that these motions set up electric currents in the metal. A magnetic field will exist around any electric current, and the earth's field is thus attributed to the currents in the liquid metal core. The presence or absence of a magnetic field thus becomes an important indicator of the presence of an iron core inside any planet.

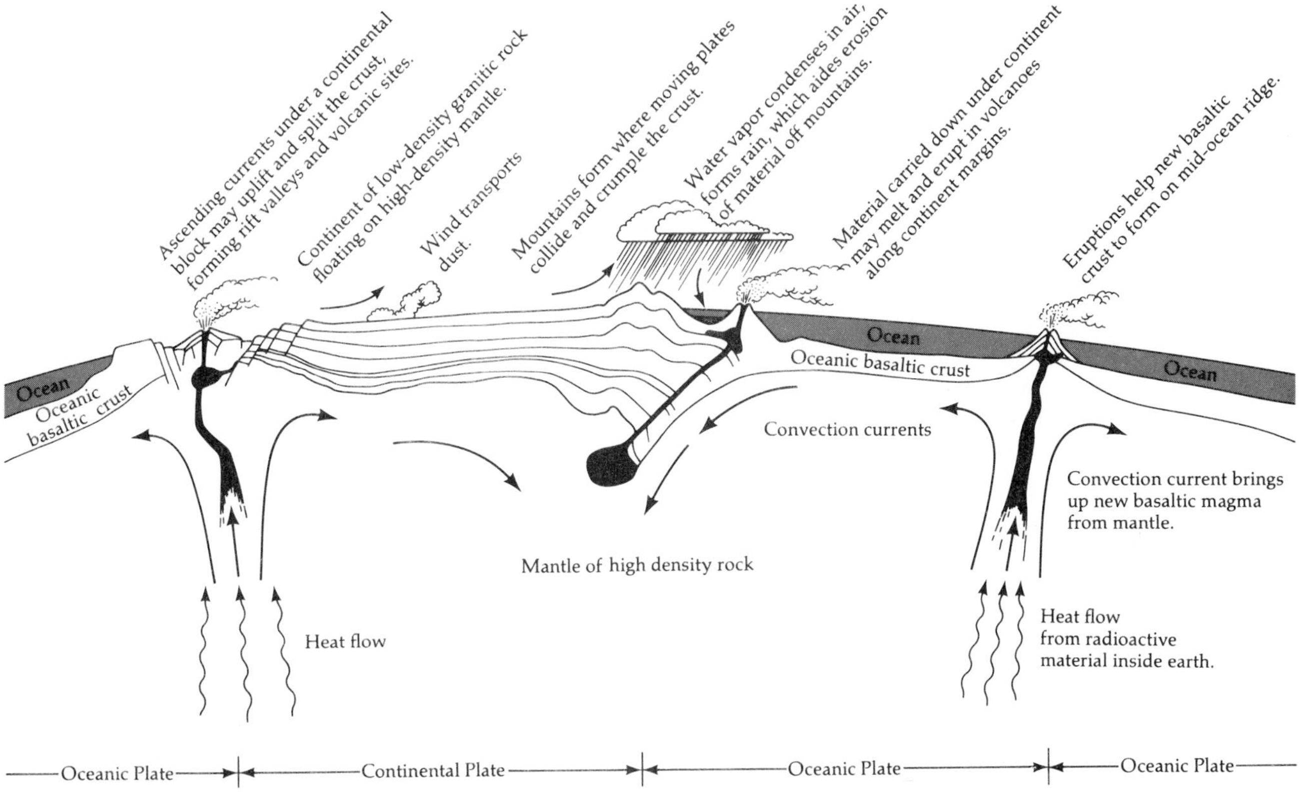

Figure 3-9. *Schematic diagram of some of the processes making the earth more geologically active than most other worlds. Convective heat flow from the interior drives convection currents in the asthenosphere. These in turn crack the more rigid surface layers and move plates. Plate movements cause earthquakes, faulting, the formation of rift valleys (left), and crumpled mountain chains (right). In addition, atmospheric processes and water flow cause rapid erosion.*

Figure 3-10. *Volcanic activity on Earth can build varied landforms. In the background of this volcanic landscape are the San Francisco peaks, near Flagstaff, Arizona. In the middle distance is a crater formed partly by collapse and partly by gas eruption; it resembles an impact crater. In the center is a cinder cone built up by an accumulation of ash and cinders over an eruptive vent. In the right foreground is a flow of dark basaltic lava from the cone's eruptive vent. (Infrared photo by author.)*

The Earth's Atmosphere and Oceans

Just as the earth's interior and surface have dramatically evolved, so has the earth's atmosphere. The original atmosphere, called the **primary atmosphere**, had less oxygen than today's atmosphere. Evidence for this is that sediments deposited more than about 2.5 billion years ago are less oxidized than modern sediments, indicating less oxygen was in the air. In addition, the earliest fossil life forms, dating from about 2.5 to 3.5 billion years ago, are types of algae found today only in oxygen-poor environments, such as salt marshes along seacoasts.

Indeed, most of the earth's present air is regarded as a **secondary atmosphere**—gas that was not originally present[2] but was added by outgassing of the earth's interior from volcanic vents and fissures, as shown in Figure 3-11. Measurements of such volcanic gases show that they are rich in water vapor (H_2O), carbon dioxide (CO_2), and nitrogen (N_2). Today's atmosphere is 76% nitrogen and 23% oxygen by weight. The proposed evolution from volcanic gases to the present atmosphere is described below.

The water vapor emitted by the earliest volcanism condensed into liquid water and eventually formed oceans. The carbon dioxide that was emitted at the same time mostly dissolved in the oceans, where it formed a weak carbonic acid solution. (Most of the carbon dioxide emitted by modern industrial processes does the same thing, with only a small portion of the total remaining in the air.) The carbonic acid attacks rocks and sediments of the seafloor, eventually forming carbonate rocks. Through this direct chemical process, as well as by biological processes, most of the carbon dioxide has thus been tied up in carbonate rocks, such as limestone. The nitrogen (N_2), being chemically inactive, remained in the air, forming the nitrogen observed today.

A small amount of water is always in the air as water vapor, and molecules of water are sometimes struck by powerful ultraviolet rays of sunlight and broken into hydrogen and oxygen atoms. The hydrogen, being the lightest element, tends to rise and eventually escape into space. The oxygen remains behind. Until about 2 or 2½ billion years ago, this was the only source of oxygen, and the atmospheric oxygen content was increasing slowly. After this time, abundant plant life evolved, consuming carbon dioxide and releasing oxygen. The atmospheric oxygen content thus increased more rapidly.

DYNAMICS OF THE EARTH'S ATMOSPHERE

Why do winds blow? Why don't they just blow themselves out, leaving a calm atmosphere? Any turbulent motions require an energy source to keep them going. The motions of our atmosphere turn out to be another astronomical application of the same physics discussed in the case of motions of materials inside the earth: Heat is transferred through the atmosphere and sets up convective

Figure 3-11. *Volcanic gases emitted by fissures (as here) and major eruptions add water vapor, carbon dioxide, nitrogen, and other gases to the atmosphere. Volcanism was a major factor in altering the earth's original atmosphere. (Kilauea Volcano, Hawaii Volcanoes National Park, Hawaii. Photo by author.)*

[2]The primary atmosphere was probably rich in hydrogen (H) compounds, e.g., methane (CH_4), ammonia (NH_3), and water (H_2O), since hydrogen dominated the gas in the early solar system (see Chapter 12).

motions. In this case the energy source is sunlight. Suppose we magically stopped all winds. Sunlight would beat down on the noontime side of the earth, warming the ground and warming the air. The air near the ground would eventually become warm enough to expand and rise. New air would flow in laterally along the ground to take its place; thus winds would arise.

If the temperature gradient in the air is steep enough (that is, very warm air near the ground and considerably cooler air aloft), strong convection currents will be set up, lifting the air to form towering thunderhead clouds.

CORIOLIS DRIFT

The earth's rotation is also involved in determining winds and weather patterns. Consider an air parcel moving northward off the equator toward a low pressure region in the atmosphere. Because the earth is turning, the equatorial air mass would make one trip around the circumference, a distance of 40 000 km, in 24 h. Thus, regardless of any other motions, it has an eastward motion of nearly 1700 km/h, whereas material to the north, near the pole, has no such eastward drift. As the mass moves northward, it moves eastward faster than the ground or local air, as shown in Figure 3-12. A northern-hemisphere air mass drifting south toward the low pressure zone would drift westward. Any such eastward and westward deflections due to the planet's rotation are called **Coriolis drift.** Coriolis drift means that rotation is set up in the vicinity of low pressure storm systems, as Figure 3-12 shows; rotation is counterclockwise in the Northern Hemisphere, clockwise in the Southern. The reader can demonstrate on Figure 3-12 that opposite rotation occurs as air moves away from high pressure systems. Coriolis drifts explain the spiral cloud patterns so prominent in space photos of earth, such as Figure 3-13 and Color Plate

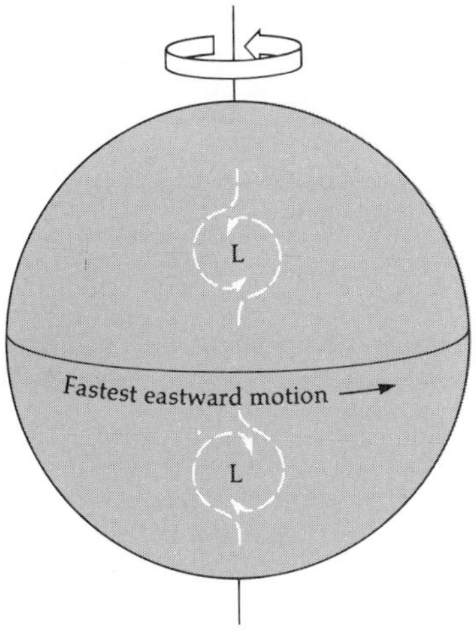

Figure 3-12. *Due to the Coriolis force, air masses moving toward or away from the equator are deflected east or west, respectively, relative to the ground. Air moving into low pressure areas (L) produces cyclonic storm patterns; the sense of rotation is opposite in the two hemispheres and is proof of the earth's rotation.*

Figure 3-13. *The spiral rotational pattern of a storm off New Zealand demonstrates the Coriolis force (see Figure 3-12) caused by the earth's rotation. (NASA photo by Skylab 4 astronauts.)*

a

Figure 3-14. Meteor Crater, in Arizona, testifies to the effects of interplanetary debris hitting Earth. A meteorite impact formed this 1.2-km crater about 20 000 years ago. (a) Road and museum buildings (left) give scale in this aerial view. (Dale P. Cruikshank.) (b) Wide-angle view from rim shows crater interior and boulders blown out onto the rim. The crater is open to the public. (Photo by author.)

4. No Coriolis drifts, hence no spiral patterns, would occur on nonrotating planets.

The Astronomical Connection

Most people think of the earth as isolated from astronomical influences, having evolved independently as a result of its own internal forces. But recent research indicates a number of ways in which external astronomical forces influence the earth's evolution in dramatic ways.

Meteorite impact craters have been created by impacts of interplanetary bodies up to 10 km or more in diameter. The eroded craters, reaching at least 140 km (90 miles) across, have been recognized on the earth only in the last century. In the last two decades, dozens have been identified. Most are thousands or millions of years old and have been damaged by erosion, but Figure 3-14 shows a young, well-preserved example. A few are important for exposing oil- or mineral-bearing strata. For example, a 40-mi-wide region in Ontario, called the Sudbury Complex, is an unusual concentration of metal ores. Its origin has been a long-standing geologic mystery, but recent work indicates it formed when a giant impact 1.8 billion years ago melted crustal rocks, allowing the native metals to concentrate (Faggart and others, 1985). The metal in a nickel in your pocket is likely to have come from this impact site.

As we will see in later chapters, impact craters formed at a much more rapid rate in the early days of the solar system. Although these ancient craters have been obliterated by erosion on Earth, they are preserved in great numbers on most other planets.

Recent evidence suggests that extraordinary climatic effects were associated with the largest ancient meteorite impacts. In strata formed 65 million years ago in various parts of the earth, researchers have found enrichment of certain elements associated with meteorites. Geologists have long known from fossil evidence that during a brief interval 65 million years ago, a major climate change occurred, causing a majority of the earth's species to die out and new species to emerge (see Table 3-1). Dinosaurs became extinct at this time. The dramatic changes were evidently

b

caused by the impact of a large (10-km-diameter?) meteorite or asteroid, as depicted in Figure 3-15 (Alvarez et al., 1980; Ganapathy, 1980). Statistics of asteroids indicate that such impacts should occur at least once every few hundred million y. Calculations indicate that such an impact would throw dust into the high atmosphere, making it too dark to see for one or two months, and too dark for plants' photosynthesis for three months to a year (Toon et al., 1982). This would wipe out the lower parts of the food chain and explain mass extinctions of most higher species. After the large reptiles were wiped out, the hitherto-obscure mammals emerged to inherit the earth.

Scientists studying climate effects of giant meteorite impacts have applied the same principles to study the climate effects of nuclear war. Just as dust ejected from an impact would block sunlight, so dust and smoke from explosions of many nuclear bombs would, according to calculations, block sunlight and cause global climate changes, dropping temperatures and damaging crops far beyond target countries. This theory of "nuclear winter" has been widely publicized in the press.

Still another example of an astronomical connection to the earth's history comes from studies of the earth's orbit. Many studies suggest that minor orbital changes are caused by gravitational forces of nearby planets, which cause changes in the earth's tilt, distance from the sun, and exposure to sunlight. These effects seem to explain most of the climatic changes associated with ice ages in the last 100 000 y or so.

As discussed in Chapter 13, certain disturbances on the surface of the sun also appear to be correlated with disturbances in the earth's climate. Thus, modern research suggests many links between astronomy and climate effects on the earth.

Table 3-1. *Geologic Time Scale*

Era	Age (millions of years)	Period	Life Forms	Events
	0			
		Quaternary	Humanity	Technological environmental modification; extraterrestrial travel; ice ages
	3			
Cenozoic		Neocene		
	22			
		Paleocene	Mammals	Building of Rocky Mountains
	65			
		Cretaceous	Extinction of many species of plants and animals, including dinosaurs	Large meteorite impact (65 million years ago); continents taking present shape
	130			
Mesozoic		Jurassic	Dinosaurs	
	180			
		Triassic	Reptiles	
	230			
		Permian	Conifers	Building of Appalachian Mountains
	280			
		Pennsylvanian	Ferns	Pangaea breaking apart
	310			
Paleozoic		Mississippian		
	340			
		Devonian	Fishes	
	405			
		Silurian	Early land plants	
	450			
		Ordovician		
	500			
		Cambrian	Trilobites	Earliest abundant fossils (trilobites, etc.)
	570			
		Ediacarian	Small soft forms	
	640			
	1000		First macroscopic life forms; sexually reproducing life forms	Growth of protocontinents; scattered fossils
Proterozoic				
	2000			
			Oxygen-producing microbes	Oxygen increasing in atmosphere
	2600			
Archeozoic	3000			Crustal and atmospheral evolution
	3600		Microscopic life	Earliest fossils (algae)
				Oldest rocks; crustal formation?
Formative era	4000			Heavy meteoritic cratering
	4500			Formation of sun and planets (see Ch. 12)
	4600			
Presolar era	12 000?			Formation of Milky Way galaxy (see Ch. 21)
	16 000?			Origin of universe (see Ch. 25)

Source: Data in part from Schopf (1975).

Figure 3-15. *Impacts by interplanetary bodies as much as 10 km across and even larger may explain some sudden environmental changes found in the ancient geologic record. These views show the impact of a 10-km asteroid as seen from an altitude of 100 km (60 mi) in views 10 s before impact, 1 s before, and 60 s after. Massive amounts of dust being ejected in the final picture will settle into the atmosphere and be blown around the world by high altitude winds. Chemical evidence suggests that such an impact 65 million y ago led to extinction of dinosaurs and many other life forms, probably due to climate changes resulting from the dust pall in the atmosphere, which blocked most sunlight for many months. (See text; paintings by author.)*

Figure 3-16. *A unique scene in the solar system, but the most common scene on Earth—an ocean of liquid water. When water degassed from the earth's interior, it encountered surface conditions warm enough to prevent it all from freezing, but not warm enough to evaporate it all. Earth is the only planet where liquid water is known to exist in substantial amounts. (Photo by author.)*

Summary: The Earth's History

Geologic studies of strata and fossils have established a chronology of events in the history of the planet earth, and radioisotopic dating of rocks has established the actual ages of these events. The sequence, known as the **geologic time scale**, is shown in Table 3-1. It is broadly divided into intervals called *eras*. Only the last 14% of the earth's history has yielded enough rock evidence, such as fossils and dates, to provide finer divisions, which geologists call *periods*.

Drawing from material presented in this chapter, we can now construct a thumbnail sketch of the earth's history.

Some 4½ billion years ago the earth formed. Giant meteorites struck much more often than today, scarring the primal landscape with great impact craters. The surface was lifeless.

Life probably originated within the first few hundred million years of the earth's history, since the earliest microscopic fossils date back about 3½ billion years ago.

The heating of the earth's interior, probably due to radioactivity, led to a prompt internal melting, within the first few hundred million years. This, in turn, caused differentiation, a draining of metal toward the center to form a nickel-iron core and a floating of lighter minerals to form a low-density crust overlying a dense rock mantle.

As the earth cooled and solidified, the surface layer formed a rigid lithosphere overlying a more plastic asthenosphere. Currents in the asthenosphere broke the relatively thin lithosphere into "plates." Continents repeatedly split and rejoined as rifts broke the lithosphere and plate motions caused crustal masses to collide.

Mountainous landscapes thus replaced the original, cratered surface, though vestiges of the old surfaces still appear in the cratered continental shields. With the coming of plants and animals, the landscape became familiar to modern eyes. Figure 3-16 (page 65) reminds us that the dominant landscape of our planet is not rolling hills but a vast expanse of water—a view unique in the solar system, as we will see in the next few chapters.

In the scale of planetary evolution, recent events such as the ice ages seem the merest of details. Larger climate changes have punctuated the past. They may have had various terrestrial and astronomical causes. Extinction of many species resulted from atmospheric dust caused by the large meteorite impact 65 million years ago. It is important to understand such events. Even small shifts in climate can produce droughts, affecting world food production. Scientists studying the 65-million-year-old event recently concluded that dust and smoke from nuclear warfare would trigger a devastating climate change, known as the "nuclear winter"—yet another argument for stopping nuclear arms proliferation!

Let us conclude our summary of the earth's history by compressing events into a single day. Life evolved sometime in the early morning, but the fossil-producing trilobites that begin the traditional geologic time scale in the Cambrian period did not appear until about 9:30 in the evening. By 10 P.M. there were fishes in the sea, and by 11 P.M., dinosaurs on the land. Mammals did not appear until about 11:40. Human beings, who have been here (depending on your definition of *human*) perhaps 2 million years, made their appearance only 30 seconds before midnight. The last few thousand years—civilization—occurred in a tenth of a second—represented by the pop of a single flashbulb at midnight. The question is, What will be here a tenth of a second after midnight?

Concepts

radioactive atom	asthenosphere
parent isotope	plate tectonics
daughter isotope	fault
half-life	volcanism
radioisotopic dating	magma
continental shield	lava
age of the earth	erosion
seismic waves	deposition
core	field
mantle	magnetic field
crust	primary atmosphere
differentiation	secondary atmosphere
basalt	Coriolis drift
granite	meteorite impact crater
lithosphere	geologic time scale

Problems

1. When were the last major earthquakes in your region? Is your region seismically active or inactive? How is it located with respect to the boundaries of tectonic plates?

2. If a rock sample can be shown to contain one-eighth of its original amount of radioactive uranium-235, how old is it?

3. In view of the preservation of ancient craters and the lack of folded mountain ranges on the moon, would you predict the moon to have more or less earthquake activity than the earth? Discuss your reasoning.

4. Fossils of apelike predecessors of the genus *Homo* (e.g., *Australopithecus*), found in Africa, are believed to date back at least 2 to 3 million years. What percent of the earth's age is this? Can you accept, philosophically, that events happened during most of the history of the earth before anyone was around to see, hear, or record them?

5. If the earth is 4.6 billion years old, about how much more radioactive uranium-238 did it have when it formed? Would the heat-production rate from radioactivity when the earth formed have been more; less, or the same as it is now?

6. The composition of the earth's atmosphere is probably much changed from what it first was.

a. What happened to the abundant hydrogen atoms initially present or produced by breakup of molecules such as methane (CH_4)?

b. Given that much ammonia (NH_3) was initially present, and that ammonia molecules break apart when struck by solar particles in the atmosphere, account for one source of the earth's now abundant nitrogen.

c. What two gases were added abundantly by volcanoes, and where did these two gases finally end up?

7. Describe how you would expect conditions on the earth to be if Earth were so close to the sun that the mean surface temperature were above 373 K (100°C). What if the earth were far enough from the sun so that the mean surface temperature were below 273 K (0°C)? Explain why a view like Figure 3-16 is probably unique to the earth.

ADVANCED PROBLEMS

8. The average near-surface temperature gradient in the earth is 20°C/km.

a. Assuming the surface temperature is about 20°C, how many kilometers would one have to drill to reach a depth where water would boil? (The temperature of boiling water is 100°C.) Compare this depth with that of the deepest mines, roughly 3.5 km.

b. If such depths could be reached economically, dual pipes could be lowered, with water pumped down one pipe, converted to steam, and blown up the other pipe. Steam-powered plants could thus tap the planetary energy source in any part of the world. Describe possible economic and political consequences of such a project.

9. As seen from Mars, the sun subtends an angle of about ⅓°. Suppose that the earth passes exactly between Mars and the sun.

a. Is the earth big enough to cause a total eclipse of the sun as seen from Mars?

b. Assuming that the human eye can resolve a disk as small as 2 minutes of arc (120 seconds of arc), could the earth be detected by the unaided eye as it crossed the sun, as seen from Mars? (*Hint:* Use the small-angle equation. The earth's diameter is 12 756 km, and its distance from Mars is 60 000 000 km.)

Project

1. Place cooking oil an inch or two deep in a flat pan over low heat and under a single strong light. Because the oil does not readily boil, heat is soon transmitted to the surface in visible convection currents. Note how the currents divide the surface into cells, or regions of ascent, lateral flow, and descent. These cells are analogous to tectonic plates in the earth's surface layers. Sprinkle a slight skin of flour on the oil's surface to simulate floating continental rocks and watch for examples of continental drift and plate collisions. Experiment with different depths of oil and different temperature gradients (by changing the heat setting), and record the results.

CHAPTER 4

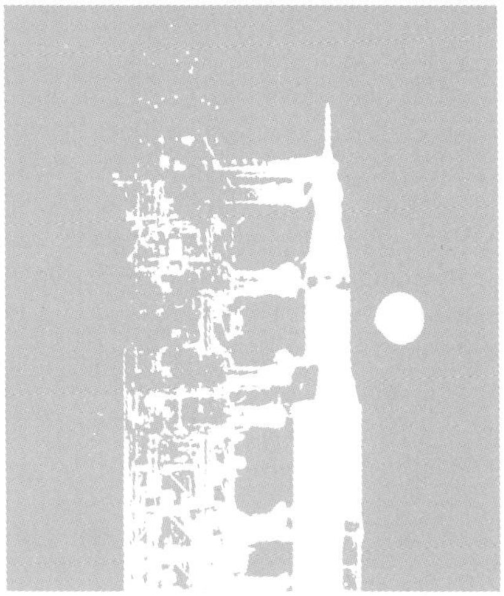

THE CONQUEST OF GRAVITY AND SPACE

The events that led us to leave our terrestrial cradle and visit the nearest planetary body were events of science, engineering, and sociology. In this chapter we will describe early dreams of space travel, then Newton's analysis of the gravitation that binds us to the earth, and finally our current capabilities to operate throughout near-earth space.

The Dream of a Flight to the Moon

Flights to the moon appear in literature as far back as Greco-Roman times and recur throughout European history. The Greek satirist Lucian (c. A.D. 190) had one of his characters put on vulture and eagle wings, take off from Mt. Olympus, and fly to the moon to learn how the stars came to be "scattered up and down the heaven carelessly." Around 1500, Leonardo da Vinci sketched devices for manned flight (Figure 4-1). In 1528 the Spaniard Eugenio Torralba confessed to the Inquisition that he had flown near the moon under the guidance of a demon.

Lucian's stories of lunar flight were translated into English in 1634 and since then literary trips to the moon have occurred often. Among them are Cyrano de Bergerac's 1656 account of *Empires of the Moon* and Daniel Defoe's *Journey to the World in the Moon* in 1705. Over a hundred fictional descriptions of voyages to the moon were published in Europe between 1493 and 1783, when the first balloon ascensions were made (Nicholson, 1949). As Figure 4-2 dramatizes, flight to the moon was not an idea conjured by twentieth-century engineers upon hearing the knock of technological opportunity! The dream is probably as old as humanity.

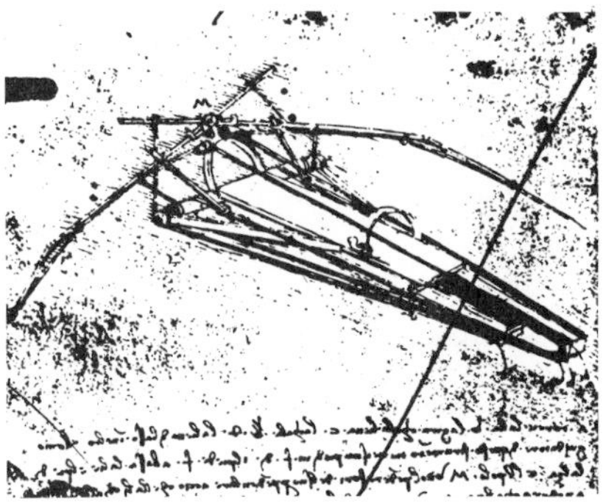

Figure 4-1. ***Leonardo da Vinci drew this engineering sketch for the framework of a flying machine in about 1500.***

Isaac Newton and Gravity

Isaac Newton was a father not only of physics and astronomy but of space travel as well. Between the ages of 23 and 25, while attending Cambridge, he almost single-handedly developed calculus, discovered the principle of gravitational attraction and certain properties of light, and invented the reflecting telescope (in which a curved mirror replaces a lens). Newton once said that he made his discoveries "by always thinking about them," a trait that no doubt contributed to his reputation for absentmindedness.

At age 41, Newton began writing his famous *Principia*, a revolutionary compendium of physics, and published it three years later in 1687. He became president of the Royal Society at 60, died at 84 in 1727, and was buried in Westminster Abbey. Of Newton, Alexander Pope wrote:

Nature and Nature's laws lay hid in night:
God said, "Let Newton be!" and all was light.

In 1726, an acquaintance gave this account of Newton's early thoughts on gravity, based on conversations with Newton (Ball, 1972):

The first thoughts [were those] he had, when he retired from Cambridge in 1666 on account of the plague. As he sat alone in a garden, he fell into a speculation on the power of gravity: that as this power is not found sensibly diminished at the remotest [height] to which we can rise, neither at the tops of the loftiest buildings, nor even on the summits of the highest mountains, it appeared to him reasonable to conclude that this power must extend much farther than was usually thought; why not as high as the moon, said he to himself?

The moon must be attracted to the earth by some force, Newton thought, because it does not travel in a straight line, as it would if no force were pulling on it. Instead, the moon follows a path curved around the earth. Newton, supposedly inspired by a falling apple, realized that he could calculate from the moon's known orbital motion how fast it "fell away" from a hypothetical straight line. He then compared this acceleration rate of a falling body at the earth's surface. After gathering accurate data, Newton showed that whereas the moon is 60 times farther than the earth's surface from the earth's center, its gravitational acceleration is about 1/3600, or $1/60^2$, of the acceleration experienced at the earth's surface. The force diminishes as the *inverse square of the distance.* Gravitational force is thus said to follow an **inverse square law**.

This result is not surprising. If anything spreads from a point in straight lines in all directions, it must become less concentrated as it gets farther from that point. Light, radio waves, and water spray from a fast-rotating sprinkler are examples.

The inverse square relation is illustrated by Figure 4-3. Imagine light from a candle shining into a pyramid. If the pyramid is cut at the point where its base has an area of 1 cm^2, then all the candlelight entering the tip of the pyramid passes through this square centimeter. Twice as far from the light, the base of the pyramid is twice as wide and so has four times the area, but it receives the same amount of light. The original radiation is now dispersed over 4 cm^2. Thus, the base receives one-fourth as much light per unit area at twice any given distance. Thus light radiation follows the inverse square law.

Newton thus knew how distance affected gravity, but he did not know what else affected it. He eventually showed that the gravitational attraction between two bodies is proportional to the amount of material in each body—that is, to its **mass.** The more mass, the more attraction. Thus, if either mass doubles, the force between them doubles; if the distance between their centers doubles, the force drops by a factor of 4. The relationship is discussed further as Optional Basic Equation II on page 70.

Mass should not be confused with *weight*. The weight of an object is merely the gravitational force with which the earth pulls the object against the earth's surface. *Weight* is merely a convenient name for gravitational force. The weight of an object would be different on different planets because the gravity of different planets differs. The mass, or amount of material, of a body remains the same regardless of the body's location.

Note an aspect of the scientific method that is illustrated here. Neither Newton nor we claim to know any "ultimate underlying cause" of gravity. We treat gravity only as a property manifested by matter, which we can confirm by experiment and by observations of nature.

***Figure 4-2.** The rocket was still not recognized as the best mechanism for space travel by the late 1800s, when French artist Gustave Doré made this illustration of a lunar voyage.*

CIRCULAR VELOCITY: HOW TO LAUNCH A SATELLITE

Newton realized that an object could be launched into orbit around the earth. His *Principia* contains a diagram of an earth satellite fired from a cannon on a mountaintop, with the barrel pointed parallel

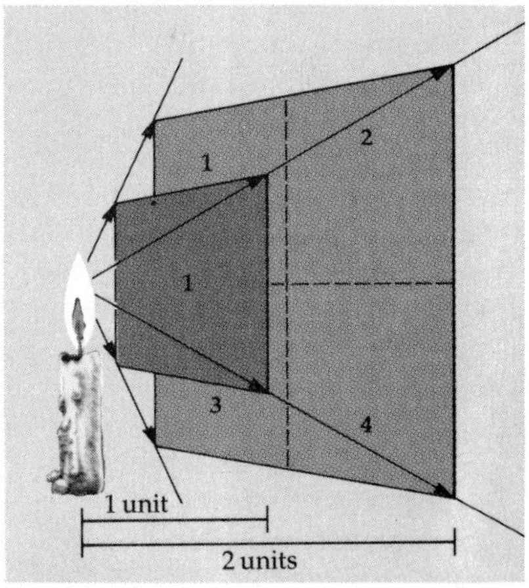

***Figure 4-3.** The inverse square law. At twice the distance from the source, the light is spread over four times the area.*

OPTIONAL BASIC EQUATION II

Newton's Universal Law of Gravitation

Once Newton realized that the gravitational force between two objects is proportional to their masses and inversely proportional to the distance between them, he could express this more simply in math than in words:

$$F = G\frac{Mm}{R^2}$$

where

F = force of gravitational attraction between two bodies
G = gravitational constant, measured to be 6.67×10^{-11} N·m²/kg², in SI units
M = mass of larger body
m = mass of smaller body
R = distance between centers of M and m

The gravitational constant, G, may require some explanation. It is called a constant of proportionality. To take another example, in the familiar equation for the area of a circle, $A = \pi R^2$, the Greek letter pi is used to stand for the number 3.1416, which is the constant of proportionality between the variables A (area) and R^2 (square of radius). Similarly, using SI units of kilograms for m and meters for R, scientists measure the numerical value of $G = 6.67 \times 10^{-11}$ for all known conditions.

A striking fact is that this equation is true throughout the known universe, according to spacecraft and telescopic measurements. The equation and the value for G are true whether measuring the gravitational force between two large lead balls in the lab, the earth and your body, the earth and the moon, the sun and Mars, Saturn and its moons, or two stars orbiting around each other. Hence the equation is known as the *Universal* Law of Gravitation. It seemed remarkable to Newton and his contemporaries in 1687 that the reasoning of a single human could reveal the elegant simplicity, $F = GMm/R^2$, behind such a profound phenomenon of nature.

Note that since we use SI in this book, we express mass in kilograms and distance in meters, and we get the SI unit for force, called the newton, which may be unfamiliar. Europeans and other users of the metric system commonly express amounts of material in terms of mass (kilograms); newtons usually appear only in physics and engineering problems. [Another slight complication is that Americans and English express amount of material in terms of weight or force units (pounds), instead of mass units. This will not affect problem solving in this book. You should remember the approximate rule of thumb that 1 kilogram has a weight of 2.2 pounds on the earth's surface. In any problem that starts with an amount of material expressed in pounds, convert to kilograms before proceeding with calculations using the SI units of this book.]

Sample problem 1. Suppose you are flying your 10-ton (10 000-kg) spaceship at a distance of 10 000 km past a planet, and you detect a gravitational force of 40 000 newtons deflecting you toward the planet. Determine the mass of the planet. *Solution*: We want to solve for M, rather than F. Rearranging the equation, we have:

$$M = \frac{FR^2}{Gm}$$

Substituting in values, using units of newtons, kg, and m, we have

$$M = \frac{40\,000(10^7)^2}{6.67(10^{-11})(10^4)} = 6.00\,(10^{24})\text{ kg}$$

Note from Table 6-1 that this planet is almost exactly the mass of earth. This problem is instructive because it shows how scientists can measure the mass of planets, moons, or stars by measuring forces on objects moving near them, such as space vehicles or natural satellites.

Sample problem 2. If the moon were moved twice as far away, how much weaker would be the earth's gravitational pull on it? *Solution*: We simply note that force is inversely proportional to the square of the distance. Thus, if R increases by a factor 2, F decreases by a factor 4. The force would be 1/4 as much as for the present moon.

to the ground (Figure 4-4). He merely applied his *first law of motion:* A projectile keeps moving forward in a straight line unless a modifying force acts on it. In the case of the cannonball, the force of gravity pulls the cannonball toward the ground. If the launch speed is too slow, the cannonball falls to the earth near the cannon (curve *A* in the figure). At a higher speed it travels farther (curve *B*). At a high enough speed, it curves toward the ground, but the earth, being round, curves away at the same rate. Thus the projectile never reaches the ground, but travels all the way around the earth and returns to the mountaintop on circular orbit *D*.[1] This speed—the speed at which an object must move parallel to the surface of a body in order to stay in circular orbit around it—is called the **circular velocity**.

The farther from the earth or other central body, the less the force of gravity that must be overcome, and therefore the lower the circular velocity. At the earth's surface, 6378 km from the earth's center, the circular velocity is 8 km/s, or nearly 18 000 mph. The moon, 384 000 km from that same center, moves at only about 1 km/s in its circular orbit.

In short, launching a satellite into earth orbit is a seventeenth-century idea! One must simply get an object above the atmosphere (so that air resistance isn't a problem), point it in a direction parallel to the ground, and accelerate it to 8 km/s.

An orbiting body's closest point to the earth is called its **perigee**; its farthest point, its **apogee**.

ESCAPE VELOCITY

As a body is launched at higher and higher speeds, the apogee point is farther and farther away. Each orbit is an **ellipse**—a type of curve describing the closed orbit of one body around a second body. A high enough launch speed would send a body from perigee near the earth to an apogee far beyond the moon. At a slightly higher speed, the body would travel infinitely far from the earth (neglecting the gravitational influence of the sun), following a curve the shape of a **parabola** (curve *F* in Figure 4-4), and it would never come back. This unique speed, which allows the object to escape the earth forever, is called the **escape velocity**, or *parabolic velocity*. At a point near the surface of the earth, the escape velocity is about 11 km/s; it is less at more distant points. Launched at a still higher speed, a body travels a similar curve called a **hyperbola** (such as curve *G*), and does not return. Thus, a launch speed exceeding escape velocity is called a *hyperbolic velocity*.

In addition to the earth, each other body in the universe, such as a planet, moon, or star, has its own gravitational field. Hence a unique escape velocity applies at the surface of each body.

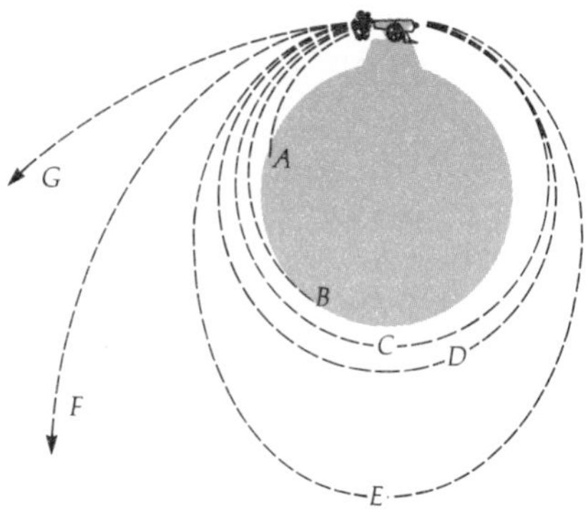

***Figure 4-4.** Newton realized that a vehicle launched parallel to the surface of the earth from a mountaintop could travel varying distances depending on its launch velocity. At low speed, it would drop along curve* A*, hitting nearby. At a higher speed, it could travel halfway around the earth on curve* B*. A slightly higher speed would put it in an elliptical orbit with a low point (or perigee) at* C*. A still higher speed, called circular velocity, would put it in a special type of elliptical orbit, the circular orbit* D*. A slightly higher speed would create an elliptical orbit with the farthest point (or apogee) at* E*. Escape velocity would create a parabolic orbit* F *that never returns to the earth. A still higher speed creates a hyperbolic orbit* G*, which also never returns.*

Rockets and Spaceships

How can a body be propelled to circular velocity or escape velocity? Jules Verne, in *From the Earth to the Moon*, imagined using a 900-ft-long cannon and 400 000 pounds of explosive. But as just noted, a cannon is impractical.

A more realistic technology has since been employed. A spacecraft must carry its own means of propulsion and operate in the vacuum of space. The second point rules out propeller or jet aircraft. Around the turn of the century, several experimenters and visionaries realized that rockets were ideal. Their use was first recorded in China and Europe in the 1200s. Rockets work essentially by **Newton's third law of motion**:

For every action, there is an equal and opposite reaction.

For example, if you sit on a wagon and throw a large mass (like a cinder block) out the back, the wagon coasts forward; the force needed to expel the mass causes an opposite force on the vehicle. In the same way, the force used to expel high-velocity gases out the back of a rocket nozzle pushes the rocket forward with equal force. This force is called **thrust**.

The Russian experimenter Konstantin Tsiolkovsky, beginning in 1898, and the American Robert Goddard, in the 1920s, studied and fired rockets. However, both were mavericks, and their work was almost ignored by their contemporaries. In the 1920s in Germany, Hermann Oberth, who remarked that Verne's book was an inspiration, published several books on rocket-powered space travel. Oberth's work attracted a group of enthusiasts, including

[1]Newton's mountaintop cannon neglects two realities: In the length of a cannon barrel, no shell could be accelerated to the necessary speed (8 km/s) without shattering; and the earth's atmosphere would retard the satellite, making if fall back to the ground. Thus, rockets must be used, and the projectile must be launched above the atmosphere.

THE UNIVERSE IN COLOR

Our picture of the universe has changed, like our television picture, from black and white to color. Recent advances in color photography, television, and computer image processing now bring us the universe's amazing variety of locales and landscapes in a color spectacular. More and more, astronomers are using color as a clue to what is happening in space.

On the planets, different minerals in the rocks reflect different colors of sunlight, creating the reddish landscapes of Mars, the grays and browns of the moon, the grays, tans, and reds of earth. Gases, dust, and condensed particles in the planets' atmospheres interact with sunlight in various ways, creating the yellowish clouds of Venus, the blue skies of Earth, the pink skies of Mars, and the yellow, brown, and red cloud formations of Jupiter and Saturn. Color photographs are thus a key to surface and atmospheric conditions on the planets.

Similarly, images of the sun and stars photographed in light of different colors (that is, different wavelengths) reveal the different properties of their surface gas layers. From these properties astronomers get clues to internal structure and composition.

Gases in space can show an incredible variety of colors because of the many ways that gases can interact with light. No one had seen these colors until recent decades because nebular gases were too dim to record on color films. Advances in color imagery intensify both images and colors, showing emissions in the red light of hydrogen, other colored emissions from other gases, and the blue light scattered by dust particles. The colors are not only beautiful; they also tell us about composition, temperature, and other conditions in the gases.

Colors can also tell us about whole star systems. Massive, hot blue stars are the brightest stars of a newly forming system. Thus, if most of a system's stars are blue, stars are probably still forming there. Many red stars in a system often mean an older system where star formation has ceased, the short-lived blue stars have burnt themselves out, and the brightest survivors are red giants near the end of their evolution. Thus color can help us to understand the evolution of systems as well as their present conditions.

Not the least of color's benefits are simple pleasure and wonder. The addition of color to the astronomer's view of the universe has increased not only our knowledge but our sense of awe of our cosmic environment. The photographs in the following section reflect both of these values. The photographs are grouped roughly in the order of topics in the text where the individual color plates are referenced. The captions that follow briefly describe the plates.

Color Plate 1 *Evidence of ancient astronomy (see Chapter 1). Top left: Aerial view of Stonehenge, built around 2200 B.C. "Avenue" in direction of summer solstice sunrise is toward lower left. (Georg Gerster.) Top middle: Mayan ruin, "Caracol," build around A.D. 1000 in Yucatan. Its window designs include astronomical alignments that may have been used in compiling extensive lists of Mayan astronomical observations. (Anthony Aveni.) Top right: Summer solstice sunset framed in a cylindrical window of Casa Grande Ruins National Monument, Arizona. The four-story structure was built by Indians around A.D. 1350. Evidence suggests careful window design to frame certain astronomical events. (Photo by author.) Bottom: Summer solstice sunrise presents a spectacular view from the center of Stonehenge. Beyond massive stone lintels, the rising sun is framed over distant Heel Stone. (Georg Gerster.)*

Color Plate 2 *The earth, photographed under nearly full phase illumination by Apollo 17 astronauts coasting outward toward the moon. Africa and Antarctica are prominent. (NASA.)*

Color Plate 3 *Los Angeles from orbit. In this view from the Landsat unmanned satellite, infrared imagery has been used, and computer processing gives vegetation a red color. The San Andreas Fault is a prominent left—right linear feature in upper center. (NASA.)*

Color Plate 4 *Cyclonic storm over the ocean near Australia during the storm's dying stages. Winds have sheared off thunderhead clouds, making cloud deck surface unusually flat. Note cloud shadows. (Photo by Apollo 11 astronauts during first successful moon landing flight; NASA.)*

Color Plate 5 *Spaceships in flight (see Chapter 4). Top: The Skylab space station in orbit. The station was manned for a total of 171 days during three intervals in 1973, allowing observations of the sun, the earth, weightless materials, and human capability during prolonged spaceflight. (NASA.) Bottom: Apollo LM landing module in lunar orbit, ascending after the first landing on the moon. (NASA.)*

Color Plates 6 and 7 *Human capabilities in space (see Chapters 4 and 5). Left: Astronaut Jack Lousma working on the exterior of Skylab space station, 1973. (NASA.) Middle: Astronaut Harrison Schmitt collecting samples with a lunar "rake" in the Taurus Mountain region of the moon during the Apollo 17 expedition, 1972. (NASA.) Right: Astronaut Charles Duke on the rim of a 40-m-diameter crater near the Apollo 16 landing site in 1972. Note the crater is old enough to be smoothed by microimpact erosion and is pocked by later craters. (NASA.)*

Color Plate 8 *Selected planets and satellites, shown to scale. Right: Jupiter, Saturn, and inset of Saturn's orangish cloud-covered moon Titan. (Voyager 2). Left (from top): Venus, the earth, the moon, Mars, and Jupiter's satellites Europa, Io, Ganymede, and Callisto. Note difference in cloud patterns on Venus and the earth in spite of similarity in size, due to Venus' slower rotation. (NASA photos.)*

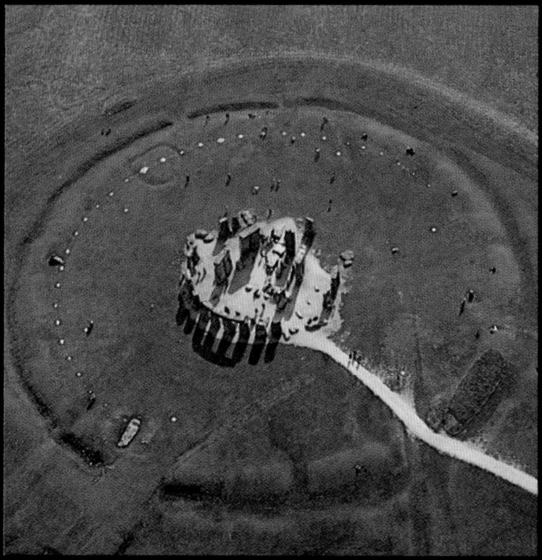

▲ Stonehenge

▲ Mayan ruin, Caracol, Yucatan

▲ Casa Grande National Monument, Arizona

▼ Stonehenge, summer solstice sunrise

Color Plate 2 ▼ The earth, photographed by Apollo 17

▼ Los Angeles from orbit

Color Plate 3

Color Plate 4

▼ Cyclonic storm over the ocean near Australia

▲ Skylab space station in orbit

▼ Apollo 11 landing module in lunar orbit

Color Plate 6

▼ Astronaut working on Skylab

▼ Astronaut collecting lunar samples

▲ Venus

▼ The earth

▼ The moon

▼ Mars

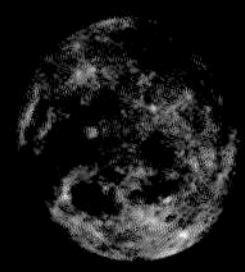

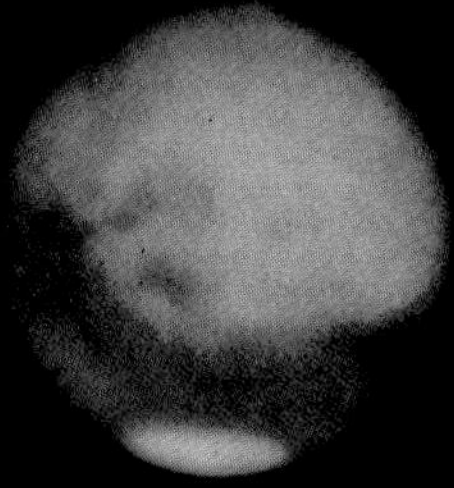

▼ Europa

▼ Io

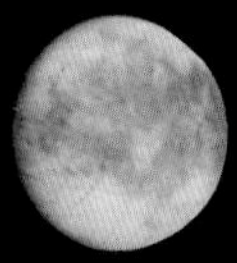

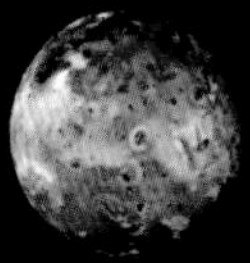

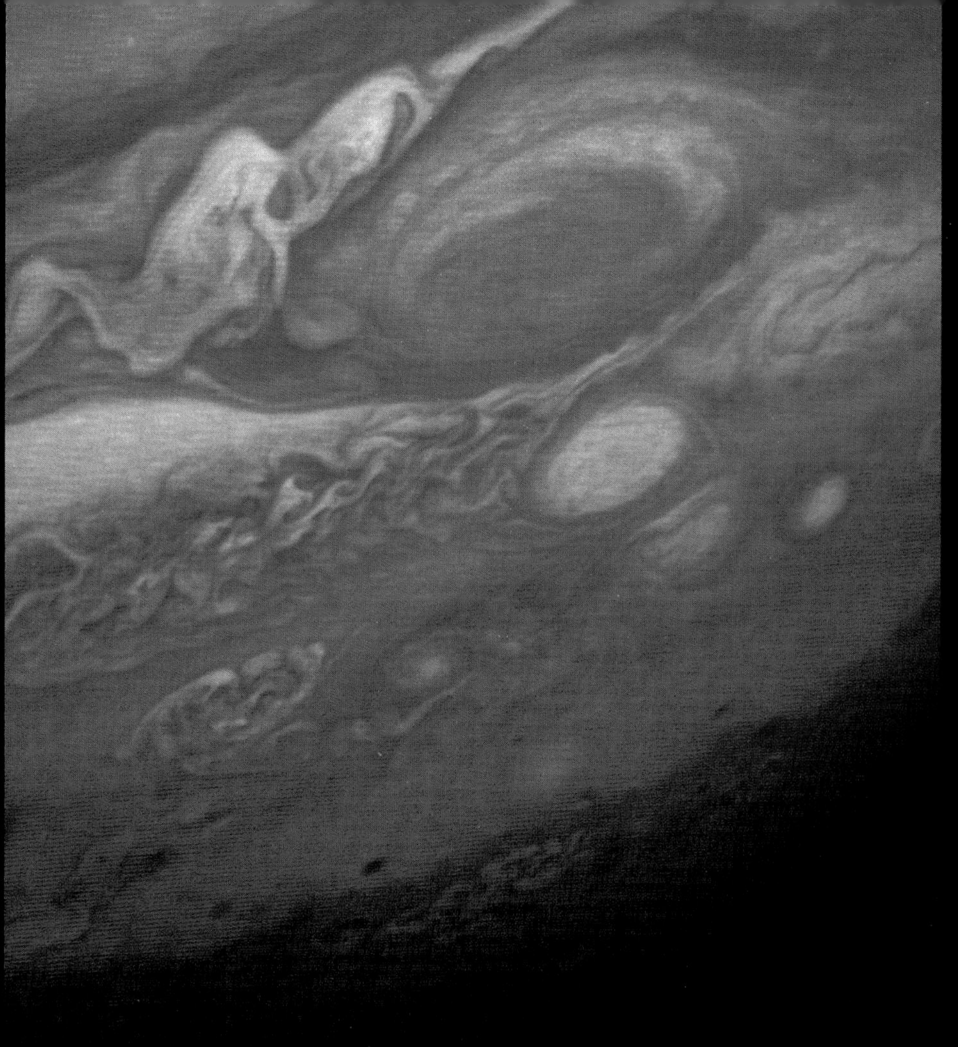

▲ Jupiter

▼ Saturn

▼ Titan

Wernher von Braun, whose astronautical experiments were converted into the V-2 guided missile program under the Nazis. At the end of World War II, about 125 German rocket experts, including von Braun, moved to the United States and continued the chain of space-travel development that stretches back to Lucian, Newton, and Verne. A remarkable footnote to our times is Oberth's survival through war and political vicissitudes to attend as a NASA guest the launch of the first successful moon flight in 1969.

THE FIRST SATELLITES

After several secret postwar studies of satellites had been made, President Eisenhower announced in 1955 that the United States would launch a satellite during the International Geophysical Year (1957–58). This was to be a civilian program using a non-military rocket called Vanguard. Within days, Soviet scientists announced their plan to launch satellites larger than the American one. This plan was not taken seriously in the West: Americans viewed the Soviets as the Soviets portrayed themselves in their poster art—as unsophisticated, shirt-sleeved tractor drivers.

On October 4, 1957, the Soviet Union astonished the world by launching the first artificial satellite, an 83-kg (184-lb), instrumented sphere shown in Figure 4-6. It was named Sputnik I (Russian for "traveling companion"). In November the half-ton Sputnik II went up, carrying a dog as a biological test. Because it was easily visible to the naked eye when it passed overhead at dusk, as shown in Figure 4-7, Sputnik II electrified the Western citizenry. In December, under hasty orders, American technicians tried to launch a small satellite in one of the Vanguard test rockets, but as millions watched on live television the rocket blew up on the launch pad.

These three months produced a crisis in Western confidence and soul searching in American education. After years of chafing at the bit, the Army team under von Braun was given the go-ahead in November; 84 days later, on January 31, 1958, the

OPTIONAL BASIC EQUATION III

Calculating Circular and Escape Velocities

One of Newton's achievements in systematizing physics was that his results, once derived for a specific case such as the earth, could be generalized. The basic physical laws, such as Newton's law of gravitation, apply to bodies orbiting around other planets or around stars. For example, for any small satellite orbiting an object of much greater mass M, the circular velocity is

$$V_{circ} = \sqrt{\frac{GM}{R}}$$

where

G = Newton's gravitational constant = $6.67 \times 10^{-11} \mathrm{N{\cdot}m^2/kg^2}$, in SI units
M = mass of central body
R = distance of orbiter from center of central body

See Figure 4-5. This is the third of nine simple but important equations that are presented in this text.

This simple equation shows that the greater the mass M, the greater the circular velocity; and the greater the distance R, the less the circular velocity. For example, we can compute the velocity required to achieve circular orbit not far about the earth's surface. Expressing units in the SI system, we start with the mass and radius of the earth:

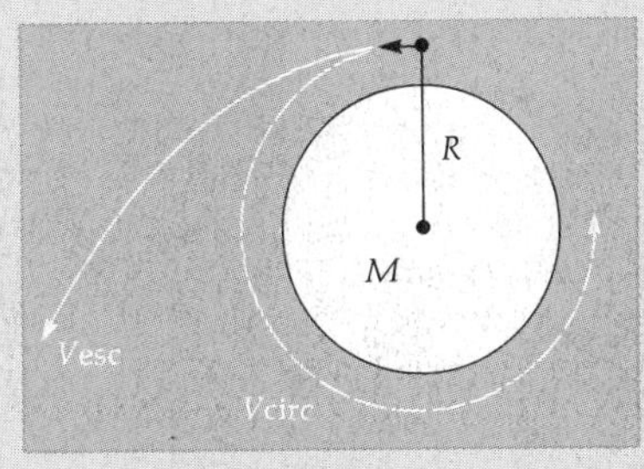

Figure 4-5. *The equation for circular velocity gives the velocity* V_{circ}*, required to place a small object in circular orbit (dashed line) around any large mass* M *from an initial position at any distance* R. *The velocity for escape* V_{esc} *from this same position is always* $\sqrt{2}\,V_{circ}$.

$M = 5.98 \times 10^{24}$ kg
$R = 6.38 \times 10^{6}$ m

Substituting these into the equation, we find that a vehicle must reach a speed of 7.91×10^3 m/s, or about 8 km/s, to stay in circular orbit, confirming the value quoted above.

Another useful fact is that the escape velocity of a small satellite is *always* $\sqrt{2}$ times the circular velocity. Thus the vehicle above would have to reach about 11 km/s in order to escape the earth on a parabolic orbit.

Figure 4-6. *The first artificial satellite. Russian artist A. Sokolov and cosmonaut A. Leonov collaborated on this painting of Sputnik I.*

Figure 4-7. *This time exposure of Sputnik II shows the trail left by the satellite (lower right) as it passed a few degrees from the moon. (Photo by Donald L. Strittmatter, January 1958.)*

first American satellite, Explorer I, was orbited. Vanguard I, a 1½-kg (3-lb) sphere, went into orbit in March, while Sputnik III went up in May. At 1½ tons,[2] Sputnik III was 56 times as massive as the three American satellites combined and intensified American anxiety during that spring.

The first satellites were designed primarily to probe the nearby environment of space. Among their discoveries were the **Van Allen belts** (doughnut-shaped zones of energetic atomic particles surrounding the earth) and the earth's slight bulge in the Southern Hemisphere, which gives the earth a slightly pear shape.

THE FIRST MANNED SPACE FLIGHTS

American engineers concentrated on miniaturizing precision instruments, while Soviet engineers concentrated on rocket power. Though either approach worked for orbiting the earth, the Soviets' large, powerful rockets gave them an edge. After putting the first probe on the moon and photographing the moon's far side for the first time in 1959, the Russians went on to test the biological possibilities of space flight with dogs, some of which they recovered from orbit. On April 12, 1961, in a 5-ton craft, the 27-year-old Russian, Yuri Gagarin, became the first person to orbit the earth, which he did in 108 min (see Figure 4-8).

The first American manned flight was Alan Shepard's 15-min suborbital flight on May 5, 1961. On August 7, Russian cosmonaut Gherman Titov made 17 orbits in a full-day flight. American's first single-orbit flight came six months later on February 20, 1962, when John Glenn piloted a Mercury capsule. One scientific result of these flights was to allay fears that the Van Allen radiation belts or meteoroids might prevent manned space flight.

The Decision to Explore the Moon: Politics and Science

Space exploration obviously has not been a purely scientific effort, divorced from politics. The pacing of the whole enterprise has been determined by social judgments and funding decisions. The technical resources needed to attack a major scientific problem have become so complex that they demand not a single genius

Figure 4-8. *The launch of Vostok I and Yuri Gagarin, the first human to circle the earth, in April 1961.*

[2]One metric ton approximately equals the familiar English ton. Specifically, 1 metric ton = 1000 kg = 2200 lb.

in a hand-built laboratory but a coordinated, well-managed program with heavy financial, political, and industrial support. The financial and political commitment must last throughout the program. Even a wealthy Isaac Newton could hardly be expected to manufacture the heavy castings, special glass, and solid state electronics necessary for sophisticated astronomical observations today, not to mention a fueled, 111-m (364-ft) Saturn rocket! Just as Newton's work might have been lost in a society that did not favor inquiry, disseminate opinions, and preserve results, exploring the moon required its own set of social conditions.

Figure 4-9 is one way of showing these social conditions. Articles on space flight and lunar exploration had a long but sparse pre-Sputnik history. The number began to rise dramatically with Sputnik in 1957, providing a strong public consciousness of space well before the Apollo program.

Verne imagined a voyage to the moon undertaken by a group of shrewd Yankees, funded by an international subscription. In the late 1930s Robert Heinlein imagined it as a commercial venture taking place in 1978, funded by an industrial tycoon. In reality it was to be a $20 billion government-inspired enterprise, conducted by an agency whose greatest problem was not the technology but the coordination of widespread resources necessary for the undertaking.

Between 1957 and 1961, planners had roughed out technical requirements and timetables for a lunar voyage. Though President Kennedy, who took office in January 1961, sought national goals that would spur creative effort, he was at first skeptical about a lunar program, particularly because of the existing Russian lead. Kennedy's science advisor recalls Kennedy remarking (Logsdon, 1970):

"If you had a scientific spectacular on this earth that would be more useful . . . or something that is just as dramatic and convincing as space, then we would do it." We talked about a lot of things . . . and the answer was that you couldn't make another choice.

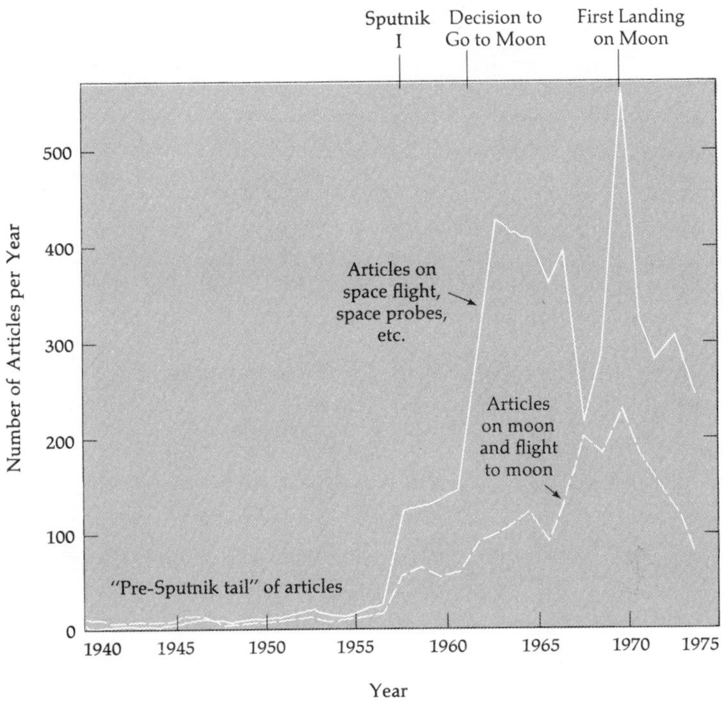

Figure 4-9. ***The evolution of public consciousness of space exploration is reflected in the number of related articles published each year in major periodicals.***

Shepard's May 5 suborbital flight gave some reason for optimism. On May 25, 1961, the goal was set in an extraordinary presidential speech before Congress:

The dramatic achievements in space which occurred in recent weeks should have made clear to us all, as did the Sputnik in 1957, the impact of this adventure on the minds of men everywhere. . . .

I believe that this nation should commit itself to achieving the goal, before this decade is out, of landing a man on the moon and returning him safely to earth.

After Kennedy's assassination in November 1963, the NASA program, especially the Apollo moon-landing program (Figure 4-10), became almost a memorial to the president, and prevented funding cutbacks until the program was achieved.

However, American manned space flight and planetary research funding declined after the last Apollo lunar flight in 1972.

The last remaining Apollo spacecraft was used in a regrettably short-lived effort at global cooperation in space exploration. In 1975, this spacecraft, carrying three American astronauts, linked together in orbit with a Russian Soyuz spacecraft carrying two cosmonauts. The explorers shared good-humored handshakes, conducted mutual experiments, and televised pictures of each others' countries to audiences below in a gesture of goodwill. The Apollo-Soyuz project was remarkable; the challenges included an initial political agreement between Presidents Nixon and Kosygin in 1972, design of a mating tunnel to dock the two vehicles, travel of engineers and astronauts between the two countries, and learning rudiments of each others' languages on the part of astronauts and technicians. The program proved that if political leaders are willing to set challenging, cooperative goals, technical communities in the involved countries can respond with enthusiasm—a more hopeful situation than

Figure 4-10. *A night view of the Apollo 17 spacecraft, the 364-ft Saturn 5 rocket. (NASA.)*

having these communities engage in bomb-building!

Meanwhile, American engineers developed a fleet of four space shuttles designed to fulfill most of the nation's launch needs for scientific, commercial, and military payloads (Figure 4-11). Its first 24 flights, from 1981 through 1985, were highly successful. Milestones included flights of the first American woman and black astronauts, launch of satellites for Indonesia, Germany, and other countries, flight of the European Space Agency's Spacelab, manned by a European-American team, and demonstration of an efficient drug-manufacturing technique utilizing weightless conditions. During the twenty-fifth shuttle launch, however, on January 28, 1986, an explosion destroyed one of the four shuttles, killing the racially mixed crew of two women and five men. Nearly all of America's spaceflight eggs had been put in the shuttle basket, to save money, and the disaster thus has crippled the U.S. space program for several years. Launches may resume in late 1987 or 1988.

Space Exploration: Costs and Results

The results of space exploration will be clearer to our grandchildren than to us, but we can at least compare some costs and benefits.

COSTS

From 1959 to 1969 (the year of the first lunar landing), the total budget of the United States was $1400 billion, whereas the NASA budget (including traditional aircraft research as well as spacecraft development) was $35 billion, or about 2.5% of the total. In the early 1970s funding for all scientific research (civilian and military) was about 6.5% of the total U.S. budget. In contrast, the Departments of Defense and Health, Education and Welfare *each* spent about one-third of the total budget.

Some writers have asked whether society might be improved by canceling space exploration and spending the money on programs to alleviate such social problems as poverty, illness, malnutrition, and the energy crisis. But as the preceding figures show, the science budget is too small to have much impact even if it were diverted to these areas.

PRACTICAL RESULTS

Seven centuries elapsed between the first use of rockets (probably in a Mongol battle in 1232) and their first scientific application (in an atmospheric research flight conducted by Goddard in 1929). Constructive benefits have come only in recent years. The first weather satellite launched in 1960 led to nearly continuous monitoring of weather conditions. TV weather reports now routinely display satellite photographs, such as the one shown in Figure 4-12. Thousands of lives and millions of dollars have been saved through the use of these photos to predict storms on land and sea.

Another important practical consequence of space flight is the communications satellite. Communications satellites, such as the Indonesian satellite launched by the Shuttle, provide the first effective communications between central governments and outlying villages in some Third World countries. Events ranging from Chinese ballet and Olympic games to outbreaks of war are now broadcast between continents. In 1983, U.S. and Russian scientists collaborated on live TV hookups between Washington and Moscow to hold an international conference on nuclear disarmament. Such events illuminate relationships among peoples. Marshall McLuhan, Buckminster Fuller, and others have predicted that this improved communication will strengthen our sense of human community on this planet, producing a "global village"—just as the bickering American colonies eventually came to accept a common identity.

Intercontinental telephone communications are made possible by a group of satellites orbiting about 42 000 km (26 000 mi) out from the earth's center. With an orbital period of 24 h, these satellites stay fixed in relation to any particular transmitting station. From an earthbound antenna, a signal beamed at the satellite is retransmitted to another part of the earth.[3]

Another use of satellites is in photography for mapping. Amazingly enough, in some remote areas of the earth—such as the Amazon jungles of Brazil and certain parts of Ethiopia—satellite photos are better than any existing map. Ultraviolet and infrared satellite sensors are being used in agriculture, forestry, and prospecting. Satellites can also spot certain crop

Figure 4-11. *The space shuttle expands human space capabilities. A successful 2-day orbital flight followed its first launch on April 12, 1981, twentieth anniversary of Yuri Gagarin's pioneer human spaceflight. (Painting by Pamela Lee.)*

Figure 4-12. *Use of space photography in weather forecasting has become familiar on the evening news. This photo, made from the Applications Technology Satellite (ATS) on September 27, 1971, shows U.S. cloud cover and two tropical storms: Ginger (right) and Olivia (lower left). Advance storm warnings reduce property damage in the United States and other countries. (NASA).*

[3]This system was first proposed in the late 1940s by science fiction writer Arthur C. Clarke. In practice today, it causes curious pauses in transcontinental phone calls. If your call goes by normal ground links from the East Coast to the West Coast, the gap between your question and your friend's answer is about 0.2 s, the normal response time for the brain to frame an answer. However, if your call is beamed up to a satellite about 36 000 km above the surface, then these radio waves must complete two round trips (144 000 km) at the speed of light (300 000 km/s) between question and answer. This takes roughly 0.5 s. Added to the 0.2 s mentioned above, it gives a characteristic awkward pause of 0.7 s between verbal exchanges.

blights by subtle color changes before the infestations are detected on the ground; detect and track pollutants in rivers, coastal waters, and the atmosphere, as seen in Figure 4-13; and locate mineral deposits by "reading" geologic features and subtle soil colorations. The weightless, clean, and airless environments in space vehicles also offer opportunities to test new manufacturing techniques in space. These applications are likely to expand in the next decade. The drug processing equipment tested on the Shuttle is being developed commercially and many observers believe it will become the first self-supporting space industry.

INTANGIBLE RESULTS

There are three important intangible results of space exploration. First, it demonstrated successful management techniques to accomplish an extremely broad technological objective.

Second, space flight creates a cosmic perspective. By no coincidence, the ecological movement emerged just at the time of the first lunar flights. Astronaut Lovell, on Christmas Eve 1968, radioed from lunar orbit that the "vast loneliness . . . of the moon . . . makes you realize just what you have back there on earth." Astronaut Anders, on the same flight, noted the earth as "the only color in the universe—very fragile . . . it reminded me of a Christmas tree ornament" (Figure 4-14). The cosmic perspective also opens the possibility of human survival even in the event of a natural or human-caused catastrophe on the earth. Rocket pioneer Wernher von Braun commented shortly after the first lunar landing (Lewis, 1969), "The ability

***Figure 4-13.** Photo made on March 27, 1973, by the Earth Resources Technology Satellite (ERTS) shows polluted waters entering the southwest corner of Lake Erie from highly urbanized areas along the Detroit (top) and Maumee (lower left) rivers. Lighter tones, brought out by contrast enhancement, reveal currents of polluted water. Similar satellite photos can be used to aid in enforcing pollution control standards. (NASA.)*

***Figure 4-14.** The scene of the earth rising over the lunar landscape provided humanity with a new view of our planet. (NASA photo by Apollo astronauts in lunar orbit.)*

for man to walk and actually live on other worlds has virtually assured mankind of immortality."

Third, space exploration provides a frontier and a sense of adventure that is important to human well-being (Figure 4-15). After several decades during which frontiers seemed to be closing all around us on the earth, we now see a new frontier opening above us. Noted astronomer Margaret Burbidge (1983) reviewed some of the opportunities and dangers of this frontier in her presidential address to the American Association for the Advancement of Science, "Adventure into Space." Princeton physicist Freeman Dyson (1969) noted:

We are historically attuned to living in small exclusive groups, and we carry in us a stubborn disinclination to treat all men as brothers. On the other hand, we live on a shrinking and vulnerable planet which our lack of foresight is rapidly turning into a slum. Never again on this planet will there be unoccupied land, cultural isolation, freedom from bureaucracy, freedom for people to get lost and be on their own. Never again on this planet. But how about somewhere else?

FUTURE PLANS

In considering space exploration, our perspective is broadening to include not just flights to the moon and planets, but exploration of the whole universe as well. The moon and nearby asteroids and planets are, of course, the easiest extraterrestrial places to reach. Every planet and satellite in the inner solar system has now been studied by spacecraft at close range several times. Jupiter, Saturn, and their amazing moons (Chapter 10) have been surveyed, and our space frontier is now somewhere beyond Saturn. A more complex U.S. mission to Jupiter, the Galileo program, is planned for the later 1980s. It will drop a probe into the atmosphere and map the Jovian satellites in detail. After American orbiters made rough radar maps of Venus' mountains and plains, and Soviet landers made surface photos, NASA scientists pushed for a more detailed radar mapping mission to understand Venus' geology, but Congress denied funds for such a program. In 1983, Russian scientists put two radar probes in Venus orbit and began mapping the planet on their own. Their images show an astonishing array of narrow ridges, valleys, and eroded impact craters.[4]

Halley's comet is being tracked on its approach to the inner solar system, and Europe, Japan, and the Soviet Union have all announced plans to launch probes to the comet's vicinity, but Congress denied funds for an American space probe to Halley's comet, though NASA will be able to redirect an old cameraless spacecraft, launched in 1978, toward a comet in 1985. The camera-carrying probes from Europe, Japan, and the Soviet Union are all scheduled to rendezvous with Halley's comet in March 1986, and should offer the first close-up details of phenomena in the mysterious heart of an active comet.

Soviet policy keeps their space goals secret, but recent evidence indicates they are working on a large permanent space station, a Shuttle-like vehicle, and a 1988–

[4]Russian scientist A. T. Basilevsky brought some slides of the Russian pictures to present to a Houston conference in March 1984, some weeks *before* release of the pictures in the Soviet Union. American scientists, excited by the myriad of new details shown in the slides, responded with vigorous applause. Later, adopting the Russian custom of exchanging small gifts, they presented the Russians with one scientist's Texas vanity license plate, which spelled CRATER, amidst handshakes and laughter. There is sad irony in sharing victories of human exploration with foreign colleagues when you know your governments are engaged in life-threatening struggles. Meanwhile, NASA has a go-ahead to proceed with a Venus radar mapping mission and is hoping to achieve wider geographic coverage and somewhat sharper images than the Russian results.

***Figure 4-15.** A Skylab astronaut working outside the space station in 1973 gave an early demonstration of our ability to work in the space environment that surrounds Earth. (NASA.)*

Figure 4-16. *Astronauts working in the cargo bay of the Space Shuttle in 1983 symbolize our growing capability to adjust to and work in space. A cloudscape on Earth's surface fills the background. (NASA.)*

89 unmanned probe to Mars' moon, Phobos. In 1984, two Soviet cosmonauts completed a record 8 months in orbit; Soviet officials have often compared their record-breaking orbital flight durations to the durations of interplanetary flights.

American manned space capabilities will leap forward as the Space Shuttle becomes routinely operational (Figure 4-16). As discussed in the preceding chapter, the earth's atmosphere strongly limits astronomical observations from the ground, and one of the Space Shuttle's early missions will be to place new astronomical telescopes into orbit above the atmosphere. A large instrument called the Space Telescope will be one of the first of these, permitting unparalleled views of distant stars and galaxies (Burbidge, 1983).

Beyond helping to launch such instruments, manned missions have a continuing role of their own, in terms of both adventure and practicality. In his 1984 State of the Union address, President Reagan called for a permanently inhabited U.S. space station. It may have international participation, and is currently planned for the early 1990s.

Space operations may ultimately help solve energy, raw material, and pollution problems on the earth. For example, large satellites, as shown in Figure 4-17, could collect pollution-free solar energy and beam it down to large (10-km) "antenna farms" that would replace power-generating plants on the earth. This could pro-

Figure 4-17. *Construction of very large structures is feasible in space because of the low stresses and lack of gravity there. This imaginary scene shows construction of a solar-powered station in earth orbit. Space Shuttle (upper right center) indicates scale. (Painting by John J. Olson, courtesy Boeing Aerospace Company.)*

vide energy with less environmental damage than that caused by coal-fired or nuclear generating plants, though some analysts have expressed concern over the effects of the microwave beams that would carry the energy to the receiving stations. While the cost of such a solar energy space program would be high (the Apollo program cost about $20 billion over 10 years), the United States now sends more than $50 billion each year to OPEC countries for oil. Once burned, the oil is gone for good, whereas each investment in space gives us new capabilities for the future (Arnold, 1980).

In a more visionary vein, we can imagine that industrial activity in space may reduce industrial and manufacturing pollution on the earth. Because of the availability of iron from asteroids, hydrogen from lunar soil, energy from the sun, and other space resources, space colonies may not only be feasible but may also pay for themselves (O'Neill, 1977). Someday the earth may be appreciated as a Hawaii in a universe of Siberias, the only place where we can enjoy the natural environment unprotected; this view may encourage us to take many industrial activities into space, where waste products cannot harm planetary ecosystems. If organized with sufficient political freedom and flexibility, space colonies might even allow experiments in new systems of government, which might defuse political tensions in the closed societies of the earth.

Summary

The idea of carrying people or instruments into space existed long before the technological possibility. Newton's theory of gravitational attraction made it possible to calculate how fast objects would have to go in order to orbit around the earth or escape from the earth. The development of rocket technology early in this century provided the means to reach these speeds. Political and social factors, including strong public support for space exploration, allowed the implementation of these means.

The technology of space travel has been used in four areas: (1) improving communication, weather prediction, and manufacturing processes; (2) helping us search for new nonrenewable resources, while helping us realize that the earth and its supply of resources is finite; (3) exploring other bodies in the solar system; and (4) improving telescopic observations of stars and galaxies far beyond the solar system. Space flight has provided a sense of adventure unmatched since the Renaissance voyages to the New World. Space travel's greatest value could be to insure human survival in other environments should warfare or environmental changes make this planet uninhabitable.

Concepts

inverse square law	parabola
mass	escape velocity
circular velocity	hyperbola
perigee	Newton's third law of motion
apogee	thrust
ellipse	Van Allen belts

Problems

1. If an astronaut is flying in circular orbit just above the atmosphere and wants to escape from the earth, how much additional velocity does he or she need?

2. Rocket engineers typically speak of the difference between one orbit and another in terms of velocity difference, for example, in meters per second. Why is this? (Consult Problem 1 if necessary.)

3. Because the earth rotates, the equatorial regions move at about 1600 km/h (1000 mph). Why is it easiest to launch a satellite in a west-to-east orbit over the equator?

4. You have just rowed a rowboat to a point at rest next to a dock. You step off the rowboat toward the dock.

a. Why does the rowboat move away from the dock?

b. In this instance, how are you analogous to the exhaust from a rocket?

5. Do you think a proposal to spend $20 billion to land a man on the moon for the first time would be endorsed by Congress or the public if it had happened this year instead of 1961? Explain your answer, accounting for similarities or differences in public attitudes during these two periods.

6. Do you believe that in the next century manned flights to other planets, satellites, or asteroids will be common? Do you think this would have positive or negative effects on social progress, intellectual stimulation, the economy, the environment, the availability of energy and materials, and other characteristics of our civilization?

ADVANCED PROBLEMS

7. The lowest point on Earth's land surface is 392 m below sea level, at the Dead Sea in Israel and Jordan. The highest point is 8847 m above sea level, on Mt. Everest. Using a mean sea level radius for the earth of 6371 km, and assuming your average weight is measured at sea level, calculate your weight near the Dead Sea and on Mt. Everest. Use SI units as needed, but convert your answer back into pounds. Does the earth's relief make much difference in gravity from place to place?

8. Suppose you landed on a planet with a mass of 0.11 earth-masses, and a radius 0.53 times that of the earth.

a. Compare your weight on that planet with your weight on the earth.

b. Using data in Table 6-1, compare this situation with that of an astronaut on Mars.

9. Suppose an astronaut is orbiting the earth at the same distance as the moon in a circular orbit, but not near the moon. How much faster would he or she have to move to escape the earth altogether?

10. Compute the velocity of an artificial satellite in a circular orbit 42 500 km from the earth's center. Show that it would be in a synchronous orbit—that is, it would revolve with a period of 24 h and thus stay above the same point on the earth's surface.

Projects

1. Observe the moon as it passes above or below a bright star. Remembering that the moon is about 3480 km in diameter, observe how many of its own diameters it moves in 1 h, and calculate its orbital velocity.

2. Observe an artificial satellite. What illuminates it? Why are artificial satellites commonly seen at dusk or dawn and not at midnight? Observe whether the satellite passes into the earth's shadow and whether it reddens slightly as it does so.

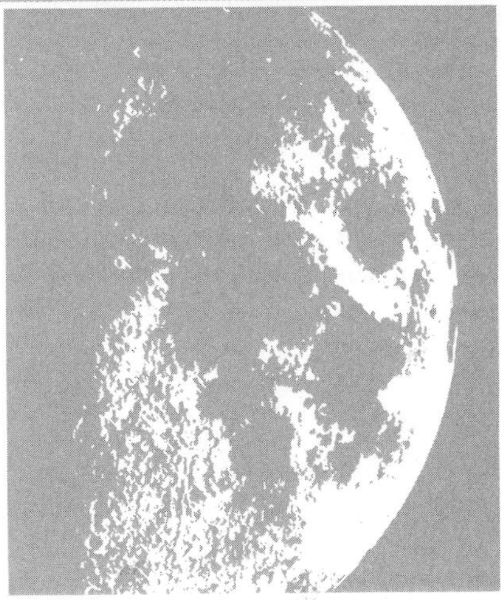

THE MOON

The same moon that we see, with almost the same features (give or take a few craters), shone down on the breakup of the continents, gleamed in the eyes of the last dinosaurs, and illuminated the antics of the first protohumans. The same moon was seen by all the historical figures we have mentioned: Stonehenge builders, Mayan eclipse observers, Aristarchus, al-Battani, Isaac Newton, Jules Verne. All those people asked what it is, what its composition is, where it came from. Today it is a little different; it has footprints on it and we know a few of the answers.

The moon is at once familiar and unfamiliar. Ours is the first generation to have seen a few of its stark landscapes. We know from experience that it has no air, no water, no weather, no blue sky, no clouds, and no life. Yet many people are still confused about its simple phases. Can you recall which way the horns of the crescent point in the evening sky? Cartoonists often draw the horns pointing down toward the horizon. Not so. Since the fully illuminated edge of the crescent must face the sun, which has just set below the horizon, the horns must point upward, away from the sun. Scoff, too, at the novelist who describes the full moon rising at midnight! The fully illuminated moon is always opposite the sun, and hence must rise as the sun sets, as seen in Figure 5-1.

The whole subject of the moon's motions around the earth, though it might at first seem mundane, provides interesting clues about the ancient history of the earth–moon planetary pair. We will thus first consider these motions, then describe the moon's surface and the astronauts' discoveries, and finally explore the puzzling problem of the moon's origin.

Synchronous Rotation of the Moon

Perhaps the idea of the earth as a spinning ball would have come centuries earlier if people could have watched the moon spin in the sky, with its gray patches carried around the lunar face and out of sight onto the back side, only to reappear again a few days later. However, the moon does not appear to spin. From day to day we see only the same familiar features of "the man in the moon"—a pattern of gray lava

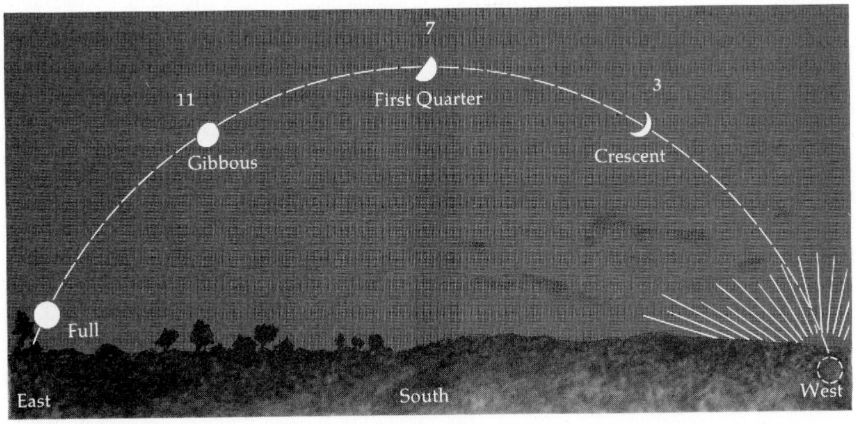

Figure 5-1. *Wide-angle view of the sky showing some of the phases of the moon as seen by a Northern-Hemisphere observer at sunset on the indicated days of the monthly lunar cycle.*

patches lit from different directions by the sun as the moon goes through its phases. As early as 1680, the French astronomer G. D. Cassini correctly explained this in a statement that is sometimes hard to grasp:

The moon rotates on its axis with a period equal to its orbital revolution period around the earth, so that the same side keeps facing the earth at all times.

The rotation of any satellite in this way is called **synchronous rotation** since it is synchronized with the satellite's own revolution, as shown in Figure 5-2. How can the moon rotate at all, you might ask, if it always keeps the same side toward the earth? The best answer is provided by an example. Put a chair in the middle of a room. The chair is the earth; the walls, the distant stars. To represent lunar orbital motion, walk around the chair. If you walk around the chair always facing it, so that an earthly observer in the chair never sees your back, you will find that you have faced all sides of the room during one circuit. In other words, you have rotated once on your axis and made one revolution around the chair.

Even writers who should know better sometimes speak of the moon's "eternally dark side." This mistake comes from a popular belief that the side eternally hidden from us must always be dark. But the far side, just like the near side, has day and night—periods of sunlight and periods of darkness. Each period lasts about two weeks, since the moon takes about four weeks to make a complete rotation. There is always a dark side, but it isn't necessarily the *far* side.

The moon takes 27.32 d to complete one revolution around the earth relative to the stars—an interval called the moon's **sidereal period**. During this period the earth moves roughly 27° around the sun, so that the moon has to move through this additional angle to complete its cycle of phases relative to the sun. Therefore, the cycle of lunar phases takes 29.53 d.

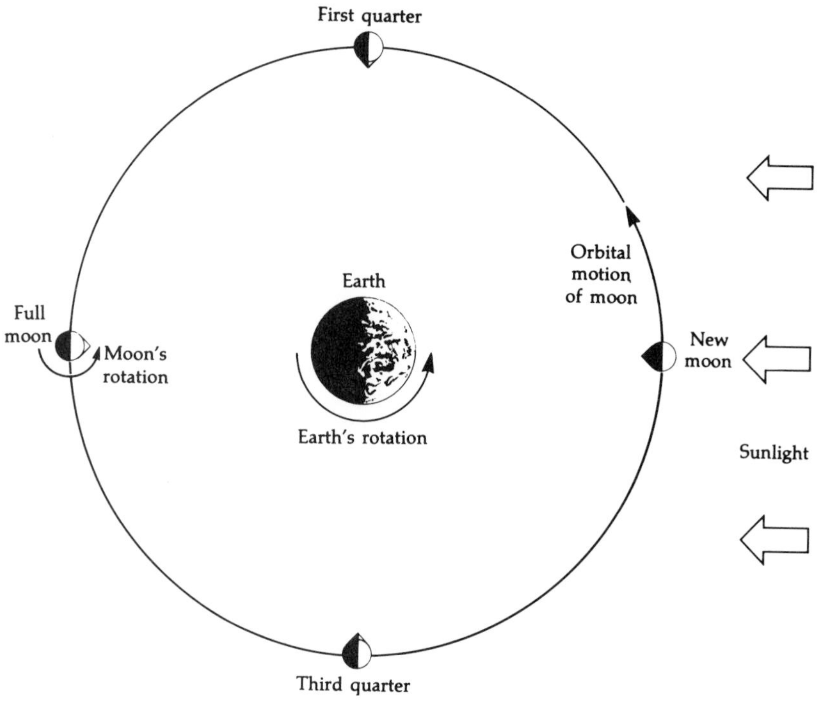

Figure 5-2. *The moon's motion around the earth. Synchronous rotation of the moon is shown by a mountain (indicated by a triangle) that always faces the earth. The moon completes one rotation during one complete revolution around the earth. Phases seen by an earthbound observer are indicated at different points in the orbit.*

Is there a reason why the moon keeps one side toward the earth? When asking for a reason, a scientist normally means to ask, "Could the observation be explained by some more fundamental properties of nature, so that it becomes a special case of a more general phenomenon?" Here the answer is yes. In the 1780s and 1790s, Joseph Louis Lagrange and Pierre Simon de Laplace used Newton's law of gravity (Chapter 4) to show that if the moon were slightly egg shaped or football shaped, gravitational forces would make the longest axis point toward the earth at all times. Confirming this, space vehicles in the last few decades have shown that one axis of the moon *is* about 2–3 km longer than the others and points steadily toward the earth.

If the moon always kept exactly the same side toward the earth, earthbound observers could never see more than exactly 50% of it. Careful mapping, however, has shown that the moon wobbles and that during a period of years 59% of the moon can be seen, with first one side and then another being turned slightly farther from the earth than its average position.

Tidal Evolution of the Earth–Moon System

Gravitational forces in the earth–moon–sun system cause **tides**, or bulges in the shape of the earth and the moon. The closer two objects are, the stronger the gravitational force each exerts on the other. Thus, the side of the moon facing the earth has a stronger force on it than the far side, because the facing side is closer. This is shown by the small arrows in Figure 5-3. As a result, the moon stretches slightly along this line, limited by the small elas-

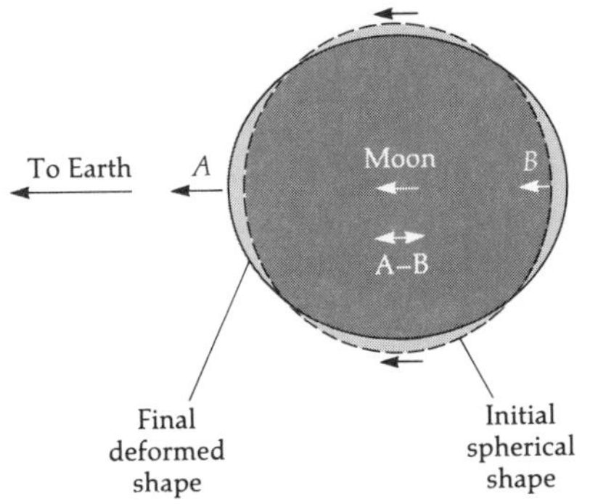

Figure 5-3. *The gravitational attraction of the earth on the moon is greater on the near side (A) than on the far side (B). The difference (A − B) acts as a stretching force deforming the moon from its unstressed spherical shape to a flattened shape with tidal bulges on the near and far side.*

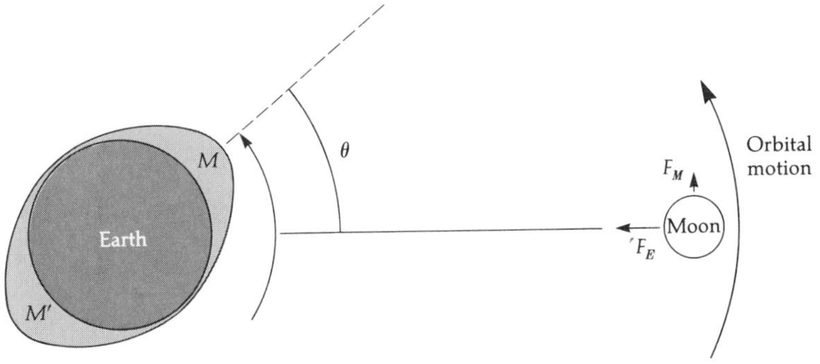

Figure 5-4. *The cause of the moon's tidal recession. Because of the earth's relatively rapid rotation, the earth's tidal bulge (MM′) gets dragged off the earth–moon line at an angle θ. Thus, in addition to the normal gravitational force of the earth (F_E), there is a net forward force (F_M) caused by the difference in attractions between the nearer bulge (M) and the farther bulge (M′). This force (F_M) pushes the moon ahead in its orbit, causing it to spiral slowly outward.*

ticity of its rock interior. This stretching is called a **body tide**. Similar forces acting on the earth produce not only a body tide but also an **ocean tide** in the fluid layer of water on the earth's surface. These tides take the form of bulges on the front and back side of both the earth and the moon, since each body comes into equilibrium with the gravitational forces by stretching along the earth–moon line.

The earth's body tides are hard to detect, but its ocean tides are obvious to anyone who visits a beach for more than an hour. They range in height from about ½ m (2 ft) to over 15 m (50 ft). One might suppose that high tide always occurs when the moon is overhead, with the highest tides at the new moon or full moon, because the tidal forces act along the earth–moon line (or, in the strongest case, along the earth–moon–sun line). However, this is only a rough tendency; complications result from motions of water around the irregularly shaped oceans and seas and because of the earth's rotation. Coastline geometry in some places can produce remarkable wave effects associated with tides, but so-called tidal waves are related not to tides but to earthquakes or volcanic activity at sea; they are properly called by their Japanese name, **tsunamis**.

Tidal bulges raised on the earth and the moon have four major effects, first described in 1898 by George Darwin, son of the famous naturalist. These effects are described in the following four subsections.

UNDERLYING CAUSE OF THE MOON'S SYNCHRONOUS ROTATION

The effect of tides explains why the moon keeps one side facing the earth. As mentioned earlier, any elongation in the moon's shape would tend to make one side always face the earth. Tides guarantee such an elongation, since they create bulges. If the moon were initially spherical or rotating at a nonsynchronous rate, gravitational forces acting on it would create tidal bulges and slow the rotation until it became synchronized and one side faced the earth at all times. This effect explains why many other satellites in the solar system, besides ours, keep one face toward *their* planet.

RECESSION OF THE MOON

The second effect is a slow **tidal recession** of the moon away from the earth because of gravitational forces on tidal bulges. As shown in Figure 5-4, the earth's rapid rotation drags its tidal bulge slightly ahead of the earth–moon line by some angle θ. The gravitational effect of the bulge *M* exceeds that of *M′*, because *M* is closer to the moon. Because *M* is in front of the moon, the net force from *M* tends to pull the moon ahead in its orbit. This accelerates the moon forward and makes it spiral out from the earth. This effect also indicates that the moon was much closer to the earth several billion years ago. Analyses do not indicate the exact date, but most researchers believe the moon was closest to the earth about 4.6 billion years ago during the formation of the two bodies.

SLOWING THE EARTH'S ROTATION

The third effect of tides is that the moon slows the earth's rotation, just as the earth has slowed the moon to synchronous rotation. The effect can be seen in Figure 5-4, where the moon pulls back on the tidal bulge *M*, acting like a brake on the earth. Studies indicate that billions of years ago, when the moon was closer, the earth's day was only about 5 or 6 h long. These theoretical results have been confirmed by an unexpected finding. Certain marine creatures create daily and monthly banded

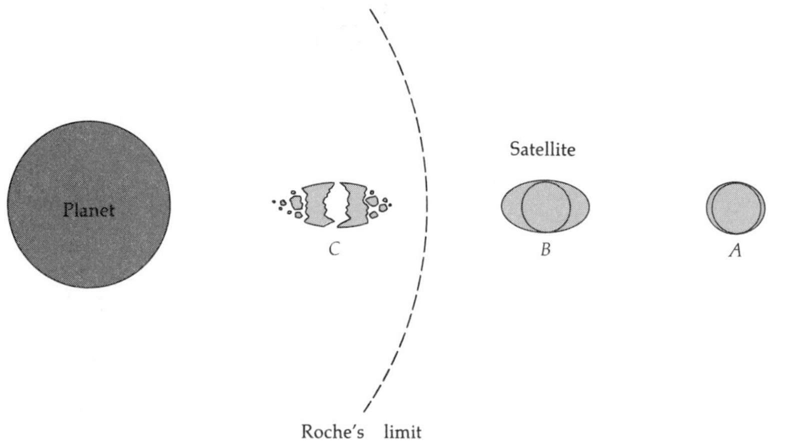

Figure 5-5. ***Roche's limit. If a satellite is located at A, a small tidal bulge develops. At closer distance B, a larger bulge develops. Within a critical distance, called Roche's limit, the stretching force differential between the near and far sides is so great that the satellite is torn apart (for example, distance C).***

structures in their shells or other hard parts, allowing biologists to count the number of day bands in a monthly cycle. Fossil evidence suggests that whereas the present month is 29.5 d long, it was only about 29.1 d long 45 million years ago. Older fossils show that tides existed as long as 2.8 billion years ago, and that the month was as little as 17 d long (Kaula and Harris, 1975). We cannot extrapolate such data further back in time because the tidal effects depend on the configuration of terrestrial oceans and continents, and these configurations were different by unknown amounts in the past.

As the earth slows and the moon recedes in the future, it might eventually recede so far that the earth's gravitational pull would be too weak to keep it in orbit. It could then escape into an orbit around the sun. However, before that point is reached, small tides raised on the earth by the sun will slow the earth's rotation so that the length of the day and the month become equal, stopping the tidal recession. The earth and the moon would then *both* be in synchronous rotation. Eventually, because of this small solar effect, the earth's rotation will slow so much that the day will exceed the month and the tidal process will reverse: The moon will begin to approach the earth again.

ROCHE'S LIMIT

The fourth application of tidal theory shows that a small body, if close enough to a large body, can be torn apart by tides. Calculated around 1850 by French mathematician Edward Roche, **Roche's limit** is the distance between any two different-sized bodies within which the tide-raising force exerted on the smaller body is sufficient to disrupt it. The effect occurs because the *difference* between the forces on the near and far sides of a satellite increases as the satellite moves closer to the primary body. This effect causes the satellite to stretch into a slight egg shape as it approaches the primary, as shown in Figure 5-5. The position of Roche's limit depends on the size, density, and strength of the satellite. For instance, a small metal spacecraft is not torn apart while orbiting near the earth. But a body as big as the moon, if similarly placed for a long period, would develop internal stresses and fractures and eventually disintegrate into a cloud of orbiting particles like Saturn's rings. The moon could not remain an intact body much closer to the earth than about 18 000 km (11 000 mi) (its present distance is 384 000 km, or 240 000 mi).

Tidal effects have a bearing on the origin of the moon. Was it ever inside Roche's limit? We will return to such questions at the end of the chapter.

Features of the Moon

Until the telescope was invented around 1608, no one knew much about the lunar surface features except for the gray patches that make up the face of "the man in the moon" and "the woman in the moon" (Figure 5-6). Some thought the moon was a polished sphere. Galileo Galilei (1564–1642), the first person known to have seen the moon's features through a telescope, reported his observations in 1610. As shown in Figure 5-7, he recorded rugged regions with prominent shadows cast by mountains and crater rims along the **terminator**, the line dividing lunar day from lunar night. Details were less prominent under high lighting (full moon) or at the edge of the disk, called the **limb**. After brief inspection, Galileo was able to report:

The moon certainly does not possess a smooth surface, but one rough and uneven, and just like the face of the earth itself, is everywhere full of vast protruberances, deep chasms, and sinuosities.

These features were later recorded in better detail by telescopic photographs (Figures 5-8 and 5-9). Most of the roughness was caused by thousands of **impact craters**, or circular depressions caused by meteorite impacts (Figure 5-10), ranging up to 1200 km in diameter. Smaller ones (up to a few kilometers) are bowl shaped, while larger ones have a more complex structure, such as central mountains or concentric rings of cliffs. Galileo found the dark gray patches that are visible to the naked eye and that form "the man in the moon" to be much smoother than the brighter, cratered areas, as seen in his sketch in Figure 5-7. He mistook these dark patches for seas. Using Latin, as sci-

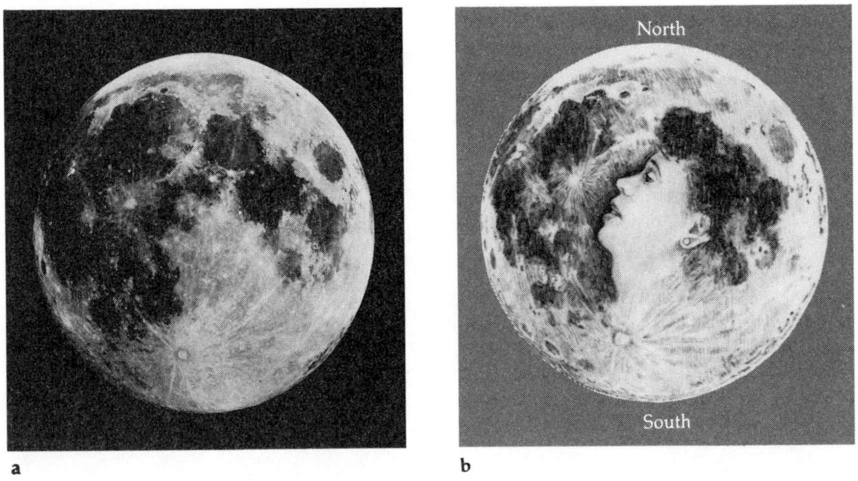

Figure 5-6. (a) *At full moon, the rugged appearance of the terrain is minimized by the absence of shadows. Compare with Figure 5-9. Full lighting emphasizes different features, such as the bright rays emanating from the crater Tycho (bottom).* (b) *Several folklore figures, such as "the woman in the moon," are formed by the pattern of dark lava plains. Squinting at* a *and* b *may help you see these features. (Photo from Lunar and Planetary Laboratory, University of Arizona.)*

Figure 5-7. *Comparison of one of Galileo's first lunar telescopic drawings (left), made in 1610, to a photograph of the moon at the same phase. Letters show corresponding features. Galileo detected mountains, plains, and craters (bottom). (Courtesy Ewen A. Whitaker, University of Arizona.)*

entists did in the 1600s, he called them *maria* (singular **mare** [MAH-ray]). These "seas," as Galileo himself probably eventually realized, are actually vast plains, which we now know are covered with dark lava. Mare surfaces cover much of the front side of the moon, but only 15% of the whole moon. The bright regions are cratered, rugged upland areas. Figure 5-11 shows an astronaut's view from above the boundary between a mare plain and the uplands.

Galileo's work greatly strengthened the conception that the moon is an earthlike, planetary body having familiar features such as mountains. He used this as a proof against those who argued that the earth was not to be included among the planets.

The first reasonably accurate lunar map was produced by the German Johannes Hevelius in 1647. (A modern lunar map with names of certain prominent features is seen in Figure 5-12.) In 1651 an Italian priest, Riccioli, started the present practice of naming craters after well-known scientists and philosophers, such as Copernicus, Tycho, and Plato. The maria were

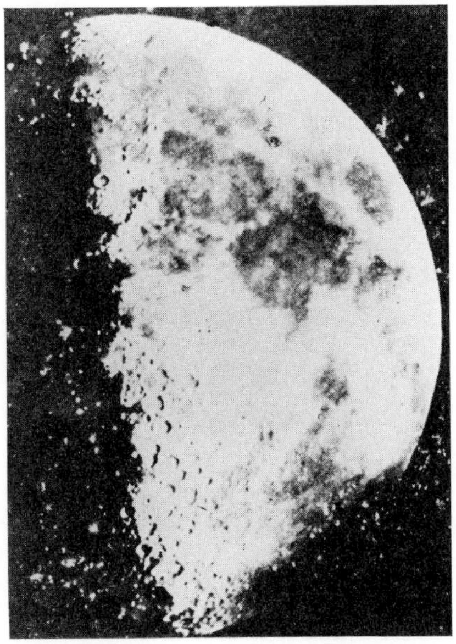

Figure 5-8. *This is one of the first known photographs of the moon, an 1851 daguerreotype, probably by the American astronomer J. W. Draper or his son Henry Draper. Compare with Figure 5-9, a modern photo of the same phase. The smallest details are about 35 km across. (NASA.)*

Figure 5-9. *A modern view of the moon at a phase similar to that in Figure 5-8, illustrating a century's advance in astronomical photography. The smallest details are about 1 km across. (Lunar and Planetary Laboratory, University of Arizona.)*

given poetic and fanciful names, such as Mare Imbrium (Sea of Rains). Lunar mountains were named for prominent terrestrial ranges, such as the Alps. These mountains, however, are unlike terrestrial folded ranges; they are the rims of vast multiringed craters, called **basins**, which in turn contain the mare plains. Figures 5-13 and 5-14 illustrate one of these huge impact features and its probable subsurface structure. Bright streaks called **rays**, which radiate from various craters but show no relief, are fine debris blasted out of the craters. Most maria are merely seas of lava that have flooded ancient basins. Meandering canyonlike valleys, probably eroded by hot lavas, are called **rilles**.

Astronomers of the 1700s and 1800s undertook a tantalizing endeavor, searching with ever larger telescopes with ever greater magnification for diagnostic details that would show how the moon formed. Were there any signs of changes—new craters forming or volcanic activity? Were there any artificial structures that might have been built by the inhabitants that popular writers continued to place on the moon?[1]

Despite years of search, no changes or artificial structures were found. Lunar photographs, first made in 1849 (see Figure 5-8), showed the same structures year after year. Because there seemed to be nothing new to discover, interest in the moon dwindled, and many astronomers saw it as a sterile nuisance that lit up the night sky and blotted out the faint objects that they wanted to observe.

Nonetheless, evidence for minor geologic activity on the present-day moon came in 1963, when on two occasions observers at Lowell Observatory saw a red glow near the crater Aristarchus. This glow may have been a volcanic eruption or gas discharge; later data from Apollo flights indicated occasional gas emission in this region.

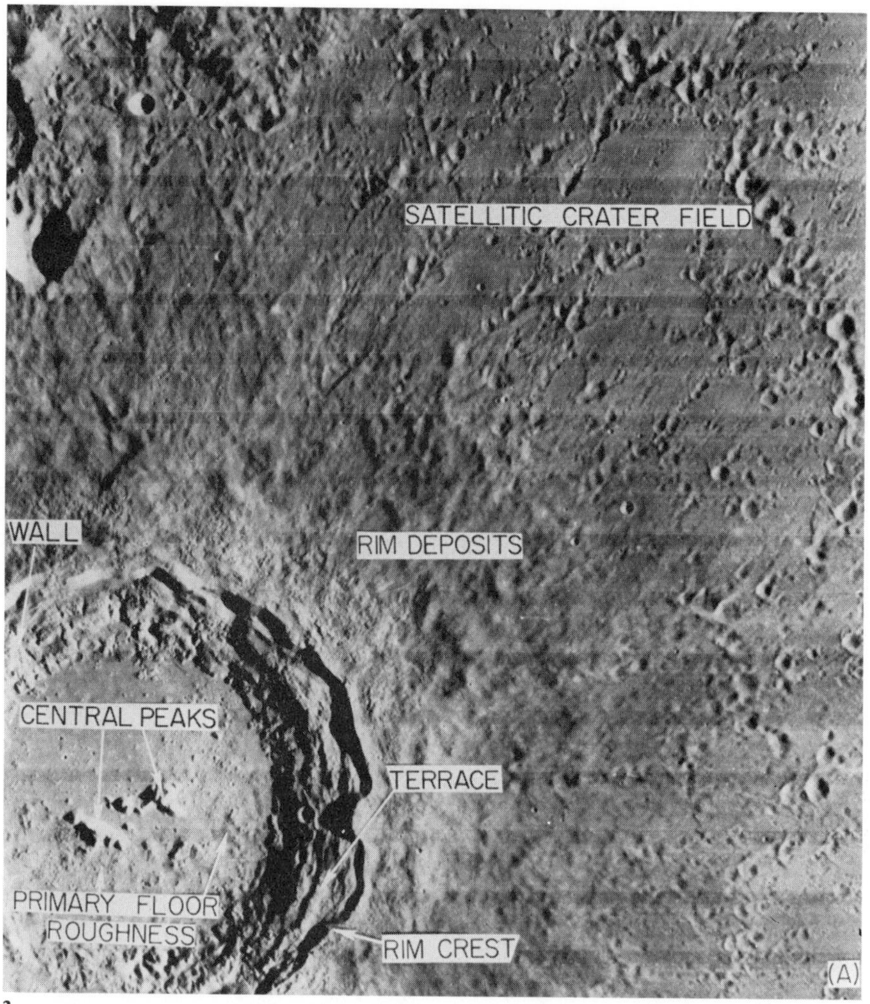

a

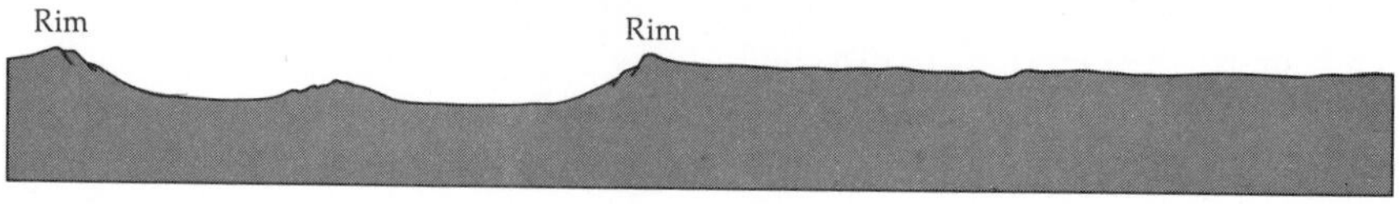

b

Figure 5-10. *(a) The typical structures in a large lunar crater are exemplified by this Orbiter photograph of the crater Copernicus, which is about 90 km across. The main bowl was excavated by the impact of a meteorite a few kilometers in diameter. Central peaks and terraces are rebound and slump features. Rim deposits and satellite craters are caused by debris blasted out of the crater during its formation. (Courtesy James W. Head III, Brown University.) (b) An approximate cross section of such a crater.*

[1] Or—an idea conceived more recently—by alien interstellar travelers who might have visited the solar system in past eras? See, for example, Arthur C. Clarke's book *2001*.

Figure 5-11. *View from an Apollo spacecraft flying over the edge of the Sea of Serenity, a 3½-billion-year-old lava-covered plain (right). Rugged area to left is lunar uplands, formed 4 to 4½ billion years ago by intensive cratering and overlapping of crater ejecta blankets. (NASA, Apollo 17, wide-angle photo.)*

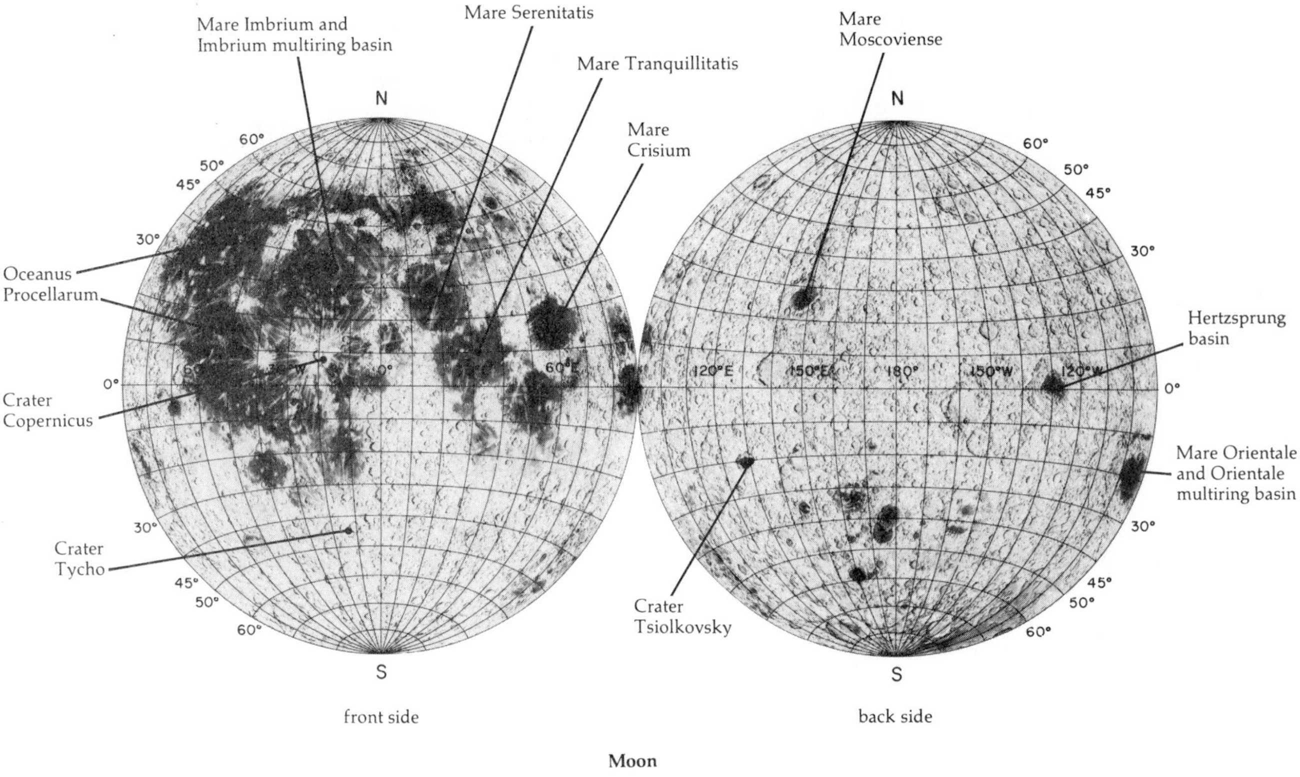

Figure 5-12. Map of the front (left) and far (right) sides of the moon, showing some prominent features. (U.S. Geological Survey, courtesy R. M. Batson.)

Why is the moon's geologic character different from the earth's? Why do craters dominate? Why are there no major mountain ranges? Where did the moon come from? What would it be like to stand on the moon? Such questions motivated actual journeys to the moon in the last decade.

Flights to the Moon

The first human device to reach the moon was a Russian spacecraft that carried little scientific equipment and crashed into the moon in 1959. The first close-up photos of the surface, from the American probe Ranger VII in 1964, showed that the surface was not craggy, but covered by a gently rolling layer of powdery soil (Figure 5-15), scattered rocks, and shallow craters of various sizes. This type of soil cover, present nearly everywhere on the moon, came to be called the **regolith** (rocky layer). The lunar regolith is typically 3 to 30 m (10 to 100 ft) deep and made primarily of debris blasted out of lunar craters as they were formed. Each well-preserved lunar crater is surrounded by a sheet of such debris, called an **ejecta blanket**. The regolith is therefore said to be composed of overlapping ejecta blankets. The powdery surface of the regolith is due to small meteorites (some microscopic), which are so abundant that they have "sand-blasted" most of the upper few meters into fine dust.

Table 5-1 lists the six Apollo lunar landings and two earlier test flights. Twelve men walked on the moon during the Apollo program from 1969 to 1972. Figures 5-16 through 5-23 and Color Plates 5 to 7 include typical scenes observed during these missions. The first two Apollo missions (Figures 5-16 and 5-17) were cautious tests of the spacecraft, touching down on flat, smooth plains. This allowed sampling of the rocks in the lava flows that make up the maria. These lavas were found to be 3½ billion years old, about as old as the oldest known earth rocks.

The subsequent four Apollo landings occurred at a variety of sites, chosen to sample the variety of lunar geology. These sites included the ejecta blanket from the giant Imbrium basin (Figure 5-18), a valley at the foot of the Imbrium basin rim (Figures 5-19–5-21), rolling terrain in the uplands (Figure 5-22, Color Plate 7), and the border of the Sea of Serenity (Color Plate 10).

The flights proved that the lunar surface features are very old. The maria—the dark regions visible to the naked eye from the earth—turned out to be vast flows of

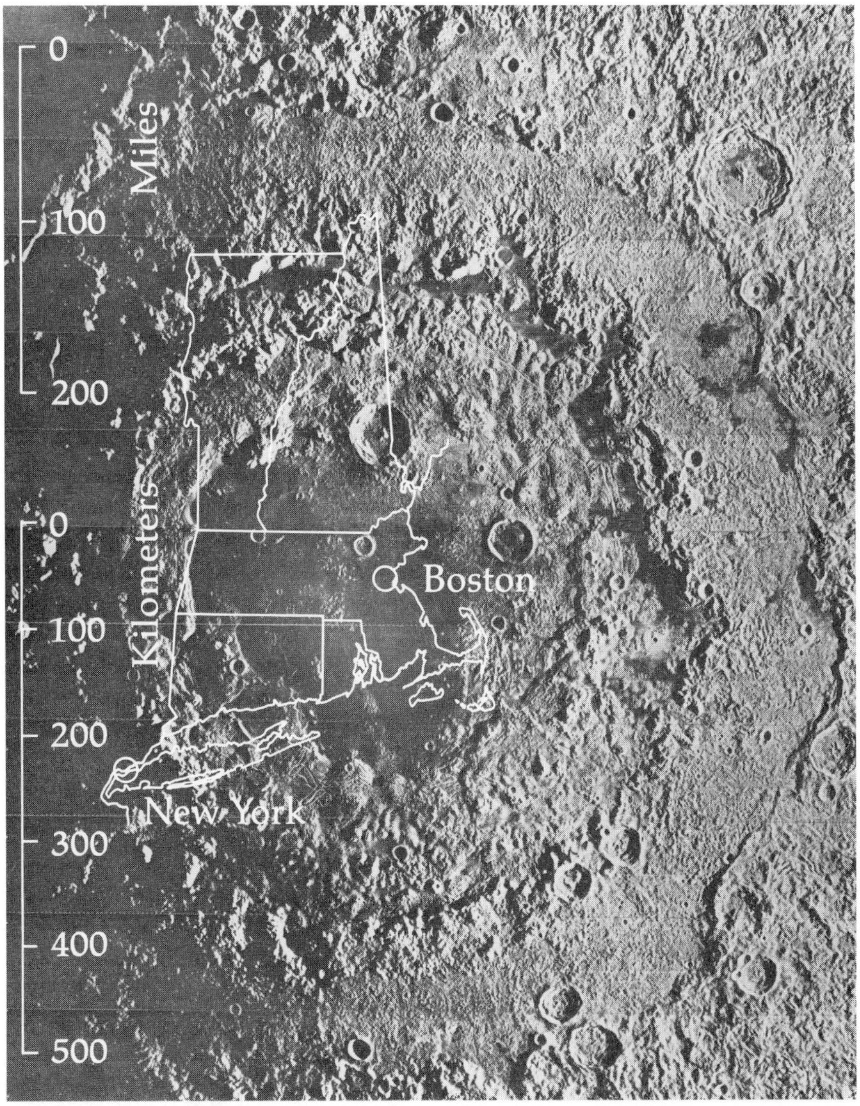

Figure 5-13. *The Orientale basin is the youngest and most dramatic multi-ring complex on the moon. The outermost ring of cliffs is nearly 1000 km in diameter. Lava erupted from fractures along the bases of some of the outer cliffs. Map shows scale relative to the northeastern United States. (NASA Orbiter photo.)*

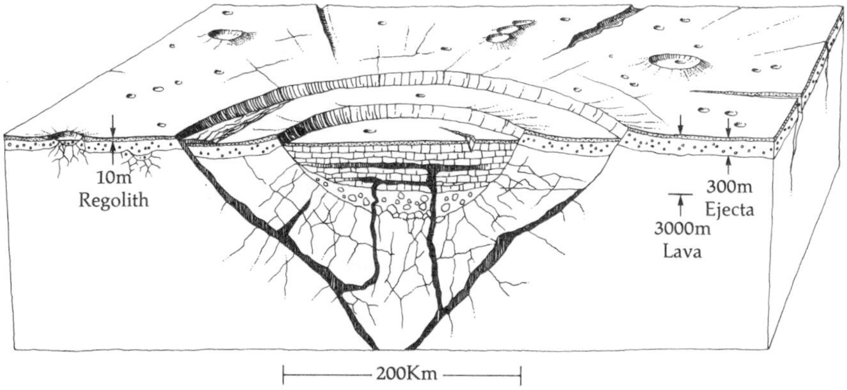

Figure 5-14. *A schematic cross section of a major lunar basin. Faulting and subsidence during or soon after the basin-forming impact created rings of cliffs. Vertical hatching indicates later intrusions of lava, typically several kilometers deep in the basin center. (Vertical scale exaggerated.)*

***Table 5-1.** Manned Apollo Explorations of the Moon*

Mission	Date	Landing Site	Results
Orbital Missions			
Apollo 8	Dec. 24, 1968[a]	———	First lunar orbit. Orbital mapping. 115-km minimum altitude.
Apollo 10	May 21, 1969[a]	———	Test of approach to approximately 17-km minimum altitude.
Landing Missions			
Apollo 11	Jul. 20, 1969	Mare Tranquillitatis	First landing. Samples of mare material.
Apollo 12	Nov. 18, 1969	Oceanus Procellarum	Sample of mare material.
Apollo 14	Feb. 5, 1971	Fra Mauro (ejecta from Imbrium basin)	Samples of ejecta from Imbrium basin.
Apollo 15	Jul. 30, 1971	Edge of Mare Imbrium at foot of Apennine Mts.	Samples of material from Apennine Mts., forming rim of Imbrium basin. Samples of mare material. First use of roving vehicle.
Apollo 16	Apr. 20, 1972	Lunar uplands near crater Descartes	First landing in uplands. Samples of upland materials.
Apollo 17	Dec. 11, 1972	Taurus Mts.; edge of Mare Serenitatis	Samples from a region suspected of recent volcanism.

[a]Date lunar orbit began.

***Figure 5-15.** The powderlike properties of the lunar regolith, or soil, are seen in this view of the imprints made by the footpad of a Surveyor unmanned spacecraft that bounced during touchdown. The photo was made by Apollo 12 astronauts who visited the Surveyor landing site. (NASA.)*

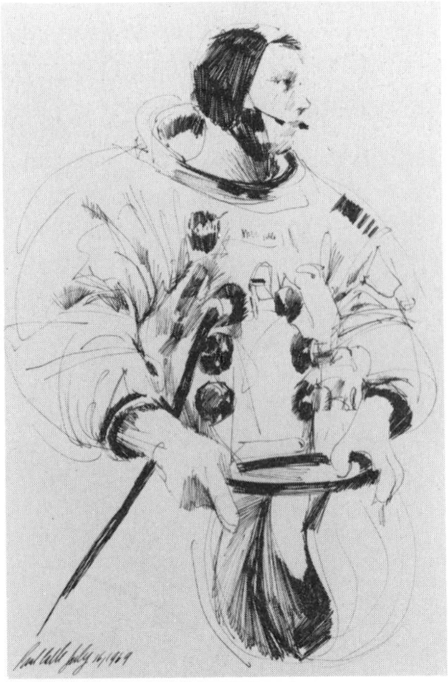

***Figure 5-16.** The first man to set foot on the moon, Neil Armstrong, was sketched on the morning of the Apollo 11 launch, July 16, 1969. A NASA-funded art program yielded many historic paintings and on-site sketches, such as this one. (NASA sketch by Paul Calle.)*

***Figure 5-17.** The Apollo 12 lunar landing module preparing to descend to the surface of the moon. (NASA.)*

basaltic lava 3 to 4 billion years old. The bright uplands are still older, formed during the early years of the solar system about 4.5 billion years ago. Chemical evidence from the rocks shows that at that time, the surface layers of the moon were molten, forming a vast sea of lava called a **magma ocean**. Low density feldspar crystals then solidified in the magma ocean. Because of their low density, these floated in the magma ocean, accumulating into an ever-thickening lithosphere of basalt-like rock called anorthosite. This is the rock type that now composes most of the uplands.

Most of the ancient anorthosite rocks did not survive intact. Numbers and ages of craters prove that the rate of meteorite impact in the first half-billion years of lunar history was thousands of times higher than today. The high cratering rate pulverized most of the primordial lunar rocks. Upland sites visited by astronauts revealed overlapping layers of ejecta blankets from many craters, composed of dust, glass droplets, rock chips, and **breccias**, or rocks composed of cemented rock fragments. These results explain why the most common lunar rocks date from about 3 to 4 billion years ago. Rocks older than 4 billion years were destroyed by impacts; after 3 billion years ago, the moon had cooled enough, and the lithosphere was thick enough, that subsequent volcanism or rock-forming activity was infrequent. Because 3 to 4 billion year old rocks are rare on the earth, lunar rocks have given geologists a welcome insight into conditions in the earth–moon system during that era. More detailed descriptions of the advances resulting from Apollo exploration are given by Taylor (1975) and El-Baz (1975).

In the years since the last Apollo landers blasted off the moon (see Figure 5-23), the moon has been deserted. In 1983 and 1984, NASA planners and scientists began holding conferences to discuss a return to the moon. They have suggested permanent lunar stations that could produce oxygen for breathing and for fuel, and could conduct lunar and astronomical research, by the year 2000.

Figure 5-18. *An astronaut approaches a shallow crater on a lunar plain formed by ejecta from the Imbrium impact basin. Light spots arise from sunlight reflection in the camera lens. (NASA, Apollo 14.)*

Figure 5-19. *The landing module at the Apollo 15 site at the foot of the Apennine Mountains. The tracks of the astronauts' battery-powered vehicle are visible in the powdery regolith. (NASA.)*

Figure 5-20. *Boulders excavated from the regolith by a modest impact crater on the Imbrium lava plain. (NASA, Apollo 15.)*

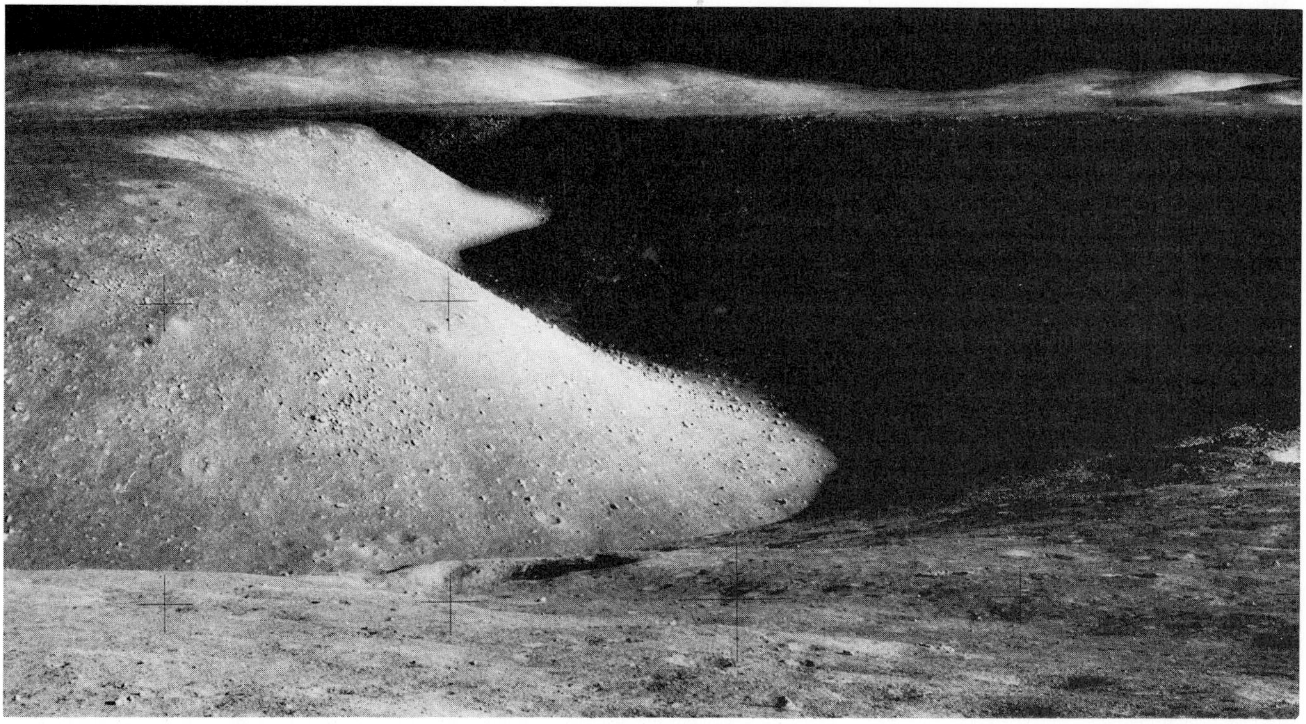

Figure 5-21. *View down a canyonlike lunar rille. Apollo 15 astronauts attempted to determine the origin of rilles by visiting this example, but found no conclusive evidence. Rilles probably are formed as channels along which lavas flowed during mare formation. (NASA, Apollo 15.)*

Lunar Rocks: Implications for the Moon and the Earth

Rocks are cosmically significant. They are solid materials that contain many clues to their long histories of crystallization, melting, recrystallization, and so on. Rocks also tell the histories of their parent planets. Rocks can be analyzed in various ways:

1. by their elements and isotopes, which indicate the material from which the parent planet originally formed;

2. by their minerals, which indicate the degree of differentiation that produced different chemical compounds inside the planet;

3. by their structure, which indicate their environments through history; and

4. by their radioisotopic ages, which indicate when the planet-forming, differentiating, and rock-altering processes occurred.

If the concepts in this list are not clear, the reader should review Chapter 3.

The elements and isotopes of lunar rocks (Figure 5-24) are similar to those of the earth's, though the moon has much less iron. No new elements have been found on the moon, and the proportions of elements suggest that the moon formed from material similar to that of the earth's mantle. However, relative to the earth's mantle, the moon is strongly depleted in **volatile elements** and compounds—substances that are driven off by heating, such as water. Similarly, the moon is enriched in **refractory elements** and compounds—substances with high boiling points, such as aluminum and titanium. These findings indicate that the lunar material may have been strongly heated before the moon formed, driving off the water and other volatiles into space.

Although the minerals of the lunar rocks are similar to those of the earth's mantle and crustal basalts, there is a very interesting difference. The moon's rocks are not as differentiated as the earth's, and there are no granitic continental blocks or plates in the lunar crust. These facts are among many lines of evidence suggesting that the moon has not been nearly so geologically active as the earth. This is because the moon is so much smaller than the earth that it cooled quicker, formed a thicker lithosphere, and never developed the plate movements or perpetual volcanic and tectonic activity that make the earth's geology so complex.

Figure 5-22. *A 900-m-diameter (½-mi-diameter) impact crater in the lunar uplands. Rock samples indicate that the crater formed about 50 million years ago. (NASA, Apollo 16.)*

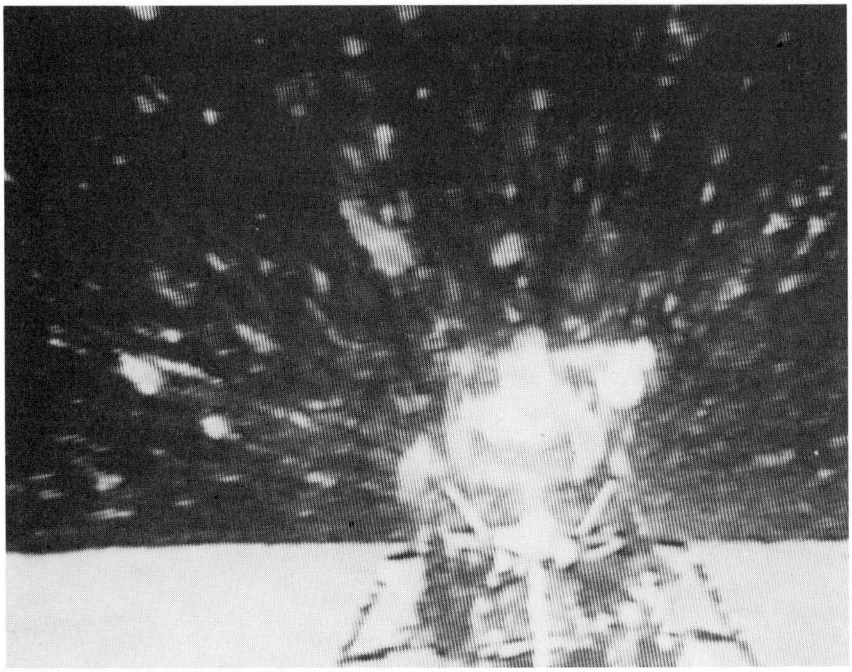

Figure 5-23. *The payoff from the Apollo missions came with the returns to the earth. This blastoff from the moon was televised to the earth by a remote camera set up by Apollo 16 astronauts. Debris flying away from the spacecraft includes material blown off the descent stage (lower) by ignition of the ascent engine. (NASA.)*

Figure 5-24. *Apollo 15 astronauts David Scott (left) and James Irwin (right) join NASA geologists in examining lunar rock samples in the sample receiving laboratory in Houston. (NASA.)*

The evidence for a lunar magma ocean has led to a new view that many or all planets may have initially had magma oceans that allowed low density minerals to concentrate near their surfaces. This influenced the rock types, minerals, ores, and so on, that we now find on the surfaces of these bodies.

Radioisotopic dating ties the whole rock story together (see Table 5-2). The moon formed about 4.5 or 4.6 billion years ago. The earliest rock fragments (often found inside breccias) show that the magma ocean finished solidifying around 4.4 billion years ago. The whole period from 4.5 to 4.0 billion years ago was marked by extremely intense cratering, as shown by the fact that most rocks older than 4 billion years are highly fractured breccias. All huge impact basins formed as part of the intense cratering that occurred before or around 4 billion years ago. The biggest impacts created fractures that allowed basaltic lavas to erupt from depths of around 300 km (a depth determined from certain mineral properties), and lunar lava flows filled many of the impact basins between 4 and 3 billion years ago. The moon has been relatively quiet since then.

It is extraordinary to realize that astronauts have walked and driven across landscapes whose mountains and craters have lain still and silent since trilobites crawled on our own seafloors and the first lizard-like creatures crawled out on our land. Perhaps the most important fact to remember is that the thick lunar lithosphere formed very early, so that its surface preserves craters and rocks formed 3 to 4 billion years ago. In contrast, the earth is so active that it has destroyed most landforms and rocks formed during this period. Thus the moon supplies some heretofore missing information about the early history of the planets and the early environment of the earth–moon system.

The Interior of the Moon

We can understand the moon better by considering its interior structure, shown in Figure 5-25.

First, the **mean density** (total mass divided by total volume) of the moon is much less than that of the earth—3300 versus 5500 kg/m^3 (3.3 versus 5.5 g/cm^3). This proves that the moon is mostly rocky, like the earth's mantle, and that it lacks a big iron core. Second, measurements indicate that the moon has virtually no magnetic field. This again suggests the lack of a large molten iron core, because scientists believe that the planets' magnetic fields originate in currents in such cores. Nonetheless, magnetic measurements on

Table 5-2. *Summary of Dated Lunar Rocks*

Mission	Rock Type	Age[a] (billions of years ago)	Notes (ages[a] based on)
Apollo 11	Mare basalts	3.48–3.72	7 rocks
Apollo 12	Mare basalts	3.15–3.37	9 rocks
Apollo 14	Fractured rocks ejected from Imbrium basin	3.85–3.96	6 rocks (date of Imbrium impact?)
Apollo 15	Mare basalts	3.28–3.44	6 rocks from 2 sites
	Upland anorthosite	4.33	2 rocks
Apollo 16	High-aluminum-content basalt	3.84	11 rock chips
	Upland breccia	4.15	
Apollo 17	Mare basalt	3.77	1 rock
	Upland breccia	4.15	3 rocks
	Fractured dunite	4.48	1 rock
Luna 16	Mare basalt	3.43	2 rock chips
Luna 20	Upland anorthosite	ca. 4.4	1 rock chip
Luna 24	Mare crisium basalt	3.34	5 rock chips
Comparison Ages for Other Rocks			
	Meteorites	4.6	Typical formation date
	"Whole moon"	4.6	Formation of moon
	Oldest terrestrial rocks	3.9	Very rare specimens
	Typical terrestrial rocks	0.01–0.5	

[a]Date of solidification of crystalline rocks, or formation of breccias, based primarily on rubidium–strontium results of Wasserburg, Papanasstasiou, Tera, and colleagues at the California Institute of Technology, and a summary by S. R. Taylor, 1982. Additional rocks and soils were collected.

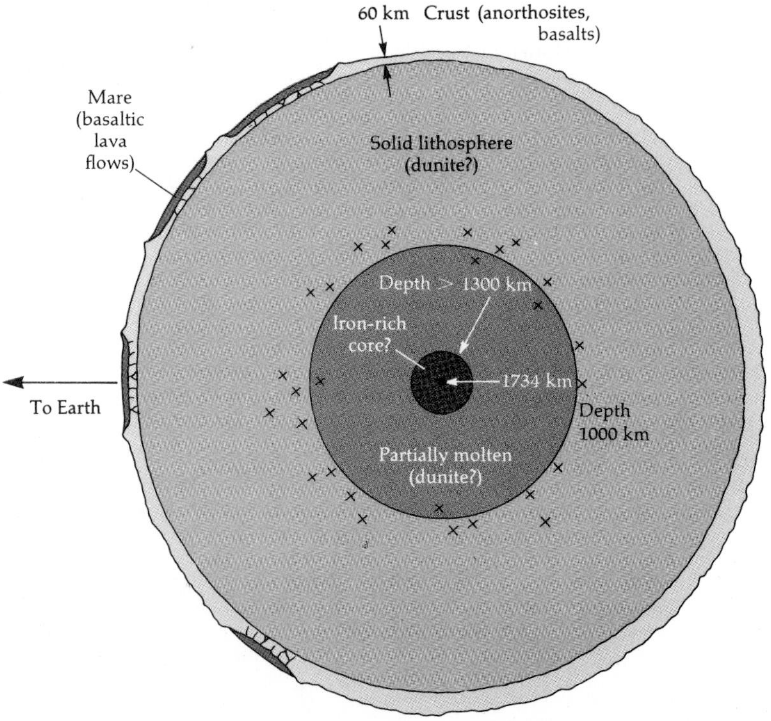

Figure 5-25. ***Cross section of the moon as revealed by Apollo and other sources. The crust is thinner on the front side than on the far side. Fractures under large impact craters on the front side allowed lava to reach the surface and create more mare lava flows than on the far side. An earthquake zone (x) marks the bottom of the lithosphere. The existence of an iron-rich core is uncertain.***

lunar rocks indicate that when they solidified billions of years ago, there was a lunar magnetic field roughly 4% the strength of the earth's field. Because no such field exists today, the implication is that the moon had a molten iron core only in the past, when its interior temperatures were higher. A small core might still exist but reach no farther than about 350 km from the moon's center, or about 20% of the radius (Lammlein et al., 1974).

Third, seismic data show much less quake activity on the moon than on the earth. Large lunar quakes rank only 0.5 to 1.3 on the Richter scale, compared with 5 to 8 for major earthquakes. In a year, the earth expends about 100 million million times as much seismic energy as the moon.

Nonetheless, moonquakes tell us something about the moon's interior. The quakes occur mostly at depths of 700 to 1200 km. Just as the earth's quakes are concentrated at the bottom of the brittle lithosphere and the top of the sluggishly-moving asthenosphere, the moon's quakes are believed to mark the bottom of a 1000-km-thick lunar lithosphere. Many moonquakes occur in monthly cycles associated with tidal flexing as the moon's slightly elliptical orbit brings it toward and then away from the earth (Toksoz et al., 1974). Some quakes detected by Apollo instruments had a different source—the impacts of modest-sized meteorites (too small to make craters visible from the earth).

Consequences of the Thick Lunar Lithosphere

The evidence from lunar rocks and seismographs indicates that the 1000-km-thick lunar lithosphere is much thicker and less geologically active than the 100-km-thick

OPTIONAL BASIC EQUATION IV

The Definition of Mean Density

When an astronomer begins to get data on a new object, such as a newly measured planet, asteroid, or star, one of the first things he or she wants to know is the mean density of the object, because this helps clarify the composition of the object. The mean density is defined as

$$\text{mean density} = \frac{\text{total mass of object}}{\text{total volume of object}}$$

This is called the *mean* density because it gives only the average density, averaging over the whole object. Parts of the object's interior might have higher or lower densities. The earth's mean density, for example, is about 5500 kg/m^3, but the iron core has much higher values, in excess of 8000 kg/m^3, while surface rocks have values closer to 2500 kg/m^3. Thus, for example, if you sighted a round object in space, it would make quite a difference to your interpretation if the mean density turned out to be 8000 kg/m^3 instead of 2000 kg/m^3!

The value of 1000 kg/m^3 is of particular interest, since this is the density of water and the approximate density of ice. (When the metric system was set up, the kilogram was defined such that 1000 kg equaled the mass of a cubic meter of water.)

Since most objects in astronomy are roughly spheroidal in shape, we can replace volume by the expression for the volume of a sphere, $4\pi R^3/3$. Thus, using R as the radius, M as the mass, and the Greek letter ρ (rho) traditionally used as the density, we get the equation

$$\text{mean density} = \rho = \frac{3M}{4\pi R^3}$$

Thus if an astronomer or astronaut can measure the mass M of an object (perhaps by its gravitational force on another object) and its radius R, he or she can quickly calculate the mean density and make some comments about the types of materials that might compose the object.

Note that in the SI system of units, M is measured in kg and R in m, so that the density is expressed as kilograms per cubic meter. While water and ice have densities of about 1000 kg/m^3, rock has densities around 2500 to 3000 kg/m^3, and gaseous objects have densities less than 1000 kg/m^3, unless they are compressed by strong gravity. In many scientific and other books, density is more commonly expressed in units of grams per cubic centimeter, which are 1/1000 of the value in kg/m^3. For example, water has a density of 1 g/cm^3.

Sample problem. Voyager spacecraft and earth-based measurements showed that the small moon of Saturn, Enceladus, has a surface of frozen water, a radius of 250 km, and a mass of 8.4×10^{19} kg. Find the mean density of Enceladus and comment on whether it might be largely icy or rocky throughout. *Solution*: Converting R to meters and inserting the given values for R and M, we compute a density of 1283 kg/m^3. Since R and M are given only to two significant figures, we should round off our answer to two significant figures, or 1300 kg/m^3. Note that this is only 30% greater than the density of ice! Enceladus thus could not be made mostly of rock, but could be mostly icy throughout!

Figure 5-26. *Explosion tests on the earth simulate many features of meteorite impacts. This explosion of 100 tons of TNT produced a crater 39 m (128 ft) across and 7 m (23 ft) deep. Secondary impact craters formed as far as 110 m (360 ft) away and the furthest ejecta went 200 m (660 ft). The jets of material may simulate material that formed lunar rays; the turbulent cloud at the base of the explosion expands across the landscape and deposits much of the ejecta blanket. (Photo by author.)*

lithosphere of the earth (see Chapter 3). We can understand why. Because the moon is a small planet, it lost its heat quickly. The initial magma ocean cooled quickly and solidified. A lithosphere was quickly established that was too thick to break into plates due to internal activity, and no granitic scum could accumulate on the surface by repeated convective churning of the interior. No folded mountain ranges, like the Rocky Mountains, could form. By 3 billion years ago, the lithosphere had grown too thick to allow much lava to reach the surface, preventing formation of large volcanic mountains. The surface therefore recorded only crater upon crater as meteorites fell onto the moon. The thick lithosphere thus largely determined the rock types and surface structures of the moon today.

Cratering of the Moon and Earth

As noted earlier, the moon underwent an intense bombardment from about 4.6 to 4.0 billion years ago, with a cratering rate thousands of times higher than today's rate. All large impact basins, such as that shown in Figure 5-13, as well as the heavily cratered uplands formed at that time. From about 4.0 to 3.0 billion years ago, the rate declined to the present level, which has been nearly constant since that time. The earth has had roughly the same cratering rate, and scientists assume that both the earth and the moon experienced similar cratering histories. The Apollo data from the moon thus help clarify the impact history of earth, paving the way for recent theories that giant impacts may have altered climates and biological evolution (see Chapter 3).

The **early intense bombardment** of the moon represents the final stages of the sweepup of debris left over after planet formation (see Chapter 12). The now visible lunar craters may represent only the last interplanetary bodies to be swept up, with still earlier craters obliterated by the ones we now see.

Craters are the most important landforms on many other planets and moons as well. The photo in Figure 5-26 shows how meteorite impact explosions may have looked. The meteorites ranged from abundant microscopic particles, through numerous kilometer-sized chunks, to objects over 100 km in diameter. The latter made craters about 1000 km across.

Repeated cratering by a rain of meteorites explains lunar topography. In the lava plains, there have been only enough craters in the last 3 billion years to cover a few percent of the surface with multikilometer craters and to grind up a layer of lava into the regolith some tens of meters thick. But in the uplands, the 4- to 4½-billion-year-old surface is virtually saturated with multikilometer craters. Small impacts and overlapping ejecta blankets have smoothed the once craggy landscape, producing the gently rolling hills photographed by astronauts.

Among the many lunar landscapes not yet seen by astronauts, large young craters would be the most rugged and dramatic, as shown by Figure 5-27.

Theories of the Moon's Origin

The moon's origin has long frustrated theorists. Prior to the Apollo missions, astronomers debated three main theories of lunar origin. The **fission theory** proposed that the moon is made of material spun off the outer layers of the earth, and that it has receded ever since, due to tidal action. The **co-accretion theory** proposed that the moon formed in orbit around the earth at the same time the earth was forming. The **capture theory** proposed that the moon formed somewhere else in the solar system, and then was later captured into an orbit around the earth.[2] Each of these theories now seems to have fatal flaws. The fission theory did not explain how enough energy could have been provided to have lifted a swarm of particles off the earth; the co-accretion theory did not explain how the moon could have formed near the earth but have received so much less iron than the earth; the capture theory did not explain how the capture might have occurred, or where the moon formed in order to receive its low-iron composition.

Apollo studies provided new information. Especially important was the finding that the moon's bulk material is generally like the earth's mantle, but with less volatiles and slightly more refractories. Another important result was that the proportions of different isotopes composing lunar oxygen exactly match those in terrestrial oxygen. Studies in the 1970s showed that oxygen in rocks from other parts of the solar system (e.g., meteorites) has different isotopic proportions. This establishes that lunar material formed at about Earth's distance from the sun and rules out, for example, capture of a body from far away.

In 1984, an international conference produced a new leading theory called the *impact-trigger hypothesis.*[3] It suggests that during the earth's formation, but after the iron core formed, the earth was hit by a very large interplanetary body—perhaps as large as Mars. (As will be clearer in Chapter 12, many secondary interplanetary bodies were growing as the planets

a

b

c

Figure 5-27. *Views of the prominent young crater Tycho, about 90 km in diameter. The age of the crater has been estimated at 100 million y to 270 million y.* (a) *Remote view of the moon centered above Tycho, showing Tycho's system of bright rays. These are caused by streamers of ejecta shot out from the crater during the impact explosion that formed it. (Rectified photo, Lunar and Planetary Laboratory, University of Arizona.)* (b) *View of Tycho showing rugged interior and surroundings and central peaks. Box shows location of photo* c. *(NASA Orbiter 5.)* (c) *Close-up of the extremely rugged floor surface just above the central peak shown in the preceding view. Smallest details are about 100 m (100 yd) across. Tycho is apparently too young for cratering to have "sandblasted" these rugged features and converted them to a smooth regolith. (NASA, Orbiter 5.)*

[2]With a bit of tongue-in-cheek male chauvinism, NASA geologist Bevan French has called them the daughter, sister, and pickup theories.

[3]This theory was first published in 1975 by D. R. Davis and myself and, essentially independently, in 1976 by A. Cameron and W. Ward, but little further work was done on it for several years. By the time of the 1984 conference, new data were available and sophisticated computer models of impact were possible, and it was interesting to see the idea emerge from relative obscurity into the status of the leading theory!

formed.) The impact blasted out hot debris from the upper mantle (Figure 5-28). This debris, lacking iron and depleted in volatiles because of the heat of the impact, aggregated to form the moon. This theory explains a moon with roughly mantlelike composition, little iron, few volatiles, and terrestrial oxygen-isotope proportions. It also explains the moon's uniqueness: Such an impact might have happened to only one of the nine planets.

Various researchers are working today on the impact-trigger hypothesis to test whether it can explain further details about the moon. The work includes chemical studies of lunar rocks and computer models of the giant impact (Hartmann, Phillips, and Taylor, 1986).

The firmest information so far obtained from lunar rocks about lunar origin is not about the mode but about the *timing* of the event. The moon formed about 4.6 billion years ago, and the process was relatively fast, in astronomical terms, taking not more than about 90 million years—the time interval estimated for formation of the entire solar system.

Summary

The moon is an ancient planetary body, little disturbed since the formative days of the solar system. Much of the information in this chapter yields a chronological history of the moon. Because the earth shared much, if not all, of this history, we summarize the history of the earth–moon system in Table 5-3.

Origin. Lunar rocks and meteorites reveal that the moon and planets were formed 4.6 billion years ago. The moon's formation may have involved a giant impact. The earth and moon were close together shortly thereafter.

Duration of Formative Process. Differences in ages among lunar and meteorite specimens indicate that the moon and planets reached approximately their present sizes a few million to 90 million years after the formative process began.

Magma Ocean. Analyses of lunar rocks indicate that the lunar surface was initially covered with a molten magma ocean several hundred kilometers deep. An initial magma ocean may also have formed on the earth, but evidence here has been destroyed.

Early Intense Bombardment. Nearly all lunar rocks that formed before about 4 billion years ago have been pulverized by repeated

Table 5-3. ***History of the Earth–Moon System***

	Time (billions of years ago): 4	3	2	1	0 (Present)
EARTH	Formation			?←Continental evolution	→
	Intense cratering[a]	Length of day increasing		Earliest substantial fossils	
	Crustal formation				
	Shorter day (due to tidal interaction with moon)[a]				
					Mammals→| |←
	Early life	Atmospheric changes (O_2 increase due to plants)			Man→||←
	Oldest surviving rocks				
MOON	Formation[a]				
	Intense cratering[a]				
	Early heating[a]	Interior cooling[a]	Sporadic cratering[a]		
	Rapid motion away from earth[a]			Occasional moonquakes and gas emissions[a]	
	Mare formation[a]				
	4	3	2	1	0 (Present)

Time (billions of years ago)

[a]Information discovered or improved through Apollo-related lunar research.

bombardment. According to Apollo data, the cratering rate 4 billion years ago was 1000 times the present rate, and before that it was still higher. The earth was presumably also bombarded at that time. Many large craters formed on the earth and the moon during this period. The cratering rate declined to the current value by 3 billion years ago.

Tidal Movement of the Moon. The moon was once much closer to the earth than it is now. Tidal analysis proves that from its moment of closest approach, it moved out quickly, reaching about half its present distance in only 100 million years, or 2% of the time since the earth's formation. If the moon formed alongside the earth as the earth formed, then it probably approached its present orbit between 4.5 and 4 billion years ago, and the earth's day approached its present length at that time.

Heating and Differentiation. Radioactive elements in the earth and the moon heated their interiors, allowing heavy metal materials to sink to the center, probably within a billion years of the formation of these bodies. The earth formed a large iron core, but the moon had less iron and formed little or no core.

Mare Lava Flows. Basaltic lavas erupted in many places on the front side of the moon, especially where the crust was thinnest because of large-scale impacts. Successive lava flows formed the mare plains 3.8 to 3.2 billion years ago. Because the moon is smaller than the earth, it cooled more quickly and most volcanism died out around 3 billion years ago.

Lithosphere Evolution. The moon cooled off and formed a 1000-km-thick lithosphere that shielded its surface from internally caused changes. Convection currents continued in the earth, breaking the 100-km-thick lithosphere into moving plates.

Sporadic Recent Cratering. Occasional large impacts, involving meteorites a few kilometers across, excavated major craters, throwing bright rays of pulverized ejecta across the dark maria and old uplands. The most prominent example is shown in Figure 5-27. Smaller meteorite impacts churned the upper soil, producing the regolith. On the earth some of the large craters formed by meteorites during the last billion years have also been preserved (as described in Chapter 3).

Terrestrial Erosion. The parts of terrestrial history studied by most geologists are only the tail end of earth–moon history. Because of plate tectonics and atmospheric erosion, most of the surface rocks of the earth represent only the last 12% of earth–moon history, whereas most lunar surface structures date back through 84% of it. The moon shows more clearly than the earth the early and middle parts of geologic history and the combined results of internal geologic processes (for example, partial melting) and external astronomical processes (impacts). The moon has given us new understanding of processes that shaped our own world, and we have left our mark on it.

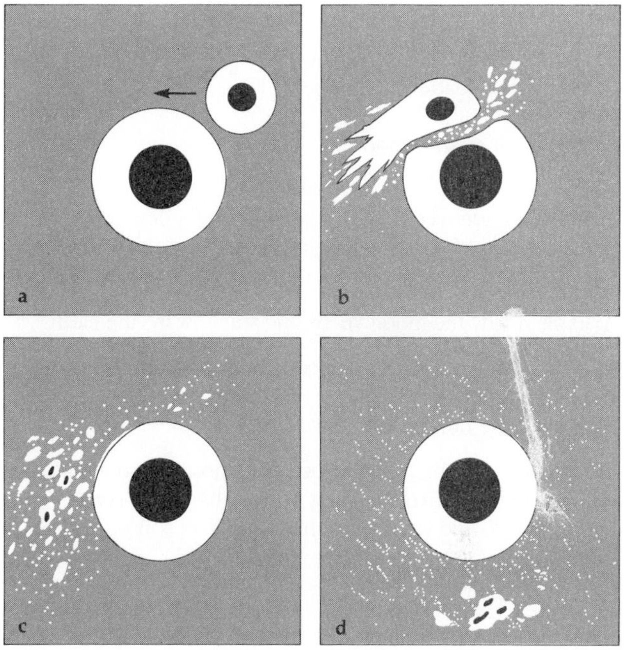

Figure 5-28. *Schematic view of the impact-trigger hypothesis for lunar origin. A large interplanetary body approaches (a) and collides with the primordial Earth (b), following formation of iron cores (dark) in both bodies. Hot gas and condensing debris from the mantles of both bodies are thrown into near-Earth space (c). Part of the debris forms an orbiting cloud around Earth, in which the moon begins to aggregate (d). (Adapted from computer models by A. Cameron, W. Benz, W. Slattery, M. Kipp, and J. Melosh.)*

Concepts

synchronous rotation	rille
sidereal period	regolith
tide	ejecta blanket
body tide	magma ocean
ocean tide	breccia
tsunami	volatile elements
tidal recession	refractory elements
Roche's limit	mean density
terminator	early intense bombardment
limb	fission theory
impact crater	co-accretion theory
mare (maria)	capture theory
basin	impact-trigger theory
ray	

Problems

1. At what time of day (or night) does
  a. the first quarter moon rise?
  b. the full moon?
  c. the last quarter moon?
  d. the new moon?

2. When a terrestrial observer is recording a new moon, what phase would the earth appear to have to an observer on the moon?

3. Explain why the moon keeps one side toward the earth.

4. Why might one naively expect the highest tides to occur at noon or midnight on the date of the new moon or the full moon? Why don't the highest tides always occur at these times?

5. If astronauts orbiting just above the atmosphere released into an orbit of their own two ping pong balls just in contact with each other, would you expect them to stay in contact with each other indefinitely? Why or why not?

6. Imagine you are selecting a lunar landing site.
  a. What type of feature might offer fresh bedrock, where regolith layers have been stripped away?
  b. Would you expect the landscape inside a young 100-km-diameter crater to be rougher or smoother than inside an old crater of the same size?
  c. Where might astronauts land to seek evidence of recent volcanic activity?

7. Imagine you are an astronaut exploring the moon.
  a. Would an ordinary compass work on the moon? Why or why not?
  b. What celestial object could serve as a navigational beacon for astronauts on the moon's front side?
  c. What property of this object's apparent motion would make it especially useful as a navigational aid during a lunar stay of several months?

8. If the ages of the earth and moon are identical, as believed, why are most rocks found on the moon so much older than earth rocks?

9. Suppose sedimentary rocks had been discovered on the moon. How would this affect our beliefs about the moon's history?

10. By comparing pictures of lunar maria (3 to 4 billion years old) and uplands (4 to 4.5 billion years old), prove that the meteoritic cratering rate was much higher during the first few hundred million years of lunar history than during the last 3 billion years.

ADVANCED PROBLEMS

11. The moon has about 0.012 of the earth's mass and 0.27 of the earth's radius.
  a. Using Newton's law of gravity, show that the moon's surface gravity is about one-sixth that of the earth.
  b. How much would an 82-kg (180-lb) person weigh on the moon? Give answer in pounds.
  c. Note that pounds are a measure of weight, while kilograms are a measure of mass. Would an 82-kg person have a different mass on the moon?

12. If you have a telescope that resolves details as small as 1 second of arc, what would be the diameter of the smallest craters and mountains you could see on the moon?

13. How fast must a projectile move to escape from the moon (mass of moon = 7.35×10^{22} kg; radius = 1.74×10^{6} m)? Compare this with the escape velocity from the earth.

14. From planetary data given in Table 6-1, confirm that the earth's mean density is roughly 5500 kg/m^3, and that the moon's is roughly 3300 kg/m^3. Explain how this result alone proves that the moon cannot contain as much iron as the earth.

Projects

1. Observe the moon at different phases with a telescope of at least 5-cm (2-in.) aperture. Locate and compare the texture of upland regions with maria (dark plains). Compare visibility of detail near the terminator and away from the terminator, and explain the difference. Sketch examples of craters, ray systems, mountains, and rilles.

2. With a telescope, find an example of a bright-ray crater, such as Tycho or Copernicus, and compare its appearance at full moon (high lighting) with its appearance near the terminator (low lighting). Why do the rays disappear under low lighting? Make a simulation of this effect by scattering a thin dusting of white flour or powder on a slightly darker, textured surface with a raylike pattern. Illuminate with a bright light bulb from above (full moon) and from the side at a low angle (low lighting), and compare the appearance.

3. Prepare a box with white flour several centimeters deep and a light dusting of darker surface powder (flush with the top edge of the box). Drop different-sized pebbles into the box to make craters. Illuminate with a bright light bulb from various angles and compare with the appearance of the lunar surface. Compare the number of craters needed to simulate a mare region and an upland region; can the surface be saturated with craters if enough stones are dropped? What physical differences exist between this experiment and lunar reality? (Example: These stones hit at a few meters per second, whereas meteorites hit the moon at several kilometers per second and cause violent explosions.)

P A R T C

The Solar System

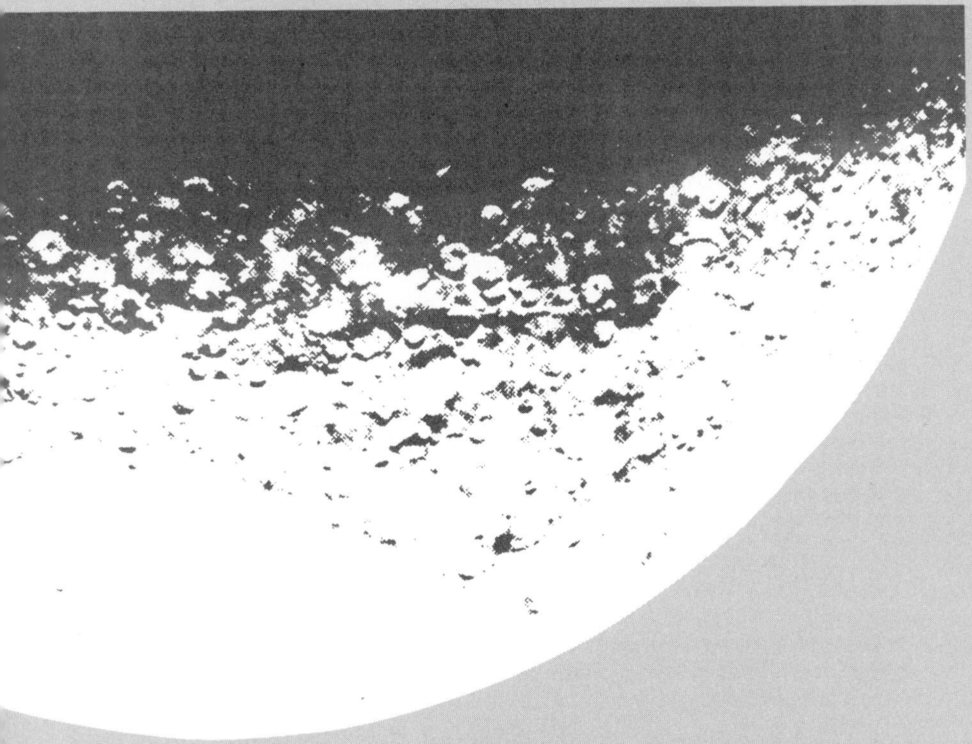

In questions of science the authority of a thousand is not worth the humble reasoning of a single individual.

Galileo

CHAPTER 6

DISCOVERING THE SOLAR SYSTEM

If we could journey far beyond the moon and look back, we would see the sun and its family of planets—the **solar system** (Figure 6-1). We would discover that the earth is only the fifth largest of many worlds that orbit around the average-sized star we call the sun. This is a far cry from the conception of twenty generations ago, when the earth was viewed as a kind of imperial capital of the universe—a unique stationary scene of human activities around which the sun, moon, planets, and stars moved. The exciting transition from the older idea to the modern conception began what astronomer Carl Sagan has called "the cosmic connection," the realization that we are only one part of a larger system of worlds—an idea still growing in our consciousness even today.

The transition began during the Renaissance, when the concept of a system of earthlike worlds seemed such an attack on traditional knowledge that it became an issue of life and death for some of the participants, as we will see. Today, the idea that we live in a system of worlds suggests a new conception of a large theater in which we may travel, investigate many new examples of geologic, meteorological, and biological processes, and perhaps exploit new sources of energy and materials.

A Survey of the Planets

The solar system is defined as the sun, its nine orbiting planets, their own satellites, and a host of small interplanetary bodies, such as asteroids and comets. Starting in the center of the solar system, the major bodies and their symbols are:

☉ **Sun**

☿ **Mercury**

♀ **Venus**

⊕ **Earth**

♂ **Mars**

♃ **Jupiter**

♄ **Saturn**

♅ **Uranus**

♆ **Neptune**

♇ **Pluto**

The symbols, mostly derived from ancient astrology, are sometimes used as convenient abbreviations today. A traditional memory aid for this outward sequence is "*M*en *V*ery *E*arly *M*ade *J*ars *S*tand *U*pright *N*icely, *P*eriod." Surely today's students can do better![1] To avoid confusion about the positions of Saturn, Uranus, and Neptune, remember that the *SUN* is a member of the system, too.

Some simple facts about the bodies of the solar system are useful to remember. For example, the sun is about 10 times the diameter of Jupiter, and Jupiter is about 10 times the diameter of the earth. Whereas the sun is a **star**, composed of gas and emitting radiation by its own internal energy sources, **planets** are bodies at least partly solid, orbiting the sun, and known to us primarily by reflected sunlight. **Satellites**, in turn, are solid bodies orbiting the planets.

The planets divide into two groups. **Terrestrial planets** are the four inner planets, Mercury through Mars. They most nearly resemble the earth in size and in rocky composition. The **giant planets** are the four large planets of the outer solar system, Jupiter through Neptune. Much bigger than the terrestrial planets, they also have a different composition, being rich in icy or gaseous hydrogen compounds such as methane (CH_4), ammonia (NH_3), and water (H_2O). Pluto falls into neither category, being a special, mysterious case.

Table 6-1 (pages 108–109) presents a

[1]A prize-winner in 1984 was created by a student at the University of Hawaii: *M*y *V*ery *E*rotic *M*ate *J*oyfully *S*atisfies *U*nusual *N*eeds *P*assionately.

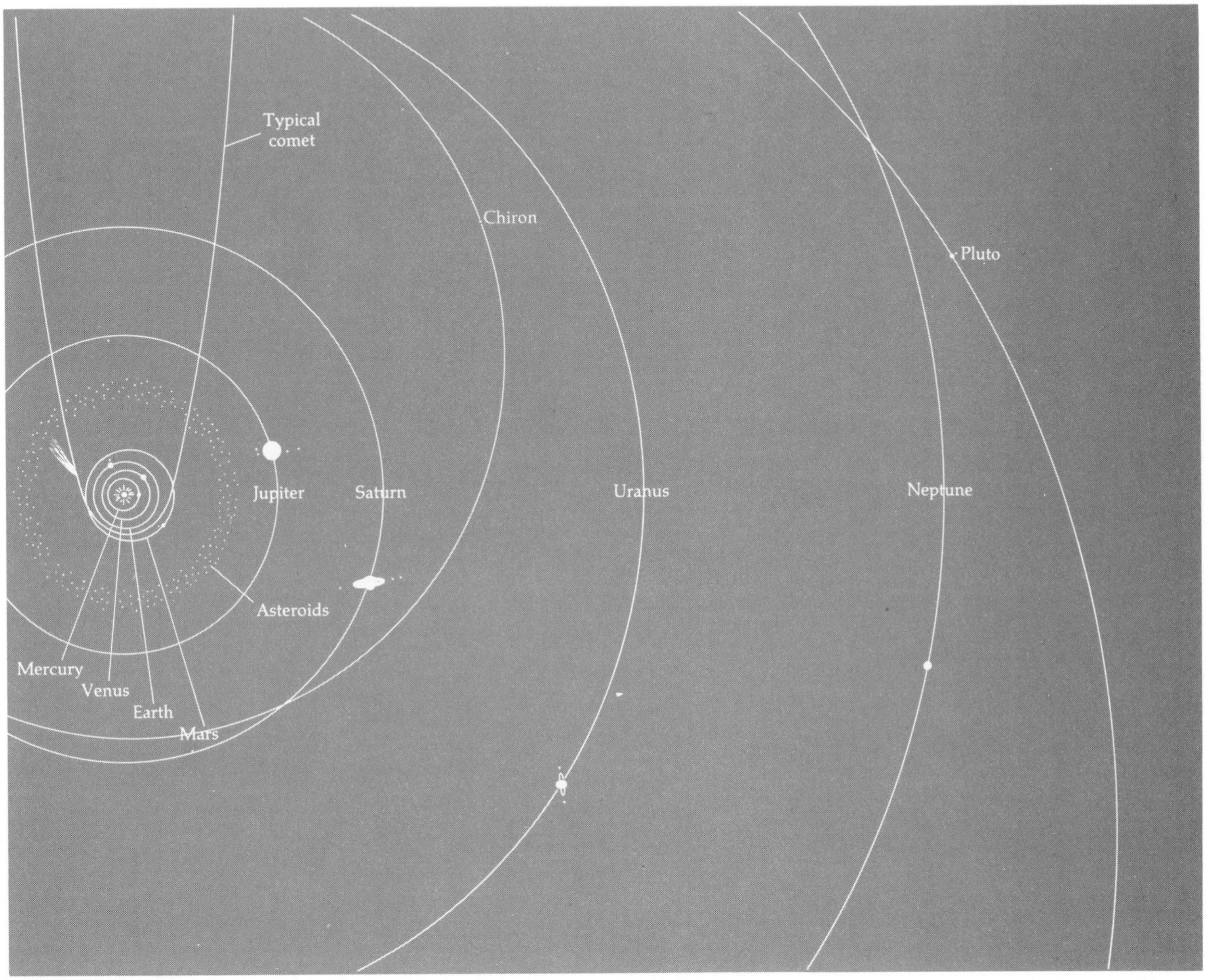

Figure 6-1. *The arrangement of the solar system as it is now known. Orbits of the nine main planets and a typical comet are shown, plus positions of typical asteroids. The orbit of the unusual asteroidlike object Chiron is also shown. Note eccentricities of the orbits of Pluto, Mercury, and Mars.*

comprehensive list of data. Notice that some satellites are bigger than some planets! The diameters of the remotest satellites are uncertain because they are hard to observe, but the largest satellites in the solar system are Jupiter's moon, Ganymede, and Saturn's moon, Titan, with diameters over 5000 km. They are both bigger than the planets Mercury and Pluto. Planets and moons are not the only large bodies in the solar system. The large asteroid Ceres (diameter 1020 km), which orbits the sun in a planetlike orbit between Mars and Jupiter, is bigger than half the known satellites of the solar system. As

Table 6-1. *Objects in the Solar System*
(Including the sun, all planets, satellites known through mid 1986, the six largest asteroids, and Chiron.)

Object	Equatorial Diameter (km)	Mass (kg)[a]	Rotation Period[c] (d)	Distance from Primary (10^3 km unless marked)	Orbit Inclination[b,c] (degrees)	Orbit Eccen-tricity	Escape Velocity (km/s unless marked)	Known or Probable Surface Material
Sun	1 391 400	1.99 (30)	25.4	0	—	—	617	Ionized gas
Mercury	4878	3.30 (23)	58.6	0.387 AU	7.0	0.206	4.2	Basaltic dust & rock
Venus	12 104	4.87 (24)	243R	0.723 AU	3.4	0.007	10.4	Basaltic & granite rock
Earth	12 756	5.98 (24)	1.00	1.00 AU	0.0	0.017	11.2	Water, granitic soil
Moon	3476	7.35 (22)	27.3	384	18–29	0.055	2.4	Basaltic dust & rock
Mars	6787	6.44 (23)	1.02	1.52 AU	1.8	0.093	5.0	Basaltic dust & rock
Phobos	27 × 19	9.6 (15)	0.32	9.4	1.0	0.015	11 m/s	Carbonaceous soil
Deimos	15 × 11	?	1.26	23	2.8	0.001	6 m/s	Carbonaceous soil
Asteroids								
1 Ceres	1020	1.2 (21)?	0.38	2.77 AU	10.6	0.08	0.6	Carbonaceous soil
4 Vesta	549	2.4 (20)?	0.22	2.36 AU	7.1	0.09	0.3	Basaltic soil
2 Pallas	538	?	0.33	2.77 AU	34.8	0.24	0.3	Meteoritic soil
10 Hygiea	443	?	0.75	3.15 AU	3.8	0.10	0.2	Carbonaceous soil
511 Davida	341	?	0.21	3.19 AU	15.8	0.17	0.2	Carbonaceous soil
704 Interamnia	338	?	0.36	3.06 AU	17.3	0.15	0.2	Unidentified soil
Jupiter	142 800	1.90 (27)	0.41	5.20 AU	1.3	0.048	60	Liquid hydrogen?
J16 Metis	40	?	0.30	126	0.0	0.0	22 m/s	Rock?
J15 Adrastea	35?	?	0.30	128	0.0	0.0	19 m/s	Rock?
5 Amalthea	270x155	?	0.50	181	0.4	0.003	0.13	Sulfur coated rock?
J14 Thebe	75?	?	0.68	223	0.0	0.0	40 m/s	Rock?
J 1 Io	3640	8.89 (22)	1.77	422	0.0	0.000	2.6	Sulfur compounds
J 2 Europa	3130	4.79 (22)	3.55	671	0.5	0.000	2.0	H_2O ice
J 3 Ganymede	5280	1.48 (23)	7.16	1070	0.2	0.001	3.6	H_2O ice, dust
J 4 Callisto	4840	1.07 (23)	16.69	1880	0.2	0.008	2.4	Dust, H_2O ice
J13 Leda	8?	?	?	11 110	26.7	0.146	4 m/s?	Carbonaceous rock?
J 6 Himalia	170?	?	?	11 470	27.6	0.158	88 m/s?	Carbonaceous rock?
J10 Lysithea	19?	?	?	11 710	29.0	0.130	10 m/s?	Carbonaceous rock?
J 7 Elara	60?	?	?	11 740	24.8	0.207	31 m/s?	Carbonaceous rock?
J12 Ananke	17?	?	?	20 700	147R	0.17	9 m/s?	Carbonaceous rock?
J11 Carme	24?	?	?	22 350	164R	0.21	12 m/s?	Carbonaceous rock?
J 8 Pasiphae	27?	?	?	23 300	145R	0.38	14 m/s?	Carbonaceous rock?
J 9 Sinope	21?	?	?	23 700	153R	0.28	11 m/s?	Carbonaceous rock?
Saturn	120 660	5.69 (26)	0.43	9.54 AU	2.49	0.056	36	Liquid hydrogen?
S15 Atlas	40 × 20	?	0.61	138	0.3	0.002	13 m/s	?
S14[d] (1980 S27)	140 × 80	?	0.61	139	0.0	0.003	90 m/s	?
S13[d] (1980 S26)	110 × 70	?	0.63	142	0.0	0.004	40 m/s	?
S11[d] (1980 S3)	140 × 100	?	0.69	151	0.3	0.009	70 m/s	?
S10[d] (1980 S1)	220 × 160	?	0.69	151	0.1	0.007	80 m/s	?

Table 6-1. *(cont.) Objects in the Solar System*
(Including the sun, all planets, satellites known through mid 1986, the six largest asteroids, and Chiron.)

Object	Equatorial Diameter (km)	Mass (kg)[a]	Rotation Period[c] (d)	Distance from Primary (10^3 km unless marked)	Orbit Inclination[b,c] (degrees)	Orbit Eccen-tricity	Escape Velocity (km/s unless marked)	Known or Probable Surface Material
S 1 Mimas	390	4.5 (19)	?	186	1.5	0.02	0.2	Mostly H_2O ice
S 2 Enceladus	500	8.4 (19)	1.37	234	0.0	0.00	0.2	Mostly H_2O ice
S 3 Tethys	1060	7.6 (20)	?	295	1.1	0.00	0.4	Mostly H_2O ice
S16 1980 S25[d]	34 × 22	?	?	295[e]	1	0	11 m/s?	?
S17 1980 S13[d]	34 × 26	?	?	295[e]	1	0	16 m/s?	?
S 4 Dione	1120	1.0 (21)	2.7	377	0.0	0.00	0.5	Mostly H_2O ice
S12 1980 S6[d]	35	?	?	377[e]	0.0	0.00	18 m/s?	Mostly H_2O ice
S 5 Rhea	1530	2.5 (21)	4.4	527	0.4	0.00	0.7	Mostly H_2O ice
S 6 Titan	5150	1.3 (23)	15.95	1222	0.3	0.03	2.7	Ices, liquid NH_3 & CH_4
S 7 Hyperion	410 × 220	?	?	1481	0.4	0.10	0.2	Ices?
S 8 Iapetus	1460	1.9 (20)	79.33	3560	14.7	0.03	0.6	Ice and soil
S 9 Phoebe	220	?	?	12 930	150R	0.16	0.1	Carbonaceous soil
Asteroid 2060 Chiron	350?	?	?	1.37 AU	7.0	0.38	0.1?	
Uranus	50 800	8.76 (25)	0.96?R	19.18 AU	0.8	0.05	21	?
1986U7	15	?	?	49.3	0	0	8 m/s?	Ice and soil
1986U8	20	?	?	53.3	0	0	10 m/s?	Ice and soil
1986U9	50	?	?	59.1	0	0	26 m/s?	Ice and soil
1986U3	70	?	?	61.75	0	0	36 m/s?	Ice and soil
1986U6	50	?	?	62.7	0	0	26 m/s?	Ice and soil
1986U2	70	?	?	64.35	0	0	36 m/s?	Ice and soil
1986U1	90	?	?	66.09	0	0	46 m/s?	Ice and soil
1986U4	50	?	?	69.92	0	0	26 m/s?	Ice and soil
1986U5	50	?	?	75.10	0	0	26 m/s?	Ice and soil
1986U1	160 × 170	?	?	85.89	0	0	85 m/s?	Ice and soil
U 5 Miranda	480	4 (20)?	?	130	3.4	0.02	0.4	H_2O ice, soil
U 1 Ariel	1170	4 (21)	?	192	0	0.00	0.7	H_2O ice, soil
U 2 Umbriel	1190	2 (21)	?	267	0	0.00	0.6	H_2O ice, soil
U 3 Titania	1590	1.1 (21)?	?	438	0	0.00	1.1	H_2O ice, soil
U 4 Oberon	1550	9 (21)	?	586	0	0.00	1.0	H_2O ice, soil
Neptune	48 600	1.03 (26)	0.92?	30.07 AU	1.8	0.01	24	?
N 1 Triton	3400	1.1 (23)	5.9	354	160R	0.00	2.5	CH_4 ice
N 2 Nereid	300	?	?	5570	27.5	0.75	0.3?	CH_4 ice
Pluto	3100?	1 (22)?	6.4	39.44 AU	17.2	0.25	0.9	CH_4 ice
P 1 Charon	1500?	1 (21)?	6.4	17	0	0	0.5?	CH_4 ice

Note: Numbers assigned to asteroids and outer planets' satellites indicate order of discovery, except for largest satellites.
[a]Numbers in parentheses are powers of 10.
[b]To ecliptic for planets; to planet equator for satellites.
[c]An R in this column indicates retrograde motion.
[d]Temporary designations until orbits are well determined. More than 20 images of faint satellites were found just outside the rings when Saturn was photographed with rings edge-on in 1979–80. More close satellites were detected during the Pioneer Saturn flyby in 1979 and the Voyager flyby in 1980. Images are designated by codes such as 1980 S3, the third satellite image of 1980.
[e]S12 in the leading Lagrangian point of Dione's orbit. S16 and S17 are in following and leading Lagrangian points of Tethy's orbit, respectively.

Sources: Data from tabulation by Chapman et al. (*Proc. Lun. Plan. Sci. Conf.* 1979, *9*, inside covers); IAU announcements 3463–3476; Voyager team report (*Science*, 1979, *204*, 964 ff.; 1979, *206*, 934 ff.; 1981, *212,* 159 ff.); Reitsema, Smith, and Larson (*Icarus,* 1980, *43*, 16); Voyager team press releases on Uranus satellites, 1986.

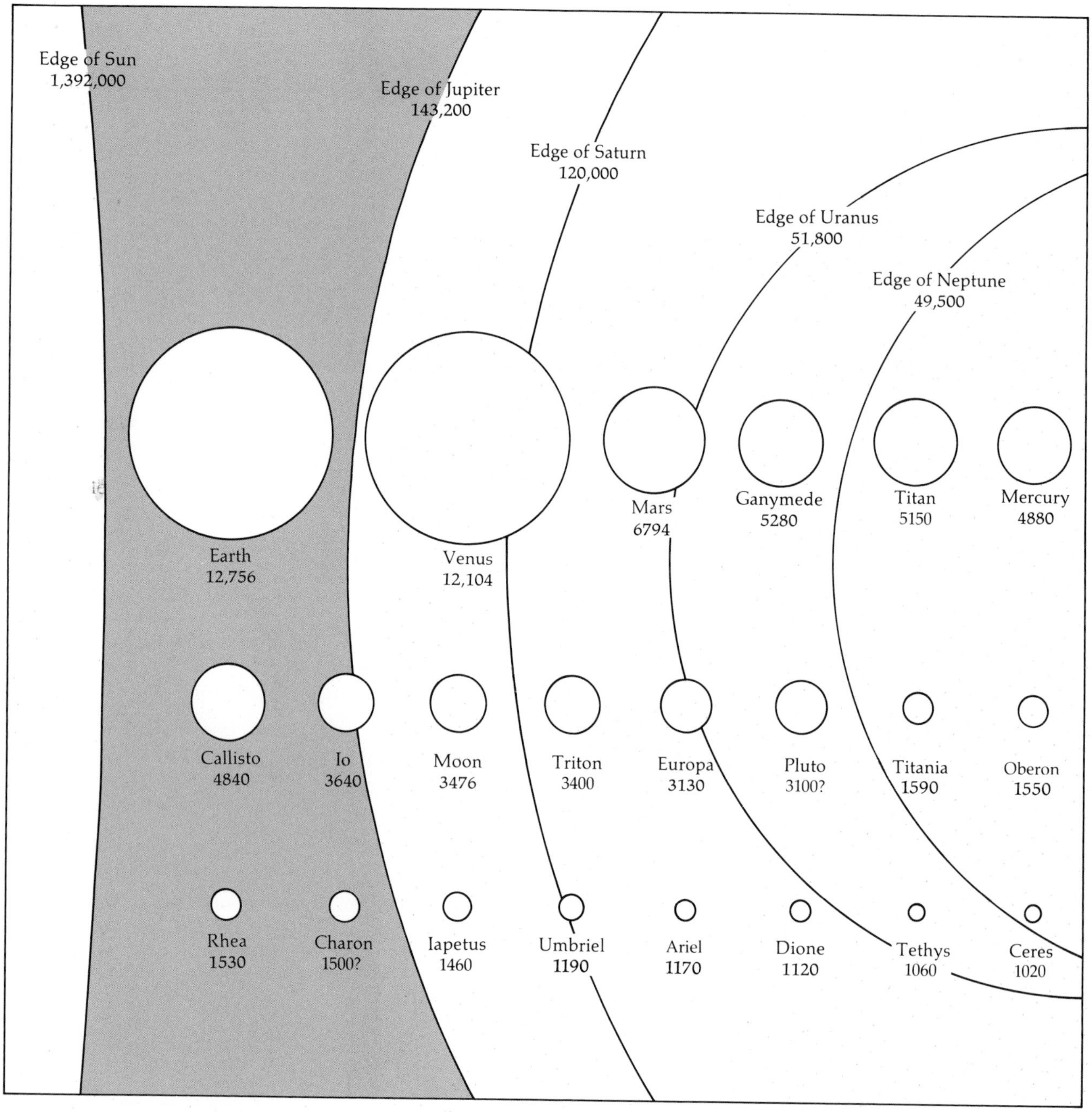

Figure 6-2. *The 27 largest bodies in the solar system, shown to true relative scale. They include 1 star, 9 planets, 16 satellites, and 1 asteroid. Diameters are in kilometers.*

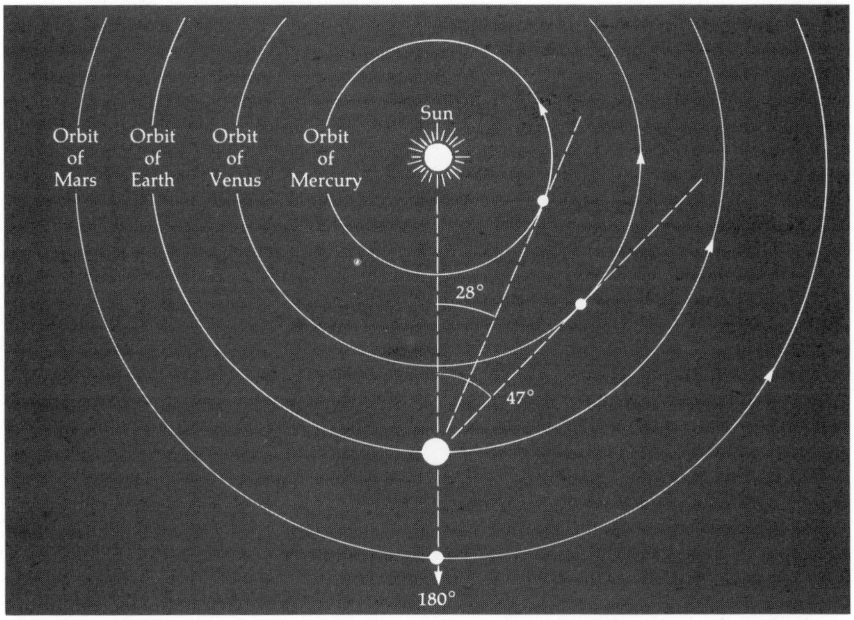

Figure 6-3. ***Observation indicates that Mercury and Venus lie between the earth and the sun. Their maximum angular separations from the sun are 28° and 47°, respectively, and they can sometimes be seen transiting in front of the sun. Exterior planets, however, can appear at opposition, 180° from the sun.***

Figure 6-2 shows, there are probably 27 worlds in the solar system larger than 1000 km across.

Until the decade of the 1970s, even the best telescopic views of Uranus, Neptune, Pluto, and all satellites beyond the moon revealed only poorly perceived disks, like a pinhead held at arm's length. Virtually nothing was known about the features on these worlds. They were anonymous—assumed to be cratered globes not much different from our own moon. Perhaps the most astonishing discovery of solar system exploration in the last decade was the unforeseen variety among these worlds. As we will see in the next chapters, distant moons include a near-featureless iceball, a sulfur-orange world with dozens of active volcanoes, a world with one blackboard-black hemisphere and an opposing snowy-white hemisphere, a cloudy world where gasolinelike compounds may rain out of the sky, and a world that may have an ocean of liquid nitrogen! The solar system is not just nine planets and a sun; it has dozens of worlds, each with its own personality.

The Solar System's Configuration

Let us now return to the story of how humans deduced the arrangement of planets in the solar system, which we left off in Chapter 2.

The most important step in discovering the solar system was a careful study of planetary motions, as described in the following section.

INFERIOR AND SUPERIOR PLANETS

An important clue about the arrangement of the solar system was that, to an earthly observer, Mercury never strays more than 28° from the sun, and Venus never more than 47°. All other planets can appear at any angular distance from the sun along the zodiac. This observation indicates that Mercury and Venus lie closer to the sun than the earth does, whereas the other six planets are outside the earth's orbit (see Figure 6-3). For this reason Mercury and Venus came to be called **inferior planets**, while more distant planets are called **superior planets**.

Final proof of this arrangement may have come when ancient astronomers observed actual **transits**, or passages of an inferior planet directly between the earth and the sun. Only a planet closer to the sun than the earth could pass between it and the sun. During a transit of Venus, a sharp observer looking through fog or a smoked glass can see Venus as a tiny black spot moving across the sun, but Mercury is too small to see without a telescope. Transits of Mercury are more common. The next two transits of these two planets are:

Mercury	Venus
Nov. 12, 1986	June 8, 2004
Nov. 14, 1999	June 6, 2012

RETROGRADE MOTION

Superior planets slowly shift position from night to night, usually west-to-east relative to the background stars. Of the superior planets, Mars most dramatically exhibits a motion that puzzled all early analysts of the solar system. Observers who plotted its position from night to night found that as Mars approaches a point opposite the sun in the midnight sky, it slows down, reverses itself, and drifts westward in so-called **retrograde motion** for some days before resuming its normal eastward (prograde) motion.

We now know that Mars does not really reverse its motion in its orbit around the sun. The appearance is an illusion caused by the motion of Mars *relative to earthly observers*, as shown in Figure 6-4. The earth moves around the sun faster than Mars does, following an "inside track," closer to the sun. Mars seems to go backward whenever we overtake and pass it.

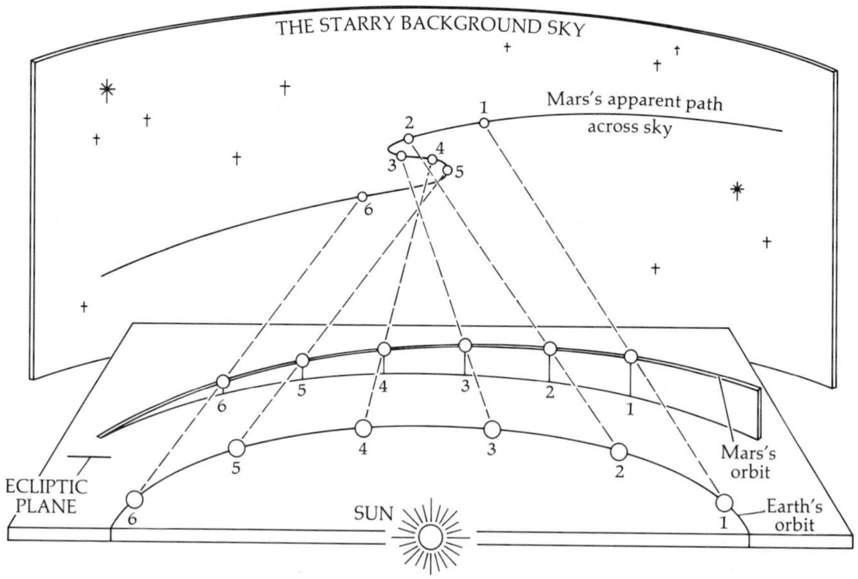

Figure 6-4. *A schematic model showing why Mars appears to have a complicated path across the sky. Positions of the earth and Mars are shown for six dates. Note that Mars moves slower than the earth and at a slight angle to the earth's orbit. As the earth overtakes Mars, Mars' apparent path relative to the background stars changes speed and direction. This phenomenon led to major problems with the Ptolemaic theory of the solar system and led to Kepler's discovery of elliptical planetary orbits around the sun.*

ized that further adjustments were needed and discussed adding small epicycles to the main epicycles.

In 1252, King Alfonso X of Castile, acting as a medieval version of the National Science Foundation, supported a 10-year project conducted by Arab and Jewish astronomers to calculate the extensive Alfonsine Tables, an almanac of predicted planetary positions based on the Ptolemaic model. These tables became the basis of planetary predictions for the next three centuries. However, the solar system now seemed very complicated because of the many overlapping cycles. (On seeing the complexity of the Ptolemaic epicycles, Alfonso is said to have remarked that had he been present at the creation, he could have suggested a simpler arrangement!) New epicycles had to be added to make positions agree with observations, and the calculations became ever more complicated.

Around 1340 the English scholar William of Occam enunciated his famous principle, called **Occam's razor**, applicable to all branches of science and philosophy:

Problems with the Ptolemaic Model

In Chapter 2, we showed how ancient scientists arrived at a solar system model with planets moving in deferent circles and epicycles around the earth, as was shown in Figure 2-11. But the relative motions of inferior and superior planets, especially Mars, were destined to cause trouble. Inferior planet motions were explained by putting Mercury's and Venus' orbits between the earth and the sun and adjusting their deferent and epicycle rates of motion so that they would never get more than 47° from the sun. Mars' motions were explained by adjusting its epicycle motion so that it would sometimes appear to be moving east-to-west.

The main goal of this model was to predict the positions of planets, often for astrological applications. The models attempted to break down the seemingly irregular motions of the planets into combinations of circular motions or uniform cycles. Eudoxus (c. 360 B.C.) reportedly represented planetary motions by a combination of motions of 27 rotating concentric spheres carrying the planets; Aristotle (c. 360 B.C.) reportedly used 55 spheres; Apollonius of Perga (c. 220 B.C.) reportedly increased the number of possible combinations of cycles by introducing epicycles; and Ptolemy perfected the **epicycle model** around A.D. 140. Ptolemy's version of the epicycle model is known as the **Ptolemaic model** of the solar system and is seen in Figure 6-5.

In the *Almagest*, Ptolemy showed how a properly chosen set of epicycles could account for all observations of planetary positions accumulated by his time, and he made relatively accurate predictions of planetary positions. Arab astronomers introduced this scheme into Europe in the early Middle Ages. However, by the 1200s and 1300s, astronomers in Damascus real-

Multiplicity ought not be posited without necessity.

In other words, among competing theories, *the best is the simplest—the one requiring the fewest assumptions and modifications in order to fit observations.* This idea may have contributed to suspicion of the tacked-on epicycles. Some European scholars, concerned about how the planets moved and remained suspended, regressed to Aristotle's more simplistic idea of spherical shells (see Figure 2-13). Purbach (c. 1460) thought the spheres were made of transparent crystal with special hollows for the epicycles; Fracastoro (c. 1550) proposed a scheme with 79 spheres!

According to astronomical historian Owen Gingerich (1973a, 1973b), the growing dissatisfaction with the Ptolemaic system was aesthetic (a result of its increasing unwieldiness) as much as intellectual (a dissatisfaction with its results). In addition, obvious errors of a few degrees had accumulated in the predicted positions. One observer remarked that a new model would be more "pleasing to the mind." This observer was Nicolaus Copernicus.

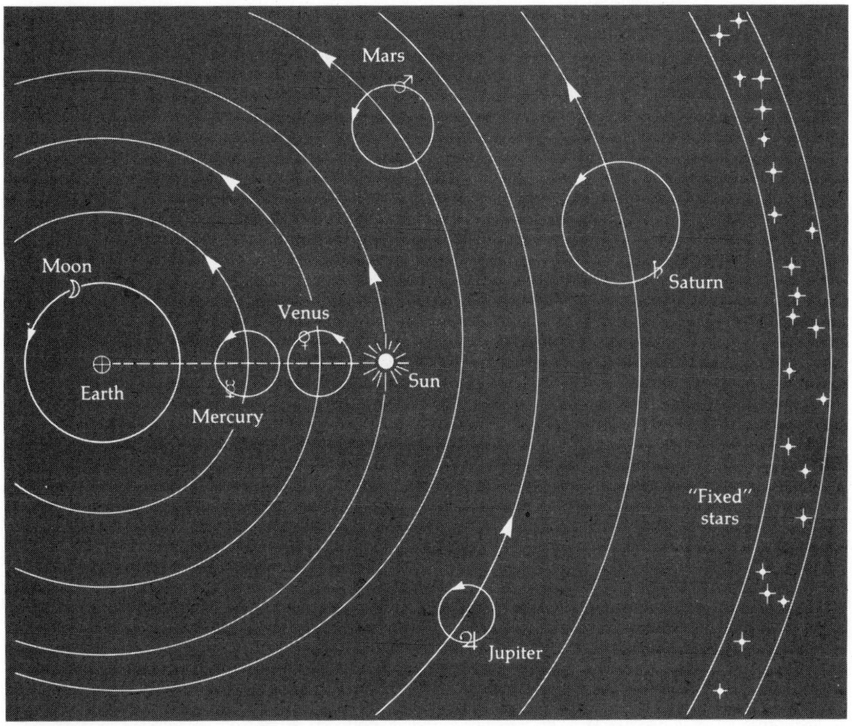

Figure 6-5. *The solar system as it might have been conceived by a Ptolemaic astronomer between* A.D. *100 and 1500.*

The Copernican Revolution

COPERNICUS' THEORY

Nicolaus Copernicus (Figure 6-6) was born February 14, 1473, the son of a Polish merchant. During his university education in Italy, he became excited by the surging scientific thought of that country. He associated with several astronomers and mathematicians and made his first astronomical observations at age 24. A few years later, a cathedral post gave him the economic security to continue his observations. At age 31 he observed a rare conjunction that brought all five known planets as well as the sun and moon into the constellation of Cancer. He found that their positions departed by several degrees from predictions in his set of Alfonsine Tables.

Familiar with several classical alternatives to Ptolemy's system, Copernicus analyzed planetary motions by various methods, including use of small "epicyclets." He soon realized that the solar system would be simpler and the prediction of planetary positions easier if the sun was placed at the center and the earth placed as one of the sun's orbiting planets. In 1512 he circulated a short comment (*Commentariolus*) containing the essence of his new thesis: The sun was the center of the solar system, the planets moved around it, and the stars were immeasurably more distant. This comment was not widely distributed, however, and few of Copernicus' acquaintances realized that the work he was pursuing would scandalize and revolutionize the medieval world.

He continued his studies but, fearing controversy, delayed publication for many years. Finally, encouraged by visiting colleagues, including some in the clergy, he allowed the *Commentariolus* to be more widely circulated. News of Copernicus' work spread rapidly. Late in life, Copernicus prepared a synthesis of all his work, *De Revolutionibus* (*On Revolutions*, 1543). In this book he laid out and explained the evidence about the solar system's arrangement:

Venus and Mercury revolve around the sun and cannot go farther away from it than the circles of their orbits permit [e.g., the 47° figure mentioned earlier for Venus]. . . .

According to this theory, then, Mercury's orbit should be included inside the orbit of Venus. . . .

If, acting upon this supposition, we connect Saturn, Jupiter, and Mars with the same center, keeping in mind the greater extent of their orbits . . . we cannot fail to see the explanation of the regular order of their motions.

This proves sufficiently that their center belongs to the sun.

Having thus laid out the correct nature of the solar system, Copernicus tackled the solar system's relation to the stars:

The extent of the universe . . . is so great that, whereas the distance of the earth from the sun is considerable in comparison with the other planetary orbits, it disappears when compared to the sphere of the fixed stars. I hold this to be more easily comprehensible than when the mind is confused by an almost endless number of circles, which is necessarily the case with those who keep the earth in the middle of the universe.

Although this may appear incomprehensible and contrary to the opinion of many, I shall, if God wills, make it clearer

Figure 6-6. ***Nicolaus Copernicus (1473–1543). Copernicus is holding a model with a central sun circled by the earth, and the earth by the moon.***

than the sun, at least to those who are not ignorant of mathematics.

Now the stage was set for turmoil. Church officials and most intellectuals held that the earth was at the center. The printer of *De Revolutionibus*, a Lutheran minister, had tried to defuse the situation by extending its title to *On the Revolutions of Celestial Orbs*, as if to imply that the earth was not necessarily included. He had also inserted a preface stating that the new theory need not be accepted as physical reality but could be seen merely as a convenient model for calculating planetary positions. From a philosophical viewpoint, we might accept this, but it did not deter medieval critics. Already Copernicus himself had come under fire from Protestant fundamentalists: In 1539 Martin Luther had called him "that fool [who would] reverse the entire art of astronomy. . . . Joshua bade *the sun* and not the earth to stand still."

In a world of strong dogmas, tampering with established ideas is dangerous. The Reformation era of ideological clashes was no exception. In the 1530s Michael Servetus had been criticized for certain writings on astrology and astronomy; in 1553 he was burned at the stake as a heretic for professing a mysterious theology that offended both Protestants and Catholics.[2]

Copernicus himself missed the height of the violent debate. He was ill in his last year and the first copies of his book were reportedly delivered to him on the day of his death, in 1543, at age 70. But the **Copernican revolution**, which put the sun instead of the earth at the center of the solar system, was under way.

About 1584, a 36-year-old Italian theologian and naturalist, Giordano Bruno, became known for tracts vigorously defending the Copernican view against academics. His message included exhortations that warring Protestant and Catholic factions in Europe could be brought together by a better appreciation of the new scientific view of earth's and humanity's place in the universe (Lerner and Gosselin, 1973).

Bruno expanded on the Copernican cosmology. The stars, he said, were all worlds like the sun. Many planets might orbit around them, offering abodes for other races. Bruno traveled in Europe, lecturing on the theological implications of astronomy and science, often with remarkable imagery that was at once poetic and scientific. For example, he discussed optics using "light" in both the physical and in the theological sense as "divine light." When he returned to Italy in 1592, he was arrested by the Inquisition, a church court established to detect and punish heresy. In 1600, after eight years of investigation of his philosophical and political views, he was burned at the stake.

In 1575, a correspondent wrote to the astronomer Tycho Brahe, "No attack on Christianity is more dangerous than the infinite size and depth of the heavens." Was the earth to be taken as merely a minor province of the universe? Where was heaven? In 1616, the Catholic church banned reading of *De Revolutionibus* "until corrected." It was corrected in 1620 by removing nine sentences asserting that it contained actual fact, not just theory.

These sad incidents illustrate the continuing problem of reconciling differences. As Will Durant wrote:

The heliocentric astronomy compelled men to reconceive God in less provincial, less anthropomorphic terms; it gave theology the strongest challenge in the history of religion. Hence the Copernican revolution was far profounder than the Reformation; it made the differences between Catholic and Protestant dogmas seem trivial.

TYCHO BRAHE'S SKY CASTLE

Tycho Brahe (Figure 6-7) was a flamboyant naturalist who wore a silver nose to cover a dueling mutilation. With funds from the king of Denmark, he built the first modern European observatory, named Uraniborg (Sky Castle), at his island home near Copenhagen. From his observations, all made with the naked eye (the telescope had not yet been invented), he made catalogs of star and planet positions. By demonstrating that stars and other bodies showed no angular shift in position as our position shifts with rotation of the earth, Tycho proved that stars and planets were many times farther away than the moon, for which he *could* detect a shift (called *parallax*; see p. 37.)

At age 16 Tycho noticed errors in predicted planetary positions in the same Alfonsine Tables that Copernicus had used. At age 25, in 1572, Tycho saw a temporarily bright exploding star (further discussion in Chapter 17), and by demonstrating that it had no parallactic shift, he disproved the popular belief that it was an object in the earth's atmosphere or near the earth–moon system. In 1577 he observed a bright comet and showed that it was also a remote object, far beyond the moon.

[2]Both Protestants and Catholics were involved in outrageous suppression. It was John Calvin himself who masterminded Servetus' execution, although, in a fit of moderation, he recommended beheading instead of burning. Servetus, a man of wide learning and varied interests, had improved geographic data on the Holy Land and discovered blood circulation in the lungs.

These discoveries were critical in overturning pre-Copernican theories. They meant that new objects could appear in the supposedly unchangeable heavens and that planets could not be attached to crystalline spheres, because such spheres would be smashed by the comets.

These observations inspired Tycho to catalog the precise positions of the stars and planets, which he did between 1576 and 1596. Unable to convince himself that the earth could move, Tycho invented a compromise solar system in which the earth was central and stationary, but other planets were placed in the correct sequence from the sun.

His pension withdrawn by the king of Denmark, Tycho moved to Prague in 1599, where he was joined in 1600 by a 30-year-old assistant named Johannes Kepler. When Tycho died in 1601, Kepler inherited the great compendium of Tycho's observations, with all its potential for fruitful analysis.

Figure 6-7. *Tycho Brahe (1546–1601), as shown in an old print. Silver plate on his nose covers a dueling scar.*

Figure 6-8. *Johannes Kepler (1571–1630).*

KEPLER'S LAWS

Devoutly religious and a believer in astrology, **Johannes Kepler** (Figure 6-8) was sure that planetary motions must be governed by hidden regularities—"the harmony of the spheres." With Tycho's material, Kepler first went to work on the orbit of Mars, which presented the most notable case of retrograde motion. This movement had plagued astronomical theorists since Ptolemy. He found something astonishing: After all the centuries of debate over the arrangement of circular orbits, the orbit that fitted Mars' motion best was not a circle at all, but an **ellipse**, a roughly egg-shaped figure that is symmetrical around two interior points called *foci* (singular, **focus**). (Recall Figure 4-4 and the discussion of elliptical orbits.) Kepler found that the sun lay exactly at one focus of Mars' orbit.

Kepler went on to discover two other related priciples, and these *three laws of planetary motion* were published in two books, *New Astronomy* (1609) and *The Harmony of the Worlds* (1619). **Kepler's laws** describe how the planets move (without attributing this motion to any more general physical laws), show that the sun is the central body, and allow accurate prediction of planetary positions:

1. Each planet moves in an ellipse with the sun at one focus.

2. The line between the sun and the planet sweeps over equal areas in equal time intervals.

3. The ratio of the cube of the semimajor axis to the square of the period (of revolution) is the same for each planet. (This is sometimes called the harmonic law.)[3]

Although the planetary orbits are ellipses, they are only slightly eccentric—that is, they are nearly circular. This is why the Ptolemaic system of circles worked as well as it did.

In the case of the earth, the semimajor axis, or average distance from the sun, is 1 **astronomical unit** (AU) and the other planets' distances are measured in multiples of this unit. Working in astronomical units and years, we can easily confirm the third law numerically, using $\boldsymbol{a}$ as the semimajor axis and P as the period. For the earth,

$$\frac{a^3}{P^2} = \frac{(1\,\text{AU})^3}{(1\,\text{y})^2} = \frac{1}{1} = 1.00$$

The same formation works for other planets.

Kepler's laws solved the mystery of the apparent retrograde motion of Mars. Kepler's laws say that because the earth is closer than Mars to the sun, the earth moves faster in its orbit than Mars (see Figure 6-4). Therefore, the earth catches up to Mars, like the faster driver on the inside track at a race.

[3] The major axis of an ellipse is its longest diameter; the semimajor axis is half that. Since most planets in the solar system have nearly circular orbits, the semimajor axes of their orbits are essentially the orbital radii.

Figure 6-9. *Galileo Galilei (1564–1642).*

GALILEO'S OBSERVATIONS

Kepler's laws might not have gone so far in establishing the Copernican model of the solar system had it not been for the contemporary invention of the telescope and for extensive observations by an Italian scientist, **Galileo Galilei** (Figure 6-9). Unlike Kepler, Galileo had a superbly practical turn of mind. For example, after reportedly watching the regular swing of a lamp in the Pisa cathedral, he applied the periodic motion of the pendulum to regulate clocks. As early as 1597, Galileo wrote to Kepler: "Like you, I accepted the Copernican position several years ago. . . . I have not dared until now to bring [my writings on the subjects] into the open."

Galileo perfected the telescope and began astronomical observations with it in late 1609. By 1610 he had made some of the most important observations until that time of the solar system. For example, he found four satellites revolving around Jupiter, *supplying proof at last that some bodies do not revolve around the earth.* These discoveries electrified European intellectuals.

Because Galileo wrote in Italian rather than Latin, he built a popular following outside the universities proper. Academics and churchmen saw in him a threat, and Galileo soon found himself being attacked from local pulpits. His invitations to reactionary academics and churchmen to look through his telescope and see for themselves led nowhere. Some looked and said they saw nothing; some refused to look; some said that if the telescope had been worth anything, the Greeks would have invented it.

From 1613 to 1633 Galileo was in frequent contact with church authorities, even in Rome. In 1616 a cardinal read Galileo an order that he must not "hold or defend" Copernican theory, though he could discuss it as a "mathematical supposition." In 1632, Galileo's great book *Dialogue of the Two Chief World Systems* appeared. It featured a fictionalized debate between Copernican and Ptolemaic advocates. In 1633, 69-year-old Galileo was ordered to Rome to stand trial before the Inquisition, where a curious episode occurred. The court produced a purported copy of the cardinal's order of 1616 telling Galileo not to "hold or defend" Copernicanism. The Inquisition's copy, still in the files today, also ordered him not to "teach" or "discuss" it in speech or in writing (contrary to the cardinal's 1616 order). The document in the Inquisition files is not the original and lacks the names of the alleged witnesses. Historians now suspect this document was a fraud created to frame Galileo.

The Inquisition jurors were inclined to be lenient only if Galileo repudiated his work. The elderly Galileo saw no point in getting himself killed; his book was already published and he had faith that intelligent people could see plain truth through telescopes or in print. So he recited a prepared recantation and was sentenced to prison, a sentence commuted by the pope to house arrest on Galileo's own estate, where he died in 1642.

NEWTON'S SYNTHESIS

In spite of the Inquistion, the evidence for a sun-centered solar system accumulated so rapidly that the Copernican revolution was almost complete. The main element still lacking was an overall theoretical scheme that would draw together Kepler's empirical laws of planetary motion into a concise physical explanation of the behavior of the solar system. To be intellectually satisfying, this theory needed to start with a few universal principles and show that the Keplerian orbits and the Galileian satellite motions *had* to exist as a consequence of these principles. The man who achieved this synthesis—the man usually deemed the greatest physicist who ever lived—was **Isaac Newton** (Figure 6-10), whose spectacular career was sketched in Chapter 4.

Once Newton discovered that masses attract each other gravitationally, he concluded that gravity is the only force involved in keeping the earth moving around the sun. But this solution led to another riddle: How could the sun influence the planets in their Keplerian orbits if it never touched them and always stayed at such a great distance from them? And how could the planets stay in the sky without spheres to hold them up?

Newton answered all these questions of "action at a distance" with three simple laws of motion and his law of gravity, the basis of most modern physics except for the corrections made necessary by work on relativity during the present century. These laws were enunciated in Newton's book *Principia* in 1687. *They are quite unlike Kepler's three laws* in that they are not merely empirical rules based on observation, but *fundamental postulates* from which Kepler's laws and many other phenomena can be predicted. **Newton's laws of motion** are:

1. A body at rest stays at rest and a body in motion moves at constant speed in a straight line unless a net force acts on it.

2. For every force acting on a body, there is a corresponding acceleration proportional to and in the direction of the force and inversely proportional to

the mass of the body. In other words, force = mass × acceleration.

3. For every force (sometimes called action) on one body, there is an equal and opposite force (called reaction) acting on another body.

Newton's laws of motion were especially powerful when combined with his universal law of gravitation, which we discussed in Chapter 4, and which explains the force acting on the planets from the sun.

Every particle in the universe attracts every other particle with a force proportional to the product of their masses and inversely proportional to the square of the distance between them.

The properties of Keplerian orbits follow from Newton's laws. An important exercise in advanced astronomy courses is to derive all three of Kepler's laws from Newton's laws. This exercise shows that if Newton's laws are true, the Copernican theory and Kepler's laws also have to be true.

Newton's laws thus tidied up the miscellaneous observations of preceding centuries and completed the Copernican revolution. If the force of gravitation did *not* obey the inverse square law, orbits could be nonelliptical. The law of gravitation, when combined with the first and second laws, shows why planets do not move in straight lines, but are always deflected toward the sun. The third law explains why rockets work and ultimately led to man-made satellites and spacecraft orbiting moons and planets in Keplerian orbits.

By the time of Newton's death, at age 84 in 1727, the solar system was conceived essentially as we see it today, lacking only the discovery of the three outer planets, Uranus, Neptune, and Pluto. Subsequent astronomical observations have shown that Newton's laws also apply in all other parts of the universe that we can see. They correctly predict properties of certain pairs of stars that orbit around each other, and properties of stars orbiting around galaxies.

Figure 6-10. *Isaac Newton (1642–1727).*

Bode's Rule

A curious relationship discovered by the German astronomer Titius and popularized by his colleague Johann Bode, in 1772, is helpful in memorizing the distances of the planets from the sun. **Bode's rule** is: Write down a row of 4s, one for each planet, and add the sequence, 0, 3, 6, 12, 24, and so on, doubling each time, as shown in Table 6-2. By dividing the sums by 10, you get the number of astronomical units between each planet and the sun.

Because Bode's rule, unlike Kepler's laws, does not necessarily follow from Newton's laws, it is considered more descriptive than explanatory, and not a law of physics. It is merely a handy way to remember planetary positions. Nonetheless, it seems to imply that the process that formed the planets, whatever it was, partitioned the solar system so that each planet formed nearly twice as far from the sun as the next inner planet. No one has fully explained why this happened.

In 1781 Bode's rule was strengthened with the discovery of Uranus at its predicted position, about twice as far from the sun as Saturn. Astronomers then noted that the rule predicted a planet between Mars and Jupiter. Telescopic searchers set out to find this missing planet, and on January 1, 1801, they found the first and largest asteroid, Ceres, just at the right distance! Ceres might have been called an ordinary but small planet had not many other small asteroids been discovered between Mars and Jupiter.

The discovery of Neptune in 1846 somewhat reduced the credibility of Bode's rule by putting a planet outside a predicted position, though the still later discovery of Pluto filled a predicted position.

Table 6-2. ***Bode's Rule: Distances of Planets from the Sun***

	Mercury	Venus	Earth	Mars	Asteroids	Jupiter	Saturn	Uranus	Neptune	Pluto
	4	4	4	4	4	4	4	4	4	4
	0	3	6	12	24	48	96	192	—	384
Predicted distance	0.4	0.7	1.0	1.6	2.8	5.2	10.0	19.6	—	38.8
Actual distance	0.4	0.7	1.0	1.5	2.8	5.2	9.5	19.2	30.0	39.4

Note: All distances are expressed in astronomical units (1 AU = average distance of the earth from the sun).

Summary

The planets are tiny specks circling the sun. If you backed off far enough to see the system as a whole, the outer giants would hardly be noticeable and the inner planets would be lost in the glare of the sun. This conception of the solar system was accepted only after one of the major intellectual upheavals in human history took place only about four centuries ago. The key to this Copernican revolution was the work of five scientists listed in Table 6-3. Of special importance were Kepler's three laws, which described how planets moved, and Newton's laws of motion and gravity, which revealed the underlying forces that explain Kepler's laws.

Concepts

solar system	Occam's razor
star	Copernican revolution
planets (names and order)	Tycho Brahe
satellite	Johannes Kepler
terrestrial planet	ellipse
giant planet	focus
inferior planet	Kepler's laws
superior planet	astronomical unit
transit	Galileo Galilei
retrograde motion	Isaac Newton
epicycle model	Newton's laws of motion
Ptolemaic model	Bode's rule

Problems

1. To an observer north of the plane of the solar system, do the planets appear to revolve around the sun clockwise or counterclockwise? Which way to an observer south of the plane? (*Hint*: All planets revolve in the same direction as the earth rotates, from west to east.)

2. Which planet can come closest to the earth (see Table 6-1)?

3. Which planet moves the largest number of degrees per day in its orbit around the sun? Which the least?

4. How many times bigger in diameter than the earth is the largest planet? How many times more massive than the earth is the most massive planet? (See Table 6-1.)

5. Can the full moon ever occult Venus (pass between the earth and Venus)? Draw a sketch to show why or why not.

6. One of Galileo's telescopic discoveries was that Venus, like the moon, goes through a complete cycle of phases, from narrow crescent to full. How did this disprove the Ptolemaic model, which restricts Venus to positions between the sun and the earth?

7. Which planet can come closer to Jupiter, the earth or Uranus?

8. Which planet can come closer to Uranus, the earth or Pluto?

9. Does Mars' actual orbital motion around the sun change in any special way during the time it exhibits retrograde motion?

10. If Venus, the earth, Mars, and Jupiter are in a straight line on the same side of the sun, what phenomenon does an observer on the earth see? What would an observer on Mars see? An observer on (or near) Jupiter?

Table 6-3. *Five Key Figures in the Copernican Revolution*

Nicolaus Copernicus	1473–1543	Proposed circular motions of planets around sun
Tycho Brahe	1546–1601	Recorded planets' positions
Johannes Kepler	1571–1630	Analyzed Tycho's records; deduced elliptical orbits and laws of planetary motion
Galileo Galilei	1564–1642	Made telescopic discoveries supporting Copernican model
Isaac Newton	1642–1727	Formulated laws of gravity and used them to explain elliptical planetary orbits

11. List systematic differences in physical properties between inferior planets and giant planets. Consider size, temperature, density, satellite numbers, and orbital properties (see Table 6-1).

12. Do you think Copernicus and other major figures in the Copernican revolution would have viewed themselves as revolutionaries? Why? Contrast the causes of scientific revolutions and political revolutions. What roles do factual discoveries, strong personalities, controversy, and publicity play in each? Which are more dangerous to human life? Which have the more lasting effects?

ADVANCED PROBLEMS

13. A small telescope will show the disks of Mars and Jupiter when they are closest to the earth. Assume that Mars approaches within 60 million kilometers and Jupiter within 630 million kilometers. Use the small-angle equation and data from Table 6-1 to calculate the angular sizes of the two planets in seconds of arc. Which appears larger in the telescope in angular size?

14. Calculate and compare the orbital velocities of the earth (1.5×10^{13} cm from the sun) and Pluto (about 5.9×10^{14} cm from the sun). The sun's mass is 2.0×10^{33} g.

15. If a comet on a parabolic orbit was closest to the sun at a point very near the earth, and if it moved in the same direction as the earth, how fast would it move (in km/s), relative to the earth?

16. Suppose the sun were replaced by a star 4 times as massive, and the earth were moved to a new orbit 2 AU from this star. Compare the gravitational force on this "new earth" to that in the present solar system.

Projects

1. On a piece of typewriter paper try to make a scale drawing of the orbits of the planets, based on orbital radii listed in Table 6-1. Which orbits are hard to show clearly? What size dots could represent planets at this scale?

2. Make a large wall chart showing the orbits of the planets out to Saturn in scale. Mark the motion of each planet in one-day or one-week intervals, as appropriate. Using the *American Ephemeris and Nautical Almanac*, an astronomy magazine, or a similar source, locate each of the planets' relative positions for the current date. Update the chart during the semester and watch for alignments that represent conjunctions, elongations, and so on. Confirm these in the night sky.

3. If a planetarium is nearby, arrange for a demonstration of planetary motions from night to night. Demonstrate the apparent retrograde motion of Mars.

4. With a telescope of at least 2-in. aperture, examine Jupiter and confirm that it is attended by four prominent satellites, as discovered by Galileo. (On any given night, one or more satellites may be obscured by Jupiter or its shadow.) By following Jupiter from night to night, confirm that the satellites move around Jupiter. This proves the Copernican dictum that not all celestial objects move in earth-centered orbits.

CHAPTER 7

LIGHT AND THE SPECTRUM: MESSAGES FROM SPACE

Now we are ready for a short pause in our cosmic journey outward from Earth. We are about to cross the interplanetary frontier beyond which humans have not ventured. Only a few of our robot space probes have sent back data from this region. To learn about still more distant environments, we humans have to rely on nature's messages from remote planets, stars, and galaxies. The messenger is light. "Light," as we will see in this chapter, means radiation of various kinds, including reflected sunlight (in the case of planets) and radiation originating inside planets and stars.

At this point in the book we begin the transition between space exploration and more traditional astronomical techniques. The traditional techniques use telescopes and related devices to analyze light. In this chapter we will give a brief introduction to the different types of light, and how they can be analyzed. You will see these techniques applied to planetary worlds in Chapters 8–11. In Chapters 12 and 13 we will begin to study the sun and other stars, and we will pause again to describe still more principles about light.[1]

The Nature of Light: Waves vs. Particles

Suppose you stand by a quiet swimming pool where a cork is floating. You disturb the cork by jiggling it. You will notice that this disturbance causes a set of waves to move out across the water. The waves have a certain spacing from one crest to the next, called the **wavelength**. They move at only one fixed speed. As they move by a given point on the water's surface, that point moves up and down; the number of these up-and-down pulsations per second at any one spot is called the **frequency** of the wave.[2]

These waves provide a useful analogy, but not a perfect description, of some **wavelike properties of light**. For instance, just as a water wave expands from its source, light spreads out in all directions from its source. **Visible light** has a tiny wavelength, around 400 to 700 nm (0.0000004–0.0000007 m). Radiation with still shorter wavelengths exists, but is too deep violet for our eyes to perceive; it is called **ultraviolet light**. Light with wavelengths longer than red light is too deep red for the eye to perceive and is called **infrared light**. Just as the speed of the water waves is constant, the **speed of light** through empty space is constant, about 300 000 km/s (186 000 mi/s). It has been measured, for example, by the time interval between the sending of a signal from a distant spacecraft and the receipt of that signal on Earth.

Now suppose you shoot a BB at something. The BB has a certain energy. Unlike a wave, which takes a while to pass a fixed object and then die out, the BB delivers all its energy to its target at the moment it hits. The BB is an analog for some **par-**

[1] Note to teachers and students: Some concepts especially relevant to planetary study are introduced here, and the student is allowed to develop familiarity with them, before more concepts, such as Kirchoff's laws, the Doppler effect, and the Stefan-Boltzmann law are introduced and applied to the sun and stars in Chapters 12 and 13. Pedagogically, this appears a useful way to introduce these physical concepts a little at a time, as needed.

[2] The velocity of a wave is equal to the wavelength (the distance between two crests) times the frequency (one divided by the time interval between crests as they pass a fixed point). Check the dimensions: Wavelength × frequency has dimensions of distance/time, that is, dimensions of velocity. Note that a light wave and its color can be specified by either its wavelength or its frequency. Some scientists tend to characterize waves by their frequency (as in radio parlance where stations are listed by cycles/s), but astronomers tend to refer to waves by their wavelength—a practice we will follow in this book.

ticlelike properties of light, but again is not a perfect description of light. For instance, light has its energy concentrated in individual units, instead of having it spread out along the wave. The energy-containing unit of light is called the **photon**. It can be visualized roughly as a microscopic, BB-like particle that moves at the speed of light, yet has a certain wavelength associated with it. An important concept is that each color of light corresponds to a photon of different wavelength and energy. *The bluer the light, the shorter the wavelength and the more energetic the photon.*

From the 1600s to the 1800s, scientists argued over whether light is "really" a wave or a particle. For example, Newton argued for a particle theory of light, while the Dutch astronomer-optician Huygens (1625–1695) argued for a wave theory. Arguments of Huygens and others established that light definitely has many wavelike properties. For example, if your swimming pool has an inward-protruding corner or wall, you can observe that the wave bends slightly as it goes past the corner, so that some energy (wave motion) reaches a target slightly behind the wall as seen from the wave's source. Light does the same thing, indicating a wavelike property. A BB would move in a straight line past the wall and not hit such a target. This process of light bending its path slightly as it passes an edge is called **diffraction**; it limits the sharpness that can be achieved in the image formed by a telescope.

On the other hand, Einstein described one of the important particlelike properties of light in 1905.[3] This is the so-called photoelectric effect—an effect in which light rays can knock electrons out of metal surfaces. The effect can be explained only if the light energy arrives in individual "packets"—that is, photons—rather than being "smeared out" along the wave.

In summary, light spreads out through space something like a wave in which energy is carried by tiny, particlelike photons. Which is light: a wave or a particle? Consider a platypus. It has some ducklike and some beaverlike properties, but is neither. Similarly, light has some wavelike and some particlelike properties, but is neither a pure wave nor a pure particle. Considered in microscopic detail, light is a phenomenon not wholly familiar in terms of analogs in our everyday world.

The Spectrum

An arrangement of all colors, in order of wavelength (or, what is the same, in order of photon energy), is called the **spectrum** (plural, spectra). Newton discovered he could see the spectrum of visible light by passing sunlight through a glass prism. The resulting band of colors, cast on a wall in a dark room, looks like the colored band in Color Plate 32. Water droplets in a rainstorm act like little prisms and allow us to see the spectrum—the same arrangement of colors from violet to red—in the rainbow. Newton's discovery proved that "white" light from the sun is really made up of all the different colors. Figure 7-1 shows the names attached to different parts of the spectrum, which extends to far shorter and longer wavelengths than the ones we can see. As this figure indicates, the radio waves we receive on our radios and TVs are simply long-wavelength versions of the light we see with our eyes.

Figure 7-2 shows an important way astronomers present this information. The bar along the bottom labels the colors. This bar, or the color band in Color Plate 32, shows the kind of picture we could get if we simply photographed a projected spectrum. But instead of photographing, suppose we used a little photocell, like a photographer's exposure meter; we could scan along the spectrum, left to right, to measure the amount of energy (that is, light intensity) at each wavelength. Then we could plot the results as the graph in the upper part of Figure 7-2. This allows us to see the amount of light of each color. For example, Figure 7-2 shows that sunlight's strongest intensity is reached at yellow wavelengths. Astronomers often present the spectrum in this way.

Astronomers' study of the properties of the spectrum of different objects is called **spectroscopy**. *Spectroscopy is the most important method for learning about remote planets, stars, and galaxies.*

Origins of Light: Electromagnetic Disturbances

Consider again the cork in the swimming pool. We had to disturb it to get a wave. If you put a second cork in the swimming pool and watch as you make a wave, you would see it bob up and down as the wave passed by. The cork would be a *detector* of the passing wave. In the same way, if you had a tiny enough electrically charged particle, and a tiny enough compass, you could detect light waves passing by. As the wave passes, the electric charge would vibrate with the frequency of the wave, and the compass needle would rapidly oscillate with this same frequency. For this reason, physicists say that light is a disturbance of electric and magnetic **fields** in space. A field is an important concept in much of physics. A field is said to exist in a certain volume of space if some physical effect can be measured and assigned a numerical value *at any point* throughout that volume. For example, at any point near Earth, Earth attracts a given "test particle" with a measurable force; therefore, a gravitational field is said to exist. Similarly, where a compass measures a certain response at any point, a magnetic field is said to exist. Also, when an electrically

[3]Interestingly, Einstein eventually got his Nobel Prize primarily for this work, not directly for his later, better-known work on relativity.

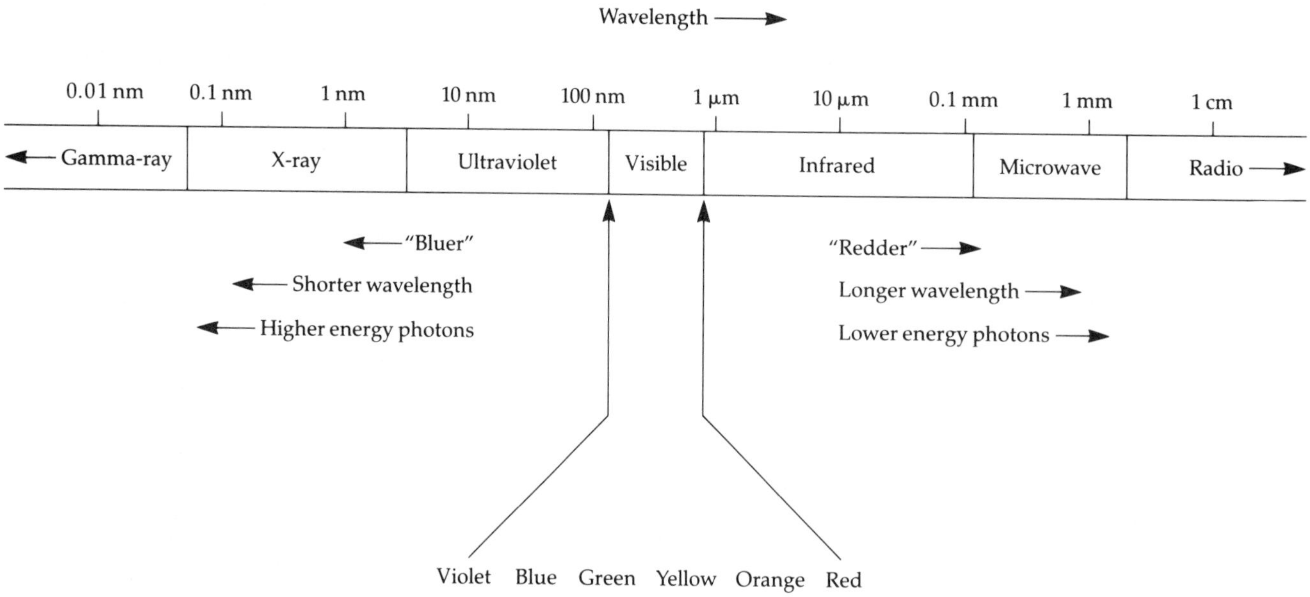

Figure 7-1. *The spectrum. Top chart shows a wide range of wavelengths with names attached to different regions. Lower chart shows the order of colors in the narrow range that is visible to the eye.*

charged particle experiences a force at any point, an electric field exists.

The charged particle and the compass just mentioned are actual tools with which we can detect electric and magnetic fields in space or in your room. They prove that light involves pulsations in these electromagnetic fields (coexisting electric and magnetic fields). For this reason, we say that all light, of whatever wavelength—gamma-ray, X-ray, ultraviolet, visible, infrared, microwave, or radio—is **electromagnetic radiation**. The word *radiation* is used for light of any wavelength.[4]

[4]However, the word must be read with caution, since physicists and engineers sometimes speak of streams of atomic particles, such as protons or electrons, as radiation. For example, charged particles escaping from radioactive material are sometimes called "radiation," or better, "particle radiation." These atomic particles are not light or electromagnetic radiation. Other examples are "radiation belts" and "cosmic rays," both of which are particles, not electromagnetic radiation.

If the second floating cork were hit by a BB or by another drifting cork, this would disturb its motion and cause it to radiate a new wave. In the same way, a drifting electron can be disturbed in such a way as to release electromagnetic radiation. Each electron has a certain amount of energy associated with it, by virtue of its motion, spin, and other properties. If an electron is disturbed in such a way as to reduce its total energy, it can give up the energy in the form of radiation—a photon that speeds away. The resulting photon has an energy just equal to the energy the electron lost. In other words, the wavelength and color of the emitted radiation is exactly controlled by the amount of energy lost by the electron. As long as this electron is unattached to an atom, it can make a change in energy by any amount, and hence can radiate photons of any wavelength. Thus, a mass of free electrons, if being disturbed, could radiate an array of photons of all wavelengths. This produces a spectrum like Figure 7-2 or the top spectrum of the three in Color Plate 32—a continuous band of colors called a **continuum**, or continuous spectrum.

THERMAL RADIATION

In fact, this type of radiation is emitted all the time, all around us. Electrons (and atoms and molecules) in all gases, liquids, and solid objects are in constant motion, jostling each other. **Temperature** is simply a measure of the average energy of these motions, which are thus called **thermal motions** (from the prefix *thermo-*, referring to heat). The higher the temperature, the faster the thermal motions, and the more disturbed the atomic particles. Because the electrons of any object are being constantly disturbed by thermal motions, all objects continuously radiate a continuum spectrum of electromagnetic radiation. The hotter the temperature, the more radiation is given off. **Thermal radiation** is the name given to the type of radiation that depends on the heat of the radiating material.

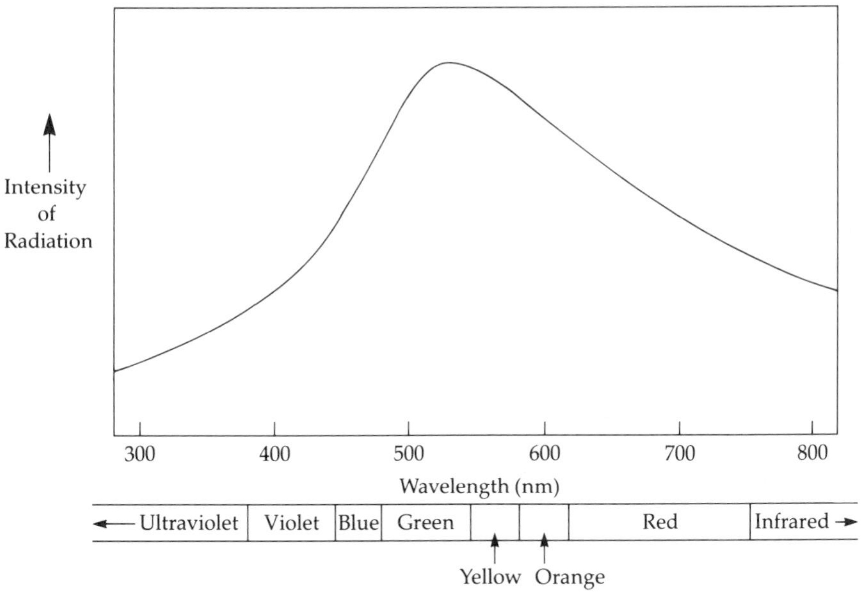

Figure 7-2. *A representation of the spectrum of sunlight. The graph gives the intensity, or amount, of light at each wavelength, and the names of the colors are inserted along the bottom. The sun's dominant radiation occurs at wavelengths we perceive as yellow.*

WIEN'S LAW

It is important to realize that the light you see from a rock or a planet in the night sky, or from this book, is *not* thermal radiation. It is **reflected radiation**—light from the sun or some other source, bouncing off the object.

Why can't we see the thermal radiation given off by a rock, planet, or book? The basic reason is that they are too cold. Cold objects give off very little radiation. More importantly, however, they give off the wrong color of radiation. Let us explain, with an important physical law discovered in 1898 by German physicist Wilhelm Wien (pronounced VEEN). It is called **Wien's law**:

The hotter an object, the bluer the radiation it emits.

This effect is familiar in everyday life. Ordinarily, a nail emits no visible radiation. However, if you heat it over a stove, it begins to radiate a dull red light. Higher temperature leads to a bright orange-red glow. At a still higher temperature, the dominant color is "white hot," or yellowish/white. If the nail could be heated enough, its radiation would become distinctly bluish. This effect is shown in terms of spectra in Figure 7-3, where we see the light change from red to blue, and increase in intensity, as the nail is heated.

Despite appearances, the nail is radiating regardless of its temperature. When it is at room temperature, the light it radiates is so red (of such long wavelength) that we cannot see it because our eyes are not sensitive to infrared radiation. For example, note that room temperature is about 300 K, and the curve labeled "300 K" on Figure 7-3 is so far to the right it is almost off the diagram. Similarly, a hot enough object radiates ultraviolet light, whose wavelength is too short for us to see. For example, note that the curve labeled "8000 K" peaks at the beginning of the ultraviolet, and hotter objects would emit a still larger fraction of their light in the ultraviolet part of the spectrum. Though we cannot see infrared and ultraviolet radiation, we can build instruments to detect them; radios, for example, detect especially long wavelengths (usually called radio waves).

MEASURING THE TEMPERATURE OF A PLANET (OR STAR)

Now we can see one of the ways that light carries messages from space. Instruments to detect infrared radiation of planets and stars were first built around World War II and are still being improved. With such infrared detectors, we can measure the temperature of a planet by measuring the color (wavelength) at which its thermal radiation peaks. As shown in Figure 7-4, an object with peak radiation at a certain infrared wavelength has one temperature, while a star with peak radiation at visible wavelengths is hotter. The peak thermal radiation lies in the infrared parts of the spectrum, for all planets and moons, from Mercury to Pluto. They are all too cold to emit appreciable visible light.

NONTHERMAL RADIATION

The thermal radiation we have been discussing occurs when electrons change energy levels at random due to disturbances arising from the *thermal* motions of atomic particles. For completeness, we should add that other types of disturbance can create a continuum spectrum of different shape than in Figure 7-3. An example occurs when an electron is accelerated by moving into a strong magnetic field. This particular type of nonthermal radiation is called synchrotron radiation, and will be encountered in later chapters.

Emission Lines and Bands

We've been describing ways to produce light as electrons are disturbed. This process, in which a particle emits a photon, is called **emission**. We started off picturing a freely drifting electron (like a cork

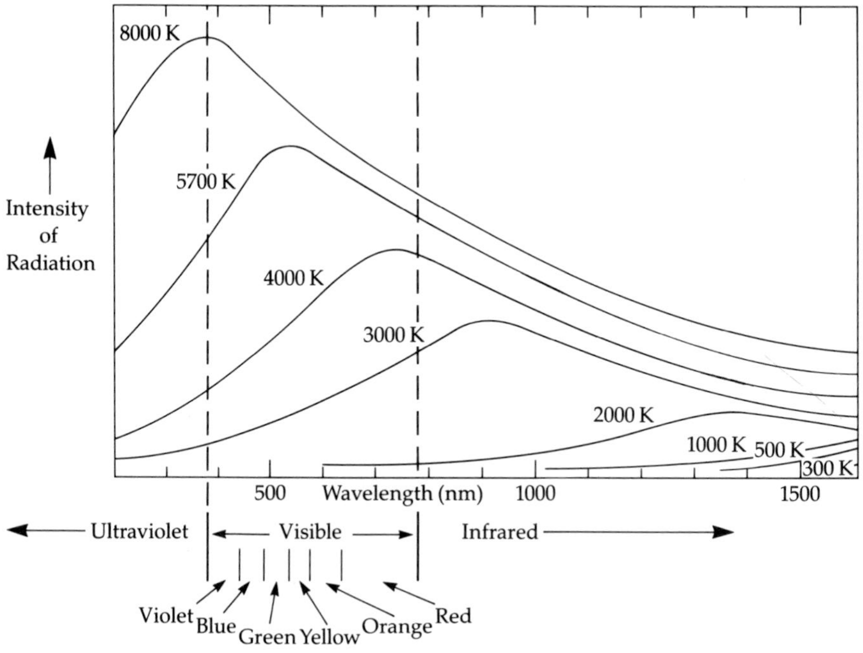

Figure 7-3. *Schematic diagrams of the spectrum of a nail (or lava or a planet) as the temperature is changed. At room temperatures (~300 K), the object glows only in the infrared, with radiation invisible to the eye. As predicted by Wien's law, the dominant color becomes more blue as the temperature increases. As it heats, the object glows first deep red, then orange, and so on. At 5700 K, the material radiates light like the sun's (cf. Figure 7-2). The total energy radiated by the object (area under the curve) also increases as the temperature grows.*

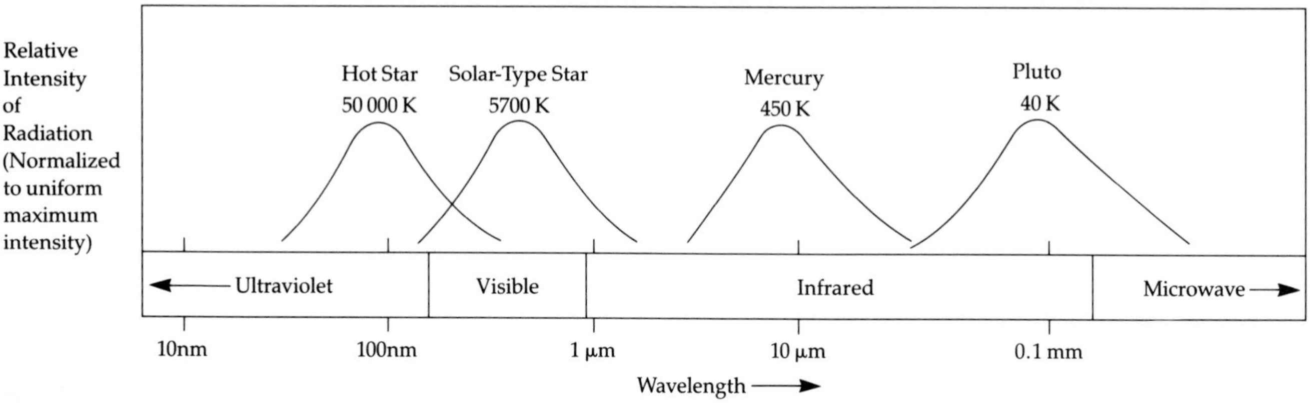

Figure 7-4. *Schematic spectra of thermal radiation from four different celestial objects, normalized to the same intensity for clarity. As predicted by Wien's law, hotter objects are bluer. The temperature can be determined by the wavelength of the dominant radiation. Planets are so cold they radiate their thermal radiation only in the infrared part of the spectrum; infrared detectors are needed to measure them. Similarly, ultraviolet detectors help study the hottest stars.*

floating in a pool), not attached to an atom. Now consider an electron orbiting around the nucleus of an atom. As soon as we consider electrons attached to atoms, extraordinary modifications occur to the emission process, and these are critical to all of astronomy.

ATOMS AND EMISSION LINES

Figure 7-5 shows the schematic structure of an atom, with its central nucleus and orbiting electrons. This may be only one of many orbiting electrons, depending on the element: hydrogen, helium, and so on. In some ways, the atom is like a tiny solar system—that is, a relatively massive central object and tiny orbiting particles. But there is an extraordinary difference. In the solar system we can put a rocket into any orbit we choose; each orbit would have

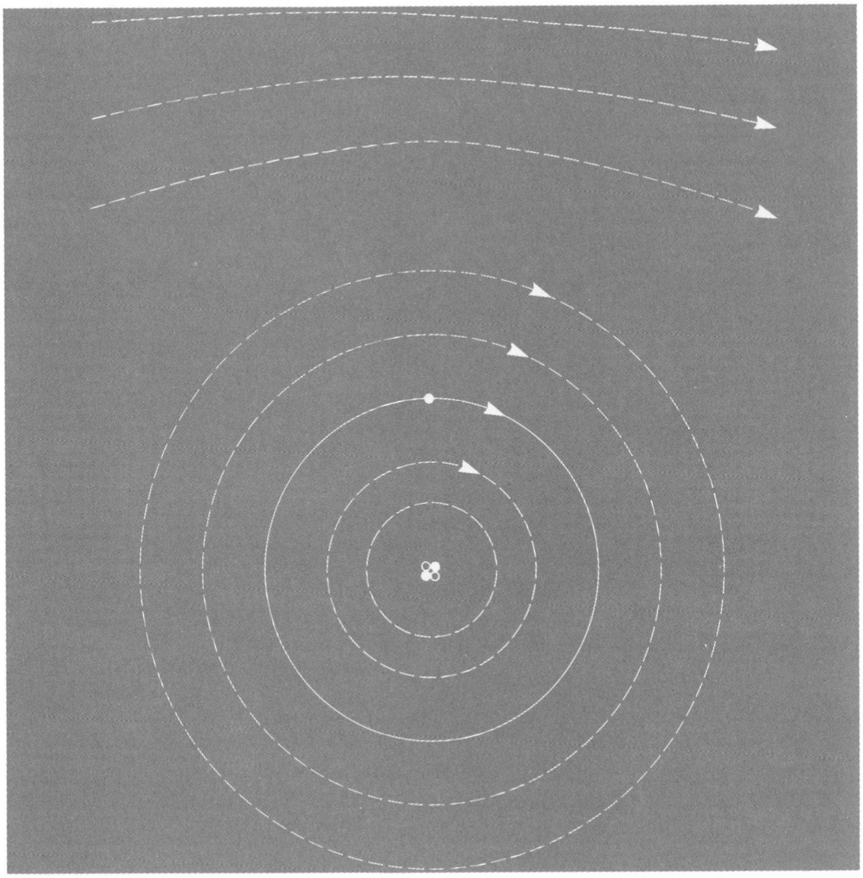

Figure 7-5. *Schematic view of an atom's structure. If an electron (dot) finds itself in an orbit (solid line) around the nucleus of an atom, there are only certain other orbits, or "energy levels," that it can occupy in the atom (dashed circles). However, outside the atom (top), virtually any orbits, or "energy levels," are possible.*

OPTIONAL BASIC EQUATION V

Measuring Temperatures of Astronomical Bodies: Wien's Law

Wien's law tells which wavelength W corresponds to the maximum amount of radiation, given the temperature T. Using SI metric units, the wavelength is given in meters and the temperature in Kelvins, abbreviated K. (These units are sometimes called "degrees Kelvin," since they define a temperature scale in degrees; but the correct SI terminology is "Kelvins.")

The law is:

$$W = \frac{0.00290}{T}$$

The number 0.00290 is a constant of proportionality and remains the same in all applications of the law. Thus, as T increases, W decreases, giving shorter wavelengths and hence bluer light.

An example is instructive. The sun has a surface temperature of about 5700 K. Therefore, we can calculate the wavelength of the strongest solar radiation. Using SI metric units and powers of 10,

$$W = \frac{2.9 \times 10^{-3}}{5.7 \times 10^{3}} = 5.1 \times 10^{-7}\text{m}$$

This, of course, is exactly in the middle of the range to which the eye is sensitive. (Otherwise we could not see sunlight!) Light of this wavelength corresponds to a yellow color.

its own velocity and hence energy. But in an atom, electrons can occupy only *certain* orbits. These are called **energy levels**, since each orbit has one specific energy. Physicists describe this by saying the atom has quantized energy levels, because only certain quantities of orbital energy are possible. This was discovered in 1913 by Danish physicist Niels Bohr. In Figure 7-5, the solid circle represents the orbit occupied by the electron, and the dashed circles are other allowed energy levels. At the top are a few of the infinite number of energy levels outside the atom, where energy levels are not discrete.

Now let us disturb the electron as before (hitting it with a photon or a neighboring atom) so that it drops to a lower energy level. It will emit a photon whose energy equals the change in energy levels. But note that being inside an atom, it can experience only certain, fixed changes in energy level, as shown in Figure 7-6. For instance, if it starts in level 3, it can drop only to levels 2 or 1. Thus it can emit only certain amounts of energy corresponding to these differences in energy level; it can emit only photons of certain wavelength or color. As a result, *each element can emit only certain wavelengths* of light. The bottom chart in Color Plate 32 shows, for example, the emissions produced by hydrogen gas. These emissions are called **emission lines** because they appear in a projected spectrum as *lines* or narrow bars of color, as seen in the color plate.

Here, then, is an astounding and useful fact! If you see a certain set of emission lines, you can match it with a certain element and infer that atoms of *that* element are present and glowing in the distant object. You can tell something about the object's composition, even without having a sample!

Note that if the electrons in an atom are in their lowest possible energy level, that atom cannot produce an emission line. This is because the electrons cannot drop to any lower energy level. An atom in which all electrons are in the lowest possible energy level is said to be in its **ground state**. An atom in which one or more electrons are in energy levels higher than the lowest available ones[5] is said to be in an **excited state**. Excited states usually last only a short time before the electrons "decay" to the lowest available energy level, producing the ground state. Atoms generally need to be disturbed to produce and maintain excited states. Thus, for example, hot gases are more apt to produce emission lines than cold gases, because atoms in hot gases collide faster.

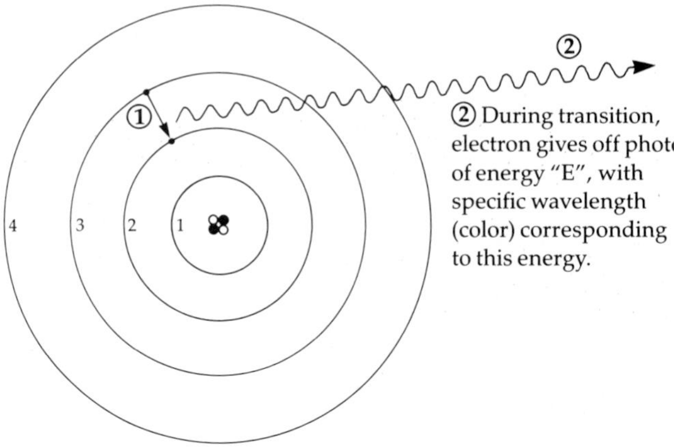

***Figure 7-6.** Schematic view of emission of a photon of light as an electron is knocked from a lower to a higher energy level in an atom.*

MOLECULES, CRYSTALS, AND EMISSION BANDS

The electron structure in a molecule is more complex than in an atom. The electron's path may take it around two or more nuclei. Thus, the emission line structure from a molecule is not so simple as that from an atom. For example, in a gas containing water molecules (H_2O), we get more complex emission lines than in a gas containing single H and O atoms. The molecule has various ways of responding to a disturbance, in addition to having its electron change energy levels; for example, it may vibrate like two balls linked with a spring. As a result, the energy levels are vastly more numerous and the resulting emission lines blend together. Instead of sharp emission lines, we get blended clusters of barely separated emission lines over a range of wavelengths, as shown in Figure 7-7. The resulting broader emission feature from a molecule is called an **emission band**. The rest of the story is the same, though: A given molecule (such as H_2O) can produce only certain emission bands. Thus we can identify the molecules glowing in a given remote source just by detecting their emission bands.

In order for atoms and molecules to produce clear emission lines and bands, they must be detached from one another, as in gases. If the atoms were linked together, they would form molecules, and if the molecules were linked, they would form a solid or liquid. In most solid or liquid substances, the electron structure is so complex that emissions are smeared out and emission features are barely discernable. This means that most emission lines and bands arise from gases.

However, an important exception occurs among rock-forming minerals, which make up most planetary surfaces and interstellar dust grains. Most rock-forming minerals are crystals, which can

[5]Each energy level can be occupied only by a certain number of electrons. For example, the lowest level can take only two.

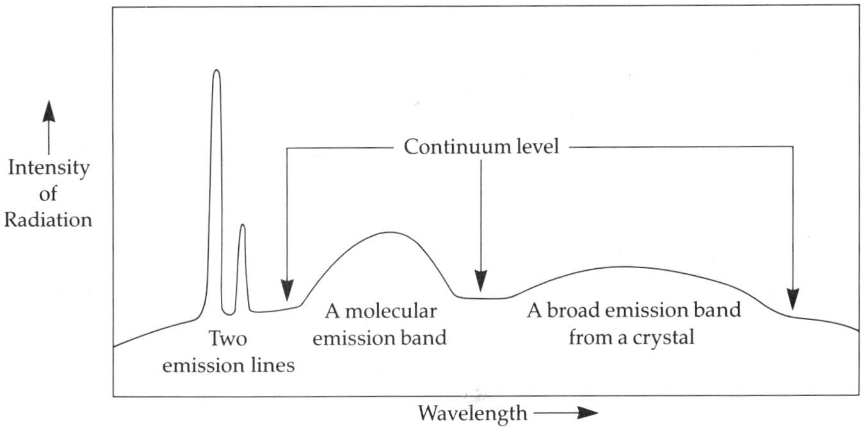

Figure 7-7. *Schematic examples of emission lines and bands as they might appear in a spectrum.*

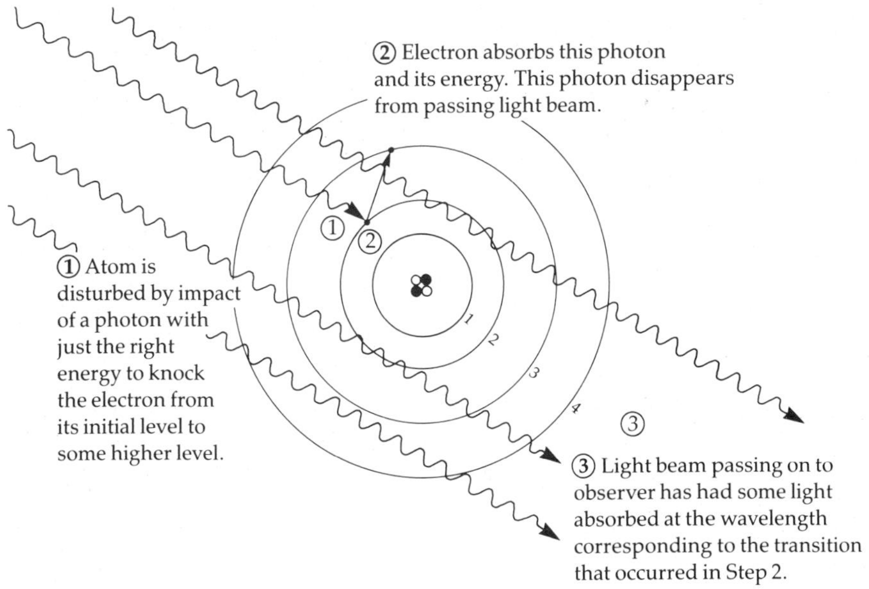

Figure 7-8. *Schematic view of absorption of a photon of light as an electron is knocked from a lower to a higher energy level in an atom.*

be thought of as giant molecules. In a crystal, atoms are joined in a fixed pattern that simply repeats as the crystal grows bigger. Light can penetrate through the outer millimeter or so of a crystal, and very broad, faint emission bands can result, as shown in Figure 7-7. Warm grains of silicate material near other stars have been identified from their emission bands, for example.

Absorption Lines and Bands

Let us return, in Figure 7-8, to an atom with an orbiting electron. This time it gets disturbed by a passing light wave (photon) that bumps it up into a higher energy level. Energy was removed from the beam to do this. This process of energy removal from a light beam is called **absorption**.

ATOMS AND ABSORPTION LINES

Only the specific energy corresponding to the specific transition (level 2 to 3 in the case of Figure 7-8) can be absorbed in an atom. Thus, only a photon of specific energy or wavelength can be removed from the beam. A beam of light passing through a cloud of such atoms will have many of these photons removed, thus absorbing some light of that color. As a result, the light in a narrow interval of the spectrum is lost, and this missing interval is called an **absorption line**. Various absorption lines can result from the various possible upward transitions—for example, level 2 to 3, 2 to 4, 2 to 5, 1 to 2, 1 to 3, and so on. Absorption lines due to hydrogen are well shown in the middle spectrum in Color Plate 32.

Since the energy level intervals possible in a given atom are the same whether absorption or emission is occurring, the pattern of emission and absorption lines for a given element is the same. An element can be identified from either its emission lines or absorption lines.

MOLECULES, CRYSTALS, AND ABSORPTION BANDS

Much of the preceding also applies to molecules and crystals. As light penetrates through a cloud of molecules (a gas) or through the upper millimeter of a crystal (in a rock surface), transitions of electrons among its numerous, closely spaced energy levels produce **absorption bands**. The molecules in the gas, or the composition of the crystal, can be identified if the absorption bands themselves can be detected and identified. Figure 7-9 shows schematic examples of absorption lines and absorption bands. The absorption bands of a substance have the same wavelength intervals as its emission bands.

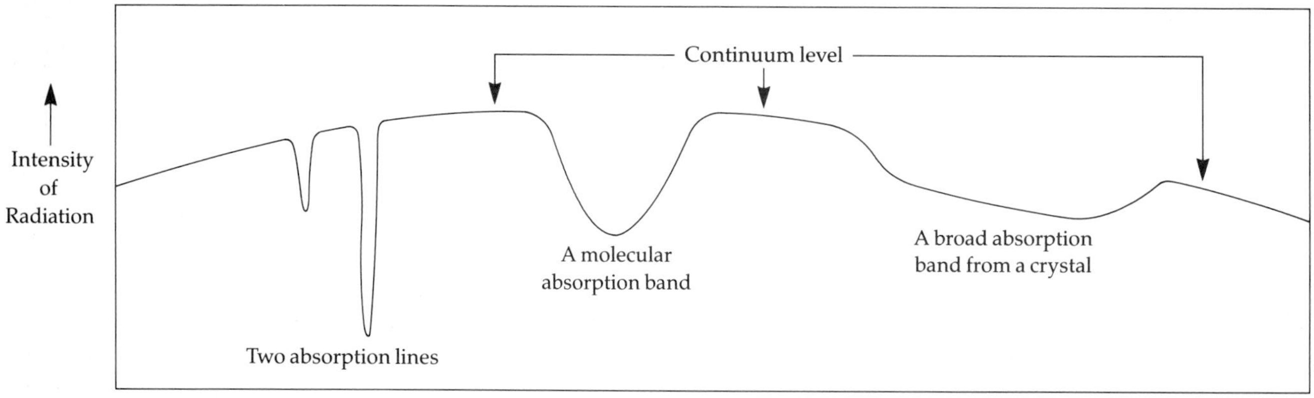

Figure 7-9. *Schematic examples of absorption lines and bands as they might appear in a spectrum.*

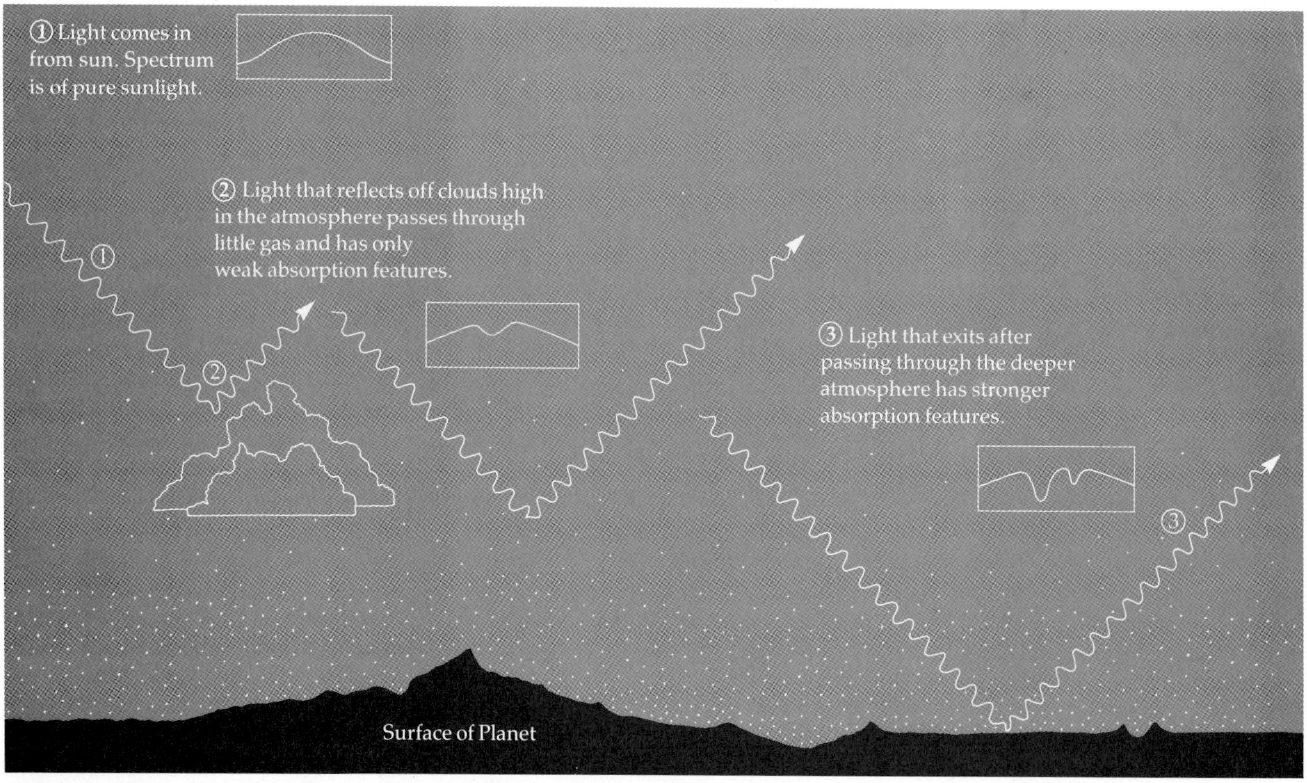

Figure 7-10. *Production of absorption lines or bands in the spectrum of a planet as light passes through the atmosphere of the planet. Identification of the lines or bands allows identification of at least some constituents in the atmosphere. The same process occurs as light radiates from a star outward through the star's surrounding gaseous atmosphere.*

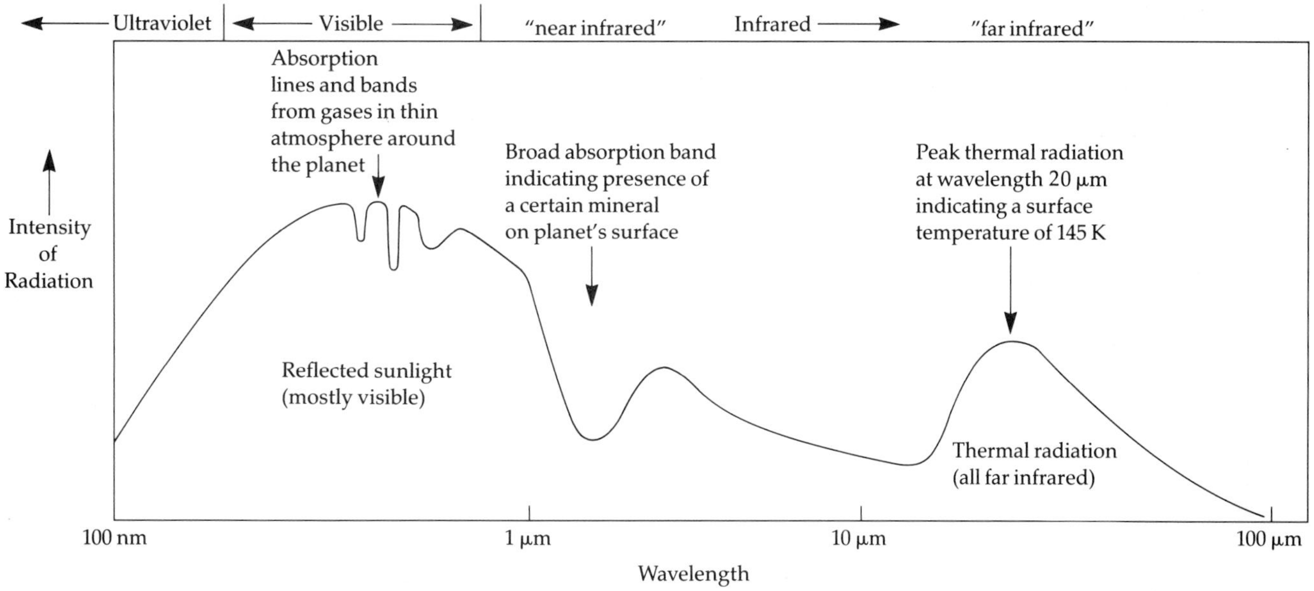

Figure 7-11. *Schematic, simplified spectrum of an imaginary planet from the ultraviolet to infrared wavelengths, showing the reflected sunlight and the thermal radiation from the planet itself. A thin atmosphere creates some absorption lines and bands, but lets enough light through to the surface that we see a broad absorption band (at about 1.5 μm) due to minerals on the planet's surface. (See text for further description.)*

MEASURING THE COMPOSITION OF AN ATMOSPHERE OF A PLANET (OR STAR)

Now we see how light carries even more messages. Suppose we look at a planet with an atmosphere. Light passes into the atmosphere and back out, as shown in Figure 7-10. As the light passes through the gas, it may acquire absorption lines and bands that allow us to identify some of the atoms and molecules in the gas. Not all atoms and molecules have equally prominent lines and bands. Some, such as carbon dioxide (CO_2) are easy to detect, and others, such as nitrogen (N_2) are hard to detect by spectroscopy. Thus, while the carbon dioxide of Venus and Mars was discovered by Earth-based spectroscopy decades ago, the abundant nitrogen of Saturn's moon, Titan, was not discovered until a Voyager spacecraft made a close pass by Titan.

How to Measure the Composition of a Planet's Surface from a Distance

As mentioned earlier, the absorption bands caused by mineral crystals are broad, often shallow, and hard to detect. For this reason, spectroscopy has only limited success in identifying planetary surface materials. There are two approaches: measuring colors and detecting actual absorption bands. Measuring the color of an object is essentially measuring the relative intensities of radiation in different parts of the spectrum. A red object looks red because it reflects more light at red wavelengths than at bluer wavelengths. A measurement of color often restricts the possible materials on a surface. For instance, most satellites of Saturn have a bluish-white color similar to ice, but inconsistent with the red rocks and minerals on Mars. If we can detect a mineral absorption band, we can probably identify the mineral. For example, the satellites of Saturn just mentioned display absorption bands caused by crystals of frozen water, in the form of ice or frost, confirming the implication of the color studies that ice is a dominant material on their surfaces.

Analyzing a Spectrum

Astronomers can spend many hours using large telescopes to obtain spectra of distant, faint objects, such as planets, moons, asteroids, and comets, as well as stars and galaxies far beyond the solar system. Figure 7-11 shows an example of some results that might be obtained if we could scan a planet's spectrum all the way from the ultraviolet to the infrared. (This is somewhat idealized, since different instruments are needed in different regions. One astronomer may obtain the spectrum only in one region. Figure 7-11 might require

a

Figure 7-12. *Views of the earth utilizing two different wavelengths of light. (a) View in visible light, showing clouds and continents. (b) View at an infrared wavelength absorbed by water vapor gas in the air. Due to absorption, water vapor haze obscures most of the earth's surface. (European Meteorological Satellite images, courtesy C. R. Chapman.)*

b

the combined work of many astronomers using different instruments at different observatories.) In the visible region we see reflected sunlight. Superimposed on it are some absorption lines and bands that tell us the planet has a gaseous atmosphere containing certain elements and compounds; these arose as shown in Figure 7-10. In the near infrared, just beyond 1 μm wavelength, we see a broad absorption feature characteristic of a rock-forming mineral. This gives us some indication of the rock type on the surface, indicating at the same time that clouds are not thick enough to block most of the light from reaching the surface. Finally, in the far infrared at 20 μm wavelength, we see the thermal emission from the planet. Noting that the dominant wavelength is at 20 μm, we can use Wien's law to estimate the surface temperature on the planet. In this case, it is a cold 145 K (−198°F).

The Spectrum and Our Atmosphere: Seeing into Space from the Earth

Figure 7-12 (pp. 130–131) illustrates several important points about the relationship between astronomy and the earth's atmosphere. First, clouds block much of our view of space. One reason for putting observatories on top of mountains is to get above the clouds. Another reason is to get above the denser, shimmery air that makes telescopic star images dance and twinkle when seen from sea level.

But a more important reason can be seen by comparing Figure 7-12(a), made in normal visible light, with Figure 7-12(b), made in infrared light. Many atmospheric molecules, such as water molecules, strongly absorb infrared light just beyond the reddest wavelengths we can see. Figure (b) is an image made with light of the infrared wavelength absorbed by water. Sure enough, it shows a water vapor haze over most of our planet, because water vapor (H_2O gas) absorbs this light, preventing it from revealing the ground clearly in this image. Conversely, a sea level astronomer who wants to look for water vapor absorptions on, say, Mars, has a problem: The infrared light that carries this message from Mars is blocked by water vapor in our own atmosphere before it ever reaches his or her telescope on the ground, as can be seen in Figure 7-13. Even the thin, high atmosphere blocks many wavelengths. For example, the **ozone (O_3) layer**, about 20 to 50 km up, absorbs nearly all ultraviolet radiation (thus protecting us from severe sunburn).

These absorptions of certain wavelengths by our atmosphere are a main reason for putting modern observatories

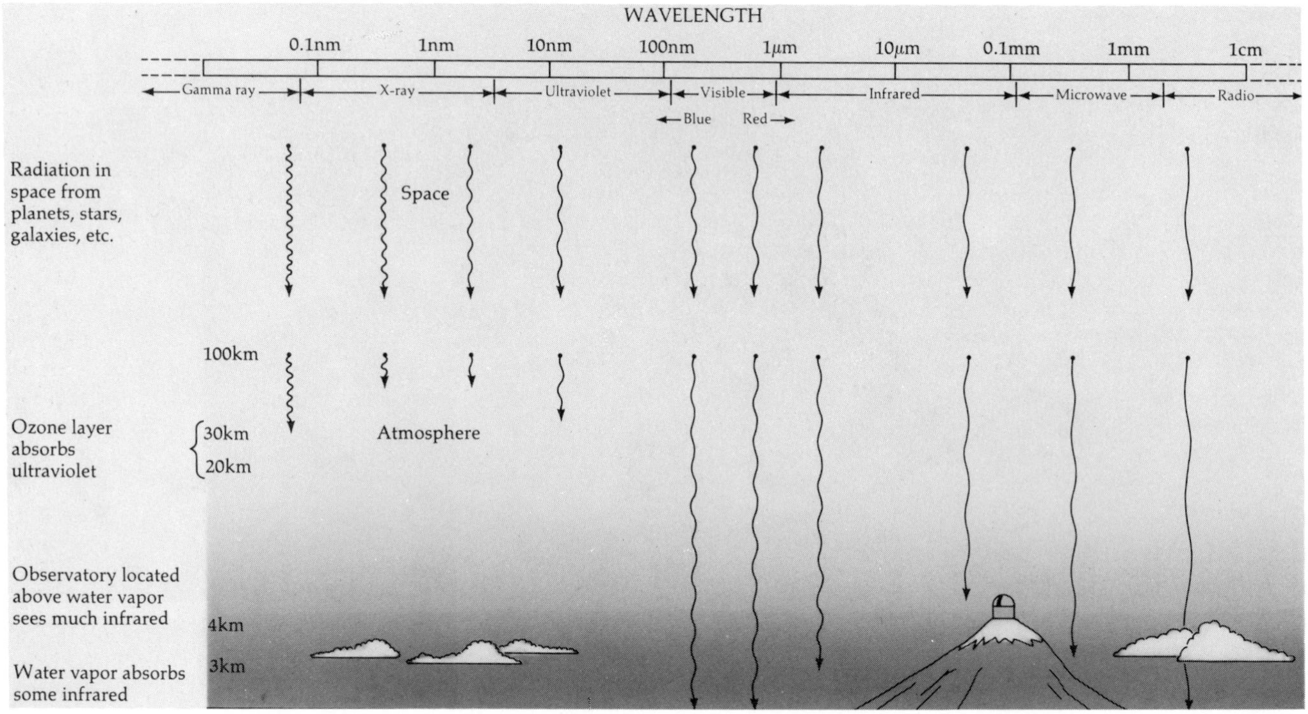

Figure 7-13. *Electromagnetic spectrum is shown at top. All these types of radiation from various astronomical sources can be detected by a telescope in space (upper middle), but many of them are absorbed by various gases in our atmosphere (bottom).*

on mountains. In particular, high observatories are above most of our atmosphere's water vapor, which is concentrated in the bottom 3 km (9000 ft) of the atmosphere, and which absorbs much of the infrared light from planets and stars. At the same time, high altitude observatories escape the clouds and the shimmering, hazy air of low elevations. Thus, for example, astronomers at 14 000-ft Mauna Kea Observatory in Hawaii can measure most of the infrared spectrum almost every night. To measure most of the ultraviolet spectrum, we must get above the ozone layer, far above even the highest mountains. Therefore, various ultraviolet, X-ray, and gamma-ray telescopes, as well as infrared and optical telescopes, have been put in orbit. These have made amazing breakthroughs that we will describe in later chapters. Of special importance is the Space Telescope project, which plans to put a large telescope into orbit in the late 1980s (Bahcall and Spitzer, 1982). The telescope will have a mirror 2.4 m (94 inches) across, making it one of the larger telescopes in operation. It is expected to produce revolutionary astronomical discoveries because of both the sharpness of its image and its space-based opportunity to sample a wide range of ultraviolet to infrared wavelengths. It may be serviced by shuttle flights or space station astronauts, so that instruments on it can be updated and replaced.

Figures 7-12 and 7-13 illustrate one of the advantages of access to a wide range of wavelengths. Photos using light of different wavelengths penetrate gaseous atmospheres of planets or stars to different depths (because of the different rate of absorption at different wavelengths). Therefore, they show different features at different depths in the atmosphere. Comparison of images at different wavelengths allows us to tell something about the vertical structure of the atmosphere we are observing.

The Function of Telescopes

A **telescope** is a device for magnifying the image of an object to a larger angular size than we perceive with our naked eye, and just as importantly, for gathering more light than the eye can. The increased amount of light allows us to produce a bright image of objects too faint to see otherwise, and to make accurate measurements of their brightness.

When light from a distant planet or star reaches the earth, a certain number of photons strikes each square centimeter of the earth each second. The pupil of the eye covers an area of less than a square centimeter and can receive only a limited number of photons per second. But in a telescope, we collect all the photons striking a larger lens or mirror. This makes a much brighter image. The larger the lens or mirror, the brighter the image. We can study the image visually through an eyepiece, photograph it, measure its brightness with a photometer, or measure its spectral properties spectroscopically. Such measurements can be used to study the temperature or composition of an object, as we described in this chapter, and still more properties we will mention in later chapters.

How Does a Telescope Work?

Two basic designs have been used for telescopes. The first to be built, the **refractor**, uses a lens to bend, or refract, light rays to a focus, as in Figure 7-14. Galileo first used this type astronomically in 1609. The second type, the **reflector**, uses a curved mirror to reflect light rays to a focus, as in Figure 7-15. Isaac Newton built the first reflector in 1668. Several reflector designs have since been constructed, but the simplest is Newton's, sometimes called the Newtonian reflector. In recent years new designs have combined lenses and mirrors. Often called **compound telescopes**, these designs are often very compact and portable.

In any telescope, the main lens or mirror is called the **objective**, as shown in Figures 7-14 and 7-15. The distance from the objective to the place where the image is focused is called the **focal length** of the objective (marked F in the figures). The larger the focal length, the bigger the image. This principle is also used in cameras, which are merely small telescope systems with film at the focus.

The diameter of the objective is called the **aperture**. This is the width of the hole through which light can pass into the instrument. The larger the aperture, the more light is gathered and the brighter the image.

RADIO TELESCOPES

Radio telescopes serve the same function as optical telescopes. But because a 50-cm radio wave is a million times bigger than a visible light wave and carries less energy per photon, radio telescopes need larger surfaces to collect and focus enough energy to give strong signals. As shown in Figure 7-16, radio telescopes are reflectors, with a large curved surface (the "dish") of metal or wire mesh, which

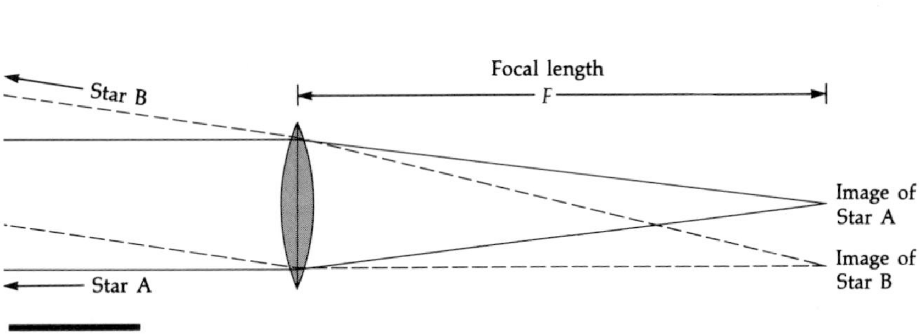

***Figure 7-14.** Cross section through a lens, showing image formation in a refracting telescope (and most cameras). Light rays from two stars are focused into two images.*

Figure 7-15. Cross section through a mirror system, showing image formation in a reflecting telescope. Light rays strike curved mirror (right) and are reflected back toward focus. The focus would normally lie in an inconvenient position in front of the mirror, but a secondary mirror is used to beam light rays to one side for easier access to the image.

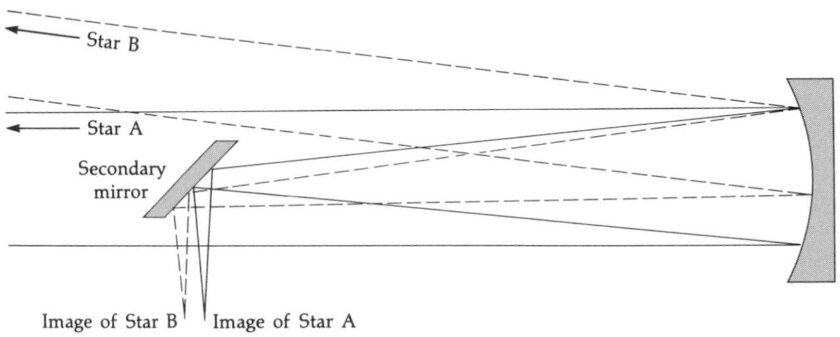

a

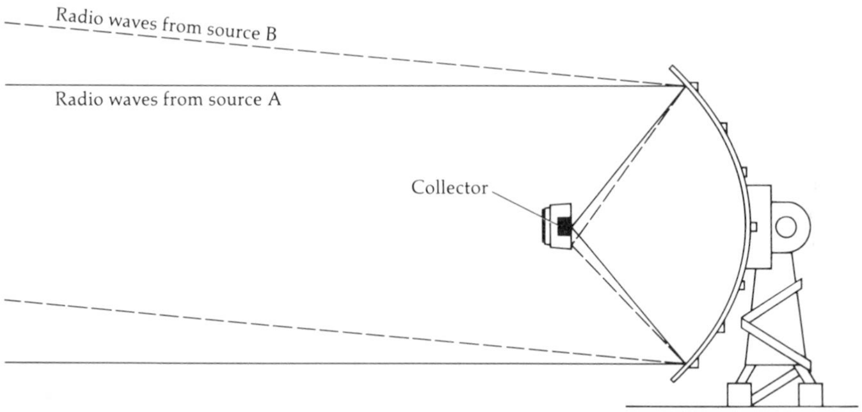

b

Figure 7-16. (a) *View of one of the radio telescopes in the Very Large Array in central New Mexico. This view shows the underside of the "dish" (top), or curved reflector, that collects the radio waves, along with the mounting assembly and figures for scale. (Photo by author.)* (b) *Schematic cross section of a radio telescope, showing the similarity to the design of the reflector-type optical telescope.*

reflects the radio waves to a focus, where a smaller radio detector picks up the signal.

Functions of Visual Telescopes

An optical telescope designed to be looked through, as opposed to a radio telescope or a camera, may be called a *visual telescope*. The optical systems of Figures 7-14 and 7-15 can be converted into visual telescopes simply by adding an eyepiece (and usually tubing to keep out stray light), as shown in Figure 7-17. The eyepiece is simply a lens or system of lenses designed to magnify the image still more and allow the eye to examine the image.

One of the first questions asked of backyard telescope enthusiasts is "How far can you see with that thing?" This is not the right question to ask, because no telescope is limited by distance. Any optical system can see as far as there is an object large enough or bright enough to detect. The naked eye as well as the 5-m (200-in.) Palomar telescope can see the Andromeda galaxy 19 billion billion kilometers away—but the telescope shows more detail and fainter regions. The two main functions of the telescope, as stated

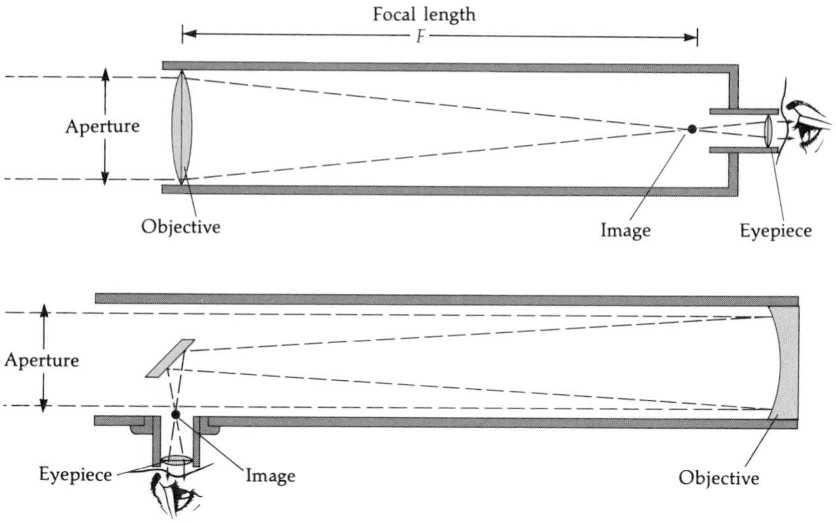

Figure 7-17. ***Cross sections showing how the systems of Figures 7-14 and 7-15 are converted to visual telescopes by the addition of an eyepiece. The eyepiece is a small magnifying lens (or several lenses mounted together) used to examine the image. Tubing from the objective to the eyepiece helps cut out stray light.***

above, are to enlarge the image in order to show small angular detail and to gather enough light to reveal objects too faint to be otherwise seen. These two functions are called **magnifying power** and **light-gathering power**.

MAGNIFYING POWER

In any given telescope, the magnifying power, or *power* as it is usually called, is controlled by the eyepiece. Suppose a distant object subtends an angle of 1 minute of arc (about the angular size of Jupiter when it is prominent). An eyepiece that makes the object appear 100 minutes across when seen through the telescope is said to give 100 power.[6]

In theory, any telescope can be made to give any magnifying power simply by inserting an eyepiece of short enough focal length. In practice, however, three effects limit the magnifying power that can be used. First, because we look into space through turbulent air, higher powers magnify air turbulence, and too high a power causes the image to shimmer hopelessly. (The air quality, called **seeing**, varies from night to night, with occasional nights of good seeing occurring when the air is still.)

Second, as the magnification is increased, the image gets fainter and fainter because the light is spread out over a larger and larger area. Too high a power gives a hopelessly faint image.

Third, due to fundamental properties of light waves, a telescope of *A*-cm aperture cannot resolve details smaller than 12/*A* seconds of arc. Thus, a 12-cm (5-in.) telescope can resolve about 1 second, but nothing smaller. Therefore, too high a power gives a hopelessly fuzzy image.

Astronomers look forward to solving the first problem by putting more telescopes into orbit in space, above the atmosphere. The last two problems can be overcome by increasing the telescope aperture, giving more light and resolution of smaller angular detail. Nonetheless, the maximum useful magnifying power for most telescopes on most nights is about 20 power per centimeter of aperture. For example, a 5-cm (2-in.) telescope might be used at 100×; a 15-cm (6-in.) at 300×; and a 30-cm (12-in.) at 600×.

LIGHT-GATHERING POWER

The other important function of a telescope, gathering light, is controlled solely by the aperture size. Many astronomical objects, such as nebulae and galaxies, are very faint, and to see their details one needs light more than magnifying power. In fact, the best visual impression of faint nebulae comes from a large telescope (lots of light) used at *low* power (to concentrate light). For this purpose, one might use only 4 power per centimeter of aperture; for instance, a 15-cm (6-in.) telescope at 60 power. If you have access to a telescope, you can experiment with these recommendations while looking at different celestial objects.

Generally, the amount of detail that can be seen on astronomical objects depends ultimately on the stillness and clarity of the air and on the telescope aperture, since these control both the useful magnifying power and the light-gathering power.

OBSERVING THE MOON AND PLANETS

Although any visual telescope can be used to observe the heavens, certain designs and techniques are best suited to certain purposes. To observe the moon and planets, magnification is the most important function, because one wants to get a big enough image to see detail. Telescopes with rather long focal lengths, about 6 to 10 times the aperture (often stated as *f* 6 to *f* 10), are desirable. Eyepieces giving magnifications of 100 to 200 power are recommended. Some of the sights you can see with such equipment, in order of increasing difficulty, are the mountains and craters on the moon, the phases of Venus, the satellites of Jupiter, the satellites of Saturn, the rings of Saturn, the cloud bands on Jupiter, divisions in the rings of Saturn, the phases of Mercury, the polar ice caps

---

[6]The magnifying power is given by the formula

$$\text{power} = \frac{\text{focal length of objective}}{\text{focal length of eyepiece}}$$

For instance, if a 1-cm eyepiece is inserted in a telescope of objective focal length 100 cm, we get 100 power, often written 100×.

of Mars, the dusky markings of Mars, the cloud bands on Saturn, and the dusky markings on Mercury. An aperture of 5 cm (2 in.) suffices to see the first five items on the list; an aperture of around 25 cm (10 in.) is recommended for all items and usually necessary for the last one.

OBSERVING STAR FIELDS, NEBULAE, AND GALAXIES

The most prominent star clusters, nebulae, and galaxies are faint but cover larger angular areas than planets; hence light-gathering ability is the most important function in observing them. Telescopes with shorter focal lengths (*f* 3 to *f* 6) are useful, and magnifications of 20 to 100 recommended. Some double stars and star clusters are interesting to study under higher power, but higher powers spread the light out too much to give good views of nebulae and galaxies, which are mostly very faint. As always, the largest possible aperture gives the best view.

OBSERVING THE SUN

A NORMAL TELESCOPE SHOULD NEVER, UNDER ANY CIRCUMSTANCES, BE POINTED AT THE SUN. An eye could be immediately burned or blinded, since the objective acts like a giant magnifying glass, concentrating solar light, heat, and UV radiation at a point near the eyepiece. Furthermore, the telescope is likely to be damaged, since the solar heat can crack the glass in eyepieces or secondary mirrors. Some telescopes can be modified for safe solar observing. One usually starts by putting an opaque screen over the objective with a new aperture as small as 1 or 2 cm, then projecting the image onto a white card so it can be viewed. Instructors or telescope owners should be consulted before attempting this procedure.

Photography with Telescopes

Instead of in an eyepiece, an image can be formed on a piece of photographic film. Development of properly exposed film

a

Figure 7-18. (a) *The 2.2-m (88-in.) reflecting telescope at Mauna Kea Observatory, Hawaii. In this telescope design, light is focused through a hole in the center of the main mirror, where the astronomer adjusts his electronic instruments to record data. Data are fed through cables to computer equipment, which performs initial analysis and stores data for further analysis. The dome (top) is open to night air in order to give the clearest observing conditions; temperatures inside the dome often drop below freezing during observing sessions. (Photo by author.)* (b) *Schematic diagram of* (a).

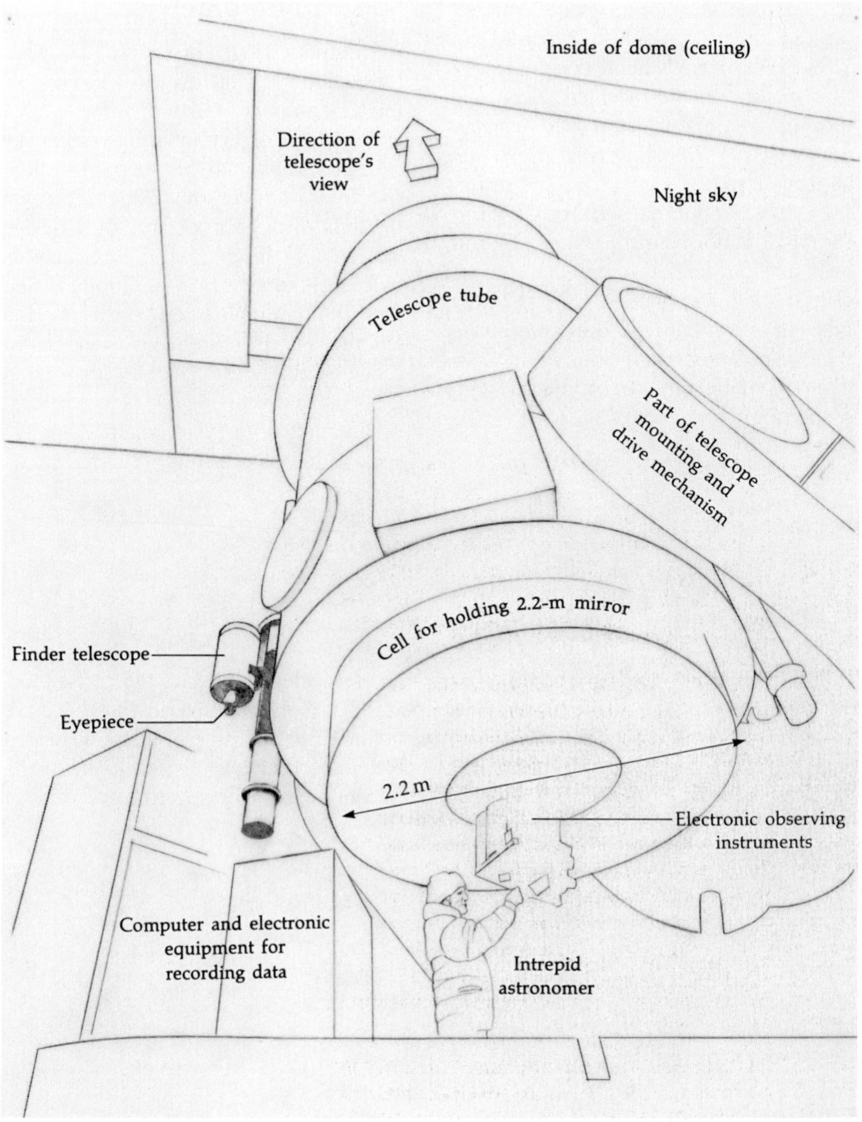

b

then provides a permanent record of the observation. Many telescopes come equipped with photographic attachments for this purpose. In the case of a bright object like the moon or a planet, the eye usually sees more than a photograph with the same telescope, because the eye can take advantage of moments of perfect seeing, whereas the photograph averages moments of poor seeing, which blurs the image. On the other hand, faint objects like nebulae and galaxies are usually better shown in photos, because light can be accumulated in long exposures, whereas the eye cannot "store" light.

Many photos of star fields and nebulae in this book were taken with small telescopes or ordinary cameras. The reader should consult picture captions, most of which describe the equipment and exposure used. In many cases, readers can duplicate or improve the results with their own equipment.

In large, modern observatories, the recording instruments are often electronic, instead of either photographic or visual, as shown in Figure 7-18. Furthermore, computer-directed machines can point the telescope almost precisely at any known celestial object. Thus, an astronomer may spend nearly the whole night observing without looking through the telescope, though he may look through a small telescope called a *finder*, mounted parallel on the side, to help find his target object, or he may view television screens that monitor the image.

Photometry

Pictures of astronomical objects are interesting, of course, but much astronomical work requires measuring an object's brightness—the amount of light coming from it at all wavelengths, or at selected ranges of wavelengths (such as blue to green). This is called **photometry**. Until a few decades ago, this was done by measuring the size and density of the image on a photograph. Today, it is done much more precisely with electronic devices. The basic device is a photomultiplier placed at the telescope's focus. Each photon of light collected by the telescope strikes the photomultiplier surface, which is designed to release a shower of electrons with each photon impact. In some designs, each electron strikes a surface and releases more electrons, thus multiplying the effect into a weak electric current that can be measured. Measurement of the current tells the number of photons arriving from the astronomical object, thus giving a measurement of its light intensity.

In recent years, these techniques have been extended into video-imaging devices that can produce TV images of objects too faint to photograph. Most important of the modern detectors are charge-coupled devices, more commonly known as CCDS. These are extremely light-sensitive detectors, made possible by micro-electronic technology. They are often used in arrays of, say, 500 × 500 light detectors, so that an image can be created from the resulting 250,000 "dots"; each "dot" corresponds to one measurement of brightness, made by one detector. Each "dot" is called a picture element, or "pixel." CCD imaging arrays can record in 10 minutes more faint detail than photographic film exposed for 100 minutes in the same telescope (for a review of CCDs see Kristian and Blouke, 1982).

Spectrophotometry

As we stressed in the first part of this chapter, it is important to measure not only the total amount of light coming from an astronomical object but also the intensity at each wavelength. This process is called **spectrophotometry**. It is done by passing the light through a spectroscopic device that breaks the light into different colors before it enters the photomultiplier. Then the photomultiplier can measure the amount in each color (that is, wavelength) range. Similarly, photos or TV images can be produced showing the object as seen in different colors (wavelengths). We will see a number of examples in this book.

Summary

Most of our knowledge of the universe comes from deciphering messages carried from extraterrestrial bodies by electromagnetic radiation. To do this we need to understand the nature of the radiation and how it is generated in atomic particles. Electromagnetic radiation is released in the form of photons, each of which has its own specific wavelength and energy. The light coming from an object, arranged in order of wavelength, is called the spectrum of the object. Each wavelength corresponds to a "color," ranging from too short (too blue) to see, through visible colors such as blue and red, to wavelengths too long (too red) to see. Photons of blue light have more energy than photons of red light.

Atoms and molecules each produce absorption and emission lines and bands in the spectrum as photons interact with them. These derive from the energy level structure of their electronic orbits. Each element and compound has its own lines and bands. Materials in a distant object can be identified by obtaining the objects' spectrum and identifying emission or absorption lines or bands, if present. Many gases in planetary atmospheres or stars can be identified in this way, but solid surfaces of planets can be only roughly characterized, since solid materials have only weak, poorly defined absorption bands.

Telescopes are devices that not only magnify the angular size of distant objects, but also just as importantly, enable us to gather much more light than gathered by the eye. Thus we can see objects much fainter than visible to the unaided eye. This chapter contains advice on viewing different astronomical objects with small telescopes. Astronomers bolt various kinds of instruments on large telescopes to analyze the light from planets, stars, and so on. One of the most important instruments is the spectrophotometer, which measures the amount of light at each wavelength over a range of wavelengths, allowing precise measurement of the spectrum of the object being observed. We are just entering an era when the largest telescopes are being placed in orbit above the atmosphere in order to give access to unblurred images throughout all wavelengths of the spectrum.

Concepts

wavelength
frequency
wavelike properties of light
visible light
ultraviolet light
infrared light
speed of light
particlelike properties of light
photon
diffraction
spectrum
spectroscopy
fields
electromagnetic radiation
continuum
temperature
thermal motion
thermal radiation
reflected radiation
Wien's law
emission
energy level
emission line
ground state
excited state
emission band
absorption
absorption line
absorption band
ozone layer
telescope
refractor
reflector
compound telescope
objective

focal length	seeing
aperture	photometry
magnifying power	spectrophotometry
light-gathering power	

Problems

1. Instruments measuring faint light sources are more efficient when measuring more energetic photons. Which would you expect to be easier to measure: blue, red, or infrared light?

2. The ground cools at night by radiating thermal infrared radiation into space. Explain why the air stays warmer on a moist, cloudy night than on a very dry, clear night.

3. A spacecraft passes close to a hitherto unknown planet. A camera system on the spacecraft photographs surface details quite clearly in all wavelengths except the absorption bands of carbon dioxide. At these wavelengths, the image is featureless. Interpret these results.

4. Why has there been more emphasis on getting ultraviolet and infrared telescopes into orbit than on getting radio telescopes into orbit?

5. You are considering buying a pair of binoculars. Two available pairs each have lenses with 50-mm aperture, but one has a magnifying power of 4× and the other, 10×. Which pair:

a. Gives the larger apparent image size?
b. Gives a greater apparent brightness in the observed image?
c. Is easier to hold still enough to observe (remember that the unsteadiness of your hands is magnified as much as the image)?

6. For your birthday, your parents offer you a choice of lenses for your 35-mm camera. One has a focal length of 24 mm and the other has a focal length of 90 mm. (Focal length numbers are marked on the inner rim of all camera lenses.)

a. Which one would give a bigger image but cover a narrower field of view in terms of degrees?
b. Which one would give a smaller image but cover a wider angular field of view?
c. Which one would be called a telephoto lens?

7. a. Why can an atom in the ground state not produce emission lines? **b.** Use this fact to explain why the colorful emission-line glows of certain nebulae occur near hot stars, but not in the cold gas of intersteller space.

ADVANCED PROBLEMS

8. Confirm that the thermal infrared radiation in Figure 7-11 could come from a planetary surface at a temperature of 145 K.

9. A planetary astronomer has an instrument that scans the spectrum from 5 to 10 μm wavelength. Looking at a certain moon, he detects thermal infrared radiation that increases from 5 to 10 μm, but does not peak in this range. Apparently it peaks at a longer wavelength than 10 μm. What can you say about this moon's surface temperature?

10. What wavelength of thermal radiation is being emitted with greatest intensity by this book?

11. Your friend's Newtonian reflector telescope has a focal length of 70 in. and he has an eyepiece with a focal length of ½ in.

a. What magnification does the telescope give?
b. With the preceding information, discuss what can be said about the brightness of the image given by this telescope.
c. If you look through the telescope at the moon, what angular size would the moon appear to be?
d. Another friend appears and asks how big a lens the telescope has. How would you answer?

12. Two telescopes each have a 4-in. aperture, but one has a short focal length and gives 30× magnification, and the other has a long focal length and gives 150× magnification.

a. Which one would give a better view of the planet Mars, which subtends an angle of about 20 seconds of arc?
b. Which one would give a better view of the faint, wispy Andromeda galaxy, which subtends as much as 3°?

Projects

1. Compare views through binoculars of different sizes. (Binoculars carry a designation such as 7 × 35, where the first number is the magnifying power and the second, the aperture.) Confirm that at fixed power, larger apertures give brighter images.

2. Observe and sketch a rainbow. Confirm that the colors appear in order of wavelength, as seen in Figure 7-1.

CHAPTER 8

MERCURY AND VENUS

Flights of unmanned space vehicles by Mercury and Venus in the 1970s and 1980s typify the surge of exploration that has opened up much of the solar system in recent years. In 1974 and 1975, the American spacecraft Mariner 10 sailed less than 6000 km from Venus' cloudtops and then made three close passes by Mercury, as Mercury and the spacecraft looped around the sun. The third encounter produced photos from only 327 km above Mercury's surface. In October 1975, two unmanned Russian probes, Venera 9 and 10, parachuted through Venus' clouds and sent back the first photos of the surface. Subsequent Russian unmanned landers sent back more pictures and other data. In 1978, the U.S. Pioneer Venus spacecraft went into orbit around Venus and mapped most of the surface with radar. In 1983, the Russian Veneras 15 and 16 began more detailed radar mapping from orbits above the clouds. Because of such flights, Mercury and Venus have been revealed as interesting worlds, astonishingly different from each other and from the earth.

The Planet Mercury

Mercury is a moonlike body with no atmosphere. It is only about 40% the size of the earth, and about 40% bigger than the moon. Figure 8-1 dramatizes Mercury's small apparent size as seen from the earth. A telescope with an aperture larger than about 15 cm (6 in.) reveals that it has phases (Figure 8-2) and very faint dusky markings on its illuminated side. These markings are probably lava plains similar to the patchy dark maria of the moon. Mercury's surface has a slight pinkish cast. No satellite has been found.

Mercury exhibits an interesting relation between its rotation and its orbital

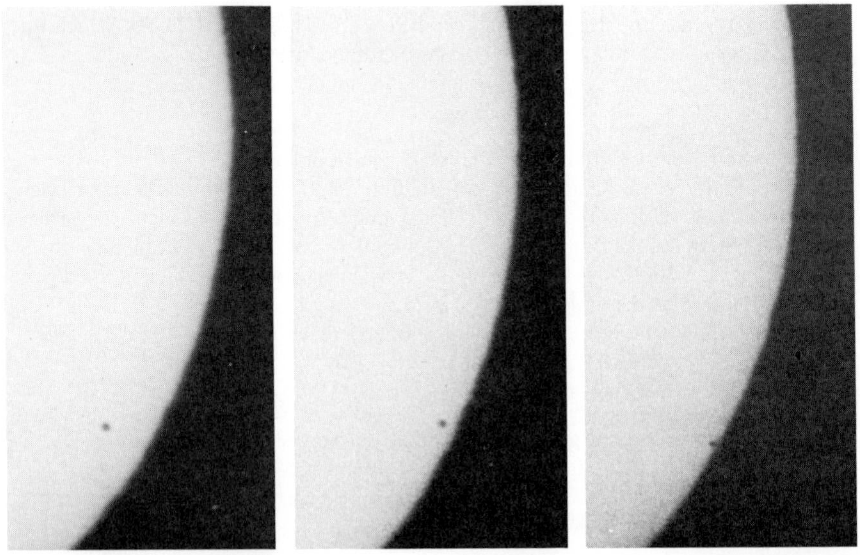

***Figure 8-1.** Three photos made of Mercury as it transited the sun on November 10, 1973. Mercury (the small black disk at bottom) is moving off the disk of the sun. These photos show the small size of Mercury's disk during its closest approach to the earth, illustrating how poor an earthbound observer's view of the planet is. (Photos with a Questar telescope by W. A. Feibelman.)*

revolution around the sun. Due to complex tidal effects (similar to those that hold the moon facing the earth), Mercury has gotten locked into a 59 d period of rotation, which is just ⅔ of its 88 d orbital revolution period. The combination of these two rates means that the sun moves very slowly across Mercury's sky, taking 176 d to go from noon until the next noon!

This can be seen better from Figure 8-3, in which we visualize Mercury at position zero, where the noontime sun is overhead at a certain mountain. At position 1, Mercury has completed a quarter of its rotation relative to the stars. By position 3, Mercury is halfway through its orbit, 44 d have elapsed, and the sun is just setting on the mountain. By position 4, Mercury is ⅔ of the way around its orbit, but it has completed one rotation, since the mountain now points in the same direction as in position 0. By position 6, Mercury has gone once around the sun in 88 d, but it is now midnight on the mountain, since the mountain faces directly away from the sun. Another 88 d are necessary to bring the mountain back to noontime, as you can confirm by sketching in the next orbital trip, positions 7 through 12.

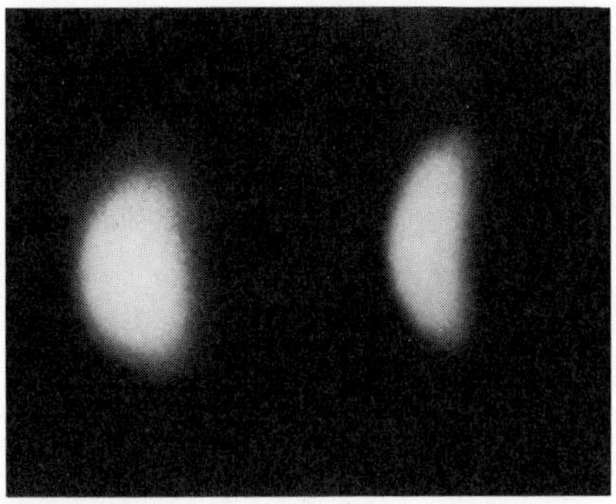

Figure 8-2. *Two photographs of Mercury (June 7 and 11, 1934) showing the changing phases of the planet. Virtually no surface detail can be seen; very few photos from the earth show reliable detail. (Lowell Observatory.)*

MERCURY AND DR. EINSTEIN

At first glance, Mercury's orbital motion seems to contradict the laws of Kepler and Newton: The *perihelion*—the point nearest to the sun—shifts in position slowly around the sun from year to year. This movement is called **orbital precession**. Some precession had been predicted from Newton's laws, but observers found an excess shift of 43 seconds of arc each century—a tiny amount, but enough to consternate orbital theorists!

The French astronomer U. J. Leverrier, around 1860, thought that the excess shift might be caused by the gravity of a small, undiscovered planet inside Mercury's orbit; they even gave the planet a name—Vulcan (Moore, 1954). Leverrier had already successfully predicted Neptune's existence from similar gravitational disturbances in the motion of Uranus, but he was not correct in the case of Mercury. Twentieth-century observations reveal no planet inside Mercury's orbit.

But how can Mercury's orbital precession be explained? The solution came in 1915, when Albert Einstein showed that the great mass of the sun disturbs the orbits of nearby planets in a way unpredicted by Newton's laws. Einstein's theory of relativity predicted almost exactly the excess precession observed—43.03 seconds of arc per century. Einstein predicted smaller excesses for Venus and the earth, and these,

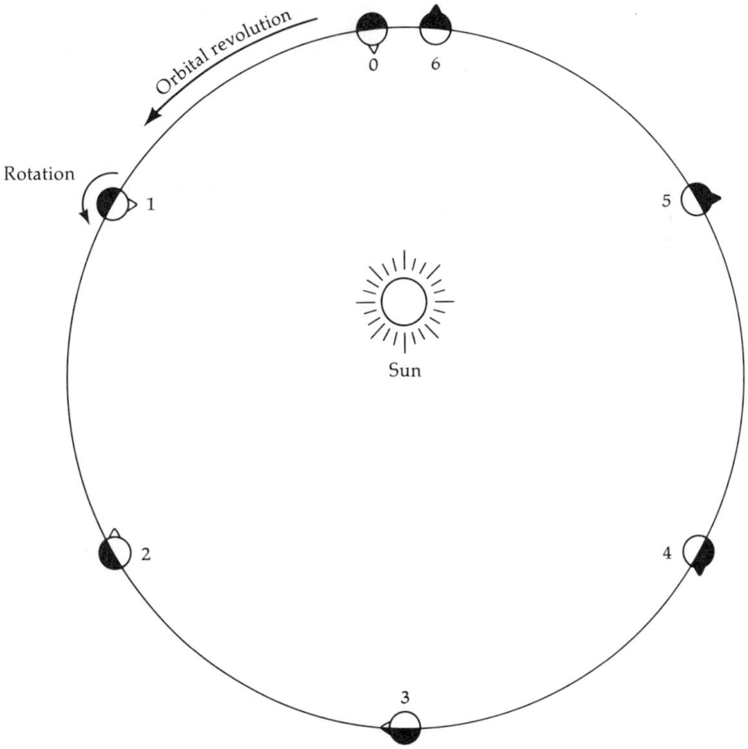

Figure 8-3. *A view of the orbit of Mercury showing the relationship between its rotation and revolution. The cartoons represent Mercury as a globe with one high mountain marking the rotation position. Shading shows the nighttime side. See the text for a discussion.*

too, were confirmed by observation. Thus, Einstein's contribution to solving the puzzle of Mercury's precession played a major role in the acceptance of his theory of relativity.

SURFACE PROPERTIES OF MERCURY

Because Mercury is so close to the sun, its daytime surface is much hotter than the earth's. Measurements by infrared detectors of the thermal radiation from both the day and night side of Mercury show that temperatures in the upper few millimeters of soil range well above 500 K (441°F) in the "early afternoon" near perihelion to lows of about 100 K (−279°F) at night. In some areas, depending on soil type, the temperature might exceed 600 K (621°F). Though science fiction of a few years ago described hypothetical pools of molten metal on the daylight side, this now seems unlikely for several reasons. First, the melting point of lead is about 600 K; of aluminum, 832 K. Second, because of the insulating effect of the overlying soil, the temperature slightly below the surface is only 314 to 446 K.

Mariner 10 produced the best available data about Mercury, including photographs of the surface (Figure 8-4) and magnetic measurements. The surface resembles the moon's in its abundance of rugged craters and in its huge, multiringed craters known as basins. A few vague features glimpsed from the earth correlate with some bright and dark features photographed by Mariner 10. The most prominent basin, shown in Figure 8-5, is named the Caloris basin (in keeping with the high Mercurian temperatures). Its concentric ring structure, more than 1200 km in diameter, strongly resembles the Orientale ringed basin on the moon (see Figure 5-13). Evidently the same impact and lava flow processes occurred on Mercury as on the moon some 3½ to 4½ billion years ago. From this discovery, most scientists believe all planets suffered an intense bombardment by interplanetary debris at the close of the planet-forming process.

Figure 8-4. *Mariner 10 spacecraft photo of Mercury shows a heavily cratered planet that resembles the moon. (Compare with Figure 5-9.)(NASA.)*

Like the moon, Mercury probably has a thick lithosphere, but some disturbances are visible. Cliffs such as those in Figure 8-6 appear to mark huge faults, suggesting that Mercury may have contracted slightly because of cooling. Parts of the ancient cratered surface appear to have been leveled by more recent melting or lava flows.

The magnetic field of Mercury is only about 1% that of the earth's. Nonetheless, Mercury's magnetic field has the form calculated for an internally generated field like the earth's, with a magnetic axis tilted about 7° to the planet's rotation axis. These results suggest that this field is caused by weak currents in an iron core like the earth's core. A large iron core also accounts

Figure 8-5. *The Caloris basin on Mercury. The center of the basin lies in shadow out of the frame to the left. The left half of the frame is dominated by curved cliffs and fractures surrounding the impact site. The photo from top to bottom covers an area of about 1300 km; the cliffs are believed to be about 2 km (6000 ft) high. (NASA.)*

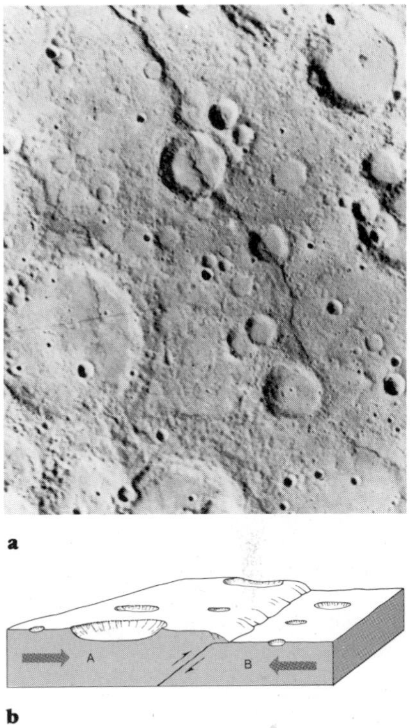

Figure 8-6. *(a) A heavily cratered portion of Mercury resembles the lunar uplands. The cliff running from the upper left to the lower right, called Discovery Scarp, may be a fault caused by contraction of the planet following early heating and volcanism. The largest crater (left center) is about 125 km across. (NASA, Mariner 10.) (b) Schematic diagram showing probable geometry of Discovery Scarp.*

for Mercury's high mean density (5500 kg/m^3), high compared to that of the moon (3300 kg/m^3), which lacks much of an iron core.

Mercury's landscape probably superficially resembles the lunar landscape: Rolling, dust-covered hills have been eroded by eons of bombarding by meteorites and covered by a regolith. Dusty lava plains and fault-cliffs remind us of violent ancient volcanism. Fresh impact craters might display rugged boulders and outcrops of craggy rocks. As shown in Figure 8-7, the sky is black, dominated during the nearly 88 d period of sunlight by a sun looking about 2½ times bigger in angular size than the sun in the earth's sky. Besides the sun, two jewel-like planets, much brighter than any stars, occasionally dominate the sky. These are yellowish-white Venus and the bluish earth.

The Planet Venus

Entirely different from Mercury is the planet Venus. Sometimes called the earth's sister because of its similarities to the earth,

Figure 8-7. *Landscapes on Mercury probably resemble those on the moon. In this imaginary view, the sun, which would appear 2½ times bigger than from the earth, is hidden behind a rock. The solar corona, or outer atmosphere of the sun (see Chapter 13), dominates the sky. In the upper right, Venus and the earth are seen in conjunction. Zodiacal light (see Chapter 11) stretches to the upper right. (Painting by the author.)*

Venus has about 95% of the earth's diameter (see Color Plate 8) and about 82% of its mass. Venus has no moons. Venus can approach the earth closer than any other planet. Nonetheless, it remained very mysterious until space probes could be sent there, because it is completely obscured by the clouds. The brilliant yellowish-white, nearly blank cloud surface is the first feature to impress a telescopic observer. The cloud layer prevented telescopic observers from seeing any surface detail or even measuring Venus' rotation.

Venus turned out to have unusual rotation, unlike that of Mercury, Earth, or Mars. Those planets all have **prograde rotation**—spin from west to east. Venus has **retrograde rotation**—spin from east to west (Figure 8-8). Venus' spin is also unusual in being very slow, taking 243 d to make a complete turn on its axis. These properties were discovered in 1962, when radar signals were first bounced off the planet. The cause of the unusual reverse spin may involve tidal forces between the earth and Venus, or an ancient collision with a body larger than our moon, of which there may have been many in the early solar system.

VENUS' INFERNAL ATMOSPHERE

The first proof that Venus has an atmosphere came as long ago as 1761, when the Russian scientist M. Lomonosov observed the backlit atmosphere extending around the disk, as shown in Figure 8-9. This phenomenon occurs when Venus is approximately between the earth and the sun, so that sunlight backlights the hazy upper atmosphere of Venus, revealing the haze layer ringing the planet.

The Venusian[1] atmosphere intrigued scientists for two centuries after it was discovered. What caused this whitish shroud? What kind of planetary surface did it hide? In 1928 the American astronomer Frank Ross photographed dusky patterns by using plates and film sensitive to ultraviolet. These patterns, shown in Figure 8-10, are formed by cloud layers differing in composition, particle size, or altitude.

In 1932 Mt. Wilson astronomers studied the spectrum of Venus and detected extraordinary amounts of carbon dioxide (CO_2, the same gas dissolved in carbonated soft drinks). Later data showed that Venus' atmosphere is about 96% CO_2, as seen in Table 8-1.

Still undiscovered was the composition of the opaque cloud layer and the air

---

[1]Adjectives for planets are controversial. Some writers, claiming that "Venusian" is an ugly word, use "Cytherean" (from a name of Aphrodite), which is merely confusing. "Venereal" is already preempted by other areas of human endeavor. While "Venusian" (ve-NOO-sian) has the sanctity of science fiction tradition, many astronomers use "Venerian" or the noun "Venus" as an adjective.

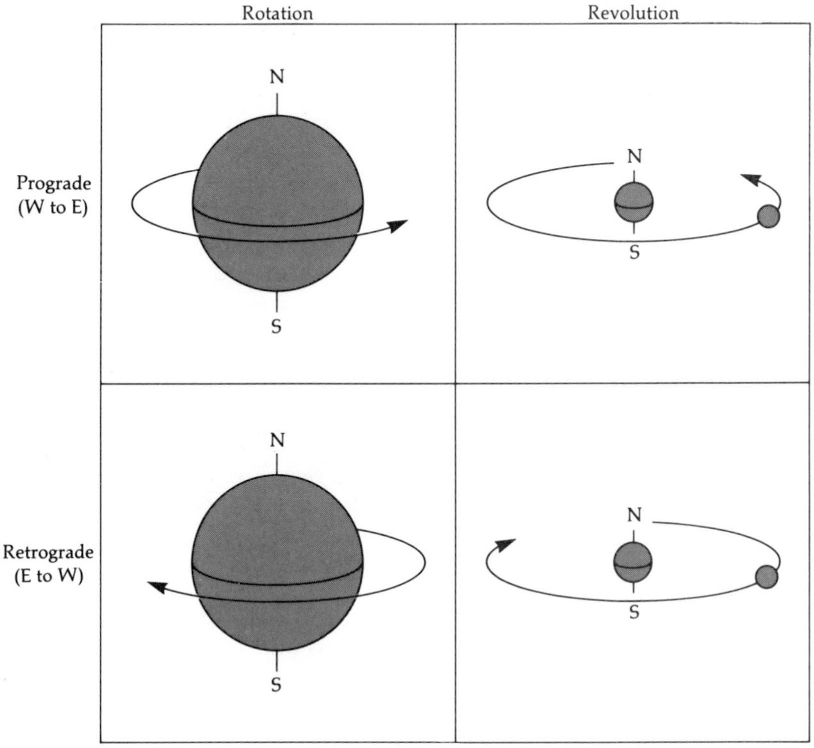

Figure 8-8. *Rotation (motion around an internal axis) and revolution (motion around an external body). Prograde motion (top) is west to east. Most planets, including the earth, have prograde rotation and revolution. The rotation of Venus is retrograde, as diagrammed in the lower left figure. Remember the difference between rotation and revolution by noting that the common handgun should be called a rotator, not a revolver.*

Figure 8-9. *A telescope view of Venus while passing nearly between the earth and the sun. Backlighting by the sun illuminates the atmosphere of Venus all the way around the disk. This effect, first recorded in 1761, not only proves the existence of an atmosphere but gives information on its structure. (New Mexico State University.)*

below it (Figure 8-11). In the 1940s and 1950s some writers imagined stormy clouds of water droplets, a surface swept by torrential rains, and vegetation like that of a Brazilian rain forest. Some supposed that the highly reflective clouds would shade the surface and moderate the climate in spite of Venus' closeness to the sun. However, in the 1960s, when the planet's thermal radiation was measured at far-infrared and radio wavelengths, Wien's law revealed that the lower atmosphere has a temperature of about 750 K (891°F)—hardly conducive to liquid water or life as we know it!

In 1970, **Venera** 7, a Soviet probe that was the first spacecraft to land successfully on another planet, transmitted data from the Venusian surface for 23 min. The data confirmed the high temperature and

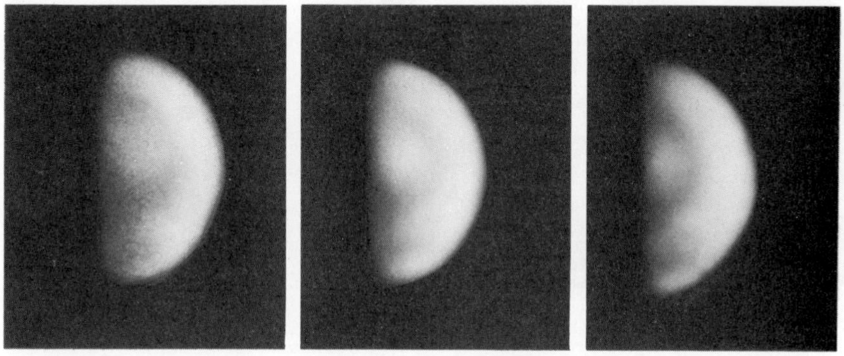

Figure 8-10. *Ultraviolet photographs of Venus made with earth-based telescopes. These pictures are among the best made with ground-based equipment and show faint cloud patterns. (Lunar and Planetary Laboratory, University of Arizona.)*

***Table 8-1.** Atmospheres of Venus and Earth*

Venus		Earth	
Gas	Percent by Volume	Gas	Percent by Volume
CO_2	96	N_2	78.1
N_2	4	O_2	20.9
CO	0.002	H_2O	0.05 to 2 (variable)
H_2O	~0.1	Ar	0.9
SO_2	0.0002	CO_2	0.03
Ar	0.000067	Ne	0.0018
CO	0.00002	He	0.0005
HCl	0.00002	CH_4	0.0002
Ne	0.000005	Kr	0.0001
HF	0.0000001	H_2	0.00005
O_2	?	N_2O	0.00005
		Xe	0.000009

Sources: Marov (1972), Oyama et al. (1979).

Note: Compositions are for near-surface conditions, with terrestrial data other than H_2O tabulated for dry conditions. CO_2 on the earth is probably increasing by 2% to 3% of the listed amount in each decade, because we are burning so much fossil fuel. This may be modifying the earth's climate.

revealed an atmospheric pressure about 90 times as great as the earth's! Instead of our 101 000 N/m^2 (14.7 pounds of force pressing on every square inch of surface) the pressure on Venus is about 9 000 000 N/m^2 (1320 lb/in.^{2}), equivalent to that endured by a diver nearly a kilometer (3000 ft) below the terrestrial ocean surface! Later Soviet spacecraft landings confirmed these results. Venus is a stranger world than most humans had imagined!

Venus was soon revealed to be stranger yet. In 1972–73 astronomers discovered that the clouds of Venus consist not of water droplets, like the earth's clouds, but rather of tiny droplets of sulfuric acid (H_2SO_4)! In 1978 probes dropped by the American spacecraft Pioneer Venus showed that the clouds lie primarily in a high layer 48 to 58 km above the surface.

In 1985, two balloons (dropped by Russian probes on their way to Halley's comet!) floated in the clouds for 46 hours, measuring hurricane-like winds (150 m.p.h.) but relatively pleasant conditions at this altitude (T = 95°F and pressure like that on Earth's surface). Venus' clouds

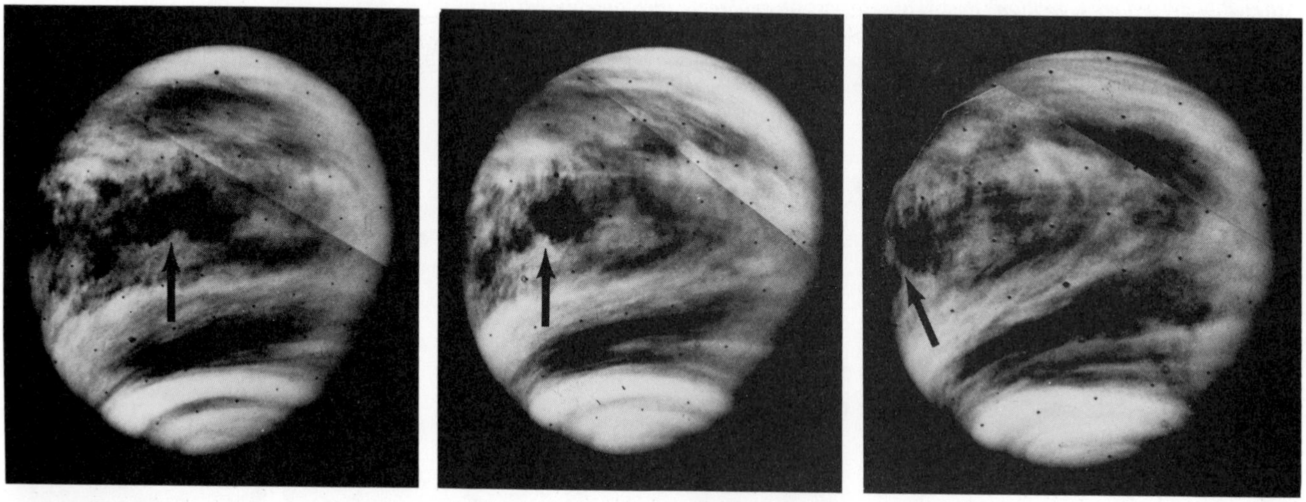

***Figure 8-11.** Three ultraviolet images of Venus made by cameras on the Mariner 10 spacecraft as it flew by Venus on February 7, 1974. The ultraviolet filters enhance cloud detail too faint to detect at visible wavelengths. The arrow pointing to a 1000-km-wide dark cloud shows motions in the atmosphere during a 14-h period. High-altitude winds make Venus' clouds circulate around the planet in about 4 d. Contrast this view with Figure 7-12, which shows more cyclonic spiral clouds on the earth, caused by Coriolis forces resulting from the earth's faster rotation (described in Chapter 3). (NASA.)*

are higher than Earth's, which are mostly less than 10 km high.

In spite of the differences, the clouds of Venus and the earth form in a similar way, in atmospheric layers where the temperature and pressure cause condensation of some relatively minor atmospheric constituent. On the earth, this constituent is H_2O, condensed either into droplets (lower clouds) or ice crystals (high cirrus clouds). On Venus, it is H_2SO_4 droplets, which begin to fall as they grow. If a droplet gets big enough, it falls out of the cloud deck, where it encounters much higher temperatures and evaporates. Venus' weird H_2SO_4 rain thus never reaches the ground. This explains why the clouds have a well-defined bottom surface, as detected by the Pioneer probes, and why the lower atmosphere and surface are clear, as found by the Russian Venera landers.

Finally, the 1978 American and Russian Venus probes made yet another startling discovery—terrific blasts of lightning play among the clouds of Venus.

THE GREENHOUSE EFFECT ON VENUS

Why is Venus so hot? Planets absorb sunlight and, following Wien's law, radiate infrared light. A planet's surface temperature is determined by the balance between the amount of visible sunlight it absorbs and the amount of infrared radiation it emits. If the solar energy absorbed each second is greater than the infrared energy radiated each second, the planet heats up. If the incoming amount is less than the outgoing amount, the planet cools. The mean, or equilibrium, temperature is reached when the two rates are equal.

The incoming rate is easily calculated as the total sunlight striking the planet minus the amount reflected. The total energy striking the planet per square meter in 1 s is called the **solar constant** for that planet. The outgoing radiation increases as the planet's surface temperature increases.

Suppose the planet has no atmosphere. Then the situation is easy to predict. The surface rocks heat up until their temperature is so high that outgoing infrared radiation equals the incoming sunlight. But if the planet has an atmosphere that absorbs some sunlight, as in Figure 7-13, the atmosphere and the surface both warm up and radiate infrared energy. Because of the nature of atmospheric gases, the outgoing infrared energy from the surface may not escape directly into space. In fact, CO_2 and H_2O gases absorb a great deal of the outgoing infrared, thus adding energy to the atmosphere and warming it even more. The warming continues until the amount of infrared escaping from the top of the atmosphere equals the amount of incoming sunlight. On Venus, the lower atmosphere has to reach about 750 K before this condition is met (Figure 8-12).

This heating is called the **greenhouse effect** because of its resemblance to the heating physics of a greenhouse. The glass panes of a greenhouse admit sunlight but block the escape of the infrared. (They also keep the warm air from escaping—an important function that, in the case of planets, is performed by gravity.) Hence the inside becomes warmer than the outside. The greenhouse effect explains why Venus is so hot: Its massive CO_2 atmosphere blocks outgoing infrared radiation. The greenhouse effect also explains why a cloudy night often stays warmer than a very clear night on the earth, since the water vapor in the cloud layer blocks outgoing infrared radiation from the cooling earth.

The greenhouse effect is of great concern to environmental scientists on Earth. Burning of fossil fuels has increased Earth's atmospheric CO_2 by perhaps 10% since 1860 (Walker, pp. 127–128). Of course, there is much less CO_2 here than on Venus, but greenhouse effects associated with this and other atmospheric changes could change climates and agricultural productivity on Earth. Venus, with its CO_2–caused greenhouse effect, thus serves as an important "natural lab" for understanding environmental change on Earth.

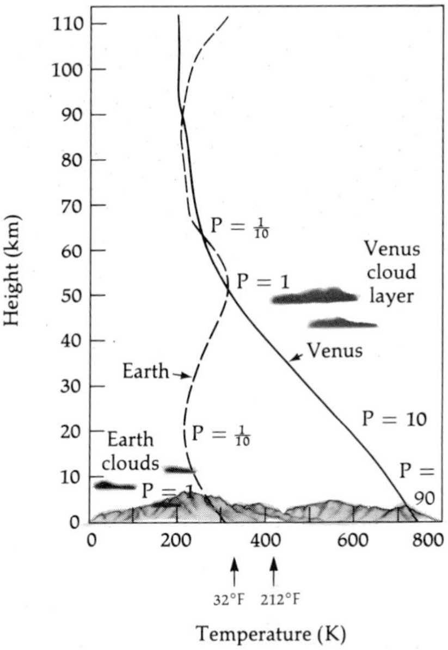

Figure 8-12. *Temperature and pressure of Venus' atmosphere (solid curve) and the earth's atmosphere (dashed). The greenhouse effect greatly heats the lower atmosphere of Venus. Pressures (P) are measured in units of the earth's surface pressure.*

Venus' Surface Conditions and Chemistry

Why should Venus' atmosphere be mostly CO_2 instead of N_2 and O_2, like the earth's? (Compare columns in Table 8-1.) The answer is clearer if we rephrase the question: Why does Earth *not* have a massive CO_2 atmosphere?

The groundwork for answering this question was laid in Chapter 3, where we saw that volcanic degassing of terrestrial planets produces secondary atmospheres. Volcanoes emit mostly H_2O and CO_2 gas. On the earth, the H_2O formed oceans, and the CO_2 dissolved in the oceans and ended up in carbonate rocks.

Venus was hotter, and most of its water evaporated and then broke down into H and O atoms. Much of the light H escaped into space, while much of the O oxidized minerals in Venus' rocks, according to 1986

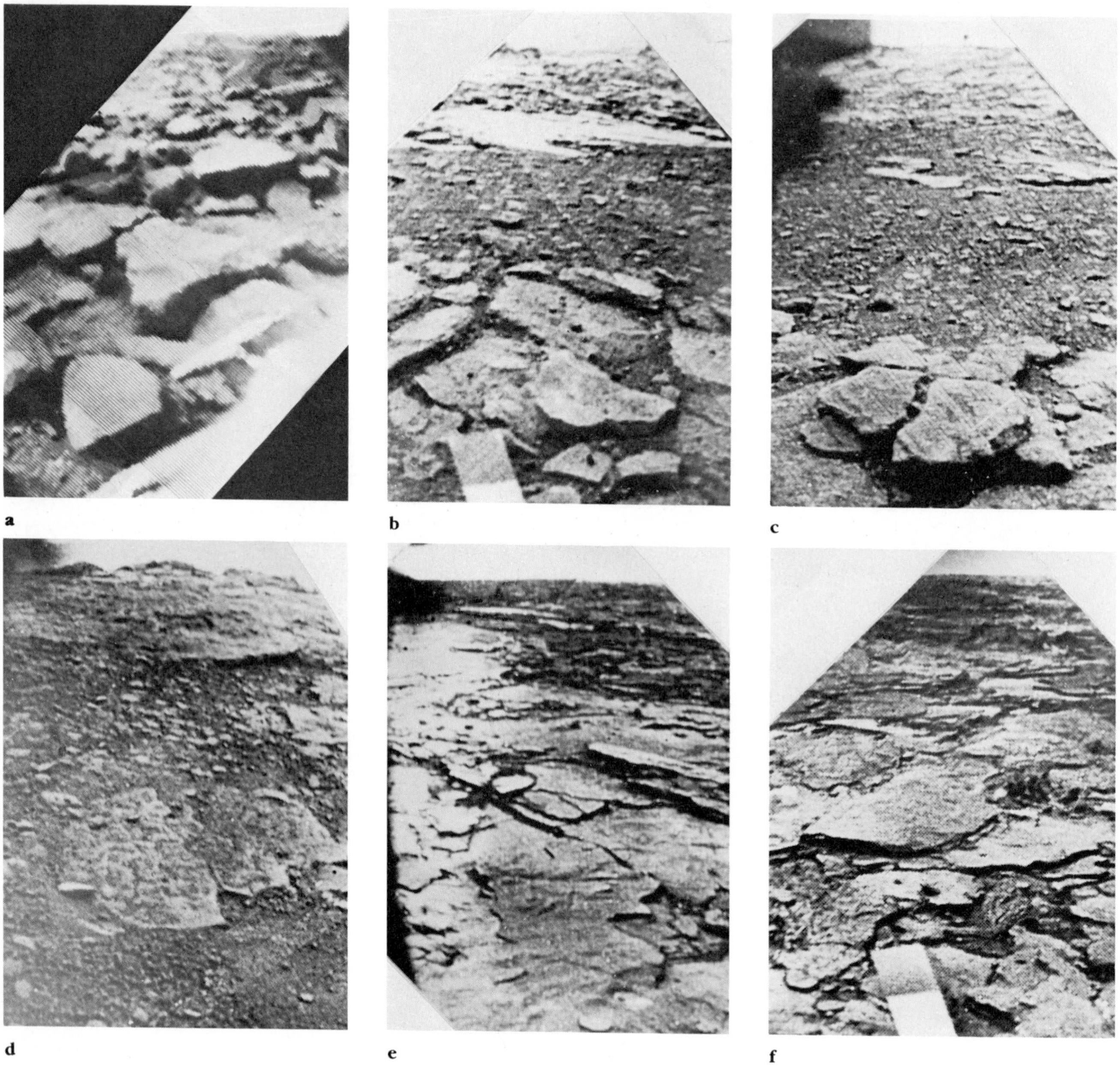

Figure 8-13. *Six landscapes on Venus: views toward the horizon (top) from Russian Venera landers. Parts of landers appear in the bottom of views* a,b, *and* f. *(*a*) First photo from the surface of Venus, showing loose boulders near Venera 9. (*b *and* c*) Boulders and gravel near Venera 13. (*d*) Platey rock outcrops and gravel near Venera 13. (*e *and* f*) Platey rock surfaces near Venera 14. (b–f Courtesy C. Florensky and A. Basilevsky, Vernadsky Institute, Moscow.)*

Figure 8-14. *Angular rocks and soil near the Venera 9 spacecraft on the surface of Venus. The white arc at bottom is part of the spacecraft. The presence of angular rocks that are neither eroded smooth nor weathered into fine soil indicates recent rock-forming or rock-exposing geologic activity, such as lava eruption or faulting.*

results by Russian and American scientists. These results suggest Venus once had enough water to make "oceans" at least 10 m (30 ft) deep; but it is not known if that much water ever actually collected on the surface at once. In the absence of oceans, Venus' CO_2 couldn't dissolve and was left as an atmospheric gas.

In support of this theory, careful inventories indicate that the earth and Venus both have about the same total content of CO_2! Venus' is in its atmosphere; the earth's is in its rocks. The unusual N_2–O_2 atmosphere that makes the higher forms of life possible on the earth seems to be a special consequence of our liquid H_2O oceans, which prevented the earth from developing the normal CO_2 atmosphere of a terrestrial planet.

LANDSCAPES ON VENUS

On October 22 and 25, 1975, Soviet spacecraft Venera 9 and 10 landed at two sites on Venus and returned the first photos of the surface. The clarity of these photos surprised many scientists who thought the surface might be obscured by murky haze. In 1982, Veneras 13 and 14 added photos from two other surface sites.

As shown in Figure 8-13, all these photos revealed barren landscapes with angular boulders, gravel, flat outcrops, and fine soil alternating at the different sites. Figure 8-14 shows a close-up of the gravel at the first site. The varied states of erosion at the sites, with angular, young-looking rocks at some sites, suggest different degrees of geologic activity. There were no signs of life. These Soviet spacecraft found only mild winds at the surface, but "jet stream" winds (up to 185 km/h) at altitudes around 40 km.

The Pioneer Venus mission in 1978 included an orbiter that mapped altitudes all over Venus by means of radar. The resulting map of Venus, shown in Figure 8-15, yields exciting scientific information that clarifies Venus' sisterlike relation to the earth. The planet is 60% covered with rolling lowland plains. The Russian Venera 8 lander indicated a granitic rock composition in this lowland, suggesting some differentiation of the crust.

The radar maps also reveal Australia-sized areas raised about 2 to 5 km above the lowland plains. These are probably continents, indicating more evolution of Venus' lithosphere than occurred on the moon, Mercury, or Mars.

In 1983, Russian scientists placed two radar-equipped probes in orbit around Venus and constructed much more detailed radar maps of certain areas on the border between the rolling plains and one of the "continents." The images showed volcanic mountains in the continental uplands, capped by clear examples of volcanic calderas on their summits. These were surrounded by a maze of complicated valleys and ridges. On the rolling plains are scattered, eroded meteorite impact craters, some exceeding 100 km diameter. From the accumulated number of these craters, the Venusian plains were inferred to be perhaps a billion years old—older and less active than most regions on the earth, but younger than most regions on the moon.

There are several indicators that volcanic activity is occurring *now* on the high continental regions of Venus! First, the Venera 9, 10, 13, and 14 landers made soil composition tests that indicated basaltic lavalike rocks and soil in these areas. Second, mountain peaks as high as 5 km (16 000 ft), and up to nearly 1000 km across, have the shapes and summit craters of active volcanoes (Figure 8-16). Third, maps of probable lightning activity suggest a greater concentration of lightning blasts around those peaks; lightning sometimes accompanies volcanic eruptions on the earth. Fourth, data show a dramatic decline by a factor ten in sulfur dioxide (SO_2) gas in Venus' atmosphere in the five years after Pioneer arrived in orbit in 1978. The 1978 value was higher than had been measured earlier from the earth. Because SO_2 is a common volcanic gas, this has been interpreted as evidence of a major volcanic eruption on Venus shortly before 1978. Similar volcanic outbursts on the earth produce increases in atmospheric SO_2, which then declines due to atmospheric reactions. These observations hint, but do not prove, that large volcanoes are active on Venus.

The data about Venus' surface tantalize scientists by suggesting geologic comparisons with the earth. Is Venus really Earth's twin in geologic evolution, as well as size?

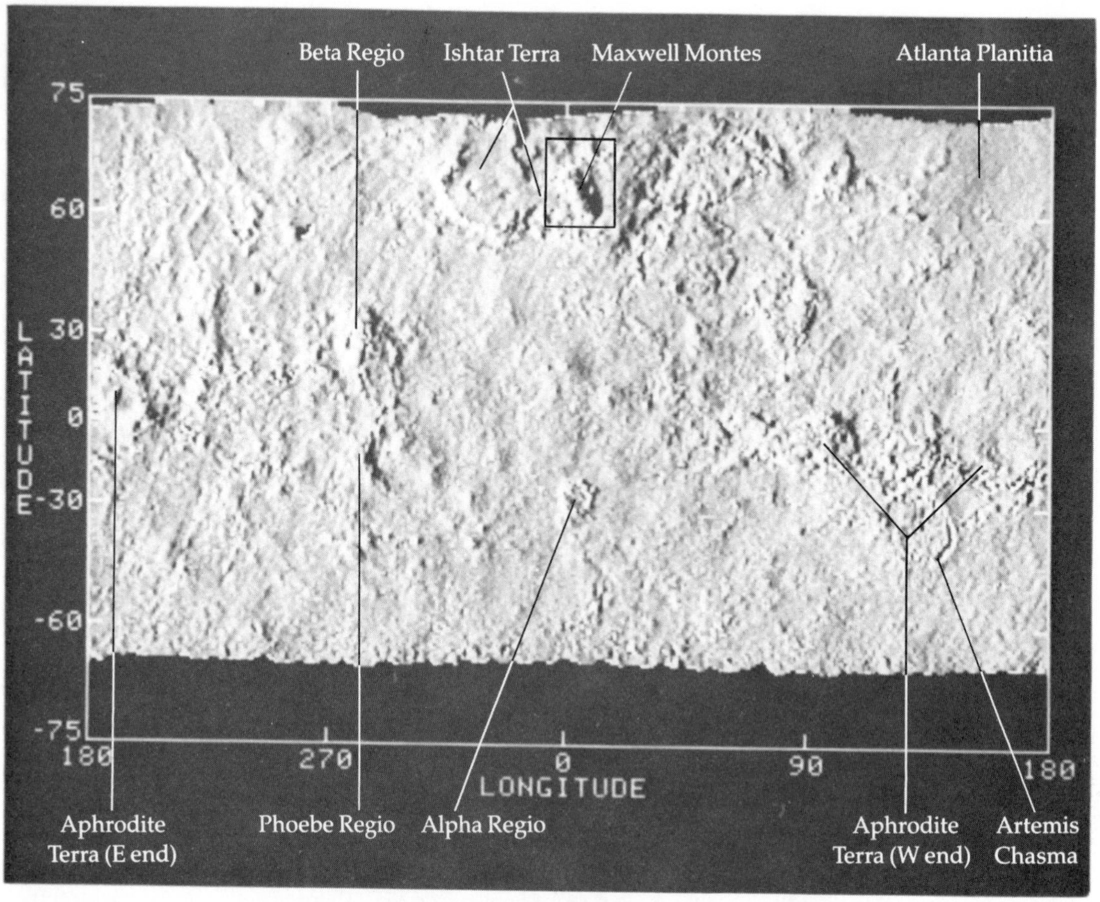

Figure 8-15. *Radar map of the surface of Venus. The relief can best be seen by imagining this as a wrinkled sheet illuminated from the upper right. Box shows location of detailed radar image in Figure 8-16. (NASA, Pioneer radar image courtesy G. Pettengill and H. Masursky.)*

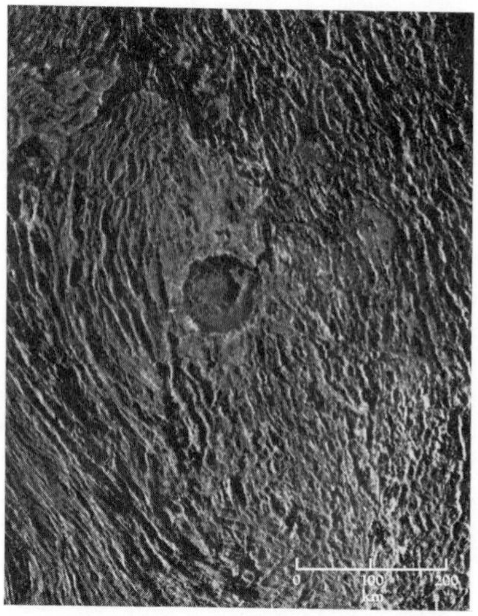

Figure 8-16. *Radar image of the highest mountain summit on Venus, the 11.5-km-high (38 000-ft) Maxwell Montes. The image shows a 100-km-wide volcanic caldera on the summit, flanked by a complex pattern of ridges interpreted as due to tectonic compression and consequent folding. Image was constructed from radar that mapped the surface below the clouds from Russian orbiters Venera 15 and 16 in 1983–84. (USSR Academy of Sciences.)*

Apparently, Venus has not produced the earth's full-fledged plate tectonics or massive granitic continents. But detailed Russian radar images, showing details as small as a few kilometers, reveal belts of folded mountains something like the Appalachians (Figure 8-16). Venus' slightly smaller interior may have provided just enough heat to start tectonic compression and stretching, without providing such active motions as on the earth.

Exploration of Venus thus provides a link between Earth and the smaller planets that may clarify geologic and atmospheric processes on Earth. The possibility of giant volcanic eruptions beneath the clouds hints at dramatic scenery we have not yet witnessed (Figure 8-17).

***Figure 8-17.** Imaginary scene on the surface of Venus, illustrating some recently discovered features. Landscape includes rocky debris, a volcanic cone (possibly active?) and background lightning. (Painting by David Egge.)*

Because Venus is named for the goddess of beauty and femininity, and is represented by the biological symbol for "female," scientists have named most of its surface features after mythical or real women. The largest continent (about half the size of Africa) is called Aphrodite Terra. Other features include Ishtar Terra, with its volcanic mountains, and a crater named Eve, which marks the zero meridian on maps.

Undiscovered Planets Near the Sun?

In the 1860s and 1870s, some observers thought they had seen a small planet, about half the moon's size, inside Mercury's orbit. They named it Vulcan. But later observations showed it did not exist. Comets sometimes pass through this region on their journeys around the sun, but no permanent worlds are known there. In the 1980s, Tucson astronomers searched for even smaller bodies (nicknamed Vulcanoids) orbiting in this region, but found none.

A concept in some science fiction and UFO literature is that an Earthlike planet could exist in Earth's orbit, but on the opposite side of the sun, where it could not be seen from Earth. Scientific study of this idea (Duncombe, 1969) showed that the hypothetical planet would be disturbed by other planets' attractions and move far enough out of the antisolar point to be detected; thus it was inferred not to exist. Subsequent spacecraft exploration also confirms that no such world exists.

Why Do Some Planets Lack Atmospheres?

Our discussion of the earth in Chapter 3 showed that the planets probably formed with an initial gas concentration called a *primitive atmosphere*, which changed to a *secondary atmosphere* as new gases were added by outgassing. Why, then, do some planets lack atmospheres while other planets have dense ones? The explanation comes from three principles that govern the motions of gas molecules in atmospheres:

1. The higher the temperature is, the higher the average speed of the molecules.

2. The lighter the molecules are, the higher their average speeds. Light gases like hydrogen and helium have faster average speeds than heavier gases such as oxygen, nitrogen, carbon dioxide, or water vapor.

3. The bigger the planet is, the higher the speed needed for a molecule to escape into space.

If you could heat a planet's atmosphere, more and more molecules would move faster than escape velocity, and fast-moving molecules moving upward near the top of the atmosphere would shoot out into space, never to return. First hydrogen, then helium, and then heavier gases would leak away into space. Cold, massive planets are most likely to retain all the gases of their primitive and secondary atmospheres; hot, small planets (with weak gravity and low escape velocity) are most likely to lose all their gases. Calculations based on these principles show that planets as small as Mercury and the moon have lost virtually all of their gases. Venus and the earth have lost most of their hydrogen and helium but have kept heavier gases.

These principles help explain why Venus could rapidly have lost its water if water molecules broke into hydrogen and oxygen atoms: The light hydrogen atoms would rapidly escape into space. (Oxygen atoms would oxidize surface rocks and thus be removed from the atmosphere as well.)

Summary

A comparison of the planets can help explain planetary phenomena, particularly on the earth. Study of planets, their origins, and their development has come to be called **planetology**. **Comparative planetology** is the comparison of different planets to understand what makes them alike or unlike.

Only two decades ago, knowledge of phenomena on the earth, such as rock chemistry, weather circulation, and mountain building, had to be derived from knowledge found on the earth alone. Today we can gain insights from other planets, treating them as special "laboratory examples" of other conditions. Venus, for instance, is an earth-sized planet nearer the sun with a different rotation rate. From it we learn that an Earth without oceans could have retained much more CO_2 gas in its atmosphere, instead of in rocks. Venus confirms that atmospheric CO_2 causes atmospheric heating through a strong greenhouse effect.

From Venus we have also learned that a slower rotating earth would have more linear weather circulation and less developed cyclonic spiral systems. The moon, Mercury, and Venus show that smaller planets apparently do not develop enough internal energy to drive the plate tectonic activity that has broken and reformed the earth's original, cratered crust.

From our studies of the earth, moon, Mercury, and Venus, we can derive three general principles that will clarify phenomena of other planets as well:

1. *The largest planets are most likely to have internal geologic activity.* Internal heat is the energy source that drives geologic activity such as tectonic faulting, earthquakes, and volcanism; the larger a planet, the more radioactive minerals it contains and the more radioactivity there is to release heat. Also, the larger a planet, the better insulated the interior and the harder it is for the heat to escape. Small planets, on the other hand, cool rapidly and lose whatever heat they may have generated. The earth, unlike Mercury and the moon, has enough internal energy to drive plate tectonics.

2. *The larger a planet is, the younger its surface features are likely to be.* This principle follows from the one above. The more internal heat, the thinner the lithosphere, and the more likely it is for the lithosphere to be broken by recent geologic activity. Small planets that cooled long ago retain very ancient surface features. The earth and probably Venus retain fewer ancient craters than Mercury and the moon.

3. *The larger and cooler a planet is, the more likely it is to have an atmosphere, and the more likely this atmosphere is to have retained its original gases.*

Concepts

orbital precession	solar constant
prograde rotation	greenhouse effect
retrograde rotation	planetology
Venera 7	comparative planetology

Problems

1. List processes of planetary evolution that have occurred on both Mercury and the moon. List processes that are unique to each, if any. Is there any indication of plate tectonic activity on Mercury? What might this indicate about heat flow and mantle convection inside Mercury?

2. Some of the best telescopic observations of Mercury and Venus are made during midday instead of after sunset or before sunrise. Why? Why are none made at midnight?

3. Which is hotter, Mercury or Venus? Why?

4. Venus and the earth are about the same size and mass, and degassing volcanoes on each probably produced both CO_2 and

H_2O gas. Why is CO_2 a major constituent of the atmosphere only on Venus, while H_2O is not a *major* constituent of the atmosphere on either planet?

5. If astronauts are to walk on both Mercury and Venus, what differences in space suit design might be needed?

6. Compare the cloud patterns on the earth and Venus (Figures 7-12, 8-11, Color Plate 8).

a. What types of features are similar?
b. These two planets have nearly the same radius, mass, and surface properties, but different rotations. What might be learned about circulation (wind patterns) of planetary atmospheres by comparing patterns of Venus and Earth?
c. Would Coriolus forces be greater or smaller on Venus? How would this affect air flow and cloud patterns?

7. Clouds of water vapor tend to absorb infrared radiation. Use this fact and the greenhouse effect to explain why desert climates on the earth have greater temperature extremes between day and night than moist climates.

8. State which of the following characteristics of Venus suggest a primitive surface (little disturbed since planet formation), and which ones an evolved surface (affected by geologic processes such as erosion, differentiation, and plate tectonics).

a. Craters
b. A large, rifted canyon
c. The lack of high mountain ranges
d. Granitic surface rocks

How would you compare the state of Venus' geologic evolution with those of the moon, Mercury, and the earth?

9. What are the chances that life as we know it exists on Venus? Why?

a. If Venus were to have a surface temperature of about 300 K and an abundance of H_2O in its clouds, how would you rate the chances for life? Why?
b. If Venus were exactly like the earth, would life necessarily exist there?

10. How close is Venus during its nearest approach to the earth (see Table 6-1)? How many times farther is this than the distance to the moon?

11. If a telescope shows Venus to be a thin crescent, where is Venus relative to the earth and the sun?

ADVANCED PROBLEMS

12. When earthbound observers can see Mercury partly illuminated, it is at a distance of about 0.9 AU, or 135 million kilometers.

a. If you used a telescope that revealed details 1 second of arc across, what would be the smallest features you could see on Mercury? (*Hint:* Use the small-angle equation.)
b. Why can virtually no detail be seen on the surface of Mercury when it is closest to the earth, about 0.61 AU away?
c. With the telescope in part a, what would be the thinnest cloud layers visible on Venus at its closest approach to the earth?

13. If Venus has a surface temperature of 750 K, at what wavelength is its strongest radiation emitted? Is this ultraviolet, visible, or infrared radiation?

14. Of the terrestrial planets, Mercury is most like the moon in size and surface character. Using the data in Table 6-1, calculate the mean density of Mercury and use this to confirm that Mercury has an iron percentage much more like the earth's than the moon's. Comment on how this might affect a lunar origin theory in which the moon is viewed as an ordinary planetary body captured into orbit around the earth.

Projects

1. Try to see Mercury. This is likely to be harder than it sounds, because Mercury is prominent only during its elongation period, which lasts only a few days and then only for an hour or less each day just after sundown or before sunrise. Determine from an almanac or an astronomy magazine when Mercury comes to evening elongation and find a site with a very clear western horizon. It is best to start a few days before the elongation so you can become familiar with the background stars in the appropriate region of the sky.

2. Determine whether Venus will be prominent in the evening or morning sky during the course, and observe its motions and brightness from day to day. Observe on which date it is farthest from the sun and estimate this angle.

3. Observe Venus in a telescope of at least 5-cm (2-in.) aperture on several dates a few weeks apart. Observe and sketch the changes in phase, and explain them in terms of Venus' motion relative to the earth and the sun.

4. If a planetarium is convenient, arrange a demonstration of the motions of Mercury and Venus.

CHAPTER 9

MARS

On a Martian summer morning in 1976, the sun came up as usual on a rock-strewn plain. The dawn temperature was around −84°C (−120°F or 189 K), but by afternoon the air warmed to about −29°C (−20°F or 244 K). The wind was light and the rust-colored rocks lay about as they had for the last few million years. The first sign of something unusual came at about 4:11 in the afternoon when a tiny, white, starlike object appeared high in the dusky red Martian sky. It was the 16-m (53-ft) diameter white parachute of the Viking 1 landing craft.

In a few moments, the contraption would have been clearly visible from the surface. At an altitude of 1.2 km (4000 ft) the lander's three engines fired with a pale, transparent flame. Within seconds the parachute cut loose and drifted away. Slowed by its rocket engines, the spacecraft dropped for another 40 s or so. As it dropped the last few meters, reddish dust swirled into the air. The first of three lander legs hit the ground about as hard as you would if you jumped off a chair, and the jolt automatically switched off the engines. As the ungainly spacecraft bumped to rest, Viking 1 became the first human-built machine to gather data on the surface of Mars.

Until the instant that Viking's cameras clicked on, no one knew what existed on the surface of Mars; many scientists expected plants or other life forms. As might be hoped for any successful exploration, Viking's voyage confirmed some existing ideas, but forced surprising revisions of some other ideas.

Mars as Seen with Earth-Based Telescopes

At its closest, Mars comes within about 56 million kilometers (35 million miles) of the earth—closer than any other planet but Venus. When it is that close, a telescope of only a few centimeters' aperture will show features on its reddish surface, including polar ice fields, clouds and dusky markings (see Color Plate 8)—features not unlike those of the earth.

THE MARKINGS OF MARS

First to see Mars through a telescope was probably Galileo, who wrote in 1610 that he could see its disk and phases. These observations indicated that it was a spherical world illuminated by the sun. When telescopes improved, observers began mapping dusky and bright patches that we can still recognize today. Dutch physicist Christian Huygens first clearly sketched these markings in 1659 (Figure 9-1), and French–Italian observer Giovanni Domenico Cassini tracked them a few years later to determine that **Mars' rotation period** is 24^h37^m, only a bit longer than the earth's.

SEASONAL CHANGES ON MARS

Mars has seasons just like the earth, though each season lasts about twice as long since the Martian year is nearly twice ours. Telescopic observations in the 1700s and 1800s (Figures 9-2 and 9-3) revealed **seasonal changes in the features of Mars**. In the Martian hemisphere experiencing summer, the bright, white polar cap shrinks away and may disappear from view, while the dusky markings darken and grow more prominent. Some early observers mistakenly thought the dark areas were oceans and called them **maria**, just as on the moon. Brighter, orange areas came to be called **deserts**, a term that proved to be more appropriate.

While retaining roughly constant shapes, the markings also change slightly from year to year, as shown in Figure 9-4. Once these changes were established, many observers thought that the dark areas were regions of vegetation, perhaps losing their leaves in winter and turning dark

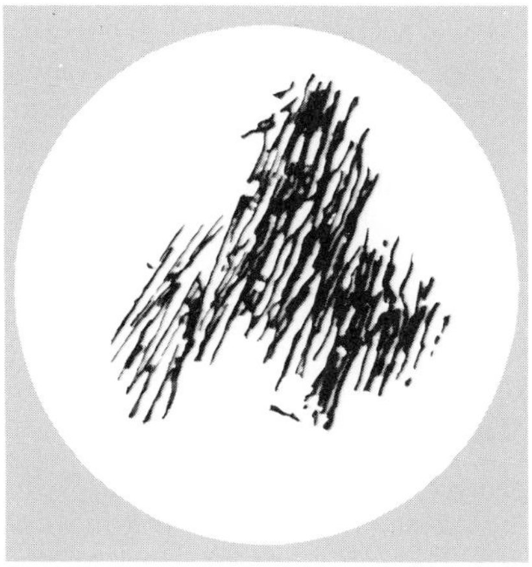

Figure 9-1. *One of the earliest known sketches of Mars was drawn by Christian Huygens on November 28, 1659. The north-extending triangle is a dark region still visible today, known as Syrtis Major. (After Huygens.)*

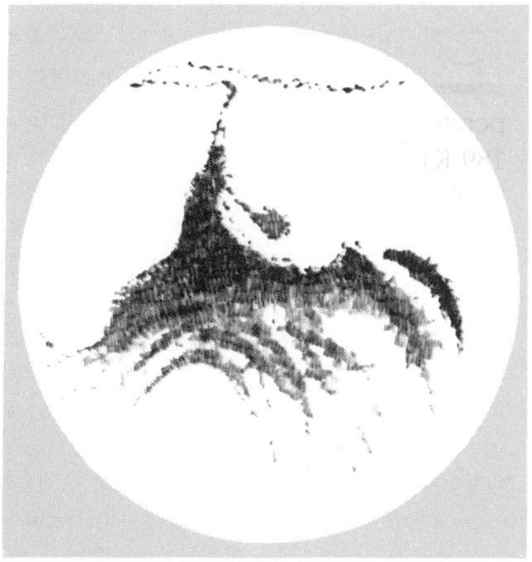

Figure 9-2. *Drawing by English observer W. R. Dawes during the 1864–65 approach of Mars. As in Figure 9-1, Syrtis Major extends northward, but Dawes recorded a streaky extension of a type later called a "canal." Clouds covering the north polar ice cap are outlined by the dotted line at the top.*

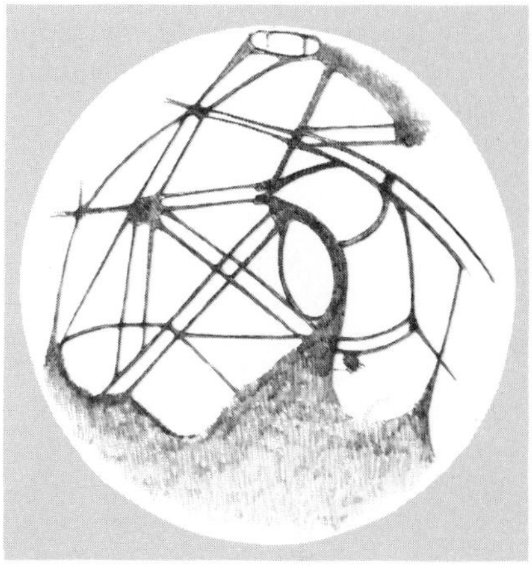

Figure 9-3. *Mars as drawn by Italian observer Giovanni Schiaparelli on June 4, 1888. As in preceding figures, Syrtis Major is the north-extending streaky dark area. Schiaparelli first popularized the "canals," shown by him here as nearly straight lines covering Mars. Spacecraft show that the "canals" do not exist in this form. The north polar ice cap is at the top.*

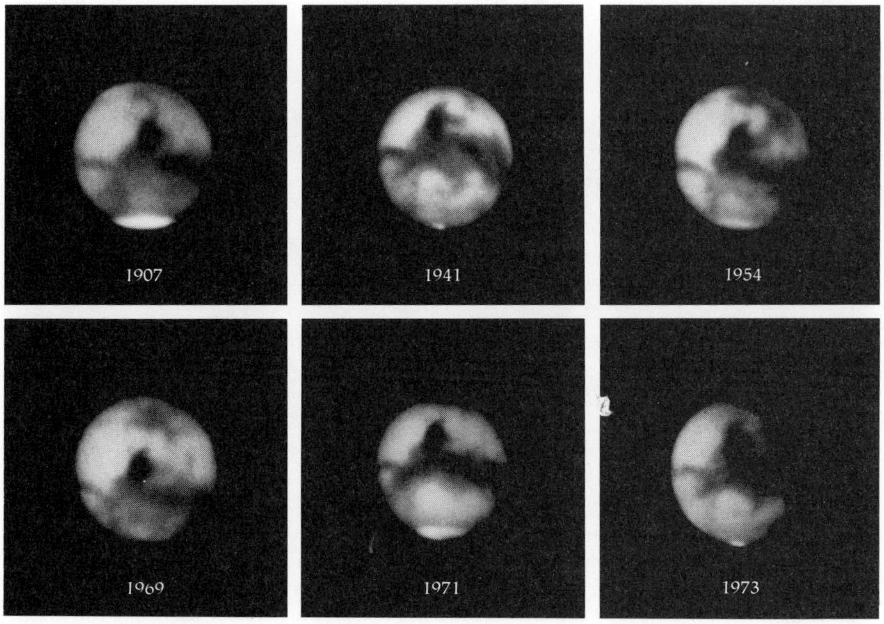

Figure 9-4. *Syrtis Major region of Mars (see Figures 9-1 to 9-3), as photographed from 1907 to 1973. Variations in the patterns of the dark markings can be seen. A dark "wing" to the right of Syrtis Major and a brightening of the nearly circular Hellas bright region south of Syrtis Major are evident in the 1941 photo. Note the large winter south polar ice cap in the 1907 view and the small summer cap in 1941 and 1973 views. (Lowell Observatory.)*

and lush in summer, as on the earth. To understand the studies of Mars, it is important to realize that many observers in the late 1800s erroneously believed that Mars had climate and vegetation like the earth. This opinion evolved even further as a result of the celebrated affair of the Martian "canals."

CANALS ON MARS?

In 1869 Father Angelo Secchi in Rome mapped streaky markings he called *canali*, maintaining the convention of naming dark areas after bodies of water. In 1877, Giovanni Schiaparelli, director of an observatory in Milan, popularized the term and drew the streaks much narrower and more linear than earlier observers had. He showed them forming a network of lines on Mars, as seen in Figure 9-3. These features came to be called **canals**.

Many other observers did not see such features. The controversy grew hotter in 1895 with a vivid description of the canals in a book, *Mars*, by Percival Lowell. Lowell was a wealthy Bostonian who founded his own observatory in the exceptionally clear air of Flagstaff, Arizona. Lowell said the canals were very sharp lines:

> *It is the systematic network of the whole that is most amazing. Each line not only goes with wonderful directness from one point to another, but at this latter spot it contrives to meet, exactly, another line which has come with like directness from quite another direction.*

Lowell concluded that the features really were canals—artificial ditches built by intelligent creatures to carry water. He pointed out that spectroscopic measurements had revealed Mars to be a dry place. He hypothesized that a once moist climate was becoming desertlike as water evaporated from the thin Martian atmosphere into space, and that a Martian civilization had turned to massive irrigation canals to carry water from their polar snow fields to the dry, warm equator.

This exciting hypothesis sparked raging debate for several decades. However, after spacecraft visits to Mars, we now know that a canal network does not exist as Lowell and some others drew it. What went wrong with the observations and the hypothesis of Martian civilization? Astronomer Carl Sagan has commented that drawings of geometric networks on Mars certainly do imply that intelligence is present, but the drawings alone don't tell at which end of the telescope the intelligence resides. Modern evidence indicates two reasons why canals existed only in observers' minds. First, streaky markings, including faulted canyons and dust deposits, do exist on Mars; seen through the shimmery atmosphere of the earth, these markings may resemble patterns of lines, as shown in Figure 9-5. Second, and more importantly, some people are more likely than others to perceive streaky patches as straight lines, especially if they already believe the lines are there. Lowell was one of these; he even drew lines on Venus.

NAMES OF MARTIAN FEATURES

Schiaparelli, a classical scholar as well as astronomer, named the larger Martian dark and light regions (as well as the illusory canals) after historical, mythological, and geographic features of his native Mediterranean area. These names, such as Hades, Arabia, and Libya, are still used for the major dark and light areas. In 1971, when the Mariner 9 spacecraft revealed actual geologic structures such as craters, mountains, and canyons (Figure 9-6, Color Plate 9)—all too small to be seen from the earth—these were assigned additional names. As on the moon, craters were named after scientists. The largest canyon complex, big enough to stretch across the United States from coast to coast, was named Valles Marineris (the Valleys of Mariner). A modern map with some of the markings and geologic structures is shown in Figure 9-7.

The Lure of Mars

On October 30, 1938, Orson Welles broadcast a realistic radio play in which listeners heard "newsmen" reporting that Martians had invaded and were laying waste to New Jersey. Thousands believed it and the resulting panic caused a national scandal. In July 1965, Mariner 4 sent back the first close-up pictures of Mars, which were widely published on front pages of newspapers; the reporting rate of UFOs shot up by a factor of 6 for several weeks. Why has Mars, of all the planets, held such a strong fascination for the public?

The answer lies in the hypotheses we have just discussed. If Copernicus knocked the earth out of the center of the universe,

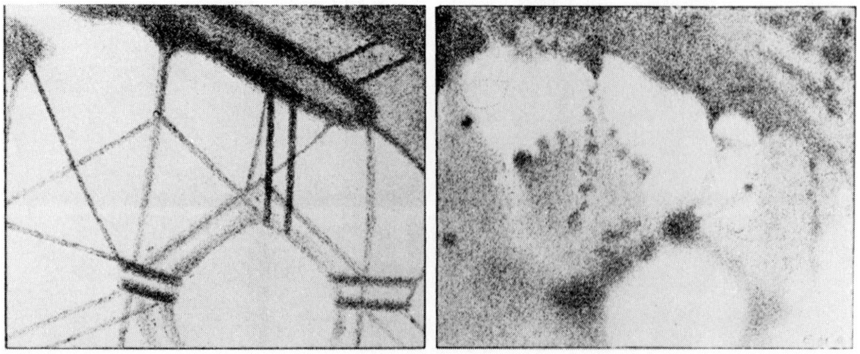

Figure 9-5. ***An explanation of the canals proposed by the French observer E. M. Antoniadi around 1930. Under mediocre observing conditions, canallike streaks may be seen in some regions of Mars, as Lowell described them (left). During moments of excellent atmospheric stability, observers with large telescopes find that these streaks break apart into a complex pattern of mottling and dark patches. The canals are a product of poor observing conditions combined with the tendency of the eye to connect dots into lines.***

Figure 9-6. *A portion of Mars, displaying a long canyon (bottom) whose length is comparable to the width of the entire United States. Its tributary canyons match our Grand Canyon in scale. Large, dark volcanic mountains with summit craters are prominent (left). (NASA Viking mosaic; Jet Propulsion Laboratory; courtesy Conway Snyder.)*

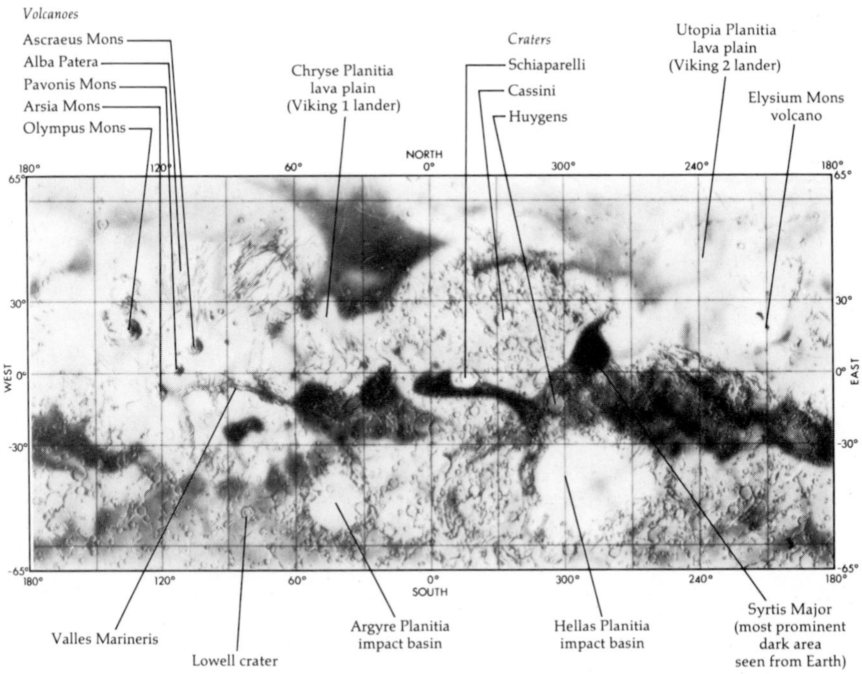

Figure 9-7. *A map of Mars. Dark shadings are the somewhat changeable markings visible from the earth, probably associated with windblown dust. Topographic features such as craters and volcanoes are too small to see from the earth. The names of some features are given. (Base map courtesy R. M. Batson, U.S. Geological Survey.)*

Lowell and the Victorians made people realize that the earth's civilization might not be the only one around. Giordano Bruno and others in the 1600s and 1700s had already suggested that other worlds might be inhabited—an idea known as the *plurality of worlds*; and Darwin's theory of evolution, published in 1859, made more plausible the idea that other planets might produce totally alien species specially adapted to their environments.

The idea electrified leading thinkers. Tennyson and other poets wrote about it. In 1898, H. G. Wells published *The War of the Worlds*, in which Martians invade the earth to escape their dying planet. Wells commented:

> *The Tasmanians, in spite of their human likeness, were entirely swept out of existence in a war of extermination waged by European immigrants. . . . Are we such apostles of mercy as to complain if the Martians warred in the same spirit?*

There was even a UFO scare in the 1890s that produced reports of Martian spaceships (described as looking like the first dirigibles, which were then flying).

Between the days of Lowell and the first space flights to Mars, several generations of readers grew up on stories by Edgar Rice Burroughs, Ray Bradbury, and others, which pictured a Lowellian Mars with remnants of a dying civilization holding out in nearly deserted cities on a dying, drying planet. While incorrect in many details, these blends of theory and fancy fostered a valid excitement about whether life might really exist on Mars and helped to encourage interest in actual voyages to the red planet. Yet, by the time of the first Martian voyages, earth-based measurements had already revealed that Mars had colder, thinner, drier air than had been thought. Could such a planet support advanced life forms? Microbes? Or no life at all?

Voyages to the Surface of Mars

The first three human-made devices to reach the surface of Mars were unsuccessful Russian probes. Mars 2 crashed in November 1971; Mars 3 landed in December 1971 but failed after 20 s on the surface; another probe sent back data while parachuting through the atmosphere in 1974 but failed moments before touchdown. Causes of the failures may have been design problems or hostile Martian conditions; a dust storm was raging during the 1971 landings.

The first successful landing on Mars was by the **Viking 1** spacecraft, which touched down July 20, 1976 (seven years to the day after the first human landing on the moon). It was followed on September 3, 1976, by a duplicate spacecraft, **Viking 2**. Both landings were on plains that looked relatively smooth from orbit but turned out to be rock strewn, as seen in Figures 9-8 and 9-9. Boulders as wide as 3 m were photographed among dunes near Lander 1 (Figure 9-10). Missing were the Martians, deserted cities, canals, or the strange vegetation imagined by early writers.

Mars turned out to be a desolate, cold, yet beautiful desert (see Color Plates 11 to 13). The reddish color of Mars was vividly shown by color photos from Viking

a

b

Figure 9-8. *Landscapes of Mars photographed by Viking landers.* (a) *Viking 1 landed in a region of rolling dunes and scattered basalt boulders.* (b) *Viking 2 landed on a plain littered with even more basalt boulders. (NASA.)*

Figure 9-9. *First close-up photo of Martian soil was sent from Viking 1 lander on July 20, 1976. Fine dust and rock chips are shown. Landing leg is at right. The landing dislodged soil, which settled in the center of the concave pad at end of the leg. The larger chips are about 5 cm (2 in.) across. The larger soil shows no signs of life. (NASA.)*

Figure 9-10. *The largest boulder photographed by Viking landers lies only 8 m (20 ft) from Viking 1. Landing on this rock would probably have destroyed the spacecraft. The boulder is about 1 by 3 m (3 by 10 ft), dark gray, probably volcanic, and covered at the top with a thin mantle of reddish dust, which may have filtered down out of the atmosphere after the last dust storm. Sand dunes surround the boulder. (NASA.)*

landers. Though some rocks appeared dark gray, like terrestrial lavas, most rocks and soil particles were covered with a coating of rustlike, reddish iron oxide minerals. Similar iron minerals are what give terrestrial deserts their familiar red-to-yellow coloration, especially when moisture is present only occasionally. Viking scientists were surprised to find the daytime sky of Mars reddish-tan instead of blue. The sky color is caused by much fine red dust stirred from the surface into the air by winds (see Color Plate 12, top).

THE MARTIAN CLIMATE

During the weeks after landing in the Martian summer, **air temperatures** at the two Viking sites ranged from nighttime lows around 187 K (−123°F) to afternoon highs around 244 K (−20°F). Temperatures of the soil, which absorbs more sunlight than the air, exceed freezing (273 K, or 32°F) on some summer afternoons, so that any frost formed at night near the surface can melt and produce moisture or water vapor. Winds at the two sites were usually less than 17 kph, with gusts exceeding 50 kph. However, much higher winds are believed to occur at certain seasons, raising clouds of dust that can be observed from the earth. **Air pressure** at each site was only about 0.7% that on the earth.

The Martian polar caps give vivid evidence that important gases freeze out of the Martian atmosphere. On the hemisphere that is having summer there is a small, permanent cap of frozen water, because even the summer temperature is below the freezing point of water. It is not known how thick the ice deposit is or how much water it contains. During Martian winter, temperatures plunge below the 146 K (−197°F) level at which carbon dioxide clouds (Color Plate 9) form, and carbon dioxide snow, or "dry ice," accumulates on the polar ground to form a large polar cap that eventually shrinks during Martian spring. As seen in Figure 9-11 and Color Plate 13, a thin and scenic snow or frost layer accumulated on the ground around the Viking 2 lander at latitude 48°N during Martian winter, but not at the Viking 1 site, at 22°N.

a

b

Figure 9-11. *Summer and winter at the Viking 2 landing site on Mars. (a) A rock-strewn plain as it appears during most of the 687-d Martian year. (b) For about 100 d during Martian winter, a partial layer of H_2O snow covers the same area. Snow apparently forms during cold nights as a mixture of CO_2 and H_2O ice, but the CO_2 component sublimes into the atmosphere when warmed by the daytime sun. (NASA, Viking 2.)*

ROCK TYPES

Martian rocks at the two sites appear to be fragments of lava flows, as judged by their textures and colors and by the chemistry of the associated soil. Proportions of elements measured in the soil resemble those for soils derived from basaltic lavas on the earth and the moon. A minor component in the soil is water, about 1%. This water exists not as a liquid or ice, but as H_2O molecules in the crystal structure of the rock particles. This water was released when the rocks were heated to around 703 K (430°C or 800°F) during automatic experiments inside the Viking spacecraft. This water, which may be a remnant of more abundant water in the past, raises the possibility that future explorers could get water by heating Martian soil. Many investigators believe that ice may be present among soil particles a few meters below the surface. This would form a layer of **permafrost**, ice-rich soil similar to that in arctic tundra regions of the earth.

If one were to search for the most Mars-like landscapes, rock types, and soil types on the earth, the best analogs would be found in extremely dry volcanic deserts. As shown in Figure 9-12, remarkably Mars-like vistas can be found in such regions as the Peruvian coastal desert, and high-altitude volcanic plains.

ATMOSPHERIC COMPOSITION

The composition of the **Martian atmosphere**, shown in Table 9-1, gives several clues about the planet's history. Like Venus, Mars has an atmosphere that is mostly carbon dioxide, probably generated chiefly by planetary degassing through volcanic activity. As noted in our discussions of the earth and Venus, volcanic gases, generated from melting of the interior rocky matter, are rich in carbon dioxide. The argon in Martian air is especially informative. Because argon is chemically inert,

Table 9-1. Composition of Martian Atmosphere

Gas	Percent Volume
CO_2 (carbon dioxide)	95
N_2 (nitrogen)	2.7
Ar (argon)	1.6
CO (carbon monoxide)	0.6
O_2 (oxygen)	0.15
H_2O (water vapor)	0.03
Kr (krypton)	Trace
Xe (xenon)	Trace
O_3 (ozone)	0.000003

Note: Amounts of gases vary slightly with season and time of day. H_2O is especially variable. Some CO_2 condenses out of the atmosphere into the winter polar cap; changing cap sizes cause small changes in the total Martian atmospheric pressure.

a

b

Figure 9-12. *The closest analogs to Martian landscapes on the earth occur in extremely arid deserts and volcanic regions. (a) Coastal desert of Peru, with weathered volcanic boulders and windblown sediments. (b) Weathered lava rubble near the arid summit of Mauna Kea volcano, Hawaii, at an altitude 4 km (13 000 ft). (c) Basalt lava boulders and windblown dust on an alluvial fan, Death Valley, California. (Photos by author)* none that was released from the planet would have combined chemically into the rocks; because it is so heavy, it could not have floated to the top of the atmosphere and escaped into space. The present amount of argon is therefore a measurement of the total amount ever degassed from the interior. Because the composition of the volcanic gases on Mars can be estimated from experience with the earth's volcanoes, the amount of argon can be used to estimate the amount of accompanying gases. These results show that a *massive* atmosphere must have been degassed in the past, perhaps enough to create a surface pressure a quarter or half that of the earth's, compared with the 0.7% of the earth's surface pressure now present. Much of this gas should have been nitrogen, carbon dioxide, oxygen, and water vapor.

In other words, Mars probably had a more extensive, earthlike atmosphere in the past. Where did the missing gases go? Meteorologists have shown that atoms and molecules of some gases, such as nitrogen, can react in the high atmosphere in such a way that atoms are accelerated out of the atmosphere or combined with the soil. This process suggests more evidence for an extensive permafrost layer under Mars' surface.

Major Geologic Structures

Data and pictures from the Viking landers have given vivid ideas of the Martian landscape. In addition, the mapping of large-scale geologic structures from orbiting spacecraft has yielded information just as intriguing.

c

MARTIAN DUST: THE MARKINGS DEMYSTIFIED

The dark markings, once thought to be vegetation, are probably caused mostly by dust deposits. Evidence for this theory includes vast dune fields (Figure 9-13), patches, and streaks associated with the leeward sides of crater rims and hills. These patches have been simulated by dust blowing over model craters and hills in wind tunnels.

Seasonal changes in contrast and shapes of markings, once thought to be evidence of vegetation, are now attributed to dust transport by season winds, especially summer dust storms. In spite of the fact that dust storms create broad changes visible from the earth, they cause only subtle changes at the local scale. Viking 1, which worked on the surface of Mars for 3.3 full Martian years (over 6 earth years) photographed only millimeter-scale deposition or erosion of dust in most areas, even though it weathered several dust storms (Arvidson et al., 1983). Some broad Martian markings, hundreds of kilometers across, may be dust deposits only millimeters thick.

MARTIAN CRATERS: CLUES TO SURFACE PROCESSES

Martian craters, such as those shown in Figure 9-14 and Color Plate 9, are interesting for three reasons. First, they are abundant and show that impacts have been common on Mars as on the other planets. Second, they show varied states of degradation, which indicate that erosive processes have been much more active on Mars than on the moon or Mercury. In some regions the craters have been obliterated by young-looking lava flows; in other regions, by massive accumulations of sediments or windblown dust hundreds of meters deep. Third, the craters give a way to estimate the ages of Martian surface features. By estimating the rate at which interplanetary debris hit Mars and created the craters, analysts have estimated that the older, heavily cratered regions are probably a few billion years, like the cratered surfaces of the moon;

but the youngest, sparsely cratered volcanoes, lava flows, and eroded surfaces may be as young as 100 million years or less.

MARTIAN VOLCANOES: CLUES TO LITHOSPHERIC STABILITY

The largest known **volcanoes** in the solar system were discovered on Mars by Mariner 9 in 1971. The highest volcano, Olympus Mons (Figure 9-15), rises about 24 km (78 000 ft) above the lower Martian deserts. In contrast, Mt. Everest is only 9 km above sea level, 13 km above the mean ocean floor, and 20 km above the greatest ocean depths. The base of Olympus Mons is about 500 km across and would nearly cover the state of Missouri. The caldera, or volcanic crater, at the summit is about 65 km across. Olympus Mons is thus several times bigger than any similar type of volcanic cone on Earth. For example, Mauna Loa reaches about 9 km above the seafloor, is about 120 km in diameter at its seafloor base, and has a summit caldera a few kilometers wide.

Nearly a dozen other large, volcanic mountains dot Mars. Most are a few hundred kilometers across at the base, and one has a summit caldera 140 km across. They are gently-sloped and clustered in

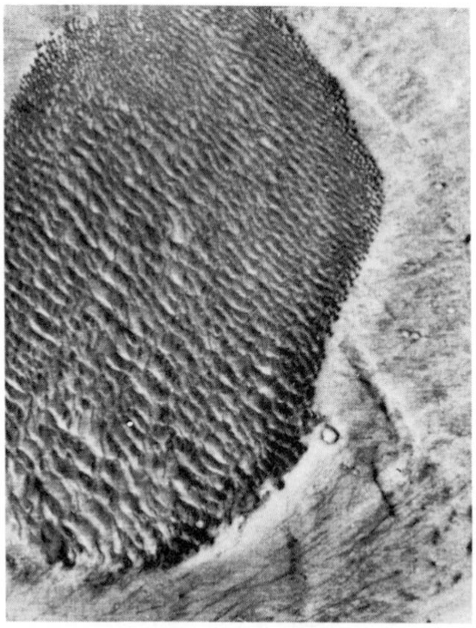

a

Figure 9-13. *Martian dunes as seen from orbit and ground level.* (a) *Telephoto orbital view reveals a dark, complex dune field on the otherwise smooth floor of a crater. Dunes are about 1–2 km apart. Such photos indicate that many of the dark and light markings may be windblown dust deposits.* (b) *A more localized dune deposit near the Viking 1 lander. (NASA.)*

b

broad volcanic regions up to 2500 km across. Geologists believe the Martian lavas were very fluid, spreading into broad flows instead of piling up in steep-sided cones, like Mt. Fuji. The volcanic regions are fresh-looking, perhaps 1 billion years old or less. These regions are surrounded by swarms of radial fractures that extend thousands of kilometers from their centers.

The largest fracture is a vast canyon system, the Valles Marineris, shown in Figure 9-16. It suggests that Mars' lithosphere is thinner than the moon's—enough so to be split by internal stresses. On the other hand, it is thicker than the earth's—enough so to resist plate tectonic crumpling.

The large volcanoes, broad domes, and fracture systems all suggest that the Martian lithosphere in these areas has been massively disturbed by lava upwelling from below. Uplift and volcanism may have been caused by rising hot currents similar to those suspected inside the earth. The Martian volcanoes have probably reached such enormous size because Mars' lithosphere was too thick to form movable plates, even though it was thin enough to split. A given spot on Mars therefore stays directly over a given ascending magma current, and the lava eruptions during many years pile up into single huge mountains. In contrast, Mauna Loa and

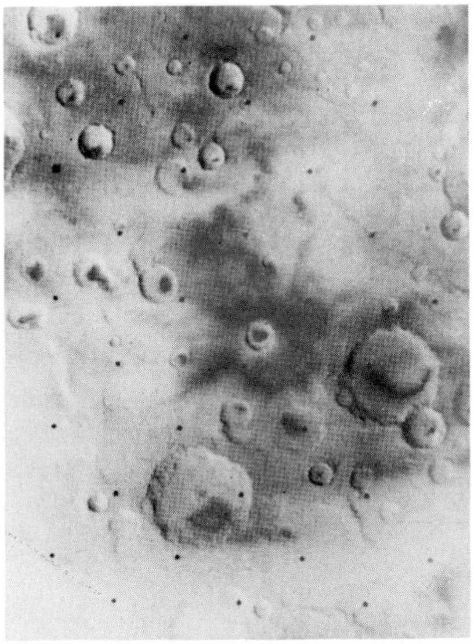

***Figure 9-14.** Older regions of Mars are fairly heavily cratered. Blowing dust often collects in dark patches in the bottoms of craters. Craters here are some tens of kilometers across. The geometric array of black dots is artificial, used for camera calibration. (NASA.)*

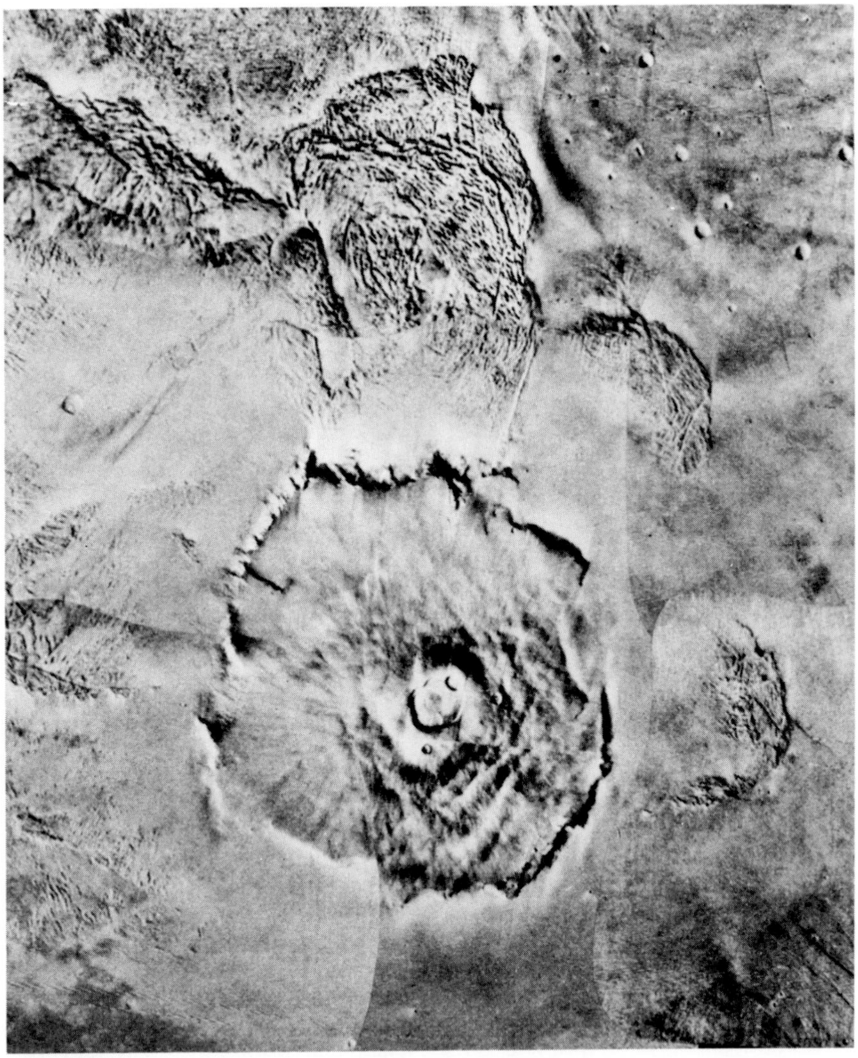

***Figure 9-15.** This mosaic of pictures shows a view looking downward onto the volcanic mountain Olympus Mons, the surrounding desert, and fractures. The mountain's base is nearly 500 km across. (NASA Viking mosaic.)*

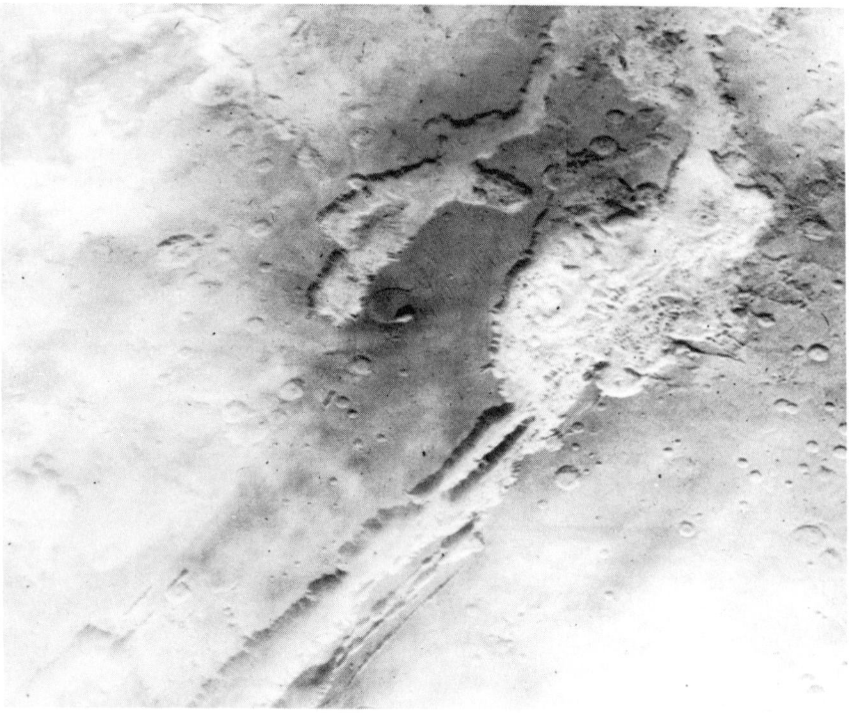

Figure 9-16. *An orbital view of a portion of the Valles Marineris canyon complex of Mars. Linear portions, comparable in scale to rift valleys like the Red Sea and the Gulf of California, may be rift valleys opened by tension in the Martian lithosphere. The canyon has been enlarged by collapses that created chaotic terrain (upper right), the source of some Martian channels. Chaotic terrain may be formed by melting of the underground permafrost and subsequent draining of water out of the region through the channels. (NASA.)*

many other terrestrial volcanoes—in fact, the whole string of Hawaiian islands—are believed to be separate lava accumulations that occurred as plates in the earth's crust drifted over a single region of intermittent mantle activity under the Pacific. A Martian volcano shows what might happen if all the Hawaiian islands were piled into one eruptive unit.

Mars confirms our general rules (Chapter 8, page 152) that large planets have more internal geologic activity and younger surfaces than small planets. Small planets lose heat faster, thus developing thicker lithospheres without the hot, convecting interiors needed to drive plate tectonics or crack lithospheres. Mars, midway in size between the earth and the moon, also falls midway between them in terms of surface geology; it has an older and less active surface than the earth, but a younger and more active surface than the moon.

MARTIAN RIVERBEDS: THE MYSTERY OF THE ANCIENT MARTIAN CLIMATE

So far Mars might sound like the moon or Mercury with a few extra volcanoes and a little air to blow the dust around. Mariner 9 shattered this conception by photographing **channels** that look like dry riverbeds, as seen in Figures 9-17 and 9-18. These channels meander in sinuous curves and often have tributaries. They get wider and deeper in the downslope direction and have sedimentary deposits on their floors. In short, they have all the features of arroyos cut by water or ice-clogged streams. Other theories of their origin—for instance, that they might have been lava flow channels—do not explain all their features.[1]

The greatest surprise of Martian exploration was to confirm that Mars is very arid, and yet at the same time to discover what appear to be riverbeds! Evidently, Mars once had flowing rivers of liquid water. The number of recent impact craters interrupting the channels indicate that the channels are perhaps 1 to 3 billion years old. The duration of flow episodes is unknown. The channels are younger than the most ancient cratered regions, but older than most volcanoes. Some channels emanate from chaotic collapsed areas believed to have formed when ice deposits melted and released water onto the surface. Others—a network of fine channels near the equator and large channels with tributaries—suggest that some water came from other sources, possibly rainfall from a once denser atmosphere, as visualized in Figure 9-19. Other signs of ancient erosion, such as degradation of craters and the buildup of layered sedimentary deposits, also suggest much more atmospheric and erosive activity in the past.

Consistent with this, Viking scientists found chemical evidence of a denser atmosphere in the past, based on chemically inert gases in the present-day Martian atmosphere. Gases such as argon are called inert because they don't combine readily with other elements. Therefore, once they are injected into the atmosphere by volcanoes, they remain there. Thus, argon and other inert gases are good indicators of total volcanic activity. By measuring the present argon content, and knowing the amounts of nitrogen and carbon dioxide that are emitted by volcanoes along with the argon, Viking scientists could estimate the total N_2 and CO_2

[1] Although a few of the major riverbeds lie near some reported positions of the once popular "canals," there is little correspondence in general. The channels do not explain the canals.

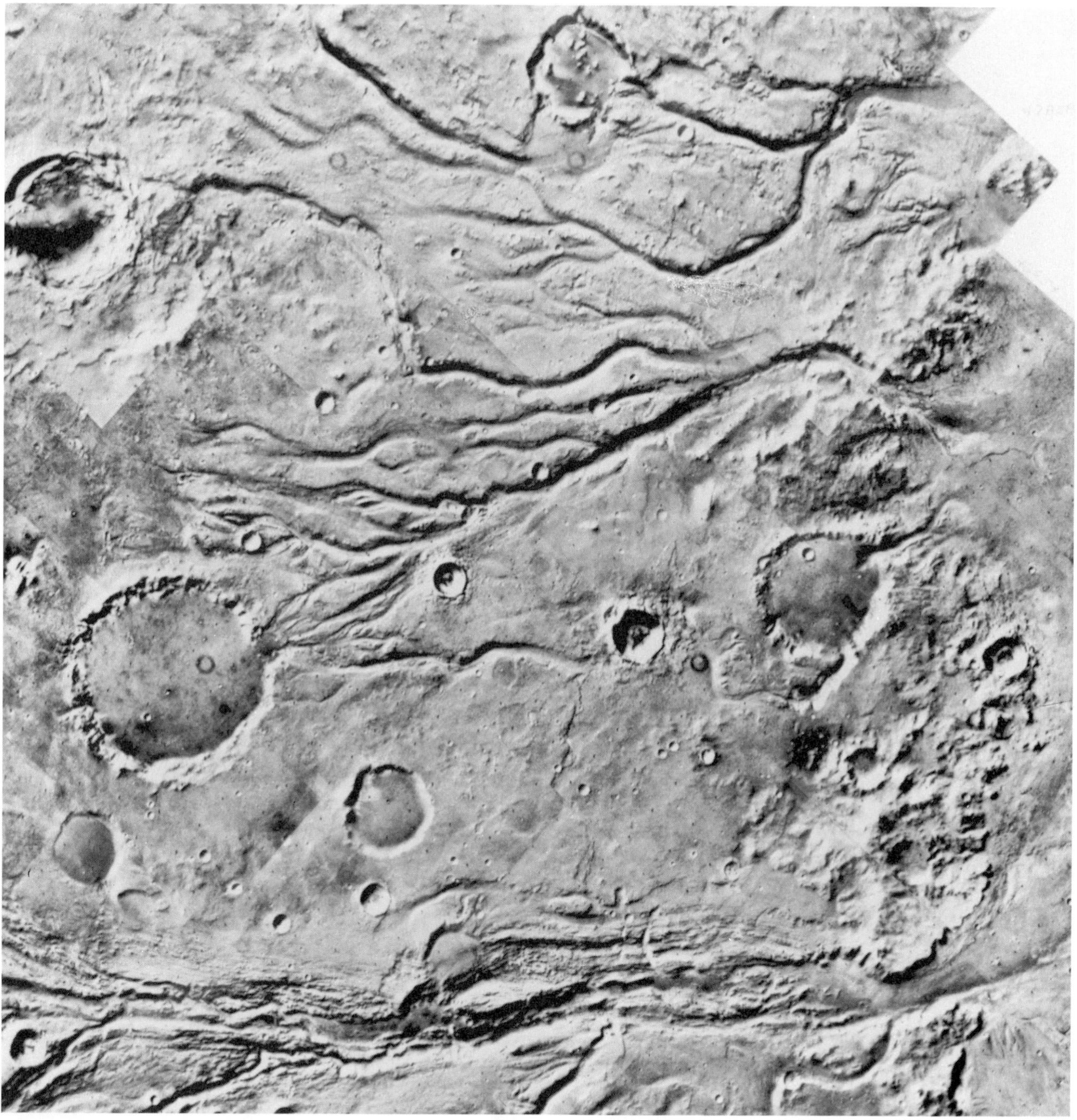

Figure 9-17. *Dramatic evidence of ancient riverbeds on Mars is found in these channels and tributary systems. The area is about 180 km (110 mi) wide and drops about 3 km in the direction of flow, from the west (left) to the east. Water apparently cut into some old craters, but predated others. (NASA, Viking photo from orbit.)*

THE UNIVERSE IN COLOR

Color Plate 9 *Orbital views of Mars. Inset: Disk of Mars showing large impact basin (lower center) and Valles Marineris canyon system (upper center). South polar cap and north polar haze are brightest features. (Viking 1 photo, 18 June 1976; NASA.) Background: Lava desert on Mars, bordered on south (bottom) by uplands cut by apparent dry riverbed erosion features. Edges of lava flows are faintly visible in the smooth desert. Largest crater has "splash" pattern believed to result from muddy ejecta involving soil and melted ice, ejected by the impact. The characteristic red color of Mars comes from rustlike iron oxide minerals in the Martian soil. Smallest craters visible are comparable to a football stadium in size. (NASA Viking mosaic, Mangala region, courtesy A. S. McEwen, U.S. Geological Survey, Flagstaff, AZ.)*

Color Plates 10 and 11 *Landscapes on four planets. Upper left: The most common view on the earth is the open ocean, covering three-fourths of the earth's surface. Upper right: Land surfaces show eroded mountains created by tectonic activity and vigorous life forms, unique in the solar system. (Photos by author; Pacific Ocean and Grand Teton National Park.) Lower left: Astronaut Harrison Schmitt on the barren lava plains of the moon. (NASA.) Lower right: Boulders in the orange deserts near the Viking 1 landing site on Mars. (NASA.)*

Color Plates 12 and 13 *Many moods of Mars. (See Chapter 9.) Left top: The Martian sky, 1977. Orange color, due to dust, gives way to bright region around the sun, out of field toward upper right. Left bottom: Sunrise over the plains of Utopia, 1978. The sun is just touching the horizon and is dimmed by haze in the lowest part of the atmosphere. Sky above the haze layer is brighter. Right: Winter on Mars, 1979. In this region of Mars, at 44°N latitude, a thin layer of mixed frozen carbon dioxide and water frost settles on the surface. The daytime sun evaporates the carbon dioxide* (CO_2) *and leaves a coat of water frost that remains on the rocks and dust for about 100 days during Martian winter. (NASA Viking 2 photos.)*

Color Plate 14 *Jupiter as seen by Voyager 1 in 1979 (see Chapter 10). Top: View from 28 million kilometers, showing the colored, turbulent cloud bands and two satellites. Orange-colored Io, covered by sulfurous lavas, is seen against Jupiter's disk. Ice-covered Europa is the bright, whitish moon to the right. Io and Europa are about the size of our moon (NASA.) Bottom: Close-up of Jupiter's Red Spot and nearby turbulent clouds. Smallest cloud details are about 95 km (55 mi) across in this view from 5 million kilometers. (NASA.)*

Color Plate 15 *A large region on Jupiter's volcanic moon, Io, computer-processed to show natural color on a Mercator-map projection. Scale can be seen by comparing features in right half with global photo, Figure 10-13. Black spots are believed to be molten sulfur lakes in volcanic calderas. Large, upside-down-heart-shaped feature (left) is created by our vertical view down through erupting plume from a major volcano. (NASA, Voyager mosaic, courtesy A. S. McEwen, U.S. Geological Survey, Flagstaff, AZ.)*

Color Plate 16 *Volcanoes of Io. (See Chapter 10.) Top: Crescent lighting of Io reveals bluish plumes being erupted out of active volcanoes on Io's surface. The plumes reach 100 km in height. These plumes were erupting when Voyagers 1 and 2 passed Io four months apart. The bluish color is probably due to scattering of sunlight in the volcanic gas, the same process that makes sunlight scattered by our atmosphere blue. (NASA Voyager 2 photo, 1979.) Bottom: Vertical view onto volcanic calderas on Io. Bright reddish sulfur flows emanate from the volcanic vents. The ribbonlike flows are 10 to 20 km wide. (NASA Voyager 1 photo, 1979.)*

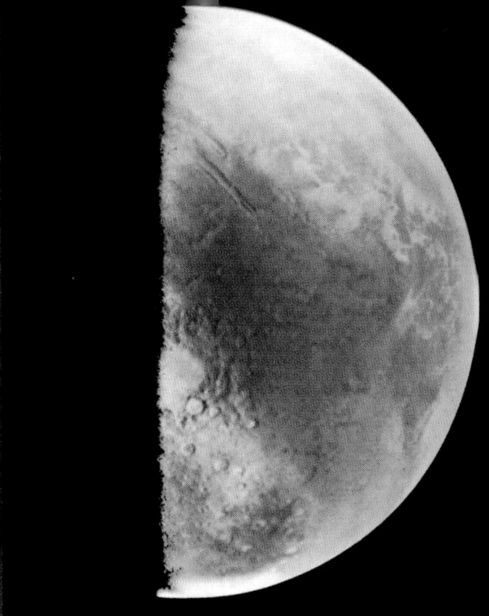

◄ Lava desert on Mars

▲ Disk of Ma

▲ The open ocean on Earth

▼ Astronaut on the lava plains of the moo

▲ Grand Teton National Park

▼ Boulders near Viking 1's landing site on Mar

▲ The Martian sky

▼ Sunrise on Mars

▲ Jupiter and two satellites

▼ Jupiter's Red Spot

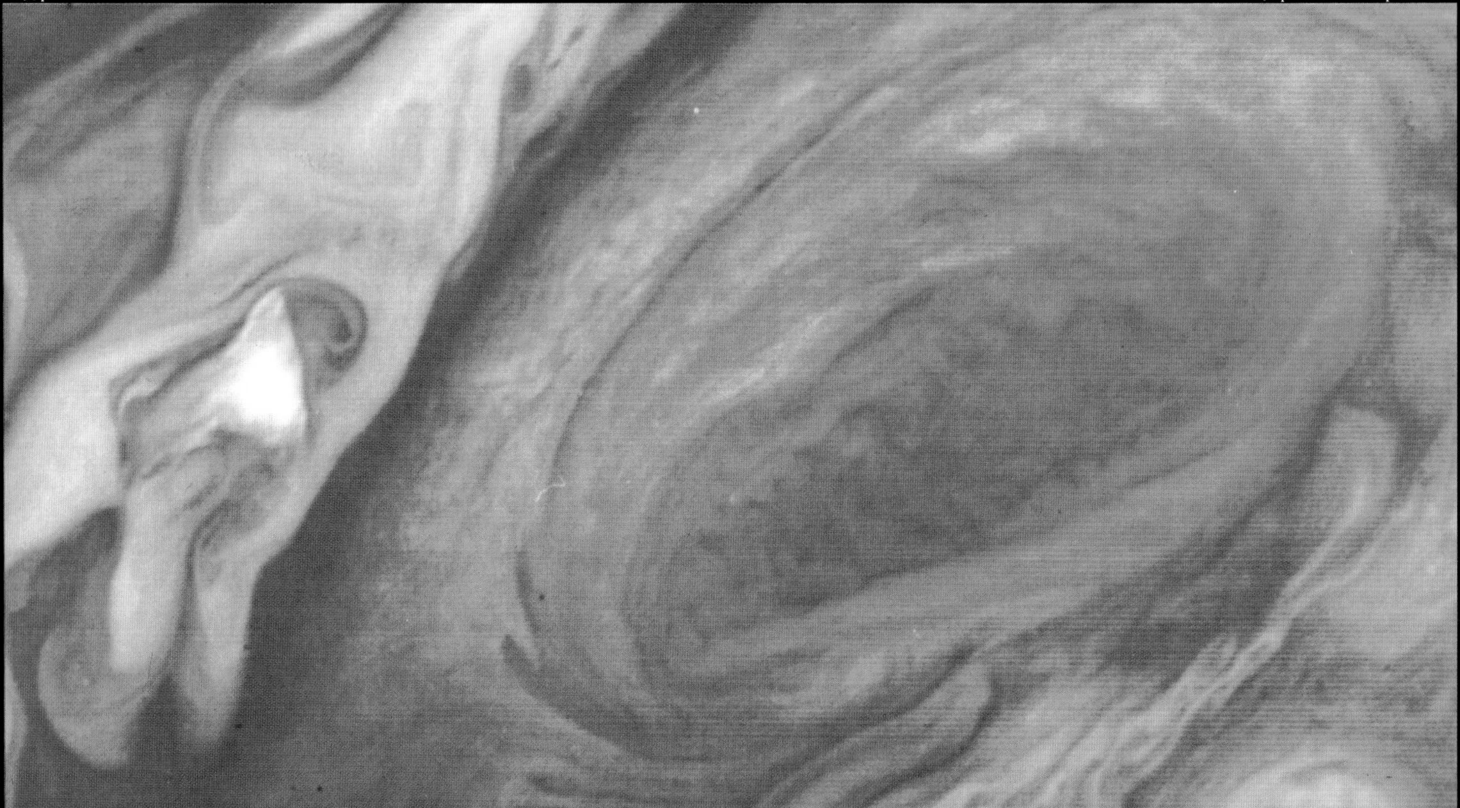

▲ Active volcanoes on Io

▼ Sulfur flows from volcanoes on Io

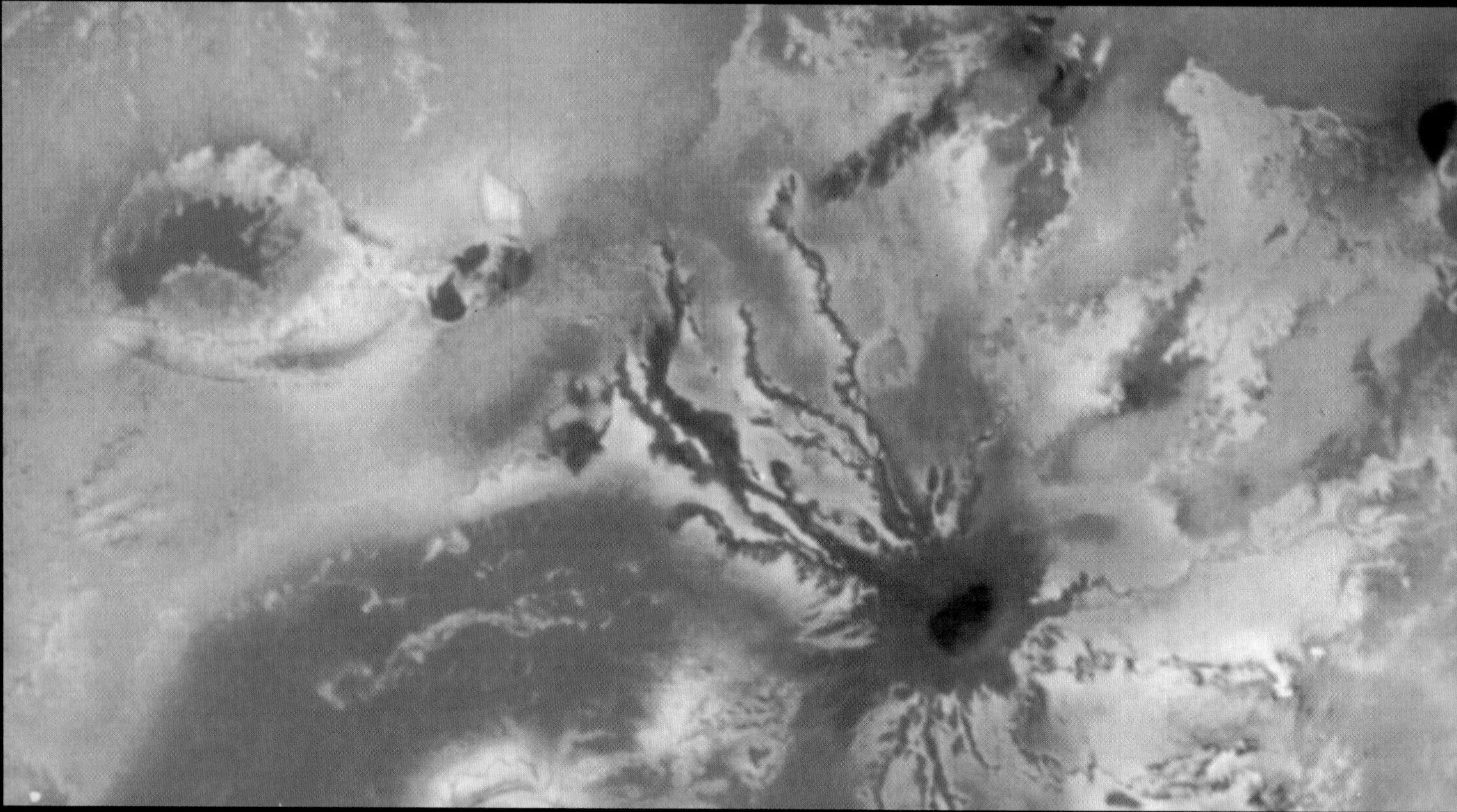

dumped into Mars' atmosphere during Martian history. They concluded that Mars once had 10 to 100 times as much of these gases as now present in the atmosphere (Owen et al., 1977; Haberle, 1986).

This leads to exciting conclusions and new questions. Although Mars' present conditions are inhospitable to evolution of life, Mars' *ancient* conditions may have been more clement.

But where have the water and atmospheric gas gone? Many Mars analysts believe that much more of the water is still on Mars. We know that much water is frozen in the permanent polar caps and that about 1% of the surface soil is chemically bound water. But much more water may still be frozen as permafrost below the surface. Some other gases may have escaped into space.

A more provocative question is, what could have caused the transition from the earlier conditions—with more liquid water, more air, and higher temperature—to the present arid, freezing conditions? Astronomers and meteorologists have combined forces to try to understand what factors could change planetary climates within intervals of a hundred million or a billion years.

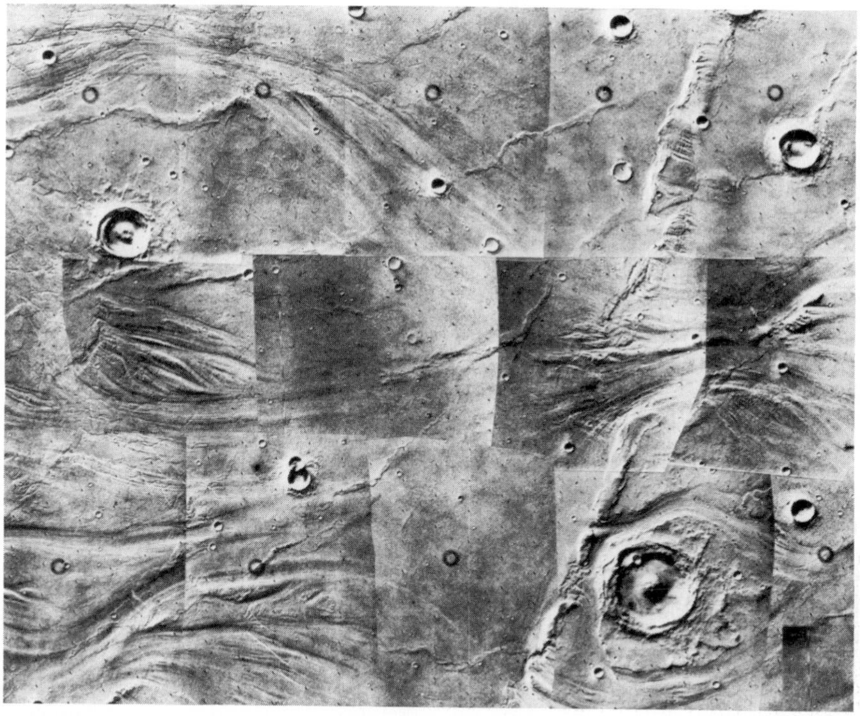

Figure 9-18. ***Viking orbiter photos were combined to show a region about 250 by 200 km (155 by 120 mi) where water has apparently flowed across the state surface. Streamlined erosion channels from left to right mark the course of flow. The background terrain is believed to be lava flows dotted by impact craters. At right, water has apparently cut through ridges formed by lava. (NASA.)***

Under present conditions, water is very unlikely to flow on Mars. It is usually frozen, and even if it warmed enough to melt, it would very rapidly evaporate into the thin Martian air (in many regions the air pressure is so low that the water would spontaneously boil away into the air). But calculations indicate that if Mars ever had more air, liquid water might have been more stable on the surface.

One possibility is that the early atmosphere (3 to 4 billion years ago) was denser and wetter but slowly dissipated into space. But some data suggest a more rapid mechanism of climate change. Cornell scientists have found that if solar radiation striking the Martian polar caps increased by only 10% to 15%, much of the frozen carbon dioxide in the cap would *sublime* (go from solid to gas), passing into the atmosphere. This denser atmosphere would make Mars warmer, carry more heat to the poles, and cause still more polar dry ice to sublime. Thus, there would be a striking feedback effect: A small initial change in polar climate could result in a large change in global climate (Sagan, Toon, and Gierasch, 1973).

These results led scientists to look for ways in which more sunlight might have reached the Martian poles in the past. One such process was revealed by calculating the changes in the rotation of Mars caused by the accumulation of lava in the volcanoes in the domed region around Olympus Mons. These calculations indicated that before that volcanism occurred, the poles of Mars might have been inclined toward the sun by as much as 45°, instead of the present 25°. This dip would mean that each summer pole would have received enough warmth to melt all the water in the permanent ice cap. The water would have circulated in the atmosphere on its way toward freezing at the winter pole. Rain might have occurred and created the observed riverbeds, or at least the water that erupted from underground sources might have lasted long enough to flow and erode the surface.

These new ideas about Martian climate history show how astronomical exploration of other planets can illuminate conditions on the earth. Everyone is familiar with the ice ages that have occurred on the earth in the last hundred thousand years, and geologic records indicate even greater climate changes in the last few hundred million years. The Martian evidence suggests that planetary climate changes are common. Some scientists have concluded that similar changes in the axial tilt of the earth, caused by known gravitational effects, fully explain the ice ages.

Figure 9-19. *Imaginary views of a Martian channel ("now" and "then"). (a) A typical channel under present conditions. (Painting by author.) (b) Reconstruction of same as it was being inundated by a catastrophic flow of water. This might have occurred during the several episodes under ancient climatic conditions, when the atmosphere was denser. Whether such scenes really occurred on Mars is still unknown. (Painting by Ron Miller.)*

Where Are the Martians?

Much of the motivation and excitement in exploring Mars has been in the search for extraterrestrial life. Discovery of life on Mars would be of major cultural importance. Just as the discoveries of the Copernican revolution showed that the earth was not the center of the solar system, **life on Mars** (or elsewhere) would show that humans are not necessarily the lords of creation. On the other hand, proof that life never evolved on Mars in spite of favorable conditions would present an exciting challenge to our present concepts of the nature of life, since biological experiments suggest (but do not prove) that life should evolve whenever conditions are suitable.

The Viking mission was specifically designed to look for life on Mars.[2] Of the five Viking experiments involved, two gave strong negative results, but three gave ambiguous results. In the first experiment, the cameras showed no signs of life. The second experiment, a soil analysis, revealed no organic molecules in the soil at either Viking site, at a sensitivity of a few parts per billion. Since **organic molecules**, or massive molecules containing carbon, are essential building blocks of life as we know it, this test strongly indicates that living organisms do not now exist in Martian soil and have not existed in the recent past.

The other three experiments were designed to look for ongoing biological processes, such as metabolism and photosynthesis, by taking soil samples (Figure 9-20), putting them in special chambers (some with nutrients), and watching for chemical changes that would indicate microscopic organisms processing the material in the chamber. All three of these tests found the type of changes that indicated life in terrestial samples!

[2]By *life* scientists generally mean "life as we know it"—the ability of carbon atoms to combine with other atoms and form very complex molecules, which in turn form organisms that grow and reproduce.

At first, investigators believed that Viking may have actually detected microbial life on Mars, but in a quantity of less than one organic molecule per billion parts of soil. However, the magnitude and rate of the changes were unlike those for terrestrial microbes. Furthermore, because most investigators believe microbes would be unlikely, if not impossible, at the measured low abundance of organic molecules, the Viking results seem to reveal unexpected but ordinary chemical reactions in the soil, *not life*. Tests with simulated Martian soil suggest that solar ultraviolet radiation reaching the surface of Mars causes different mineral properties than those familiar on the earth; in their altered states, Martian minerals may react in unanticipated ways in the test chambers.

Thus, Viking has shown that Mars "is not teeming with life from pole to pole" (to use the memorable understatement by astronomer Carl Sagan). It probably has no present or recent microbial life. Why? Ultraviolet light irradiates the surface because Mars lacks an ozone layer. It would break down organic molecules, preventing current life on exposed soil or rock surfaces. Conceivably, primitive life might have evolved in the past, but its traces may have been destroyed as ultraviolet light sterilized the windblown soil. A less likely possibility is that Martian life exists, but the Vikings missed it. Researchers since 1979 have reported microorganisms in antarctic soils where Viking-type instruments did not detect organic molecules, reopening the question of whether Martian life is too subtle for Viking to find. Also, scientists in 1977 discovered algae and lichens in microscopic pore spaces *inside* rocks in barren, Marslike Antarctic valleys where no life was previously thought to exist. Perhaps we need to seek life inside Martian rocks!

A third possibility is that life never evolved on Mars. Liquid water may never have existed long enough to allow even microbes to evolve. Future exploration may clarify whether Martian life ever existed. The dusty plains of Mars have not given up their last secrets about the origins of life in the solar system. Perhaps a Mars expedition will someday ferret out the answer (Figure 9-21). Lunar astronaut (and later U.S. senator) Harrison Schmidt has remarked that today's students will be the parents of the first Martians!

Figure 9-20. ***Sampler arms on both Viking landers scooped up soil samples for chemical analysis inside the spacecraft. Here a Viking 2 sampler arm pushes aside a frothy-textured lava rock to gather a soil sample from under it. Protected soil in such locations would have less irradiation by solar ultraviolet light, expected to be harmful to any possible Martian life forms. This experiment occurred on October 9. 1976. (NASA.)***

Martian Satellites: Phobos and Deimos

Not all of the mysteries of Mars are on its surface. In 1877, the American astronomer Asaph Hall became the first human to see a satellite of Mars. Shortly afterward, he charted the positions of two Martian moons, naming the inner satellite Phobos ("fear") and the outer one Deimos ("terror") after the chariot horses of Mars in Greek mythology. Close-up photographs by spacecraft have revealed these moons to be strange, potato-shaped, cratered chunks of rock (Figures 9-22 to 9-25). Their dimensions are given in Table 9-2.

The craters of Phobos and Deimos were caused by collisions with small bits of meteoritic debris. The largest crater on Phobos is 8 km (5 mi) across and can be seen at the top of Figure 9-22. It is named Stickney, the maiden name of Mrs. Hall,

Figure 9-21. *Manned expeditions to the surface of Mars may eventually clarify Mars' secrets. Astronauts could set up a base similar to Antarctic research stations and probe Martian geology and chemistry more effectively than unmanned probes. Such expeditions are within our technical capability; Russian cosmonauts have occupied their Salyut space station for durations equivalent to a flight to Mars. (Painting by Paul Hudson.)*

who encouraged her husband's successful search for the satellite. A collision violent enough to create such a crater would release as much energy as 100 000 atom bombs of the Hiroshima size, or about 1000 hydrogen bombs of megaton size.

A peculiar feature of Phobos was revealed by close-up photos from Viking orbiters. Networks of grooves reach widths of around 100 m and lengths of as much as 10 km. Some are rows of adjoining craters. The exact nature of the grooves is unclear, but they radiate roughly from the large crater Stickney and may be fractures (somewhat masked by surface dust) caused by the mighty Stickney impact, which nearly shattered Phobos.

The origins of Phobos and Deimos are unclear. Their surfaces are dark in color and apparently of a composition resembling a carbon-rich type of meteorite (called carbonaceous chondrite) known to be common in the nearby asteroid belt. Are Phobos and Deimos merely captured asteroids? Were they formed in orbit around Mars? Are they fragments of a much larger satellite blown apart in a collision with a large meteorite? No one is certain. Further exploration of the red planet may provide us with some of the answers to our questions.

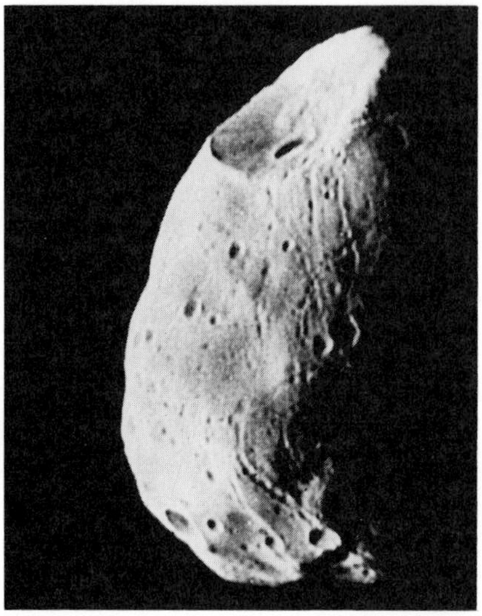

Figure 9-22. *Phobos, a 28-km-long moon of Mars. The largest crater Stickney (top) marks an impact that fractured and almost shattered Phobos. (NASA.)*

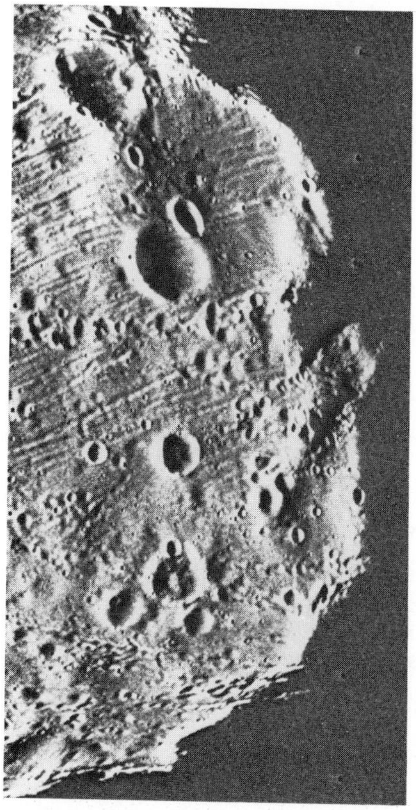

Figure 9-23. *Close-up of Mars' satellite Phobos by Viking orbiter revealed surprising grooves covering the asteroid-sized moon. The photo, made from 880 km (545 mi) away, shows objects as small as 40 m (130 ft) across. Grooves might be traces of fractures occuring on Phobos (or its parent body if Phobos is a fragment) during impacts that made the larger craters. (NASA.)*

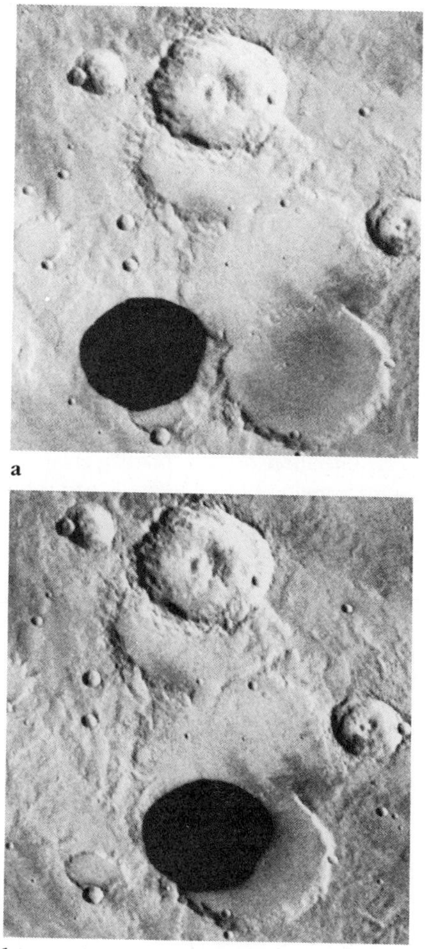

Figure 9-24. *Sequential photos from Viking orbiter show 28-km-wide Phobos passing in front of cratered Martian terrain, 9400 km further away in the background. Phobos is so much darker than Mars that these normal contrast photos, adjusted for Martian tones, underexpose the moon and render it black. (NASA.)*

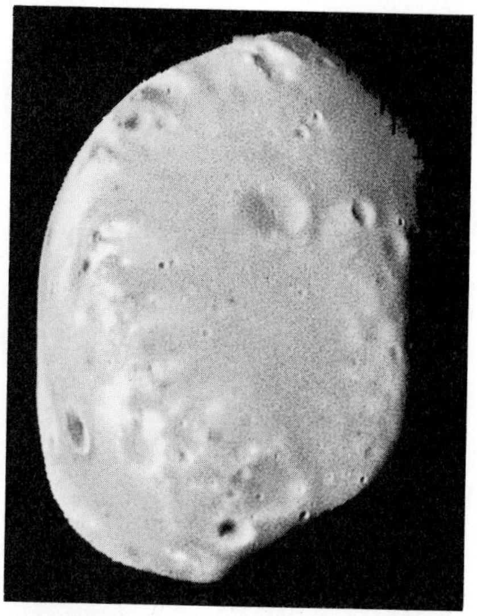

Figure 9-25. *Deimos, a 16-km-long moon of Mars. Its surface texture is smoother than that of Phobos, perhaps because of a different duration of microcratering since the most recent large impacts. Streaky markings are believed to be due to slipping of loose material in locally downhill directions in Deimos' weak, asymmetric gravity field. (NASA.)*

Table 9-2. *Dimensions of Phobos and Deimos*

Diameter	Phobos		Deimos	
	km	mi	km	mi
Longest	28	17	16	10
Intermediate	23	14	12	7½
Shortest	20	12	10	6

Expeditions to Mars

Decades of interest in the comparative planetology of Mars and Earth, as well as the potential habitability of Mars, have led to emphasis on Mars as the major target of planetary exploration. In terms of a long time scale, various advocates, including the President's National Commission on Space in 1986, have urged construction of a permanently inhabited Martian base as a long-term space goal.[3] The project could be international—serving as a model for international technical cooperation as well as an impetus for scientific and engineering growth. The National Space Commission Report targets the base for the decade of the 2020s.

The Soviet Union has already announced plans for an unmanned mission to Phobos that will maneuver within about 50 yd of that moon and then send over a lander, which can hop from site to site,

[3]The commission included Neil Armstrong, astronaut Kathryn Sullivan, test pilot Chuck Yeager, U.N. ambassador Jeane Kirkpatrick, as well as several well-known scientists and engineers.

analyzing surface chemistry. A duplicate craft sent at the same time may go on to Deimos if the Phobos mission works. NASA has announced plans for a "Mars Observer" that would orbit Mars and gather new data on surface mineral compositions, atmospheric dust, and geologic structures. American and Soviet researchers hope to coordinate the work of the vehicles near Mars in the 1990s. Meanwhile, the Soviet Union has proceeded with its manned program near Earth, where its astronauts have stayed in orbit for periods comparable in time to a Martian expedition. It is rumored they may be building toward manned expeditions around Mars or even to its surface, perhaps in the 1990s.

Summary

Mars, the planet once thought to have fields of vegetation or even a dying civilization, has been revealed by spacecraft to be a barren but beautiful desert lacking any advanced life forms. Orbital photos reveal a wide variety of landscapes including lava flows, grand canyons, landslides, polar snowfields, eroded strata, dunes, impact craters, and arroyos. Biological experiments aboard Viking landers indicate that interesting chemical reactions take place in Martian soil, and the seemingly lifeless Martian environment may yet yield interesting clues about the evolution of biochemical reactions and their dependence on climate.

Evidence about Mars' ancient climatic history seems to contradict its present-day barrenness. Nearly all water on Mars today is locked in polar ice, frozen in the soil, or chemically bound in the soil; virtually no liquid water exists. But liquid water apparently once flowed and eroded the surface, indicating different past climates. Current research indicates that the climates of both Mars and the earth may have varied over hundreds of millions of years, possibly due to dynamical wobbles of the rotation axis.

Concepts

Mars' rotation period	Martian air pressure
seasonal changes on Mars	permafrost
mare	Martian atmosphere
desert	volcanoes
canal	channels
Viking 1	life on Mars
Viking 2	organic molecules
Martian air temperatures	

Problems

1. Describe a day on Mars as a future astronaut might experience it. Include the length of the day, the appearance of the landscape, possible clouds and winds, possible hazards, and objects visible in the sky.

2. What scientific opportunities for long-term exploration does Mars offer, as compared with the moon? What qualities of the Martian environment might make operating a long-term base or colony easier on Mars than on the moon once initial materials were delivered to the site?

3. Compare photos of craters on Mars and on the moon.
- **a.** Assuming that all craters had similar sharp rims when fresh, which craters have suffered most from erosion?
- **b.** What does this say about lunar versus Martian environments?

4. Give examples of how the Martian environment and geology are midway between those of the smaller planet Mercury and the larger planet Earth. Include comments on atmosphere, craters, volcanism, and plate tectonics.

5. What scientific knowledge might be gained from close-up investigation of Phobos and Deimos? What measurements would be of interest if rocks from Phobos and Deimos were available for study?

6. Imagine you are a visitor from outer space exploring the solar system.
- **a.** If two Viking-type spacecraft landed at random places on the earth, took photos, measured the climate, and took soil samples, what might they reveal about the earth?
- **b.** How many landings might be needed to characterize the earth adequately?
- **c.** To characterize Mars to the same degree, how many might be needed?

ADVANCED PROBLEMS

7. What is the weight of an astronaut on Phobos, if he or she weighs 130 pounds on Earth?

8. In a gravitational field, objects fall through a height h in a time $t = \sqrt{2h/a}$, where a is the acceleration due to gravity. Suppose you are an astronaut exploring Deimos, where a is about 0.0042 m/s^2. Suppose you drop a tool from eye level.
- **a.** How long does it take to reach the ground?

b. How long would it take on the earth where *a* is about 10 m/s^2? (*Hint*: Because we are using the SI system, express *h* in meters.)

9. With a telescope that can resolve details as small as 1 second of arc, what is the smallest detail you can see on Mars during its closest approach to the earth, at a distance of about 56 million kilometers?

10. Calculate the velocity needed to launch an object into circular orbit around Deimos from a point on the highest "mountain" on Deimos, assumed to be 8 km from the center. Assume the mass of Deimos is 4×10^{15} kg. Many people can throw an object at about 30 m/s. Would you need a rocket to launch a satellite from Deimos?

Projects

1. Observe Mars with a telescope, preferably within a few weeks of an opposition and with a telescope having an aperture of at least 15 cm (6 in.). Magnification around 250 to 300 is useful. Sketch the planet. Can you see any surface details? Usually the most prominent detail is one of the polar caps, a small, brilliant white area at the north or south limb contrasting with the orangish disk. Can you see any dark regions? Compare the view on different nights and at different times of the night. (Because Mars turns about once in 24 h, the same side of Mars will be turned toward the earth on successive evenings at about the same hour.) If no markings can be seen, three explanations are possible: (1) Observing conditions are too poor; (2) the hemisphere of Mars with very few markings may be turned toward the earth; (3) a major dust storm may be raging on Mars, obscuring the markings.

2. For the previous observations, determine which side of Mars you were looking at. (Your instructor may need to assist you.) First determine the date and Universal Time of your observations (UT = EST + 5 h = PST + 8 h. Thus, 10 P.M. EST on April 2 = 03^h00^m on April 3, Universal Time). In *The American Ephemeris and Nautical Almanac*, the table "Mars: Ephemeris for Physical Observations" gives the central meridian (or longitude on Mars of the center of the side facing Earth) at 0^h00^m Universal Time on each date. From these tables you can find the Martian central meridian for the time of your observation. (Mars turns about 14.7°/h.) Compare your observations with a map of Mars, locating the part of Mars that you observed.

CHAPTER 10

THE OUTER PLANETS

The outer solar system contains four **giant planets**—Jupiter, Saturn, Uranus, and Neptune—and a small planet, Pluto. The name of the monarch of the Roman gods is fitting for Jupiter. The biggest planet, it contains 71% of the total planetary mass—nearly 2½ times as much as all other planets combined. The giant planets also tend to have large families of satellites—at least 41 in all. The four giant planets together contain 99½% of the total planetary mass, and harbor about 91% of the known satellites.

The four giant planets have much lower mean densities than the terrestrial planets—700 to 1600 kg/m^3 as compared with 3900 to 5500 kg/m^3. Saturn, at 700, would float like an ice cube if we could find a big enough ocean (water's density is 1000 kg/m^3). This simile is significant. The giant planets evidently *are* made largely of ices, as well as low-density liquids such as liquid hydrogen.

Jupiter's diameter measures a little over 10 times the earth's diameter, and Saturn just under 10 times. Uranus and Neptune have diameters about 4 times the earth's. Placed on the face of Jupiter, the earth would look like a dime on a dinner plate (see Color Plates 8 and 14, top).

Perhaps the most important principle to remember about the outer solar system is that because it is further from the sun and much colder than the inner solar system, the outer solar system contains much more ice than the inner solar system. Because the gases that formed the sun and its planet-spawning surroundings were mostly hydrogen, the ices that formed in the outer solar system are frozen compounds of hydrogen, such as water (H_2O), methane (CH_4), and ammonia (NH_3). Thus, instead of forming worlds of rock like the terrestrial planets, the outer solar system formed worlds of rock *plus* ice, often in roughly 50–50 mixtures. This explains many properties of worlds in the outer solar system. The giant planets are giant for two reasons: (1) They had ices in addition to rock, and (2) once they reached a large size during their formation, their gravity was so great that they began to pull in the surrounding hydrogen-rich gases. (The gravity of the earth and smaller planets was too weak to hold these light gases.) Hence, the giants are huge, hydrogen-rich worlds. Their surfaces are hidden by colored clouds. Their thick atmospheres may bear some resemblance to the primordial atmosphere of the earth.

Similarly, most of the moons in the outer solar system have icy surfaces or dirty-ice surfaces. On some of them, geologic processes tended to evaporate the ice off the surface, leaving darker, soil-rich surfaces. On others, internal heating melted the ice, producing watery "lava" that erupted and formed bright ice patches. These moons are fascinating worlds—some larger than the planets Mercury and Pluto, as seen in Figure 6-2. Some of the larger ones are compared in Color Plate 8.

The Planet Jupiter

Even a week's observations with a backyard telescope reveal the swirling cloud patterns of Jupiter, prominent in Figure 10-1. The most obvious pattern is the system of dark and light cloud bands parallel to Jupiter's equator, as sketched in Figure 10-2. The dark ones are called **belts**, and the bright ones **zones**. Within these bands, wispy spots and streaks arise, develop, and die out. These features look small, but some of them are larger than the earth! Though the smaller ones evolve in days, the larger ones may last for months or years. Dark clouds may grow and darken an entire bright zone for months at a time (Figure 10-3).

The belts and zones have distinct colors as shown vividly in Color Plate 14. Usually, belts are brown, reddish, or even greenish, whereas zones are light tan or yellowish. These colors are caused by smoglike compounds formed by photo-

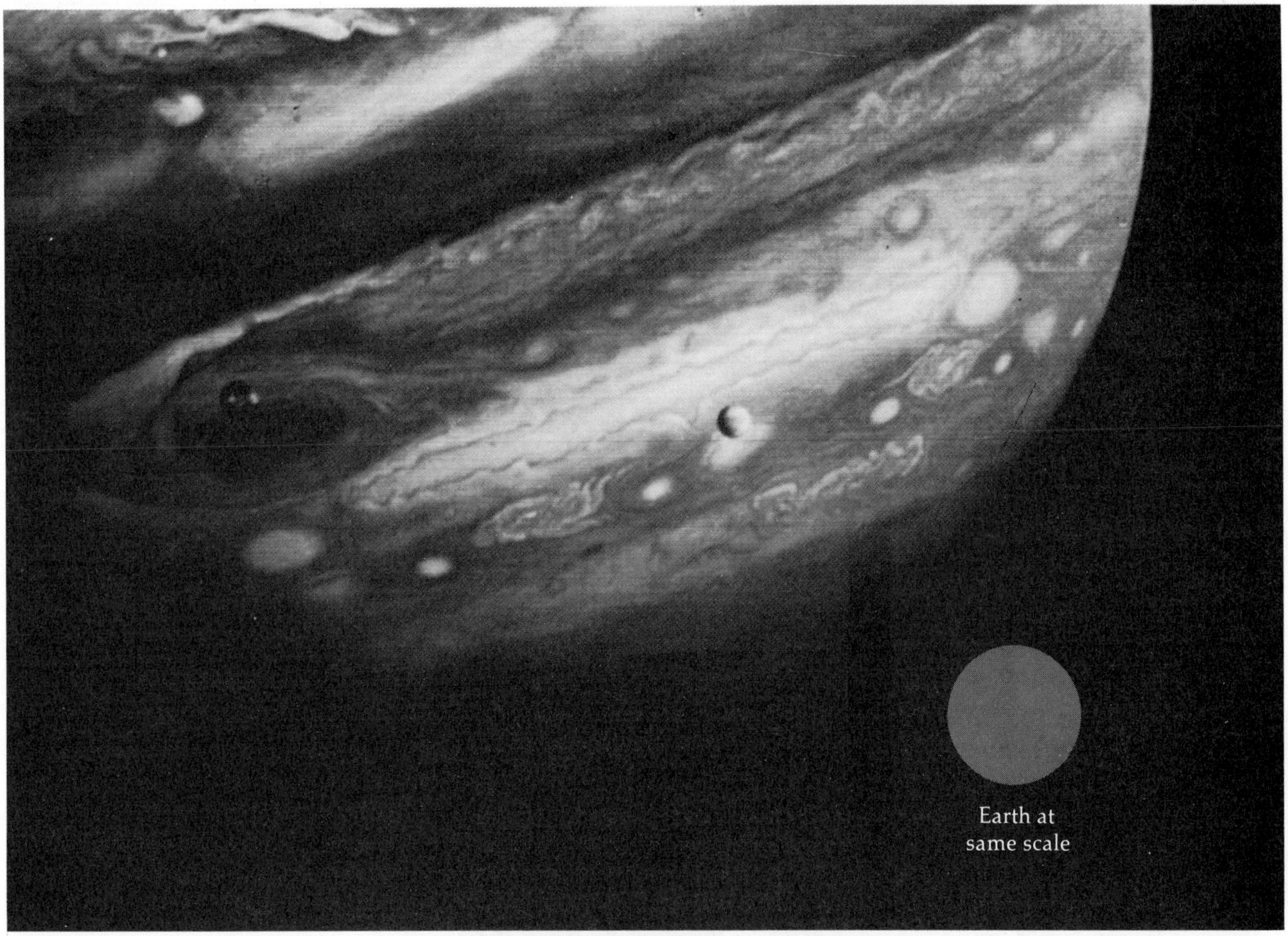

Figure 10-1. *Massive Jupiter is mottled by swirling cloud formations, including the Great Red Spot (oval at left center). In the foreground are two moons, Io (left, in front of the Great Red Spot) and Europa (right center); they are about the size of our moon. The circle in the lower right shows the relative size of the earth. (NASA, Voyager photo from a distance of 20 million kilometers.)*

chemical reactions among minor constituents of the clouds, perhaps involving hydrogen sulfide, organic molecules, or metallic sodium particles. The clouds are believed to be primarily ice crystals of ammonia, ammonium hydrosulfide, and frozen water.

Jupiter's atmospheric composition is mostly hydrogen and helium, as indicated in Table 10-1. Water, ammonia, and methane are also present. Voyager 1 and 2 in 1979 revealed additional trace gases.

Infrared radiation reveals **Jupiter's temperature** in the upper atmosphere to be very cold because of Jupiter's great distance from the sun—about 133 K (−220°F), both on the sunlit and nighttime sides of the planet. At a lower level, in the poisonous upper clouds (where the air pressure equals the earth's), the temperature is closer to 150 K (−189°F).

The lower atmosphere of Jupiter may be more interesting. Gaps in the highest clouds have revealed lower haze layers

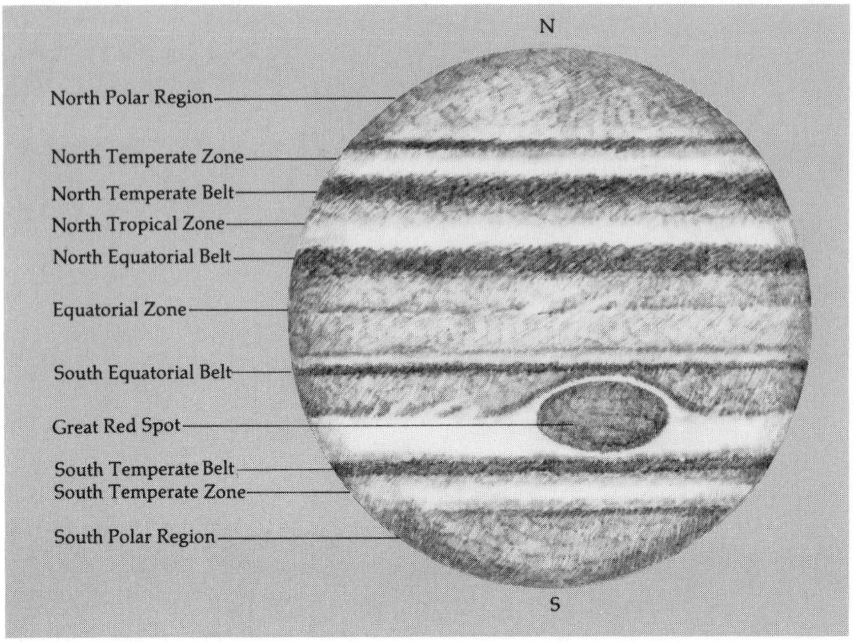

Figure 10-2. *Semipermanent cloud formations of Jupiter, visible from the earth when viewed through small telescopes.*

Table 10-1. *Atmospheric Compositions of Jupiter and Saturn*

Gas	Estimated Percent by Mass	
	Jupiter	Saturn
H_2 (hydrogen)	79	88
He (helium)	19	11
Ne (neon)	1?	?
H_2O (water)	Trace	?
NH_3 (ammonia)	0.5?	0.2
Ar (argon)	0.3?	?
CH_4 (methane)	0.2?	0.6
C_2H_6 (ethane)	Trace	0.02
PH_3 (phosphine)	Trace	Trace
C_2H_2 (acetylene)	Trace	Trace

Note: Data derived mainly from Voyager 1 and 2 infrared observations (*Science 204:* 972; *206:* 952; *212:* 192). "Trace" indicates a small fraction detected; "?" indicates a likely constituent that has not been measured. The hydrogen and helium abundances are very close to those in the sun, 78% and 20%, respectively.

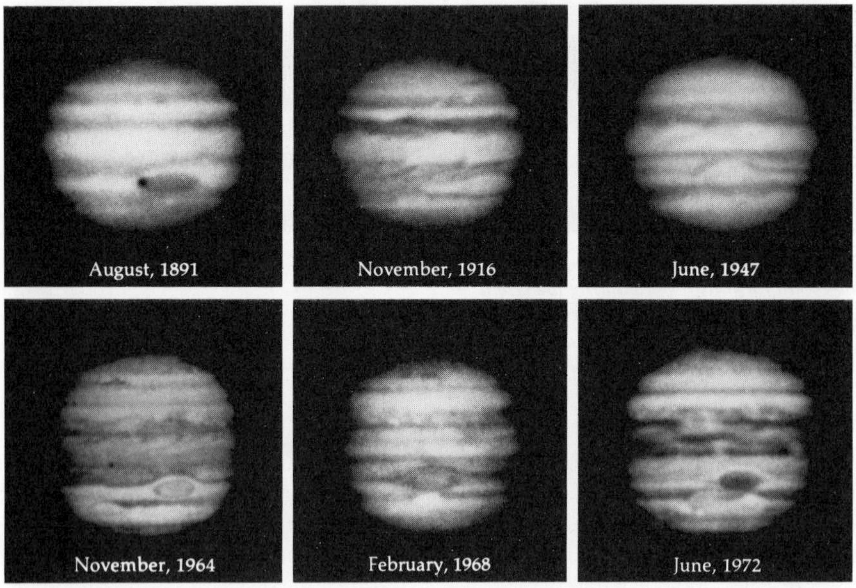

Figure 10-3. *Photographs of Jupiter spanning 81 years, showing the changing array of Jupiter's belts and zones. The dark north temperate belt is relatively permanent, but the Equatorial Zone changes from bright (1891) to dark (1964). Many of these images show the Great Red Spot. (Lowell Observatory)*

with temperatures of around 250 K (−9°F). Lower regions are even warmer. According to some scientists, the lower regions resemble the hydrogen-compound-rich primordial atmosphere of the earth when terrestrial life originated.

A few scientists speculate that some level may even be warm enough for complex organic molecules to evolve into simple organisms that could float in the atmosphere. A recent model of Jupiter's atmosphere calls for temperatures similar to those at the earth's surface at a level of about 60 km below the Jovian cloud tops, where the pressure would be about 10 times the earth's surface pressure. Such conditions might be hospitable to primitive life, but strong updrafts and downdrafts would probably draw hypothetical organisms to levels that would be too cold or hot for life. Thus, most scientists doubt that any advanced life forms exist on Jupiter.

Voyagers' cameras revealed mighty blasts of lightning playing among the clouds, and enormous auroral displays flickering high above the clouds in the polar regions.

For at least 300 years a storm three times bigger than the earth has been raging on Jupiter. In the telescope it appears as an enormous reddish oval in the south tropical zone, probably first studied by G. D. Cassini in 1665. It became very prominent in 1887, when it was rediscovered and called the **Great Red Spot** (see Figure 10-4 and Color Plate 14, bottom). The Red Spot has reached diameters of 40 000 km. It and other, smaller transient spots are probably vast, hurricanelike storm systems in Jupiter's atmosphere. Small clouds approaching the Red Spot get caught in a counterclockwise circulation like leaves in a great whirlpool.

The Voyagers revealed eastward jet streams blowing along the Equatorial and Temperate Zones at wind speeds 150 m/s (338 mph) faster than in the dark Equatorial Belts. Figure 10-5 shows an imaginary view of the awesome cloud vistas near the top of the main cloud deck of this stormy world.

Figure 10-4. *Close-up of Jupiter's Great Red Spot, a giant storm system in Jupiter's clouds. The planet Earth is about three-quarters as large as the spot's long axis. For a color view 4 d later, see Color Plate 14. (NASA, Voyager 1 photo.)*

Figure 10-5. *Enormous "cloudscapes" would greet a visitor to the upper atmospheres of the giant planets. This imaginary view shows cloud decks at several levels. Note that breaks in the clouds allow views of lower levels, as in the lower left. Here lightning flashes are visible, as discovered on Jupiter by Voyager spacecraft. (Painting by Ron Miller.)*

JUPITER'S ROTATION

Cloud belts and zones have a rotation period of about 9^h50^m near the equator, but several belts and zones at higher latitudes average 9^h56^m. The Red Spot has its own rate, sometimes lagging behind or drifting ahead of nearby clouds. The best estimate of the underlying planet's rotation is $9^h55\frac{1}{2}^m$, based on radio radiations from deep-atmosphere electrical storms. However, no one is sure if any of these periods marks the true rotation of a well-defined solid or liquid surface beneath the clouds—or whether such a surface exists.

JUPITER'S RADIATION AND MAGNETISM

Besides reflected sunlight, Jupiter emits three other types of radiation. In order of increasing wavelength, they are infrared thermal radiation, shortwave radio radiation, and longwave radio radiation.

The first is called **Jupiter's infrared thermal radiation** because it is due to the heat of the planet itself (review Figure 7-4). From measurements of the total amount of this radiation, scientists know the total amount of energy being radiated by Jupiter. Surprisingly, this figure turns out to be about twice as much energy as Jupiter absorbs from the sun! This is very different from the case of the earth or other terrestrial planets, where the heat radiated from inside is negligible compared to the heat received from the sun. Jupiter's extra internal heat must be coming from somewhere, but where?

Theorists believe Jupiter is slowly contracting, releasing gravitational energy as heat and radiation. This radiation was most intense when Jupiter formed and has declined ever since to the low level observed today.

Although Jupiter is radiating its own energy, it is not a true star, because its energy is not produced by thermonuclear fusion, the way a star's energy is. Jupiter's mass is not great enough to create the central pressure and heat necessary for starlike fusion reactions in its interior.

The second kind of radiation emitted by Jupiter is shortwave radio radiation, often called Jupiter's decimeter radiation because its wavelength is around $\frac{1}{10}$ m (1 dm). The distribution of this radiation with respect to wavelength shows that it is **synchrotron radiation**, a type of radio wave emitted when very fast electrons move through a magnetic field. (These electrons move at nearly the maximum possible speed for any object: the speed of light. The same radiation is found in synchrotron atom smashers on the earth.)

Synchrotron radiation implies that Jupiter must have a magnetic field. The field was confirmed in 1973 and 1974 when Pioneers 10 and 11 flew through it and measured it. As shown in Table 10-2, it is the strongest planetary field known. Like the earth's field, it is not perfectly aligned with the rotation axis, but is about 15° off. The polarity is opposite the earth's, with the north *magnetic* pole on the south side of the planet; a compass would thus point south on Jupiter. Because such fields are believed to result from the circulation of conductive material deep inside the planets, Jupiter's field indicates such a core. Further studies of Jupiter's field will help us understand the earth's magnetic field.

Another similarity to the earth is that charged particles emitted by the sun have been trapped by the magnetic field in doughnut-shaped rings around the planet, like the earth's Van Allen belts. The particles are 100 000 to a million times more concentrated than near the earth, making a zone of quite hazardous radiation near Jupiter's inner satellites.

The third kind of radiation, the strong, erratic decameter radio radiation (wavelength ~10 m), was discovered in the mid-1950s. Its sources are located on Jupiter itself rather than in nearby space. They may be localized storm areas. The strangest property was discovered by radio astronomer E. K. Bigg in 1964: The earth receives the bursts of radio noise primarily when Jupiter's satellite Io is in certain positions *with respect to the earth.*

Why should radiation emitted from Jupiter depend on the orbital position of *one* of its moons—particularly on its position relative to the earth? Io moves through the magnetic field of Jupiter, disturbing this field and its particle motions in such as way as to produce beams of radiation that, like searchlight beams, sweep through space as Io moves along. Intermittently they point toward the earth.

Table 10-2. *Magnetic Fields in the Solar System*

Location	Approximate Field Strength Near Planet (nT)[a]	Source of Data
Sun (average surface)	200 000	Spectra
Sun (sunspot)	100 000 000	Spectra
Interplanetary space	4	Various spacecraft
Mercury	600	Mariner 10
Venus	Low[b]	Mariner 10
Earth (surface)	40 000	Field measures
Mars	Low[b]	Mariner 9
Jupiter	400 000	Pioneer 10, 11
Saturn	20 000	Voyager 1
Uranus	25 000	Voyager 2

[a]The tesla (T) is a unit of magnetic field strength. Think of it as measuring the strength of response of a compass placed in the field. $1 \text{nT} = 10^{-9}\text{T}$.

[b]Spacecraft near Venus and Mars detected little or no magnetic field increases over the interplanetary field strengths.

JUPITER'S INTERNAL STRUCTURE AND SURFACE

What is Jupiter like under its clouds? Its mean density is too low for it to be a rock-and-iron planet like the earth. Instead, **Jupiter's interior** is believed to consist of 60% hydrogen, the rest being helium with small amounts of silicates and other "impurities." The heavier elements have sunk, so that a core of silicate or iron–silicate material—resembling terrestrial planets—may exist near the center. According to some theoretical models, the core may be over twice the earth's size and have a temperature around 30 000 K.

Under the high pressure within Jupiter's interior, hydrogen takes on an unfamiliar form called **metallic hydrogen**, which can conduct electric currents. As shown in Figure 10-6, much of Jupiter's interior may be taken up by a mantle of liquid metallic hydrogen, in which convection currents carry heat to the surface and electric currents generate Jupiter's magnetic field. Much of the outer half may be a sea of ordinary liquid molecular hydrogen, H_2. There may be no solid surface at all, only a slushy mixture of liquid hydrogen and crystals of various compounds.

The gravity at this ill-defined "surface" is stronger than that of any other planet. A person weighing 150 pounds on the earth would weigh about 400 pounds on Jupiter! Similarly, the pressure of the thick atmosphere is roughly 100 times the air pressure on the earth. Various data suggest that the cloud-obscured, high-pressure ocean of liquid hydrogen begins as much as 1000 km or more below the cloud tops.

Jupiter's Satellites

Jupiter has an impressive array of at least 16 moons, ranging from 8-km bodies to four large worlds, including two slightly larger than the planet Mercury. These moons are numbered in order of discovery. The four largest, shown together in Color Plate 8, were discovered independently by Galileo and the German astronomer Marius on two consecutive nights in 1610 as they viewed Jupiter through the newly invented telescope.[1] These bodies are called the **Galilean satellites.**

Until the 1979 flights of Voyagers 1 and 2 through the satellite system, we knew the large satellites only as pinhead disks in even the largest telescopes. Some observers noted vague dusky markings. Since they are in the size range of the moon and Mercury, these moons were assumed to be cratered, dead worlds with little geologic activity. Spectra showed some differences among the surface materials of these worlds, but no one realized how unique each of these moons really is.

A clue came during the months just before Voyager sailed by Jupiter. California dynamicists S. J. Peale, P. M. Cassen, and R. T. Reynolds (1979) calculated that although Io's orbit is nearly circular, gravitational pulls of neighboring satellites cause Io's distance from Jupiter to vary slightly. As the distance varies, the tidal stretching of Io changes. This flexing heats Io's interior, just as rapid flexing of a tennis ball makes it heat up from friction. The calculations showed that this heating effect is stronger in Io than in any other satellite and causes Io's interior to be molten. In a beautiful example of the scientific method at work, Peale and his colleagues published their prediction that volcanoes might be found on Io in a journal that came out just a few days *before* Voyager 1 reached Io. A few days later, Voyager discovered the solar system's most active volcanoes, spewing 100-km-high plumes of debris (Color Plate 16, top)!

Another heating effect on Jupiter's moons was radiant heat from primordial Jupiter. As Jupiter formed 4½ billion years ago, it acted like a miniature sun in its own miniature solar system of moons, heating the surfaces of the inner moons most and the outer moons least. As we describe moons in the next paragraphs, you will note that the outer ones seem least disturbed by heating effects or geologic activity.

Of the four Galilean moons, Callisto (Figure 10-7 and Color Plate 18) is the outermost. It is 5000 km across, 2% bigger than the planet Mercury. Here we have a primitive world with little sign of internal geologic activity. Its surface is covered with craters caused by meteorite impacts. Compared with water's density of 1000

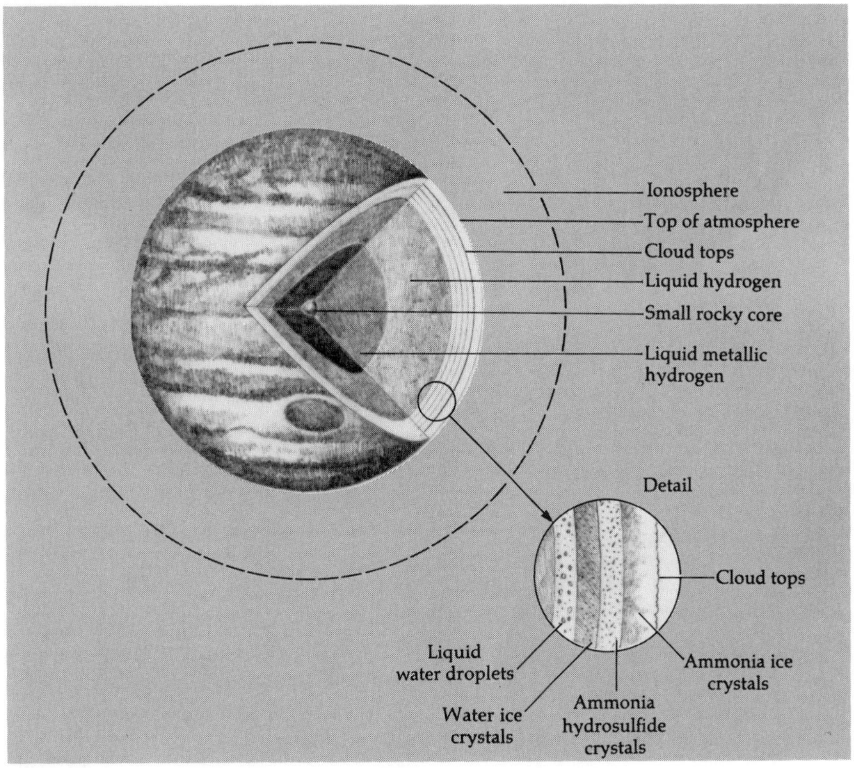

***Figure 10-6.** A view of the interior and atmosphere of Jupiter, based on telescopic and spacecraft data. The atmospheric depth to the liquid zone is perhaps 100 km. (NASA.)*

[1]Keen-sighted individuals can sometimes see at least one of Jupiter's large moons with the naked eye; hence, it may have been spotted long before 1610. In 1982, Chinese historians noted such a report of a faint Jupiter companion by ancient Chinese astronomer Gan De in 364 B.C. It will remain hard to prove such sightings; in any case, the significance—moons orbiting *around* Jupiter—was not known until 1610.

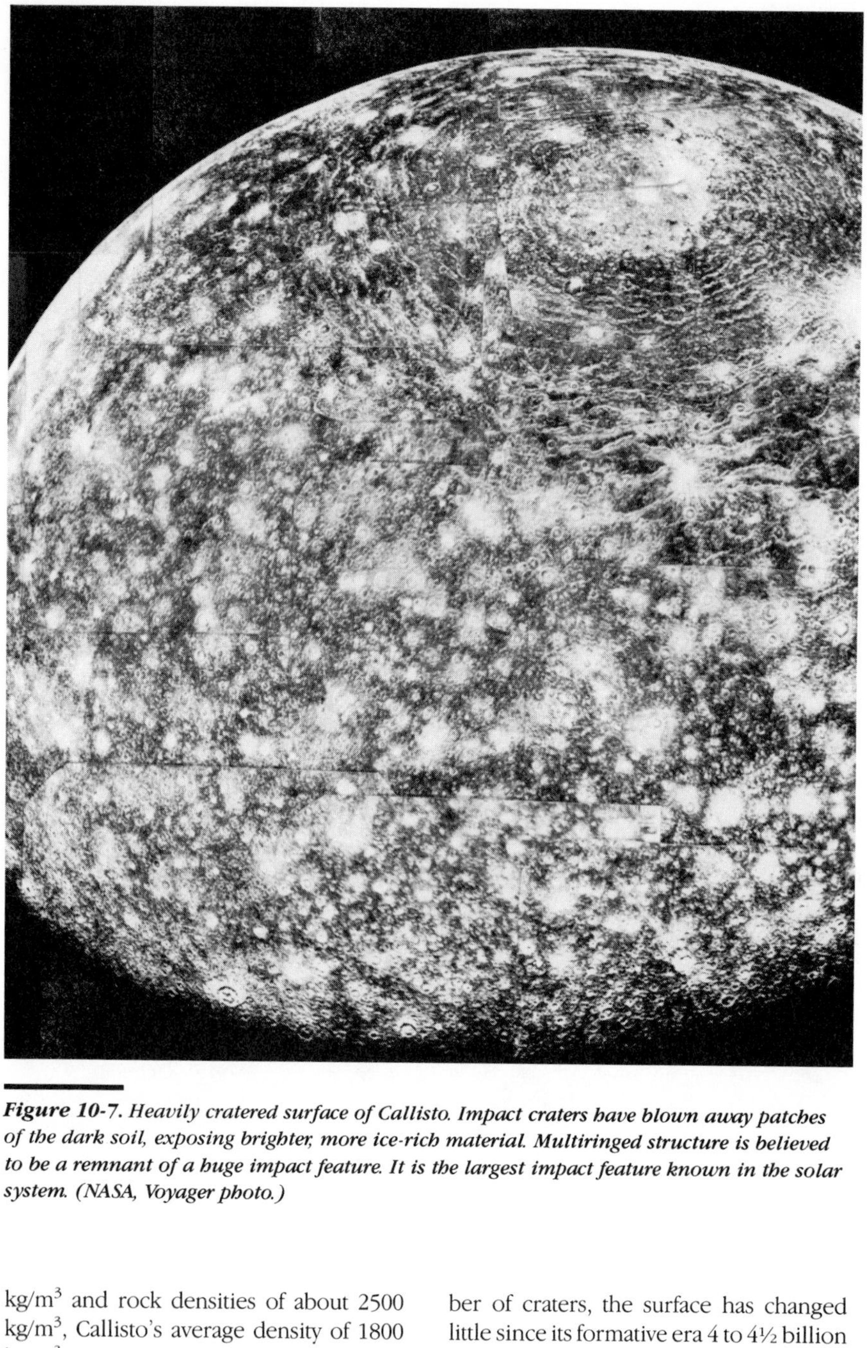

Figure 10-7. *Heavily cratered surface of Callisto. Impact craters have blown away patches of the dark soil, exposing brighter, more ice-rich material. Multiringed structure is believed to be a remnant of a huge impact feature. It is the largest impact feature known in the solar system. (NASA, Voyager photo.)*

Figure 10-8. *The dark regions of Ganymede are cratered crust resembling Callisto's. These are cut by swaths of brighter, younger, icy material. Note the polar cap of frost across the top of the disk. (NASA, Voyager photo.)*

kg/m^3 and rock densities of about 2500 kg/m^3, Callisto's average density of 1800 kg/m^3 indicates a composition of about half ice and half rock. The surface is mostly covered by a layer of dark, icy soil, probably left behind as impacts evaporated the surface ices. As seen in Figure 10-7, the larger craters have blown away the dark surface soil and exposed bright ice underneath. Judging from the high number of craters, the surface has changed little since its formative era 4 to 4½ billion years ago.

Ganymede, with a 5270-km diameter, is the largest moon of Jupiter and probably the largest moon in the solar system. It is shown in Figure 10-8 and Color Plate 18. It is 8% bigger than Mercury and nearly twice as big as Pluto. Its average density of 1900 kg/m^3 is barely higher than Callisto's, implying a similar ice–rock composition. A thin polar cap, visible at the top of Figure 10-8, is believed to be water frost. This would indicate that water vapor has been emitted from the interior, circulated poleward in an almost negligible atmosphere, and frozen onto the ground at the poles.

The ancient parts of Ganymede's surface are very similar to Callisto's surface, being heavily cratered and covered with dark, dusty soil. Contrary to Callisto, though, these areas are divided by swaths of light-toned, grooved terrain. The ancient crust seems to have split, allowing fresher icy material to erupt.

Such splitting may imply internal heating that melted at least the icy component and caused expansion and cracking. Water erupted, forming ice flows—the Galilean satellites' equivalent of the earth's lava flows. The grooves appear to be fractures along which water erupted (Figure 10-9). Scattered impact craters on the grooved terrain suggest that it formed 2 to 4 billion years ago, after which Ganymede's geologic activity declined. The source of heat that drove Ganymede's geologic activity

***Figure 10-9.** Imaginary panorama of Jupiter and the icy surface of its moon, Ganymede. Ganymede's surface is split by parallel fissures, where water may have erupted and refrozen. This view, from such terrain, shows a ridge of jagged ice blocks thrust up during movements of Ganymede's crust. In the sky are Jupiter's inner moons, Europa (left) and Io (in front of Jupiter). Jupiter would look about 15 times as big as our moon in our sky, and Europa would look slightly smaller than our moon. (Painting by Ron Miller.)*

may have been radiant heat from primordial Jupiter or radioactivity in the rocky component of Ganymede's interior.

Ganymede is telling us an important story that may help us understand the earth better. Ganymede offers a case of incipient plate tectonics, since its brittle "lithosphere" of ice split into platelike blocks as it floated on a warmer, watery, subsurface. Although Ganymede's interior was not active enough (hot enough?) to cause full-scale continental drift among the plates, offsets do show that the plates have moved slightly. Ganymede thus provides a transitional example from cold, inactive planets to larger, geologically active worlds.

Europa, the next inner moon, is only as big as our moon. It is dramatically different from Callisto or Ganymede. As seen in Color Plate 17, it is an icy billiard ball. There are no mountains and scarcely any craters. The only markings are delicately shaded, shallow grooves crisscrossing the surface. Spectra show that the surface is mostly frozen water, but the relatively high mean density of 3000 km/m^3 for Europa as a whole shows that Europa's interior consists mostly of rock. Perhaps the rocks' radioactivity, together with tidal flexing, heated the interior, melted the ices, and allowed water to erupt onto the surface, obliterating the original crust. That water froze into the ice layer we see, which has been cracked by subsequent heating or cooling. The ice layer may be 100 km (60 mi) thick, perhaps overlying a liquid water layer (see Figures 10–10 and 10–11). The near absence of impact craters shows this ice surface is younger than Callisto's or Ganymede's. It may be continuing to form by intermittent eruptions of water, even in modern times. Some Voyager team members believe they have identified a bright transient cloud marking a possible eruption on the final Voyager 2 photo of Europa. Europa may occasionally have giant geysers! In his novel *2010*, science fiction writer Arthur C. Clarke speculated that life might have evolved in Europa's buried oceans. Confirmation of such ideas will have to await new spacecraft flights.

The inner Galilean moon, Io, which is the size of our moon, is probably the most bizarre world in the solar system. Voyagers 1 and 2 discovered not only active volcanoes (see Figure 10-12 and Color Plates 15 and 16), but also a strangely mottled surface of orange, red, and white sulfur deposits (Figure 10-13 and Color Plate 15). One Voyager team member commented that it looked like the kind of worlds we used to laugh at when they appeared outside the spaceship windows in grade B science fiction movies with cheap special effects.[2] The heating by tidal flexing has kept the interior of Io molten for a long time, and the ensuing eruptions have driven off into space any water that Io may have had in its upper mantle, leaving sulfur compounds instead of ice as the main constituents of Io's "lavas." As exemplified in Figure 10-14, at least eight active volcanoes were found, shooting plumes of material in graceful arcs as much as 250 km into the near vacuum over Io's surface.

Although Jupiter's satellites generally have very cold surfaces (measured dawn

[2]Another analyst commented that he didn't know what was wrong with Io, but it looked as though it might be fixed with a shot of penicillin. Still another wag called it the Pizza Moon, complete with cheese, tomato, and pepperoni strata.

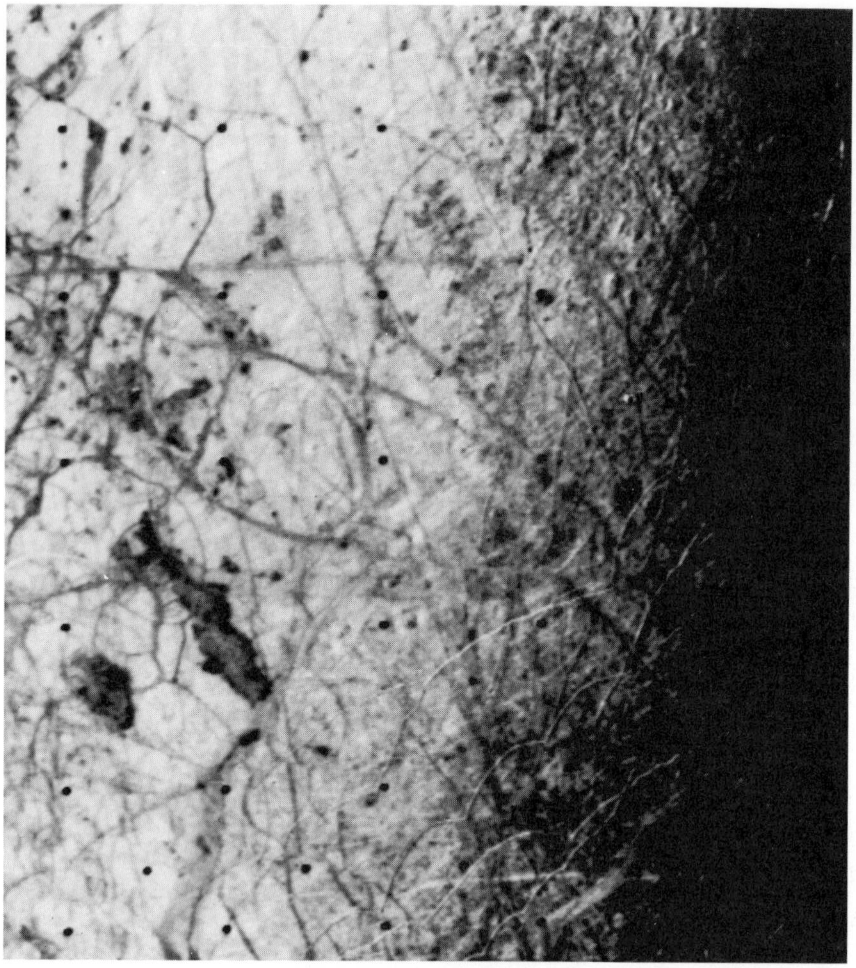

Figure 10-10. *This contrast-enhanced close-up of Europa under low lighting shows a filigree of fine grooves and ridges, possibly related to fractures in the icy surface. The area is 600 × 800 km. (NASA, Voyager photo.)*

and afternoon temperatures range between 80 and 155 K, or −315°F and −180°F), Io's volcanoes are local hot spots. Their measured lava temperatures are as high as 600 to 700 K (621°F to 800°F). Infrared observations from the earth have detected explosive outbursts of Io's volcanoes, confirming the continuous volcanic activity there.

If we could survive the earthquakes, volcanic outbursts, and intense irradiation by ions trapped in Jupiter's nearby magnetic field, a day on Io would be an interesting experience. This experience is depicted in Figure 10-15. All Galilean moons display synchronous rotation—they keep one side toward Jupiter just as tidal locking keeps one side of the moon toward the earth (see Chapter 5). As a result, Jupiter hangs in one spot in Io's sky. It covers an angle of 20°, looking 40 times bigger than our moon in our sky. In Figure 10-15(a), we see the day starting with an eclipse, as the sun emerges from behind Jupiter above a dimly lit landscape. About 15 min later (Figure 10-15[b]), the sun has emerged and the ground is covered by a whitish frost of sulfur dioxide (SO_2), condensed from volcanic gases during the cold darkness of the eclipse. This occasional phenomenon has been glimpsed from the earth.

Io's day—one rotation and one trip around Jupiter—is about 42 h long. In Figure 10-15(c), 10½ h after Figure 10-15(b), we have traveled a quarter of the way around Jupiter. The sulfur dioxide frost has burned off and we see the colored sulfur deposits. A strange phenomenon has happened in the sky. Sulfur and sodium atoms knocked off Io form a permanent cloud around it, and when Io moves toward or away from the sun, these atoms give off a visible yellow glow due to a process involving the absorption of Doppler-shifted sunlight. (The Doppler shift will be explained in Chapter 14.) Thus a faint yellow aurora lights the sky in Figure 10-15(c). In Figure 10-15(d) we have moved another quarter of the way around. The sun is behind us, shadows stretch toward the horizon, and Jupiter is full. In Figure 10-15(e), the sun has set, the landscape is lit by Jupiter, and the yellow glow is back in the sky. Each time Io moves ¼ of the way around Jupiter, incidentally, Jupiter turns once with respect to the stars, and so the same cloud features can be seen in Figures 10-15(c), 10-15(d), and 10-15(e) from different directions. In 10-15(f), we witness a predawn volcanic eruption. The distant volcano spews curtains of ash up into the sunlight (just as high clouds may be lit by sun just before dawn on Earth). The shadow of Io darkens the lower part of the eruptive cloud.

The fifth largest moon of Jupiter is Amalthea (sometimes designated J5), which orbits inside Io's orbit. Close-up photos by Voyager showed it to be an irregular, heavily cratered, rocky object 270 km long and 155 km wide, with the long axis pointed at Jupiter. From its surface, Jupiter would look even more formidable than from Io, covering 41° as shown in Figure 10-16.

Voyager discovered three additional moons, each around 40 km across, close to Jupiter. One orbits between Amalthea and Io, and two are near the outer edge of Jupiter's faint ring of dust.

The discovery of Jupiter's ring, which

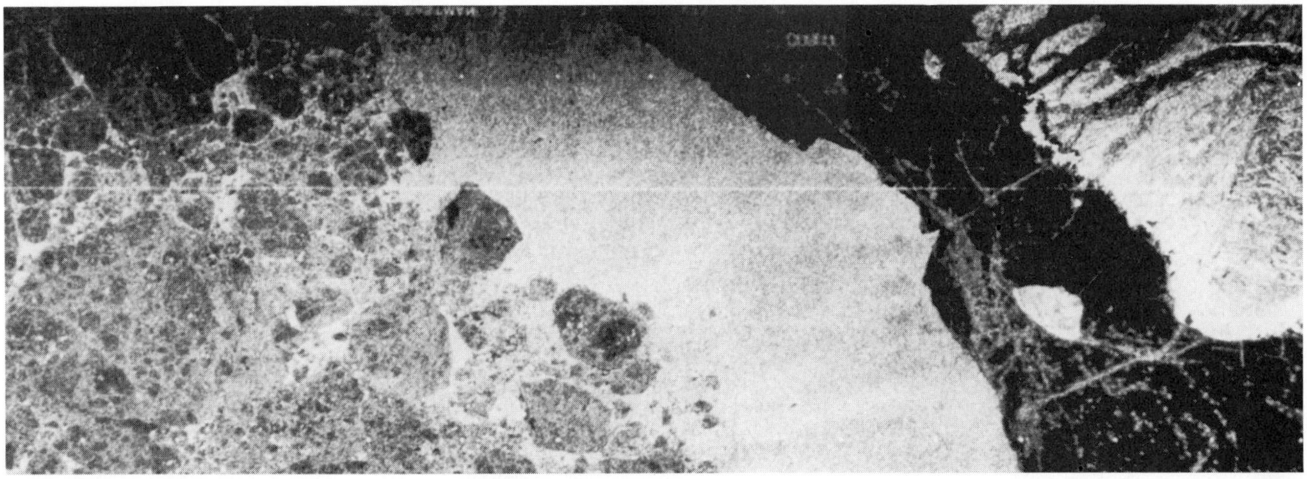

Figure 10-11. *An orbital view of the fractured Arctic Ocean ice pack (left) reveals a fracture pattern similar to those seen on Ganymede and Europa, which may also involve ice slabs floating on water. The dark lane (right) is open sea, which is bordered by an island at right. The length of the image is 120 km. (NASA, Seasat satellite radar image.)*

is too faint and edge-on to be seen from the earth, was an additional surprise of the Voyager flights (see Figure 10-17). It probably consists mostly of microscopic particles being knocked off the inner moons at its edge. Forces acting on such particles make them spiral inward toward Jupiter, thus creating a ring through which material slowly flows over millions of years.

Far outside the orbits of the Galilean moons are two curious groups of small satellites. Satellites J6, J7, J10, and J13 have orbits around 12 million kilometers from Jupiter and inclined 27° to Jupiter's equator. All these moons, as well as the other moons mentioned so far, orbit in a **prograde** direction, like our own moon. But satellites J8, J9, J11, and J12 all orbit in a **retrograde** direction 23 million kilometers from Jupiter with inclinations of around 52°. These moons have dark surfaces of rocky material, resembling surfaces of nearby asteroids. How did the outer satellites come to be clumped into two such groups? Each group may be fragments of an ancient moon smashed by an asteroid, or they may be captured asteroids.

NASA's Galileo program, planned to send more probes to Jupiter in the late

Figure 10-12. *The photo on which Io's volcanoes were first recognized. While analyzing the position of Io in order to help navigate the Voyager spacecraft, researcher Linda Morabito recognized the plume extending off the limb at right. A second eruption creates a bright plume near the terminator. (NASA, Voyager photo.)*

1980s, should tell us more about this mighty planet, its ring, and its strange moons.

The Planet Saturn

The next planet beyond Jupiter is Saturn, famous for its ring system (Color Plates 8, 19, and 20). The globe itself is less interesting than Jupiter's, lacking such striking features as the Great Red Spot. Yellowish and tan cloud zones and belts, diagrammed in Figure 10-18, hide the surface.

As on Jupiter, atmospheric circulation varies with latitude. Rotation periods near the equator are about 10^h14^m, but cloud formations near the poles show longer periods, such as 10^h40^m.

That the mean density of Saturn is the lowest of any planet and less than that of water indicates that Saturn has only a small core of rocky material. Its layer of metallic hydrogen is smaller than Jupiter's. Most of its bulk is molecular hydrogen in either

Figure 10-13. *The surface of Io is mottled by sulfur flows and volcanoes. The doughnut at the center is the outer part of a plume from the erupting caldera at the center. (NASA, Voyager photo.)*

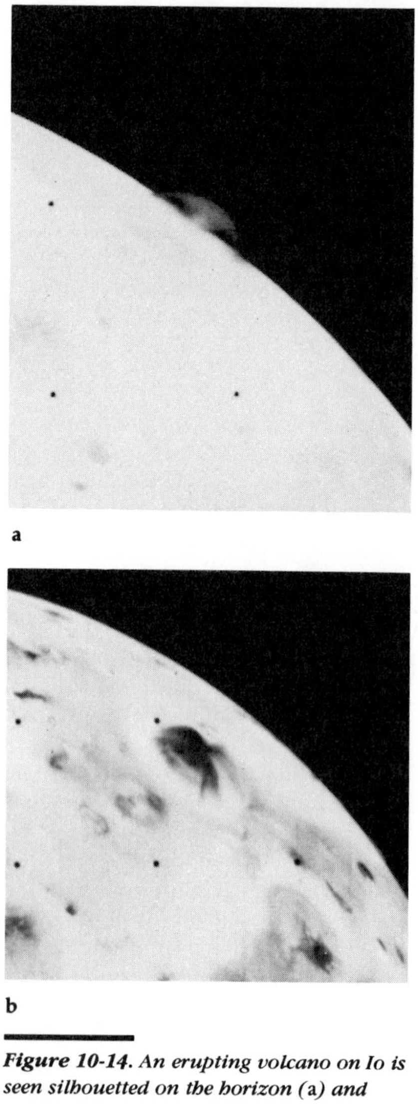

Figure 10-14. *An erupting volcano on Io is seen silhouetted on the horizon (a) and against the background landscape (b). Because there is no air, the cloud does not billow as a terrestrial eruption does; rather, the debris follows curved ballistic paths and falls back to the ground. (NASA, Voyager photo.)*

liquid or highly compressed gaseous form.

Saturn's atmosphere resembles Jupiter's in composition and structure (Table 10-1). Figure 10-19 dramatizes the strong absorption of certain light wavelengths by methane (compare with the absorption by water vapor in the earth's atmosphere, pp. 130–131). Saturn's cloud markings have less contrast than Jupiter's. Nonetheless, small-scale cloud features may be seen in Color Plate 21 (top). Voyager photos, such as Figure 10-20, revealed turbulent structures and stronger eastward jet streams than on Jupiter, with wind speeds up to 450 m/s (1000 mph) eastward along the Equatorial Zone. Only the larger disturbances can be seen from the earth; they can be tracked for many weeks.

SATURN'S RINGS

Galileo's first sight of Saturn, in 1610, was a strange one: a blurry disk with blurry objects on either side. He drew it as a

***Figure 10-15.** A day on Io. Sequence shows events as seen from one spot on Io as Io moves around Jupiter in 42 h. See the text for further description. (Paintings by author.)*

triple planet. Not until 1655 did Christian Huygens discover that a **ring system** encircled the planet.[3] A small modern telescope easily shows the astonishingly beautiful rings, largest in the solar system. The rings' proportions are surprising. They stretch about 274 000 km (171 000 mi) tip to tip, but they are probably less than 100 m thick! The system is so thin that terrestrial observers lose sight of it altogether when it appears edge-on for a few days once every 15 y as the earth passes through the ring system's plane (see Figure 10-21).

The rings are composed of billions of separate particles, too small to be seen individually from earth or in Voyager photos. This was proved by a stepwise combination of theory and observation. In 1895, Scottish physicist James Clerk Maxwell showed that the rings could not be a solid disk because they are inside Roche's limit, where a disk would be pulled apart into myriads of particles. In 1895, American astronomer James Keeler confirmed spectroscopically that different parts of the rings move at different speeds, each in its own orbit around Saturn. Modern proof that the rings are not solid comes from spacecraft photos showing that we can see through them, as seen in Figure 10-22. In 1970, American astronomers proved spectroscopically that the ring particles are composed of, or covered by, frozen water (Lebofsky, Johnson, and McCord, 1970). Various measurements from the earth and from the Voyagers show that common particles in the rings range from ping pong ball to house sizes. Some particles of larger or smaller size may also

[3]Huygens announced this discovery in the form of a famous anagram (succession of apparently meaningless letters). Huygens later explained that the anagram was to represent the sentence "Annulo cingitur, tenui, plano, nusquam cohaerente, ad eclipticam inclinato" ("It is surrounded by a thin, flat ring, nowhere touching, inclined to the ecliptic"). In the days before copyrights, such anagrams were a common form in which scientists published initial results, thus establishing priority while giving themselves time for more observations. Not until about 1661 were most observers convinced of Huygens's discovery (Alexander, 1962).

Figure 10-16. *A spectacular view of Jupiter could be seen from its closest satellite, the small moon Amalthea. Jupiter dominates the sky, subtending an angle of about 41°. (Painting by James Hervat.)*

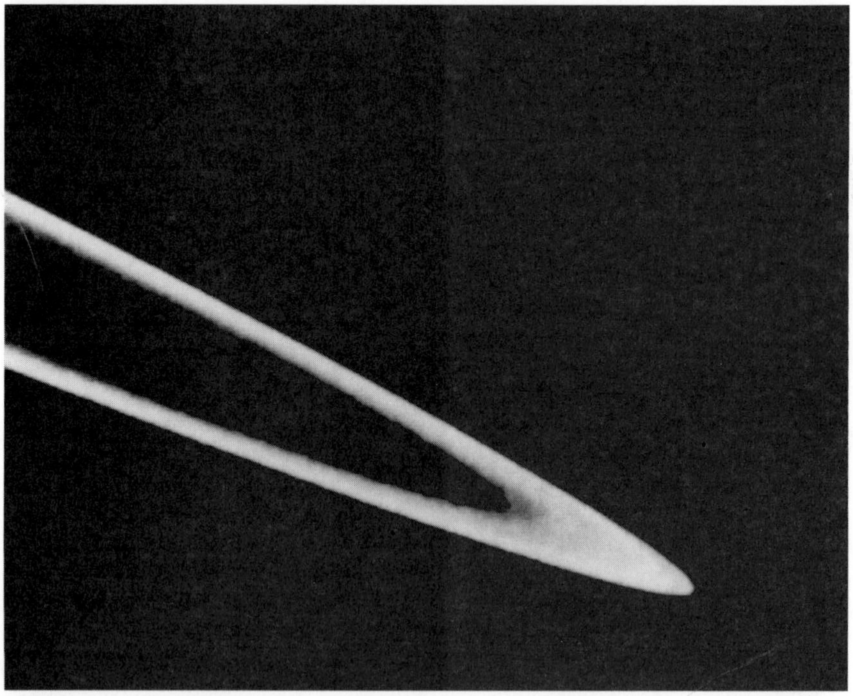

Figure 10-17. *Jupiter's ring is believed to be fine material spiraling in toward Jupiter from small moons near the outer edge of the ring. Jupiter is out of the picture at left. (NASA, Voyager photo.)*

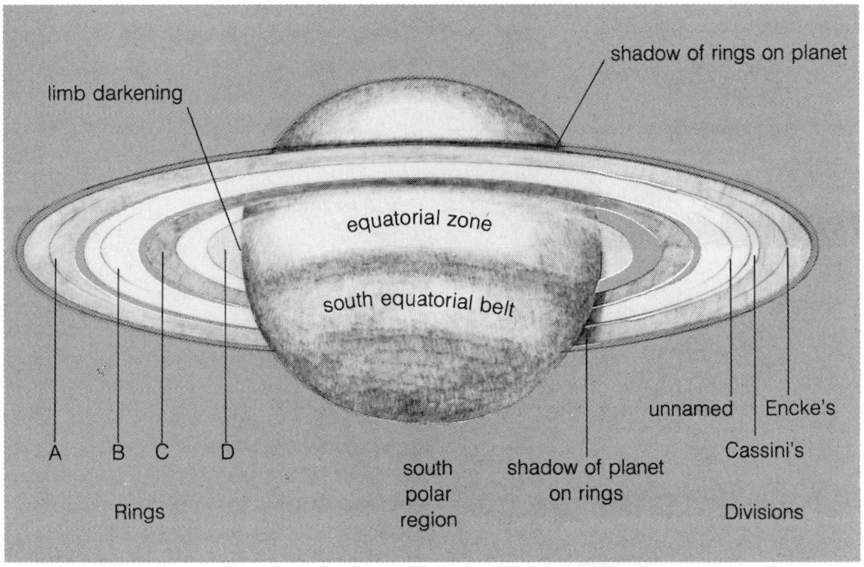

Figure 10-18. *Features of Saturn. Telescopes with apertures as small as 5 cm (2 in.) will show the rings; telescopes larger than about 25 cm (10 in.) will sometimes show all these features.*

Figure 10-19. *Photographs of Saturn with films and filters sensitive to different colors of light. Top = blue (450 nm), the most common view; bottom = wavelength absorbed by methane gas (898 nm). Abundant methane in the atmosphere absorbs most of the light, while the rings remain bright, being covered with ice that reflects sunlight at this wavelength. (Lunar and Planetary Laboratory, University of Arizona.)*

Figure 10-20. *Close-up of Saturn reveals cloud bands with turbulent structure, but fewer prominent structures than on Jupiter. (NASA, Voyager 1 photo.)*

Figure 10-21. *Varying aspects of Saturn during its 29-y orbit around the sun, as photographed with Earth-based telescopes. Because the rings maintain a fixed relation to the ecliptic, the Earth-based observer sometimes sees from "above" and sometimes "below" the ring plane. This series shows half the 29-y cycle, including two views that show the disappearance of the rings as the earth passes through the ring plane. (Lowell Observatory.)*

exist. The remarkable view from within the rings is suggested by Figure 10-23.

The rings are divided by several gaps. The biggest is called **Cassini's division,** after G. D. Cassini who discovered it in 1675 (see Figures 10-18, 10-24, and Color Plate 20). A dusky ring region outside Cassini's division is called Ring A; the brighter region just inside it is called Ring B. Fainter inner ring regions are C and D (see Figure 10-18). Ring A is cut by Encke's division; other finer divisions have been mapped by telescopic observers.

Scientists were astonished when Voyager 1 and 2 photos showed not just a few major divisions, but thousands of fine-scale divisions and ringlets, such as seen in Figure 10-24. Wavy ring-edges, gentle twists in ringlets, and transient radial shadings were also found, as seen in Figure 10-25. What maintains such intricate structure? One theory was that unseen moonlets clear swaths among the ring particles, create wavy edges, and maintain ringlets by their gravitational effects. Supporting this theory was the Voyager 1 discovery of two 200-km moonlets straddling the narrow F ring just outside the A ring, confining the F ring to a narrow zone. These two moons came to be called "shepherd"satellites. Voyager 2 searches failed to turn up other "shepherd" moons among the rings, down to a diameter of a few kilometers. Nonetheless, ring structure may result from a combination of poorly understood gravitational effects caused by known moons and still-unseen moonlets.

Voyager radios picked up distinct static signals caused by electrical discharges in the rings. This discovery suggests that electrostatic forces may be more important in the ring particle interactions than hitherto suspected.

How did such a remarkable ring system form? A clue is that any particles in the location of Saturn's rings would have to form a flat ring system. Mutual collisions plus gravitational forces from Saturn's oblate mass would force the particles into Saturn's equatorial plane. Since they are inside Roche's limit, tidal forces prevent them from coalescing into a large satellite.

The problem remains of where the particles came from in the first place. They may have been:

1. particles condensed from gas as Saturn formed,

2. part of a satellite blown apart by collision with a comet or asteroid, or

3. part of a comet or asteroid broken apart by tidal forces after approaching too close to Saturn.

In addition, new ring particles may be added by erosion of small moons in or near the rings as they are hit by meteorites. Close-up study of the individual particles might reveal whether one of these theories is correct.

Figure 10-22. *Seeing* through *Saturn's rings. In this close-up view from Voyager 2, we are looking through the rings (diagonal grey bands). The curved edge of Saturn's brightly lit globe is in the distance at right. The A ring fills the middle part of the picture and is partially transparent. Above it (upper right) is Cassini's division, through which we can see clearly. The B ring (upper right corner) is thickly crowded with particles and is opaque. Other thin gaps and ringlets are visible. (NASA.)*

Figure 10-23. *In this imaginary view we are floating among the icy particles within Saturn's rings. Saturn's globe is at left, and the distant sun (right) backlights the scene. The rings, perhaps less than 100 m thick, stretch off into the distance, lower center. (Painting by author.)*

Saturn's Satellites

Figure 10-24. ***Detailed photo of Saturn's rings shows hundreds of narrow ringlets divided by narrow divisions of uncertain origin. Cassini's division and A ring beyond it are at top. Spokelike dusky shadings are possibly clouds of fine particles disturbed by electromagnetic forces. These "spokes" were unexpected before the Voyager pictures; their origin remains unclear. (NASA, Voyager 2 photo.)***

Figure 10-25. ***Orbital motion of ring material around Saturn is shown in this sequence of photos, taken half an hour apart. Note the movement of the dusky "spokes" from one position to another. (NASA, Voyager 1 sequence.)***

Saturn's satellite system, like Jupiter's, is extensive and includes exotic worlds. There are at least 17 moons (see Table 6-1, page 108), including a host of small moons near the rings, a small outer moon, and a grouping of larger moons at intermediate distance, including the giant moon Titan.

If we start on the outskirts of the rings, we find five small moons with diameters between 20 and 220 km, mostly discovered in 1979 and 1980 by Voyager 1 and by intense earth-based observation during a brief period when the rings were seen edge-on and thus did not obscure the satellites by their glare. Voyager photos have shown some of these small moons to be lumpy and cratered (Figure 10-26). Bright white tones of at least some of them suggest icy surfaces, but their interior composition is uncertain. They seem to be battered iceballs.

The next moon is Mimas, 390 km across, round, icy, and heavily cratered with one very large, fresh-looking crater (Figure 10-27). Mimas's bright surface (60% reflectivity) and low bulk density (1200 kg/m^3 compared to 1000 for pure ice and around 3000 for rock) indicate that Mimas is composed mostly of ice. Color Plate 20 shows the pale creamy color of Mimas.

The next satellite, Enceladus, is especially fascinating as a "missing link" between Jupiter's moons Europa and Ganymede: Enceladus (Figure 10-28) has the linear grooves and bright, sparsely cratered, icy surface of Europa, but it also has some moderately cratered, fractured regions that resemble Ganymede. It has the brightest surface known in the solar system. Scientists suspect that Enceladus may have a young surface created by eruptions of water and ice crystals from an interior heated by tidal interactions with other Saturnian moons. Calculations like those used to predict Io's volcanoes show that Enceladus would be the Saturnian moon most strongly heated by this mechanism, although the calculations suggest tidal heating is barely adequate to produce such activity. The mean density of Enceladus, 1100 kg/m^3, indicates that Enceladus is mostly icy throughout, so that radioactivity from rock minerals is an unlikely heat source. Nonetheless, researchers have discovered a so-called E ring of tiny particles extending far beyond the main rings of Saturn, and having a concentration in Enceladus' orbit (see Figure 10-29). Together with the cracks, brightness of the ice, and presence of sparsely cratered young regions, the particles suggest that occasional eruptions of water may recoat the surface with fresh ice and blow some ice grains clear off the surface. Enceladus bears further study!

The next three moons—Tethys, Dione, and Rhea—are around 1000 to 1500 km across, and have bright, icy surfaces. Tethys has a bulk density of only 1000 kg/m^3, and hence must be nearly pure ice. Dione and Rhea have densities of around 1300–1400 kg/m^3 and probably have some rocky materials mixed with their ice. These

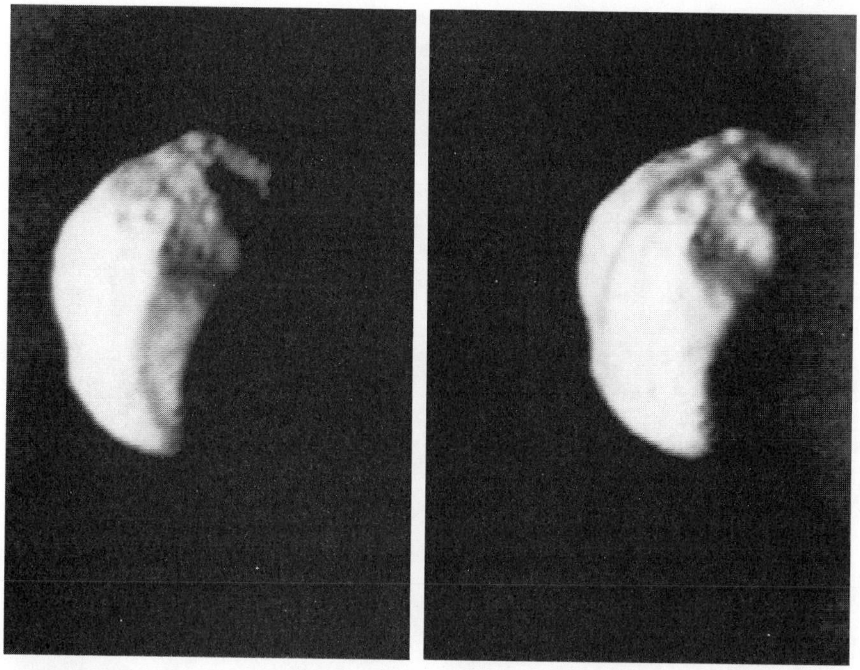

Figure 10-26. *Saturn's satellite S11 is a 70 × 135 km object, probably eroded into its irregular shape by intense cratering. These two views, taken 13 min. apart, show the shadow of Saturn's nearby rings moving across the satellite. (NASA, Voyager 1 photo.)*

Figure 10-27. *Saturn's small satellite Mimas, 390 km across, is marked by a crater that is unusually large relative to the satellite itself. This crater is about 130 km (81 mi) across and has a large central peak that rises about 6 km (20 000 ft) from the crater floor. The impact that made this crater was almost large enough to disrupt Mimas. Voyager scientists nicknamed this the "Darth Vader Deathstar picture" for its fancied resemblance to the* Star Wars *villain's space station with its large, circular landing port. (NASA, Voyager 1 photo.)*

worlds are moderately to heavily cratered, as seen in Figures 10-29 to 10-31. However, Tethys's surface is broken by a huge canyon system and Dione's by swaths of light-toned material, reminiscent again of Ganymede. Rifting and fracturing of the surfaces may have occurred due to some source of internal heat.

In 1980, earth-based observers discovered interesting additional small moons roughly 20 to 50 km across in the orbits of Tethys and Dione. One orbits 60° behind Tethys; one, 60° ahead of Tethys; and the third, 60° ahead of Dione. Gravitational forces make this 60° point a stable location for small objects. Certain asteroids are known in similar 60° points ahead of and behind Jupiter. These discoveries expand the list of similarities between satellite families and the solar system as a whole—supporting the concept of satellite families as miniature solar systems.

We called Io the most bizarre moon, but Saturn's largest moon, Titan, is a close runner-up. With a diameter of 5140 km (3200 mi) it is probably the second-largest moon in the solar system, and the only moon with a thick atmosphere. Methane (CH_4) was discovered from spectra of Titan in 1944. Observations in 1973 showed that Titan's sky is not clear, but rather filled with reddish haze, seen in Color Plates 8 and 21 bottom right. Later observations showed that this haze is photochemical smog produced by reactions of the methane and other compounds when they are exposed to sunlight—like the smog produced by the action of sunlight on hydrocarbons over Los Angeles. Titan is the smoggiest world in the solar system.

The Voyagers showed that the methane and smog are not more than 10% of the atmosphere. The main constituent is nitrogen, meaning that the atmosphere is denser that realized before the Voyager flights. Voyager 1 discovered that the surface air pressure is 1.6 times that on the earth! Since the earth's air is also mostly nitrogen, Titan's atmosphere offers fascinating comparisons to present or primeval conditions on earth. The main difference is that Titan is very cold, around 93 K (−292°F). Minor constituents detected in Titan's air include ethane, acetylene, ethylene, and hydrogen cyanide. The abundance of these organic molecules suggests that Titan offers a good natural laboratory for research on the origins of life.

Based on the measurements of temperature and pressure at Titan's surface, researchers visualize dramatic weather conditions, illustrated in Figure 10-32. Methane may exist not only as a gas, but may be able to rain out of the clouds and exist as snow or ice, playing the same triple role of gas, liquid, and icy solid as water does on the earth. In addition to methane (CH_4) meteorologists conclude

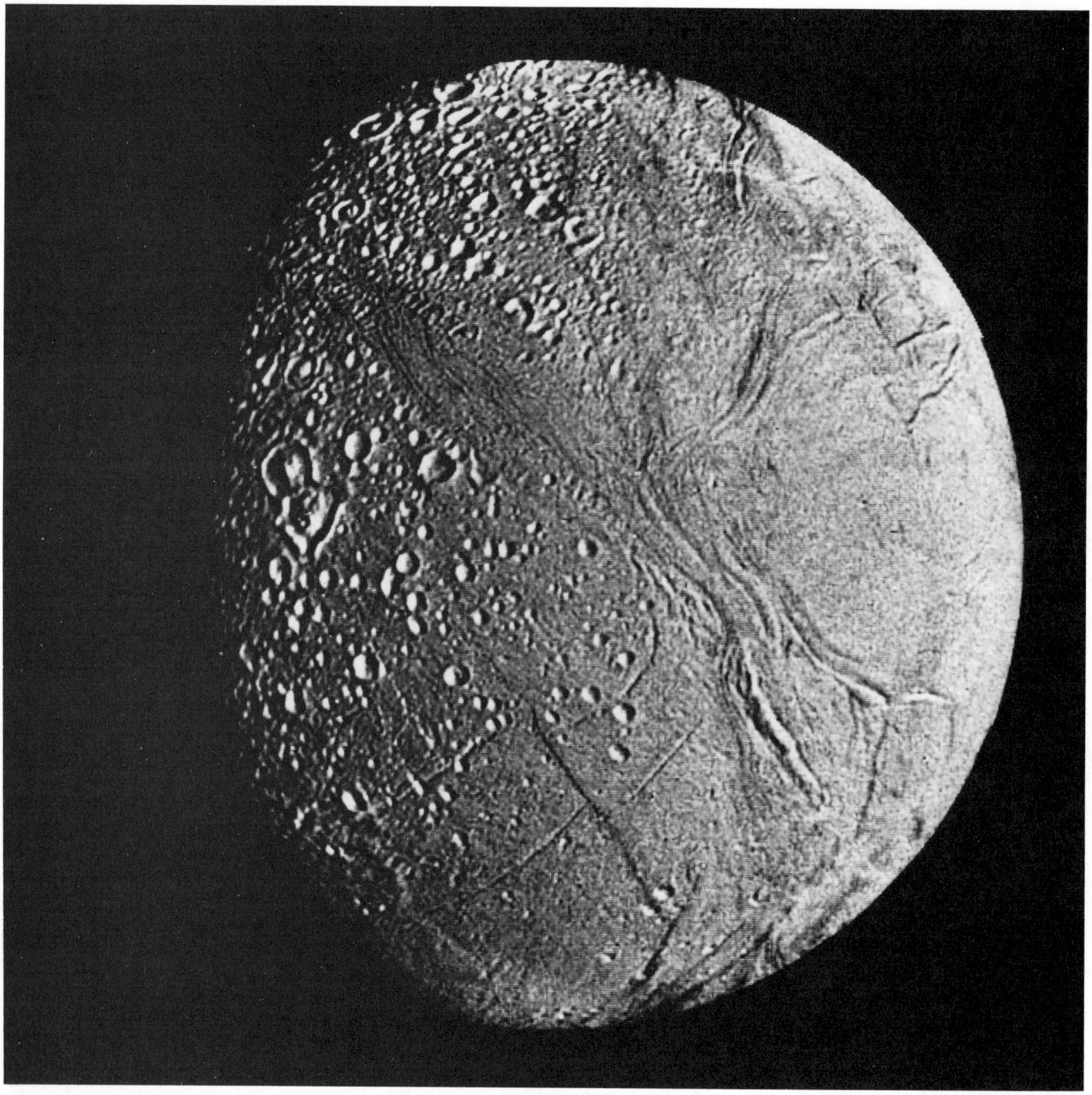

Figure 10-28. *Saturn's moon Enceladus offers a missing link between the "cracked billiard ball" surface of Jupiter's moon Europa, and the cratered, fissured surface of Jupiter's moon Ganymede—though Enceladus is only one-sixth the size of Europa. Portions of the ancient cratered surface seem to have been cracked as water erupted and froze into smooth ice plains, obliterating older craters— for example, bottom central region. This suggests that Enceladus may have been heated enough to melt portions of its icy interior at some time in its history. (NASA, Voyager 2 photo.)*

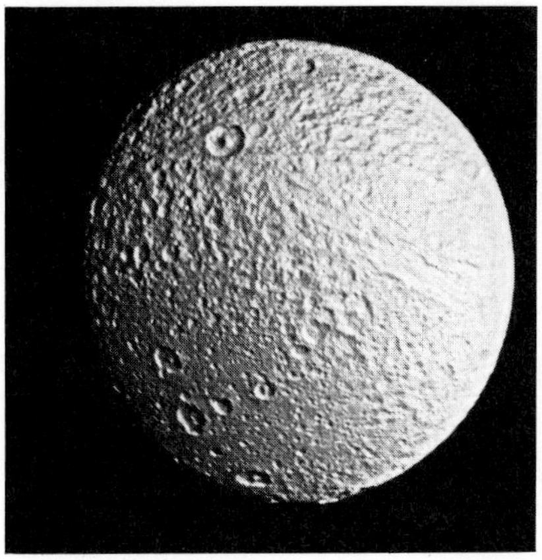

a

b

***Figure 10-29.** Views of Saturn's moon Tethys. (a) The heavily cratered globe of Tethys, about 1050 km across, is cut in places by* broad valleys *or* rift systems *(left center). These may indicate a period of internal activity that split the original rigid surface. (NASA, Voyager 2 photo.) (b) This imaginary view on the surface of Tethys shows the beginning of an eclipse of the sun by the globe of Saturn. The fuzzy extension of the rings (upper left) is the E ring, backlit by the sun. It is composed of tiny particles believed to have been ejected from the neighboring moon Enceladus, visible at upper left. (Painting by author.)*

***Figure 10-30.** The trailing hemisphere of Dione is covered by darkish material broken by an array of bright swaths, possibly frost deposits. The leading hemisphere is brighter, more uniform in tone, and heavily cratered. (NASA, Voyager 1 photo.)*

***Figure 10-31.** The intensely cratered surface of Saturn's moon Rhea, believed to be composed primarily of ice. (NASA, Voyager 1 photo mosaic.)*

that ethane (C_2H_6) would form in the atmosphere, slowly drizzle out, and accumulate on the ground. Thus, the surface of Titan may be largely covered by a cold ocean of liquid methane and liquid ethane estimated to be a kilometer deep (Lunine et al., 1983). Still more complicated, gasolinelike compounds may form in the smog and rain out of the hazy clouds. As these facts unfolded during the Voyager mission, one Voyager scientist characterized Titan as "a bizarre murky swamp," shown in Figure 10-32(b). More interesting to scientists are the possible biochemical reactions among these organic molecules. Life is unlikely in the cold temperatures of Titan, but biochemists would be very excited to learn what complex organic compounds have formed on Titan. Titan may be a natural laboratory for studying primordial biochemical evolution on the earth! Titan remains a fascinating target for future exploration and NASA has considered long-range plans to parachute probes beneath the mysterious clouds to see what is really on the surface.

As if to reemphasize the rule that each world is unique, the next satellite is completely different. Hyperion is a biscuit-shaped chunk roughly 360 km across and 210 km thick. As seen in Figure 10-33, Voyager 2 photos showed an irregular, cratered shape. Hyperion's long axis is not quite lined up with Saturn, as would be

a b

Figure 10-32. *Scenes on Titan.* (a) *From above the upper cloud layer visitors would see Saturn hanging in a blue sky with an appearance perhaps not unlike Earth's sky. (Painting by author.)* (b) *In the dusky light below the clouds, Saturn would be obscured. Measurements of temperature and pressure at the surface suggest snowstorms and rainstorms of solid and liquid methane compounds, with possible rivers, "waterfalls," and oceans of liquid methane or liquid ethane, or a mixture of the two. (Painting by Ron Miller.)*

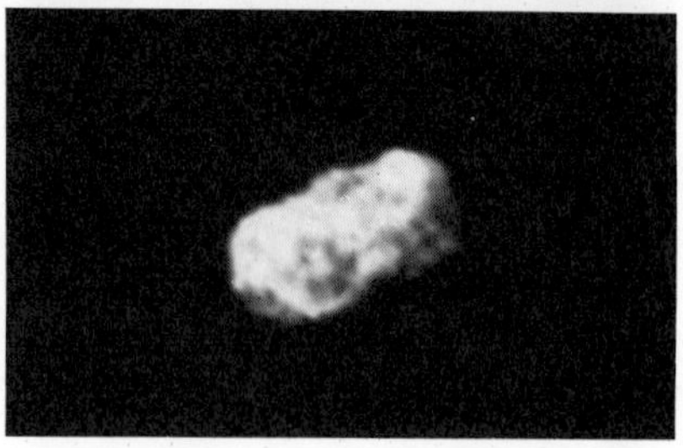

Figure 10-33. *Top view and side-on view of Saturn's biscuit-shaped satellite, Hyperion. The irregular shape may have been created by a major meteorite impact. Hyperion is about 360 km long. The side-on view (bottom) was made from a greater distance and is less sharp. (NASA, Voyager 2 photo.)*

expected if tidal forces had acted over a long period (see page 85). Its rotation rate varies irregularly, apparently due to forces associated with neighboring moons. Some scientists believe Hyperion may be an oddly shaped, wobbling piece of a former moon broken by an asteroid impact.

The next outer moon, 1440-km (900-mi) Iapetus, is another unique world. The hemisphere on the leading side in Iapetus' orbital motion around Saturn is covered with dark, frosty soil as black as tar, but the trailing hemisphere is covered with bright, white ice. As seen in Color Plate 21 (at the bottom left), the difference between the two sides is striking. What could create this black-and-white moon? The answer almost certainly involves Iapetus' neighbor moon, Saturn's outermost satellite Phoebe, a 220-km (140-mi) black-surfaced moon moving around Saturn in a retrograde orbit. Black dust knocked off Phoebe by meteorites would spiral inward toward Saturn and hit the leading side of Iapetus at high speed. Darkening of Iapetus' leading-side ice layer by black dust from Phoebe may partly explain Iapetus' dark leading side. But Iapetus' dark side is more reddish-black than Phoebe; some scientists think the story must be more complex. They theorize that reddish-black organic minerals form from methane in regions of fresh Iapetus ice disturbed by dust from Phoebe. Moons can be complicated places!

The Uranus System

In 1781 William Herschel became the first known person to discover a new planet when he detected Uranus during an ambitious star-mapping project. Uranus is so far from Earth that an ordinary telescope reveals almost no markings on it, as indicated by the view in Figure 10-34. But in 1986, Voyager 2 flew past Uranus, giving us our first close-ups of its bluish haze, its dark rings, and its unexpectedly varied moons.

The planet's name was suggested by J. Bode (of Bode's law) because in mythology Uranus was the father of Saturn, who in turn was the father of Jupiter. Because of Uranus' long, 84-y period of revolution around the sun, it has made only two orbital circuits since its discovery.

Herschel found several satellites of Uranus as well. The orbits of these satellites lie in the plane of the planet's equator, but are steeply inclined to the planet's *orbital plane* by 98°. As the French dynamicist Laplace noted in 1829, steep inclination to the orbital plane can arise only if the planet has a substantial equatorial bulge, whose gravitational attraction holds the satellites over the planet's equator. Thus, Uranus itself must have an equatorial bulge, and its equator must be inclined 98° to its orbital plane, an unusual planetary configuration shown in Figure 10-35. This **obliquity** (deviation of the equator from the plane of orbit) being greater than 90° means that the **rotation of Uranus** is *retrograde,* as illustrated in Figure 10-35.

Thus Uranus has a unique seasonal sequence. When the north pole points almost directly toward the sun, the southern hemisphere is plunged into a long, dark winter, lasting for about a quarter of the planet's 84-y revolution. After this 21-y south polar winter and north polar summer, the sun shines on the equatorial regions. Each point on the planet now goes from day to night during the planet's 17-h rotation. After 21 more years the south pole points approximately toward the sun and the southern hemisphere experiences a 21-y summer.

The 1986 Voyager flyby of Uranus revealed that Uranus' atmosphere has a deep layer of almost featureless haze overlying deeper clouds. Unlike Jupiter and Saturn, Uranus thus reveals very few cloud markings (Figure 10-36). The haze gives Uranus a sky-blue color, similar to the color of Earth but tinged with green. The haze layer is very cold—Voyager recorded a temperature minimum of 51 K (−368°F). However, the temperature increases down into the clouds. Observations from Earth, as early as the 1930s, revealed methane gas (CH_4) in Uranus' atmosphere, and in the 1960s an even greater abundance of hydrogen (H_2) was found. Voyager gave much more detail. Preliminary results indicate a composition matching that of Jupiter and Saturn, dominated by hydrogen and helium.

Calculations suggest that the inner quarter of the planet's diameter is occupied by a central rocky core, perhaps molten, similar to Earth in size. This core is surrounded by a deep ocean of hot, high-pressure, liquid water. Above this is the deep atmosphere comprising the outer half of the planet.

***Figure 10-34.** Uranus (overexposed) and its five satellites, photographed with the McDonald Observatory 2-m telescope. (NASA photo, Lunar and Planetary Laboratory, University of Arizona.)*

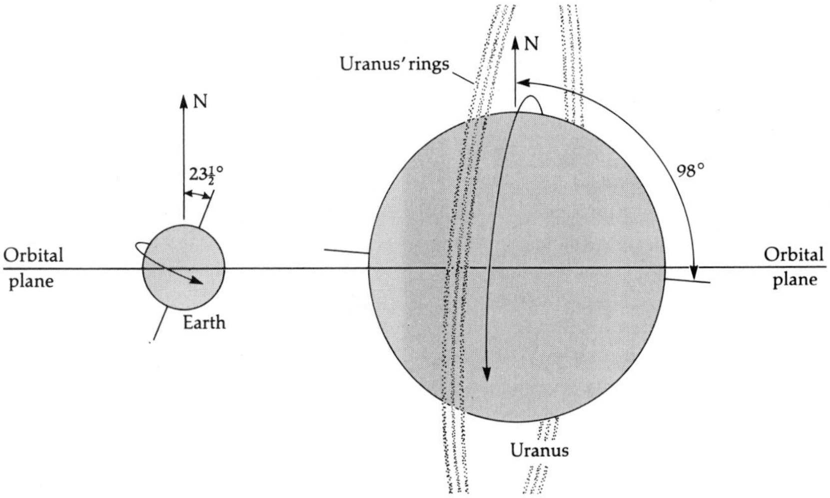

***Figure 10-35.** Comparison of the sizes and rotations of the earth and Uranus. The earth has an obliquity (or axial tilt) of 23½° and a prograde (west-to-east) rotation. Uranus has much steeper obliquity and retrograde rotation.*

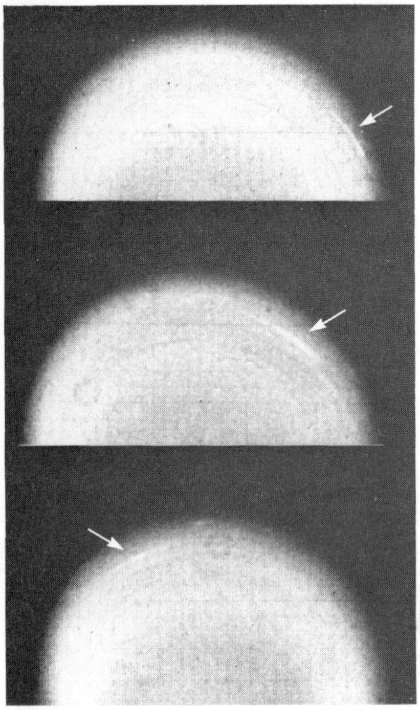

***Figure 10-36.** Three Voyager photos of Uranus, a few hours apart, show only one or two bright cloud features in the bland, hazy atmosphere. Arrows in this nearly pole-on view show the movement of a bright cloud at low latitude as the planet rotates. (NASA Voyager 2 contrast-enhanced photos; faint dark "doughnuts" in images 2 and 3 are camera defects.)*

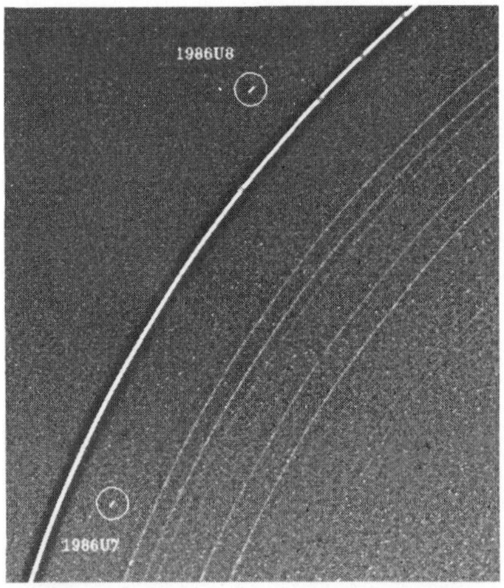

***Figure 10-37.** (Left) Portion of the rings of Uranus and two "shepherd moons" (circled), whose gravitational pulls help keep ring particles confined to narrow widths. Width of widest, outer ring is only 100 km (60 mi). All nine rings appear; inner three may be glimpsed by holding up book and sighting along the line of the rings. (NASA Voyager 2 photo.)*

***Figure 10-38.** (Bottom) Uranus' moon Ariel is about one-third the size of our moon. Its cratered surface has been intensely fractured, possibly due to an ancient episode of internal heating and expansion. (NASA photo mosaic from Voyager 2.)*

In 1977 Uranus passed in front of a relatively bright star. A number of astronomers watched, expecting to see the star dim as it passed behind Uranus' upper atmosphere and thus to learn about the haze layer's structure. They were astonished to see, in addition, the star dim several times at some distance from the Uranian disk, but symmetrically on each side of the disk. These observations marked the discovery of narrow rings of dark material around the planet. The rings blocked the starlight as Uranus moved. Voyager 2 obtained the first clear pictures of these rings (Figure 10-37), which are much narrower than the rings of Jupiter or Saturn. Scientists do not fully understand what confines Uranus' rings into such narrow bands, although "shepherd moons," like those on each side of Saturn's narrow F ring, are believed to be involved. Voyager cameras discovered two shepherd moons" straddling the thickest Uranian ring (Figure 10-37). Gravitational forces of other undiscovered moonlets may help confine the rings. Further research is under way to clarify these curious "shepherding" effects.

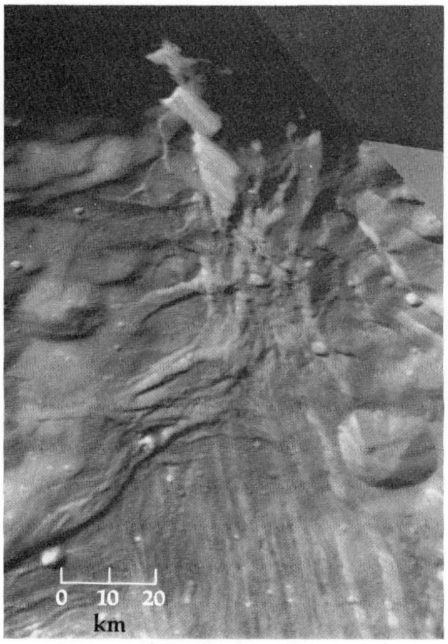

Figure 10-39. *Detailed view of a system of fractures, faults, and valleys on Uranus' moon Miranda. Compare with the full-disk view of Miranda (facing page 1), where this valley appears near the top. At the top, extending onto the night side, is an enormous cliff with an estimated height of 5 km (16 000 ft) and a slope of about 30°. Much of the region to its right, and in the foreground, has slumped along parallel fractures. (NASA Voyager 2 photo; upper right corner, outside original frame, printed as uniform gray here.)*

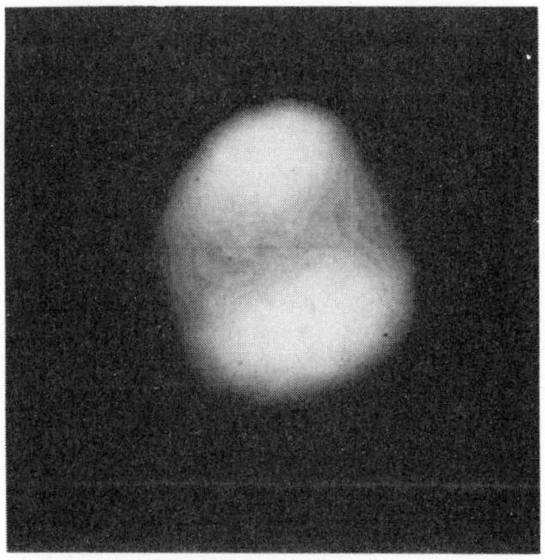

Figure 10-40. *The cloud belt pattern on Neptune, detected in 1979 by photographing through a filter centered on a color absorbed by methane gas. High clouds reflect light at high latitudes and appear bright. A cloud-free band along the equator reveals deeper methane layers that absorb this light. Images at other visible wavelengths are blank, suggesting that the markings are not as prominent as on Jupiter. (Lunar and Planetary Laboratory, University of Arizona, courtesy H. J. Reitsema.)*

Before Voyager, five major moons of Uranus were known and named after characters in Shakespeare's *Midsummer Night's Dream* (see Table 6-1, p. 109). The largest, Titania, was a mid-sized moon about the size of Saturn's Rhea. Voyager discovered ten more smaller, dark moons near the outer edge of the rings, including the two "shepherds" mentioned above. The five large moons, as revealed by Voyager, turned out to be much more complicated than expected. They are composed of ice (primarily frozen H_2O) and black soil. Umbriel, middle of the five, is darker than the others, but the reason for its excess of surface soil is unknown. All the moons are heavily cratered; on several, the ancient regions are cut by systems of younger cracks and canyons (Figures 10-38 and 10-39, frontispiece facing page 1). These cracks indicate strong internal activity, perhaps due to internal heating, but the causes are still unknown.

The Planet Neptune

The discovery of Uranus led to the discovery of Neptune. After 1800, theorists tried unsuccessfully to fit observations of Uranus' position into Kepler's laws of planetary motion. Uranus seemed to have its own somewhat irregular motions. A few scientists thought this might signal a breakdown of Newton's law of gravity at large distances from the sun. Others correctly suggested that Uranus was being attracted by a still more distant planet. In the 1840s an English astronomer and a French mathematician set out independently to predict where the new planet should be.

Both men tried to shorten their laborious calculations by following Bode's rule, assuming that the new planet would be 38 AU from the sun.[4] Unfortunately, Neptune lies at 30 AU—a serious exception to Bode's rule. The English astronomer J. C. Adams finished his prediction in 1845 after two years of calculations, but had trouble getting his senior professors at Cambridge interested in searching for the new planet with a telescope. In 1846 English astronomers began a desultory search. Several times they actually charted the new planet but failed to realize what they had seen.

Meanwhile, the French mathematician Leverrier, who had finished his work soon after Adams, interested two young German astronomers in the search. Armed with Leverrier's predictions, they located the new planet within half an hour of starting their search on September 12, 1846. Although Adams and Leverrier are both now credited with the discovery, the incident became an international scandal at the time because of the Britons' failure to grasp their opportunity.

Methane has been found in Neptune's atmosphere, and the planet is believed to resemble Uranus in general. As seen in Figure 10-40, faint cloud patterns resemble the bands of Jupiter and Saturn. It apparently rotates in about 18 h.

Observers have attempted to repeat the successful discovery of Uranus' ring by watching Neptune pass in front of stars. One blockage of starlight was seen during a 1981 event, none during a 1983 event, and several during a 1984 event. The 1984 observations indicated an object 10 to 20 km wide but at least 90 km long, so some observers now think Neptune has a set of strange, incomplete, arc-shaped rings that

[4] Today the necessary calculations could be done quickly on computers, but in the 1800s they took months or years.

will be revealed when Voyager passes by Neptune in August 1989. If so, theorists will again be challenged to explain the forces that maintain particles in arc-shaped segments instead of complete rings.

Neptune has a satellite system that is peculiar: the largest satellite is moving in a circular but highly inclined retrograde orbit, instead of the low-inclination prograde orbit common for large moons. This satellite, Triton, has an estimated diameter of 3400 km or somewhat larger, making it one of the largest moons in the solar system (see Figure 6-2). As with several other moons we have discussed, the more observations that are obtained, the more unique Triton becomes. Spectroscopic observations in the late 1970s indicate that Triton has a thin atmosphere of methane (CH_4) gas and may also have frozen methane ice on its surface (Cruikshank and Silvaggio, 1979). More recent spectroscopic studies suggest the presence of liquid nitrogen. If these studies are correct, Triton may be partially covered by an ocean of liquid nitrogen at least 10 cm deep, and may have a nitrogen atmosphere that is thicker (and much colder) than that of Mars (Cruikshank et al., 1983). Perhaps there are three solar system worlds with oceans: Triton with its liquid nitrogen, Titan with its liquid ethane–methane, and Earth with its unique ocean of liquid water!

Triton is strange in another respect. Due to tidal forces it is slowly spiralling in toward Neptune. Analysts predict it will eventually crash into Neptune. Neptune's only other known moon is only a tenth as large as Triton and moves in a highly elliptical prograde orbit, contrasting with Triton's circular orbit. The two orbits seem so unusual that many theorists think they may have been disturbed at some time in the past by passage of a large body (a giant comet or Chiron-like body?) through the system.

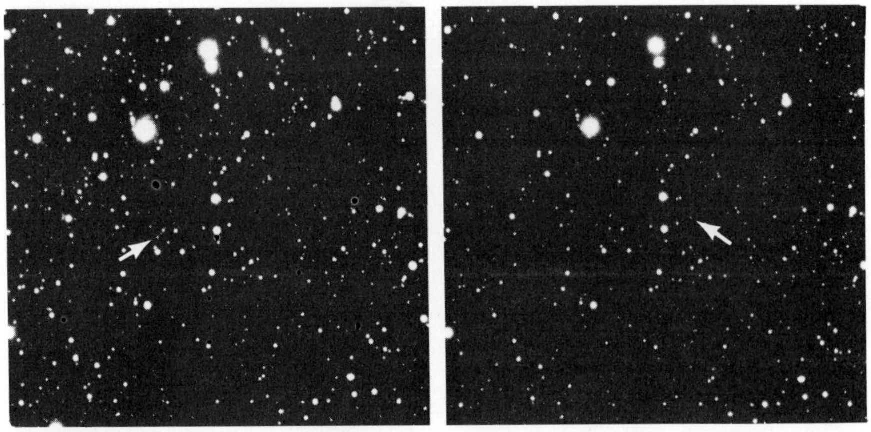

Figure 10-41. *Portions of the photographs on which Pluto (arrow) was discovered. The planet was distinguished from stars by its motion. (Lowell Observatory.)*

The Planet Pluto

Because the motions of Uranus and Neptune were still unaccountably irregular, Percival Lowell began a search for a ninth planet in 1905. As shown in Figure 10-41, this search led to discovery of Pluto in 1930, when Clyde Tombaugh discovered the faintly moving object on Lowell Observatory photos. Pluto lies close to the position predicted by Bode's rule for a planet, about 39 AU from the sun. After it was named for a god of the underworld, a new planetary symbol (♇) was created from the first two letters, which were also Lowell's initials.

Pluto is almost too small to qualify as a planet. It is probably a bit smaller than our moon and has only 0.2% the mass of the earth. Its light variations indicate a rotation every 6.39 d on an axis tipped about 120° to its orbit. The density is roughly measured to be about 500 to 1300 kg/m^3 (½ to 1.3 g/cm^3), with a composition estimated to be 74% water ice, 21% rock, and 5% methane ice by mass, according to a 1980 model. Spectroscopic observations in the late 1970s indicated a very thin atmosphere of methane (CH_4) gas, probably derived from methane ice on parts of the surface, and possibly accompanied by additional, unknown gases.

Astronomers in 1978 discovered that Pluto has a satellite about half as big as the planet—the largest size ratio of any satellite/planet pair. The satellite has been named Charon, after a mythological figure associated with Pluto. Charon goes around Pluto in 6.39 d, the same interval in which Pluto rotates, so the two are believed to keep the same faces pointing toward each other.

Pluto's orbit around the sun is inclined 17° to the plane of the solar system—more than twice the inclination of the next most inclined orbit. The orbit is so elliptical that Pluto is the only planet to cross inside another planet's orbit; it crossed inside Neptune's orbit in 1979 and will remain closer to the sun than Neptune until 1999! Because of these unusual orbit properties, some astronomers have suggested that tiny Pluto should not be considered a true planet. Some have suggested that the Pluto–Charon pair might have originated when an asteroid passed through the Neptune satellite system and disrupted it. Such an idea, though unproved, might help explain why Neptune's satellites' motions are also unusual.

On cold, remote Pluto, the surface temperature is only about 50 K (−369°F), and the sun would be too far away to be perceived as a disk. From this outpost of the solar system, the sun would look like an intensely bright streetlight across the street, reminding us that it is one star out of many.

Planet X?

Several astronomers have sought dynamical or photographic evidence of a planet beyond Pluto, sometimes called **Planet X**. Clyde Tombaugh, the discoverer of Pluto, conducted a long search of the region beyond Pluto and ruled out any planet as large as Neptune near the plane of the solar system out to a distance of around 100 AU.

However, the 1978 discovery of Pluto's satellite allowed a better mass determination than had been possible earlier for Pluto, and the mass turned out to be too small to account for all of the perturbations on the motions of Neptune. This has prompted a number of astronomers to suspect that other objects may exist, perhaps well out of the plane of the solar system. Naval Observatory astronomer Thomas Van Flandern (1981) has proposed a search for an object with as much as three earth masses in a highly inclined eccentric orbit. There is growing suspicion that the outer solar system may harbor additional small bodies. Comets traverse this region; they may grade upward in size, if not to three earth masses, then at least to other objects resembling Pluto.

Why Giant Planets Have Massive Atmospheres

At the end of Chapter 8 we gave three principles that explain why some planets lack atmospheres. Let us apply them here to explain in more detail why giant planets have massive atmospheres rich in hydro-

OPTIONAL BASIC EQUATION VI

Typical Velocities of Atoms and Molecules in a Gas

A useful result about the motions of atoms or molecules in a gas was worked out around 1860 by physicists, particularly the Scottish researcher James Clerk Maxwell and the Austrian researcher Ludwig Boltzmann. Like our other Basic Equations, it is a simple law that gives us many insights throughout astronomy. It comes from the recognition that any gas is made of particles (atoms or molecules or even microscopic dust grains), all moving at various speeds determined only by the temperature T. In fact, T is defined as a measurement of the average velocity or energy of a particle in the gas. The higher T, the higher the mean velocity (or energy) of any representative particle, as noted as the end of Chapter 8. Now recall from basic physics that the kinetic energy of any moving particle is $\frac{1}{2}mv^2$, where m is the particle mass and v, its velocity. The discovery of Maxwell and Boltzmann is that the mean kinetic energy is proportional to T; this statement is usually written

$$\tfrac{1}{2}mv^2 = \tfrac{3}{2}\,kT$$

where

$$k = \text{Boltzmann constant}$$
$$= 1.38\,(10^{-23})\ \text{joules/degree}$$

The joule (abbreviated J) is the SI unit of energy. Solving for v, we have a value that may be thought of as the typical velocity of an average particle:

$$\text{typical velocity} = v = \sqrt{3kT/m}$$

(To be more precise, this value is the square root of the mean squared velocity and is about 10% higher than the true mean velocity. Physicists widely use this so-called "root mean square" velocity as an all-purpose indicator of gas atom speeds, however.)

Note that the mass of an atom or molecule is conveniently figured from

$$\begin{aligned} m &= (\text{atomic weight of atom or molecule}) \\ &\quad \times (\text{mass of a hydrogen atom}) \\ &= (\text{atomic weight}) \times 1.67\,(10^{-27})\ \text{kg} \end{aligned}$$

Sample problem. What is the typical velocity of a molecule of air in your room? *Solution.* Assume $T = 68°\text{F} = 20°\text{C} = 293\ \text{K}$. (We have to convert to SI units, which for T are kelvin degrees.) Assume the air molecule is the nitrogen molecule, N_2. A check in a chemistry book shows that these two nitrogen atoms have total atomic weight 28 (i.e., 28 × the weight of a hydrogen atom), so that $m = 4.68(10^{-26})$. Therefore,

$$\begin{aligned} v &= \sqrt{\frac{3(1.38)(10^{-23})(293)}{4.68(10^{-26})}} \\ &= 509\ \text{m/s} \end{aligned}$$

Thus, the atoms and molecules around you are typically hitting your skin at about ½ km/s! You don't feel the individual hits because they are too small, but you do feel the sustained pressure exerted by the ensemble of them.

Sample problem. Prove that a hydrogen atom would move much faster than a nitrogen atom in the same atmosphere. *Solution.* Since v is proportional to $\sqrt{1/m}$, we confirm that the smaller the mass the larger the velocity. Since an N atom is 14 times more massive than an H atom, it would move $1/\sqrt{14}$ as fast. Or, to say the same thing, the H atom moves 3.7 times faster. The student should also show that an H *atom* would move 5.3 times faster than an N_2 *molecule* in the same gas.

As illustrated in the advanced problems at the end of this chapter, this basic equation can easily be used to show why hydrogen has escaped from the earth and not from Jupiter.

gen gas. The first principle said that the higher the temperature, the higher the molecules' speed. (For more detail, see accompanying Optional Basic Equation VI.) This means that the cold, outer planets will generally have low molecular speeds in their atmospheres. The second principle said that the lighter the molecule, the higher its speed. This means that on any given planet, hydrogen will be the fastest moving molecule. The third principle said that the bigger the planet, the higher the speed needed for a molecule to escape the planet's gravity and shoot off into space. This means that giant planets will hold whatever gases they have much better than smaller planets like the earth.

Combining these principles, we can compare the earth and a giant. On small, warm Earth, the fastest moving hydrogen molecules will move more than fast enough to escape into space. Thus, the earth has lost virtually all its original hydrogen, which was by far the most abundant primordial gas. The slower nitrogen and oxygen did not escape. Conversely, on a cold giant like Jupiter, the fast-moving hydrogen and all other gases have been retained. Since the original hydrogen was more abundant than all other gases put together, today's hydrogen-rich Jupiter atmosphere is very massive and very deep.

Summary

The four giant planets are the dominant planets in the solar system in terms of diameter, mass, and numbers of satellites. They have such strong gravity that they have retained light atmospheric gases such as hydrogen, helium, and hydrogen compounds that were once common in the solar system but have escaped from terrestrial planets and moons due to their low gravity. The giant planets thus have larger proportions of these gases. They may also have high-density cores of solid or liquid materials that may contain silicate minerals and ices.

Each of the four giant planets has at least one large satellite (ranging from lunar size to the size of Mercury) and smaller satellites a few hundred kilometers across. These satellites are much richer in ices than terrestrial planets, because the outer solar system is so cold. Since the giant planets and their satellites form miniature solar systems, they may contain clues about the origin of the solar system as a whole. The larger satellites of the outer solar system, neglected until recent years, include exotic worlds. Io has active volcanoes; Europa is an ice covered "billiard ball"; Ganymede shows incipient plate tectonics; Titan has a dense, cold atmosphere where nitrogen or methane compounds may create strange rains, snows, and liquid ethane–methane oceans. Triton also has a cold atmosphere and may have liquid nitrogen oceans. Pluto is anomalously small for the outer solar system, resembling one of the larger satellites; it may have had a different origin than the four giants.

Concepts

giant planet
belts
zones
Jupiter's atmospheric composition
Jupiter's temperature
Great Red Spot
Jupiter's infrared thermal radiation
synchrotron radiation
Jupiter's interior
metallic hydrogen
Galilean satellites
prograde and retrograde satellite orbits
Saturn's atmosphere
Saturn's ring system
Cassini's division
Saturn's satellite system
obliquity
Uranus' rotation axis
Planet X

Problems

1. Answer the following problems:
  a. How might studying cloud patterns on Jupiter, Saturn, Mars, and Venus help us understand terrestrial meteorological theory?
  b. What planetary or environmental characteristics might figure in such a theory?
  c. How do these factors vary among these planets?
  d. Why are other planets not included on the list?

2. Which planet has
  a. the highest surface gravity?
  b. the lowest surface temperature?
  c. the highest atmospheric pressure?
  d. the most atmospheric hydrogen?
  e. the highest percentage of atmospheric oxygen?

3. List places in the solar system where you might expect to find droplets, pools, or oceans of liquid water.

4. Describe the appearance of the sky to an observer flying above the clouds in the upper atmosphere of Saturn at dusk at low latitudes. Do the same for the polar regions.

5. How would you expect gravity on the Galilean satellites to compare with gravity on the moon (see data in Table 6-1)? Which three bodies in the solar system would you expect to have general environments most like the moon's?

6. Describe a landscape on Io.

7. What difficulties might be met in sending a spacecraft through the rings of Saturn to examine the ring environment? Consider a pass *perpendicular* to the ring plane versus a pass *in* the ring plane. (Note that without elaborate retrorockets, such a spacecraft would probably travel about 20 to 30 km/s relative to the rings.)

8. Describe the seasons and other effects that would occur if the earth had the same obliquity as Uranus.

9. Do orbits of planets ever cross, as seen from far north or south of the plane of the solar system?

10. Where would Bode's rule, if extended, predict a tenth planet? Would it be easier to detect if covered with snow or rock? Why?

11. Two large planets have the same size and mass, but different orbits.

a. Which would you expect to have more hydrogen? Why?
b. If they have had different amounts of volcanism, which would you expect to have more carbon dioxide? Why?

ADVANCED PROBLEMS

12. Jupiter's four Galilean satellites revolve in orbits about 400 000 to 2 million kilometers from the planet.

a. What would be their maximum angular separation from the planet when Jupiter is at its closest distance of about 4 AU from the earth?
b. These satellites have a brightness of about the sixth magnitude, equal to the faintest stars visible to the unaided eye. The eye can normally distinguish details as little as 2 minutes of arc apart. Comment on the feasibility of seeing the Galilean satellites with the naked eye.

13. If you had a telescope that could reveal details as small as ½ second of arc, would you be able to see dark markings on Ganymede (diameter of 5270 km)?

14. If a spacecraft with an infrared sensing device flew by Jupiter or Saturn and measured a cloud formation whose strongest radiation came at a wavelength of 2×10^{-5} m, what would be the temperature of the material in the cloud?

15. Assume the mass of Saturn is 5.7×10^{26} kg.

a. What is the orbital velocity of a particle orbiting around Saturn in Saturn's rings, about 250 000 km from the center of Saturn?
b. Suppose this particle hits another one that orbits 1 km farther away from Saturn. Estimate how fast they come together, using whatever mathematical techniques you know.

16. Calculate the weight of a 140-lb. person on the "surface" of Jupiter. Use Table 6-1.

17. Calculate the mean density of Saturn (lowest of any known planetary body) and discuss its implications for a dense core of rock or metal inside Saturn. Use Table 6-1.

18. What is the typical velocity of an H atom:

a. In the 1500 K gas at the top of the earth's atmosphere?
b. In a thin 1500 K gas at the surface of the moon?
c. In a 200 K gas at the top of Jupiter's atmosphere?
d. The fastest atoms in a gas may move 3 or more times faster than the typical velocity calculated above, due to random collisions in the gas. Calculate the escape velocities of the earth, moon, and Jupiter, and use these facts to explain the absence of hydrogen on the earth or moon, but its retention on Jupiter.

Projects

1. Observe Jupiter with a telescope with at least an 8-cm (3-in.) aperture. Sketch the pattern of belts and zones. Which are the most prominent belts? Which zones are brightest? Compare these results with photos in this book. Is the Red Spot or other dark or bright spots visible on the side of Jupiter being observed?

2. Observe Jupiter with large binoculars or a telescope with at least a 2½-cm (1-in.) aperture. How many satellites are visible? Observe the satellite system at different hours over a period of several days and try to identify the satellites. (This could be done as a class project, with different students making sketches at different hours.)

3. With a telescope having at least a 15-cm (6-in.) aperture, determine the rotation period of Jupiter by recording the time when the Red Spot is centered on the disk. Note that intervals between appearances must be an integral number (1, 2, 3, etc.) of rotation periods.

4. With a telescope having at least an 8-cm (3-in.) aperture, observe Saturn. Sketch the rings. Estimate the angle by which the rings are tilted toward the earth during your observation. Does this angle change much from day to day? From year to year? (Your teacher might save drawings by students from past years for comparison.) Can you see any belts or zones? (Usually they are less prominent than on Jupiter.) Can you see Cassini's divisions? Sketch nearby starlike objects that may be satellites and track them from night to night. Identify Titan, the brightest satellite.

5. With a telescope having at least a 15-cm (6-in.) aperture, observe Uranus and Neptune. What color are they? Can you see any detail? Any satellites?

CHAPTER 11

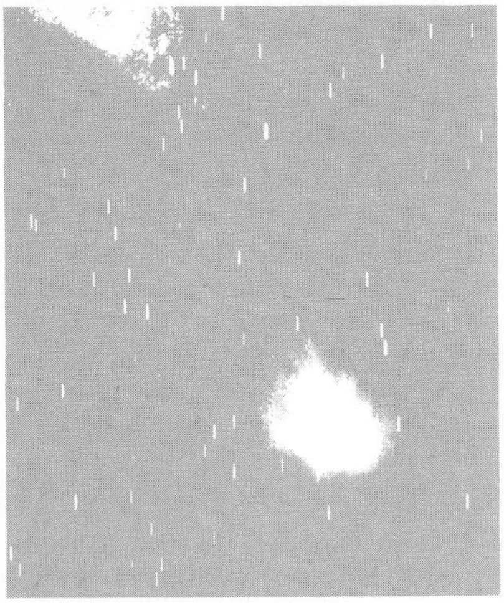

COMETS, METEORS, ASTEROIDS, AND METEORITES

On June 30, 1908, a mysterious explosion occurred in Siberia. English observatories, 3600 km (2200 mi) away, noted unusual air pressure waves. Seismic vibrations were recorded 1000 km (600 mi) away. At 500 km, observers reported "deafening bangs" and a fiery cloud. The explosion was caused by an unknown object that struck the earth's atmosphere from space.

Some 200 km (120 mi) from the explosion, the object was seen as "an irregularly-shaped, brilliantly white, somewhat elongated mass . . . with [angular] diameter far greater than the moon's." Carpenters were thrown from a building and crockery knocked off shelves. An eyewitness 110 km from the blast reported that

the whole northern part of the sky appeared to be covered with fire. . . . I felt great heat as if my shirt had caught fire . . . there was a . . . mighty crash. . . . I was thrown onto the ground about [7 m] from the porch. . . . A hot wind, as from a cannon, blew past the huts from the north. . . . Many panes in the windows [were] blown out, and the iron hasp in the door of the barn [was] broken.

Probably the closest observers were some reindeer herders asleep in their tents about 80 km (49 mi) from the site. They and their tents were blown into the air and several of the herders lost consciousness momentarily. "Everything around was shrouded in smoke and fog from the burning fallen trees."

The cause of this remarkable explosion was a collision between the earth and a relatively modest bit of interplanetary debris. Many such objects circle the sun. The Siberian event was merely the fall-in of the largest object to hit the earth in the last century or so. Even larger objects have hit the earth in earlier eras, as proven by the 20 000-year-old Arizona crater, shown in Figure 3-14.

Smaller impacts have been recorded more often; interplanetary stones fall from the sky in various locations every year (sometimes remote and sometimes inhabited—several houses have been hit in recent decades). Tiny dust grains that burn up in the atmosphere before hitting the ground can be seen every night; these are sometimes called "shooting stars."

Two conclusions are apparent. First, events ranging from small meteorite impacts to large, rare impact explosions may have occurred in the ancient past, perhaps influencing the beliefs of ancient people about forces in the sky.

The second conclusion is that the solar system contains more than just the planets and satellites we have already discussed. Floating between the planets are debris of many different sizes. They have been divided into different types depending on observed or inferred properties.

Comets are icy worlds a few kilometers across. **Asteroids** are rocky–metallic worlds ranging from a few hundred meters or less up to 1000 km in diameter. **Meteoroids** are still smaller bits of debris, probably dislodged from comets and asteroids, floating in their own orbits in space. Table 11-1 shows some orbital and other data on selected examples of all of these classes.

Some of these bodies have planetlike orbits that are nearly circular, prograde, and lying in the plane of the solar system. Others have very different orbits—highly elliptical, retrograde, or inclined at steep angles to the solar system's plane. Some of these orbits are shown in Figure 11-1, which emphasizes the diversity of interplanetary objects.

Objects whose orbits cross the orbits of the earth and other planets can collide with the planets. When they collide with the earth, they are heated by friction with the atmosphere (just as returning space vehicles are). **Meteors** are the smaller ones (usually up to a few centimeters in diameter) that burn out completely before striking the ground. **Meteorites** are the larger ones that survive and fall to the ground: They are made of rock or metal. They are free samples of other planetary worlds. Their properties testify that they broke off of larger bodies, called **parent bodies**, in the remote past. One unsolved mystery is the question of what objects were the parent bodies of different meteorite groups.

Comets, asteroids, and meteoroids of every size are all debris left over from the origin of the solar system 4.6 billion years ago. As the present planets formed, the solar system was filled with innumerable small, preplanetary bodies, ranging up to 1000 km across, known as **planetesimals.**

Comets, asteroids, and meteorites are all leftover planetesimals and planetesimal fragments. Thus, they provide a direct link to the conditions under which the planets and the sun itself formed. By implication, they provide clues to the formative conditions of other stars and hence provide a link with the most distant stellar regions, which will be a topic of much of the rest of this book.

Table 11-1. *Properties of Selected Small Bodies in the Solar System*

Class	Example	Diameter (km)	Orbit: Semimajor Axis (AU)	Orbit: Eccentricity	Orbit: Inclination	Remarks
Comets						
Short-period	Encke	1–8	2.2	0.85	12°	Probably dirty ice
	Halley	Few?	18	0.97	162°	Probably dirty ice
Long-period	Kohoutek	Few?	Very large	1.0	14°	Probably dirty ice
Meteors						
Shower	Perseid	10^{-6}	40	0.97	114°	Cometary debris
	Taurid	10^{-6}	2.2	0.80	2°	
Fireballs	July 31, 1966	?	32	0.98	42°	May be related to comets or asteroids
	May 31, 1966	?	3.0	0.80	9°	
Asteroids						
Belt	1 Ceres	1020	2.8	0.08	11°	Carbonaceous rock surface
	2 Pallas	538	2.8	0.23	35°	Rocky surface
	3 Juno	248	2.7	0.26	13°	Rocky surface
	4 Vesta	549	2.4	0.09	7°	Lava-like surface
	14 Irene	170	2.6	0.16	9°	Rocky surface
Trojan	624 Hektor	100 × 300	5.1	0.02	18°	Unusual shape
Apollo	433 Eros	7 × 19 × 30	1.5	0.22	11°	Elongated
	Apollo	?	1.5	0.56	6°	Lost before orbit precisely measured
Unusual "asteroid"	2060 Chiron	100–320?	13.7	0.38	7°	
Meteorites (chondrites)	Pribram	10^{-4}	2.5	0.68	10°	
	Lost City	10^{-4}	1.7	0.42	12°	
	Leutkirch[a]	10^{-4}?	1.6	0.40	2.5°	

Source: Data from Chapman and Morrison (1974); Hartmann (1972, 1975); Gehrels (1972). The diameter of Comet Encke was measured with radar by Kamoun et al. (1982).

[a]An object photographed over Europe in 1974. No fragments recovered as of late 1974, but believed to be a stone meteorite.

Comets

Comets are the most spectacular of the small bodies in the solar system. When they pass through the inner solar system near the earth, they can be seen drifting slowly from night to night among the stars. (Writers sometimes incorrectly describe comets as "flashing cross the sky," like shooting stars. They do not. They seem to hang motionless and ghostly among the stars. Their motion relative to the stars can be detected by the naked eye only after a few hours.)

Comets have several parts as seen in Figures 11-2 to 11-4. The brightest part is the **comet head.** The **comet tail** is a fainter glow extending out of the head, usually pointing away from the sun. Although a typical comet tail can be traced for only a few degrees by the naked eye, binoculars or long-exposure photos may reveal fainter extensions of the tail extending tens of degrees, or even extending clear across the night sky. A telescope reveals a brilliant, starlike point at the center of the comet head. This is the **comet nucleus,**

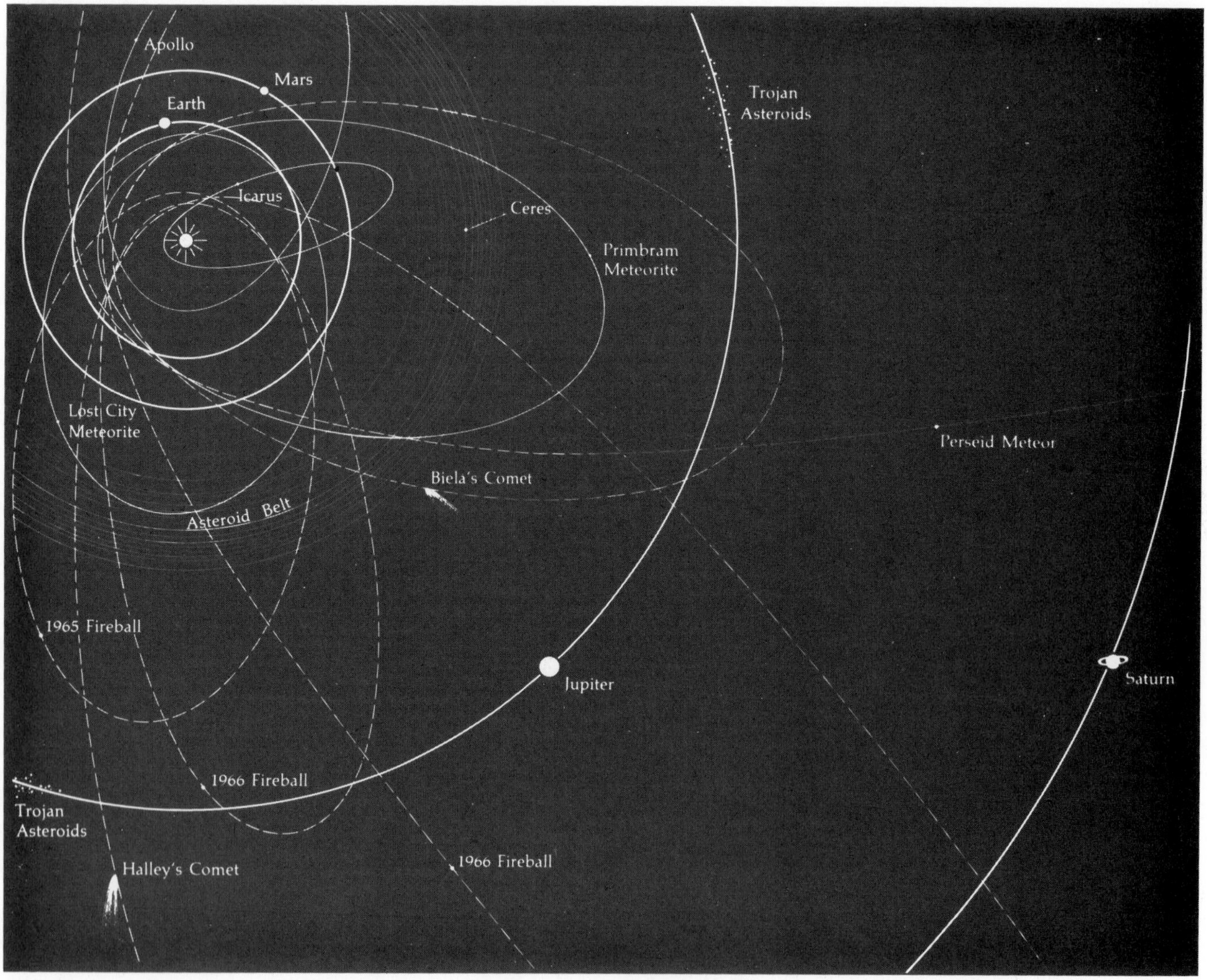

Figure 11-1. *A portion of the solar system showing orbits of a few selected interplanetary bodies including comets, main-belt asteroids, Trojan asteroids, Apollo asteroids, meteorites, and large meteors, or fireballs.*

which is the only substantial, solid part of the comet. Telescopic studies have revealed that a typical comet nucleus is a worldlet of dirty ice only about 1 to 20 km (a few miles) across—tiny compared to most planets and moons! The gas and dust that make up the rest of the comet's head and tail are material emitted from the nucleus. As the comet nucleus moves through the inner solar system, the sunlight warms it and causes the ice to evaporate[1] into the form of gas. This gas, together with dislodged dust grains from the dirt in the nucleus, is then carried away from the nucleus by the pressure of radiation and thin gas rushing outward from the sun. This forms the comet's tail, often stretching more than an astronomical unit. Caught in this "wind" of outrushing solar gas and radiation, the tail streams out behind the comet as the comet approaches the sun, but leads as the comet recedes

[1] A more correct technical word is *sublime,* which means to change from solid form directly into gaseous form. *Evaporate* technically means to change from liquid form into gaseous form.

Figure 11-2. *Comets are an occasional spectacle in the sky. This view of Comet Ikeya-Seki was made in 1965 with a stationary 35-mm camera; 20-s exposure at f1.9 on Tri-X film. (Steven M. Larson.)*

from the sun (Figure 11-5). It is like the long hair of a woman walking in a strong wind, streaming out behind as the woman walks into the wind, but in front of her if she reverses direction.

In 1704 the English astronomer Edmund Halley applied newly developed orbit computation methods and discovered that comets travel on *long, elliptical orbits* around the sun and that certain comets reappear. Calculating the orbits of 24 well-recorded comets by Newton's methods, Halley found that four comets (seen in 1456, 1531, 1607, and 1682) had the same orbit and a periodicity near 75 years. Halley correctly inferred that these appearances were by a single comet and that the slight irregularities in periodicity

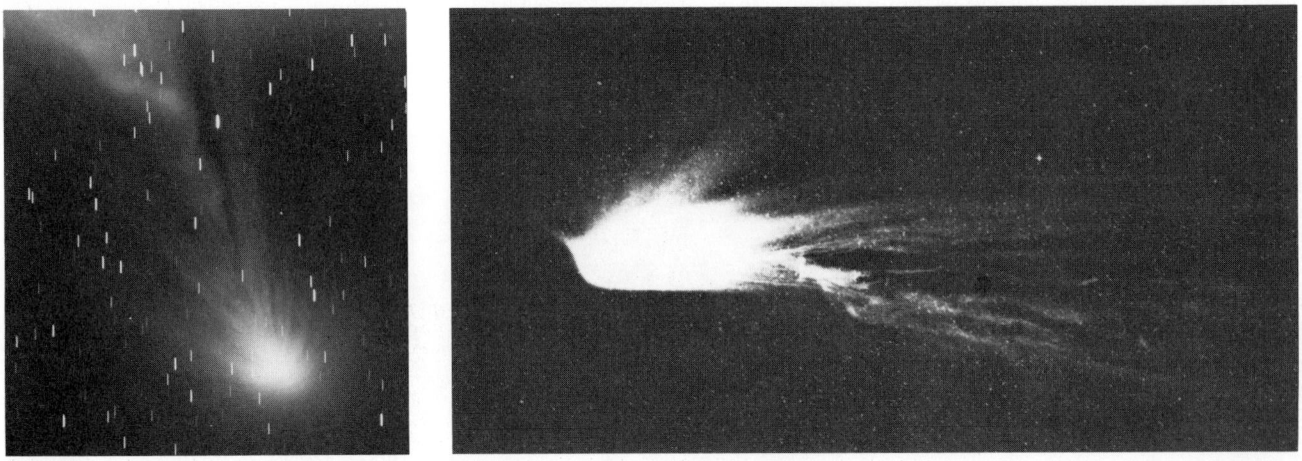

Figure 11-3. *Two comets photographed while the earth was between the comet and the sun, so that we are looking past the head, along the line of the tail. This perspective shows the fan-shaped structure of the tail. (a) Comet Humason. The telescope tracked the comet for 60 min, causing background star images to appear as streaks. (Naval Observatory; official U.S. Navy photograph.) (b) Comet Halley on February 22, 1986, photographed in Australia, where the comet was higher in the sky and more prominent than for Northern Hemisphere observers. The comet's head is overexposed in order to show the faint tail structure. (Copyright Royal Observatory, Edinburgh, courtesy International Halley Watch program.)*

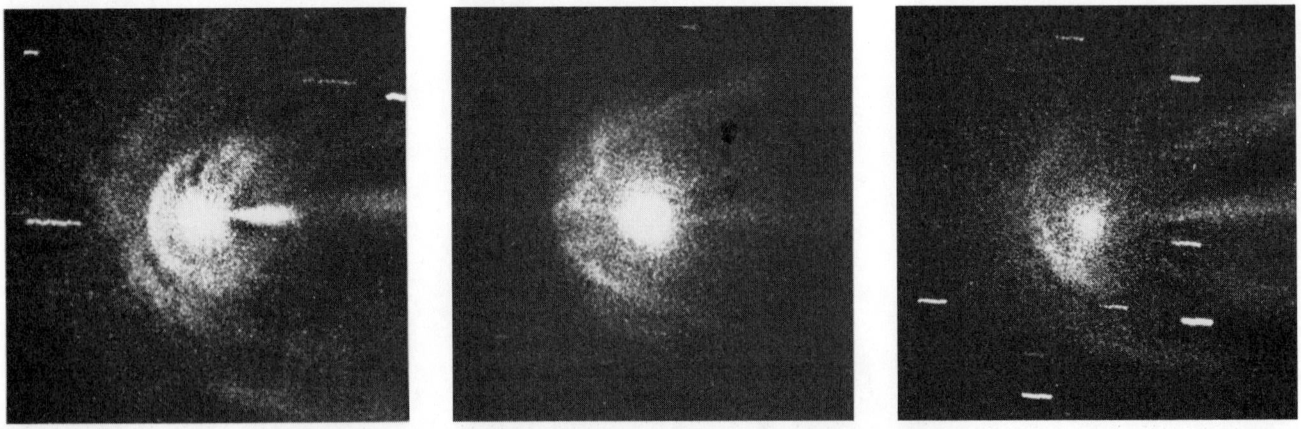

Figure 11-4. *A new look at old photos of Halley's comet. The head of Comet Halley was photographed on different days as it passed unusually close to the earth in June 1910, but photographic prints made by standard techniques of that era showed only vague structure. In 1984, astronomers Z. Sekanina and S. M. Larson reprocessed some of the old plates with modern equipment that digitizes the image and enhances low-contrast detail. These images clearly show individual curved jets of dusty debris blown off the nucleus by individual outbursts. Sun is on left; tail is to right. Streaks are star images trailed by long-exposure tracking on comet. (Photos courtesy S. M. Larson, University of Arizona.)*

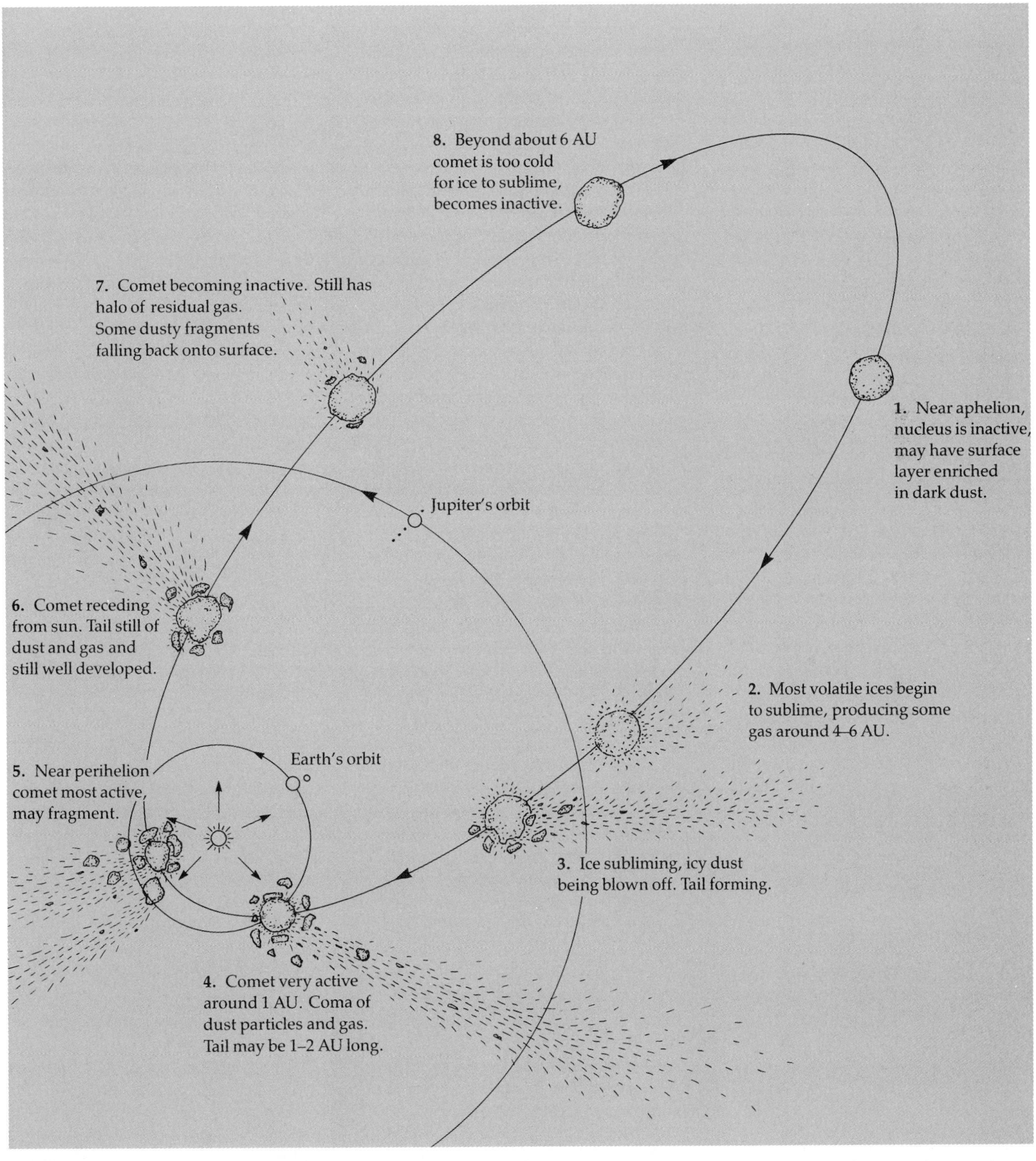

Figure 11-5. *Stages in the development of a typical comet as it travels from the outer solar system into the inner solar system, where it loops rapidly around the sun and moves outward again. The tail develops only within a few astronomical units of the sun, where sunlight is warm enough to sublime the ice in the nucleus. In order to show comet detail, the figure is not drawn to scale.*

were caused by gravitational disturbances from the planets, especially Jupiter. Halley predicted that the comet would return about 1758. It did so on Christmas night. It was named Halley's comet.[2] In 1982, astronomers rediscovered the dim nucleus of Halley's comet beyond the orbit of Saturn, on its way to its current swing around the sun.

Halley's comet became very famous due to its prominence in 1910, when the earth passed through its tail. It was not so prominent in 1986, because it did not pass as close to the earth.

COMETS AS OMENS

Before Halley's discoveries, comets were often regarded as evil omens. For example, the appearance of Halley's comet in A.D. 66 was said to have heralded the destruction of Jerusalem in A.D. 70. Five circuits later it was said to mark the defeat of Attila the Hun in 451. In 1066 it presided over the Norman conquest of England. In 1456, its appearance coincided with a threatened invasion of Europe by the Turks, who had already taken Constantinople three years before. Pope Calixtus III ordered prayers for deliverance "from the devil, the Turk, and the comet." Of course, we now recognize that some noteworthy events are likely in most cultures every decade or so, and comets are no longer regarded as ominous portents by most people.

WHERE COMETS COME FROM: THE OORT CLOUD

In ancient times no one knew how far away comets were, and many people thought comets were phenomena in our own atmosphere. Seneca, the Roman contemporary of Jesus, wrote, "Some day there will arise a man who will demonstrate in what regions of the heavens the comets take their way." That man was Tycho Brahe, who in 1577 arranged observations of a bright comet from two different locations. Finding no parallactic shift in the comet's angular position relative to the stars (see p. 114). Tycho correctly concluded that it was more distant than the moon and thus not terrestrial.

Comet orbits have several important characteristics. First, orbits are inclined at nearly random angles to the plane of the solar system. Thus, comets populate a roughly spherical volume of space around the sun, rather than being confined to a disklike volume as the planets are.

From statistics of comets' orbits, Dutch astronomer Jan Oort discovered around 1950 that this spherical swarm of comets is very extensive, and that most comets spend most of their time about 50 000 AU from the sun. This swarm, containing millions of comets, is called the **Oort cloud.** Here, comets are so far from the sun that gravitational attractions of nearby stars, as well as the sun, are important. Following Kepler's and Newton's laws, comet nuclei drift slowly at these great distances, being disturbed by stellar forces, until they eventually fall back into the inner solar system. Still following Kepler's laws, they then move rapidly through the inner solar system on elliptical, nearly parabolic orbits (see page 115).

Most comets pass among the planets in only a few years. Only then can they be seen by earthbound observers. Then they swing slowly back toward the Oort cloud, where they may spend hundreds or thousands of years. Occasionally some of these comets pass near Jupiter or other planets and are redirected by gravitational forces into orbits that remain in the inner solar system. Many of the best-known comets, which return every few decades, are of this type. These comets are called **short-period comets.** In Figure 11-1, Biela's comet can be seen to be such an object, while Halley's comet can be seen to have a much larger orbit reaching toward the outer solar system.

No comets have been observed to reach or leave the solar system with speeds much greater than escape velocity. Therefore, they are all believed to be true members of the sun's family (probably formed along with the planets) and not to come from or escape to regions as distant as the stars.

PROBING COMETS BY SPECTROSCOPE

As explained in Figure 7-11, measurement of the spectrum of an astronomical object is our most important technique for determining its nature, short of visiting it. What can the spectrum tell us about comets? We can observe the spectrum of sunlight reflected off dust particles in the comet's head and tail. Results suggest the particles are dark-colored dust or soil grains, rich in carbon compounds. The nucleus itself is obscured by the glow of the head, so that we cannot get the spectrum of the nucleus alone. If comets were made up only of solid dust grains, the spectrum would merely show (a) sunlight reflected off the dust and (b) thermal infrared radiation (right portion, Figure 7-11). But if the diffuse material includes gas, then the atoms and molecules of gas would be struck by solar radiation, which would either create **excited atoms and molecules** (molecules with altered electron structures) or **ions** (atoms or molecules in which at least one electron has been knocked out entirely). Just as excited gas in a fluorescent tube glows, excited gas in a comet's head and tail glows with certain colors, called **emission bands.** Different gases give off emission bands of different wavelength. Thus, the spectrum of a comet can be studied not only to learn about its dust, but also to learn if gases are present, and if so, what kind.

This test was first carried out by the English astronomer William Huggins as early as 1868. With his spectroscope he could see three bright emission bands, which he could identify with gaseous carbon. Modern astronomers have added other gaseous atoms, molecules, and ions to the list, including H (hydrogen), O (oxygen), CN, CH, OH, H_2O^+, CN^+, CH^+, OH^+, $N_2{}^+$, CO^+, and $CO_2{}^+$. Recent studies have made exciting additions to the list: more complex organic molecules such as CH_3CN, H_2CO, and HCO. This discovery proves that organic molecules—the

---

[2]Comets discovered today are first named for their discoverers but are later given scientific names in order of their passage around the sun, for example, Comet 1982 I.

building blocks of life—form elsewhere in the universe besides on the earth.

Note that all these atoms (H, O, C, N) and molecules are just the ones that would be expected if the gas were formed by the sublimation of common solar system ices, such as H_2O (water), CH_4 (methane), NH_3 (ammonia), and CO_2 (carbon dioxide).

From such spectroscopic evidence scientists have long inferred that comet nuclei are dirty, drifting icebergs—mixtures of ice and sooty dirt. As the ices are warmed by the sun, they sublime into space, dislodging dust grains and gas molecules that break into fragments such as H, O, and CN.

COMETS' NUCLEI

The idea of a comet nucleus as a solid, "dirty iceberg" was first developed by Harvard astronomer Fred Whipple around 1950. All the spectacular gas and dust in the coma and tail must come from this icy nucleus in the heart of the coma. The secret of a comet's nature is hidden in this body. What is it really like?

Five important clues were developed over the last decades through ground-based observations. The first clue is the spectroscopic work mentioned above. The gases must come from solid ice that composes much of the nucleus.

The second clue is that the coma and tail develop only within about 3 AU of the sun. Calculation shows that this is about the distance at which frozen water sublimes rapidly when exposed to sunlight. Hence frozen water is a main constituent of the nucleus. Outbursts at larger distances suggest that other, less stable ices, such as frozen CO_2 or CH_4, are also present.

The third clue is that some comets spontaneously break apart, as shown in Figure 11-6, often when they pass close to the sun. The breakup is apparently due to solar tidal forces (see Chapter 5) and the pressure of expanding gas from cracks in the nucleus. These forces are small, indicating that the nucleus is a weak agglomeration, not a strong, solid body. Indeed, radar signals bounced off a nearby comet in 1983 revealed not a single large nucleus but a cloud of millions of pebble-sized fragments, perhaps surrounding the real nucleus.

A fourth clue is that slightly irregular motions of comets can be attributed to jetlike outbursts of gas from individual spots on the nucleus, perhaps individual fissures that expose fresh ice.

A fifth clue is similarity in spectrum and color between comets' dark dust grains and the surface dust of dark asteroids in the outer solar system. In both cases, the dust is believed to be carbon-rich blackish soil, sometimes with a reddish tinge due to organic compounds. Certain dust grains collected by aircraft in the high atmosphere and believed to be cometary (see next section) are typically black, carbon-rich, and fragile.

From these clues, a comet nucleus was visualized, prior to Halley's 1986 arrival, as a weakly consolidated dirty iceberg, perhaps fractured or chunky due to reaggregation after previous breakups. Diameters of a few kilometers were estimated, and observers argued whether the nucleus was light-colored (due to the ice) or very dark (due to efficient coloration by the black dirt).

A NUCLEUS AT CLOSE RANGE: SPACECRAFT GREET HALLEY'S COMET

An international fleet of five spacecraft (two Japanese, two Russian, and one European) probed Halley in 1986. The first close-up data and pictures supported the earlier work.[3] After the first four craft tested the environment at moderate distance, the European probe flew by closest, only about 600 km from the nucleus. It was named Giotto, after the Italian artist whose 1304 painting of the comet may have been the first to show it.

The heart of Halley's comet was a hazy environment, where the probes found more fine dust than expected. Giotto was hit by a dust grain about 960 km from the nucleus, knocking it partially out of commission for 32 min during the closest approach! As seen in Figures 11-7 and 11-8, the photos revealed bright jets of illuminated gas and dust coming off "active" spots on the black, peanut- or potato-shaped nucleus, about 15 km long and 7 to 10 km wide. The haze made the pictures fuzzy. The coma gas was found to be about 80% water vapor by volume. The dust particles were found to be rich in H, C, N, and O and somewhat resembling

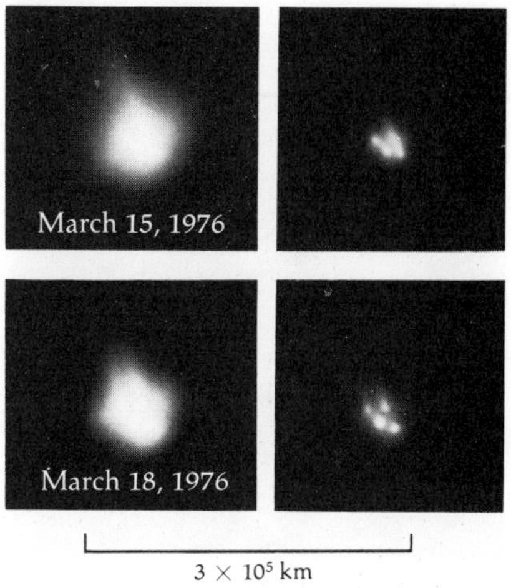

Figure 11-6. *Several comets have been seen to break apart. These pictures show the breakup of Comet West (1975n) into four pieces in March 1976. The left column photos are longer exposures; the right column photos are shorter exposures, revealing four nucleus fragments, each with its own stubby tail. The pieces separated at relative speeds of about 1 to 5 m/s (2 to 11 miles per hour). One piece was seen for only two weeks; others were followed for over six months. (Lunar and Planetary Laboratory 1.5-m telescope photo, University of Arizona, courtesy S. M. Larson.)*

[3] The encounter dramatized that while our news media may be adept at reporting tragedies, they fail in dealing with the adventure of human exploration. During the night of the first Russian encounter, a TV hookup with the USSR allowed live coverage of humanity's first close look at a comet; but my local affiliate carried instead an old Evel Knievel movie. The next day the papers ran a front-page picture of the comet, reportedly from the Russian spacecraft. A few days later, however, on an inner page they reported sheepishly that this was not the new Russian picture at all but a computer-processed 1910 photo that had been misidentified by a news agency.

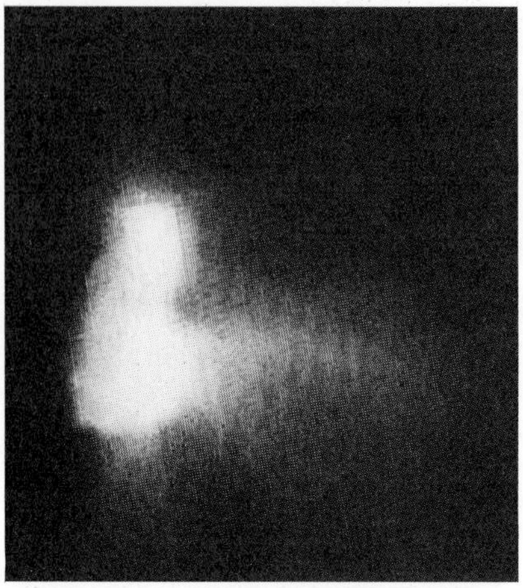

***Figure 11-7.** Nuclear region of Halley's comet imaged by Russian spacecraft Vega 2 from a distance of about 8000 km. Sun is behind observer and to the right, and jets can be seen blowing off the sunlit side of the nucleus. However, it is unclear whether the brightest part of the image is the illuminated surface of the nucleus or the most intense part of the gas jets. (Courtesy USSR Academy of Sciences.)*

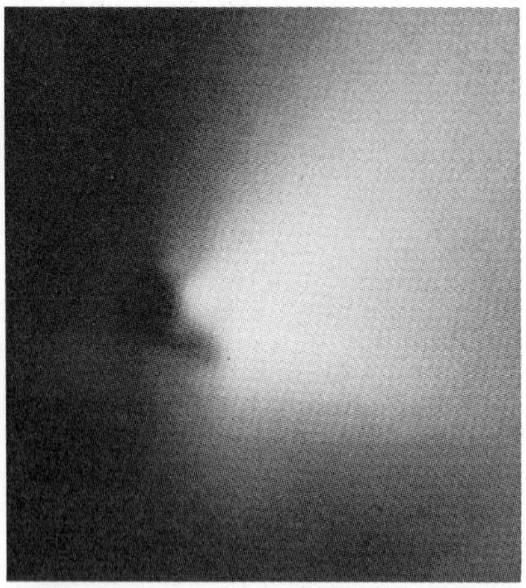

***Figure 11-8.** Nucleus of Halley's comet imaged by European spacecraft Giotto from a distance of about 600 km. Sun is on the right but slightly on the far side of the comet (unlike Figure 11-7). The dark, peanut-shaped nucleus is silhouetted against the luminous coma, with jets extending in the sunward direction. (Copyright Max Planck Institut für Astronomie.)*

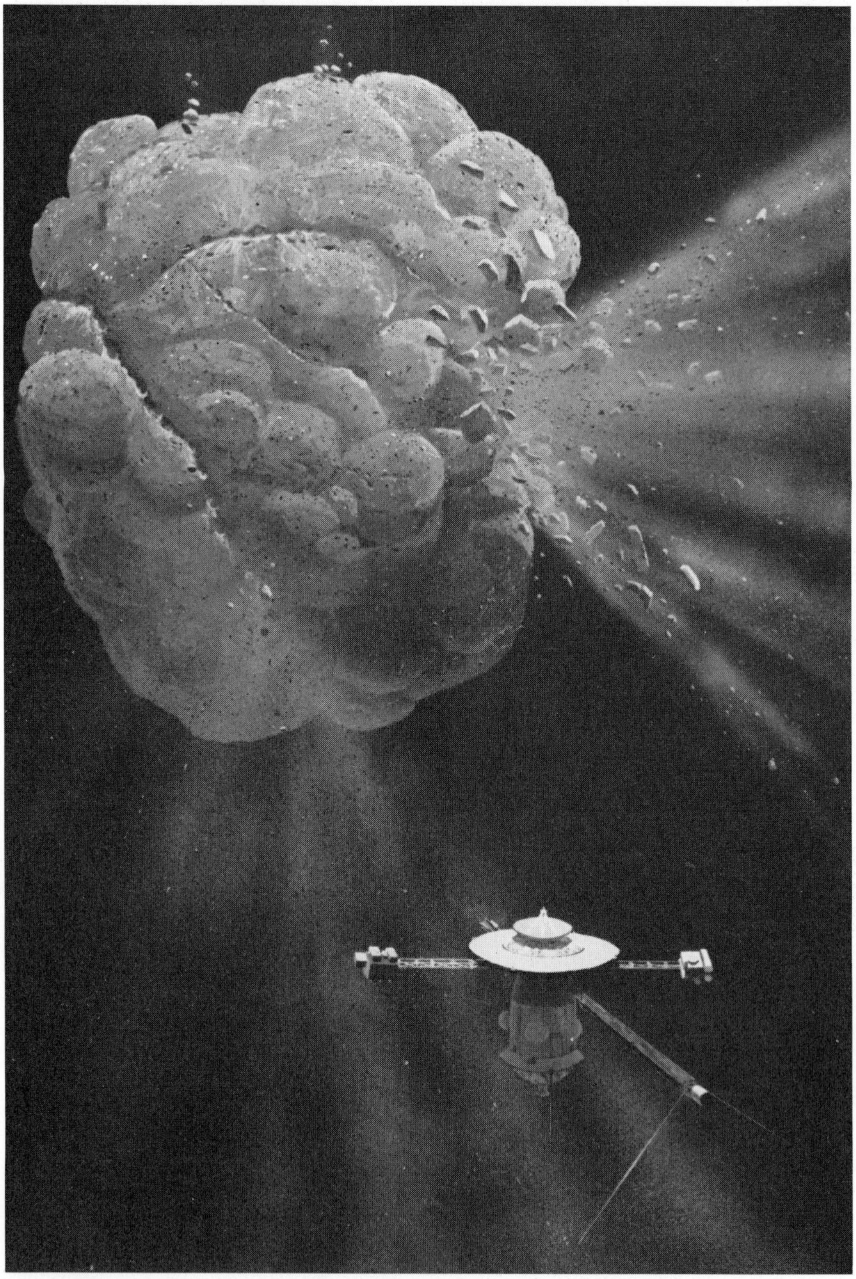

***Figure 11-9.** A proposed future comet mission would park a probe in orbit alongside a comet and watch it "turn on" as it approached the sun. This painting was made two months before the Halley encounters and was a final attempt to predict some features of the nucleus. It shows jetting similar to that seen in spacecraft photos, Figures 11-7 and 11-8. (Painting by author.)*

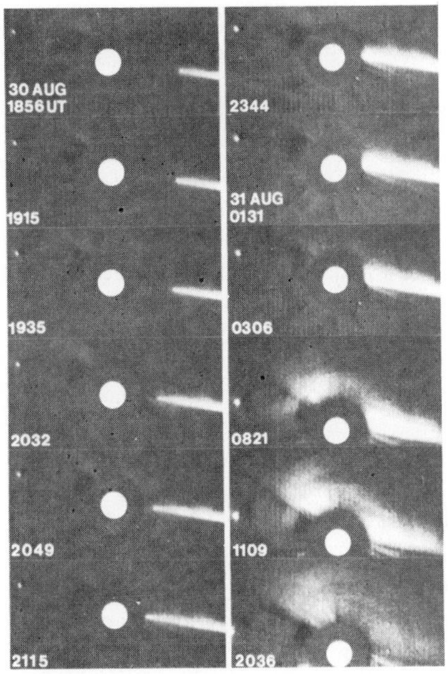

Figure 11-10. *The collision of comet 1979 XI with the sun on August 30, 1979, photographed with an orbiting telescope. The white disk shows the size of the sun, which was covered by a larger disk (making a black ring around the sun). In the post-impact series (right) the contrast has been enhanced; rapid sublimation of ices in the nucleus has added material to the tail. Material from impact or breakup of the nucleus streams outward in several directions. The comet did not reemerge on the left side of the sun. Venus is at left. (LOLWIND satellite photos from Naval Research Laboratory, courtesy D. J. Michels.)*

carbon-rich meteoritic particles collected on Earth. Analysis of the data from Halley probes is still continuing (Sagdeev et al., 1986).

For many years, astronomers have known of a dramatic group of comets called the sun-grazers; they have perihelia very close to the sun. Among these are several that have fragmented or disappeared during their close approach to the sun! In 1979 an even more exceptional subgroup of this class was discovered in photographs from an orbiting telescope in space—comets that actually collide with the sun. An example is seen in Figure 11-10. Furiously streaming a tail behind it, it moved behind the solar disk but never reemerged. Calculations showed that it collided with the upper gaseous layers of the sun. At least two additional sun-colliding comets were discovered by space telescopes in the following three years.

THE ORIGIN OF COMETS

From the evidence described so far, astronomers have theorized a four-stage history that explains the origin of comets. First, comets must have formed by aggregations of ice crystals and dust grains in the cold, outer regions of the solar system while the planets were forming. We know from studying the gases in the atmospheres of Jupiter and other giant planets that hydrogen, carbon, oxygen, and other elements were available as the outermost planets formed. In the environment of the outer solar system these would have condensed into ice grains (such as H_2O) that could accumulate into icy planetesimals—the comet nuclei. Second, many of the nuclei that were still orbiting around the sun in the outer solar system by the time the giant planets had formed must have approached close to these planets and been flung by the planets' gravity onto highly elliptical orbits that took them out of the solar system, just as Voyager and Pioneer space vehicles were flung out of the solar system by close encounters with Jupiter. This process randomized many comet orbits and formed the Oort cloud of objects circling the sun beyond Pluto. Third, comets were "stored" in the Oort cloud for long periods of time. Fourth, comets in the Oort cloud occasionally found themselves on trajectories back into the inner solar system, where they once again passed among the planets. Only during these brief trips were the ices warmed enough by the sun to evaporate and form the bright heads and tails of gas and dust for which comets are known.

Meteors and Meteor Showers

Meteors that flash momentarily across the sky might seem totally unrelated to comets. However, a study of their frequency on different nights reveals a direct connection. On an average night you may see about 3 meteors per hour before midnight and about 15 meteors per hour after midnight. (You see more meteors after midnight because you are then located on the leading edge of the earth as it moves forward in its orbit, sweeping up interplanetary debris.) But on certain dates each year you may see **meteor showers** of 60 meteors or more per hour, all radiating from one direction in the sky (Table 11-2).

The best-known example is the Perseid shower, which occurs every year around August 12, when bright meteors streak across the sky every few minutes from the direction of the constellation Perseus. (A shower is named for the constellation most prominent in the area of the sky from which the shower radiates. Occasionally the showers are so intense that meteors fall too fast to count. During the Leonid shower of November 17, 1966, meteors fell like snowflakes in a blizzard for about half an hour, at a rate estimated to be more than 2000 meteors per minute (see Figure 11-11).

What is the connection between the showers and the comets? In 1866 G. V. Schiaparelli (of Martian canal fame) discovered that the Perseid meteor shower occurred whenever the earth crossed the orbit of Comet 1862 III. The Perseids, then, must be spread out along the orbit of that comet. Other relationships were soon found between specific meteor showers and specific comets, as Table 11-2 shows. In 1983, an infrared astronomical telescope in orbit (called IRAS) discovered the thermal infrared emission from a swarm of meteor dust spread along the orbit of comet Tempel 2. Here, then, was a direct detection of the dust scattered by a comet. Even the orbits of individual sporadic meteors, tracked photographically, often resemble long- or short-period cometary orbits. Therefore, *most meteors must be small bits of debris scattered from comets.*

Since the 1960s, rockets and balloons have collected microscopic fragments believed to be meteoroids, shown in Fig-

Table 11-2. *Dates of Prominent Meteor Showers*

Shower Name (After Source Constellation)	Date of Maximum Activity[a]	Associated Comet
Lyrid	April 21, morning	1861 I
Perseid	August 12, morning	1862 III
Draconid[b]	October 10, evening	Giacobini–Zinner
Orionid	October 21, morning	Halley
Taurid	November 7, midnight	Encke
Leonid	November 16, morning	1866 I
Geminid	December 12, morning	"Asteroid" 1983 TB[c]

[a]Showers can last several days before and after the peak activity on the listed date. Observations are best when the constellation in question is high above the horizon, usually just before dawn.

[b]The Draconids are now weak because their orbits have been disturbed by the gravity of planets, but further disturbances may again strengthen the shower in the future.

[c]This object was discovered in 1983 by the IRAS satellite and is probably a "burnt-out" comet nucleus.

ure 11-12. These fragments are irregular glassy silicate and metallic particles, supporting the theory that comet nuclei are dirty icebergs with bits of entrapped grit. Their compositions are similar to those of certain types of meteorites. Most meteors are far too small to reach the ground, "burning" at altitudes around 75 to 100 km. Occasional large ones, called **fireballs**, are very bright and spectacular. They often explode before hitting the ground, again indicating that they are too fragile to survive atmospheric entry.

Asteroids

Asteroids are the largest of the interplanetary bodies, ranging from the largest, Ceres (about 1000 km in diameter), down to a few hundred meters across and probably less. Spectral studies show that they are rocky, often with metal embedded in them. They appear in various parts of the solar system, but are most abundant in the **asteroid belt**—a region between Mars and Jupiter.

DISCOVERY OF ASTEROIDS

Asteroids have played an interesting role in the mapping of the solar system. Normally too small to be seen by the naked eye, they were unknown before 1800. Bode's rule, confirmed by the discovery of Uranus in 1781, called for a planet at 2.8 AU from the sun, in the large space between Mars and Jupiter. Therefore, astronomers set out to find the "missing planet" in 1800; success came on the first night of 1801 when Ceres was discovered, just at 2.8 AU.[4]

Between 1802 and 1807, three more small, planetlike bodies turned up between 2.3 and 2.8 AU from the sun. Because of their small size, they came to be called minor planets, or *asteroids,* a name that now applies to any interplanetary body that is not a comet (that is, is not known to emit gas). By 1890, 300 asteroids were known. In 1891, German astronomer Max Wolf began searching for them photographically by time exposures, detecting many new asteroids by their telltale motion among the stars, as shown in Figure 11-13.

Asteroids are known by numbers (assigned in order of discovery, but only after the orbit has been accurately identified) and a name (chosen by the discoverer); examples are 1 Ceres and 2 Pallas. The names cover a wide range of human interests, including cities, mythology (1915 Quetzalcoatl), politicians (1932 Hooveria!), spouses and other lovers.[5] Over 2000 are now cataloged, and perhaps 100 000 observable ones remain uncharted.

ASTEROIDS OUTSIDE THE BELT

Several interesting subgroups of asteroids have orbits outside the main asteroid belt. **Apollo asteroids**, for example, come into the inner solar system and cross the earth's orbit. They are named after Apollo, first of the group to be discovered. Dozens are known, including one as small as 800 m across. Most are irregularly shaped, and many approach much closer to Earth than any planet. Indeed, meteorites that hit the earth can be thought of as the smallest examples of Apollo objects!

Apollos generally last only a few hundred million years before crashing into the earth or other planets, creating impact craters. New Apollo asteroids are constantly created, probably as asteroids are thrown out of the belt by the gravitational force of massive Jupiter. Other Apollos may be burnt-out stony remnants of short-period comets that have exhausted most of their ices after spending too much time in the inner solar system.

Trojan asteroids[6] lie in two swarms in Jupiter's orbit, 60° ahead of and 60° behind the planet, called **Lagrangian points**. The astronomer Joseph Louis Lagrange dis-

[4]As noted by Arthur C. Clarke, this discovery came just when the philosopher Hegel had "proved" philosophically that there could be no more than the seven then-known planetary bodies, Mercury through Uranus—which suggests that going out and looking is worth more than sitting at home speculating.

[5]Recently I learned of a new astronomical amusement—making sentences using only asteroid names. My favorite (not my own creation, I'm sorry to say) is "Rockefellia Neva Edda McDonalda Hamburga."

[6]The name "Trojan" comes from the tradition of naming asteroids after heroes in the Homeric epics.

Figure 11-11. *A rare meteor shower: the Leonids of November 17, 1966. The rate of meteors visible to the naked eye was estimated to exceed 2000 per minute. The brightest star (upper left) is Rigel, in the constellation Orion. This exposure of a few minutes' duration was made with a 35-mm camera. (D. R. McLean.)*

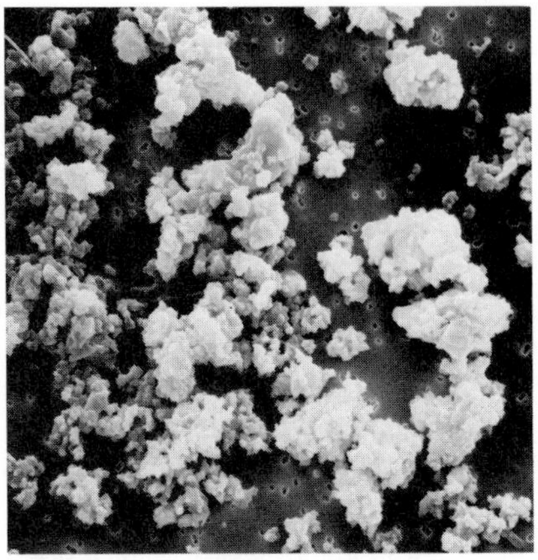

Figure 11-12. *Clumps of microscopic meteoroidal particles collected by high-flying aircraft. They may have originated in comets. The largest clump is about 60 μm across. (Courtesy D. E. Brownlee, University of Washington, Seattle.)*

covered that particles can be held in the two swarms by Jovian and solar gravitational forces. At least 45 Trojans have been found at one Lagrangian point, and observers estimate about 700 observable Trojans. The largest Trojan is Hektor, estimated to be 100 km wide and 300 km long, tumbling end over end in Jupiter's orbit. Trojan asteroids and their neighbors, the outer satellites of Jupiter (J6 through J13), are similar in having a very dark color and a different composition (more carbon minerals) than that of belt asteroids. They probably formed near Jupiter's orbit. Jupiter's outer satellites may even be Trojan-like asteroids that were captured by Jupiter's strong gravity into orbits around the giant planet.

A unique asteroid even further from the sun is 2060 Chiron, discovered in 1977. It orbits the sun, ranging from a point just inside Saturn's orbit out to a point just inside Uranus' orbit. It is estimated to be 300 km across and is the only asteroid known beyond Saturn's orbit. It may contain much dirty ice, and might turn into a super comet if it ever passes close enough to Saturn to get deflected into the inner solar system. Possibly Chiron, the small planet Pluto, and Pluto's satellite Charon are only the biggest examples of a whole host of ice-rich worldlets in elliptical orbits on the outskirts of the solar system, with others waiting to be discovered.

THE ROCKY MATERIAL OF ASTEROIDS

As indicated in Figure 7-11, the spectra of asteroids can be used to obtain rough estimates of their mineral properties. The asteroids divide into several classes with different spectra, and the spectra of these classes have been compared with spectra of minerals and rocks from Earth, the moon, and many meteorites. The spectra of some asteroid classes closely match those of some types of meteorites. This is strong evidence that certain meteorites are fragments from certain classes of asteroids.

Astronomers discovered in the 1980s that these different asteroid compositional classes are concentrated at different distances from the sun, so that the asteroid belt has a rough zonal structure. Thus, in the inner[7] and central parts of the belt, we find many asteroids with a rock or mixed rock-and-iron composition similar to two important classes of meteorites called chondrites and stony-irons (see next section). In some asteroids, such as 1 Ceres, the largest (1020 km across), the surface materials include water-bearing minerals; the water molecules are chemically bound in the mineral grains (Lebofsky, 1982). Also in this region are some asteroids (such as 4 Vesta, the second largest, 549 km across) that did melt—they have surface rock similar to basaltic lavas of the earth and the moon. In the outer belt are many asteroids with very black surface rocky material composed of carbon-rich and water-rich minerals. These materials (also found in certain meteorites) formed in a region of the solar system colder than the earth's environs. Still more remote asteroids, such as some of the Trojans, have reddish-black surfaces probably consisting of carbonaceous minerals colored by organic compounds. Some of the remote asteroids may also contain some ices, mostly hidden by the black soil. Near or beyond Jupiter may be stony-icy asteroids that are transitions between normal asteroids and comet nuclei of the outermost solar system.

Asteroids that come near the earth, such as Apollo asteroids, are a mixture of these various types, apparently perturbed onto earth-crossing orbits from various regions of the solar system.

Although no asteroid has been photographed at close range, asteroids are known to have rounded and irregular shapes; the close-up appearance may be suggested by Figure 11-14.

MINES IN THE SKY

Because some asteroids probably contain metals and other useful minerals, and because some come very near the earth, asteroids may someday be economically exploited. Studies are already underway to examine the feasibility of flying to Apollo asteroids for scientific as well as economic exploration, as suggested by Figure 11-15. Some close asteroids are actually easier to reach and return from than the moon, in terms of energy expenditure. Asteroids of pure nickel–iron might be mined in space, brought to space stations for processing, or shaped into crude entry vehicles and landed on the earth in remote areas. Based on meteorite samples, other Apollo asteroids are believed to contain good ores of economically important platinum-group metals. Masses worth billions of dollars—enough to supply the earth's needs of certain metals for decades—might be obtainable (Gaffey and McCord, 1977; Hartmann, 1982). Other

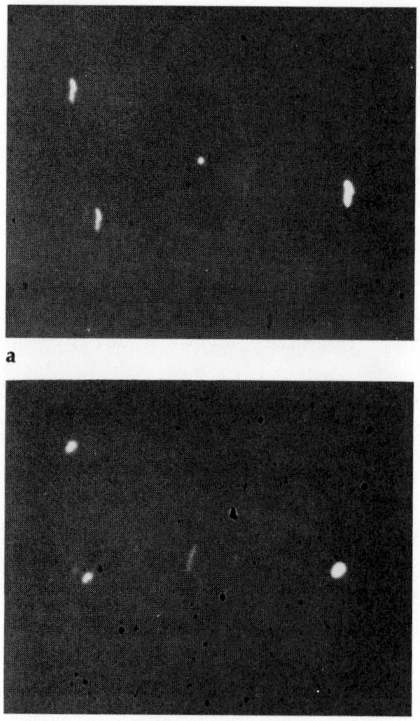

a

b

Figure 11-13. *Two photos of asteroid 887 Alinda (center).* (a) *The telescope was guided on the asteroid; star images trailed.* (b) *The telescope was guided on the stars, showing the movement of the asteroid during the exposure. (New Mexico State University Observatory.)*

Figure 11-14. *No one has observed an asteroid from close range but the view might resemble this close-up of Mars' 16-km-wide, irregularly shaped satellite Deimos, seen under crescent lighting. (NASA.)*

[7]"Inner" asteroid belt is the portion nearest Mars.

asteroid materials such as rock, ice, and hydrogen may be used for constructing and maintaining space stations.

ORIGIN OF ASTEROIDS

Asteroids are probably planetesimals that never finished accumulating into a planet. Possibly disturbed by nearby Jupiter, the planetesimals in the belt region collided too fast to coalesce into a planet. Instead, they collided and broke into thousands of fragments, as depicted in Figure 11-16. Most smaller asteroids in the main belt are such fragments. In 1983, the Infrared Astronomy Satellite ("IRAS") discovered distinct rings of dust circling the sun in the asteroid belt, each interpreted as debris from a relatively recent, individual asteroid collision. Thus, the collisions continue, sporadically, even today.

Meteorites

Meteorites are stony and metallic objects that fall from the sky. Modern evidence is that they are fragments of Apollo-type asteroids that cross the earth's orbit and eventually collide with the earth.

METEORITES AS VENERATED OBJECTS

Stones from the sky have long been a source of awe. Meteorites have been buried with American Indian artifacts (Figure 11-17), wrapped in "mummy cloth" (Casas Grandes, Mexico), and kept as a sacred possession of an Alaskan tribe. A stone venerated in the Temple of Diana at Ephesis (one of the seven wonders of the ancient world) reportedly fell from the sky and was probably a meteorite. A black stone, enshrined around A.D. 600 or earlier in the sacred Moslem shrine at the

Figure 11-15. *Beyond the moon, the closest planetary bodies are Apollo asteroids. Small Apollos, a few hundred meters across, may approach close enough to the earth–moon system (lower right) to be visited by post-shuttle-era spaceships. They could become the next planetary bodies to be visited by astronauts, shown here in transit from a "parked" ship to the asteroid surface. Some NASA planners believe their materials could be economically exploited for use on or near the earth. (Painting by author.)*

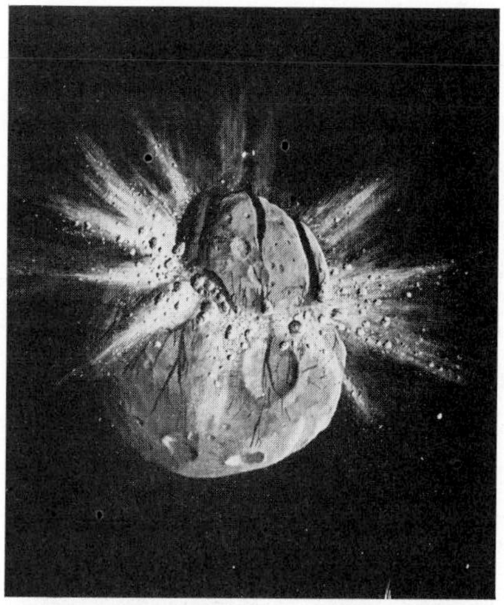

Figure 11-16. *An imaginary view of the collision of two asteroids. The diameter of the larger body is about 50 km; the debris includes pulverized dust and rocks of various sizes that may ultimately fall on the earth and planets as meteorites. Fractures through the bodies may cause them to fall apart during this or later collisions. (Painting by author.)*

Ka'aba in Mecca is also believed to be a meteorite (Sagan, 1975). When a meteorite fell at Ensisheim, France, in 1492 (Figure 11-18), the emperor Maximilian, in residence nearby, decided he should go on a Crusade. The main mass of the meteorite, suspended in the Ensisheim church, was regarded with reverent awe by the peasants.

SCIENTIFIC DISCOVERY OF METEORITES

In 1794 a German physicist, E. F. F. Chladni, reported that the Ensisheim and other supposed celestial stones seemed similar to each other and different from normal terrestrial stones. He concluded that these "meteorites did indeed fall from the sky." This conclusion started a controversy among naturalists. Upon hearing that a meteorite had fallen in Connecticut, Thomas Jefferson, himself an accomplished naturalist, is supposed to have said, "It is easier to believe that Yankee professors would lie than that stones would fall from heaven."

The French Academy, the scientific establishment of the day, dismissed as superstition the notion of stones from the sky. As luck would have it, a meteorite exploded over a French town in 1803, pelting the area with stones. Cautiously, the academy sent the noted physicist J. B. Biot to investigate. His report, one of the historic documents of science, methodically constructed an irrefutable chain of evidence from eyewitness accounts, measurements of the 2 × 6 km area of impacts, and specimens of the meteorites themselves. This report established that stones can fall from the sky.

METEORITE IMPACTS ON EARTH

Interplanetary debris collide with the earth at very high speeds, usually 11 to 60 km/s (24 000 to 135 000 mph). At such speeds, material is heated by friction with the air. Dust grains and pea-sized pieces burn up before striking the ground. Larger pieces are usually slowed by drag, although at least two grapefruit-scale specimens have punched through house roofs, one bruising a woman occupant! They pass through the atmosphere too fast for their interiors to be strongly heated, and stories of meteorites remaining red-hot for hours after falling are untrue.

Large meteorites are rare. Only a few brick-sized meteorites are recovered each year. In 1972 an object weighing perhaps 1000 tons just missed the earth, skipping off the outer atmosphere; it was filmed from the ground and detected by Air Force reconnaissance satellites. Objects weighing 10 000 tons, like the Siberian object of 1908, which are large enough to cause nuclear-scale blasts, fall every few centuries. Larger blasts, thousands of years apart, may form multikilometer-scale craters, such as that in Figure 3-14.

Since meteorites are "free samples" of planetary matter, recovering and reporting a meteorite is a rare honor. Meteorite discoveries should be presented, or at least lent, to research institutions, for example, the Smithsonian Institution in Cambridge, Massachusetts, and Washington, D.C., and the Center for Meteorite Studies, Arizona State University, Tempe, Arizona.

Figure 11-17. *Burial of the Winona stony meteorite in a crypt constructed by prehistoric Indians in northern Arizona. The Indians probably saw it fall and thus attached special significance to it. (Museum of Northern Arizona, Anthropological Collections; photo by author.)*

Figure 11-18. *An old woodcut showing the fall of a meteorite near Ensisheim, France, in 1492.*

TYPES AND ORIGIN OF METEORITES

Meteorites are among the most complex rocks studied by geologists. There are many different types, and sometimes they appear as **brecciated meteorites**—meteorites made of mixed fragments of different meteorite types all jumbled into one rock! After decades of study, scientists have begun to understand how meteorites formed and what they have to tell us about the ancient history of the solar system. The key to the story is that meteorites are fragments of asteroids (and sometimes mixtures of fragments of dissimilar asteroids), created when asteroids collided and blew apart at various times during the history of the solar system. Their fragments were ejected in various directions, often getting thrown onto orbits that eventually intersected the earth.

The different types of meteorites thus tell us about conditions inside the asteroidal parent bodies. Some are samples of surface rock layers, and other are samples of deep cores inside these bodies. Table

11-3 summarizes the major types and their abundances. The first two types are called **chondrites** (KON-drites) because they contain peculiar BB-like glassy spherules called **chondrules** (from the Greek term for seedgrains), as shown in Figure 11-19. These chondrules are believed to be spherules that solidified from molten droplets sprayed out during planetesimal collisions as the planets formed. The preservation of the chondrules in these meteorites proves that they were never melted or severely altered since their formation long ago. This is also shown in the case of **carbonaceous chondrites**, blackish meteorites with water chemically bound in their minerals; this water can be driven off by mild heating to only a few hundred degrees. The carbonaceous chondrites were probably formed in the cold, outermost asteroid belt or outer solar system where ices were abundant. They are undoubtedly fragments of the blackish carbonaceous types of asteroids found in these regions.

Achondrite meteorites, on the other hand, are rocky meteorites with no chondrules. They resemble basaltic lavas found on the earth, moon, and Mars, and are fragments of regions that were once melted and then resolidified inside or on parent asteroids.

Stony–iron meteorites (mixtures of stone and iron alloy) and **iron meteorites** (pure iron alloy metal) are probably samples from the central regions of asteroids that once melted, probably due to heat-producing radioactive minerals incorporated in them when they formed (see Figures 11-20 and 11-21). As they melted, the heavy molten metals sank to the center while the lighter silicate magmas floated. As the asteroids later cooled, the central regions formed nickel–iron cores (like that believed to exist in the earth) while the surface regions formed rocky lithospheres. When the asteroids were smashed in collisions, fragments of the metal regions, or mixed fragments of metal and achondrite rock, were released.

Studies of minerals in several types of meteorites show that they formed inside parent bodies with diameters of tens to hundreds of kilometers—fitting the theory that they formed inside asteroids.

AGES OF METEORITES

One of the most important aspects of meteorites is that they can be dated by the techniques discussed in Chapter 3, and this gives us a history of events in the early solar system. Dating by various techniques shows that the most primitive meteorites formed during a "brief" interval of only 20 million years, 4.6 billion years ago. "Formed" means that they made the transition from dispersed dust grain and gas to solid, rocklike objects. This

Table 11-3. *Types of Meteorites*

Meteorite Type	Percentage of All Falls[a]	Remarks
Stony		
Carbonaceous chondrite	5.7	Most primitive, least-altered material available from the early solar system.
Chondrite	80.0	Commonest type. Defined by millimeter-scale spherical silicate inclusions, sometimes glassy, called *chondrules*.
Achondrite	7.1	Most nearly like terrestrial rocks. Defined by lack of chondrules, which have been destroyed by a heating process. Some resemblance to terrestrial lunar igneous rocks of basaltic type.
Stony–iron	1.5	Contain stony and metallic sections in contact with each other.
Iron	5.7	Nickel–iron material. Museum specimens are often cut, polished, and etched to show interlocking crystal structure.
Total	100	

[a]This table is based on meteorites called *falls:* those actually seen to fall. Meteorites found by chance in the soil, called *finds,* are more numerous but less valuable statistically because they are biased toward iron meteorites, which attract attention whenever found in the ground, while stony meteorites eventually weather to resemble ordinary stones.

Figure 11-19. *A 3-cm portion of the chondrite meteorite Bjurbole showing many chondrules of various sizes, including a large one protruding at lower left. (Center for Meteorite Studies, Arizona State University)*

interval, then, marked the birth of planetary material. Other types of dating show that many major collisions happened at this time, 4.6 billion years ago, and sporadic additional collisions smashed some meteorites' parent bodies at other times scattered through solar system history. We will discuss some of these important events in more detail in the next chapter.

METEORITES FROM THE MOON AND MARS

One of many meteorites found in Antarctica in 1982 was noted to resemble rocks brought back from the moon. By 1983, studies showed conclusively that this rock matched lunar anorthositic composition (see Chapter 5) and indicated that this rock had been blasted off the moon. By 1984, two more were found, also in Antarctica. These findings confirm the theory that crater-forming collisions blow rocks off other planetary bodies, and that some reach the earth. The reason that moon rocks are rare on the earth, even though the moon is closer than asteroids, is that the moon has much larger gravity than asteroids; hence only a few rocks blown out of impacts at the fastest speeds can escape from the moon and have any chance of hitting the earth.

If rocks can be blown off the moon, what about even larger neighboring worlds, such as Mars? Scientists have identified nine other achondrite meteorites that share odd characteristics. They apparently formed in lava flows 1.3 billion years ago (unusually young for meteorites), and then were blasted loose about 180 million years ago. Further study showed that some of these contain gases matching certain traits of the atmosphere of Mars, measured by the Viking lander. Therefore many scientists now think these meteorites may be "free samples" of rocks from 1.3 billion-year-old lava flows on the plains of Mars! This finding, if correct, would prove that Martian volcanoes are much younger than the moon's, and that they were active well within the most recent third of Mars' history.

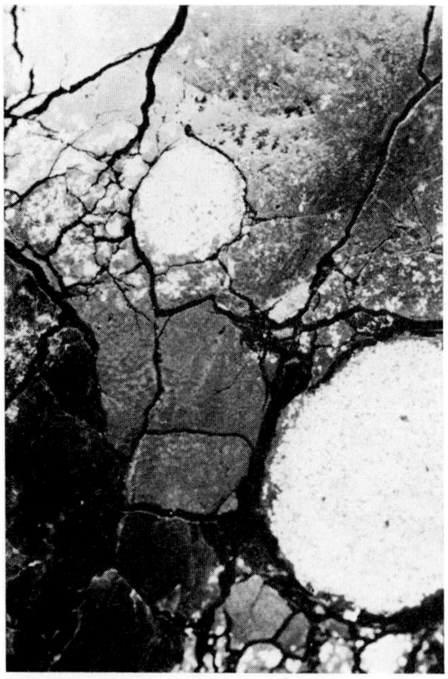

Figure 11-20. ***Cut and polished section of a stony–iron meteorite, from the Bondoc Peninsula, Philippines. The dark fractured matrix is silicate rock, possibly fractured in collisions. The bright nodules are nickel–iron nuggets, a few centimeters across. (Photo by author.)***

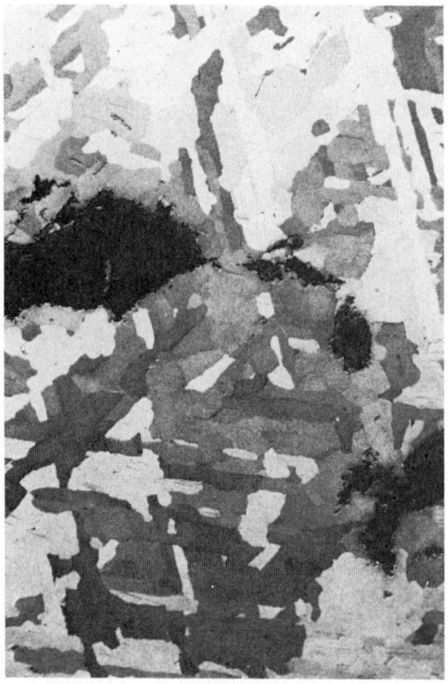

Figure 11-21. ***A cut and polished section of the Campo del Cielo (Argentina) iron meteorite. Intersecting metal crystals can be studied to reveal environmental conditions inside the parent body. The dark inclusions are stony silicate bodies, the largest being about 4.5 cm across. This pattern of metal crystals is called the Widmanstätten pattern, after its discoverer. (Photo by author.)***

Zodiacal Light

The smallest interplanetary particles are microscopic dust grains and individual molecules and atoms, spread out along the plane of the solar system and concentrated toward the sun. If you look west in a very clear rural sky as the last glow of evening twilight disappears (or look east before sunrise), you can detect sunlight shining off the cloud of these particles. It appears as the **zodiacal light**—a faint glowing band of light extending up from the horizon and along the ecliptic plane, shown in Figure 11-22 and 11-23. It is brightest at the horizon. Measurements made by astronauts indicate that it merges with the bright glow of the sun's atmosphere.

Siberia Revisited: Asteroid or Comet Impact?

We now find the mysterious Siberian explosion of 1908, which we described at the beginning of this chapter, easier to understand. Interplanetary space contains debris of many sizes. Rocky and icy chunks up to a few kilometers in size cross the earth's orbit and are likely to collide with the earth from time to time. One of these objects apparently struck the atmosphere over Siberia on June 30, 1908.

Other dramatic events in Russia at this time kept Russian scientists from visiting the site until 1927. During the 1927 expedition and later expeditions, scientists found that, surprisingly, the object did not

a

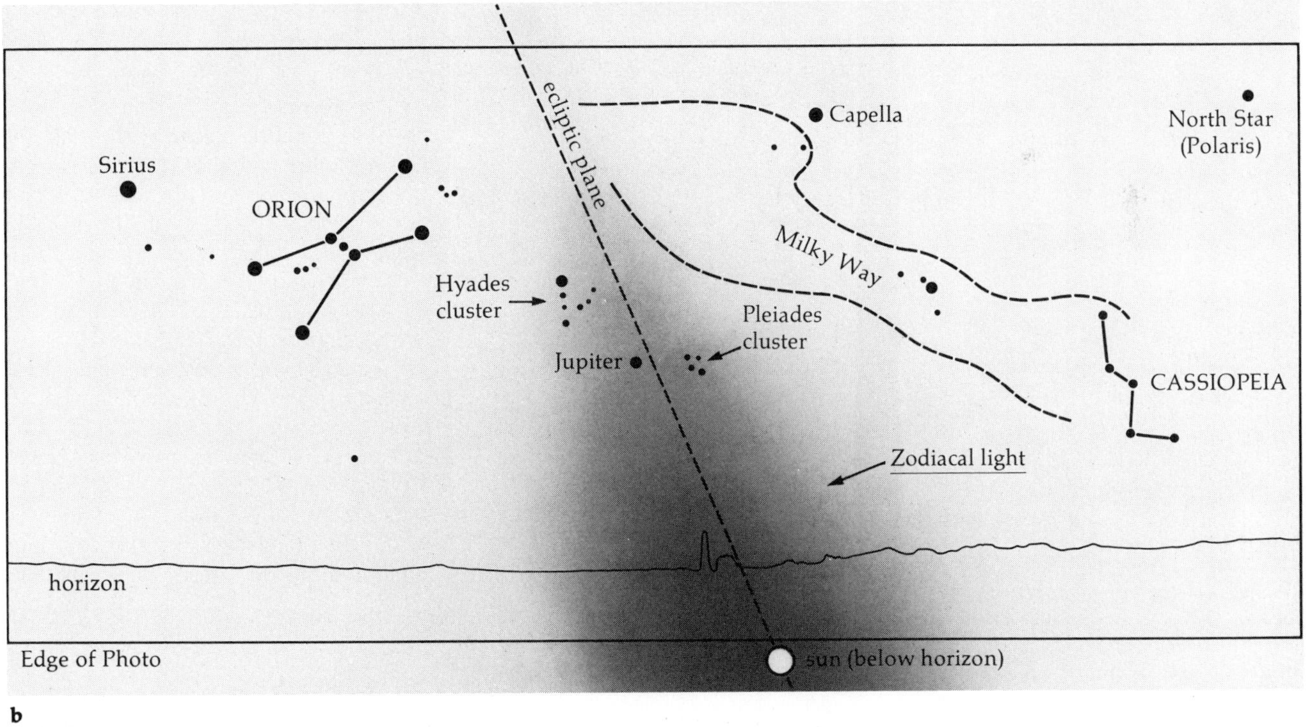

b

***Figure 11-22.** The zodiacal light. (a) Wide-angle panorama of the western sky sunset, about 120° × 50°, shows the glowing band of zodiacal light. The phenomenon is best seen from a rural setting about an hour after sunset, as seen in Figure 11-23. (Exposure 23 min, f2.8, Tri-X; photo by author.) (b) Key to features. Zodiacal light extends about 40° along plane of solar system (dashed line).*

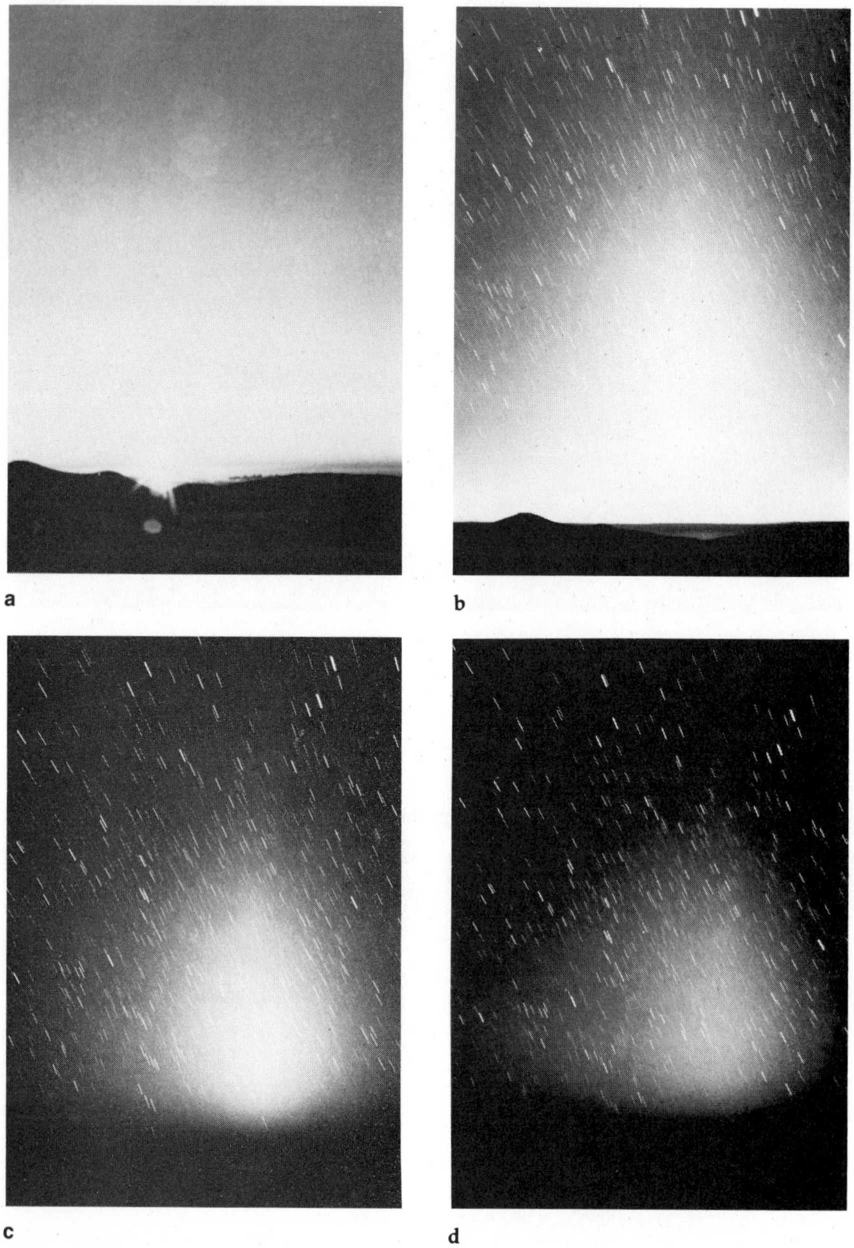

Figure 11-23. *Emergence and setting of the zodiacal light after sunset. Vertical height of these wide-angle views is 72°. Camera remains in a fixed position.* (a) *Sunset.* (b) *1 h after sunset, with zodiacal light extending 49° above horizon.* (c) *1½ h after sunset.* (d) *2 h after sunset. The brightest stars in this view are in the constellations Aries and Cetus. (Exposures* b–d *6 min,* f2.8, *Tri-X; photos by author.)*

Figure 11-24. *Devastation caused by the great Tunguska meteorite fall of 1908. This view, more than a decade later, shows trees some miles from "ground zero" blown over by the force of the explosion. (Photo from E. L. Krinov, 1966.)*

reach the ground and form a crater. It apparently exploded in the atmosphere. Trees at "ground zero" were still standing but had their branches stripped by the blast forces in a downward direction. Trees had been knocked over by the blast out to 30 km from ground zero, as shown in Figure 11-24. A forest fire was started and trees were scorched by the blast out to about 14 km. Researchers found carbonaceous chondrite dust but no meteorites.

Most investigators believe the object was composed of weak material that disintegrated in the atmosphere. It may have been a small asteroid resembling a stony meteorite some 90 to 190 m across (Sekanina, 1983) or a similar-sized icy comet fragment. It injected much dust high into the stratosphere. On June 30 and July 1, sunlight shining over the North Pole illuminated this dust during the night, and newspapers could be read at midnight in western Siberia and Europe.

Such was the effect of a relatively small bit of interplanetary debris striking the earth. To put this event in perspective, if the same object had exploded over New York City, the scorched area would have reached nearly to Newark, New Jersey. Trees would have been felled beyond Newark and over a third of Long Island.

The man knocked off his porch could have been in suburban Philadelphia. "Deafening bangs" might have been heard in Pittsburgh, Washington, D.C., and Montreal.

New reconnaissance satellites may provide early warning against such objects, which may hit the earth every century or so, allowing us to take action, or even intercept such objects to use their materials for more constructive purposes.

An interesting sidelight on large ancient impacts is that the distortions of geologic strata on the earth have been economically valuable in some cases. Rich iron ore deposits at Sudbury, Ontario, occur in a probable impact crater more than 100 km across and about 1.8 billion years old; the impact exposed native iron-bearing strata. Several oil deposits in North America and the Soviet Union occur in sediments collected in ancient craters.

Summary

All of the solar system's small bodies—comets, meteors, asteroids, and meteorites—are examples of the planetesimals or their fragments. Planetesimals were the preplanetary bodies that formed in the solar system 4.6 billion years ago, during an interval of less than 100 million years.

Most asteroids are now in a belt between Mars and Jupiter; the planetesimals in other regions interacted with planets and either were ejected from the solar system after near misses or crashed onto planet and satellite surfaces, making craters. Asteroids in the inner belt are dominated by metal-rich rock materials; asteroids from the outer belt to Jupiter have carbon- and water-rich minerals. Planetesimals formed farther out were ice rich and evolved to what we know as comets. The relation between the various asteroids, comets, meteors, meteorites, and craters that we see today is summarized in Figure 11-25.

The most detailed information about these bodies comes from meteorites, whose chemistry reveals that they came from asteroidlike parent bodies a few hundred kilometers across. Many of these were melted, causing minerals to separate into iron and silicate phases. Others survive with only minimal heating, providing samples nearly unaltered since the solar system's earliest days.

Thus the small bodies of the solar system provide some of the best clues about the origin of planets, both in our solar system and perhaps near other stars. Recent studies also indicate that they may be exploited as source of raw materials for use on the earth and for space exploration.

Concepts

comet
asteroid
meteoroid
meteor
meteorite
parent body
planetesimal
comet head
comet tail
comet nucleus
Oort cloud
short-period comet
excited atoms and molecules
ions
emission bands
dirty iceberg
meteor shower
fireball
asteroid belt
Apollo asteroids
Trojan asteroids
Lagrangian points
brecciated meteorite
chondrite
chondrule
carbonaceous chondrite
achondrite
stony–iron meteorite
iron meteorite
zodiacal light

Problems

1. If a comet should happen to pass through Saturn's satellite system, why would it probably not be detected from the earth?

2. Kepler's third law states that $a^3 = P^2$, where a is the semimajor axis of a body orbiting the sun (expressed in astronomical units) and P is the period (expressed in years). Should this result apply to comets? If a typical comet in Oort's cloud has a semimajor axis of about 100 000 AU (10^5 AU), how often would it return to the inner solar system?

3. In terms of measuring and reporting useful scientific information, what actions would be appropriate if you observed an extraordinarily bright meteor or fireball? In two columns, list examples of useful and nonuseful descriptions of the fireball's speed, brightness, and apparent size. What actions would be appropriate if you saw a meteorite strike the ground?

4. Summarize relations between the bodies and particles responsible for meteors, comets, the zodiacal light, asteroids, and meteorites.

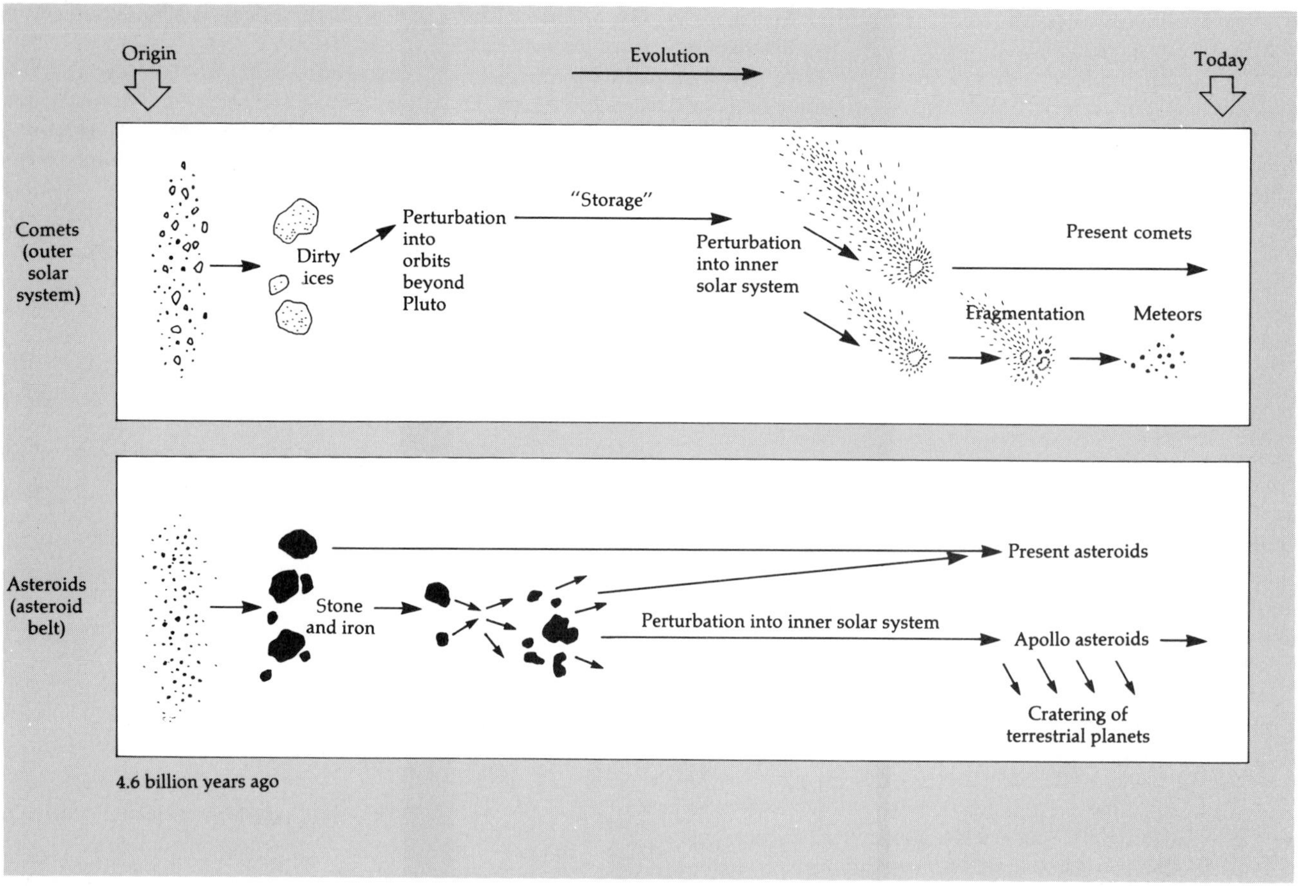

Figure 11-25. *Schematic histories of cometary and asteroidal material, showing condensation into multikilometer bodies (left) and subsequent perturbation and fragmentation to explain phenomena now observable.*

5. Suppose future astronauts could match orbits with a comet and reach its nucleus. Describe the possible surface appearance of a comet. Consider gravity, surface materials, sky appearance, and so on. Which would be easier to match orbits with, a long-period or short-period comet?

6. Typical interplanetary material may move at about 15 km/s relative to the earth–moon system.

a. If a kilometer-scale asteroid was discovered on a collision course with the earth when it was 15 million kilometers away, how much warning time would we have?

b. What would be the potential dangers?

c. If a much smaller asteroid was similarly discovered at the distance of the moon, how much warning time would we have?

d. How long would the objects take to pass through the 100-km thickness of the atmosphere?

7. Based on everyday experience, what is the danger of an event such as in Problem 6, compared with the danger of other natural disasters, such as earthquakes? Is a large-scale meteorite disaster a plausible source of myths during the 10 000-year history of humanity? Defend your answer.

ADVANCED PROBLEMS

8. Use the small-angle equation to calculate how close a 2-km-diameter asteroid would have to come to the earth to allow its shape to be resolved by a telescope that can reveal angular details 1 second of arc across. How does this distance compare with the moon's distance?

9. Suppose a small asteroid of pure nickel–iron, with radius r = 100 m, could be located and exploited. If the value of the alloy were 90¢ per kilogram, what would be the potential economic value of the asteroid? Compare this with the $20 billion dollar cost of the Apollo program. Assume that the density of the material is 8000 kg/m^3.

10. How many asteroids would be required to provide enough material to make one planet the size of the earth? Assume a typical asteroid is 120 km across and the earth is about 12 000 km across.

Projects

1. Use a large piece of cardboard to make a model of the inner solar system out to the orbit of Jupiter. Assume that the planets travel approximately in the plane of the cardboard. Use orbital properties listed in Table 11-1 to cut out scale models of the orbits of various interplanetary bodies. (A slit through the first cardboard could be used to show how comet or asteroid orbits penetrate through the ecliptic plane. Note that the sun must always occupy one focus of each orbit.)

a. Show how the geometry of passage of a comet (or other body) through the ecliptic plane, especially for highly eccentric orbits, depends on the angle between the perihelion point and the ecliptic plane, measured in the orbit. (This angle is fixed for each body, but is omitted from Table 11-1 for simplicity.)

b. Show how the prominence of a given comet may depend strongly on where the earth is in its orbit as the comet passes through the inner solar system.

2. Visit Meteor Crater, Arizona. Why is this feature misnamed? Observe the blocks of ejecta and deformation of rock strata, as explained in museum signs and tapes. What would prehistoric observers, if any (estimated impact date was 20 000 years ago), have witnessed at various distances from the blast that formed this crater, nearly a mile across? (Other impact sites are known in various states, but are eroded or undeveloped.)

3. Examine meteorite specimens in a local museum. Compare the appearance of stones and irons. Are chondrules visible in any stones? Heat damage? In iron samples that have been cut, etched, and polished, look for the crystal patterns that give information about cooling rate and environment when the meteorite formed inside its parent body.

CHAPTER 12

THE ORIGIN OF THE SOLAR SYSTEM

The planets, satellites, meteoritic material, and sun did not exist before about 4.6 billion years ago. Samples from the earth, the moon, and meteorites suggest that the solar system formed from preexisting material within an interval of about 20 to 100 million years. Prior to that time, the atoms in the earth, in this book, and in your own body were floating in clouds of thin gas in interstellar space. It is remarkable to realize that this interstellar material somehow aggregated not only into a dazzling star but also into a family of surrounding planets, as symbolized in Figure 12-1. How did the sun and planetary system form?[1]

Facts to Be Explained by a Theory of Origin

Nobel laureate Hannes Alfvén, who has spent years researching the solar system's origin, once said, "To trace the origin of the solar system is archaeology, not physics." He meant that our ignorance of the initial conditions forces us to work backward through time, reasoning from whatever clues we can find. The most important clues are *facts about the solar system that have no obvious explanation from present-day conditions,* but must have arisen from initial conditions as the solar system formed. Table 12-1 lists some of these clues. In this chapter, we will account for them one by one. This process leads to an interesting realization. As we sift through clues found in meteorites, lunar rocks, and orbits in our own solar system, we will learn about the formation of not only our own system, but also the stars themselves. Thus, this chapter makes a link to the ensuing chapters.

Catastrophic Versus Evolutionary Theories

Throughout history, theories about the origins of things have ranged from so-called catastrophic theories to evolutionary ones. **Catastrophic theories** tend to explain things by sudden events, rare accidents, unusual circumstances, or even the intervention of a Creator whose actions happened once and then ceased. **Evolutionary theories** tend to explain things as relatively gradual developments from relatively normal earlier conditions. For example, catastrophists (especially in the 1700s) argued that the earth's mountains were created in violent, sudden upheavals and that all biological species were created at once. Most modern naturalists, however, refute this with overwhelming evidence derived from rock ages, fossils, and observations of ongoing changes in mountains and species today. This evidence indicates that mountains and species evolved over very long periods by processes that included many slow changes as well as occasional rapid changes. Modern data strongly indicate that many natural systems, such as clouds, ocean basins, planets, stars, galaxies, and biological species evolved.

However, this does not deny sudden catastrophic events. The physical matter of the universe seems to have come into being in such an event—the "big bang" discussed in Chapter 25. On a more provincial scale, the amazing change in the earth's climate that resulted in the extinction of many terrestrial species 65 million years ago (see Chapter 3) probably involved a sudden asteroidal impact.

Unfortunately, because of the controversy accompanying the acceptance of

[1]Note that this question is not addressing the origin of the whole universe, which probably occurred around 14 billion years ago, or the origin of our galaxy, which probably occurred around 12–14 billion years ago. These topics will be taken up in later chapters.

Darwin's theory of the evolution of species, many people think that science defends only evolutionary theories. Many state legislators, scientists, and theologians have become embroiled in controversies over this subject. Even today proposals have been made that any presentation of evolutionary theories (at least in biology) be accompanied by an "equal-time" presentation of a catastrophist (sometimes called creationist) theory. Usually the creationist alternative is drawn from a fundamentalist interpretation of Judeo-Christian tradition with 3000-year-old roots in the Middle East, but other alternatives could be proposed. Advocates of such proposals don't realize that science doesn't work by the advocacy system, such as is used by two lawyers paid to defend two prechosen positions. Instead, it works by laying all available *evidence* out on the table and then seeing if any picture emerges on which all observers can agree. Scientists gain the regard of their colleagues not so much by "winning" arguments as by presenting new pieces of evidence. The evidence, in a sense, argues its own case.

In this chapter we will show how *evidence,* much of it from the preceding chapters, is used to reconstruct the ancient processes that formed planets and explain the items in Table 12-1.

Historically, both catastrophic and evolutionary theories of solar system origin have been proposed. The first evolutionary theory was proposed by the French philosopher René Descartes in 1644. He assumed that, regardless of how matter was created "in the beginning," it was free to evolve according to its incorporated laws of nature. He proposed that space was initially filled with swirling gas in which local eddies, or dense regions, evolved into individual stars. Smaller eddies around them made planets.

The first catastrophic theory was also French. The naturalist Georges Buffon suggested in about 1745 that the sun had been accidentally hit by a passing star and the debris formed planets. Calculations today show that such collisions would be rare. As we will see, the evidence favors an evolutionary view for formation of planetary systems around stars.

Figure 12-1. *This photo taken by Apollo astronauts symbolizes the problem of solar system origin: Why does a typical star like the sun have planets nearby? The photograph shows a solar eclipse caused as the earth (the large crescent) passed between the sun and the spacecraft. Bright spots in the solar glare are reflections in the camera lens. (NASA.)*

The Protosun

Before the sun existed, its material must have been distributed in interstellar space in a large cloud like the interstellar clouds we see today. The sun began to form a little more than 4.6 billion years ago when

***Table 12-1.** Solar System Characteristics to Be Explained by a Theory of Origin*

1. All the planets' orbits lie roughly in a single plane.
2. The sun's rotational equator lies nearly in this plane.
3. Planetary orbits are nearly circular.
4. The planets and the sun all revolve in the same west-to-east direction, called prograde, or direct, revolution.
5. Planets differ in composition.
6. The composition of planets varies roughly with distance from the sun: dense, metal-rich planets lie in the inner system, whereas giant, hydrogen-rich planets lie in the outer system.
7. Meteorites differ in chemical and geologic properties from all known planetary and lunar rocks.
8. The sun and all the planets except Venus and Uranus rotate on their axis in the same direction (prograde rotation) as well. Obliquity (tilt between equatorial and orbital planes) is generally small.
9. Planets and most asteroids rotate with rather similar periods, about 5 to 10 hours, unless obvious tidal forces slow them (as in the earth's case).
10. Distances between planets usually obey the simple Bode's rule.
11. Planet–satellite systems resemble the solar system.
12. As a group, comets' orbits define a large, almost spherical cloud around the solar system.
13. The planets have much more angular momentum (a measure relating orbital speed, size, and mass) than the sun. (Failure to explain this was the great flaw of the early evolutionary theories.)

this cloud (or part of it) became so dense that its own inward-pulling gravitational forces became stronger than the outward force of the pressure generated by the motions of its atoms and molecules. The cloud thus began to shrink.

EARLY CONTRACTION AND FLATTENING

Even if gas in the cloud had been randomly circulating initially, the contracting cloud would have developed a net rotation in one direction or another. An experiment to show this can be done with a cup of coffee or a pan of water. If you stir the liquid vigorously but as randomly as you can and then wait a moment and put a drop of cream in it, the cream will usually reveal a smooth rotation in one direction or the other, because the sum of the random motions will usually be a small net angular motion in one direction or the other. Because the interstellar cloud was relatively isolated from the rest of the universe (as such clouds are today), this rotating motion, or angular momentum, could not be easily transferred anywhere, and thus was *conserved* as the cloud shrank. (The term *angular momentum* refers to a quantity used by physicists to measure the total rotary motion of a system—it increases with the amount of rotating material and with the speed and diameter of the rotating system.)

This principle of **conservation of angular momentum** predicts that the cloud would have rotated faster as its mass contracted toward its center, just as a figure skater spins faster when she pulls in her arms. Mathematical studies show that centrifugal effects associated with the faster spin caused the outer parts of the cloud to flatten into a disk, while material in the center contracted fastest, forming the sun, as shown in Figure 12-2. The planets formed in this disk, accounting for item 1 in Table 12-1, the *single plane of all planets' orbits.* Because the sun itself was an integral part of this disk, item 2, *the sun's rotation in the same plane,* is also explained.

At first, when the cloud was large, its atoms were far apart and could have fallen more or less freely toward the center of gravity in the middle of the cloud. This state is called **free-fall contraction.** If free-fall persisted, the cloud could have completely contracted into the sun in only a few thousand years. But eventually, because of their growing concentration and random motions, the atoms in the cloud began to collide with each other. Atoms interacting in a gas cause outward *pressure.* A buildup of pressure in the cloud would have slowed the collapse. Thus, the inward force of gravity and the outward force of gas pressure competed. Because the properties of gases are well understood, astrophysicists can analyze the later contraction and evolution of the cloud under these competing forces.

HELMHOLTZ CONTRACTION

Gravitational contraction in which the shrinkage is slowed by outward pressure is called **Helmholtz contraction** (after Hermann von Helmholtz, the German astrophysicist who first studied it). In 1871, Helmholtz showed how contraction would have caused heat to accumulate in the contracting protosolar cloud. Helmholtz noted the following (quoted in Shapley and Howarth, 1929):

If a weight falls from a height and strikes the ground, its mass loses . . . the visible motion which it had as a whole—in fact, however, this is not lost; it is transferred to the smallest elementary particles of the mass, and this invisible vibration of the molecules is [what we call] heat.

In the contracting protosun, atoms or swarms of atoms would have fallen toward the center until they collided with other parts of the gas cloud. Temperature would have increased inside the cloud. Wien's law guarantees that the protosun would have radiated energy and thus warmed the surrounding cloud.

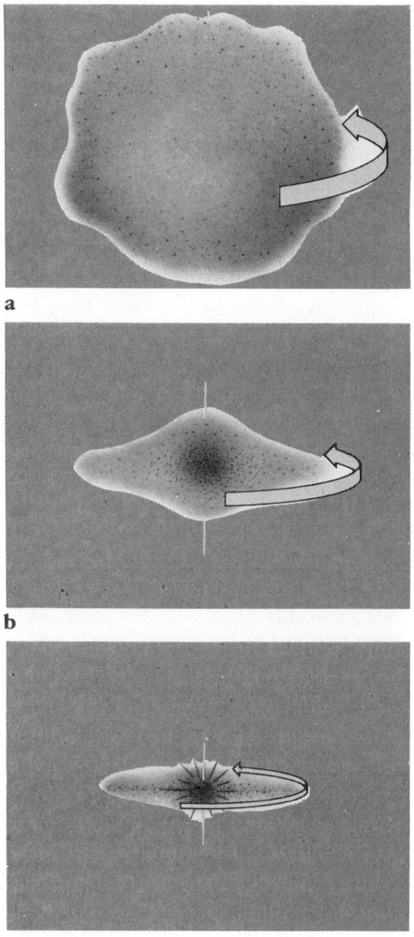

Figure 12-2. *Three stages in the evolution of the protosun. (a) A slowly rotating interstellar gas cloud begins to contract because of its own gravity. (b) A central condensation forms and the cloud rotates faster and flattens. (c) The sun forms in the cloud center, surrounded by a rotating disk of gas.*

Table 12-2. ***Condensation Sequence in the Solar Nebula***

Approximate Temperature (K)	Element Condensing	Form of Condensate (with examples)	Comments
2000	None		Gaseous nebula
1600	Al, Ti, Ca	Oxides (Al_2O_3, CaO)	
1400	Fe, Ni	Nickel–iron grains	Parent material of planetary cores, iron meteorites?
1300	Si	Silicate and ferrosilicate minerals (enstatite, $MgSiO_3$; pyroxene, $CaMgSi_2O_6$; olivine, $(Mg, Fe)_2SiO_4$) in form of microscopic grains	First stony material, combined to form meteorites; some still preserved in primitive meteorites
300 ↓ 100	H, N, C	Ice particles (water, H_2O; ammonia, NH_3; methane, CH_4)	Large amounts of ices; still preserved in outer planets and comets

Source: Data from Lewis (1974), Grossman (1975), and others.

Calculations based on the Helmholtz theory indicate the conditions: Eventually the cloud's central temperature rose to 10 million Kelvin or more, starting the nuclear reactions that made it a star and not just a ball of inert gas. Meanwhile, and more important for the formation of planets, the outer parts of the cloud, shown in Figure 12-2, formed a disk of gas as big as the solar system, at temperatures of a few thousand Kelvin. *These theorized steps have been confirmed by actual observations of newly formed stars, surrounded by clouds of gas and dust.*

The Solar Nebula

A cloud of gas and dust in space is called a **nebula** (plural, *nebulae*), from the Latin term for mist. The disk-shaped nebula that surrounded the contracting sun is called the **solar nebula.** Molecules of gas or grains of dust must have moved in circular orbits, because noncircular orbits would have crossed the paths of other particles, leading to collisions that would have damped out the noncircular motions, in the same way that noncircular eddies get damped, or cancelled, in the cup of coffee mentioned earlier. Thus, neglecting small-scale eddies in the gas, broad-scale motions in the cloud were in parallel circular orbits, accounting for items 3 and 4 in Table 12-1. As the nebula stabilized, its gas began to cool.

CONDENSATION OF DUST IN THE SOLAR NEBULA

The solar nebula was initially gas that Helmholtz contraction heated to at least 2000 K. At such a temperature, virtually all elements were in gaseous form. As is the case with other cosmic gas, most solar nebula atoms were hydrogen, but a few percent were heavier atoms such as silicon, iron, and other planet-forming material.

How did solid particles form in this gas? The answer can be seen on the earth. When air masses cool, their condensable constituents form particles: snowflakes, raindrops, hailstones, or the ice crystals in cirrus clouds. Similarly, as the solar nebula cooled, condensable constituents formed tiny solid particles of dust. Various mineral compounds appeared in a sequence known as the **condensation sequence.**

Chemical studies show that as the temperature in any part of the nebula dropped toward 1600 K, certain metallic elements such as aluminum and titanium condensed to form metallic oxides in the form of **grains,** or microscopic solid particles, as shown in Table 12-2. At about 1400 K a more important constituent, iron, condensed. Microscopic bits of nickel–iron alloy formed as grains, or perhaps coated existing grains. Still more important, at about 1300 K abundant silicates began to appear in solid form. For instance, a magnesium silicate mineral, enstatite ($MgSiO_3$), formed at about 1200 K (Lewis, 1974). These silicate minerals are the common rock-forming materials, so that the solar nebula at this point was acquiring a large quantity of fine dust with rocky composition.

Complicated mixtures of magnesium-, calcium-, and iron-rich silicates condensed, depending on the temperature, pressure, and composition of the gas at various points in the solar nebula. Since local conditions were determined by the distance from the newly formed sun, different compositions of mineral particles may have dominated at different locations, explaining items 5 and 6 in Table 12-1. Exploration of the planets has helped clarify conditions in the nebula. For example, comparison of Pioneer probe results on the atmosphere of Venus in 1979 with earlier Viking results on Mars shows that the amount of primordial argon gas incorporated from the solar nebula into the planets decreased by a few hundred times from Venus, past the earth, to Mars. This suggests that the nebular gas further out from the sun had lower pressure (Pollack and Black, 1979). Similarly, the outer nebula must have been cooler than the inner nebula, being further from the sun and more shaded by the inner dusty nebula (Cameron, 1975). In these regions, at about 300 K, water was trapped in many minerals. Beyond the asteroid belt, snowflakes of H_2O ice condensed. At 100 to 200 K, in the outermost nebula, ammonia and methane ices also condensed. In the outer solar system, these ices survive even direct sunlight; they survive today on comets and icy satellites of the giant planets.

The compositions of meteorites strongly support this theory of condensation. Carbonaceous chondrites contain inclusions of material believed to be among the earliest solid particles in the solar system. These inclusions, shown in Figure 12-3, are rich in elements that would have condensed first (at the highest temperatures), such as osmium and tungsten. Their minerals formed at temperatures of about 1450–1840 K, as is consistent with item 7 in Table 12-1. Yet the bulk of carbonaceous chondrites contain microscopic grains formed at low temperatures, showing that different kinds of primeval dust aggregated into individual bodies. Water is a common constituent in minerals in parts of carbonaceous chondrites. Also as predicted by the theory, enstatite is a common mineral in meteorites.

Figure 12-3. *A piece of the Allende carbonaceous chondrite, showing white inclusions. These demonstrate how material that formed in one environment was later trapped in other material. The black matrix is composed of microscopic dust grains condensed at a few hundred Kelvin. The white inclusions contain aluminum-rich minerals condensed at high temperatures. As recognized in the 1970s, the inclusions shed light on the earliest formative conditions in the solar system. (Courtesy R. S. Clarke, Smithsonian Institution.)*

A Presolar Explosion?

As meteorite compositions were being studied in the 1970s, an interesting mystery emerged. Light-colored inclusions of high-temperature minerals in certain carbonaceous chondrites, pictured in Figure 12-3, were found to have queer abundances of certain isotopes. Most elements occur in one main, stable isotopic form and several other forms, both unstable (radioactive) and stable. In the light-colored inclusions, researchers found certain isotopes that could only have been created very shortly before the solar system itself.

For instance, they found xenon 129, a form of xenon that arises from the decay of radioactive iodine 129. This decay process is very fast, geologically speaking, once the iodine is created—the iodine 129 half-life is only 17 million years. Therefore, the parent iodine must have been trapped in the meteorite inclusions in the brief interval (1 to 20 million years?) between the creation of the iodine and its decay into xenon. Remember that the inclusions were among the first minerals condensed in the solar nebula, and they are the main ones that got a dose of the mysterious iodine.

The mystery is how a batch of radioactive iodine and other isotopes was created just *before* the first planetary material formed, and how it got trapped in that material. The types of isotopes involved are now believed to be created by nuclear reactions inside certain massive stars, which burn their nuclear fuel quickly and then explode. Thus, the evidence indicates that such a star was located near the presolar nebula and exploded, spewing short-lived radioactive isotopes and other debris into the cloud that was becoming the solar nebula. Indeed, the blast from the explosion probably helped compress the presolar cloud, helping to initiate the collapse that produced the sun!

This may sound like a very unlikely catastrophic theory—a giant stellar explosion close enough to play a chance role in forming the sun. Not so. As will become clearer in Chapter 16, stars form in close groups, which usually include massive exploding stars. Explosions initiating new sequences of gas cloud compression, star formation, and silicate dust condensation now seem to be a normal aspect of the slow, somewhat sporadic evolution of matter in our galaxy over time spans of billions of years (Herbst and Assousa, 1979).

From Planetesimals to Planets

Although the preplanetary particles may have formed as microscopic grains, they clearly grew bigger (otherwise there would be no planets!). The hypothetical intermediate bodies, from millimeters to many kilometers in size, are usually called **planetesimals**, as noted in Chapter 11. Evidence that they existed includes the following:

1. Craters on planets and satellites indicate impacts of planetesimals with diameters of at least 100 km.

2. Meteorites are their surviving fragments, and their microstructures reveal how the dust grains clumped together.

3. Asteroids and comet nuclei reaching more than 100 km in diameter survived until today throughout the solar system.

But exactly how did microscopic grains aggregate to produce 100-km-size planetesimals? If they had circled the sun in paths comparable to present asteroidal and cometary orbits, they would have collided with one another at speeds much faster than rifle bullets. They would have shattered, and the solar system would still be a nebula of dust and grit.

But did dust particles have such high speeds in the early solar nebula? According to dynamical analyses, they collected in a swarm of particles with nearly parallel, circular orbits in the central plane of the disk. Because the orbits were nearly parallel, collision velocities were low. In low-velocity collisions, some dust grains simply stuck together, held by weak adhesive forces such as gravity and electrostatic attraction. Mutual gravity probably caused groups of planetesimals to clump together, growing to multikilometer size in only a few thousand years—a mere moment in cosmic time. At this point, the solar nebula would have resembled the scene in Figure 12-4, with rocky planetesimals orbiting in dusty clouds that dimmed the central sun.

Dynamical studies indicate that coalescing particles tend to form bodies rotating in a prograde motion (item 8) with similar rotation periods (item 9). No one is sure why the planetary spacings are regular (item 10), but studies suggest that gravitational forces tended to divide the solar nebula into ring-shaped zones, each favoring the formation of one planet.

The larger planetesimals collided with smaller ones, knocking off debris. Some small particles fell back on their surfaces, forming a powdery soil layer that, like the lunar regolith, was effective in trapping other small fragments. Thus the largest planetesimals grew fastest, sweeping up

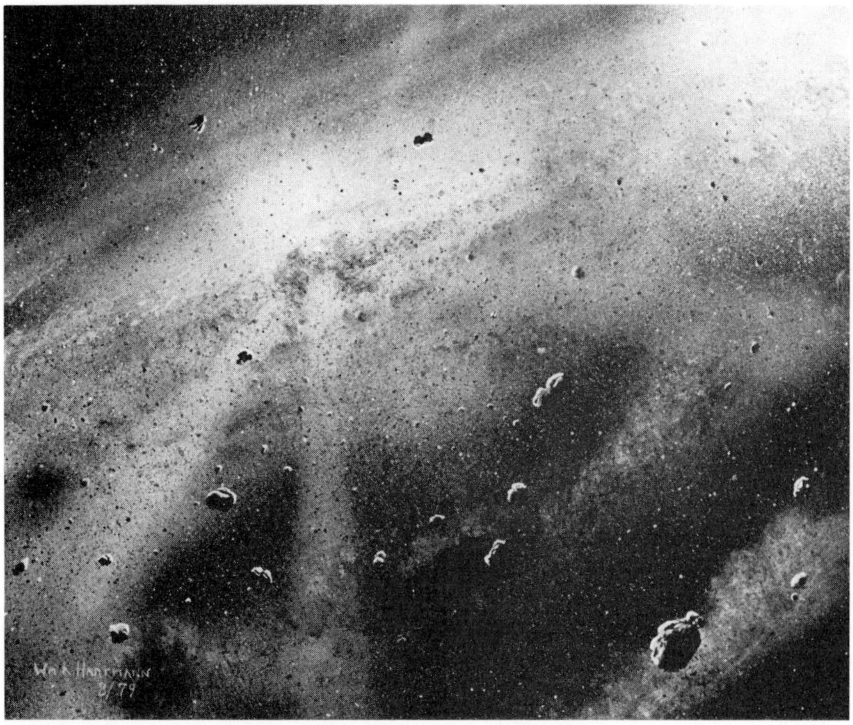

Figure 12-4. *A scene in the early solar nebula. Planetesimals of rocky and icy materials orbit in the foreground. The sun is partially obscured by dust and gas in the inner nebula (background). (Painting by author.)*

Color Plate 17 *A portion of Jupiter's satellite Europa, about the size of our moon. The surface is covered with smooth deposits of frozen water, laced by a network of linear features believed to be cracks. Some of the features are very shallow furrows, but most of the surface is very flat and uncratered. (NASA Voyager 2 photo, 1979.)*

Color Plate 18 *Views of Jupiter's moons, Ganymede and Callisto, which are about the size of the planet Mercury. Upper left: The globe of Ganymede consists of old, dark, cratered regions broken by zones of younger, lighter deposits. The surface is a mixture of dark soil and ice. A polar frost cap of ice is visible at the upper left. (NASA Voyager 2 photo, 1979.) Right: Close-up of Ganymede's dark, cratered terrain broken by lighter ice-rich remnants of ancient impact craters. Like glaciers, the icy crust may flow to fill in craters. Arc-shaped cracks may be remnants from a very ancient, giant impact. (NASA Voyager 2 photo, 1979.) Lower left: The globe of Callisto, the darkest of the four large Jovian moons. The surface is a dark soil with some ice mixed in. This picture uses a UV image for the blue-color component, emphasizing color contrasts between craters (fresher ice?) and background. (NASA Voyager 2 photo, 1979.)*

Color Plate 19 *Saturn and two of its moons, Tethys (above) and Dione. The moons are about 1000 km across, about a third the size of our moon. The photo dramatically shows the gap in the rings known as Cassini's division. Faint cloud bands are visible. One moon casts a shadow on Saturn's surface (upper right). (NASA Voyager 1 photo, 1980.)*

Color Plate 20 *Views of Saturn and its moons. Top: Saturn and its rings in crescent-phase lighting, a view never seen from the earth because we are always closer to the sun than to Saturn. Variations in density of the rings are prominent, and Saturn can clearly be seen through part of the rings. The shadows of the rings are prominent on Saturn, uper left. Lower left: The 1100-km moon Dione, transiting in front of Saturn's clouds. Dione has an icy surface and keeps this face permanently away from Saturn. The leading hemisphere (right) is brighter; the trailing hemisphere (left) is dark and mottled by bright streaks of unknown origin. Lower right: The heavily cratered 390-km moon Mimas. (NASA Voyager 1 photos, 1980.)*

Color Plate 21. *Top left: Saturn's 1440-km moon Iapetus has one face (right) surfaced with clean bright ice, and an opposing hemisphere (left) covered with dark material—probably a mixture of ice and dark soil rich in carbon minerals. The central crater has exposed a brighter subsurface material (cleaner ice?). Top right: Saturn's giant 5140-km moon Titan in a crescent phase backlit by the sun. Sunlight filters through the hazy atmosphere, creating a ring of light visible around the disk. Reddish haze is a smog of hydrocarbon compounds in a dominantly nitrogen atmosphere. Bottom left: Backlit crescent view of the planet Uranus. The atmosphere is bluish colored due to light scattered from the thick layer of clear gas above the lower haze, a process that creates a similar color of blue sky on Earth. Lower right: Uranus' 1600-km moon Titania. The heavily cratered, dirty-ice surface is reminiscent of that of Saturn's moon Mimas. Cliffs and fractures revealed by shadows along the right edge suggest tectonic activity. (NASA Voyager 2 images, 1981 and 1986.)*

Color Plates 22 and 23 *Views of the sun in different wavelengths. Upper left: Skylab astronauts' photo of a total solar eclipse showing the pink chromosphere, visible just as the moon covers the last parts of the bright solar surface. (NASA.) Lower left: X-ray view of the sun, showing flares as sites of intense x-radiation. Since X rays are not visible to the eye, picture has been processed with orange tones usually associated with sunlight. (NASA photo by Skylab astronauts, 1973.) Upper right: Researcher Joseph Sutorik planned this spectroheliograph photo to silhouette hiker Galen McFadyen against the setting sun, as seen in the red light of Hα emission. (National Oceanic and Atmospheric Administration/SESC 13-cm telescope photo.) Lower right: A major eruption on the sun, photographed in extreme UV light at 30-nm wavelength, emitted by ionized helium. (NASA photo by Skylab astronauts, 1973.)*

Color Plate 24 *Top: The Orion nebula is the heart of a star-forming complex. The nebula is about 5 pc across and 460 pc away. It is a mass of glowing gas and dark, obscuring dust clouds. As early as two centuries ago, William Herschel examined this region with his pioneering large reflecting telescope and prophetically described it as "an unformed fiery mist, the chaotic material of future suns." (Copyright Anglo-Australian Telescope Board.) Bottom: The constellation of Orion shows stars of different colors. Note red Betelgeuse (upper left) and blue Rigel (lower right), whose colors can be seen by the naked eye. The nebula lies in the middle of Orion's sword. "Star effect" filter gives added emphasis to the brighter stars. (Photo by author; 35 mm camera, 55 mm lens, f2.8, 20 minute exposure 3M ASA1000 slide film.)*

Color Plate 25 *(Found after p. 392) The spectacular young "Eagle nebula" and associated star cluster NGC 6611 dominate a region of star formation about 2000 pc away. (See Chapter 16.) The nebula, about 6 pc across, contains dark dust clouds silhouetted against red-glowing hydrogen. The complex is rich in hot, bluish-white O stars and is about 3 million years old. (Copyright Association of Universities for Research in Astronomy, Inc., Kitt Peak National Observatory.)*

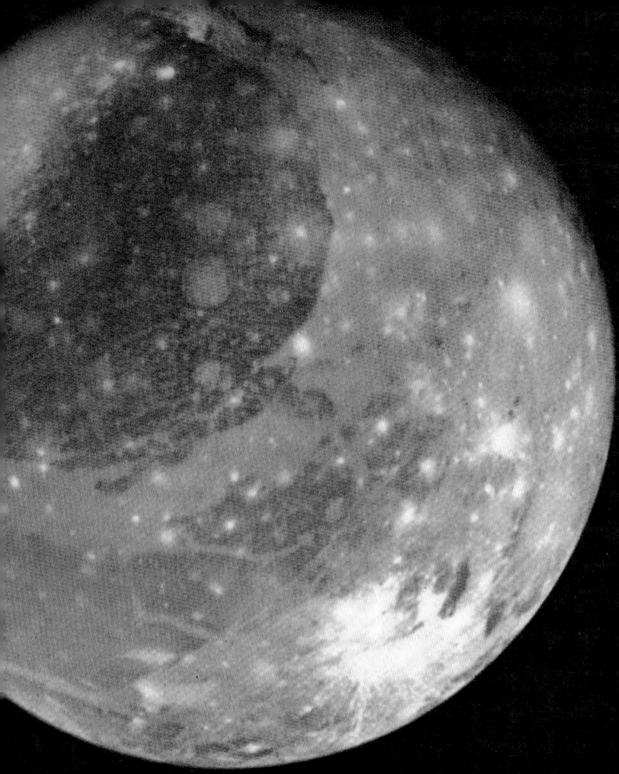

Ganymede

▼ Callisto

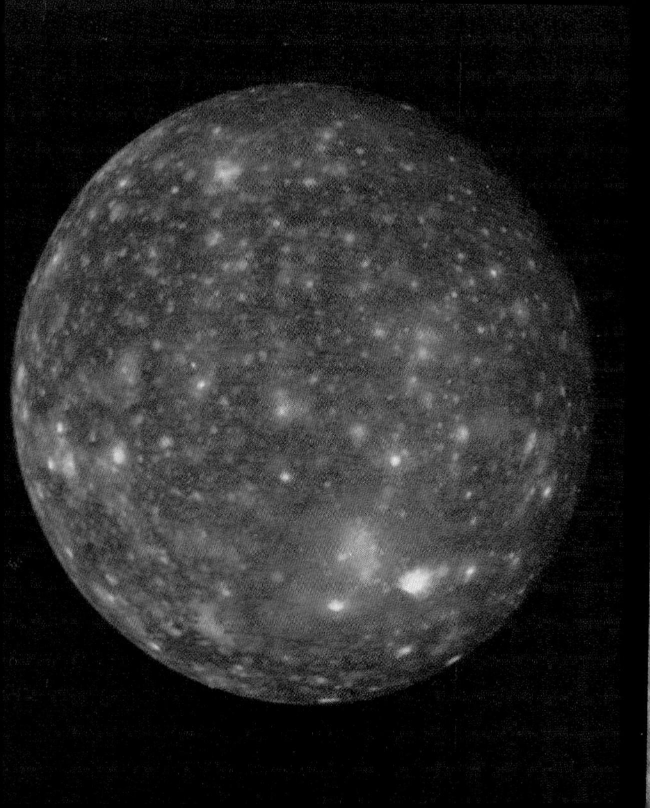

Ganymede's cratered terrain ▶

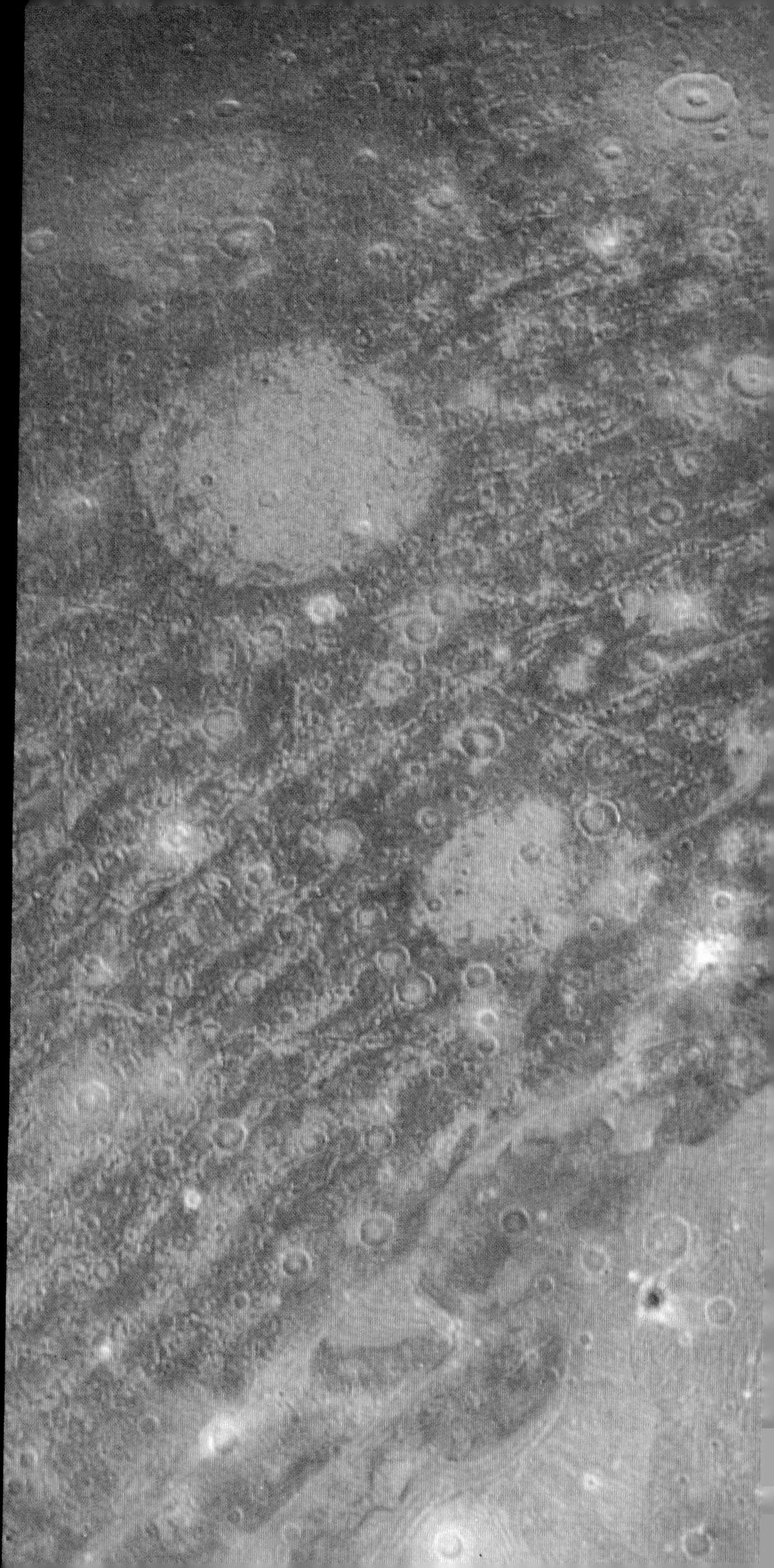

▲ Saturn and its rings

▼ Dione

▼ Mimas

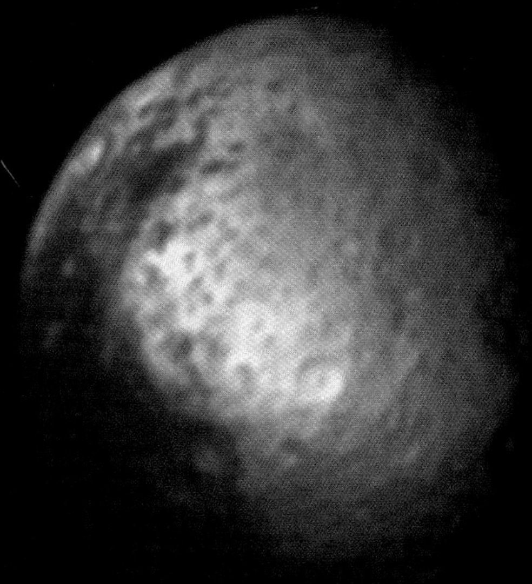

▲ Iapetus

▼ Uranus

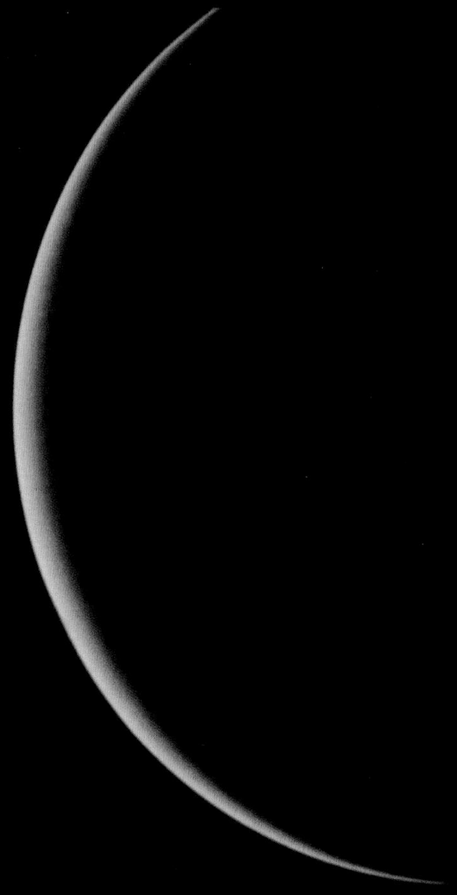

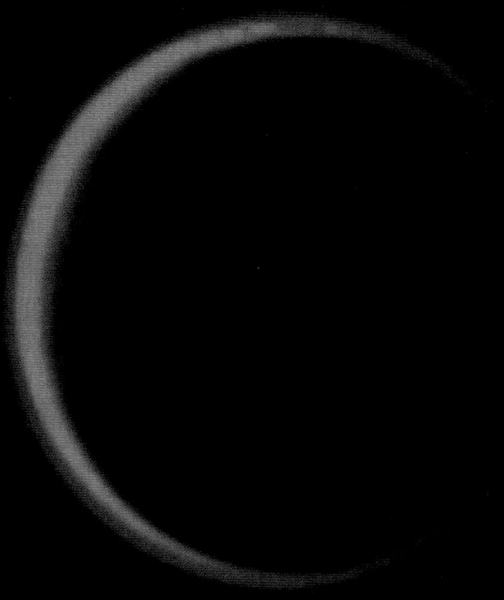

▲ Titan

▼ Titania

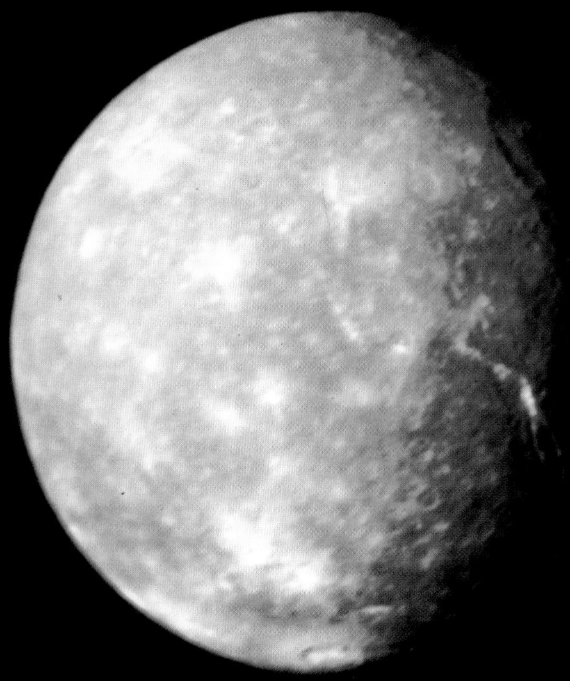

Color Plate 22

▲ Total solar eclipse

▼ X-ray photograph of the sun

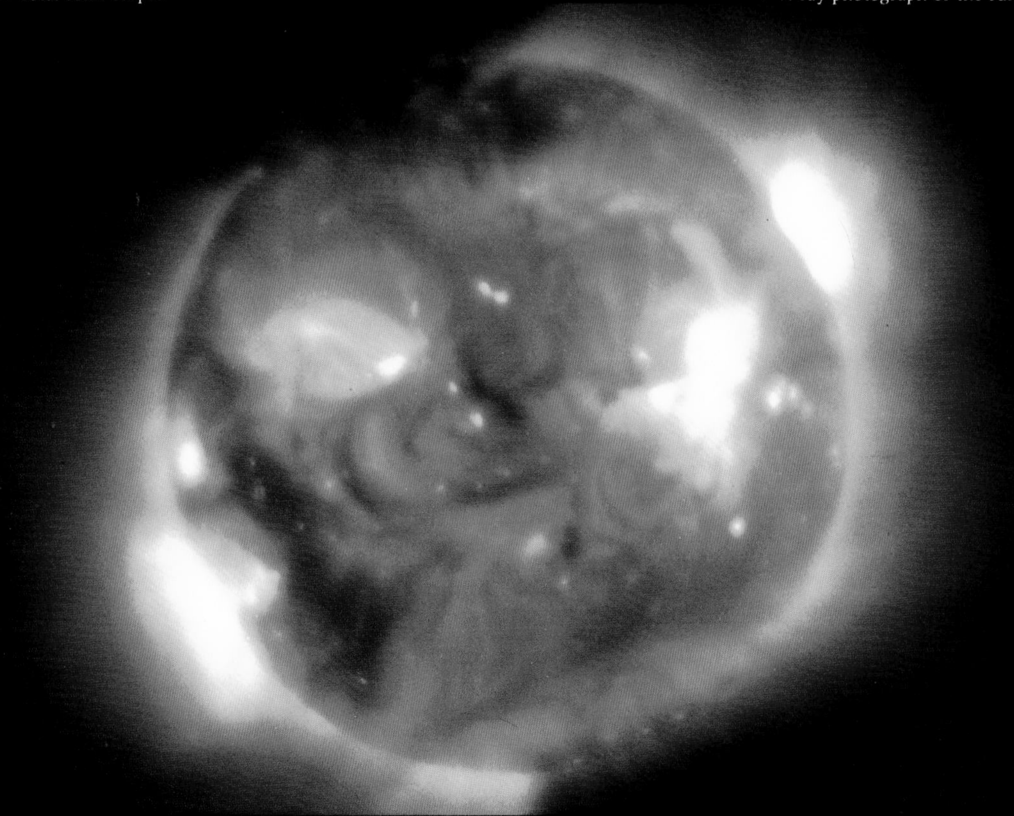

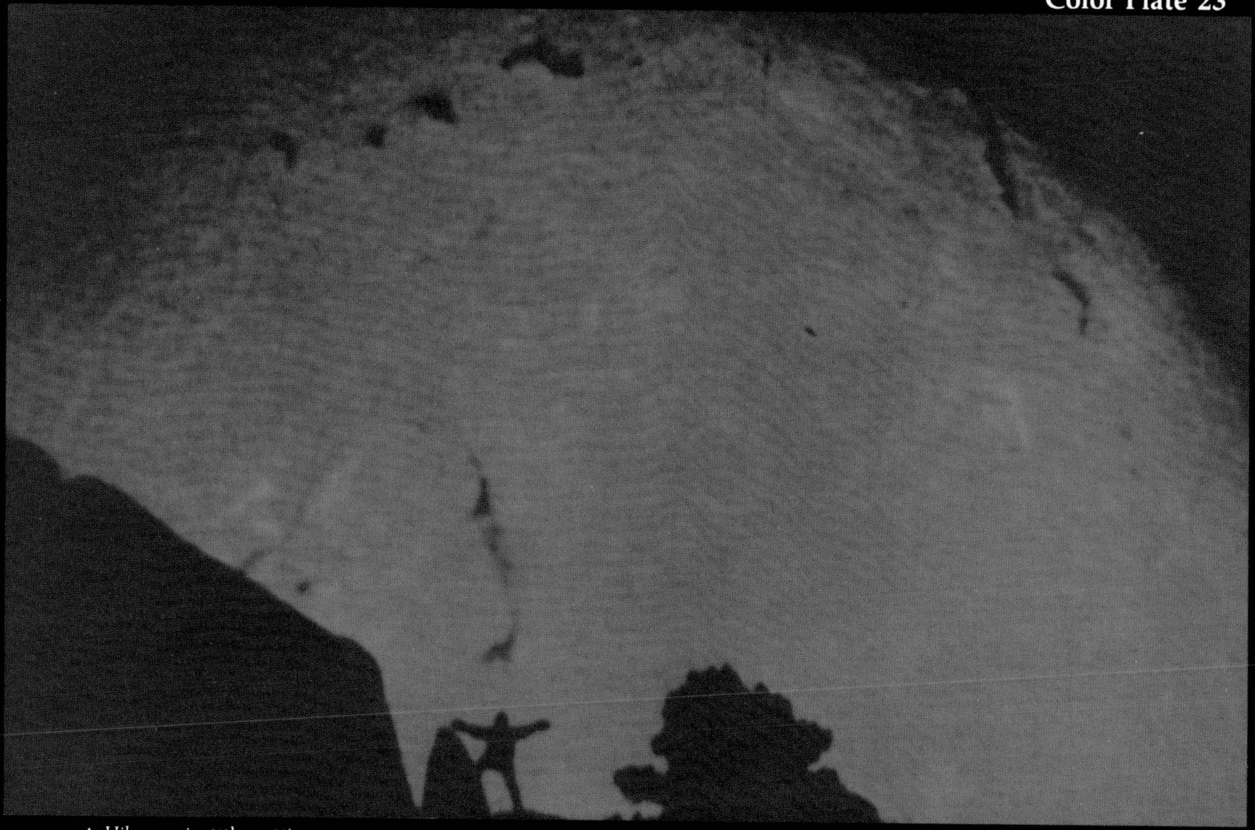

▲ Hiker against the setting sun

▼ A solar eruption

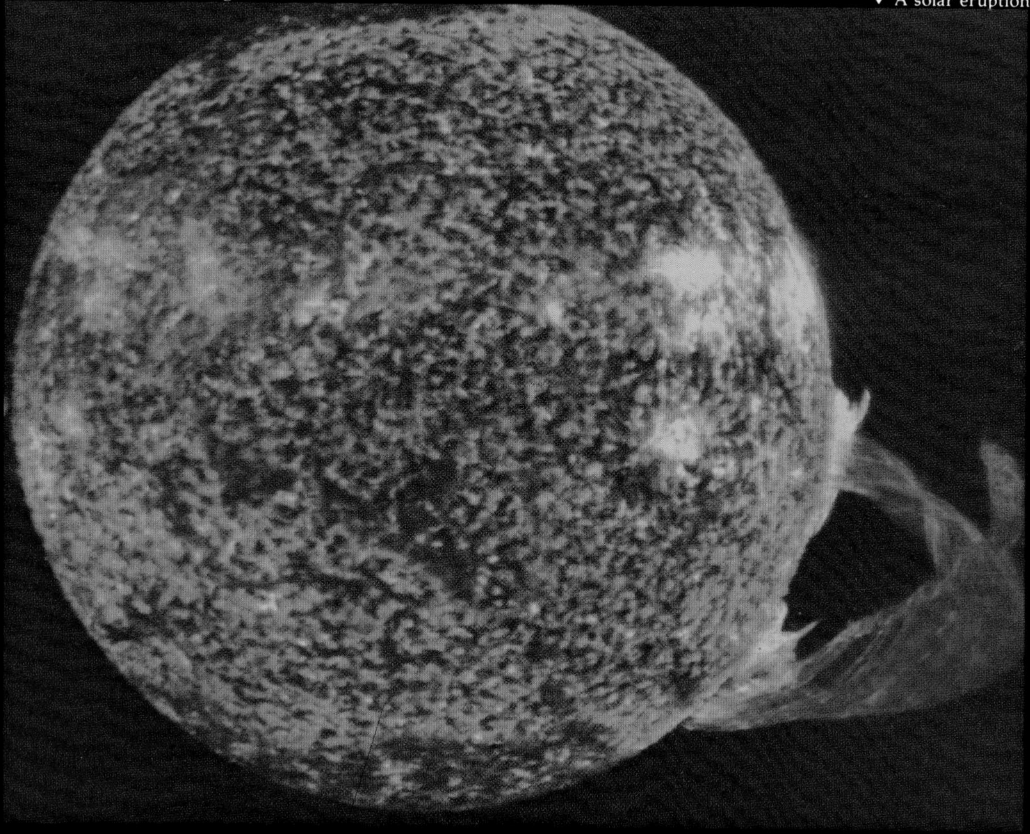

Color Plate 24

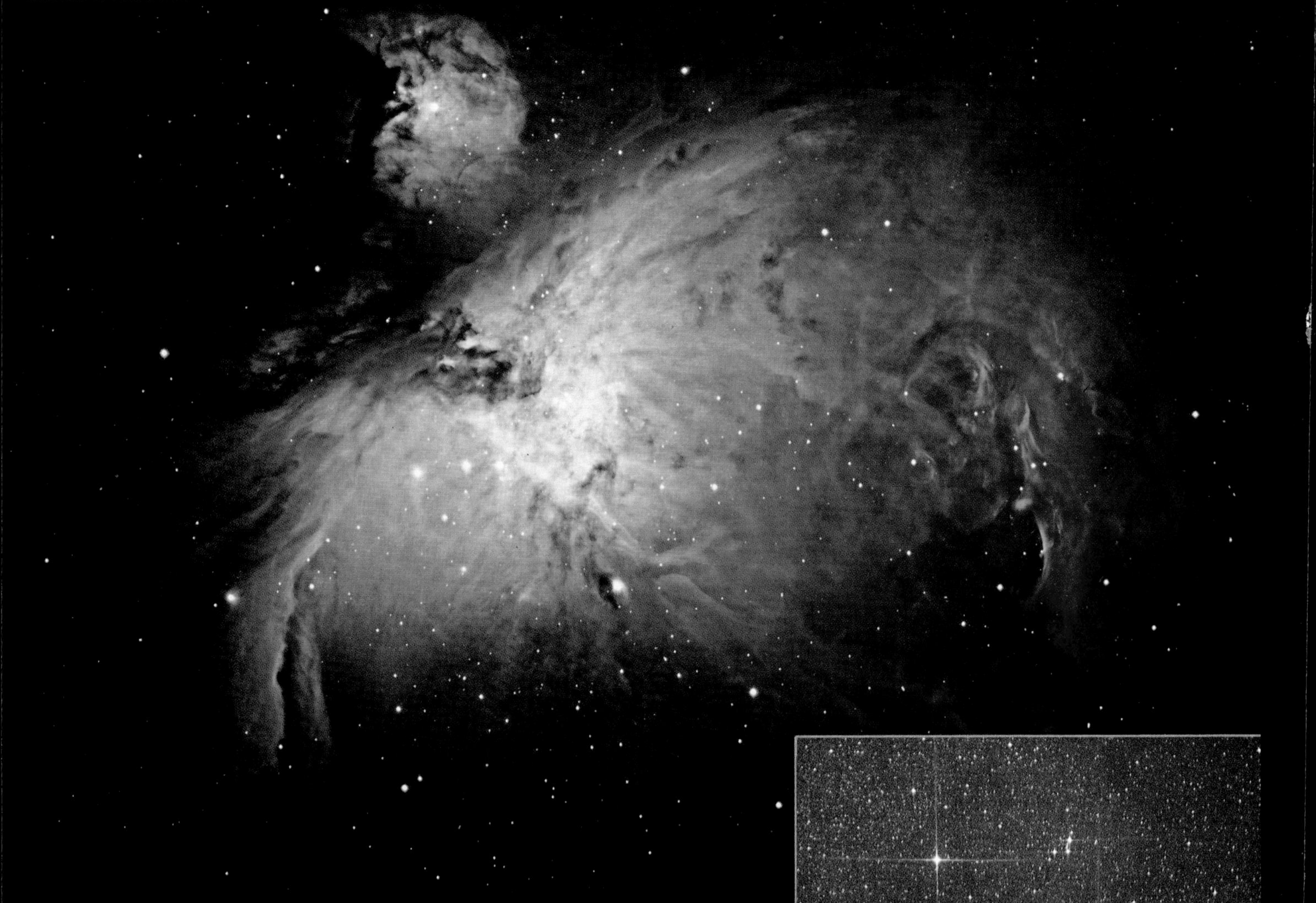

▲ The Orion nebula

Constellation of Orion (nebula marked) ▶

the others. These bodies may have grown to 100-km-size in a few million years. Some planetesimals, the *parent bodies* of meteorites, were heated, melted, and differentiated into metal and rock portions. Some were shattered by collisions with other large or fast neighbors, freeing iron, stony–iron, and achondrite meteorites from their interiors, as was illustrated in Figure 11-16. But as some shattered, others grew to replace them.

Once the largest planetesimals reached a few hundred kilometers in breadth, they had strong gravitational fields of their own and began to pull in the remaining planetesimals. Giant planets formed when some of the growing planets acquired enough gravity (that is, mass) to pull in the surrounding nebular gas, forming massive, hydrogen-rich atmospheres. They were probably massive enough to form miniature "solar nebulae" of their own, in which the process was repeated to form families of satellites, accounting for item 11.

As remaining small planetesimals passed near the planets, they were often deflected into a different orbit (just as Voyagers 1 and 2 were deflected into the outer solar system by near collisions with massive Jupiter). Icy planetesimals among the outer planets passed near Jupiter and were flung into wide-ranging orbits that account for item 12 in Table 12-1.

WHAT BECAME OF THE PLANETESIMALS AND THE NEBULA?

Most of the planetesimals were eventually accumulated by the planets, probably within 100 million years of the solar system's beginning. For another 500 million years, as lunar rocks show, some of the remaining planetesimals rained down onto planetary surfaces, creating still more craters on all the planetary bodies, as shown in Figure 12-5. Others were thrown into the outer solar system, forming comets. Some may have been ejected by the gravity of massive Jupiter. Between Mars and Jupiter, the planet-forming process was aborted, and a swarm of asteroids was left as an example of the early solar system.

A few planetesimals may have been captured into orbits around the larger planets, perhaps explaining the fact that some outermost satellites are retrograde and have compositions resembling asteroids' compositions. The few largest planetesimals to strike each planet may have significantly affected properties such as the tilt of the rotation axis, the speed and direction of rotation, the character of the satellite systems, and geologic differences between hemispheres. Eventually the smallest particles, such as dust grains, molecules, and atoms of gas, were carried away from the solar system by the outward forces of solar radiation and outrushing solar gas. The solar system had then reached essentially its present configuration.

The Chemical Compositions of Planets

Why do the planets vary in their chemical composition? Why do meteorites differ from lunar and terrestrial materials? The answers to these questions now become quite clear. The condensation sequence shows that different groups of minerals existed at different *temperatures* in the nebula. In the cold outer parts of the nebula, hydrogen-rich ices formed, but in the hot inner parts, only metals and silicates. Massive bodies such as the sun and Jupiter retained their original gases; studies of these bodies show that the nebula was originally mostly hydrogen and helium, with only a few percent of heavier atoms such as silicon, oxygen, and iron, as shown in Table 12-3.

Table 12-3 shows what happened after the gas condensed into solid planetary materials. Virtually none of the hydrogen or helium was trapped in the solid materials of the inner solar system. Primitive meteorites of the carbonaceous chondrite and chondrite types show that primitive silicates were composed mostly of oxygen, iron, silicon, and magnesium. However, as soon as these materials melted, probably due to heat from radioactivity in certain minerals, the iron drained to form central cores resembling iron meteorites. This differentiation process meant that the surface materials of planets became more iron-poor and silicon-rich than the primitive materials. This sequence can be seen by following Table 12-3 from left to right. Note that *surface* rocks of lunar and terrestrial uplands have only 2%–4% iron, compared with 24% in the original solid planetary material. On the other hand, they have increased percentages of elements like silicon and oxygen, which combine into lightweight minerals. (Review also the discussion of differentiation on page 54.)

Magnetic Effects and the Sun's Spin

As noted before, the early sun would have been spinning fast enough after its contraction to deform into a disk, but today it is rotating slowly and is spherical. It has evidently slowed down. What happened to the missing angular momentum? For many years this unanswered question was the fatal flaw in evolutionary theories of planet formation. However, in the 1950s and 1960s Swedish–American astronomer H. Alfvén and others realized that the early sun probably had a strong magnetic field. The hot surface of the early sun shot out ionized gas, probably at a greater rate than it does today. Because ions must move with any local magnetic field, the rotating magnetic field of the sun tried to drag the ions of the inner solar nebula with it. The gas would thus have exerted a braking force on the rotating sun, just as water does on a spinning tennis ball dropped into a pool. The ball's rotation slows.

This **magnetic braking** probably slowed the sun's rotation, transferring angular momentum to the planetary material in the solar nebula. This theory solves the angular momentum problem of item 13 in Table 12-1.

Stellar Evidence for Other Planetary Systems

Since no unusually rare processes seem to have been necessary to form the solar system, other stars may also have spawned planetary systems. All stars should form

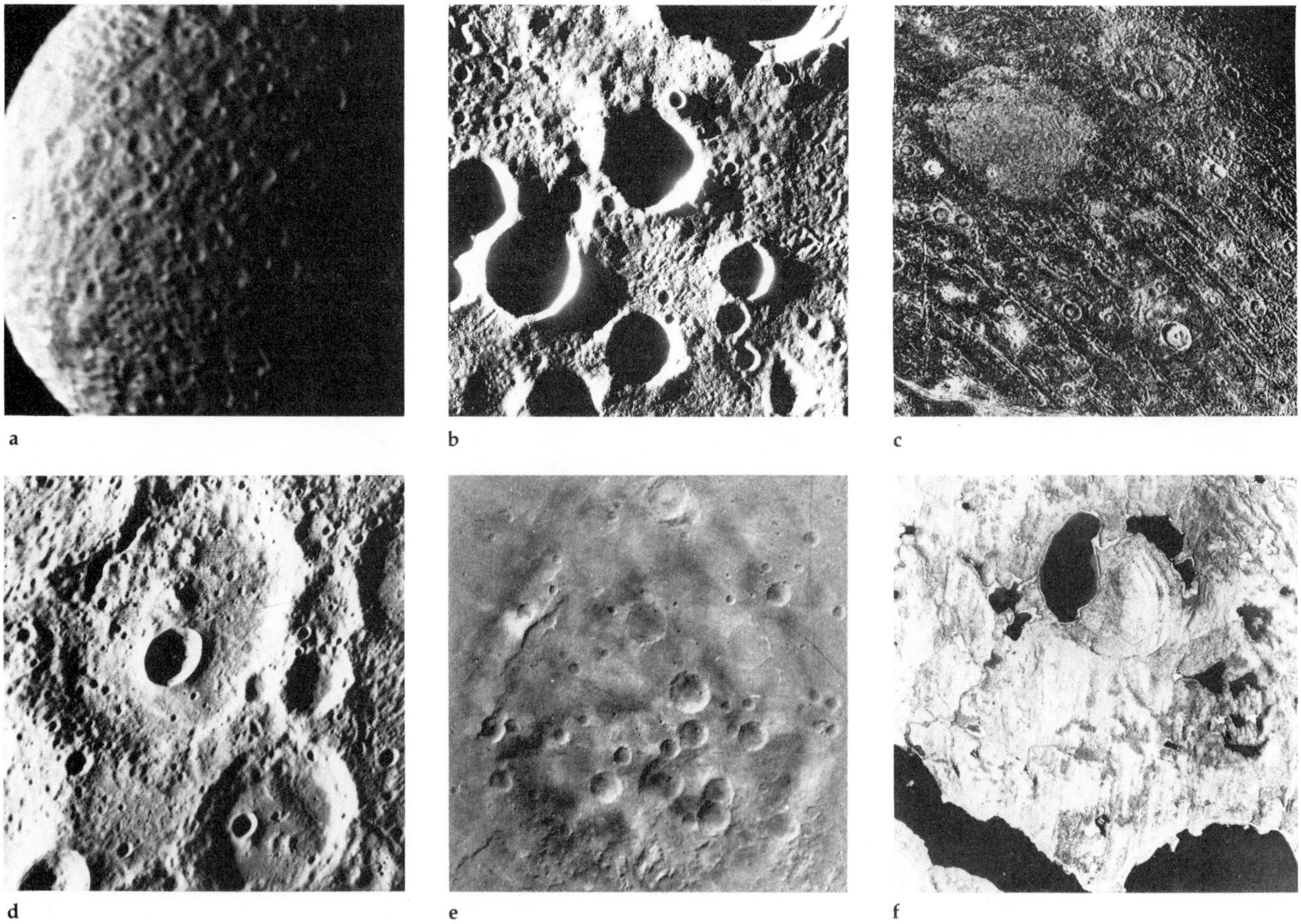

Figure 12-5. *Impact craters throughout the solar system: (*a*) Saturn's moon Mimas, (*b*) our moon, (*c*) Callisto, (*d*) Mercury, (*e*) Mars, and (*f*) the earth. The arrangement in order of planetary size reveals that the larger planets have enough erosional activity to obliterate or degrade the more ancient craters. In most of these views, larger craters are a few tens of kilometers across. (*a–d*: NASA. e: Soviet Mars 5 orbiter photo, courtesy C. P. Florensky, Vernadsky Institute, Moscow. f: 45-million-year-old eroded Canadian crater; Earth Physics Branch of the Department of Energy, Mines, and Resources, Ottawa, Canada.)*

from gravitationally unstable, collapsing nebulae. All stars should have some angular momentum and thus produce rotating, flattened clouds around themselves. This cloud should cool as its heat radiates away into space, and dust grains of varied compositions should therefore appear. Many observations of stars indicate that these processes do occur elsewhere.

But would the subordinate material always accumulate into planet-sized bodies? Probably not, according to simple observations of stars. Many if not most stars are double or triple star systems. In such a system the second biggest object is not a planet, like Jupiter, but a full-fledged star with nuclear reactions and an incandescent gaseous surface. Nonetheless, the evolutionary sequence of events just described indicates that, among the remaining stars, dust particles may very well have accumulated into planetlike bodies. Thus, many astronomers suspect that a sizable number of other stars may have planets orbiting around them. If so, favorable environments for living creatures might not be restricted to our solar system. Even if only 0.1% of all stars had planets on which liquid water could exist and life as we know it could form, 100 000 000 habitable planets could plausibly exist in our galaxy! This thought, with its exciting implications, will be deferred until Chapter 25, after we have more critically examined the nature of stars, their formation, history, and multiplicity.

Table 12-3. *Elemental Abundances in Cosmic Matter (Selected Elements)*

		Percent by Weight								
Element	Designation[a]	Sun ("Cosmic" Composition)[b]	Carbonaceous Chondrite (Most Primitive) Meteorite	Chondrite Meteorite	Earth: Ultra-basic Rock	Mars: Basaltic Rock	Moon: Mare Basalt	Earth: Basalt	Moon: Upland	Earth: Granite (Continental Crust)
Hydrogen (H)	V	78	—	—	—	—	—	—	—	—
Helium (He)	V	20	—	—	—	—	—	—	—	—
Oxygen (O)	L	0.8	34	35	43	—	43	44	45	48
Iron (Fe)	S	0.04	24	26	9.4	13	11	8.6	4.6	2.1
Silicon (Si)	L	—	15	18.5	20	21	21	23	23	33
Magnesium (Mg)	L	—	13	15	20	5	5.5	4.6	4.4	0.6
Sulfur (S)	—	—	8.6	2.3	0.03	3	0.2	0.3	—	0.3
Calcium (Ca)	L	—	1.5	1.3	2.5	4	8	7.6	7	1.5
Nickel (Ni)	S	—	1.4	1.4	0.2	—	0.0001	0.01	0.0001–0.01	0.001
Aluminum (Al)	L	—	1.2	1.1	2	3	8	7.8	13	8
Sodium (Na)	L	—	0.46	0.70	0.4	—	0.4	1.8	0.4	2.7
Titanium (Ti)	L	—	0.08	0.08	0.003	0.5	2.1	1.4	0.3	0.2
Potassium (K)	L	—	0.05	0.09	0.004	0.25	0.1	0.83	0.01–0.1	3

Source: Data from Fairbridge (1972), Taylor (1973), Mason and Melson (1970), Page (1973), and Clark et al. (1977).

[a]V = volatile (driven off by heating).
L = lithophile (concentrated in silicaceous rocks; sulfur usually classified separately).
S = siderophile (iron-affinity elements; concentrated in basic rocks and metallic minerals, as in the earth's core and mantle).

[b]This composition is called cosmic because it is believed to be the same gas that formed not only the sun but all nearby stars.

Summary

Information about the origin of the solar system has been culled from meteorites, lunar samples, the oldest terrestrial rocks, and chemical and dynamical analysis of planets and satellites. This information indicates that the sun formed from a contracting cloud of gas about 4.6 billion years ago. As outer parts of this cloud cooled, solid grains of various minerals and ices condensed and accumulated into planetesimals during a relatively brief interval, lasting from a few million to 100 million years.

During this interval, neighboring planetesimals collided, often gently enough to allow them to hold together by gravity. In this way small planetesimals aggregated into a few larger bodies. Sometimes these bodies collided at high enough speeds to shatter each other, producing meteoritelike fragments. The largest bodies survived and grew into planets. Moons formed near planets by a similar process. Many planetesimals crashed into planets and moons, forming craters. When icy planetesimals in the outer solar system made near-miss encounters with giant planets, the planets' gravity flung many of them almost out of the solar system, forming the Oort cloud reservoir of comets (as described in Chapter 11).

The sun and the outer planets, with their high masses and strong gravity, retained the light, hydrogen-rich gases of the original cloud. Lower-gravity terrestrial planets lost these gases.

No data about the early solar system restrict these processes to the solar system; thus, many scientists suspect that planetary debris or full-fledged planets may have formed near some other stars besides our sun.

Concepts

catastrophic theories
evolutionary theories
conservation of angular momentum
free-fall contraction
Helmholtz contraction
nebula
solar nebula
condensation sequence
grain
planetesimal
magnetic braking

Problems

1. The gas in the early solar system was about 78% hydrogen (Table 12-3). Considering the theory of thermal escape of gases from planetary atmospheres (see Chapter 7), explain the absence of abundant hydrogen in the terrestrial planets' atmospheres. Why does this not need to be listed as a fact to be explained by theories of solar system *origin,* in Table 12-1?

2. Why do meteorites have different chemical and geologic properties than rocks you might find in your own yard, since all planetary material condensed from the same nebula?

3. Because of heating by the sun and by the contraction process, gases in the inner solar system were probably warmer than gases in the outer solar system when planetary solid matter formed. In terms of the condensation sequence, relate this to the estimated or observed composition of the planets.

4. Judging from planetary composition, where was the inner boundary of the part of the solar nebula where ices condensed?

5. If planets orbited the sun in randomly inclined orbits, both prograde and retrograde, how might theories of such a solar system's origin differ from the theory described in this chapter?

6. List some observations that support the theory of solar system origin described in this chapter.

7. How are theories of solar system origin different in principle from theories of the origin of the universe?

8. What reasons do we have for believing that the entire solar system formed during a single, relatively short interval?

9. If a 5-year-old member of your family asked where the world came from, how would you answer?

ADVANCED PROBLEMS

10. If a planetary system was forming around a nearby star, and the star was obscured by a "solar nebula" of dust grains at a temperature of 1000 K, explain how such a dust nebula might be detected. (*Hint:* Apply Wien's law and assume that a large telescope is available with infrared detectors.)

11. If dust grains in circular orbits in the early solar nebula collided at a speed equaling 0.1% of their orbital velocity, how fast would they have collided near the present orbit of the earth? How fast near the present orbit of Pluto?

12. A rule of thumb is that a planet can retain a gaseous atmosphere for a geologically long time if the typical velocities of the molecules are no more than about ¼ the escape velocity of the planet. Assume that icy bodies with density 1000 kg/m^3 were growing in the outer solar system in a nebula of hydrogen at temperature 500 K.

a. Use this rule of thumb to show that if any primitive planets grew to around 38 000 km across they would form "cores" that would begin to retain hydrogen atoms and form massive gaseous envelopes, thus beginning to form giant planets.

b. Analysts have shown that the giant planets all have cores of ice and rock that have about 8 to 30 times the mass of the earth. Prove that the mass of the "cores" described in problem 12a would fall between Earth's mass and the mass of these observed giant planet cores.

c. Comment on how these simple calculations help explain how the giant planets may have formed.

P A R T D

Stars and Their Evolution

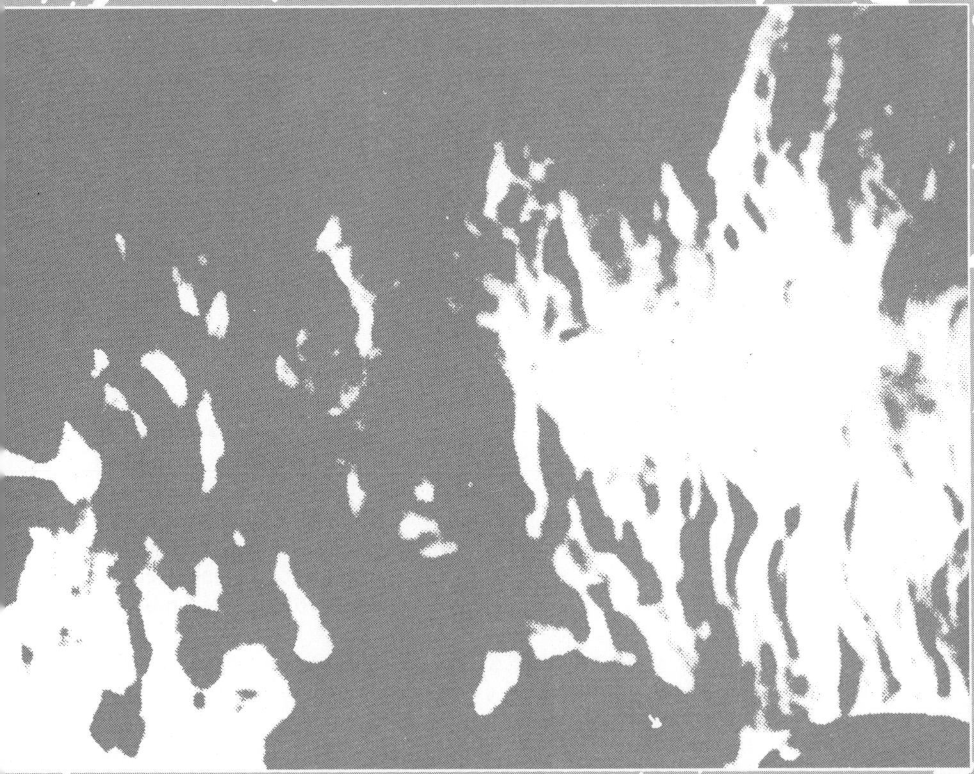

This is Outer Space
Where the atom changes mass
Where the dynamos are fast
Where the Cosmic Hoard is stashed
Where the Carbon Torch is passed
To the Oxygen Orbiters
Humid in gas
From the Hydrogen Past . . .

Burgert Roberts, from Spacewalks

CHAPTER 13

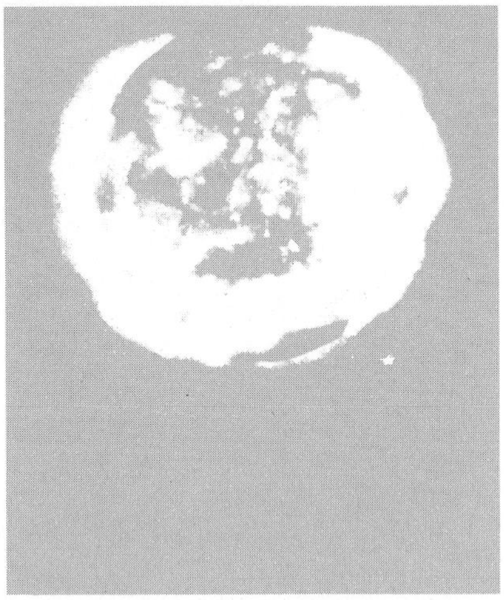

THE SUN AND ITS RADIATION

How can we study such remote objects as the stars? We can study the closest one, which is nearer to the earth than five of the nine planets. Light from its surface reaches us in only eight minutes. Our eyes are dazzled by it. The earth is bathed in its flow of radiation, washed by the winds of its outer atmosphere, blasted by seething swarms of atoms blown out of it, bombarded by bursts of X-rays and radio waves emitted by it. It is our sun, a million-km ball of hydrogen and helium in the center of the solar system.

The sun's million-km diameter is roughly ten times Jupiter's size, and roughly one hundred times the earth's size.

Humans pondered the stars for many centuries before they realized that the sun is just another of the pinpoint lights of the nighttime sky. The sun is the only star on which we can see surface details and processes. It is so close that we can watch storms develop on its surface and track them as they are carried around it by **solar rotation**, as shown in Figure 13-1. This rotation takes about 25.4 d relative to the stars (and 27.3 d relative to the earth, since the earth's orbital motion must be added in). Like Jupiter's atmosphere, the equatorial region rotates faster than the polar regions—proof that the sun has a gaseous, not solid, surface.

Spectroscopic Discoveries

We cannot send space vehicles to land on the sun, as we have done for planets. To study the sun or stars we must rely on interpreting their light, sampled by means of our telescopes.

Chapter 7 described certain properties of light, such as the radiation maximum occurring at certain colors (wavelengths), depending on the temperature of the radiating body (Wien's law, p. 123). Here, we will touch on a few of these most basic concepts again as we explain how they apply to the sun.

THE SOLAR SPECTRUM

As Newton showed, sunlight or any **white light** is a mixture of all colors. Sunlight passing through Newton's prism revealed a **spectrum**, or array of colors in order of wavelength, as shown in Color Plate 32. Review pages 132–133 (especially Figure 7-13) to recall how parts of the ultraviolet and infrared spectrum of sunlight and starlight are absorbed by our atmosphere before reaching the ground.

In 1817, German physicist Joseph Fraunhofer found that certain wavelengths were missing from the sun's spectrum, so that the spectrum appeared to be crossed by narrow, dark lines. As explained in Chapter 7, these are called **absorption lines**, shown in the middle part of Color Plate 32. Fraunhofer named them A, B, C, and so on, from red to blue.

What were these lines? Let us review and expand on what we learned in Chapter 7. By the mid-1800s, scientists discovered that when a given element is burned, it emits glows of certain colors and no others. These very narrow wavelength intervals, unique to each element and as unmistakable as a set of fingerprints, are called **emission lines**. Researchers soon found that some of the emission lines exactly matched the position of Fraunhofer's solar absorption lines. For instance, Fraunhofer's D absorption line (actually a close pair) matched an emission line from sodium; his H and K lines, calcium. Did this mean that the D line in the solar spectrum was caused by sodium in the sun, and the H and K lines by calcium?

KIRCHHOFF'S LAWS

The answer proved to be yes. In the 1850s, German physicist Gustav Kirchhoff discovered the conditions that produce the

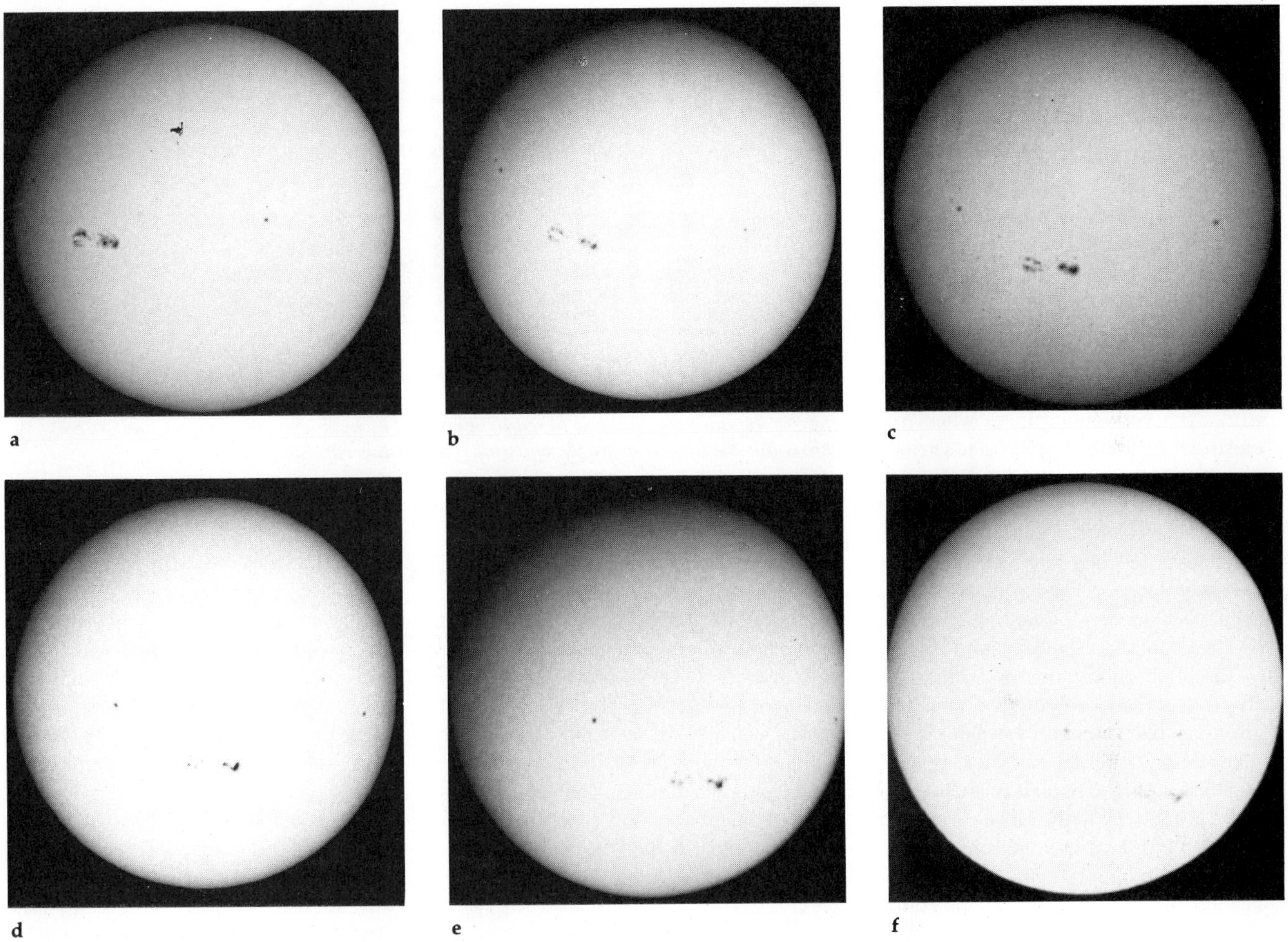

Figure 13-1. *Six photos of the sun in sequence, August 21–26, 1971, showing the rotation of the sun by the movement of sunspot groups. (W. A. Feibelman.)*

three different kinds of spectra described in Chapter 7: the glow consisting of all colors, called the **continuum**; absorption lines; and emission lines. For instance, when Kirchhoff looked through his spectroscope toward a sodium flame against a dark background, he saw the sodium D line in emission, like the emission lines in the bottom of Color Plate 32. But when he changed the background to a brilliant beam of sunlight passing through the same flame, he saw a strong D absorption line, like the lines in the middle of Color Plate 32. In each case, the D lines came from the gaseous sodium atoms in the flame.

Thus, in simple terms, the absorption lines found by Fraunhofer in the solar spectrum occur when photons of solar white light pass outward through the cooler gas of the solar atmosphere. Photons of certain wavelengths (such as the D line), when striking certain atoms (such as sodium), are absorbed from the outgoing beam, causing an absence of that color in the beam.

Kirchhoff reduced such observations to three statements called **Kirchhoff's laws:**

***1.** A gas at high pressure, a liquid, or a solid, if heated to incandescence, will glow with a continuous spectrum, or continuum.*

***2.** A hot gas under low pressure will produce only certain bright colors, called emission lines.*

***3.** A cool gas at low pressure, if placed between the observer and a hot continuous-spectrum source, absorbs certain colors, causing absorption lines in the observed spectrum.*

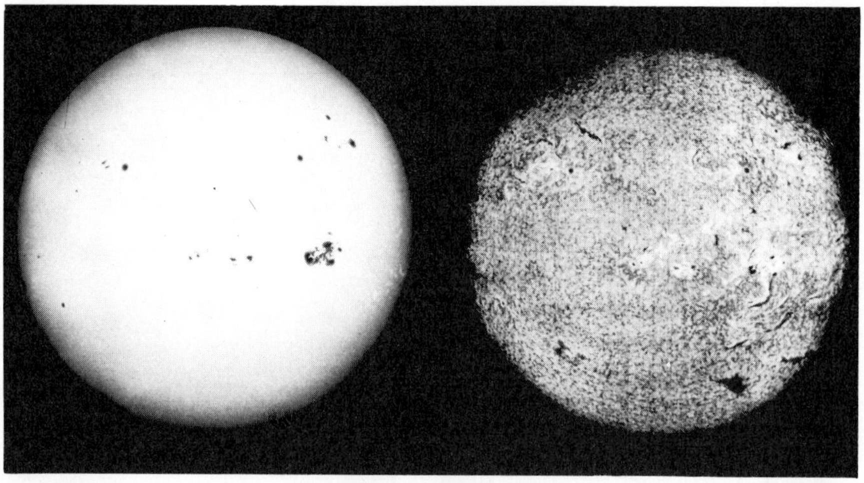

***Figure 13-2.** The left photo shows the sun in normal visible light, with sunspots and limb darkening caused by solar atmospheric absorption of light. The right photo shows the same face of the sun photographed in red light emitted by hydrogen (Hα spectral line); bright areas involve intense hydrogen emission. (Hale Observatories.)*

Chapter 7 explained why these laws work. The continuum arises when free electrons are available; this occurs in high pressure gases, liquids, or solids. Emission lines arise from electrons inside atoms in an excited state, as in a hot gas. Absorption lines arise when atoms are in, or near, the ground state, as in a cooler gas. If these principles seem unclear, you should review Chapter 7, especially pp. 121–127.

Kirchhoff himself found that the absorption lines and emission lines of a given gas have identical wavelengths, as seen in Color Plate 32. What is seen depends on the temperature and density of the gas relative to the radiation coming from behind it, as indicated in the third law. Later, an important modification was made to Kirchhoff's laws: An absorption spectrum need not originate *in front of,* or in a cooler gas than, the continuous spectrum. This is because within a single gas, electrons may be jumping upward in some atoms (making absorption lines), and downward from the free state into other atoms (forming the continuum). Indeed, in the sun, some absorption lines originate in the same surface layers that produce the continuous spectrum we call sunlight. These layers form the well-defined visible surface of the sun, and are called the **photosphere**.

In summary, spectroscopy allows us to determine many things about the properties of the sun. Most importantly, *identification of the spectral lines allows identification of the elements in the sun.* Application of Kirchhoff's laws proves that the photosphere is a layer of hot gas. Application of Wien's law (p. 123) allows us to use the wavelength of the strongest solar radiation to determine the photosphere temperature. Other spectroscopic principles enable astronomers to measure temperatures and pressures at different depths in the gas near the sun's surface.

IMAGES OF THE SUN

Normal photographs of the sun (or any other scene) are made from light of many colors, since normal photographic films are sensitive to many wavelengths. However, color filters and other devices can restrict the wavelength range. By using a narrow enough range, we can see the sun in only the light emitted by a chosen gas (hydrogen, for example) in a specific atomic state, thus tracing the distribution of that gas. The **spectroheliograph** is an ingenious instrument that allows this to be done. It spreads the light into different colors, from which one color alone is selected to form an image. When hydrogen is in certain states of temperature and pressure, it emits an especially useful red color—the well-known **hydrogen alpha line**, or *Hα emission,* shown in the lower right part of Color Plate 32. In an image using Hα light, such as Figure 13-2 (right side), the only bright regions are those where hydrogen of a certain temperature is emitting this specific red glow. Dark parts of the image are regions where there is still much hydrogen but little of it is in this Hα-emitting condition. Figure 13-3 and Color Plate 23 show additional striking Hα images of the sun.

Many techniques illustrated in this chapter have been used to obtain solar images in the light of other specific gases. These images allow analysis of the distribution of gases and temperatures on the sun.

Composition of the Sun

After more than a century of spectroscopic study, the **sun's composition** is accurately known. The most abundant

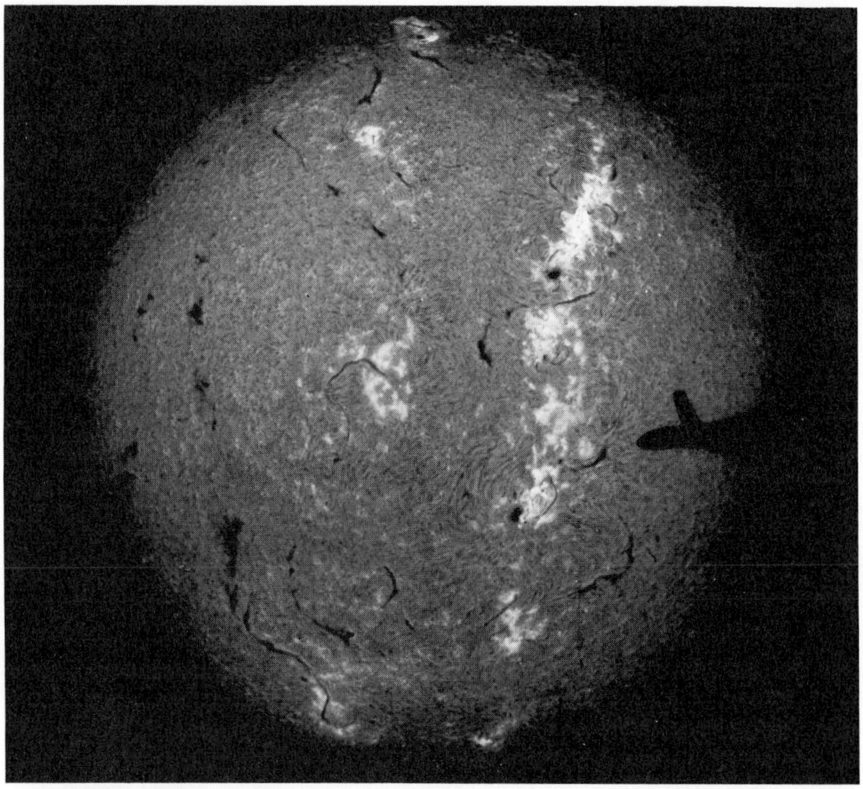

Figure 13-3. *A spectroheliograph view of the sun in the light of Hα emission happened to catch the silhouette of a commercial jet crossing the solar disk. The solar equator is nearly vertical, with parallel bands of dark and bright hydrogen clouds. (Pennsylvania State University.)*

Table 13-1. *Composition of the Sun*

Element	Percent Mass of the Sun	Atomic Number
Hydrogen (H)	76.4	1
Helium (He)	21.8	2
Oxygen (O)	0.8	8
Carbon (C)	0.4	6
Neon (Ne)	0.2	10
Iron (Fe)	0.1	26
Nitrogen (N)	0.1	7
Silicon (Si)	0.08	14
Magnesium (Mg)	0.07	12
Sulfur (S)	0.05	16
Nickel (Ni)	0.01	28

Source: Adapted from Anders and Ebihara (1982).
Note: Based on spectroscopic measurements of the sun and measurements of meteorites and other samples.

elements are listed in Table 13-1, which is believed to reflect the bulk composition of material from which the solar system formed. An interesting episode occurred in 1868, when the French astronomer Pierre Janssen and the English astronomer Norman Lockyer independently found solar spectral lines corresponding to an unknown element. This element, named *helium* (from the Greek *helios,* "sun"), was the first to be discovered in space instead of on the earth (where it was not observed until 1891). It is the second most abundant element in the sun, after hydrogen.

The sun is about three-quarters hydrogen and one-fifth helium. The heavy elements common in the planets comprise only 2% of the sun by mass.

SOLAR ENERGY FROM NUCLEAR REACTIONS

Hermann von Helmholtz calculated in 1871 that the energy output of the sun corresponds to the burning of 1500 lb of coal every hour on every *square foot* of the sun's surface. No ordinary chemical reactions can produce energy at this rate. The sun is not "burning" in the normal sense.

In the 1920s, astrophysicists realized that the energy of the sun and other stars comes from **nuclear reactions**, interactions of atomic nuclei. Normally, nuclei are protected from interacting by their surrounding clouds of electrons. Familiar **chemical reactions**, such as coal burning, involve interactions only between *electrons* of different atoms, far outside the central nucleus. However, if the temperature and pressure are high enough, atoms collide fast enough to knock away electrons, allowing the nuclei to interact. This happens in the sun's central core.

Nuclear reactions in stars do two important things: They change the star's composition, and they generate energy. The most important reactions in the sun are believed to be a three-part fusion sequence that fuses four hydrogen atoms into a helium atom.[1] Because the sequence

[1]This set of reactions will be discussed in more detail when we describe the details of stellar evolution in Chapters 16 and 17.

starts with two hydrogen nuclei, that is, two single protons, it is called the **proton–proton cycle.** The total amount of mass left at the end of the cycle is slightly less than the mass of the initial hydrogen atoms. During the fusion, a small amount of mass m is converted to an amount of energy E, according to Einstein's famous equation $E = mc^2$. (The constant c is the velocity of light, 3×10^8 m/s. When units are in the SI metric system, m is in kilograms and E is in joules.) In the sun's fusion sequence, about 0.007 kg of matter is converted into energy for each kilogram of hydrogen processed. This liberates 4×10^{26} J/s inside the sun, and the sun radiates this much every second to maintain its equilibrium. This corresponds to 400 trillion trillion watts—which equals a lot of light bulbs!

Every second, the sun converts 4 million tons of hydrogen into energy and radiates it into space. Long before the sun can use up all its mass, the solar core will have converted so much hydrogen to helium that there will not be enough hydrogen left in the core to fuel further reactions and the reactions will stop. According to a recent calculation, the sun won't run out of hydrogen for about 4 billion years. This proves the important point that stars are not permanent, but must evolve and run down.

Interior Structure: How Energy Gets to the Surface

The theory of Helmholtz contraction (see Chapter 12) predicts that the temperature at the sun's center reached 15 million Kelvin as the proton–proton reactions became established. This remains the current temperature in the **solar core,** where nuclear energy is generated.

Approximate conditions in any layer between the core and the surface can be calculated by using equations that describe the pressure at any depth, the properties of gas under different pressures, and energy generation rates at points inside the sun; results appear in Figure 13-4.

These calculations indicate that the gas pressure at the sun's core is about 250 billion times the air pressure at the earth's surface. This high pressure compresses the gas in the core to a density of about 158 000 kg/m^3—158 times denser than water and about 20 times denser than iron. One cubic inch of this gas would weigh nearly 6 lb! The core of the sun occupies about the inner quarter of the sun's radius. This 1/64 of the sun's volume contains about half the solar mass and generates 99% of the solar energy.

As heat energy always flows from hot to cool regions, solar energy travels outward from the hot core, through a cooler zone of mixed hydrogen and helium, toward the surface. Throughout most of the sun's volume, this energy moves primarily by **radiation.** That is, the energy radiates through the gas, just as light travels through our atmosphere. Very little moves by **conduction,** the mechanism by which a pan on a stove becomes hot.

In the outer part of the sun we find a third mechanism of energy transport—**convection** (p. 54). Convection occurs when the temperature difference per unit length between the hot and cold regions is so great that neither radiation nor conduction can carry off the outward-bound energy fast enough. So-called "cells" of gas, having become heated enough to expand, become less dense than their surroundings and rise toward the surface (Figure 13-5).

The solar interior still harbors mysteries. For instance, any object (such as a bell or a star) vibrates with a certain natural frequency if disturbed, but the sun's surface layers have been observed to vibrate up and down with unexpected frequencies. Some theorists infer from this that the solar core rotates two to nine times faster than the surface (Newkirk and Frazier, 1982). This result is controversial; if right, it could require changes in our ideas about the core.

Another mystery involves neutrinos—tiny, subatomic particles that should be produced by the proton–proton reac-

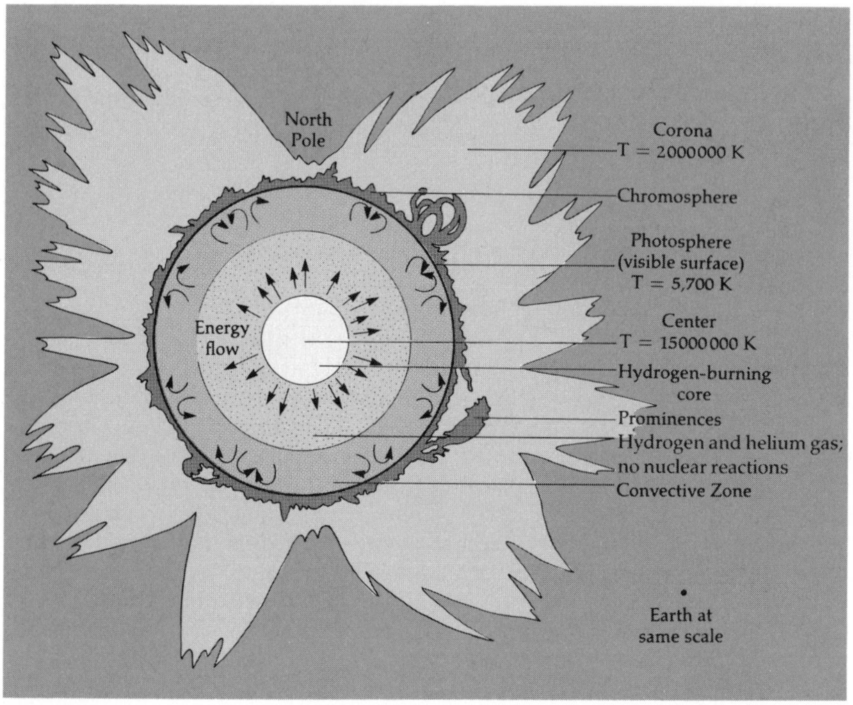

Figure 13-4. *A cross section of the sun and its atmosphere, showing (to approximate scale) the energy-producing core, the outer convective zone, and the thin, inner corona.*

tions in the core. A flood of these nearly massless particles should escape from the interior, according to predictions. Experiments to detect them began in the 1970s but succeeded in detecting only about one-third as many solar neutrino units (abbreviated SNU)[2] as predicted. Theorists have worried for several years whether the neutrino shortfall implies a fatal flaw in our understanding of the sun's interior. In the mid-1980s, two new theories were advanced to explain the shortfall without radical changes in older models. The first suggests that the sun contains certain subatomic particles (charmingly called WIMPs by the physicists—for weakly interacting massive particles) that transport energy out of the core, cool the core, and thus retard the thermonuclear reaction rate and neutrino production. This theory also claims to explain the solar surface vibration rate better than other theories. A second theory stems from the work of two Russian physicists who pointed out in 1985 that certain types of neutrinos would transmute into particles not detectable by the current experiments, thus explaining the deficiency. Further work is needed, but the two theories illustrate the close tie between modern astronomy and modern physics.

The Photosphere: The Solar Surface

Energy ascending from inside the sun heats the photosphere, or bright surface layer of gas that radiates the visible light of the sun. The photosphere's temperature can be found quickly from Wien's law. In the solar spectrum the maximum solar radiation is yellow, with a wavelength of about 510 nm. Application of Wien's law shows that the temperature of the photosphere is about 5700 K.

If the sun is a giant ball of gas, why does it appear to have a sharply defined surface? The answer involves the **opacity** of the gas—its ability to obscure light passing through it. Air, for example, has low opacity. Gas in the phospheric layer has many **negative hydrogen ions** (hydrogen atoms with an extra electron, designated H^-), which obstruct light and cause high opacity. They produce an opaque layer, beyond which we cannot see. Only a few hundred kilometers above this layer, at the top of the photosphere, there are few H^- ions and the gas is clear. Most of the sun's light comes from a layer about 400 km thick, giving the appearance of a sharply defined surface. There is also hydrogen in the solar atmosphere above the photosphere, but no abrupt change in the gas density; a solid probe, if it could survive the 5700-K temperature, could drop directly through the photospheric "surface" and plunge into the sun, like an airplane passing through the surface of a cloud.

The convective motions that bring mass and energy from the interior disturb the photospheric surface. A photograph of the sun shows pronounced **granules** in the photosphere, as shown in Figure 13-5 (compare with Figure 13-6). Each bright granule is a convection cell 1000 to 2000 km across, rising from the subphotospheric layers. Each granule rises at a speed of 2 to 3 km/s and lasts for a few minutes. Slightly darker regions between granules mark areas where cooled gas descends again into the sun.

Surging wave motions have also been observed in the gas, with wavelengths of about 5000 km and periods of about 5 min. Such sizes may seem abstract, but keep in mind that many moving masses being churned on the surface of the sun are as large as the entire earth!

Chromosphere and Corona: The Solar Atmosphere

Shooting up into the solar atmosphere are **spicules**: columns of incandescent gas 10 000 km long. Essentially giant flames, they fade in 2 to 5 min and are a transition from the photosphere to the overlying layer called the chromosphere. They are shown in Figures 13-7 and 13-8.

The **chromosphere** (which means "color layer") is a pink-glowing region of gas just above the photosphere. Its light is mainly the red Hα emission line, mentioned earlier. E. G. Gibson (1973) has called the chromosphere "froth on top of

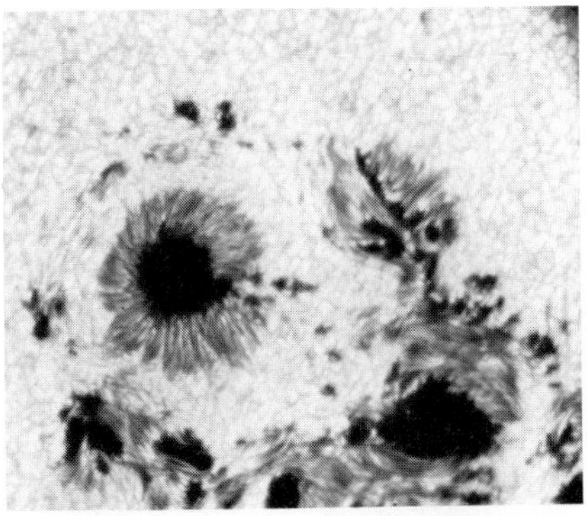

Figure 13-5. *A detailed photo of the solar surface in the region of a sunspot. Outside the sunspot, the normal solar surface is mottled by granules believed to be convection cells in the solar gas. The main sunspots are comparable to the earth in size, and the large granules are comparable to continents. (Balloon-borne telescope photo, Princeton University, Project Stratoscope, supported by NSF, ONR, and NASA.)*

Figure 13-6. *Oblique downward view of convection cells in cumulus clouds from about 10 km over Indiana, showing a convective pattern similar to that found on a larger scale in the solar gas (see upper right part of Figure 13-5). (Photo by author.)*

***Figure 13-7.** Spicules, or clustered jets of solar gas, photographed in red light of hydrogen Hα emission. Though the spicules are luminous gas, they appear darker than the bright solar background when silhouetted against the photosphere. (Hale Observatories.)*

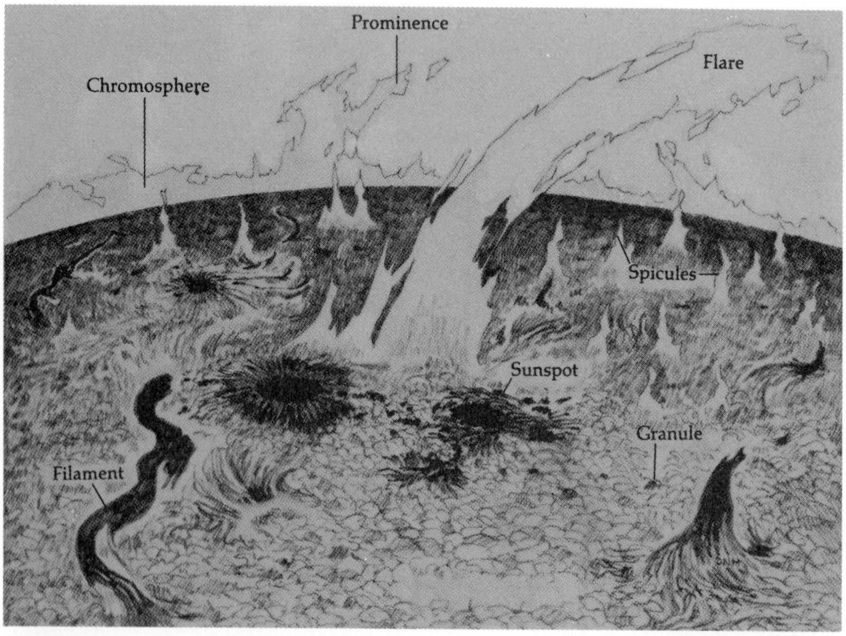

***Figure 13-8.** A schematic, oblique view of the solar surface, showing the types of features mentioned in the text.*

the turbulent and relatively dense photosphere." The chromosphere can be seen by the naked eye during a solar eclipse. When the moon covers the rest of the solar disk, this thin outer layer is visible as a ring of small, intense red flames, just visible in Color Plate 22 (top). As drawn in Figure 13-4, the chromosphere is a thin layer, about 2500 km thick. In its upper regions the temperature exceeds 10 000 K.

Above the chromosphere is the rarefied, hot gas of the **corona.** Gas in the corona reaches the amazing temperature of 2 million K, due to heating by violent convective motion in the photosphere and chromosphere. As a result of the extreme heat, it expands rapidly into space. Whereas the gas density is about 0.001 kg/m^3 within a few hundred kilometers of the photosphere, it drops to 10^{-7} kg/m^3 in the middle chromosphere and to less than 10^{-11} kg/m^3 in the lower corona. Clearly, the corona is only the outermost, tenuous atmosphere of the sun.

During eclipses, the corona is visible to the naked eye as a pearly, glowing gas around the sun, as seen in Figures 1-19 and 1-21. Although Plutarch recorded this, some early astronomers thought it was an optical illusion. Early photographs finally proved its existence. Spectra were obtained in 1869, eventually revealing that the coronal gas has extremely high temperatures of 1 to 2 million Kelvin! In 1930 French astronomer Bernard Lyot built an instrument called the **coronograph,** which artificially eclipses the sun and allows the intriguing gases of the corona and chromosphere to be studied at will (Figure 13-9).

Why are both the chromosphere and the corona hotter than the photosphere, since they are farther from the sun's internal energy source? Magnetic effects and shock waves from the violent subsurface convection transfer large amounts of energy to this gas. The sun's magnetic field controls motions of gas in the corona, creating delicate streamers (see review by Wolfson, 1983). The coronal gas expands and merges with the interplanetary gas and dust. While the inner glow of the corona comes from solar gas, the outer glow is really the inner zodiacal light—sunlight reflected from interplanetary dust.

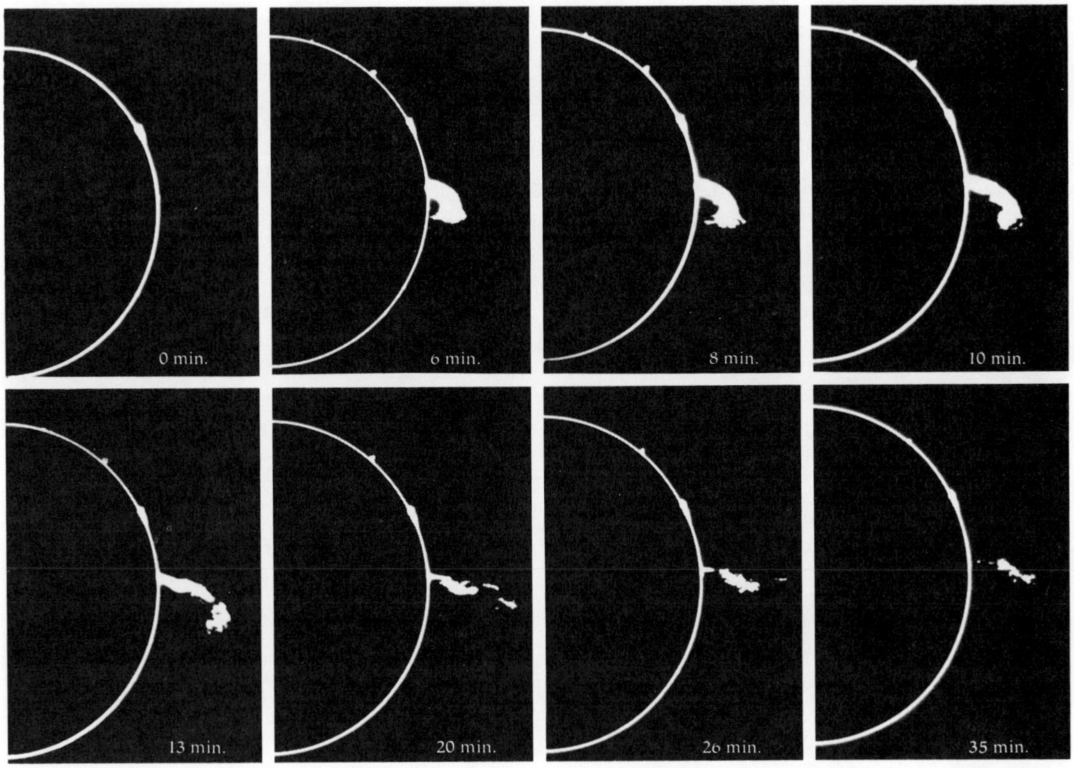

Figure 13-9. *This sequence of coronograph photos shows a jet of gas blasting off the sun over a period of 35 min. The photos were made with a coronograph, which obscures the brighter solar disk and allows solar atmospheric activity to be monitored. (National Center for Atmospheric Research.)*

Sunspots and Sunspot Activity

Although sunspots sometimes can be seen with the naked eye when the sun is dimmed by fog or a dark glass,[2] their nature was not realized until 1613, when Galileo studied them and concluded that they are located on the solar surface and are carried around the sun by solar rotation, as seen in Figure 13-1.

A **sunspot** is a magnetically disturbed region that is cooler than its surroundings. A sunspot looks dark only because its gases, at 4000–4500 K, radiate less than the surrounding gas at about 5700 K. Motions of solar gas near sunspots are controlled not by atmospheric forces, as with terrestrial storms, but by magnetic fields of the sun. *Ions* (charged atoms or molecules), which are common in the sun, cannot move freely in a magnetic field, but must stream in the direction of the field—for example, from the north magnetic pole to the south.

A gas with many ions is called a **plasma**, and unlike a neutral gas, its motions are strongly influenced by magnetic fields. For this reason, plasmas in the sunspots and elsewhere in the solar atmosphere move in peculiar patterns that indicate the twisted patterns of the solar magnetic field.

Huge clouds of gas, larger than the whole earth, erupt from the disturbed regions of sunspots. These are called **prominences**, which can be seen when silhouetted above the solar limb, or edge. *Eruptive prominences,* as seen in Figure 13-9, shoot out at speeds averaging about 1000 km/s. *Quiescent prominences,* as seen in Figure 13-10, are masses of flowing gas held in relatively fixed positions above the sun for hours or days by magnetic fields. The largest blasts of material and their very active sunspot sites are called **flares**. Figures 13-9 and 13-11 show flare material blasting off the sun and a vertical view of a flare near a sunspot pair, respectively.

THE 22-Y SOLAR CYCLE

Around 1830 an obscure German amateur astronomer, H. Schwabe, began observing sunspots as a hobby. After years of tabulating his counts, he announced in 1851 a **solar cycle**: the *number* and *positions* of sunspots vary in a cycle, as shown

[2] ***Never point a telescope at the sun without professional aid!*** Unmodified telescope optics collect enough solar light and heat to blind the observer and damage the telescope.

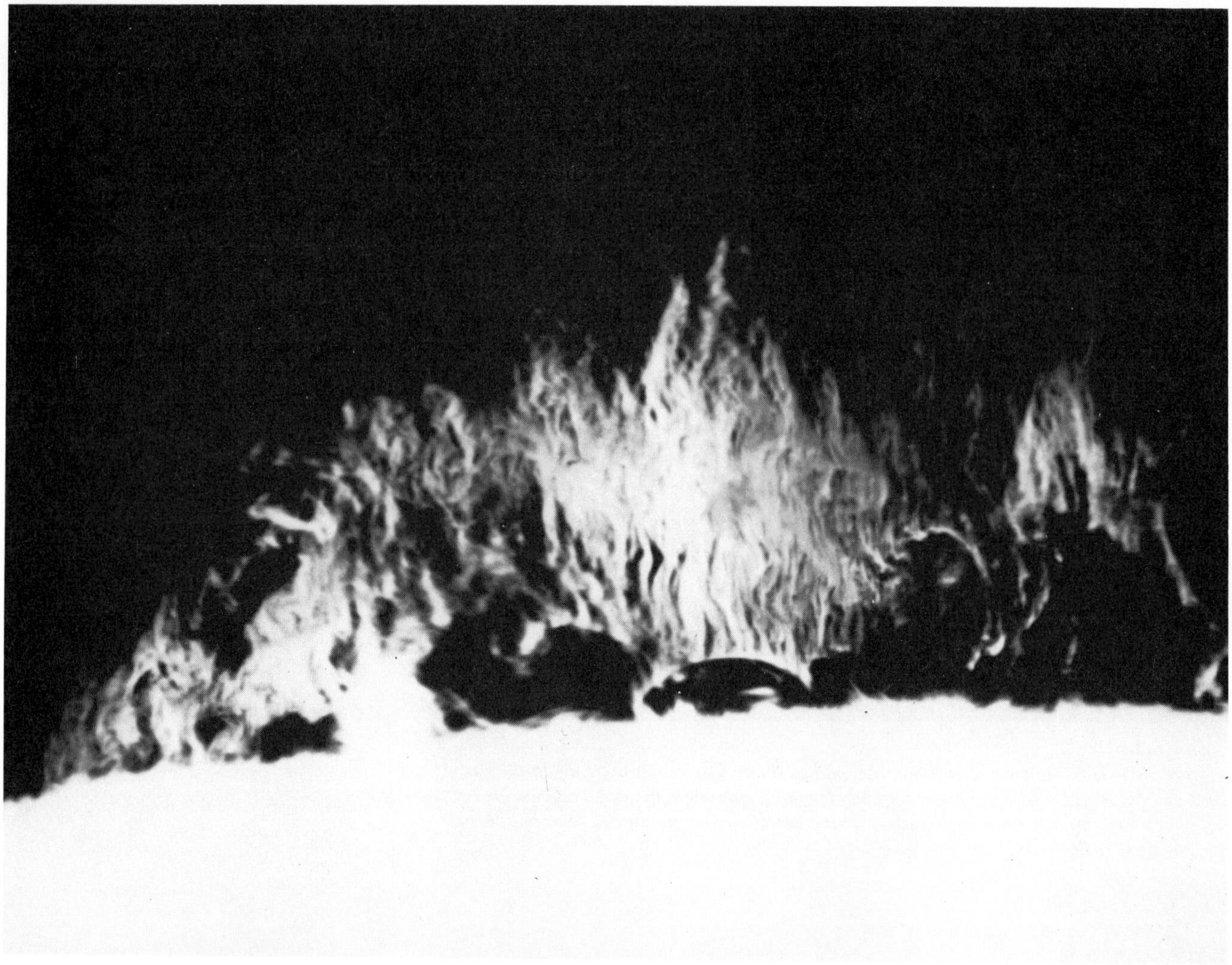

Figure 13-10. *A quiescent prominence, about 50 000 km high, hanging like a sheet of flame above the solar surface. The earth's diameter is less than a quarter the height of this plasma cloud. (Sacramento Peak Observatory.)*

in Figure 13-12. This discovery, followed a year later by the discovery that terrestrial magnetic compass deviations exactly follow the same cycle, was a key step in understanding the sun and its effects on the earth.

The cycle duration averages 22 y and consists of two 11-y subcycles, shown in Figure 13-13. At a time of "sunspot minimum" (when there is a minimum of sunspots), the few visible spots are grouped within about 10° of the solar equator. When a new cycle begins in a year or so, groups of new spots appear at high latitudes, about 30° from the solar equator. The spots often appear in pairs, and the eastern spot in each northern hemisphere pair is of specific polarity, for example, a north magnetic pole. In the southern hemisphere the polarity is reversed. After a few years, the sunspot number reaches a maximum and the spots are at intermediate latitudes, about 20° from the solar equator. After about 11 y, the spots appear mostly about 10° from the equator, and a sunspot minimum occurs again.

Now the cycle begins to repeat, except for a noticeable difference. The new spots forming at ±30° have reversed polarity. The eastern spots in the northern hemisphere are now south magnetic poles! Thus, it takes another 11 y to complete the full cycle, when all features resume their initial pattern. A sunspot maximum should occur around 1990.

Still more remarkable is the fact that

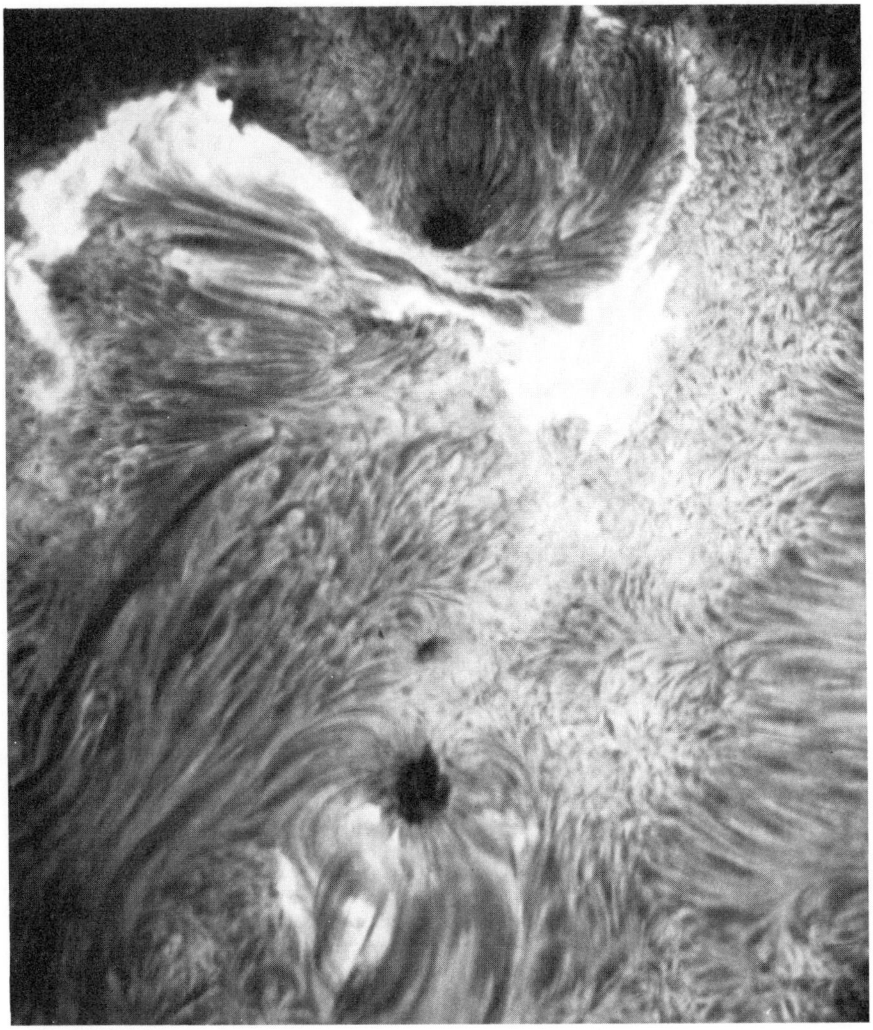

Figure 13-11. *Vertical view of gaseous streamers and solar flare activity (bright region, top) around a pair of sunspots (dark areas at top and bottom). (Air Force Cambridge Research Laboratories.)*

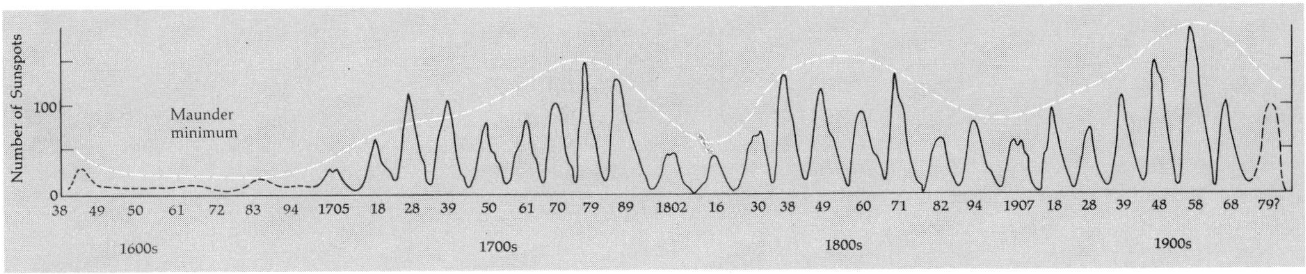

Figure 13-12. *Sunspot counts since 1600s show the cycle averaging 11 y for sunspot numbers (half the 22-y magnetic cycle), with evidence for a longer 80-y cycle (white dashed line). The extensive period of low sunspot activity in 1600s is believed to correlate with climate changes at that time. (After data of M. Waldmeier in Gibson, 1973; Pasachoff, 1980.)*

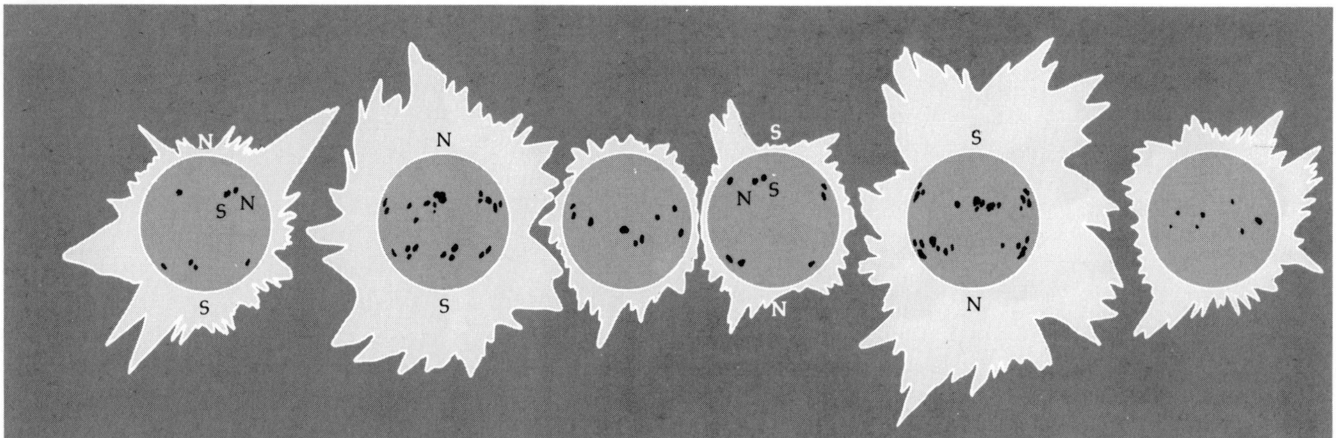

Years	0	2	4	6	8	10	12	14	16	18	20	22
Sunspot number	Minimum		Maximum			Minimum			Maximum			Minimum
Sunspot latitude	High		Medium		Low		High		Medium		Low	High
Flares, storms, prominences	Minimum		Maximum			Minimum			Maximum			Minimum

Figure 13-13. *A schematic sequence of solar changes during the 22-y cycle. Typical coronal appearances are shown for the minimum and maximum of the cycle in the two drawings at left.*

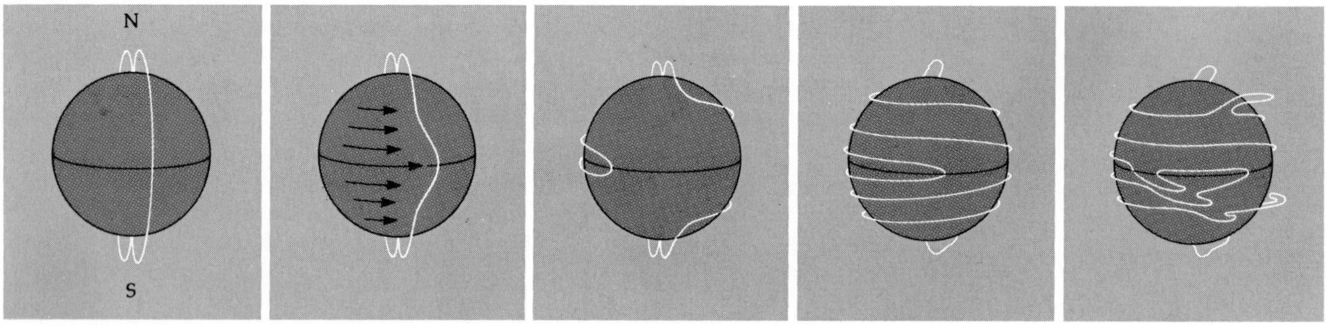

a b c d e

Figure 13-14. *Some features of the solar cycle are explained by this simple model following the history of a "magnetic field line," which defines the direction of the solar magnetic field (a). Because of differential rotation of the sun (b), and because the field lines must move with the ionized gas, field lines tend to distort (c), becoming wrapped around the sun (d). Convection and eruptions further distort the field (e). After about 11 y the field becomes unrecognizable and neutralized, and a new field forms.*

the magnetic field of *the entire sun* reverses during each 11-y subcycle; thus the entire sun participates in the full 22-y cycle. Imaginary observers on the sun would find their compasses pointing north in one direction for 11 y (subject to disturbances by frequent magnetic storms), and in exactly the opposite direction for the next 11 y. This behavior is not entirely unknown: The earth's field reverses every few hundred thousand to few million years. Both patterns of reversal may involve cyclic flow patterns in the deep fluid cores of the two bodies. The sunspot cycle is important to us because during years of maximum sunspot activity, solar particles shooting off the sun affect the magnetic field and upper atmosphere of Earth, disturbing radio communications and causing aurorae. The next such period should occur around 1990.

According to American astronomers H. Babcock (1961) and R. Leighton (1969), the sunspot cycle occurs because the sun rotates faster at its equator than near its poles. This causes a shearing and twisting of the magnetic field that controls motions of the ionized solar gas, as shown in Figure 13-14. The field is further distorted by convection, until the twisted pattern breaks and a new pattern forms, starting a new cycle (see reviews by Newkirk and Frazier, 1982; Parker, 1983).

Solar Wind

The importance of sunspots to us on the earth becomes clearer when we realize that particles blasted out of flares and spots rush outward through interplanetary space. The solar coronal plasma, having been heated to nearly 2 million Kelvin by the violence of photosphere convection, expands rapidly into space (limited only by magnetic forces acting on charged particles). Together these effects cause the **solar wind:** an outrush of gas past the earth and beyond the outer planets. Near the earth, the solar wind travels at velocities near 600 km/s, and sometimes reaches 1000 km/s. The gas has cooled only to 200 000 K, but is so thin that it transmits no appreciable heat to the earth. The solar wind extends at least as far as Saturn's orbit, according to spacecraft data.

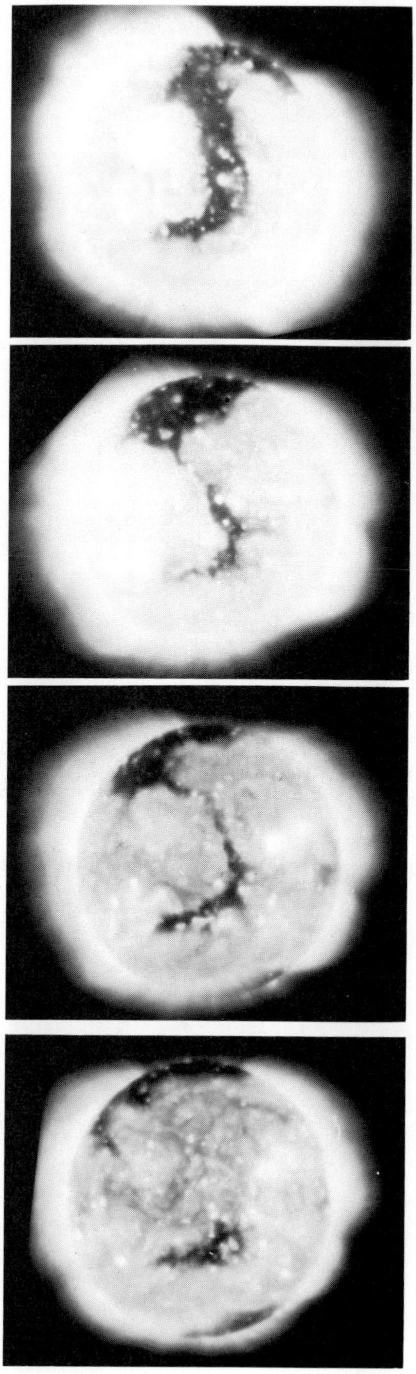

Figure 13-15. *The coronal structure of the sun's atmosphere, photographed by Skylab astronauts using an X-ray-sensitive telescope. This technique revealed rifts in the coronal structure, through which the cooler (darker) surface gases can be seen. The evolution of such a rift is shown in the sequence. (NASA; American Science and Engineering, Inc.)*

In addition to the solar wind, solar radiation itself exerts an outward force on small dust particles. This effect, which is greater on smaller particles, is called **radiation pressure**. Together these two forces blow micrometeorites out of the solar system soon after they are dislodged from comets and asteroids.

Aurorae and Solar–Terrestrial Relations

Solar radiation and particles blasted out of the sun strike the earth. Because 99.98% of all energy passing through the earth's atmosphere comes from the sun,[3] it is not surprising that small "flickers" in solar output cause major effects on the earth.

Effects come from changes in both solar radiation and solar particles. During a solar flare, the total visible radiation from the sun changes by much less than 1%, but the X-ray radiation may increase by a hundredfold. The X-ray photons are energetic because they have very short wavelength. When they strike the earth's upper atmosphere, they change its distribution of ions and affect radio transmission on the ground. Imaging of the sun at X-ray wavelengths from above the atmosphere has permitted us to visualize the remarkable appearance of the X-ray sun, as shown in Figure 13-15. Color Plate 22 (bottom) vividly shows flares and active areas emitting X rays. The flares shoot material upward into the corona, affecting the coronal structure, as seen in Figures 13-16 and 13-17.

[3]Based on a tabulation by Hubbert (1971); the rest is in the form of tidal energy, starlight, cosmic rays (atomic particles from space), and subterranean geothermal energy.

Figure 13-16. *Surface and atmospheric activity of the sun. During the June 30, 1973, solar eclipse in Kenya, the outer corona of the sun was photographed in white light (outer image). About an hour earlier, astronauts in the orbiting Skylab photographed the solar surface (circular inset) using X-ray radiation, showing centers of flare activity. Major streamers in the outer corona are shown to be aligned with X-ray flares on the surface. (National Center for Atmospheric Research; American Science and Engineering, Inc.; NASA.)*

Figure 13-17. *The same eclipse shown in Figure 13-16 was photographed with special techniques to reveal the great length of glowing gas streamers extending into the outer corona. (Los Alamos Scientific Laboratory, J-Division, courtesy Charles F Keller.)*

a

b

c

Figure 13-18. *Typical forms of the aurora. (a) Vertical rays. (W. A. Feibelman, Pittsburgh.) (b) Overhead "corona" photographed from State College, Pa. (Photo by author.) (c) Curtain form. (From NASA aircraft over Canada.)*

Solar flares emit not only radiation, such as X rays, but also streams of atomic particles, such as protons and electrons. The solar particles that strike the upper atmosphere, after a few days' travel from the sun, change its chemistry and create vivid glows, called **aurorae** (singular, *aurora*). As shown in Figure 13-18, these are fantastic, colored, swirling patterns of light visible in the night sky occasionally (more often during sunspot maxima) from locations near the north and south poles.

Discovering the cause of the aurora took 250 years. As early as the 1700s, the English astronomer Edmund Halley, of Halley's comet fame, discovered that auroral forms lie along lines of the earth's magnetic field. Around 1920, Norwegian observer F. C. Störmer showed that aurorae occurred at an altitude usually exceeding 100 km (65 mi).

In 1950 American astronomer Aden Meinel showed that the solar atomic particles enter the atmosphere at several thousand kilometers per second, disturbing atmospheric atoms at heights of about 100–300 km. Artificial satellites revealed in 1958 that many of these particles are trapped temporarily in the earth's magnetic field before they hit the earth. Here they form the **Van Allen belts** of high-energy ions, shown in Figure 13-19. Aurorae are produced when swarms of these particles are funneled by the earth's magnetic field and dumped from the belts into the regions where the belts touch the atmosphere—around the north and south magnetic poles. The ions collide with air molecules, causing spectral emission bands whose colors are determined by the elements and collisional energies involved. Even under ordinary conditions, when no

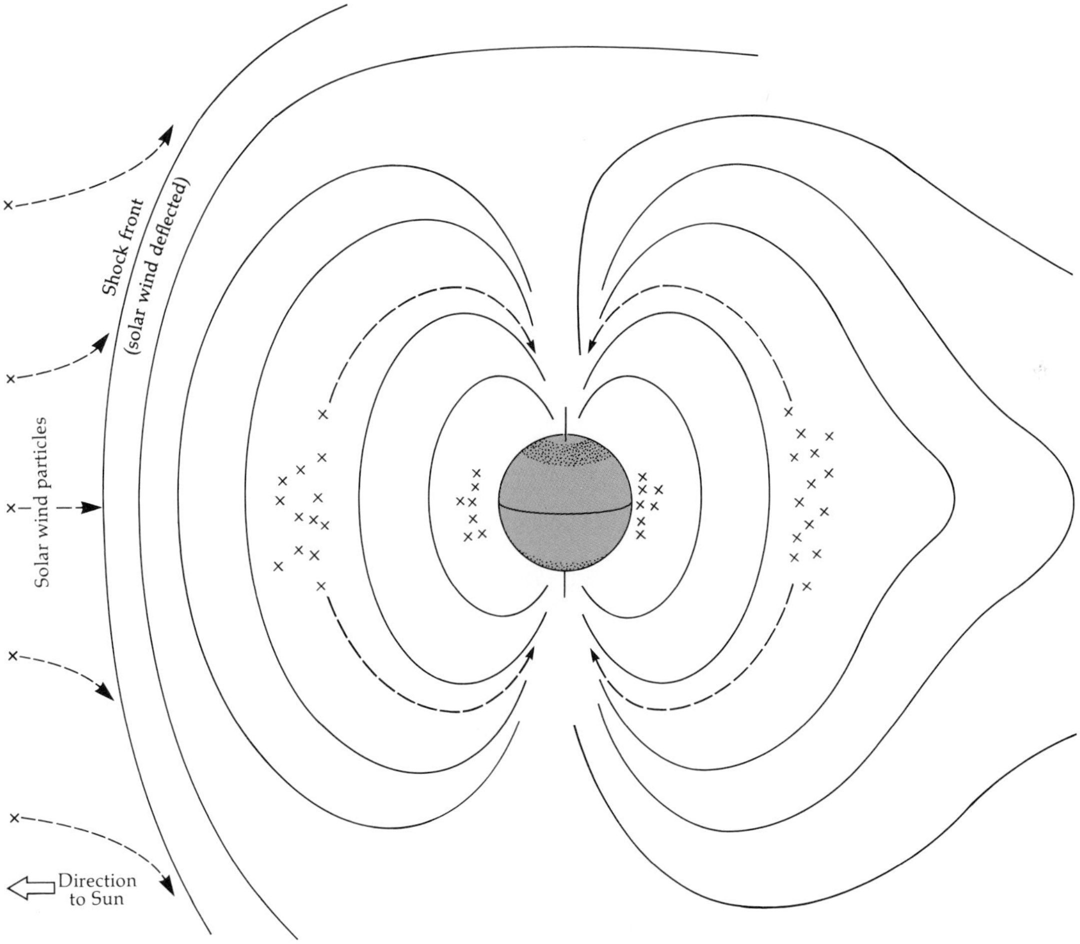

Figure 13-19. *Interaction of solar wind ions (from left) with the magnetic field of the earth. The shock front is analogous to the bow wave cut by a moving boat. Dashed lines show the typical paths of solar ions (**x's**). Those ions that penetrate the earth's field accumulate in radiation belts, eventually emptying into the earth's polar atmosphere, colliding with air molecules, and forming auroral zones near the North and South Poles.*

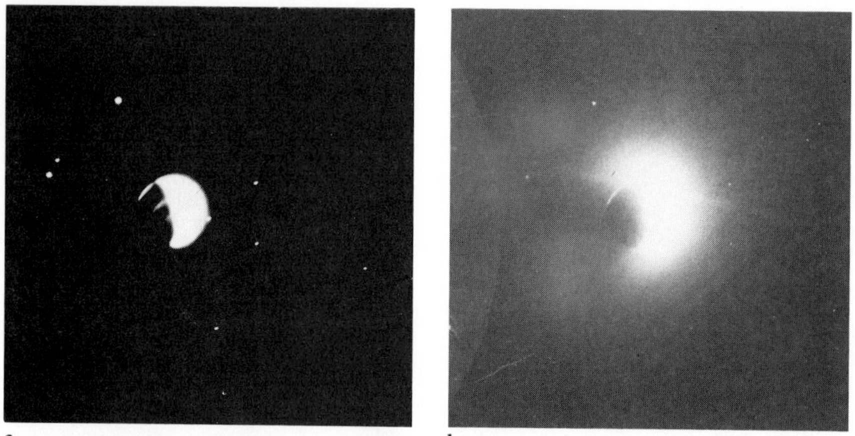

a b

Figure 13-20. Ultraviolet photos of the earth from Apollo 16 show auroral-related features. (a) A true auroral glow on the night side near the North Pole and unexplained glowing bands near the equator. (NASA.) (b) A longer exposure of the same view shows a glowing cloud of hydrogen accumulated near the earth from the solar wind. (NASA.)

aurora is visible, atomic interactions in the upper atmosphere cause a faint glow called **airglow** (see Figures 13-20 and 13-21).

Major solar flares often cause major aurorae and magnetic disturbances a few days later, when the solar particles reach the earth. During these disturbances, compass needles may deviate by a degree from normal, and aurorae may be seen from areas with low latitudes, such as Mexico and India. Electromagnetic effects may overload electric power lines, burn out transformers, and disrupt telephone, telegraph, radio, and power transmission.

Voyager space vehicles discovered tremendous polar aurorae on Jupiter in 1979. These are probably similar in origin to the earth's aurorae, involving solar particles interacting with Jupiter's magnetic field, and Van Allen belts.

Figure 13-21. Airglow, the emission of light from the earth's high atmosphere by atoms excited by sunlight, lightens the sky even on dark nights. Auroralike bands are seen in this infrared photo of emissions from OH molecules at wavelengths around 800 nm. The horizon is at the bottom. The W-shaped constellation Cassiopeia dominates this exposure, made with a 35-mm camera with infrared film and a red filter in New Mexico. (Alan Peterson and Lois Kieffaber.)

Is the Sun Constant?

Solar radiation reaches the earth at a rate of 1.370 kW/m^2, called the **solar constant**. That is, 1370 J of energy reach every sunward-facing square meter of Earth every second.

The rate is called "solar constant" because it was once assumed not to vary. But is it truly constant? Apparently not. Measurements from satellites show a drop in brightness of 0.02% each year from 1979 to 1985. One explanation is a possible solar expansion by 100 km! (See review by Kerr, 1986.) Such changes could be crucial in understanding the earth's climate fluctuations. During the few days of a single flare, X-ray radiation may change dramatically, though the total solar output changes by less than a percent. During the 22-y solar cycle, the number of these flares changes. Moreover, the 22-y cycle itself is erratic and solar activity varies from one cycle to another. During the period 1645–1715, for example, sunspot numbers were probably unusually low (see Figure 13-12), solar flares were virtually absent, and coronal light was pale during eclipses (Eddy, 1976). This period is called the **Maunder minimum**, after the astronomer who recorded sunspots during this time.

Tree rings, radiocarbon dating, and other evidence shows that climatic conditions were measurably altered during the Maunder minimum, and that climatic trends correlate with solar changes (Schneider and Mass, 1975). Evidence from tree rings indicates a 22-y drought cycle in the western United States and a striking 22.4-y cycle in the isotope chemistry of trees over a 1000-y period (Dicke, 1979). Subtle solar changes could drastically affect food production, because rainfall and crop yields respond to average temperature changes of a few degrees or less. For example, Figure 13-22 shows a correlation between the sunspot cycle and murderous African droughts, which seem to concentrate during periods of sunspot minimum.

The whole question of solar influence on biology and climate is attracting new research. A Soviet–American agreement has increased the international exchange of information about this subject (Wilcox, 1976), and solar studies may have increasing importance for astronomy, meteorology, biology, geology, agriculture, and world economics. Astronomers trying to understand basic causes of solar variability have focused more and more on processes in the outer convective envelope of the sun, where instabilities distort the magnetic field, create flares, and temporarily change the total radiation output (Gilliland, 1982).

Solar Energy and Other Cosmic Fuels

Solar energy originally stored in organisms has accumulated in the ground as coal and petroleum. Industrial civilization has been built with this fossil solar energy, and a region's standard of living is correlated with its rate of consumption of such energy. Unfortunately, while it took about 300 million years to accumulate the fossil fuels, we have burned through a significant fraction of them in only 100 years. If today's trends continue, fossil fuels will be largely gone in a few decades.

A cosmic perspective helps us recognize a hurdle we face: As a planetary culture, we must be sure that the dwindling reserves of fossil fuel are used to ensure the production of alternative fuels, because *after* we run out of fossil fuels, it won't be easy to develop the technology to produce alternative sources of energy.

With the sun providing every daylit square meter in the earth's vicinity with 1.37 kW, solar energy seems a likely alternative to fossil fuel. A 70% efficient 1-m^2 solar collector could run a 1000-W appliance. The largest solar-generating station built by 1983 was Solar One, near Barstow, California. It uses 1818 mirrors to focus sunlight on a central collector, generating 10 000 kW of electricity (see review by Kreith and Meyer, 1983). The *total* energy needs of the United States (at the use rate of the 1970s) could be met by collectors operating at only 10% efficiency, spread in giant solar farms, perhaps in desert areas covering only a tenth the area of Arizona. Greater collection efficiency would be possible in the permanent sunlight of space. Engineering studies (Arnold, 1980; Lewis, 1977) have indicated the economic feasibility of building orbiting space colonies from lunar or asteroidal materials, with solar collectors that convert solar energy to radio waves and beam it back to the earth (Figure 13-23).

More down-to-earth is the fact that the roof area of a typical home intercepts roughly 1000 to 2000 kW·h of energy per day, equaling the average daily energy consumption of a U.S. household (Snell et al., 1976). Rooftop solar collectors are a means of using this energy to create heat and electricity. Many people are just realizing that American homes built during the 1950s and 1960s were designed when energy costs were being kept artificially low through government policies. Insulation was skimpy because it was more expensive than the electricity to heat or cool the building. Today, as energy runs out and energy costs rise correspondingly, energy costs in many American homes exceed $1000 per year and may rise further, so that self-sufficient, solar-energized homes are increasingly attractive.

In contrast to fossil fuels, energy from basic planetary or astronomical sources might be called **cosmic fuels.** In addition to solar energy, cosmic fuels include nuclear fission energy (used in present-day reactors), nuclear fusion energy (use of solar-type fusion reactions, possibly using water as a source of hydrogen; still in the experimental stage), geothermal energy (heat from the interior of the earth), and energy from tides and winds. Although nuclear fission energy is the current favorite to get us over the short-term energy hurdle, it has the overwhelming disadvantage of involving and creating deadly radioactive wastes that could wreak health havoc if they escaped into the environment, either by accident or terrorism.

The other cosmic fuels are cleaner than nuclear or fossil fuels. They have an additional subtle advantage: Since they can be collected anywhere in the world, they could make countries energy independent, allowing a more stable planetary culture. As long as the planet depends on fossil fuels, it can be dominated by the economic power of whatever nation controls the cheapest reserves (the United States once did, but has now burned through them). Thus, our generation must respond to the coming exhaustion of fossil fuels not by inaction or a panic-stricken rush to exploit the last reserves, but by a systematic conversion from fossil fuels to safe cosmic fuels.

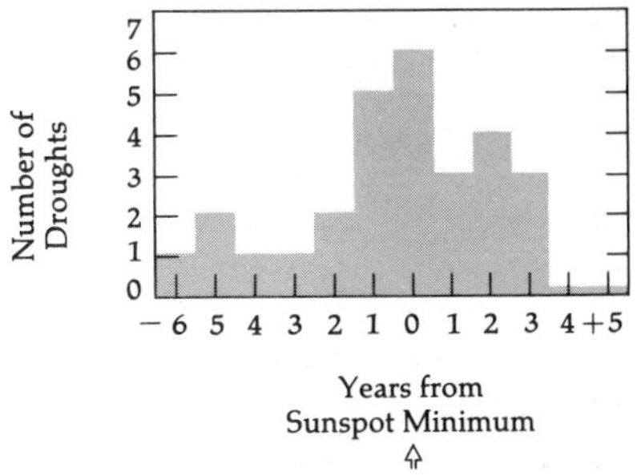

Figure 13-22. *A graph of suspected correlation between terrestrial weather and the solar sunspot cycle. Major droughts in Ethiopia, recorded from 1540 to 1974, appear to correlate with sunspots minima. Such studies may benefit agriculture by helping to predict weather cycles. (Data from Wood and Lovett, 1974.)*

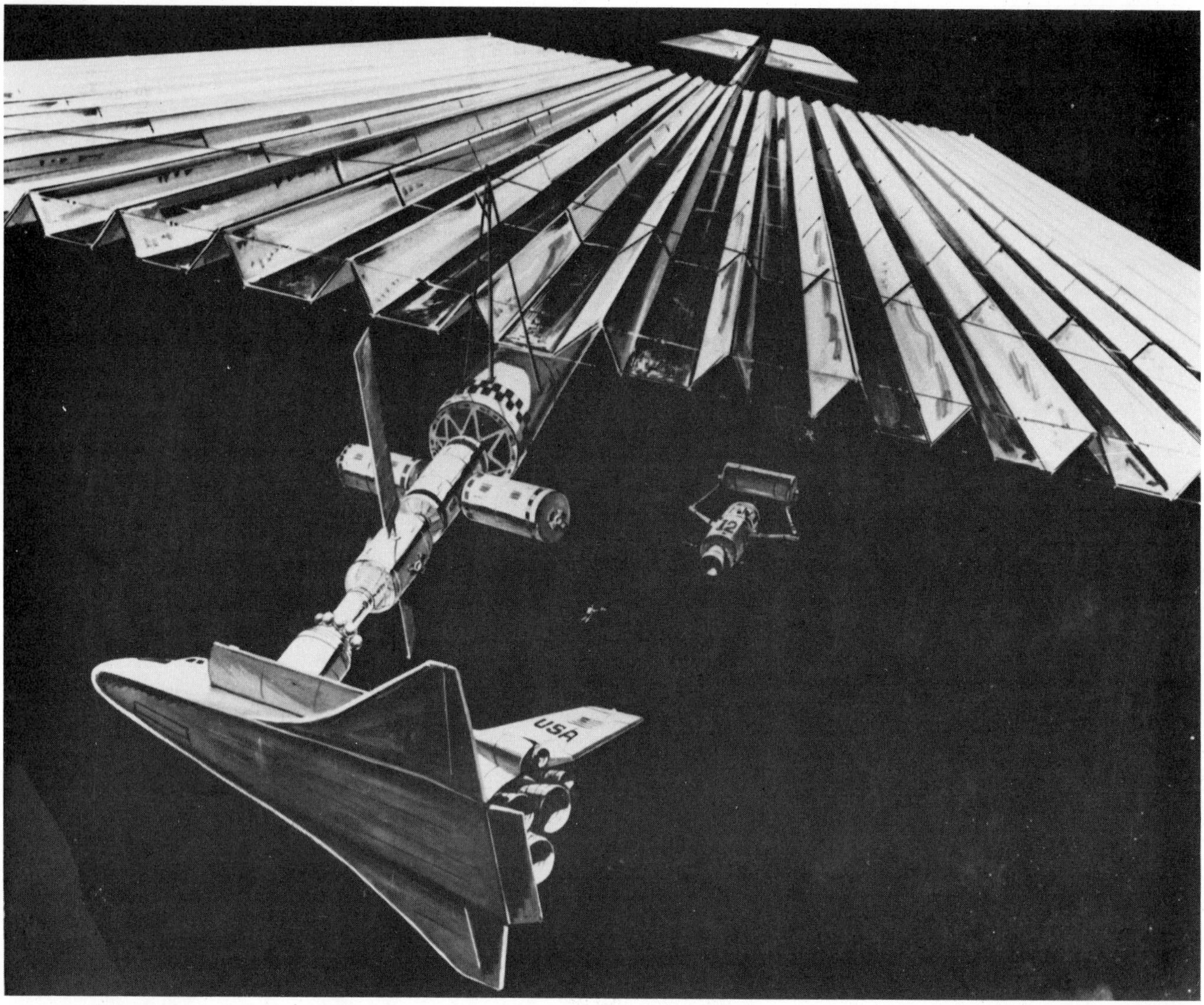

Figure 13-23. *A preliminary design study of earth-orbiting solar power station is shown in this NASA artist's conception. Giant solar collector panels could be fabricated from materials brought from the earth (or possibly from lunar or Apollo asteroid material). Energy derived could be beamed to the earth by microwave and converted to nearly pollution-free electricity.*

Summary

The sun can be studied both observationally and theoretically. Observational studies have yielded information about solar gas revealed by spectral absorption and emission lines created in the solar spectrum as sunlight from the sun's interior passes outward through the layers of solar gas. These studies reveal, among other things, that the sun is about three-quarters hydrogen; most of the rest is helium, and a few percent is composed of heavier elements.

Theoretical studies reveal that as the sun formed by contraction of an interstellar gas cloud, it got so hot that atoms at the center collided at high speeds. These collisions cause nuclear reactions in which hydrogen atoms are fused into helium atoms, releasing energy. This nuclear fusion is the source of the sun's light and heat. The most important reactions in the present-day sun are a series called the proton–proton cycle.

Transport of this energy from the sun's center to the outer layers violently disturbs the surface, producing phenomena such as granules, prominences, flares, and sunspots. Particles are shot off the sun in outward-moving gas called the solar wind, which interacts with the earth, causing aurorae and other phenomena.

Comparing the sun with other energy sources used on the earth shows that we are very rapidly consuming our planetary budget of fossil fuels and will soon have to convert to cosmic energy sources, such as solar energy, to maintain our present rates of energy consumption.

Concepts

solar rotation	negative hydrogen ions
white light	granule
spectrum	spicule
absorption line	chromosphere
emission line	corona
continuum	coronograph
Kirchhoff's laws	sunspot
photosphere	plasma
spectroheliograph	prominence
hydrogen alpha line	flare
sun's composition	solar cycle
nuclear reaction	solar wind
chemical reaction	radiation pressure
proton–proton cycle	aurora
solar core	Van Allen belts
radiation	airglow
conduction	solar constant
convection	Maunder minimum
opacity	cosmic fuels

Problems

1. By what logic do astronomers infer that the sun's energy comes from nuclear fusion reactions of the proton–proton cycle? How do we know it does not come from chemical burning?

2. If you see a cluster of sunspots in the center of the sun's disk, how long would the spots take to reach the limb, carried by the sun's rotation? How long would it take for the cluster to appear again at the center of the disk?

3. Why is the solar cycle said to be 22 y long, even though the number of sunspots rises and falls every 11 y?

4. Suppose you could make detailed comparisons of the appearance of the sun at different moments of time. What variations in appearance would you see if the intervals were:

- **a.** 10 min?
- **b.** 1 wk?
- **c.** 5 y?
- **d.** 10 y?
- **e.** 100 million years?
- **f.** 9 billion years?

5. Why is a radio disturbance on the earth likely to occur within minutes of a solar flare near the center of the sun's disk, whereas an aurora occurs a day or two later, if at all?

6. How much more massive is the sun than the total of all planetary mass (see Table 6-1 for data)?

7. What is the most abundant element in the solar system? The second most abundant element?

8. If the earth formed in the same gas cloud as the sun, why is the earth made from different material than the sun?

9. Why will the sun change drastically in several billion years?

10. Why is the sun's energy generated mostly at its center, and not near its surface?

ADVANCED PROBLEMS

11. Using the small-angle equation, calculate the following:

- **a.** The angular size of a sunspot that has the same diameter as the earth.
- **b.** The angular size of the earth as seen by an imaginary observer on the sun.

12. At what velocity must particles move to escape from a region of the corona 2 solar radii out from the center of the sun?

13. Use Wien's law to confirm the temperature estimate for the sun's surface, based on a maximum energy emission at wavelength 510 nm, as quoted in the text.

14. a. Calculate the speed of a typical hydrogen ion in a 2-million degree portion of the solar corona.

b. Compare this with the escape velocity in the corona (from problem 12). Using the rule of thumb that the fastest atoms move more than three times faster than the typical thermal velocity, comment on the possibility of solar atoms escaping from the corona and thus supplying gas for the solar wind.

Projects

1. According to the principle of the pinhole camera, light passing through a small hole will cast an image if projected onto a screen many hole diameters away. Confirm this by cutting a 1-cm hole of any shape in a large cardboard sheet and allowing the sunlight to pass through the hole onto a white sheet in a dark room or enclosure several feet away (3 m, more if possible). Confirm that the projected image is round, an actual image of the sun's disk. Are any sunspots visible?

2. Cut a 1-cm hole in a sheet of cardboard and use it as a mask over the end of a small telescope. *After* masking the telescope, point it toward the sun and project an image of the sun through an eyepiece onto a white card. *Under no circumstances should anyone ever look through the eyepiece, since all the light entering the telescope is concentrated at that point and can burn the retina!* Professional supervision of the telescope is suggested. Note also that an unmasked large telescope may concentrate enough light and heat to crack eyepiece lenses.

Are sunspots visible? If so, trace the image on a piece of paper and reobserve on the next day. Confirm the rotation of the sun by following sunspot positions for several days. Class records kept from year to year can be used to record the cyclic variations of the numbers of sunspots.

MEASURING THE BASIC PROPERTIES OF STARS

Now we are ready to take the great leap out of the solar system into the realm of the stars. It is quite a leap; the next star beyond the sun is 260 000 times as far as the sun and 6800 times as far as Pluto. If we make a model solar system the size of a half dollar, the neighbor stars would be dots smaller than a period in this book, scattered about half a block apart.

Because of optical limitations and turbulence in the earth's atmosphere, the world's largest telescope cannot distinguish details smaller than a few hundredths of a second of arc (about 10 millionths of a degree). But the stars with the largest apparent angular size are no larger than about 0.04 seconds. So the disks and surface details of nearly all distant stars are hidden from us, in contrast to the great detail we can study on the sun.

In spite of this, we know that there are giant stars bigger than the whole orbit of Mars, stars the size of the earth, and stars the size of an asteroid. There are red stars and blue stars. There are stars of gas so thin you can see through parts of them, and stars with rocklike crusts that may contain diamonds. There are stars that are isolated spheres, stars with disklike rings around them, and stars that are exploding.

All of these bodies fit the definition of stars: **Stars** are objects with so much central heat and pressure that energy is (or has been) generated in their interiors by nuclear reactions. The most familiar stars visible in the night sky are balls of gas with solar composition and sizes usually a few times smaller or larger than the sun.

How can we know all these details about stars if we cannot even see the disks of stars in telescopes? In the next chapters we will describe the details of familiar and unfamiliar types of stars, but first we will describe *how* we learn about them.

Names of Stars

Stars are named and cataloged by several systems. Because Ptolemy's *Almagest* was passed on by Arab astronomers, many of the brightest stars ended up with Arabic names. Since *al-* is the common Arabic article, many star names start with *al:* Algol, Aldebaran, Altair, and Alcor. Other scientific "*al* words" also have Arab origins: *algebra, alchemy, alkali,* and *almanac.* Stars in constellations are cataloged in approximate order of brightness using Greek letters. Thus the brightest star in the constellation of the Centaur is called Alpha Centauri (α Centauri). Fainter stars or stars with unusual properties are often known by English letters or catalog numbers, such as T Tauri or B.D. 4° + 4048.

Images of Stars

Figures 14-1 and 14-2 show typical telescopic photographs of bright stars. The strange patterns of light, such as rays or circles, have nothing to do with the actual shapes of the stars, which would be tiny pinpoints buried in these overexposed images, but are caused by optical effects in the telescope and film system.

Visual examination of a bright star in a modest telescope shows what appears to be a tiny disk surrounded by faint rings, but the disk and rings are caused by an optical effect called **diffraction**. The disk is sometimes called the **Airy disk**, after the English astronomer who explained it. It again has nothing to do with the real star, and in fact prevents the much smaller real star image from being seen.

In 1975 astronomers at Kitt Peak National Observatory picked one of the stars with the largest apparent angular size, the bright star Betelgeuse in the constellation Orion, and used computer techniques to remove the blurring due to atmospheric turbulence and diffraction. Their analysis produced an image of a fuzzy disk, shown in Figure 14-3, which is believed to be the actual disk of Betelgeuse. Analysis showed that this star's actual linear diameter is 500 to 750 times larger than the sun's. If Betelgeuse replaced the sun, it would engulf the sun and the terrestrial planets and extend out into the asteroid belt! This alone proves that not all stars are alike.

Defining a Stellar Distance Scale: Light-Years and Parsecs

The vast distances that separate stars and make them so hard to observe are awkward to express in ordinary units. Astronomers use units appropriate to these distances. The easiest to understand is the **light-year** (abbreviated "ly"), the distance light travels in 1 year, which is about 6 million million miles, or 10^{16} m.

Remember that the light-year is a unit of distance, not of time. The common mistake of using light-year as if it were a unit of time is like saying that the ball game lasted for 2 miles.[1]

The nearest star beyond the sun, Proxima Centauri (which is in orbit around Alpha Centauri), is about 4.3 light-years away. The sun could be said to be 8 light-minutes away. The North Star, Polaris, is about 650 light-years away. Polaris' light takes about 650 years to reach us, so we are seeing it now as it was in the 1300s. If Polaris had suddenly exploded in 1950, we would not know it until about A.D. 2600.

Astronomers more commonly use a still larger unit of distance called the **parsec** (abbreviated "pc"):[2]

$$1 \text{ parsec} = 3.26 \text{ light-years} = 3 \times 10^{16} \text{ m}$$

In this book we will use the parsec and its multiples (kiloparsecs and megaparsecs) to express cosmic distances as we move to more remote parts of the universe. You can convert parsecs to light years approximately by multiplying by 3. Another convenient fact to remember is that near the sun, stars are roughly a parsec apart. For instance, Alpha and Proxima Centauri, the closest stars to the solar system, are about 1.3 parsecs away.

[1] Even noted space pilot Han Solo confused distance and time units when he claimed in *Star Wars* that the Millenium Falcon could "make the Kessel run" in less than 12 parsecs.

[2] The reason for defining this unit will become clearer on p. 267.

Figure 14-1. *This overexposed image of the star Altair illustrates the great differences in brightness among the stars. The true disk of Altair would be a pinpoint at the center of the overexposed image. Raylike structures are optical effects produced in the 76-cm (30-in.) telescope used to make the 30-min. exposure. Field of view about 6 × 9 minutes of arc. (Walter Feibelman, Allegheny Observatory.)*

Figure 14-2. *A typical overexposed and enlarged stellar image from a reflecting telescope shows a prominent "cross hair" pattern caused by optical diffraction effects of the secondary mirror's supporting structure. Also visible is a halation ring caused by reflection off the back of the glass photographic plate. Both effects occur inside the telescope. Neither effect relates to the true star, which would be a microscopic dot at this scale. Fainter stars provide too little light to show either effect. (Mauna Kea Observatory, University of Hawaii.)*

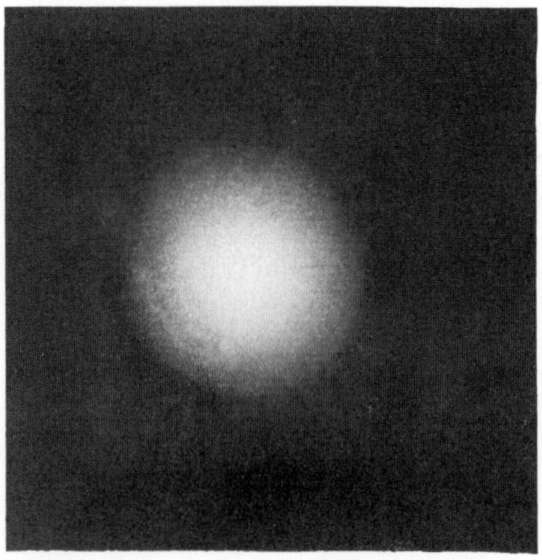

Figure 14-3. *Special techniques were used to make this image of the actual disk of the star Betelgeuse. Though only a fraction of a second of arc, this star's angular size is one of the largest of stars seen from Earth. The star's diameter approximates the diameter of Jupiter's orbit; it is classed as a supergiant. The image suggests limb (edge) darkening, also observed on the sun. (Kitt Peak National Observatory.)*

Defining a Brightness Scale: Apparent Magnitude

A nearby candle may *appear* to be brighter than a distant streetlight, whereas in *absolute* terms the candle is much dimmer. This statement contains the essence of the problem of stellar brightness. A casual glance at a star does not reveal whether it is a nearby glowing ember or a distant great beacon. Hence astronomers distinguish between **apparent brightness** (the brightness perceived by an observer on the earth) and **absolute brightness** (the brightness that would be perceived if all stars were magically placed at some identical, standard distance).

Let us look first at apparent brightness. Because scientists generally use the decimal system, we might expect astronomers to use a brightness scale in which each unit of brightness difference corresponds to a factor of 10. No such luck. Astronomers, too, are victims of history. When Hipparchus cataloged 1000 stars in about 130 B.C., he ranked their **apparent magnitude**, or apparent brightness, on a scale of 1 to 6, with 1st-magnitude stars the brightest and 6th-magnitude stars the faintest visible to the naked eye. This system stuck.

For more precision, nineteenth- and twentieth-century astronomers extended that scale to cover objects brighter than 1st magnitude and dimmer than 6th magnitude. Because Hipparchus' 1st-magnitude stars are about 100 times brighter than his 6th-magnitude stars, they used this as the definition of the brightness system. A difference of five magnitudes is *defined* as a factor of 100 in brightness. Therefore, any given star is about $2\frac{1}{2}$ times brighter than a star of the next fainter magnitude.[3]

The system is illustrated in Figure 14-4, which shows a photo of the constellation Orion the Hunter, with some stars labeled with their magnitudes (0 to 9) on the right and, in the case of brighter stars, their Greek letter names on the left.

Table 14-1 shows how the apparent magnitude scale has been extended, including fractional magnitudes. Extending the scale beyond the brightest objects included by Hipparchus meant using negative numbers. For example, the planet Venus at its brightest ranks more than minus 4th magnitude (-4.4), whereas the sun reaches beyond minus 26th (-26.5). As the famous astrophysicist Arthur Eddington remarked: "You have to remember that stellar magnitude is like a golfer's handicap—the bigger the number, the worse the performance."

Our accuracy in measuring stellar brightness is only 30 times better than Ptolemy's was 1800 years ago! One problem is that stars have different colors, and light detectors (the eye, photographic films, and photoelectric devices) have different sensitivities to different colors. For this reason, astronomers specify exactly what color any set of measurements refers to. Standards have been derived to express apparent magnitudes measured in blue light, red light, infrared light, and so on. Here, we will usually refer to a system having the same color sensitivity as the human eye, sometimes called **visual apparent magnitude** (often abbreviated m_v).

A Magnitude Scale for Expressing "True" Brightness of Stars

The apparent magnitudes that we have just discussed describe only the apparent brightness of each star as seen from the earth. This depends on the star's distance and thus doesn't express the star's true energy output. This true energy output can be expressed using a magnitude system called the star's **absolute magnitude**, which is simply the apparent magnitude the star would have if it were moved to a standard distance of 10 pc. In this system, a star with the same energy output as the sun has an absolute magnitude of about $+5$. A star with 100 times that energy output is said to have an absolute magnitude of 0. A star 100 times fainter has an absolute magnitude of $+10$.

Basic Principles of Stellar Spectra

The last chapter showed how the sun's **spectrum**, or distribution of light into different colors (**wavelengths**), gives information about the sun's atmosphere, surface layers, and interior. The same is true of other stars. So critical is **spectroscopy**—the study of spectra—to astronomy that many astronomers devote their entire careers to it. Chapter 7 gives more details on the principles of spectra, and the student may wish to review that material while studying this chapter. Let us review the nature of spectra to see how they can be used to reveal the nature of stars.

As mentioned briefly in Chapter 7, spectra can be presented in two ways, photographically (Figure 14-5) or as a chart (Figure 14-6, bottom), depending on the instrument used. A **spectrograph** produces a photographic image of the spectrum, usually placed with blue to the left and red to the right, as in Figure 14-5. A color image of such a photo would appear as in Color Plate 32 (bottom). The intensities of the photographic print represent intensities of light, or radiant energy. The vertical width of the spectrum is arbitrary; sometimes such spectra are artificially enlarged vertically to make an image more readily measurable.

A **spectrometer** produces a spectrum in the form of a chart or graph, as in the lower portion of Figure 14-6. Wavelength is shown along the bottom, again with red usually to the right; the vertical direction represents the intensity of the light. **Absorption lines** appear as dark vertical lines on a photographic spectrum (see Color Plate 32) and as notches or valleys on a graph. **Emission lines** appear as bright vertical lines or as sharp peaks. The gen-

[3]The actual factor is $\sqrt[5]{100}$, or 2.512. This unusual factor as a unit of brightness difference is used in no other science. Physicists are heard to mutter obscenities when they first try to analyze astronomical data expressed in magnitudes. But the 2100-year-old magnitude system is so ingrained in astronomy that we continue to use it here.

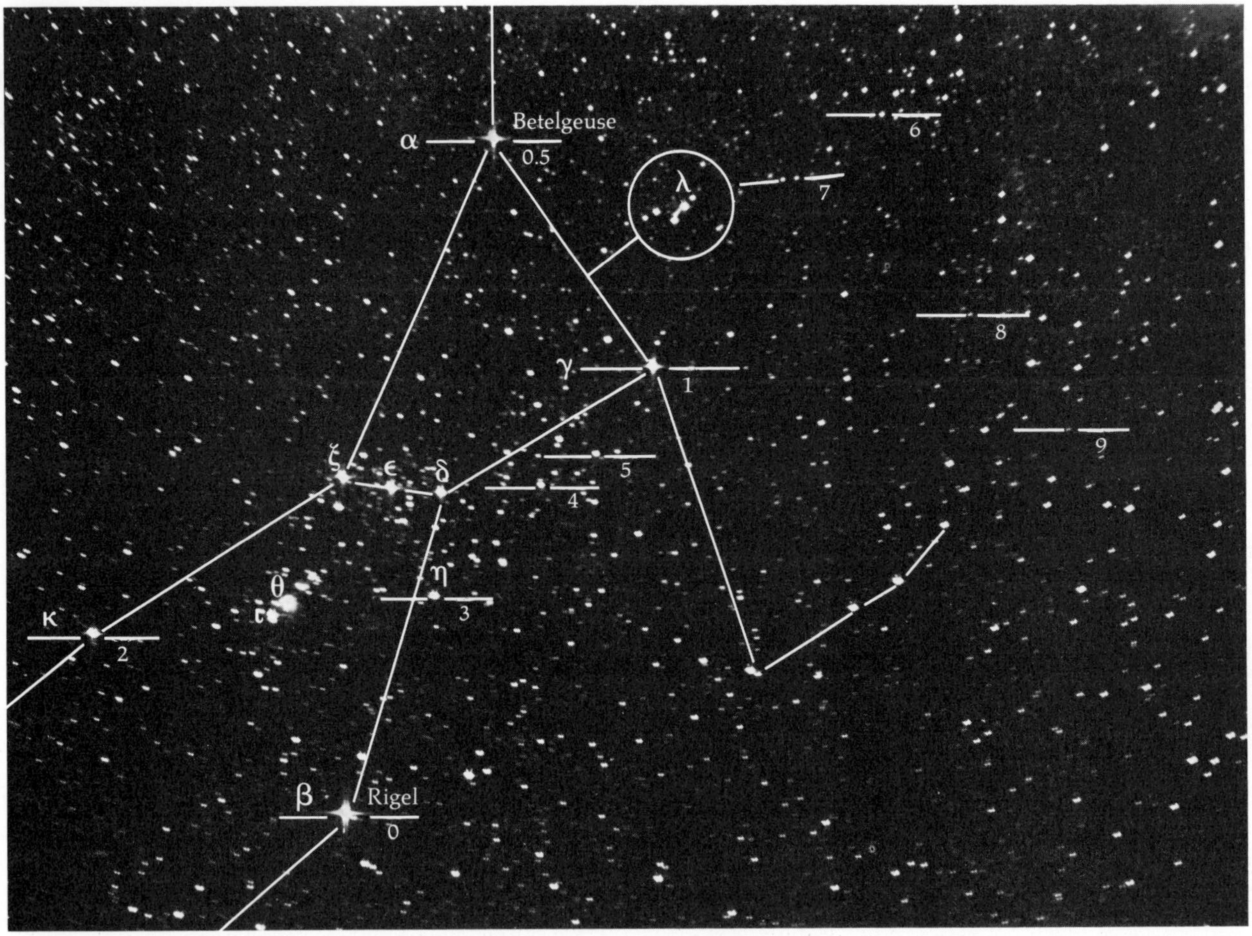

Figure 14-4. *The constellation of Orion the Hunter (stick figure) showing Greek letter designations and approximate visual magnitudes of selected stars. This exposure (on Tri-X film with a 35-mm camera) records stars to about the 10th magnitude. (Photo by author.)*

eral level of brightness between absorption or emission lines is the **continuum**. For color renditions of absorption lines, emission lines, and continuum, see the lower part of Color Plate 32, middle, bottom, and upper images, respectively.

SPECTRA OF STARS

In 1872, Henry Draper, a pioneer in astronomical photography, first photographed stellar spectra. This represented a tremendous advance; instead of sketching or verbally describing spectra, astronomers could directly record, compare, and measure them. Spectra of thousands of stars became available for precise analysis.

Such massive amounts of data required a classification scheme. This was begun in the 1880s by Harvard astronomer Edward C. Pickering, and completed by Annie J. Cannon and a group of young woman assistants who invented a system of **spectral classes** based on the number and appearance of spectral lines. The classes—A, B, C, and so on—started with spectra with strong hydrogen lines. When Annie Cannon published the Henry Draper Catalog (1918–1924), it contained spectral data on 225 320 stars and became the basis for all modern astronomical spectroscopy.

Further work showed that the classes had to be rearranged to bring them into a true physical sequence based on temperature, as shown in Figure 14-7. The sequence finally adopted begins with the hottest stars—class O—which show ionized helium lines in their spectra. The sequence of classes is O, B, A, F, G, K, and

Table 14-1. ***Apparent Magnitudes of Selected Objects***

Object	Apparent Magnitude
Sun	−26.5
Full moon	−12.5
Venus (at brightest)	−4.4
Mars (at brightest)	−2.7
Jupiter (at brightest)	−2.6
Sirius (brightest star)	−1.4
Canopus (second brightest star)[a]	−0.7
Vega	0.0
Spica	1.0
Naked eye limit in urban areas	3–4
Uranus	5.5
Naked eye limit in rural areas	6–6.5
Bright asteroid	6
Neptune	7.8
Limit for typical binoculars	9–10
Limit for 15-cm (6-in.) telescope	13
Pluto	15
Limit for visual observation with largest telescopes	19.5
Limit for photographs with largest telescopes	23.5

Source: Data from Allen (1973)

[a]This lesser-known Southern Hemisphere star is used as a prime orientation point for space craft. A small light detector on spacecraft is called "Canopus senor."

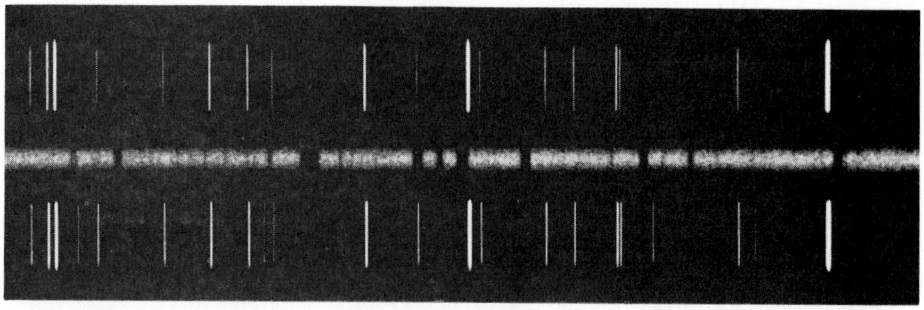

Figure 14-5. *High-resolution spectra photographed with a large telescope usually take this form. The stellar spectrum is the fuzzy horizontal luminous band in the center, crossed by vertical dark absorption lines. It is flanked at top and bottom by emission lines of a known element (usually iron) produced artificially by a spark inside the instrument. These lines, whose positions are known, permit measurement of line wavelengths in the stellar spectrum. (Lick Observatory.)*

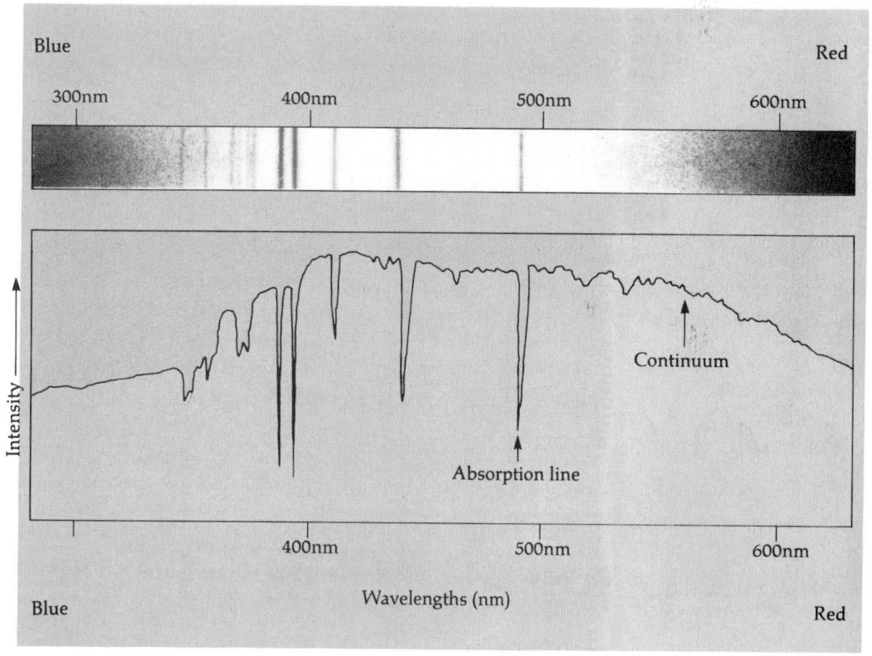

Figure 14-6. *The visual portion of a typical stellar spectrogram of a solar-type star, as shown on a photographic plate by a spectrograph (top) and in a scan by a spectrophotometer (bottom).*

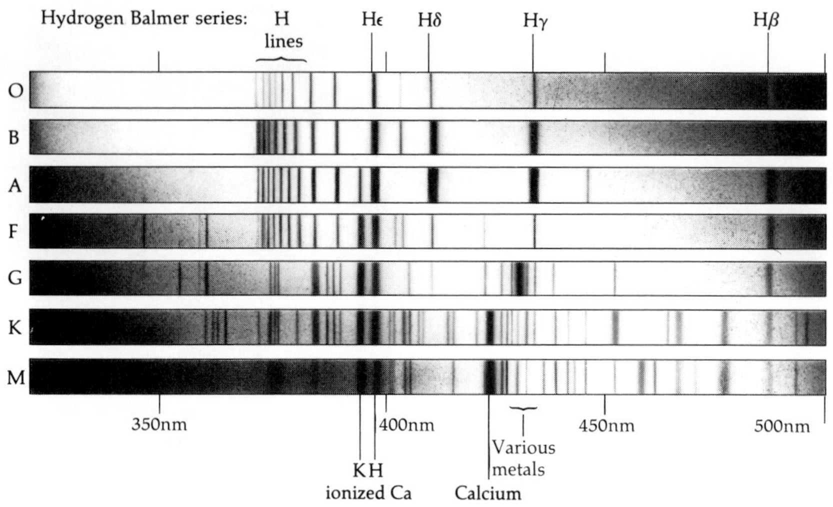

***Figure 14-7.** Representations of spectra of stars in the seven spectral classes. Classes A and B show especially well the Balmer hydrogen lines (compare with Table 14-2).*

M.[4] The M stars are the coolest. About 99% of all stars can be classified into these groups. For finer discrimination the classes are sometimes divided from 0 to 9; the sun, for example, is said to be a G2 star.

SPECTRAL LINES: INDICATORS OF ATOMIC STRUCTURES

In Chapter 7 we gave a simplified description of how emission and absorption lines depend on the orbital structure of electrons in atoms. Let us now consider the process in more detail by considering the simplest and most abundant type of atom in the universe, hydrogen. As shown in the schematic diagram, Figure 14-8, hydrogen atoms' orbits can be numbered. Since a neutral hydrogen atom has just one electron, an atom in the ground state would have its electron in the orbit $n = 1$. If the atom had been bumped by other atoms or if it had absorbed radiation, the electron might be in the orbit $n = 2, 3,$ and so on. Further absorption of energy might cause it to jump from $n = 3$ to $n = 4$, creating an absorption line, or it might spontaneously revert from $n = 3$ to $n = 2$, creating an emission line. Each possible transition (1 to 2, 2 to 3, 4 to 2, and so on) creates a different line. As Figure 14-8 shows, hydrogen lines ending at the $n = 2$ level are the lines prominent in the visible part of the spectrum, called the **hydrogen Balmer series** of lines. The famous red line called **hydrogen alpha** (Hα, shown in Color Plate 32) or **Balmer alpha**, involves transitions between $n = 2$ and $n = 3$ in the hydrogen atom. This line lends its brilliant red color to many astronomical gases, including the solar chromosphere and many nebulae.

A hydrogen atom needs to absorb only about 10^{-19} to 10^{-18} joules to bump its electrons from one level to another. About 2×10^{-18} joules is needed to ionize a hydrogen atom in its ground state. These are small amounts of energy, but sources of such energy are important in determining the state of the gas. In a hydrogen-rich environment where either the photons from surrounding radiation or jostling by neighboring atoms provides this much energy, many hydrogen atoms will be ionized.

The astute reader will have noticed that the strengths of different hydrogen lines in the spectrum will depend on the fraction of atoms with the $n = 1$ level occupied by an electron, the fraction with the $n = 2$ level occupied, and so on. This in turn is controlled by the temperature, density, and pressure in the gas. The higher the temperature, the harder the atoms bump together and the higher the fraction with electrons excited into higher energy levels. For any given element, how can we tell exactly what fractions of the atoms are in the ground state and each excited state and thus what the *relative strengths of different spectral lines will be?*

This crucial question began to be answered by the work of the Indian physicist Megh Nad Saha soon after the stellar spectral classes were recognized. In 1920 he derived an important result, now called the **Saha equation**, which allows us to tell the relative numbers of atoms in each excited state, given the conditions in the gas, or *vice versa*.

Note then the power of the knowledge we have discussed so far. *Identification* of the spectral lines tells us what elements are present in a star. *Measurement of the relative line strengths,* interpreted with the Saha equation, tells us what fraction of the atoms is in each excited state, and hence tells us something about the temperature, density, and pressure conditions in the photospheric layer where the spectrum arises in the surface of the star. This method supplements the temperature information gained from other techniques, such as Wien's law, discussed in Chapter 7.

---

[4]As Annie Cannon noted, the rearrangement of letters made the sequence harder to remember. American astronomer Henry Norris Russell soon (ca. 1920s) solved this problem with Yankee ingenuity by proposing a sentence as a memory aid: Oh, Be A Fine Girl, Kiss Me! Two later classes, R and N, were easily accommodated by adding the words *Right Now*. However, when class S was added, a controversy broke out between Harvard and Cal Tech over whether it signified the word *Sweetie* or *Smack*.

SPECTRAL CLASSES AS A TEMPERATURE SEQUENCE

An important principle is that each spectral class corresponds to a different *temperature* in the gases in the light-emitting layers of the stars. **Temperature** is merely a measure of the average velocity of the atoms or molecules of a substance. The hotter a gas, the faster its atoms and molecules move. If the temperature quadruples, the velocities of the atoms double. The faster the atoms collide, the more they disturb or dislodge each other's electrons. Furthermore, the hotter the gas, the more radiation it emits; the resulting photons also disturb electron structures of atoms. Because spectral lines depend on electron structures, and because spectral classes depend on spectral lines, spectral classes therefore form a temperature sequence.

Consider what happens as we decrease the temperature of matter. The hottest stars, the O stars, have temperatures of 40 000 K or more, as shown in Table 14-2. Here, the atoms and radiation are so energetic that even the most tightly bound atoms, such as helium, have had their electrons knocked off, as the table indicates. By class B, the temperature has dropped to around 18 000 K, and helium atoms keep their electrons. By classes A and F, even relatively weakly bound hydrogen keeps its electrons. Many metals have at least one electron that can be very easily knocked off the outer part of the atom, and in G-type stars like the sun, at around 5500 K, lines of ionized metals are prominent along with Balmer lines of hydrogen. By class K, at 4000 K, even many metals are neutral. By class M, at 3000 K, energies are so low that different atoms can stick together into molecules; even water molecules have been identified in spectra of such cool stars, as shown in Figure 14-9.

Stars cooler than M stars are faint and rarely seen, but the concept may be extended to lower temperatures. In Chapter 12, we discussed how grains of minerals such as silicates can form at low temperatures of around 2000 to 1300 K; such grains probably exist in the atmospheres of the coolest stars.

Planets, of course, have even cooler temperatures, such as 300 K for the earth, where liquid water coexists with solid rocks. At cool enough temperatures, near absolute zero, nearly all matter would assume solid form.

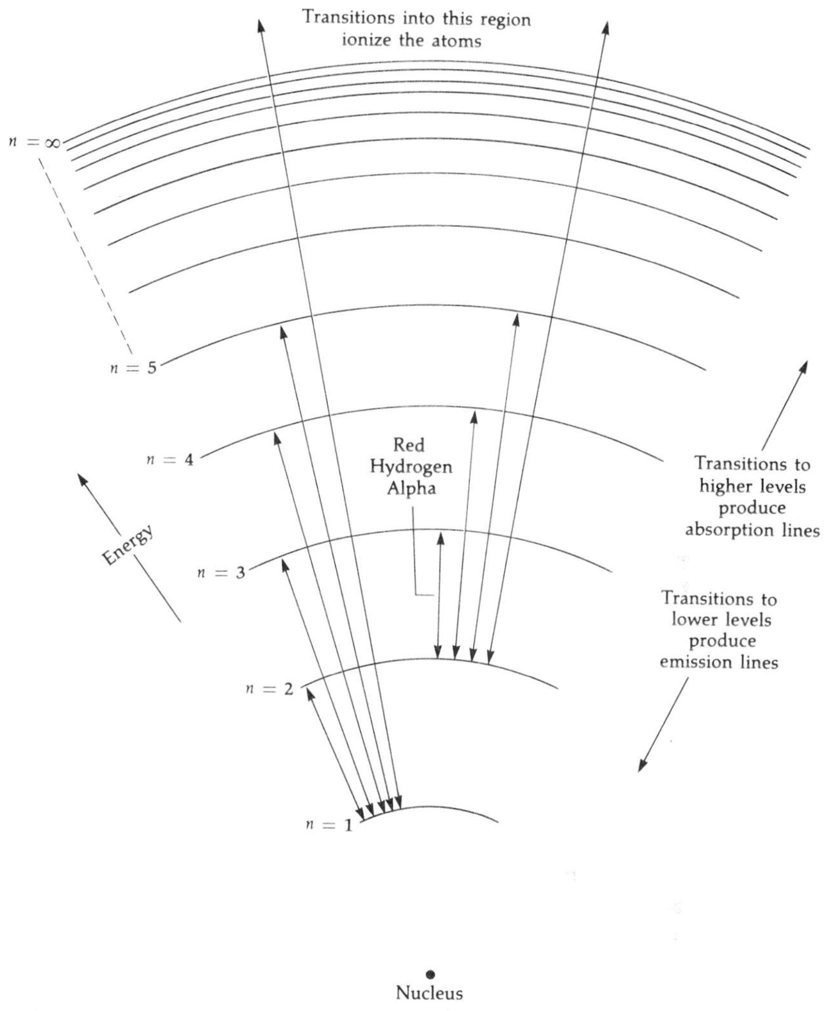

Figure 14-8. *Schematic diagram of a hydrogen atom, showing different possible electron orbits or energy levels (n = 1, 2, 3, . . .). Each change of orbit by an electron produces a spectral line, because energy is removed or added to the light beam. The series of lines starting from or ending on n = 1 occur in the ultraviolet part of the spectrum and are called the Lyman series. Lines starting from or ending on n = 2 occur in the visible part of the spectrum and are called the Balmer series. The well-known red Hα line of the Balmer series involves transitions between n = 2 and n = 3.*

TWO IMPORTANT LAWS OF SPECTROSCOPY

We now introduce two important effects that will help us understand many astronomical phenomena. Both are described in mathematical detail in the accompanying optional boxes, but they need to be

Table 14-2. *Principal Spectral Classes of Stars*

Type	Spectral Class	Typical Temperature (K)	Sources of Some Prominent Spectral Lines	Representative Stars
Hottest, bluest	O	40 000	Ionized helium atoms	Alnitak (ζ Orionis)
Bluish	B	18 000	Neutral helium atoms	Spica (α Virginis)
Bluish-white	A	10 000	Neutral hydrogen atoms	Sirius (α Canis Majoris)
White	F	7000	Neutral hydrogen atoms	Procyon (α Canis Minoris)
Yellowish-white	G	5500	Neutral hydrogen, ionized calcium	Sun
Orangish	K	4000	Neutral metal atoms	Arcturus (α Bootes)
Coolest, reddest	M	3000	Molecules and neutral metals	Antares (α Scorpii)

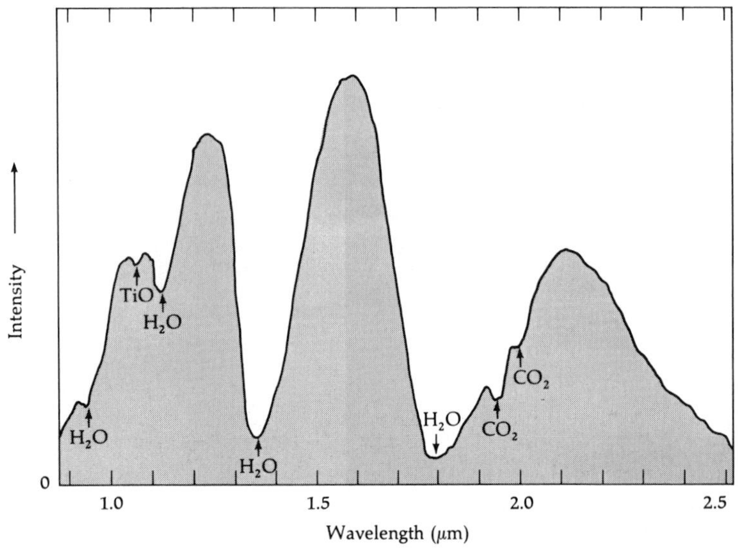

Figure 14-9. *A portion of the infrared spectrum of the giant star* o *Ceti (Mira), showing identification of various molecules, including the first identification of water vapor in a star. (After a diagram by G. P. Kuiper.)*

briefly described to all readers, because they are important in interpreting distant stars.

The Doppler Effect.

The* Doppler effect *is a shift in wavelength of a spectral absorption or emission line away from its normal wavelength, caused by motion of the light source toward or away from the observer.

If the source approaches, there is a **blue shift** toward shorter wavelengths, or bluer light. If the source recedes, a **red shift** occurs toward longer wavelengths, or redder light. The amount of the shift is just proportional to the approach or recession speed of the source (as long as that speed is well below the speed of light). In the shorthand of mathematics,

$$\frac{\text{shift in wavelength}}{\text{normal wavelength}} = \frac{\text{approach or recession speed}}{\text{velocity of light}}$$

If the light source is receding at 10% the speed of light, the light will be red-shifted by 10% of its normal wavelength; a line normally found at wavelength 500 nm would appear at 550 nm. By measuring such wavelengths, velocities of stars can be studied.[5]

The Stefan–Boltzmann Law.

The* Stefan–Boltzmann law, *discovered by two Austrian physicists about 1880, states that the higher the temperature of a surface, the more energy radiated by each square centimeter in each second.

[5]The same shift occurs for sound. This explains why the noise of a rapidly traveling automobile, truck, or train changes from higher pitch (shorter wavelength) to lower pitch (longer wavelength) as the vehicle rushes by. The sound a child makes to imitate a race car rushing by is a representation of the Doppler shift.

The mathematical form of the law gives the total energy E or luminosity L (in joules) radiated per second by a surface of area A (square meters) and temperature T (Kelvin). The law states that L is proportional to AT^4. If the temperature of a source doubles, the amount of energy radiated increases by 2^4, or 16. Thus, while doubling the area of a star would increase its output twice, doubling its temperature would increase its output 16 times! Therefore, as shown in Figure 14-10, the hotter stars not only radiate bluer light than cooler stars (a result that was predicted by Wien's law), but also radiate *more* light per unit area. The Stefan–Boltzmann law thus provides us with a way to learn about the sizes of areas of stars once we measure their temperatures and total radiation output.

OPTIONAL BASIC EQUATION VII

The Doppler Effect: Approach and Recession Velocities

Probably the most important physical phenomenon in astronomical spectroscopy is the Doppler effect (named after the Austrian physicist Christian Doppler, who discovered it in 1842). *The Doppler effect is a change in wavelength proportional to any line-of-sight velocity between observer and source.* (That is, the wavelength of a spectral line from a fast-approaching body will differ from that of a fast-receding body *and* that of one slowly approaching.)

Doppler discussed the effect in the light from distant stars orbiting around each other, but his effect also applies to any signal transmitted by waves, including familiar sound waves on earth. The most familiar example of the Doppler effect is the shift in pitch of sound from an approaching or receding source. As a car or train passes by and recedes in the distance, the pitch of the sound dramatically decreases. As the source approaches, the sound waves rush past the observer apparently more closely spaced than normal, decreasing the observed wavelength. As the source recedes, the waves are perceived to be more spread out, increasing the wavelength. The effect modifies whatever property of the signal is determined by wavelength—the color of light or the pitch of sound.

The equation that describes the Doppler effect is the seventh of the nine major equations we will use in this book. Since λ is used to designate wavelength, Δ to designate change, v to designate velocity, and c to designate the speed of light, the equation cited in the text is commonly written $\Delta\lambda/\lambda = v/c$. Read this as "change in wavelength divided by the original wavelength equals v divided by c." Thus, if the source of light is not approaching or receding, the Doppler shift is zero. The faster the source approaches or recedes, the greater the Doppler shift in wavelength of the light.

For example, suppose a certain infrared spectral line normally has a wavelength of 1000 nm. If the light source is receding from the observer at 1/1000 the speed of light, the line would appear at the slightly longer wavelength of 1001 nm, a little toward the red end of the spectrum. If the source approached at 1/1000 the speed of light, the line would appear at 999 nm.

When source and observer are getting closer together, the wavelength of all light coming from the source is *shifted toward the blue* (shorter wavelengths). But when source and observer are getting farther apart, the spectrum of light from the source is *shifted toward the red* (longer wavelengths).

The two kinds of shifts are loosely referred to as red shifts and blue shifts, and their sources as red-shifted or blue-shifted. As a mnemonic aid, remember that *re*cession produces *re*d shifts.

Note that the observer of a Doppler shift cannot say whether it is the source or the observer who is moving relative to any absolute external frame of reference. According to the principle of relativity, only recession or approach of one body with respect to another can be measured.

Example. The hydrogen alpha absorption line in a certain star is carefully measured and found to lie at 656.4 nm instead of the normal 656.3. Describe the motion of the star in the line of sight direction toward or away from the observer. Does this observation tell us anything about motion perpendicular to this direction?

Solution. Solving the equation for v, and looking up c in Appendix 2 (Table A2-2), we have

$$v = \frac{\Delta\lambda}{\lambda}c = \frac{0.1}{656.3} \times 3(10^8) = 4.6(10^4)\,\text{m/s}$$

The Doppler shift is toward longer wavelength, or redward, and so the star must be moving away. The Doppler shift is sensitive only to motion toward or away from us, so we cannot say anything about a possible additional component of motion laterally. We can only say it is moving away from us at 46 km/s in the line of sight.

Measuring 12 Important Stellar Properties

We now can apply the Stefan–Boltzmann law, the Doppler effect, and other principles to explain how 12 important physical properties of remote stars can actually be measured. These applications, to be discussed in the following sections, are summarized in Table 14-3.

HOW TO MEASURE A STAR'S DISTANCE

From what you have read so far, can you think of a way to measure the distance of a star? The question is not far-fetched, since star distances were first measured more than a century ago. A crude method may be applied to the information in Table 14-1. The brightest stars have *apparent* magnitudes about 25 magnitudes fainter than the sun. In 1829, the English scientist William Wollaston used this simple fact to estimate that most typical stars must be at least 100 000 times more distant than the sun, since dimming by 25 magnitudes corresponds to increasing distance 100 000 times.

Because this gives only a typical distance, we need a better technique to measure distances of individual stars. The most important such technique is **parallax** measurement. You will recall that parallax is the apparent shift in the position of an object caused by a shift in the observer's position (p. 37).

The principle is easy to understand, as shown in Figure 14-11. Hold your finger in front of your face and look past it toward some distant objects. Your finger represents a nearby star; the objects, distant stars. Your right eye represents the view on one side of the sun, your left eye represents the view 3 months later, after the earth has traveled to a point lined up with the sun (a shift in the earth's position by 1 AU as seen from the star). First wink one eye

---

OPTIONAL BASIC EQUATION VIII

The Stefan–Boltzmann Law: Rate of Energy Radiation

Before describing how to measure specific properties of stars, we need the eighth of the nine basic equations to be used in this book. This is the Stefan–Boltzmann law, which describes the total amount of energy radiated in each second from any hot surface. The total energy radiated per second by a star is called its luminosity, abbreviated "*L*."

In studying Wien's law, we saw that any warm body radiates. The higher the temperature is, the bluer the radiation. About a century ago the Austrian physicists Josef Stefan and Ludwig Boltzmann discovered another characteristic: *The higher the temperature, the more energy is radiated each second,* as shown in Figure 14-10. Although the subjects of their study were radiating objects in the laboratory, the law applies to stars and all other bodies in the universe.

The law gives the total energy L radiated per second from a body with temperature T and surface area A. For bodies that radiate efficiently, like stars, the law is:

$$L = \sigma T^4 A$$

Sigma (σ), called the Stefan–Boltzmann constant, equals 5.67×10^{-8} W/m²·K⁴. T is given in Kelvin, A in square meters, and L in the SI metric units of watts.

This equation shows that if the temperature or area of a star increases, the total energy radiated every second will increase. Suppose the star has some radius R; then its area is $A = 4\pi R^2$. Thus

$$L = 4\pi\sigma T^4 R^2$$

This variation of the Stefan–Boltzmann law allows us to calculate a star's radius if the luminosity and temperature are known.

Example. Suppose a certain B-type star is found to have a temperature of 18 000 K and a luminosity of 10 000 times that of the sun. What is its radius, relative to the sun's radius?

Solution. The straightforward way to do the problem is to solve the equation for R,

$$R = \sqrt{\frac{L}{4\pi\sigma T^4}}$$

look up the luminosity of the sun, $L_\odot$ in Appendix 2 (Table A2.2), multiply it by 10 000 to get L, and plug in the appropriate numbers. The answer, in meters, can be compared to the sun's radius, using the diameter listed in solar system data Table 6-1. Solve the problem this way for experience. However, also try the following simpler way. Since we asked for R relative to the sun, write the equation again for the sun, with solar subscripts on R, L, and T. Then divide it into the preceding equation to get the ratio of stellar R to solar R:

$$\frac{R}{R_\odot} = \sqrt{\frac{L}{L_\odot}\left(\frac{T_\odot}{T}\right)^4} = \sqrt{10\,000\left(\frac{5700}{18\,000}\right)^4} = 10.0$$

In this method the bothersome constants simply cancel out, since we are solving for the ratio, and we quickly see that the star is 10 times bigger than the sun.

and then the other. Your finger (the nearby star) seems to shift back and forth. Hold your finger only a few centimeters from your eyes; the shift is large. Hold your finger at arm's length; the shift is smaller. Likewise, the farther away the nearby star, the smaller the parallax. The parallax in this experiment may be measured in degrees, but the parallaxes of actual stars are all less than a second of arc.

Chapter 2 described how Aristotle correctly realized, around 350 B.C., that if the earth moved through space, nearby stars ought to show parallax. Seeing none, he concluded that the earth stood still. Actually, the shift was there, but too small for him to measure. Likewise, around 1600, Tycho looked for stellar parallax with the naked eye and found none. But he concluded (correctly) that annual movements of the earth were of small scale compared with stellar distances.

In 1837 the German–Russian astronomer Friedrich Struve published an analysis of stars that were expected to be closest to the earth (and therefore those most likely to have the largest and most easily detected parallaxes). He chose the brightest stars with the fastest angular motions across the sky, just as on a dark night one might sensibly infer that the closest fireflies are the ones that are brightest and move with the highest angular speeds. (These angular motions, though not apparent to the naked eye, can easily be measured from year to year with the telescope.)

Now the race was on to see who could measure the first true parallax. Credit is usually given to the German astronomer Friedrich Bessel, who in 1838 published the parallax and distance of 61 Cygni (star 61 in the constellation of Cygnus, the swan). Bessel found the parallax of 61 Cygni to be $\frac{1}{3}$ second of arc, and the distance to be about 3 pc. Most stars are farther than 3 pc away, and their parallaxes correspondingly smaller. Parallaxes as small as $\frac{1}{20}$ second of arc have been reliably measured, and the distance of such stars is 20 pc.

As the preceding figures show, the distance of a star in parsecs is simply the inverse of its parallax angle in seconds of arc. This explains the origin of the term parsec: It is the distance corresponding to a PARallax of one SECond of arc.

The more distant a star, the smaller its parallax. Thus, if a star is too distant, its parallax is too small to be measured. Parallaxes smaller than about $\frac{1}{20}$ second of arc are difficult to measure accurately. Therefore *stars farther away than about 20 pc are beyond the* **distance limit for reliable parallaxes**, although distances from 20 pc to 100 pc can be roughly estimated by the parallax technique.

Knowing accurate distances is the prime requirement of measuring most other properties of stars. Thus, the estimated 1000 to 2000 stars that lie within 20 pc are our main statistical sample for measuring stellar properties. Other techniques have been devised for estimating distances to more remote stars, but they all depend ultimately on the accuracy of the parallax measures of nearby stars. (For example, see Figure 14-12, which applies the concept of absolute magnitude, or luminosity, as discussed further in the following.)

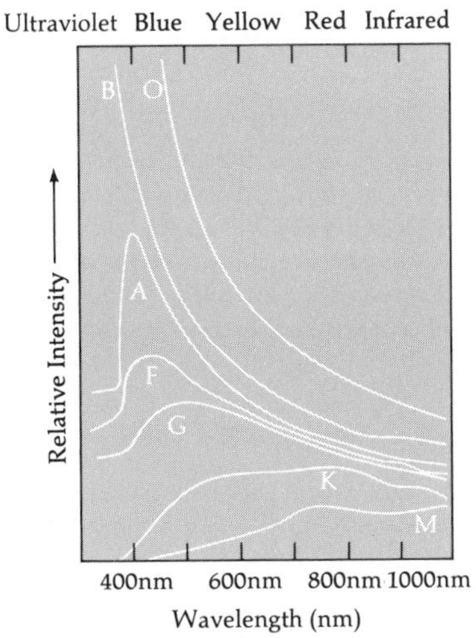

Figure 14-10. ***A schematic comparison of the energies emitted by various stars. Curves show the intensity of radiation in the continuous spectrum of each star, neglecting absorption lines. The hottest and brightest are O stars, which emit most of their energy as blue and ultraviolet light. G stars, like the sun, emit most of their radiation in the visible part of the spectrum, especially as yellow light. The faintest stars, class M, emit most light in the red and infrared part of the spectrum. Absorptions cause some irregular structure in the curves. Compare similar diagram for idealized radiator at different temperatures, Figure 7-3.***

HOW TO MEASURE A STAR'S LUMINOSITY

The **luminosity** of a star is its *absolute brightness*—the total amount of energy it radiates each second. Unlike apparent

Table 14-3. *Basic Properties of Stars and How They Are Measured*

Property	Techniques of Measurement
Distance	Trigonometric parallax
Luminosity (absolute magnitude)	Distance combined with apparent brightness
Temperature	Color or spectra
Diameter	Luminosity and temperature
Mass	Measures of binary stars and use of Kepler's laws
Composition	Spectra
Magnetic field	Spectra, using Zeeman effect
Rotation	Spectra, using Doppler effect
Atmospheric motions	Spectra, using Doppler effect
Atmospheric structure	Spectra, using opacity effects
Circumstellar material	Spectra, using absorption lines and Doppler effect
Motion	Astrometry or spectra, using Doppler effect

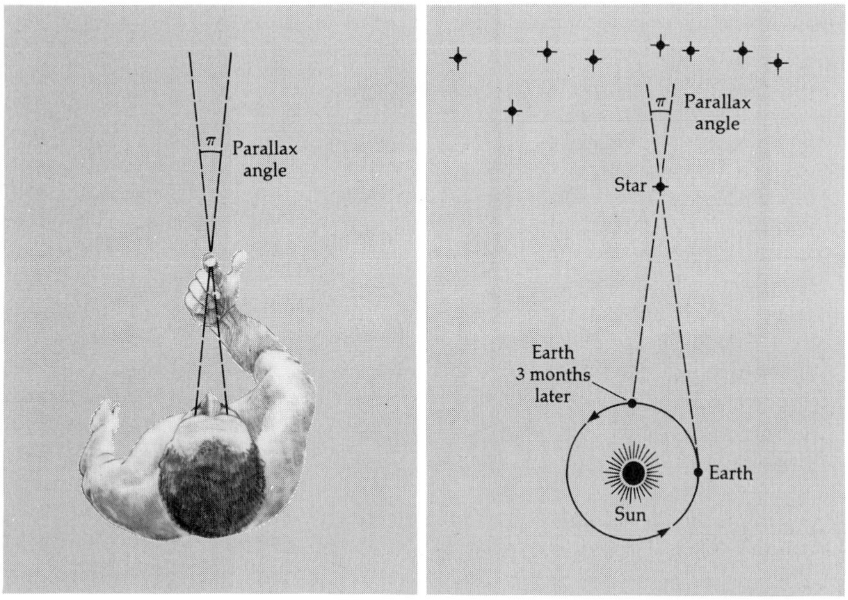

a b

Figure 14-11. *The principle of parallax determination, applied to one's finger (*a*) and to stars (*b*). Nearby objects observed from two positions appear to shift by a parallactic angle π compared with background objects; measurement of π gives the distance of the object, once the separation of the two positions is known. (By astronomical convention, the angle cataloged as a star's parallax corresponds to the angle diagrammed here; but astronomers actually observe from two positions 6 months apart. This yields a larger parallactic shift, which is easier to measure.)*

brightness (discussed earlier in this chapter), luminosity is intrinsic to a star. The most useful concept of luminosity is **bolometric luminosity**—*the total amount of energy radiated each second in all forms at all wavelengths.*

A second concept sometimes used is *visual luminosity*—the energy radiated each second in the visible part of the spectrum (the wavelengths to which the human eye is sensitive). Since many stars radiate primarily visible light, the two kinds of luminosity, bolometric and visual, are often roughly the same. In this book *luminosity* will generally mean *bolometric luminosity,* abbreviated as *"L."* Similarly, our use of the term *absolute magnitude* will generally refer to *absolute bolometric magnitude,* a measure of the bolometric luminosity expressed in the magnitude system.

The bolometric luminosity of the sun ($L_\odot$) *is* 4×10^{26} *watts (note: one watt = 1 joule of energy radiated each second).* Many stars, of course, are much brighter than the sun. Thus, a star twice as luminous as the sun, radiating 8×10^{26} watts, would be said to have a luminosity of $2\ L_\odot$.

The most basic method of estimating luminosity derives from the measurement of distance. A faint light in the night may be a candle a hundred meters away, a streetlight a few kilometers away, or a brilliant lighthouse beacon 100 km away. Once we know the distance to an object, we can determine its absolute brightness. For example, Figure 14-12 shows that a 1st magnitude star (that is, apparent magnitude $m = 1$) 10 pc away would have an absolute magnitude of +1 and a luminosity of about $40\ L_\odot$.

This method becomes less accurate for very distant stars. For one thing, as mentioned earlier, parallax distance measures become unreliable beyond about 20 pc. More important, there is some interstellar haze that dims starlight over larger distances. Just as a fog bank might keep you from telling whether a distant light was a streetlight or a more distant lighthouse, variable interstellar haze in uncertain amounts complicates estimates of stars' luminosities.

However, other methods are available. For example, very luminous stars have slightly different spectral characteristics than faint stars of the same spectral class. Spectra can therefore be used to measure luminosities as shown in Figure 14-13.

Note that the luminosity, the distance, and the apparent magnitude of any object are all interrelated. As shown in Figure 14-12, if we know any two of these quantities, we can estimate the third.

HOW TO MEASURE A STAR'S TEMPERATURE

Here again spectra come to our aid. Stellar temperatures are measured merely by applying Wien's law. By measuring the spectrum, we can see which color is the most strongly radiated (see Figure 14-10). Application of Wien's law gives the temperature. This is often called the **effective temperature**, because it is the effective average temperature of the various hotter and cooler layers of gas from which light reaches us.

As mentioned earlier, the temperature is also indicated by the relative numbers of atoms in different excited states, derived by the relative strengths of lines and use of the Saha equation.

Temperatures range all the way from about 2500 K for the cool M-type stars to 40 000 K or more for hot O-type stars.

HOW TO MEASURE A STAR'S DIAMETER

The diameters of only a handful of very large stars can be measured directly from images such as Figure 14-3. Most diameters are measured from temperature and luminosity by applying the Stefan–Boltzmann law. By this law, a star radiating a certain number of watts and having a certain temperature must have a certain area *A.* The values can be inserted in the Stefan–Boltzmann equation and the equation solved for *A.* The star's diameter can

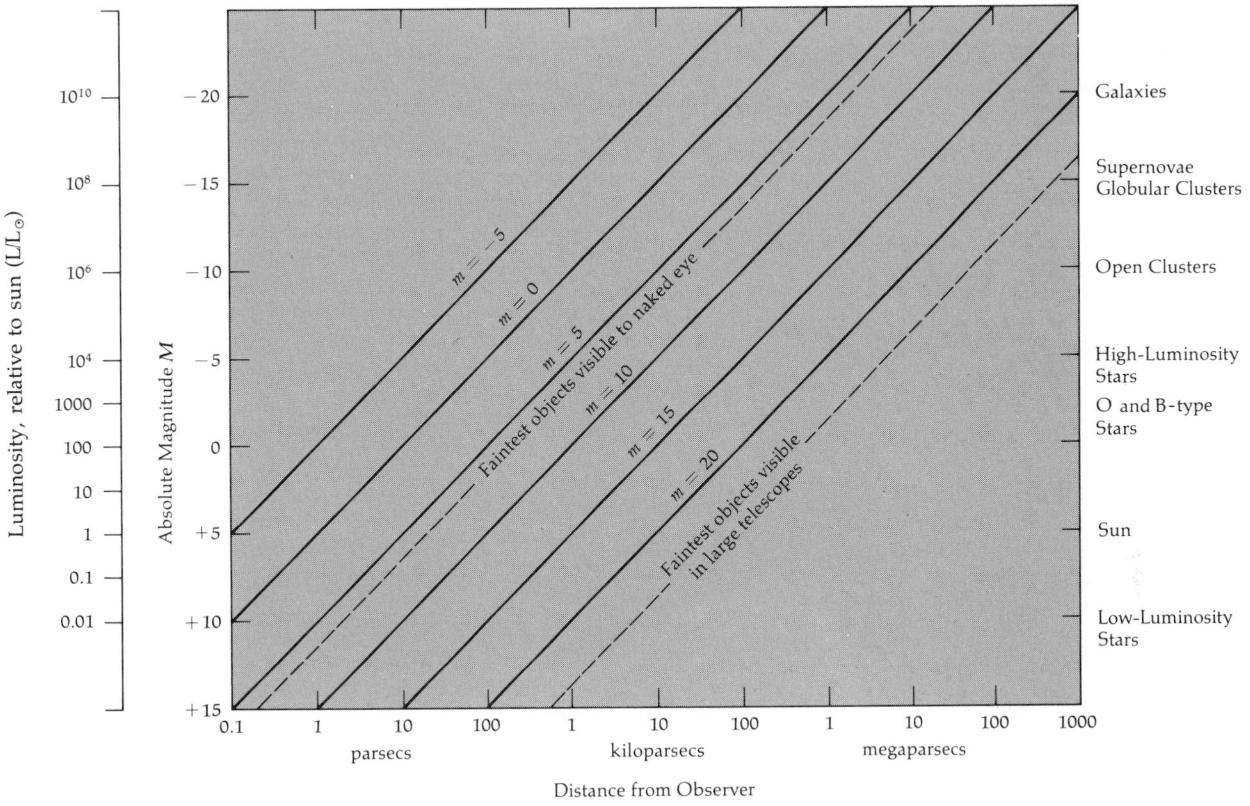

Figure 14-12. *If a certain star has a known absolute magnitude* M *(vertical scale, possibly determined from spectra or other means), astronomers can measure its apparent magnitude* m *and estimate its distance from a chart like this one. For example, a high-luminosity star of* M = −5 *could be visible to the naked eye with visual magnitude* m = 5 *at a distance of around 1000 pc. Note that galaxies (see right end of graph) are so luminous that some can be visible to the naked eye at distances of 1 Mpc, or 1 000 000 pc. The values shown here assume clear space; see Chapter 18 for a discussion of dimming due to interstellar "smog" in many regions of space.*

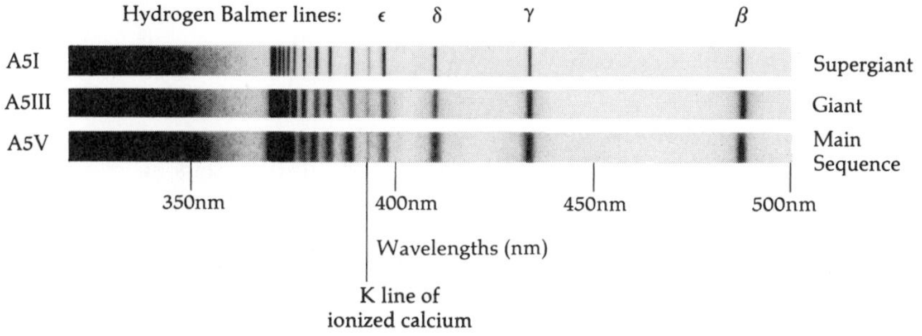

Figure 14-13. *Comparison of spectra of a supergiant, giant, and main-sequence star of constant spectral class. In the larger stars, pressures are less and atoms less disturbed by collisions, resulting in slightly narrower, better-defined absorption lines.*

be found from its area. Such measures indicate a vast range of stellar diameters, from less than earth size, through sun size, to huge stars bigger than the diameter of Mars' orbit!

HOW TO MEASURE A STAR'S MASS

According to Kepler's laws as modified by Newton, when any two cosmic objects are in orbit around each other, the period of revolution (the time it takes to complete an orbit) increases as the distance between the objects increases and as the sum of the masses of the two objects decreases. Many pairs of stars orbit around each other. For many of these *binary stars* we can measure both the period of revolution and the distance between the two stars (binary stars are discussed in further detail in Chapter 19). Thus we can calculate the sum of the masses, which is sometimes designated $m_A + m_B$, where A and B designate the two stars.

But we want to know each individual mass, not the sum of the two. Newton showed that in a system of orbiting bodies, each body orbits around an imaginary point called the **center of mass**, and that by measuring the distance of each star from the center of mass, we can measure the ratio of the masses.

Now we have both the sum of the masses and the ratio of the masses. From this information, we can get each individual mass. For example, take the double star Sirius, consisting of the bright star Sirius A (which has the greatest apparent brightness of any star in the night sky) and the faint star Sirius B. The preceding procedure shows that the sum of the masses of A and B is 3 solar masses ($3\ M_\odot$). The ratio indicates that A is twice as massive as B. From these facts, the only possible solution is that Sirius A has a mass of 2 $M_\odot$, and Sirius B, 1 $M_\odot$.

Study of many stars by this technique reveals an important fact: *Stars with nearly identical spectra usually have nearly identical masses.* This fact allows the masses of many stars to be estimated roughly from their spectral properties alone.

HOW TO MEASURE A STAR'S COMPOSITION

As described earlier, compositions are revealed by spectra. There are two problems: detecting the *presence* of an element, and measuring its *amount*. The presence of an element is detected by identifying at least one—or preferably several—of its absorption lines or emission lines in the spectrum of the star.

The amount of the element is indicated by the appearance of the spectral line. Generally, the wider and darker the absorption lines, the more atoms of the element are present. Likewise, the wider and brighter the emission lines, the more atoms. These properties are said to measure **spectral line strength**: The stronger the line, the more of the element is present. The *relative* strengths of different lines also depend on temperature, pressure, and other conditions in the photospheres of stars, as described by the Saha equation.

Astrophysicists have developed techniques to untangle the effects of composition, pressure, etc. These techniques allow astronomers to gain information from spectra not only about composition but also about the temperature, pressure, and other properties of stars. Studies of this type around 1925 by Cecilia Payne-Gaposchkin and others showed that most nearby stars are approximately sunlike in composition.

HOW TO MEASURE A STAR'S MAGNETIC FIELD

Spectral absorption and emission lines divide into two or more close lines in the presence of a magnetic field, a phenomenon called the *Zeeman effect*. The stronger the field, the greater the splitting. Measurements of this effect reveal that stars like the sun typically have fields about as strong as the sun's, but many stars have fields thousands of times stronger.

HOW TO MEASURE A STAR'S ROTATION

If a rotating star is seen from any direction except along the rotation axis, one edge will be approaching the observer and one edge will be receding (Figure 14-14). Light emitted or absorbed at the approaching edge will be blue-shifted. Light from the other edge will be red-shifted. Consider an absorption line being formed in all parts of the star's atmosphere. If the star were not rotating, neither shift would occur and the line would be very narrow. Because the star rotates, the line is broadened. The faster the rotation and the closer our line of sight to the equatorial plane, the more line broadening occurs.

The rotations of stars can thus be inferred to some extent from measurements of the broadening of spectral lines. In general, the fastest rotators are the hot, bluish, massive stars of class O, which have equatorial rotation speeds of around 300 km/s. Slightly cooler stars rotate more slowly, with a sharp break in the trend occurring at stars a little hotter and more massive than the sun (class F). Stars cooler than F stars rotate much more slowly. The sun's equatorial speed is only 2 km/s.

HOW TO MEASURE A STAR'S ATMOSPHERIC MOTIONS

If masses of stellar gas rise and fall in convection cells (as they do in the sun), the various cells will display a range of approach and recession velocities. Therefore, blue and red Doppler shifts (respectively) would occur, broadening the stars' spectral lines, as shown in Figure 14-15. In practice, astronomers have trouble distinguishing this effect from rotational line broadening, though some discrimination is possible. Especially strong turbulence has been found in the atmospheres of certain cool, large-diameter stars known as red giants. In these stars, masses of gas rise and subside with speeds as high as 40 km/s.

We have seen how the Saha equation allows us to derive conditions in a gas producing a given set of spectral lines. Now suppose that we could find different sets of lines arising in different layers at different depths in the star. Then we could analyze the conditions at these different levels.

HOW TO MEASURE A STAR'S ATMOSPHERIC STRUCTURE

As noted in the last chapter, the **opacity** of an atmosphere determines how far light of a specified wavelength can penetrate into that atmosphere. Generally, a stellar atmosphere has a different opacity at each wavelength. If the opacity at a certain wavelength is low, we can see deep into the star's atmosphere at that wavelength. If the opacity at another wavelength is high, we can see only into the uppermost atmosphere. (An example of this principle is Figure 7–12b, where we look at Earth in the high-opacity wavelengths absorbed by water vapor, and hence see mostly haze layers well above the surface.) Applying this principle by sophisticated methods, astronomers gain information on pressure and other conditions at all levels of stellar atmospheres.

HOW TO DETECT CIRCUMSTELLAR MATERIAL

Unstable stars, such as newly forming or dying stars, may throw out clouds of gas and dust (see Figure 14-16). These surround the stars and are called **circumstellar nebulae**. Because many of these clouds are nearly transparent (as with flame), we can see light arising from both the near side and far side of the cloud, as shown in Figure 14-17. As such a cloud expands away from the star, absorption lines arising on the near side (where the starlight passes through the gas) are blue-shifted compared with lines in the star because the gas is approaching the observer. Emission lines from glowing gas on the near side are also blue-shifted, but emission lines from the far side of the cloud are red-shifted.[6] These shifted lines

[6]For a discussion of how such spectral lines are produced, see Kirchhoff's third law, Chapter 13.

***Figure 14-14.** Light from a rotating star would be blue-shifted if from region A, red-shifted if from region B, or unshifted if from the central region C. The net result is a broadening of spectral lines.*

***Figure 14-15.** Currents of descending and ascending gas in a star's atmosphere cause Doppler shifts that broaden spectral lines.*

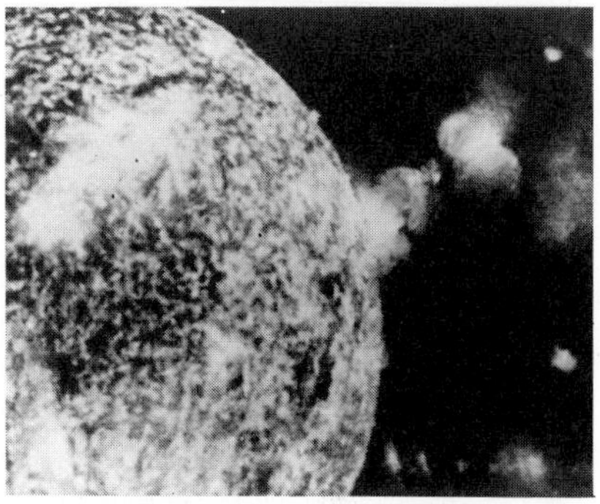

***Figure 14-16.** Spectroheliograph photo in Hα light shows a cloud of material blasted off the sun. Some other types of stars blast off even larger clouds that can be detected spectroscopically, as described in the text. (NASA.)*

are telltale clues to expanding circumstellar nebulae.

In gas around certain dying stars, expansion at speeds of 1000 to 3000 km/s has been observed. This gas is believed to have been blown off the star in explosions related to the exhaustion of the star's energy supplies.

In certain newly formed stars circumstellar material has been found not in the form of glowing gas, but in the form of dust grains. These are warmed by the star and detected by the thermal infrared radiation they give off (similar to that shown in Figures 7–3 and 7–4).

HOW TO DETECT A STAR'S MOTION

The Doppler shift is a very simple way to detect *part* of a star's motion—its **radial velocity**, or *motion along the line of sight.* A consistent blue or red shift of all of a star's spectral lines proves that the star is moving toward or away from us. Among the 50 nearest stars, about half the radial velocities measured are more than 20 km/s toward or away from us.

The other component of a star's motion is its **tangential velocity**—the *motion perpendicular to the line of sight.* It cannot be measured as simply as radial velocity. In order to measure tangential velocity, we must measure the distance of the star and its rate of angular motion across the sky, called **proper motion.** (Values of a few seconds of arc per year are common for nearby stars.) These kinds of measurements are often lumped together in a special branch of astronomy called **astrometry.**

If both radial and tangential velocities are known, they can be combined to give the star's **space velocity**: its true speed and direction of motion in three-dimensional space *relative to the sun* (Figure 14-18). Space velocities of most stars near the sun are a few tens of kilometers per second and are nearly random in direction.

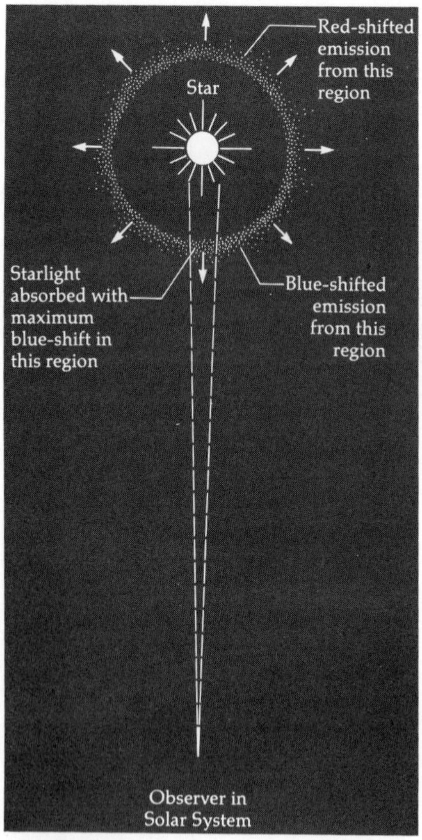

Figure 14-17. *Gaseous envelopes expanding (as shown here) or rotating around stars produce characteristic spectral effects, allowing their detection even if they cannot be seen separately.*

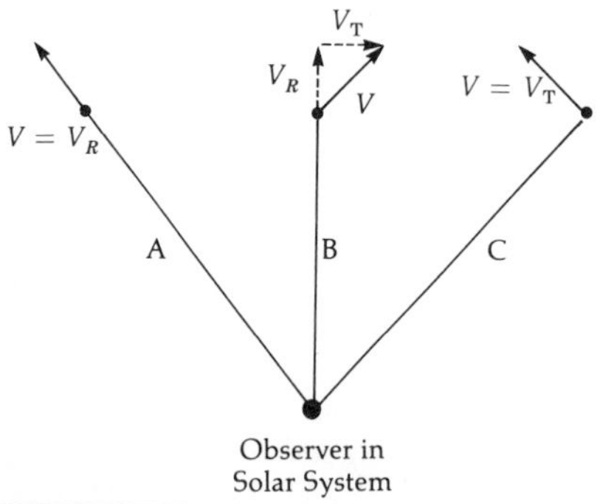

Figure 14-18. *Three examples of a star's space velocity (V) compared with its radial velocity (V_R) and tangential velocity (V_T). In A, the star's space velocity is aligned with the line of sight; no proper motion would be seen. B is the most common case, in which both V_R and V_T are appreciable. In C, the space velocity is perpendicular to the line of sight; no radial velocity or Doppler shift would be seen.*

The Importance of Telescopes in Space

Now we can see more clearly the tremendous advantage of spaceborne telescopes, as opposed to ground-based telescopes. Figure 14-19 is a schematic diagram of several stars' spectra. The top box, which portrays the solar spectrum, shows that although most solar radiation falls in the part of the spectrum that can reach the ground, parts lie in the ultraviolet and infrared regions that can be studied only from above the atmosphere. But now look at the case of the extremely hot and cool stars in the middle box. Most of their radiation, especially the important regions of most intense emission, can be studied only by a telescope above the atmosphere. The bottom box represents an exciting 1983 discovery about the star Vega, made with the orbiting IRAS (infrared astronomical satellite) telescope. Before 1983, Vega seemed a solitary star somewhat hotter and more massive than the sun. The IRAS telescope revealed far infrared radiation that (according to Wien's law) must be emitted by cool material at only 85 K (−306°F). This is too cold for a star and is believed to be a ring of dust about 80 AU from Vega (Neugebauer, et al., 1984). From the intensity of the infrared radiation, the total mass of dust is believed

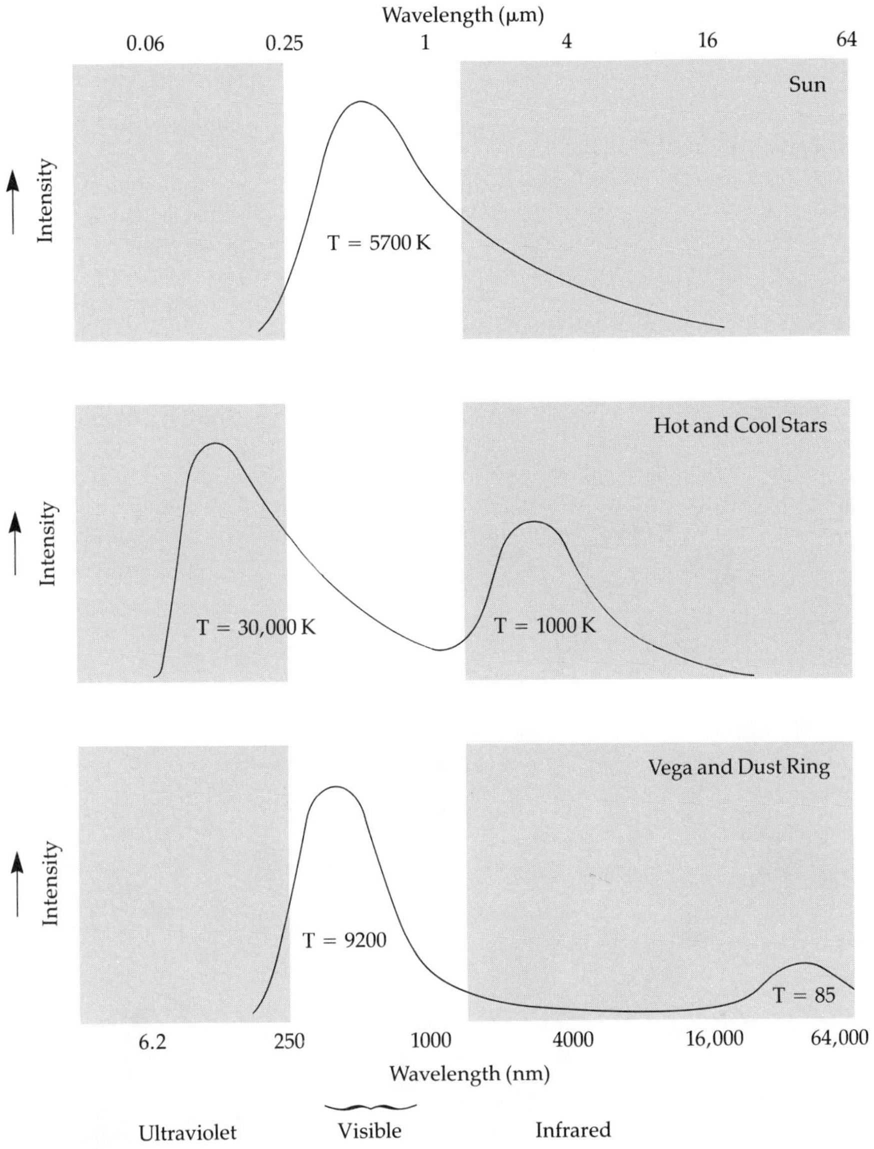

Figure 14-19. *Schematic spectra of several objects, showing advantages of telescopes above the atmosphere. Grey-shaded wavelengths are absorbed by the atmosphere and inaccessible to ground-based telescopes. See text for further discussion.*

comparable to the mass of asteroids or comets in our own solar system. Some astronomers believe this discovery is a new line of evidence of planetary material around other stars (Weissman, 1984).

These examples show why existing and future space telescopes will create new advances in our understanding of hot material, cold material, extremely energetic events, planet formation, and a host of other cosmic phenomena associated with other stars. Many astronomers expect a revolution in our understanding during the next decade as a result of their use.

Summary

This chapter emphasizes methods of studying stars, rather than the nature of the stars themselves. The chapter has two basic sections. The first section describes concepts: the system of naming stars; the photographic images of stars; the definitions of *light-year* and *parsec;* the system of magnitudes for measuring brightness; stellar spectra; the Doppler effect; and the Stefan–Boltzmann law, which describes the amount of stellar radiation.

The second section describes methods for determining twelve basic stellar properties, as summarized in Table 14-3.

Clearly, our journey outward from the earth has taken us far beyond the places that we can investigate in the near future by manned or instrumented flights. Even at the speed of light (which is 10 000 times greater than speeds our instruments have achieved), it would take years to reach the nearest stars. Thus, instead of sending instruments to the stars, we must rely on the messages in the starlight coming to us. Our abilities to interpret starlight are rapidly increasing, not only with construction of new large telescopes, but also with planned launching of such telescopes into orbit by means of the Space Shuttle (as shown in Figure 14-20). As mentioned in Chapter 7, the Hubble Space Telescope will be the largest such instrument, a 2.4-m (94-in.) telescope. Its launch was planned in 1986 but was delayed by the Challenger disaster. The Hubble Space Telescope is now ready to go and at this writing is scheduled to be launched in late 1988. Many astronomers expect it to revolutionize our view of the universe, gathering starlights' messages with unprecedented clarity (Bahcall and Spitzer, 1982; see also *Physics Today,* November 1982, for reviews of new astronomical instruments). In the coming chapters, we will interpret what those messages to date have told us about the character and histories of the stars that surround us.

Figure 14-20. *A new technique for studying stars and other remote objects is to put large telescopes into space. Two major advantages over earth-based telescopes would be freedom from the blurring caused by atmospheric shimmering, and freedom from absorption of ultraviolet and infrared wavelengths by atmospheric gases. (NASA artist's conception.)*

Concepts

star	Saha equation
diffraction	temperature
Airy disk	Doppler effect
light-year	blue shift
parsec	red shift
apparent brightness	Stefan–Boltzmann law
absolute brightness	parallax
apparent magnitude	distance limit for reliable parallaxes
visual apparent magnitude	luminosity
absolute magnitude	bolometric luminosity
spectrum	effective temperature
wavelength	center of mass
spectroscopy	spectral line strength
spectrograph	opacity
spectrometer	circumstellar nebula
absorption line	radial velocity
emission line	tangential velocity
continuum	proper motion
spectral class	astrometry
hydrogen Balmer series	space velocity
hydrogen alpha (Balmer alpha)	

Problems

1. If a telescope of 25-m (1000-in.) aperture could be put in orbit or on the moon, it could resolve an angle of only about 0.005 seconds of arc. Compare this resolution with the maximum angular size known for stars (0.03″). Why would this telescope perform better on the moon than on the earth? Why do astronomers never use the highest magnifications theoretically possible with earthbound telescopes?

2. A star is 20 pc away. How many years has its light taken to reach us?

3. What is the chemical composition of most stars?

4. A reddish star and a bluish star have the same radius. Which is hotter? Which has higher luminosity? Describe your reasoning.

5. A reddish star and a bluish star have the same luminosity. Which is bigger? Describe your reasoning.

6. How does the magnitude scale differ from the scales by which we measure lengths (centimeters or inches), temperature (degrees Celsius or Kelvin), weight (kilograms or pounds), or other familiar properties?

7. A certain star has exactly the same spectrum as the sun, but is 30 magnitudes fainter in apparent magnitude. List some conclusions you would draw about it and describe your reasoning.

8. A star has a parallax of 0.05 seconds of arc. How far away is it?

9. In a binary pair of stars, the sum of the masses is found to be 5 $M_{\odot}$; the stars are equidistant from their center of mass. What are their individual masses?

10. The spectral lines of metals, such as calcium, are prominent in the solar spectrum. Hydrogen lines are less prominent. Why does this not indicate that the sun consists mostly of these elements instead of hydrogen?

11. Each of a certain star's spectral lines is found to be smeared out over a wide range of wavelength. What might cause this?

ADVANCED PROBLEMS

12. How many times brighter is a daylit landscape than the same landscape lit at night by the full moon (*Hint:* See Table 14-1; note that 5 magnitudes equals a factor of 100 in brightness; 1 magnitude, a factor of 2½.) If your camera needs a 1/100-s exposure in daylight, what exposure might record the moonlit scene? (Actually about twice the calculated exposure will be better, because film sensitivity falls off in weak light.)

13. A certain star has a spectrum similar to the sun's, but all the spectral lines are shifted 0.1% of their wavelength toward the red.

a. What do you conclude about this star's motion toward or away from the observer?

b. Is its speed in this direction unusually fast, average, or slow?

14. A star 10 pc away is observed to have a proper motion of 1 second of arc per year. Use the small-angle equation to derive its tangential velocity.

15. If the stars in Problems 13 and 14 were the same, what would the space motion be like?

16. A star has the sun's luminosity (4×10^{26} J/s), but a temperature of only 2850 K. How big is it?

17.a. Using Optional Basic Equation VI, prove that the mean energy of atoms and ions in a star will increase as the temperature increases. (*Hint:* See discussion of Optional Basic Equation VI.)

b. Discuss how this explains the pattern of increasing ionization as temperature increases in the spectral sequence of stars.
c. In previous problems we have used a rule of thumb that the fastest atoms of any element move more than three times faster than the typical velocity of the same element in the same gas. Note that in this case they would have nine times as much kinetic energy. Assume that these fastest atoms dominate in ionizing a mass of hydrogen gas and calculate the temperature at which these fastest atoms would have enough energy to ionize adjacent hydrogen atoms. As mentioned in the text, the ionization energy for ground state hydrogen is about 2×10^{-18} joules.
d. Comment on where a star of such temperature would fall in the sequence of spectral classes, and show that hydrogen is indeed ionized in hotter stars. (Note: The model for ionization in problems c and d is overly simplistic. First, ionization in a star is also caused by photons of ultraviolet radiation, as well as thermal jostling by neighboring atoms. Second, the percent of atoms ionized depends on gas density and pressure, as well as temperature, as worked out by Saha in 1920. Nonetheless, these problems give a rough idea of how ionization sets in as temperature increases in a gas, due to increasing energy of the gas particles.)

Project

1. Using a telescope of fairly high magnification (such as 300x or 400x), examine the image of a bright star on several different nights and sketch it. Are there differences from night to night? Can you identify the Airy disk, diffraction rings, or diffraction spikes? If so, label them on your sketches. Note any shimmering due to atmospheric turbulence or air currents in and around the telescope. Run the eyepiece inside or outside the focus point, making the image a round blob. Shimmering and other turbulent effects are often more evident in this way.

CHAPTER 15

THE SYSTEMATICS OF NEARBY STARS: THE H–R DIAGRAM

As soon as astronomers could begin to measure different properties of stars, they began to categorize the stars. This chapter will outline three interesting discoveries that resulted. First, stars come in a wide variety of forms: massive and not so massive, large and small, bright and faint. Second, the reason for the different forms is that stars evolve from one form to another. Third, stars of different initial mass evolve at different rates and into different final forms.

In this sense, the population of stars can be likened to that of people. If you travel from country to country and observe people, you tend to see mostly ordinary people—youngsters, middle-aged people, and elderly people. The creation of new people—the birth of babies—is secluded in special places such as hospitals and is not readily visible to the tourist. In addition, the tourist may see tombstones, but the actual deaths of people are usually brief events not commonly encountered. This metaphor explains the approach of this chapter and the next two chapters. In this chapter we will tour the local population of stars. We will find that the great majority of these stars are ordinary stars—that is, youthful, middle-aged, or growing elderly. The next chapter will show how stars come into existence and cite some observed examples of star birth. Chapter 17 will examine some of the extraordinary phenomena of stellar old age.

Classifying Star Types: The H–R Diagram

When astronomers began to determine properties of stars by using the techniques described in the last chapter, they found a puzzling variety of star forms. Masses range from about 0.1 to 60 $M_{\odot}$. One star in a billion is even more massive, and a few dazzling stars hundreds of times more massive than the sun are believed to exist (Humphreys and Davidson, 1984). Luminosities range from around one-millionth to a million times that of the sun. Temperatures of surfaces range from about ⅓ to nearly 10 times the solar temperature. Sizes range from less than 1/100 to more than 100 times the sun's size.

When astronomers realized that there is such a variety of star forms, their first step was to devise a sensible way to arrange and study the data about them, in hopes of finding relationships among the various forms of stars. This could have been done in various ways, but one method has become traditional. This method was introduced around 1905 to 1915 by Danish astronomer Ejnar Hertzsprung and American astronomer Henry Norris Russell. They constructed diagrams plotting the spectral class of stars versus their luminosity, as shown in Russell's version of the diagram (Figure 15-1), published in 1914. This type of plot is usually called the **H–R diagram**, in honor of Hertzsprung and Russell.[1]

Even the earliest H–R diagram revealed an important discovery about stars: *Among most stars there is a smooth relation between spectral class and luminosity,* indicated by the dashed lines on Figure 15-1. This band came to be called the *main sequence* of stars. Hertzsprung called stars obeying this relation (falling on the diagonal band of the H–R diagram) **main-sequence stars.** Russell correctly argued that the smooth relation between spectral type and luminosity meant that *the principal differences in stellar spectra, however they may originate, arise mainly from variations in a single condition in the stellar atmosphere.*

As we saw in the last chapter, *temperature* is the principal factor that governs

[1] Sometimes this graph is called the spectrum–luminosity diagram or color–magnitude diagram, because plots of spectral type versus luminosity or color versus absolute magnitude produce equivalent information.

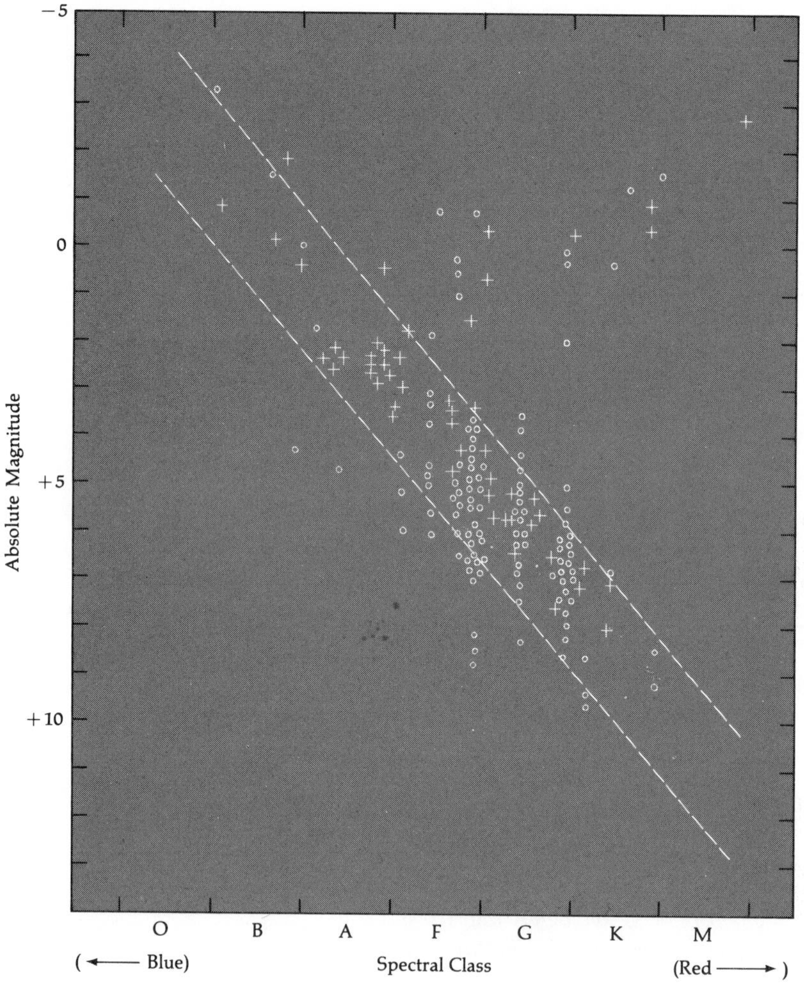

Figure 15-1. *The first H–R diagram, redrawn from Russell's original 1914 publication. Dots represent stars with directly measured parallaxes; crosses represent estimated data from stars in four clusters. Dashed lines enclose Russell's identification of the main sequence.*

the differences between spectra of O, B, A, F, G, K, and M stars. For this reason, many astrophysicists make H–R diagrams by plotting luminosity against temperature. (Following Russell's original version, the spectral classes are arranged with *cooler* stars to the right.) In this book, we identify temperature, spectral class, luminosity, and absolute magnitude along the edges of the diagram, as an additional aid to interpretation.

As shown in Color Plate 32, different locations on the H–R diagram correspond to different types of stars. It is important to understand why this is true. First, remember from Wien's law that stars of different temperature have different color. Thus, as shown schematically in Color Plate 32, the right part of the diagram corresponds to redder, cooler stars, and the left part, to bluer, hotter stars.

Next, note that the sun is near the middle of the diagram (luminosity 1, spectral type G). If we move upward from the sun's position, we encounter stars more luminous than the sun, even though they have the same temperature. Thus, these stars must be bigger than the sun. (The reasoning here is as follows: The Stefan–Boltzmann law, p. 266, assures us that each square meter on a star at the same temperature as the sun will radiate as much energy as a square meter on the sun. Thus, to get more total luminosity, we must have more square meters of surface area and hence larger size.) Similarly, a star below the sun's position must be smaller than the sun.

Do stars appear *throughout* the diagram? Or are there only certain types such as main-sequence stars, giants, and dwarfs? If so, why? The H–R diagram allowed astronomers to investigate these questions.

The Nearby Stars as a Representative Sample of All Stars

In keeping with the pattern of expanding horizons in the cosmic journey of this book, let's start our discussion of stars with the stars nearest the sun. Figure 15-2 shows a modern H–R diagram for the 100 stars closest to the sun. Stars at this range are close enough for us to measure accurate distances and to detect even very faint examples. They are also numerous enough to provide a good sample. Stars in such a random sample are believed to be **representative stars**, that is, a representative sample of all stars in our general neighborhood of the galaxy. The figure indicates that about 93% of these stars fall on the main sequence. This shows the importance of the main sequence: *It includes the great majority of stars.* The sun symbol plotted at spectral class G, luminosity 1, indicates that the sun is brighter than most representative stars.

Table 15-1 lists more detailed properties of the 26 stellar systems within 4 pc of the sun. This table reveals another interesting fact about representative stars: Most of them are inhabitants of systems in which two or three stars orbit around one another. Stars without stellar companions, like the sun, are in the minority.

The closest star system to the sun is the triple system of **Alpha Centauri,** brightest star in the southern constellation of Centaurus. Shown in Figure 15-3, Alpha Centauri is easily visible from Hawaii, but is too far south to be prominent from most of the rest of the United States. Because Alpha Centauri is so close and so bright, we can accurately measure the distances, spectral classes, temperatures, luminosities, and sizes of the components. By measuring the orbital size and period, we can use Kepler's laws of orbital motion and Newton's universal law of gravitation to deduce the masses of the components. The main two components are Alpha Centauri A and B, which are very similar to the sun and located 1.33 ± 0.01 pc away. These two components orbit around each other in 80 y and are easily visible as a pair—yellowish and orange—in a small telescope. In the 1800s, telescopic observer Sir John Herschel called them "beyond all comparison the most striking object of the kind in the heavens." They reached greatest separation as seen from the solar system (22 seconds of arc) in 1980 but will be only 4 seconds of arc apart in 2015.

Orbiting around A and B once every 1½ million years is a small faint star with only ⅒ the sun's mass. It happens to be on our side of the system at a distance of only 1.29 pc (Kamper and Wesselink, 1978). Because it is the sun's closest neighboring star, it is called Proxima Centauri (from the same root as *proximity*).

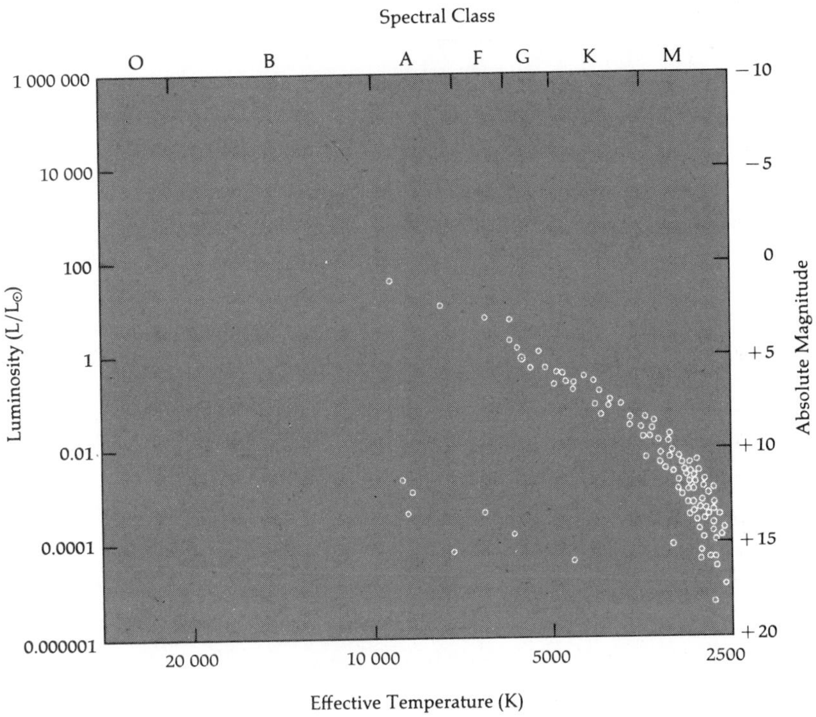

Figure 15-2. *An H–R diagram for the 100 nearest stars—a representative sample from a volume of space around the sun. This diagram shows that most representative stars are on the main sequence. Also note that most are fainter and cooler than the sun. Therefore they fall in the lower right corner.*

Representative Stars Versus Prominent Stars

If we try to expand our statistical sample of stars by tabulating more distant ones, we run into a problem. At great distances the less luminous stars, like those at the bottom in Color Plate 32, are too faint to see. The only distant stars we see are unusual, luminous ones. In fact, the **prominent stars** in our night sky are mostly distant, unusually luminous stars. As we learned from Figure 15-2, these stars are quite rare in three-dimensional space. Hertzsprung aptly called them the "whales among the fishes." Thus, the statistics of distant stars do not give us a sample of representative stars. Rather, they are biased toward the prominent "whales." Nonetheless, they are important because they reveal that some of the "whales" don't belong to the main sequence.

NON-MAIN-SEQUENCE STARS

Figure 15-4 adds the 100 most prominent stars in our sky (as ranked by apparent magnitude) to those shown in Figure 15-2. Now we see quite a different pattern because the prominent stars that *look* brightest in our sky are a nonrepresentative sample. They include not only nearby ones but also the very luminous ones—"whales" that are far away but intrinsically very bright. Thus the upper part of the H–R diagram in Figure 15-4 is much more populated than the upper part of Figure 15-2.

As shown in Figures 15-1 and 15-4, certain K and M stars have much higher luminosity than the main-sequence K and M stars. That is, these stars radiate more light than others of the same temperature. According to the Stefan–Boltzmann law, they must have more area. Hertzsprung named them **giant stars.** Most are called *red giants,* since they are reddish K and M stars. Color Plate 32 (top) makes the situation clearer by cartooning the relative brightnesses and sizes of these stars.

As more data came in, other stars appeared on the H–R diagram below the main sequence. By the same logic used for giants, these stars were recognized as unusually small. They came to be called

Table 15-1. *The Nearest 26 Stellar Systems*

Distance[a] (pc)	Star Name	Component	Apparent Magnitude (m_v)	Absolute Magnitude (M_v)	Spectral Type	Mass ($M_\odot$)	Radius ($R_\odot$)	Semi-major Axis in Multiple Systems (AU)
0.0	Sun (and Jupiter	A	−27	5	G2	1.0	1.0	5.2
		B	—	—	—	0.001	0.1	
1.3	Alpha Centauri	A	0	4	G2	1.1	1.2	23.6 (AB)
		B	1	6	K0	0.9	0.9	
		C	11	15	M5	0.1	?	13 000 (AC)
1.8	Barnard's Star	—	10	13	M5	?		1.3
2.3	Wolf 359	—	14	17	M8	?	?	—
2.5	BD +36°2147	A	8	10	M2	0.35	?	0.07
		B	?	?	?	0.02	?	
2.6	L726-8	A	12	15	M6	0.044	?	10.9
		B	13	16	M6	0.035	?	
2.7	Sirius	A	−2	1	A1	2.3	1.8	19.9
		B	8	11	Wh. dwarf	1.0	0.02	
2.9	Ross 154	—	11	13	M5	?	?	—
3.1	Ross 248	—	12	15	M6	?	?	—
3.3	ϵ Eridani	—	4	6	K2	0.98	?	—
3.4	Ross 128	—	11	14	M5	?	?	—
3.4	L789-6	—	12	15	M6	?	?	—
3.4	61 Cygni	A	5	8	K5	0.63	?	85 (AB)
		B	6	8	K7	0.6	?	
		C	?	?	?	0.008	?	
3.4	ϵ Indi	—	5	7	K5	?	?	—
3.5	τ Ceti	—	4	6	G8	?	1.0	—
3.5	Procyon	A	0	3	F5	1.8	1.7	15.7
		B	11	13	Wh. dwarf	0.6	0.01	
3.5	Σ 2398	A	9	11	M4	0.4	?	60
		B	10	12	M5	0.4	?	
3.6	BD +43°44	A	8	10	M1	?	?	156 (AB)
		B	11	13	M6	?		
		C	?	?	K?			
3.6	CD −36°15693	—	7	10	M2	?	?	—
3.7	BD +5°1668	A	10	12	M4	?	?	?
		B	?	?	?	?	?	
3.7	G51-15	—	15	17	?	?	?	—
3.8	L725-32	—	12	14	M5	?	?	—
3.8	CD −39°14192	—	7	9	M0	?	?	—
3.9	Kapteyn's Star	—	9	11	M0	?	?	—
4.0	Krüger 60	A	10	12	M4	0.27	0.51	9.5 (AB)
		B	11	13	M6	0.16	?	
		C	?	?	?	0.01	?	
4.0	Ross 614	A	11	13	M5	0.14	?	3.9
		B	15	17	?	0.08	?	

Source: Data from van de Kamp, 1981. A companion to Barnard's Star, listed by van de Kamp, is regarded as dubious and deleted here.

[a]Stars within 4 pc.

a

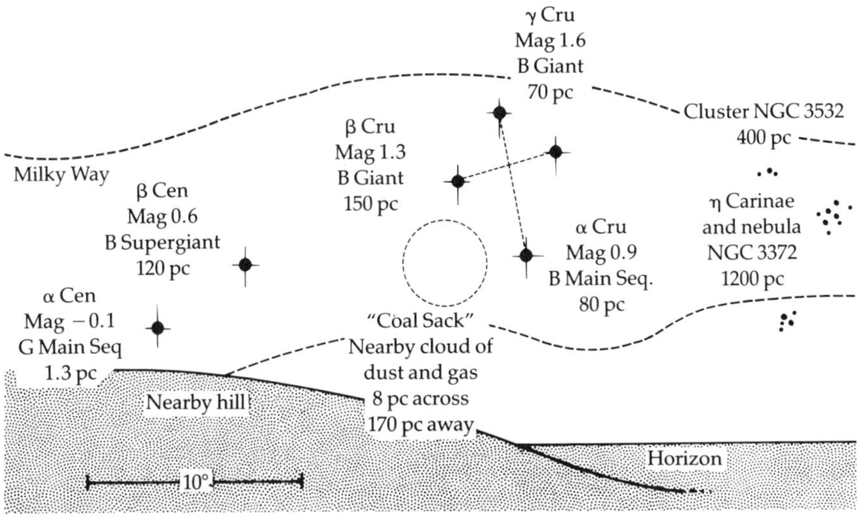

b

Figure 15-3. *The nearest star system, Alpha Centauri. Neither of two companion stars orbiting around Alpha is seen here—one (Alpha B) because it is too close to Alpha A, and the other (Proxima) because it is too faint. Map* b *identifies prominent stars, most being at least 100 times further away than Alpha Centauri. Other prominent features include the constellation of the Southern Cross and a remote dust cloud known as the Coal Sack, which blots out part of the background Milky Way. (Photo by author; 55-min. exposure on commercially available 2475 recording film; 55-mm lens, f2.8; "star effect" filter emphasizes brightest stars by diffracting light into a cross pattern around their images.)*

white dwarf stars,[2] because their colors were intermediate, like the sun's, being neither strongly blue nor red. Still other stars were found to be brighter than the giants or the main-sequence O stars; they came to be called **supergiant stars.**

The work of Hertzsprung, Russell, and their followers was important because it showed systematic groupings of stars: one group fitting on the main sequence, a distinctly different group of giants, a group of dwarfs, and a group of supergiants. Further work showed that these groups occur in all known regions of space. This meant that universal physical processes could be sought to explain the groups.

Table 15-2 tabulates the 17 most prominent stars in our sky and reveals that nearly half are giants or supergiants. Although the whales among the fishes are rare, they account for many of the familiar stars, as illustrated by Figure 15-5.

RADII ON THE H–R DIAGRAM

To emphasize the distinctness of these stellar groups, Figure 15-5 shows the same H–R diagram as Figure 15-4 but with lines added to show positions of constant radius. The important concept is that *any point on the H–R diagram corresponds to a star of a certain radius.* The reasoning should be clear to you already. Any point on the diagram corresponds to a certain temperature T. According to the Stefan–Boltzmann law, any surface at temperature T must radiate each second a certain amount of energy per square meter. But any point on the diagram also corresponds to a certain luminosity L, which is the total amount of energy radiated each second. This luminosity fixes the number of square meters involved; hence the total area; hence the radius.

The diagram confirms that the names of the types are apt. Supergiants can be 500 to 1000 times the diameter of the sun;

[2]Unfortunately, main-sequence stars are sometimes called dwarf stars to distinguish them from giants, but for clarity we reserve the term *dwarf* for white dwarfs.

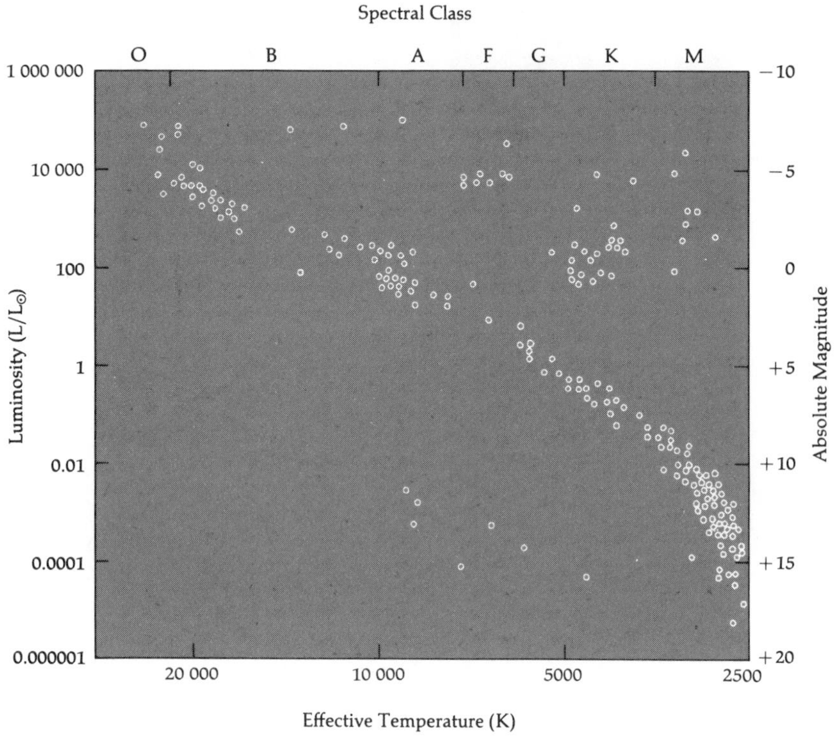

Figure 15-4. *A modern H–R diagram with a sample of well-studied stars (the 100 of greatest apparent brightness and the 100 closest). Spectral class (top) is equivalent to temperature (bottom), while luminosity (left) is equivalent to absolute magnitude (right). The main sequence (upper left to lower right), giants (upper right), and dwarfs (lower left) are prominent. (Data from Allen, 1973.)*

giants, 20 times; and dwarfs, only 1/100. Some well-known examples of these various types are marked on Figure 15-5. We can see that the sun and the brightest star, Sirius, are main-sequence stars; the North Star, Polaris, is a supergiant; and faint companions to Sirius and Procyon are white dwarfs. The enormous size of the supergiant Antares can be visualized in the imaginary close-up view shown in Figure 15-6.

Explaining the Types of Stars: Different Masses and Different Ages

What causes the distinctive types of stars found on the H–R diagram, such as main-sequence stars, the giants, and the white dwarfs? First, stars have different masses, as shown in Figure 15-7. *High-mass* main-sequence stars are hotter and brighter and bigger than *low-mass* main-sequence stars. Second, stars evolve from one type to another, and since we see stars with different ages, we see stars with different forms.

How did we discover the principles that control the structure and evolution of stars? The English astrophysicist Arthur Eddington (1926) once commented:

At first sight it would seem that the deep interior of the sun and stars is less accessible to scientific investigation than any other region of the universe.... What appliance can pierce through the outer layers of a star and test the conditions within?

The problem does not appear so hopeless when misleading metaphor is discarded. It is not our task to "probe"; we learn what we do learn by awaiting and interpreting the messages dispatched to us by the objects of nature. And the interior of a star is not wholly cut off from such communication. A gravitational field emanates from it.... Radiant energy from the hot interior after many deflections and transformations manages to struggle to the surface and begins its journey across space. From these two clues alone a chain of deduction can start which is perhaps the more trustworthy because it [employs] only the most universal rules of nature—the conservation of energy and momentum, the laws of chance and averages, the second law of thermodynamics, the fundamental properties of the atom, and so on.

Eddington showed why stars are the way they are. The same two major, opposing influences at work in the formation of the sun are competing in any star: Gravity pulls *inward* on stellar gas while gas pressure and radiation pressure push *outward*.[3] In a stable star, these forces are just balanced.

As we have already seen in the discussion of Helmholtz contraction (p. 228), heat is produced during gravitational contraction of a star. A stable main-sequence star is one that has contracted until the inside is hot enough to start nuclear reactions among hydrogen atoms. At this point the interior becomes a stable heat source, radiating light and creating enough outward pressure to counterbalance the inward force of gravity. What had been a contracting ball of gas now becomes a star with a stable size, governed by the interior energy-generating conditions.

[3]As mentioned in Chapter 13, radiation exerts a distinct pressure on material through which it passes. In stars, especially massive ones, this pressure is important.

Table 15-2. *The 17 Stars of Greatest Apparent Brightness*

Star Name	Apparent Magnitude (m_v)	Bolometric Luminosity ($L_\odot$)	Type[a]	Radius[b] ($R_\odot$)	Distance (pc)
Sun	−26.7	1.0	Main sequence	1.0	0.0
Sirius (α Canis Majoris)	− 1.4	23	Main sequence (primary)	1.8	2.7
Canopus (α Carinae)	− 0.7	(1400)	Supergiant	30	34
Arcturus (α Boötis)	− 0.1	115	Red giant	(25)	11
Rigel Kent (α Centauri)	0.0	1.5	Main sequence (primary)	1.1	1.33
Vega (α Lyrae)	0.0	(58)	Main sequence	(3)	8.3
Capella (α Aurigae)	0.1	(90)	Red giant (primary)	13	14
Rigel (β Orionis)	0.1	(60 000)	Supergiant (primary)	(40)	(280)
Procyon (α Canis Minoris)	0.4	6	Main sequence (primary)	2.2	3.5
Achernar (α Eridani)	0.5	(650)	Main sequence	(7)	37
Hadar (β Centauri)	0.7	(10 000)	Supergiant (primary)	(10)	150
Betelgeuse (α Orionis)	0.7	10 000	Supergiant	800	160
Altair (α Aquilae)	0.8	(9)	Main sequence	1.5	5
Aldebaran (α Tauri)	0.9	125	Red giant (primary)	(40)	21
Acrux (α Crucis)	0.9	(2500)	Main sequence (primary)	(3)	(110)
Antares (α Scorpii)	0.9	(9000)	Supergiant (primary)	(600)	(160)
Spica (α Virginis)	1.0	(2300)	Main sequence (primary)	8	84

Source: Data from Burnham (1978).

Note: These are stars brighter than the first apparent visual magnitude.

[a]The designation "primary" indicates data for the brighter companion in binary pairs.

[b]Parentheses indicate estimates.

EXPLAINING THE MAIN SEQUENCE

Although Eddington didn't know it, the main source of energy in a main-sequence star's interior is nuclear reactions in which hydrogen is consumed. Because stars form with a huge supply of hydrogen, stars remain stable on the main sequence for a relatively long time at a fixed size. If a stable star were magically expanded, its gas would cool and the reactions would decline, reducing the outward pressure, and the outer layers would fall back to their original state. If it were magically compressed, the inside would get denser and the reactions would increase, raising the outward pressure and expanding the star. The star has to stay in its stable state *as long as its internal chemistry and energy production rate stay the same.*

Although Eddington didn't know the details of the hydrogen-burning process, he was able to explain the main sequence by calculating the structures of stars with fixed compositions and energy-producing processes. His calculations revealed a **mass–luminosity relation:** A hydrogen-burning star more massive than the sun has higher luminosity and surface temperature than the sun. Hence, on the H–R diagram, it would lie to the upper left of the sun. Similarly, a star of lower mass would lie to the lower right of the sun. These stars therefore fall along a line in the H–R diagram—the main sequence. In other words, *the main sequence is explained as the group of stars of different masses that have reached stable configurations and are generating energy by consuming hydrogen in nuclear reactions.* Any hydrogen-"burning"[4] star with 1 $M_\odot$ and solar composition must look like the sun.

Figure 15-7 shows this trend more clearly by charting masses of some stars on the H–R diagram. Note the smooth trend of increasing mass and luminosity as we

[4]Burning technically refers to chemical reactions involving only electrons. But astrophysicists use the term informally for nuclear reactions that convert small fractions of the mass of atomic nuclei into energy.

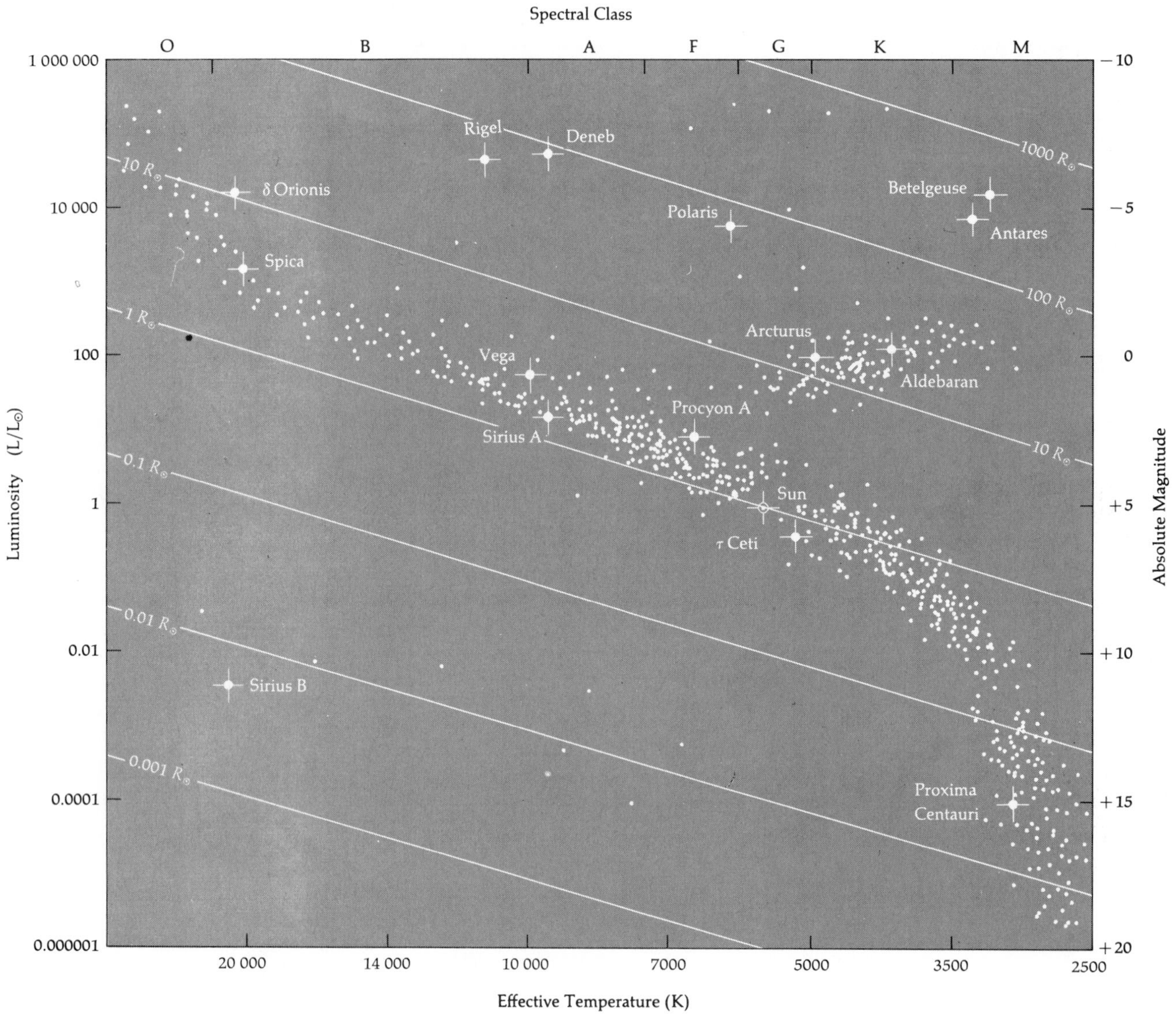

Figure 15-5. *An H–R diagram with lines of constant radius, showing the dimensions of stars in different parts of the diagram. Selected well-known stars are marked.*

ascend the main sequence. The O-type supergiants may have 40 or even 100 times the sun's mass.

Eddington's explanation of the main sequence helped explain why some stars lie off the main sequence. The same assumptions simply do not apply to them. In other words, giants, supergiants, and dwarfs are stars that do *not* have the same energy generation process, composition, or energy transport processes as the sun.

THE RUSSELL–VOGT THEOREM

As Eddington reached these conclusions, the astrophysicists H. N. Russell and H. Vogt independently derived a related result in 1926 known as the **Russell–Vogt theorem**:

The equilibrium structure of an ordinary star is determined uniquely by its mass and chemical composition.

***Figure 15-6.** Imaginary view of the reddish M-type supergiant star, Antares, as seen from a hypothetical planet $9\frac{1}{2}$ AU away, at the same distance as Saturn is from our sun. Although the sun would resemble only a brilliant dot from such a planet, Antares is so big that it would cover an angle of almost 40°. It would look 80 times bigger than our sun in Earth's sky, and have a reddish-orange color. The fainter star in the upper right is a B-type star that orbits around Antares at a distance of several hundred AU. A color version of this picture appears on the endpapers after the front cover of this book. (Painting by author.)*

This says that a certain mass of material with fixed composition—for example, one solar mass of solar composition—can reach only one stable configuration. This is the point on the H–R diagram actually occupied by the sun. But if the composition were altered to ½ hydrogen and ½ helium, the configuration of the star and its location on the H–R diagram would be different. In other words, the Russell–Vogt theorem says, "Give me the mass and composition of any star at any given moment, and I can tell you what the star looks like." Using these principles, astronomers have actually calculated what the stars look like—they have made tabulations of the pressures, temperatures, and other characteristics of the interiors, surfaces, and atmospheres of stars.

The Russell–Vogt theorem's effect is indicated in Figure 15-8; main-sequence stars of a given mass tend to have the same radius. But giants, supergiants, and white dwarfs of the same mass, because they have evolved and have different compositions, have different equilibrium structures and radii.

WHY MAIN-SEQUENCE STARS MUST EVOLVE OFF THE MAIN SEQUENCE

The Russell–Vogt theorem explains the cause of **stellar evolution** from one form to another form: A star burning hydrogen converts it to helium and changes its composition; therefore, the star must change to a new equilibrium structure. All nuclear reactions cause changes in composition, and all changes in composition cause evolution.

Philosophical Implications of Theoretical Astrophysics

Philosophically, these achievements were profound. Eddington pointed out that humans (or other intelligent creatures), reasoning purely from elementary principles even without telescopic observation, could show that the universe must be populated by objects like stars because gravity causes gas to "clump" into star-sized masses. Smaller or larger objects have too little gravity to assemble themselves from interstellar gas, and larger objects would produce so much radiation that they would blow themselves apart. These suppositions by Eddington have been proved correct by further observation and theory: The most common size of star in the universe has a mass around 0.1 to 1.0 $M_{\odot}$, as shown in Figure 15-9. Stars smaller than 0.1 $M_{\odot}$ are less common, and stars bigger than 60 $M_{\odot}$ are very rare.

Eddington (1926) wrote:

> *We can imagine physicists working on a cloud-bound planet such as Jupiter who have never seen the stars. They should be able to deduce by [these methods] that if there is a universe existing beyond the clouds, it is likely to aggregate primarily into masses of the order of [10^{27}] tons. They could then predict that these aggregations will be globes pouring out light and heat and that their brightness will depend on the mass in the way given by the [mass–luminosity relation].*

Eddington became fascinated by the idea that intelligent creatures could mentally derive the appearance of the universe without actually observing more than a few fundamental relations. He asked whether a superscientist, locked forever in a closed physics lab with certain minimal equipment, could discover the nature and arrangement of the universe. In other words, is the arrangement we observe the only possible arrangement? No one knows, but we can be sure that if some single physical constant—such as Newton's constant of gravitation—suddenly changed, the structure of the universe would dramatically change. For example, if gravity suddenly decreased, the earth would expand, causing earthquakes. The earth's rotation and orbit would change, and the moon's orbit would change. The sun would expand, causing a change in luminosity and a change in the earth's temperature.

It is thus plausible that a single complex theory, using the measured values of various physical constants, would allow prediction of everything that can be measured about the universe. Such a theory would allow treatment of matter, gravity fields, electromagnetic fields, and radiation from a single set of equations. Toward the end of his life, Eddington believed that he was approaching a theory that could predict the values of certain physical constants that had previously been thought to be underivable, intrinsic properties of the universe. His last work, however, has not been widely accepted.

An Overview of Stellar Evolution

Because of the work of Eddington, Russell, Vogt, and others, modern astrophysicists can calculate models of stars of any given mass and chemical composition. Thus, we can calculate not only what the sun's interior structure is like now, but what its future structure will be when 1% of the present hydrogen is gone, 2% is gone, and so forth. Each different composition corresponds to a certain (temporary) structure, and to a certain position in the H–R diagram. Knowing the rates of nuclear reactions, we can calculate the time scale associated with this evolution from one state to another. Using computer modeling, we can estimate what happens during unstable transition states, such as the transition from main-sequence to giant and post-giant stages.

Because a star's temperature and luminosity, as well as other properties, can be calculated for each state of its evolution, the stages of that evolution can be plotted on an H–R diagram. The sequence of such points is called an **evolutionary track**, the

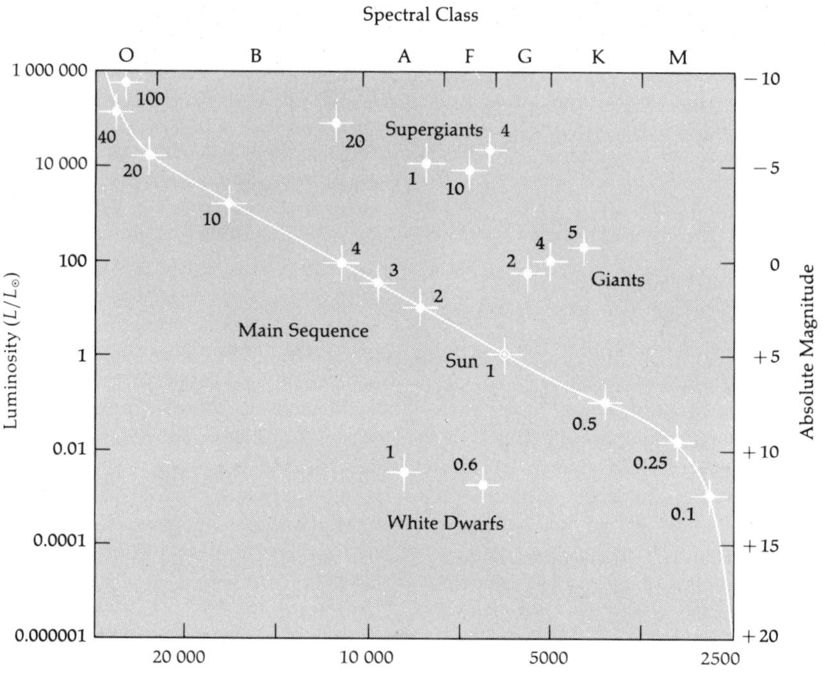

Figure 15-7. *An H–R diagram with measured masses of stars (in solar masses) marked in different parts of the diagram. A smooth progression of masses is found along the main sequence, but masses in other parts of the diagram are mixed because stars of different mass evolve into the same regions—for example, the giant branch.*

set of footprints a star leaves on an H–R diagram as it evolves.

EVOLUTION OF A ONE-SOLAR-MASS STAR (SUCH AS THE SUN)

Figure 15-10 shows the approximate evolutionary track followed by the sun or any star of solar mass. Some parts of the track are more certain than other parts. Consideration of different parts of the track shows clearly why different groups of stars appear on the H–R diagram. As we saw in Chapters 12 and 13, the sun began its life as a cool, dim interstellar cloud. Therefore, a sunlike star first appears on the H–R diagram at the right, with low temperature, low luminosity, and large radius, perhaps exceeding 100 $R_{\odot}$. It contracts rapidly, becoming quite bright, and then reaches a few solar radii in about a million years (only an instant of cosmic time). At this point it approaches the main sequence. Nuclear reactions begin in the interior, because the central temperature reaches a very high value.

Once the nuclear reactions start, the huge reservoir of hydrogen begins to be converted into helium, and a star reaches a relatively stable, main-sequence configuration. A star takes only about 100 million years to reach the main sequence, but it is destined to sit on the main sequence for roughly 9 billion years.

Finally, the supply of hydrogen begins to be exhausted. The central core is now mostly helium. The star's composition having changed, its structure must change, according to the Russell–Vogt theorem. Because there is less hydrogen in the core to fuel the reactions, not enough energy is produced to maintain the outward pressure and keep the star in its main-sequence state. The interior contracts, keeping the temperature high enough to maintain hydrogen burning in a shell around the helium core. Meanwhile, the outer layers of the star adjust by expanding. The star rather rapidly moves off the main sequence to a higher luminosity and much larger radius. It becomes a giant.

Theoretical calculations can trace a star quite accurately to the giant state, but subsequent states are less well understood, as indicated by the dashed line in Figure 15-10. Nuclear reactions begin to "burn" heavier elements, such as helium, fusing them into still heavier nuclei, as indicated in Table 15-3. Generally, after one element is consumed, slight core contractions raise the temperature to a point where still heavier elements begin to burn in fusion reactions. Evolution now takes the star somewhere to the left. Stars in this region are unstable, often variable,

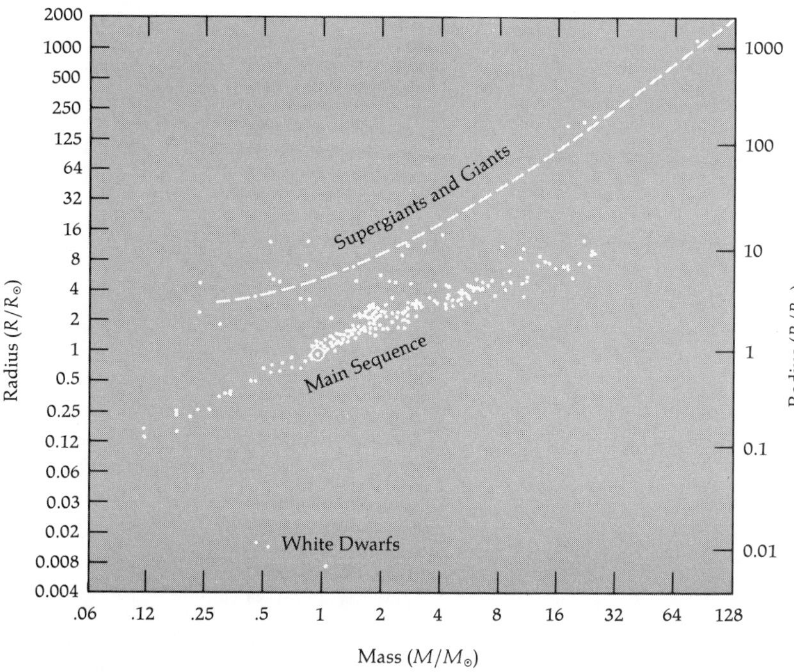

Figure 15-8. *This figure reveals that distinctly different radii exist among stars of a given mass. This is a result of evolutionary changes. Most stars in the mass range shown here remain at nearly constant mass during most of their life, but can evolve into states with different radii. (Data are derived from double stars' orbits, taken from Popper, 1980; Aller, 1971; and Liebert, 1980.)*

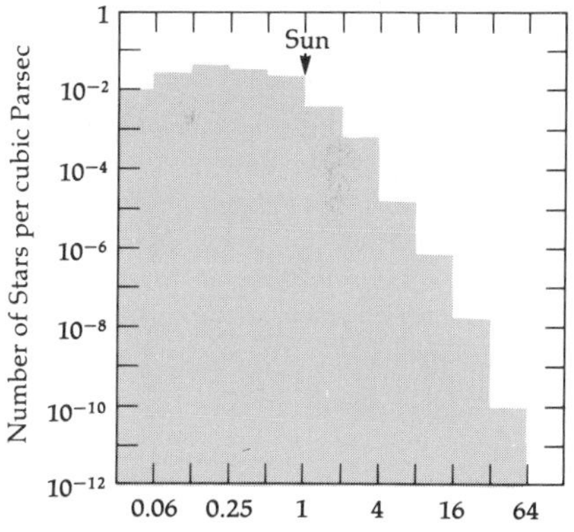

Figure 15-9. *A histogram showing the frequency of occurrence of different stellar masses in the solar neighborhood. The most common stars have about* $\frac{1}{4}$ $M_{\odot}$.

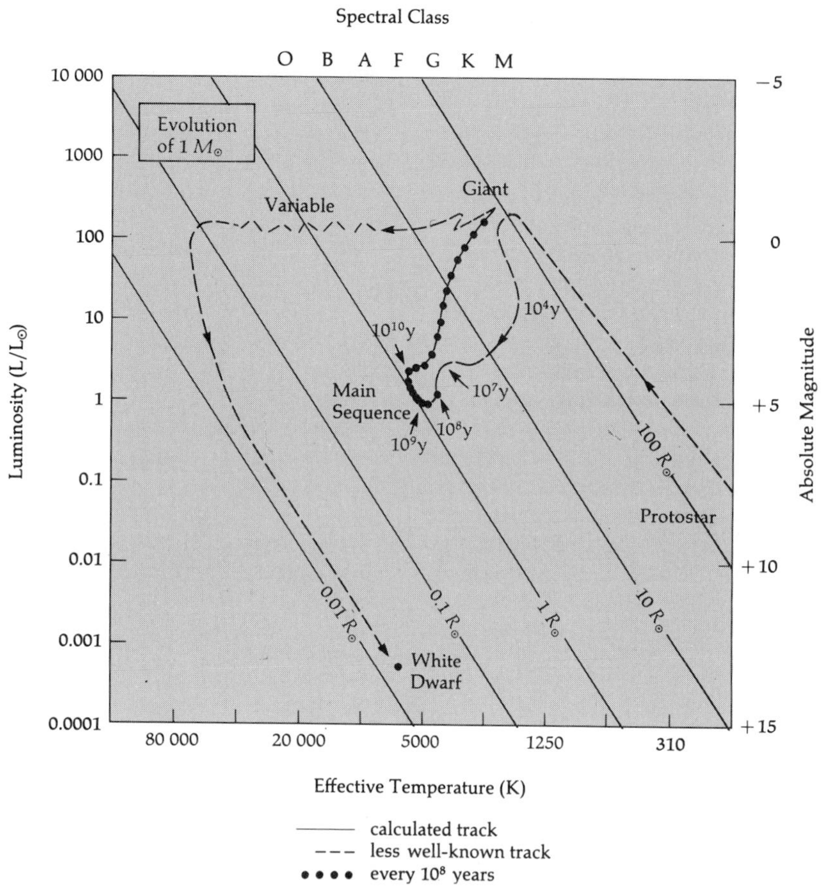

Figure 15-10. *An H–R diagram showing the calculated evolutionary track for the sun (or any star of 1 M$_\odot$). Dots are placed about 100 million years apart. The track shows the evolution from protostar state to main sequence, then to giant, variable, and white dwarf states. (Based on calculations by Westbrook and Tarter, 1975; Iben, 1967; Larson, 1969; and others.)*

and rapidly evolving. Some very massive stars in this region explode, throwing off masses of gas.

Ultimately there are no more "combustible" elements in the star's interior. No matter how much it contracts, no matter how hot the inside gets, no new reactions produce energy. So the star collapses to a very small size—the white dwarf has been created (or one of several other strange, small star forms, to be discussed in Chapter 17).

The following sequence summarizes the evolution of stars according to the groups found on the H–R diagrams (multiple arrows indicate faster evolution):

Protostar
↘ ↘ ↘
Pre-main sequence
↘ ↘
Main sequence
↘
Giant
↘ ↘
Variable or unstable
↘ ↘ ↘
White dwarf or other small star

You will get a better idea of how these stages relate if you trace the dashed line in Figure 15-10 with your pencil. Start with the pencil point in the lower right and count seconds, moving forward one round dot each second. Each second equals 100 million years. As you can see, the star almost reaches the main sequence in just one of these "cosmic seconds" but then spends nearly 100 cosmic seconds on the main sequence. Then there is a rapid evolution to the giant stage, followed by a rapid transformation (in another cosmic second or so) to a small, faint, defunct state below the main sequence. No wonder we see so many main-sequence stars: Most stars spend nearly their whole lives in this state.

MASSIVE STARS EVOLVE FASTEST

Stars evolve at different rates. The more massive a star, the higher its interior temperatures, and the faster it uses its nuclear fuel. A sunlike star of 1 $M_\odot$ stays on the main sequence about 9 billion years (9×10^9 y), but a star of 10 $M_\odot$ stays there only about 20 million years (2×10^7 y). The entire history of a star of 1 $M_\odot$ (protostar to white dwarf) takes about 11 billion years, whereas a 10-$M_\odot$ star lasts only about 24 million years. In spite of the differences in total time, the largest fraction of each star's life is still spent on or near the main sequence.

The most massive stars stay on the main sequence for only the twinkling of a cosmic eye. Some of them evolve into the supergiant region, and some less massive ones become ordinary giants. All of them quickly evolve to unstable configurations; many may explode; and all disappear from visual prominence.

DETERMINING THE AGES OF STARS

Consider a group of stars that formed at the same time. After about 10 million years, stars larger than 20 $M_\odot$ will have disappeared from the main sequence. That is, the H–R diagram of the group will contain no main-sequence O stars. After about 100 million years, stars more massive than 4 $M_\odot$ will have evolved off the main sequence, and the H–R diagram will contain scarcely any main-sequence B-class stars. The older the cluster, the more of the main sequence will be gone. The missing stars will have been transformed into giants, white dwarfs, or even fainter terminal objects.

Thus, we reach the important conclusion that *the H–R diagram can serve as a tool for dating groups of stars that formed together*. This principle will be applied often in later chapters as we probe the **ages of stars** in our galaxy.

Isolated individual stars are harder to date. The sun's age was measured at 4.6 billion years by dating planetary matter, unavailable to us in the case of other stars. Certain indicators, such as the amount of "unburned" light elements (lithium, for example) in a star's atmosphere, can also be used to estimate a star's age. By this latter method the main-sequence stars closest to the solar system—the Alpha Centauri system (Figure 15-3)—are estimated to be at least 3 billion years old (Boesgaard and Hagen, 1974).

Table 15-3. ***Elements Important in Nuclear Reactions at Different Stages in Stellar Evolution***

	Elements	Comments
	Lithium and other light elements	Relatively unimportant source of energy during formation of star before main-sequence stages (see Chap. 16)
	Hydrogen	Primary energy source; defines main sequence of stars; proton–proton chain; carbon cycle (see Chap. 16)
Time ↓	Helium	Important after hydrogen is depleted; involved in reactions during rapid evolution off main sequence and into giant branch of H–R diagram (see Chap. 17)
	Heavy elements, including carbon and metals	Energy sources during final period on giant branch and during rapid evolution off giant branch, including supergiant states; s-process, r-process (see Chap. 17)

Summary

The nearby stars, within about a hundred parsecs of the sun, display a range of masses, luminosities, temperatures, and compositions. These properties are not randomly distributed, but grouped into distinct types. Some stars are on the main sequence, some are giants, some supergiants, and some dwarfs. Differences among these types are conveniently displayed on the H–R diagram, which plots stellar luminosity against temperature.

To explain these different types of stars, this chapter developed three important principles, mentioned in the introduction. First, stars form with a variety of masses, covering a range from as low as 0.06 to 60 $M_\odot$ and more. As the stars form and begin to burn hydrogen in nuclear reactions, they settle onto the main sequence, but the more massive the star, the more luminous its main-sequence position in the H–R diagram.

Second, stars evolve from one form to another. They begin with pre-main-sequence configurations, settle onto the main sequence for a relatively long time to burn their hydrogen, and then evolve off the main sequence. Post-main-sequence forms include giants, variables, and dwarfs.

Third, stars evolve at different rates. The more massive a star, the faster it goes through its sequence of life stages. This is because more massive stars reach higher central temperatures and therefore consume their hydrogen faster.

The great majority of stars, including most of those in the sun's neighborhood, have masses in the range of 0.1 to 1 $M_\odot$. They are relatively faint and cool; they lie in the lower right part of the H–R diagram. The most massive stars are much brighter and can be seen from much further away. They lie in the upper part of the H–R diagram. Thus, many of the stars prominent in the night sky are prominent not because they are close, but because they are the very luminous "whales among the fishes," far away among myriads of distant fainter stars.

Concepts

H–R diagram
main-sequence star
representative star
Alpha Centauri
prominent star
giant star
white dwarf star
supergiant star
mass–luminosity relation
Russell–Vogt theorem
stellar evolution
evolutionary track
ages of stars

Problems

1. Suppose you could fly around interstellar space, encountering stars at random.

a. Describe the stars you would encounter most often. Mention masses and H–R diagram positions.

b. About what percentage of stars would be as massive as or more massive than the sun? (*Hint:* Use statistics from either Table 15-1 or 15-2, as appropriate.)

c. Would the stars encountered be similar to bright stars picked at random in our night sky?

2. Do giant stars necessarily have more mass than main-sequence stars? Why are they called giants?

3. Four stars occupy the four corners of an H–R diagram. Which one has

a. the highest temperature?

b. the greatest luminosity?

c. the greatest radius?

4. Two stars lie on the main sequence in different parts of the H–R diagram. Which is

a. larger?

b. more luminous?

c. more massive?

d. hotter?

5. Two stars form at the same time from the same cloud in interstellar space, but one is more massive than the other. Describe differences or similarities at later moments in time.

6. What would an imaginary terrestrial observer see as the sun runs out of hydrogen in the future? If life is confined to the earth when this happens, would life perish from heat or cold?

7. Which part of Problem 4 could not be answered just from location in the H–R diagram if both stars were not on the main sequence?

8. Consider newly forming cosmic objects of stellar composition. What would you expect to happen in terms of stars' life cycles as the masses decrease from 1 $M_\odot$ toward values comparable to Jupiter's mass?

9. Show that astrophysical estimates of the sun's lifetimes are consistent with meteoritic and lunar data on the age of the solar system. Why would the data be inconsistent if astrophysicists calculated that solar-type stars stay on the main sequence only 1 billion years?

10. Why don't normal stars all collapse at once to the size of asteroids or tennis balls under their own weight, since gravity always pulls their material toward the stars' center?

11. How was the chemical composition of the sun different 3 billion years ago from what it is now?

12. Red giants have proceeded further in the evolutionary sequence than main-sequence stars, but is it correct to make the blanket statement that red giants are older than main-sequence stars? Explain. Which star is older, the sun or a red giant of 10 $M_\odot$?

ADVANCED PROBLEMS

13. Use Wien's law to show that an M star with temperature about 2900 K is strongly red in color, and that an O star with temperature about 29 000 K would be strongly blue. Would the strongest radiation emitted by each star be visible to the human eye?

14. For a group of stars of equal radius but different temperatures T, the Stefan–Boltzmann law shows that the total luminosity will be proportional to T^4.

a. Among such stars, how many times brighter would a 20 000-K star be than a 2000-K star?

b. Using Figure 15-5, confirm your result by reading luminosities on one of the lines of constant radius.

15. a. As shown in Table 15-1, a typical white dwarf could have a radius of 0.015 $R_\odot$ and a mass of 0.8 $M_\odot$. Calculate the mean density of matter in such a star, and compare it with the density of familiar material such as water (1000 kg/m^3), rock (about 3000 kg/m^3), and lead (11 350 kg/m^3).

b. How many pounds would a matchbox full of such material weigh?

Project

1. Locate or construct spherical objects that have the same proportional sizes as selected different types of stars, such as an Antares-type supergiant, the sun, and a white dwarf.

CHAPTER 16

STELLAR EVOLUTION I: BIRTH AND MIDDLE AGE

Just as new living individuals are being formed from chemical materials against a background of slowly changing genetic pools, new stars are being formed from interstellar gas and dust against a background of slowly changing elemental abundances in the cosmic gas. Locally — and by this we now mean a substantial part of our galaxy—there is a constantly changing environment, with new stars forming and old stars blasting material into interstellar space. From a philosophical point of view, identifying and understanding newly forming stars are challenges in our long quest to comprehend the role of stars in our universe.

Three Proofs of Present-Day Star Formation

It is easy to provide **evidence for ongoing star formation**:

Youth of the Solar System. The solar system is only about 4.6 billion years old, whereas the entire Milky Way galaxy—our system of 100 billion stars—is at least 10 or 12 billion years old. Thus, our own sun formed much more recently than the galaxy, and stars did not all form in one burst at the beginning.

Short-Lived Star Clusters. Many young, massive stars are grouped in *open star clusters.* Because of tidal disruptive forces and the tendency of each star in the cluster to follow its own orbit around the galactic center, most clusters dissociate into isolated stars in only a few hundred million years. This is confirmed by analyses of H–R diagrams of clusters, which show that few of these clusters are much older than this. The famous Pleiades cluster, shown in Color Plate 28, is only about 50 million years old.

Short-Lived Massive Stars. Chapter 15 showed that massive stars evolve fastest. Calculations show that stars of 20 to 40 $M_\odot$ can last only a few million years in a visible state, yet we see them shining. They must have formed less than a few million years ago.

Thus, star formation has been a continuing process during the whole history of the galaxy, including the last million years. There are stars in the sky younger than the species *Homo sapiens.*

The Protostar Stage

Stars form from tenuous material between other stars. The average conditions between stars are a near vacuum, with very thin, cold gas. Atoms and molecules number only about 1 to 5 per cubic centimeter. About three-quarters of them are hydrogen (H atoms or H_2 molecules), and most of the rest are helium—as is the case with the atoms in most nearby stars. There are scattered dust particles. The temperature is typically around 100 K (−279°F).

MOLECULAR CLOUDS

In certain regions of space, the density of gas and dust may be 10 000 times greater. With some 10 000 atoms per cubic centimeter, the atoms are close enough together to collide and form molecules, rather than single atoms. These regions are called *molecular clouds.* They are rich in atoms, dust grains, and diverse molecules such as molecular hydrogen (H_2), water (H_2O), and more complex forms such as ethyl alcohol (C_2H_5OH) and the amino acid glycine (NH_2CH_2COOH). Such complex molecules are virtually absent in most parts of interstellar space. Although a molecular cloud is denser than most interstellar gas, it is still a near vacuum compared with ordinary room air, where there are about 20 billion billion (2×10^{19}) molecules per cubic centimeter.

Molecular clouds are important because they are the most active regions of star formation. In these regions the atoms and

Figure 16-1. *The young star cluster NGC 6611 (upper right) and associated nebulosity (center) show many features of star-forming regions, including hot, massive, blue stars and dark globules rich in dust. See the same region in Color Plate 25. (Steward Observatory.)*

molecules and dust grains are crowded close enough to begin to attract each other gravitationally—a key step in forming stars from gas (Scoville and Young, 1984). An example of the chaotic forms of gas and dust in such regions is shown in Figure 16-1 (a broader view of this dramatic region is shown in Color Plate 25).

TOWARD AN ASTROPHYSICAL THEORY OF STAR FORMATION

Chapter 12 already sketched how the sun formed when one of these diffuse interstellar clouds contracted and produced a central star surrounded by a dusty nebula. Now we need a more general theory of this process that can explain other stars with different masses.

When scientists use the word *theory,* they usually mean a well-tested body of related ideas often with a mathematical formulation that can be applied to a variety of cases and often backed up by observations.[1] Such a theory of star formation has been developed. This theory can be approached by asking: What causes some clouds to contract and not others? Why don't all interstellar clouds contract into stars and be done with it, once and for all? The answer involves the same two opposing forces we have considered before: gravity versus thermal pressure. Gravity pulls all the atoms in a cloud inward. But even at only 100 K the atoms are dancing, striking each other, and creating an outward pressure that opposes the tendency to collapse.

Gravity increases if the density of the cloud increases, since this gets more mass into the same amount of space. Outward pressure increases if the gas heats up, since this makes the atoms and molecules move faster. We can already say, then, that the reason not all clouds contract is that not

[1]A less complete or less tested idea is called a *hypothesis,* or a *working hypothesis.* An untested idea is often called *speculation.* Unfortunately, newspapers and magazines often publish scientists' speculations, but label them with the more imposing term *theory.*

all clouds are dense enough or cool enough. Furthermore, a noncontracting cloud can be turned into a contracting cloud if it suddenly experiences some turbulence (caused by a nearby stellar explosion, for example) that compresses it.

Astronomers want a more quantitative **theory of star formation** than this description. For interstellar material of any given density, they have calculated what combination of mass and temperature a cloud would need to begin its **gravitational contraction**, or shrinkage toward stellar dimensions. The calculations are based on principles sometimes called the *Jeans theory*, after English astrophysicist Sir James Jeans, who first made the calculations in 1902. (It is also sometimes called the *virial theorem*, after certain principles in mechanics.)

The results are shown in Figure 16-2, where the two diagonal lines give the typical mass that would form at 10 K or 1000 K from a gas of specified density. Along the bottom, the figure shows densities ranging all the way from 10^{-32} kg/m^3 (perhaps equal to densities between galaxies) to 1 kg/m^3 (about the density of air). This brackets the interstellar density within enormous margins, but the theory need not apply only to interstellar gas.

First, let us apply it to the interstellar medium in star-forming regions. What masses will contract there? The figure shows that in the denser interstellar medium, at about 10^{-17} kg/m and 100 K, masses contracting would be around 10^{34} kg, or several thousand solar masses, and about 100 times the mass of the largest stars. But our theory was supposed to predict the formation of stars! Is something wrong? In fact, the theory agrees with observations, because, as we noted in the second of the three proofs of present-day star formation, stars apparently form in star clusters that have masses thousands of times greater than the sun. In other words, conditions in the interstellar medium are such that *the gas subdivides into enormous chunks containing enough mass for thousands of stars.* These huge clouds can be thought of as protoclusters.

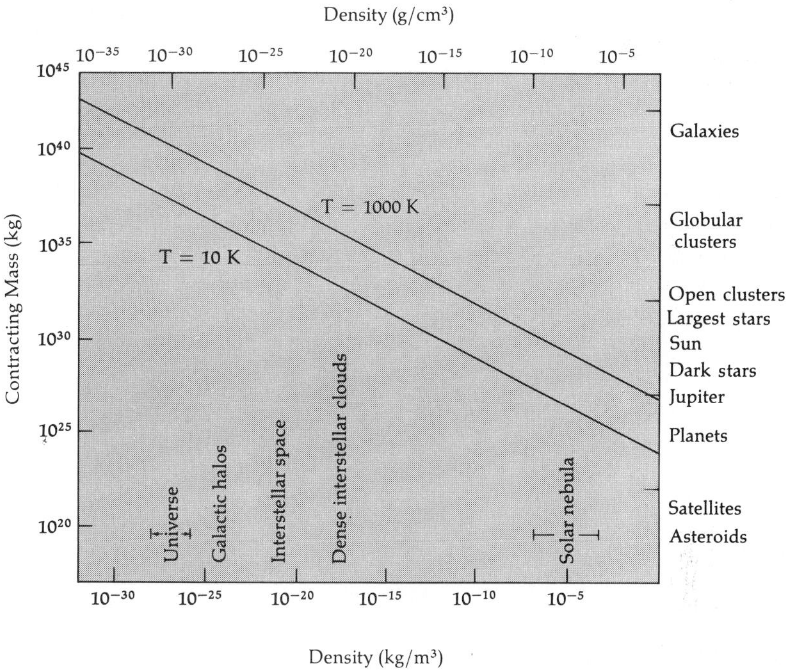

Figure 16-2. ***This diagram, based on the theory of star formation, shows the mass likely to be involved in self-gravitating contraction for different gas densities and two gas temperatures. Comparison of different environmental densities (bottom) with observed types of masses (right) shows that many features of the universe can be roughly explained with this simple theory. (See text.)***

But how do *individual* stars form? Figure 16-2 shows that as a protocluster cloud continues to contract to still higher densities, smaller, star-sized condensations can form inside it. In other words, shrinking protoclusters **subfragment** into individual stars. The Canadian researcher A. Wright (1970), for example, followed the contraction of a 6 000 000-$M_\odot$, 120-pc-wide cloud and a 500-$M_\odot$, 12-pc-wide cloud. He found that

both large and small density perturbations will grow in size, given the conditions found throughout most of a gas cloud. It would appear that fragmentation is an almost inevitable consequence of the cloud collapse process.

The new fragments become stars, and the whole mass turns into a star cluster. It is not hard to see why the interstellar gas in the disk of the galaxy might begin to contract. The material is not uniform; clots of dust and gas exist and are constantly being stirred by the galaxy's rotation, the expansion of hot nebulae, and other influences. Naturally, some clouds accumulate material or become compressed until their density exceeds the critical density that allows contraction to begin.

COMMENTS ON THE FORMATION OF JUST ABOUT EVERYTHING

It is provocative to consider the evolution of the whole universe from this point of view. For example, if all the material in the universe were spread out uniformly, a very low density of roughly 10^{-26} to 10^{-28} kg/m^3 would be obtained. The diagram shows that in a more or less uniform, relatively hot gas of this density, the condensations would have the mass of galaxies, about $10^{11}\ M_\odot$. In other words, if there was ever a time when no galaxies existed but hot gas filled space, galaxies should have formed. We'll return to this intriguing clue about galaxy formation in Chapter 25.

Galaxies' outer regions, or halos, have densities of around 10^{-24} kg/m^3. At these densities and at temperatures of 100 K, according to the diagram, masses of 10^6 $M_\odot$ could condense by their own gravity. Such objects actually exist in halolike regions around galaxies (as noted in the Prologue). Called *globular star clusters,* these objects consist of hundreds of thousands of stars. They are second in size only to galaxies, and our theory makes it quite plausible that they formed where they did through the subfragmentation of contracting galactic masses.

We have already explained that inside galaxies, where densities are higher, open star clusters form from clouds of about $10^3\ M_\odot$. We also noted that inside those clouds, where it is still denser, individual stars form by further subfragmentation involving masses around 1 $M_\odot$.

The same hierarchy of subfragmentation occurs with smaller masses. Inside the solar nebula, densities may have reached as high as 10^{-6} to 10^{-3} kg/m^3. Under these conditions, according to the same theory, masses about the size of Jupiter might have condensed. Some theorists believe that the miniature solar systems of Jupiter and its satellites or Saturn and its satellites formed in this way.

Finally, Chapter 12 described a current theory that concentrations of dust grains in such nebulae led to even higher densities, where small masses resembling asteroids formed by self-gravitation of dust grains and then accumulated into larger planets.

Thus we see how many loose ends from earlier chapters come together into one simplified theory that accounts for many objects in the universe. This illustrates how science can help us perceive a unity in the phenomena around us. The theory as described here is simplified in that it ignores such complications as magnetism, rotation, and the effects of dust opacity. But by so simplifying, astrophysicists can concentrate on the dominant influences to rough out the origin of various objects.

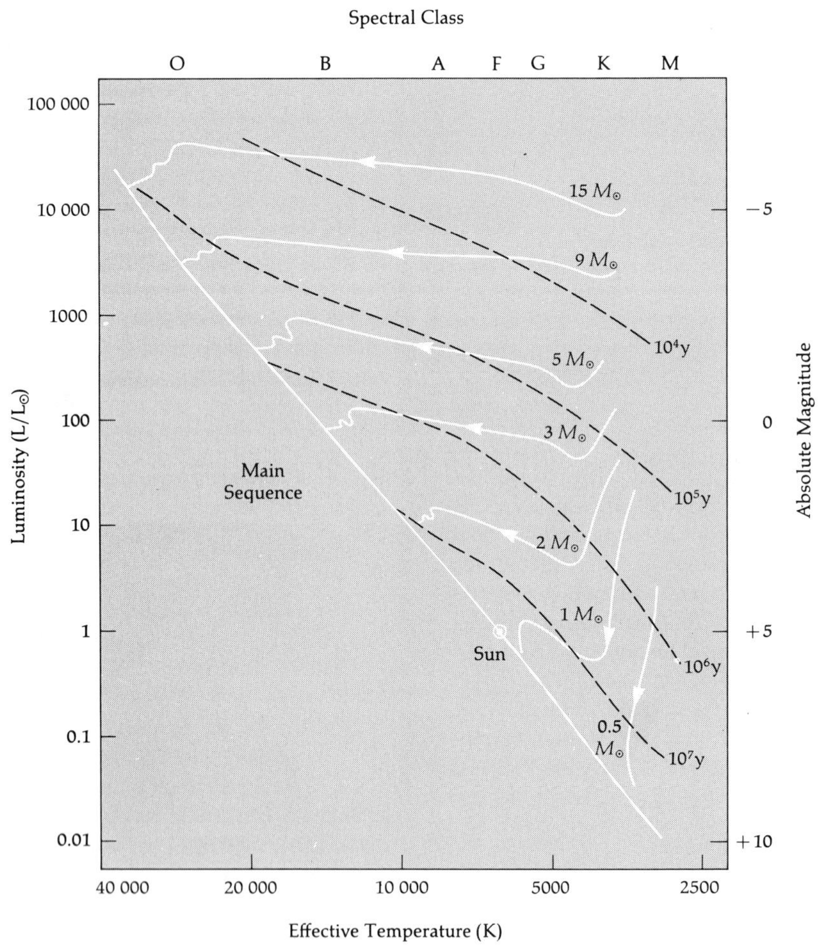

Figure 16-3. *H–R diagram showing simplified pre-main-sequence tracks for seven stars of different mass. Note that the high-mass stars settle on the upper end of the main sequence, while low-mass stars are on the lower end. Dashed lines show the states reached after the indicated number of years. More detailed recent calculations, taking into account the nebula around the star and other details, show more complex "squiggles" in the evolutionary tracks, but these details are not firmly established. (Data from Iben, 1965.)*

THE PROTOSTAR'S COLLAPSE

The preceding discussion allows us to define a **protostar** as a cloud of interstellar dust and gas that is dense and cool enough to begin contracting gravitationally into a star. An interstellar cloud may hover on the verge of this state for millions of years. A nearby exploding star or some other disturbance might compress a gas cloud and thus trigger the collapse, which then proceeds very rapidly. For instance, typical protostars contract to stellar dimensions in only a few thousand years—a wink of the cosmic eye. This explains why astronomers use the term **collapse**: The initial contraction is very rapid in astronomical time. Of course, once the cloud begins to reach stellar dimensions (a few astronomical units, or 0.00001 pc), the atoms of gas bump into each other frequently enough to produce substantial outward pressure, so that the collapse is slowed. At this point we can speak of the object as a *pre-main-sequence star*.

The Pre-Main-Sequence Stage

What are the forms of pre-main-sequence stars? Where do they lie on the H–R diagram? Can they actually be observed among the stars in space? Can we actually see stars being born?

The **pre-main-sequence star** stage covers the evolution from the end of the protostar stage to the main sequence. Astrophysicists such as the Japanese theorist C. Hayashi (1961) and his American colleague L. Henyey (Henyey, LeLevier, and Levée, 1955) pioneered in calculating the evolutionary tracks and appearance of stars contracting toward their main-sequence configurations. Results of such calculations are shown in Figure 16-3, which clearly shows that stars of all masses approach the main sequence from the right in the H–R diagram. Newly formed stars would be cooler than main-sequence stars.

The energy during most of the pre-main-sequence period does *not* come from nuclear reactions, which have not yet started inside the star; instead, heat is generated by the Helmholtz contraction (p. 228). After only a few thousand years of protostellar Helmholtz collapse, surface temperatures reach a few thousand Kelvin, thus causing visible radiation. Hayashi's work showed that convection would transport large amounts of energy from the interiors of most newly forming stars, making them very bright for short periods known as **high-luminosity phases** or *Hayashi phases.* A star of solar mass, for example, contracts in less than 1000 y from a huge cloud to a size about 20 times bigger than the sun with a luminosity about 100 times greater than the present sun's (see Figure 15-10). Figure 16-3 shows such a star entering the H–R diagram in its pre-main-sequence stage on a rapidly descending track as it fades from its high-luminosity phase.

These calculations have three important consequences. First, they show that stars have complicated, if short-lived, evolutionary histories even before nuclear reactions start. Second, they show that *newly forming stars must lie above and to the right of the main sequence.* (This is important in identifying new stars by direct observation.) Third, the calculations indicate that stars spend only a small fraction of their lifetimes in the pre-main-sequence stage.

More massive stars evolve faster to the main sequence than less massive stars. Note in Figure 16-3 that a 15-$M_{\odot}$ star reaches the main sequence in only 100 000 y. A 5-$M_{\odot}$ star takes about 1 million years, while a sunlike star takes several million years.

PREDICTION OF COCOON NEBULAE AND INFRARED STARS

While some theorists were calculating evolving conditions in newly forming stars, others, such as the Mexican astronomer A. Poveda, hypothesized that as young stars form from collapsing clouds, remnants of the clouds might surround and obscure these stars. As in the early solar nebula, dust would form in the cooling cloud and

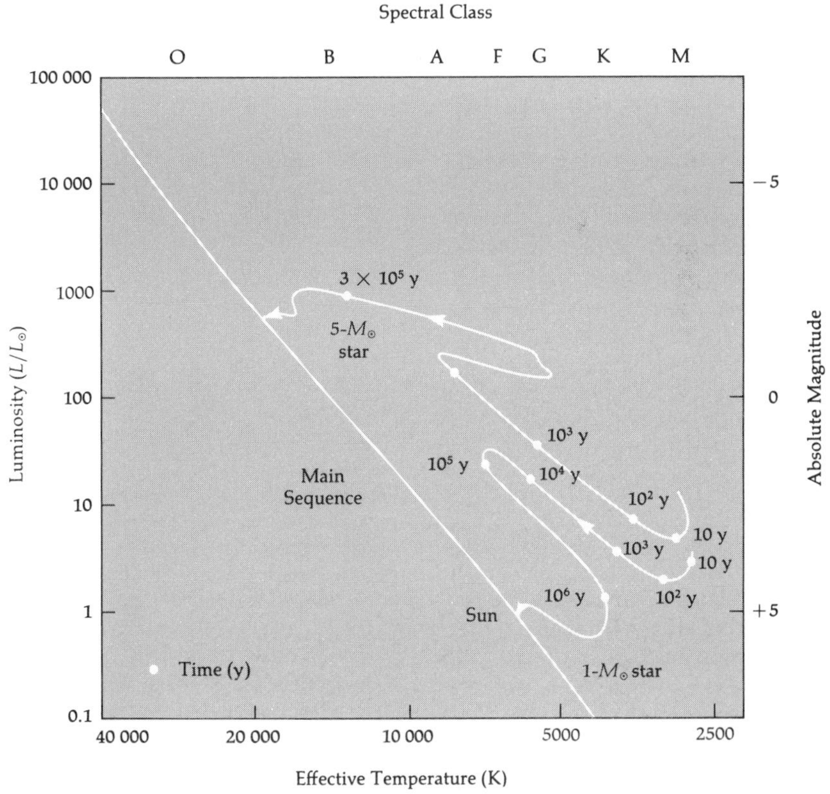

Figure 16-4. *H–R diagram showing pre-main-sequence evolutionary tracks and complex luminosity changes based on the theory illustrated by Figure 16-3, modified to take into account a surrounding nebula much larger than the star. Tracks represent only the central contracting star, or "core," of the nebula. (Data from Larson, 1969.)*

block the outgoing starlight. The radiation from the new star would heat the nebular dust grains to a few hundred Kelvin, and, according to Wien's law, the dust grains would then radiate infrared light. The nebula would eventually dissipate, revealing the star.

The term **cocoon nebula** was coined in 1967 to describe the nebula that hides a star from view during its earliest formative period (perhaps a few million years for a solar-sized star). The nebula is cast off later, just as a cocoon is cast off by an emerging butterfly.

These concepts showed the need for new theoretical models, because the earlier calculations had not included the effects of surrounding clouds.

Astronomer R. Larson (1969) calculated the evolutionary tracks of stars forming as central condensations inside larger clouds. The results are shown in Figure 16-4, page 295. The track is more complicated than before, but still approaches the main sequence from the right.

What would the newly forming star and cocoon nebula look like to an observer before the nebula dissipated? Could examples ever be observed? The outermost dust of the opaque cocoon would have a temperature of only a few hundred Kelvin. Therefore, the nebula would radiate not visible light, but rather infrared light (review Wien's law, p. 123), and the star–nebula complex, as seen from outside, would lie far to the right of the main sequence. Such an object could be detected as an **infrared star** at wavelengths of a few micrometers. As shown in Figure 16-5, it would evolve across this region of the H–R diagram in less than a million years and thus be detectable as an infrared star for a relatively brief period of its life. Stars of this type should therefore represent only a small fraction of all stars.

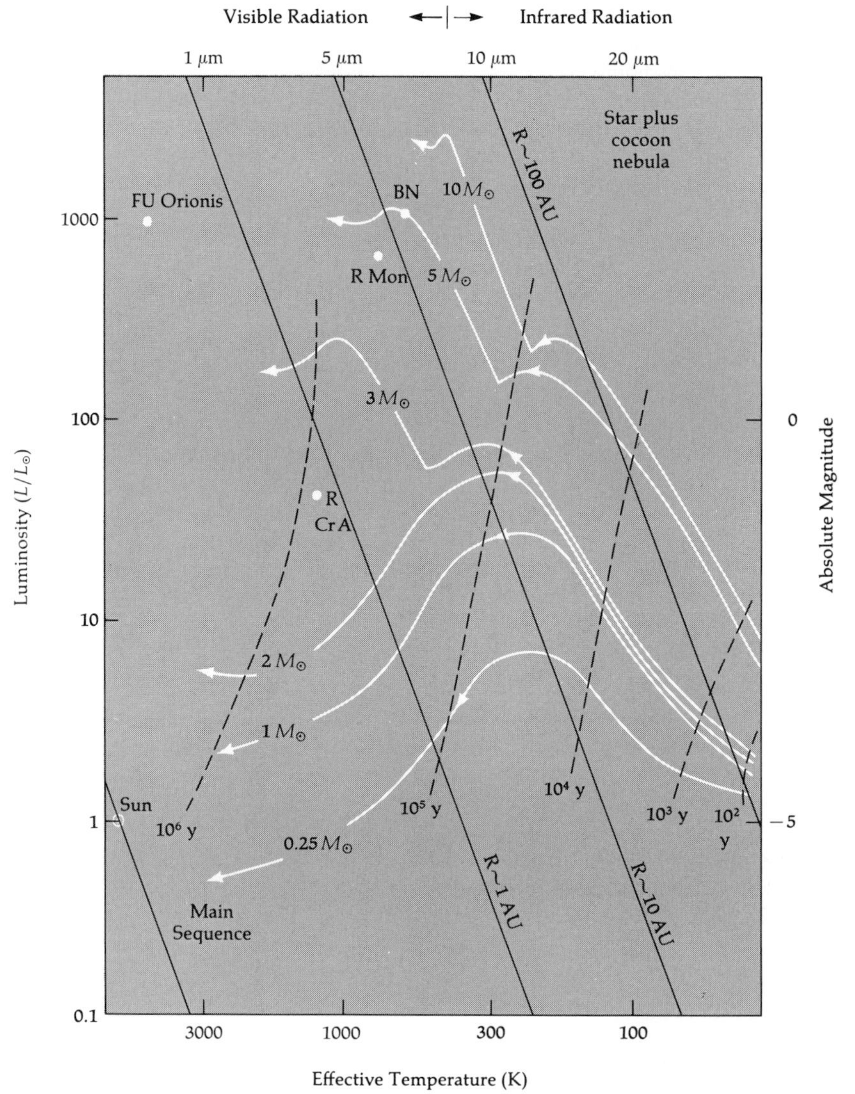

Figure 16-5. *H–R diagram showing apparent tracks followed by pre-main-sequence stars of various masses surrounded by opaque cocoon nebulae, as viewed from the outside. Unlike the central core stars (Figure 16-4), the configurations as a whole would be visible only in the infrared for some 10^5 y. (Data from Larson, 1972.)*

WHAT HAPPENS IF THE MASS IS TOO SMALL?

If a cloud is dense and cool enough to contract but has less than about 8% of a solar mass, it will contract but never develop a high enough central pressure and temperature to reach a full-fledged main-sequence state. Stars from about 1% to 8% $M_\odot$ may develop a few feeble nuclear reactions and glow for a while, while smaller objects bypass all nuclear reactions and evolve directly toward a planetlike state, as shown in Figure 16-6. The smallest known luminous object may be a temporarily glowing star of the first type. Cataloged as RG 0050-2722, it has an estimated mass of about 2.3% $M_\odot$ (only 23 × Jupiter's mass!) and a temperature of 2600 K. It is about 25 pc from us.

Since stars are usually defined as stable objects, steadily generating energy by nuclear reactions, objects with less than about 8% $M_\odot$ are not true stars. Objects too small to become stars are sometimes called **black dwarfs** (or **brown dwarfs**);

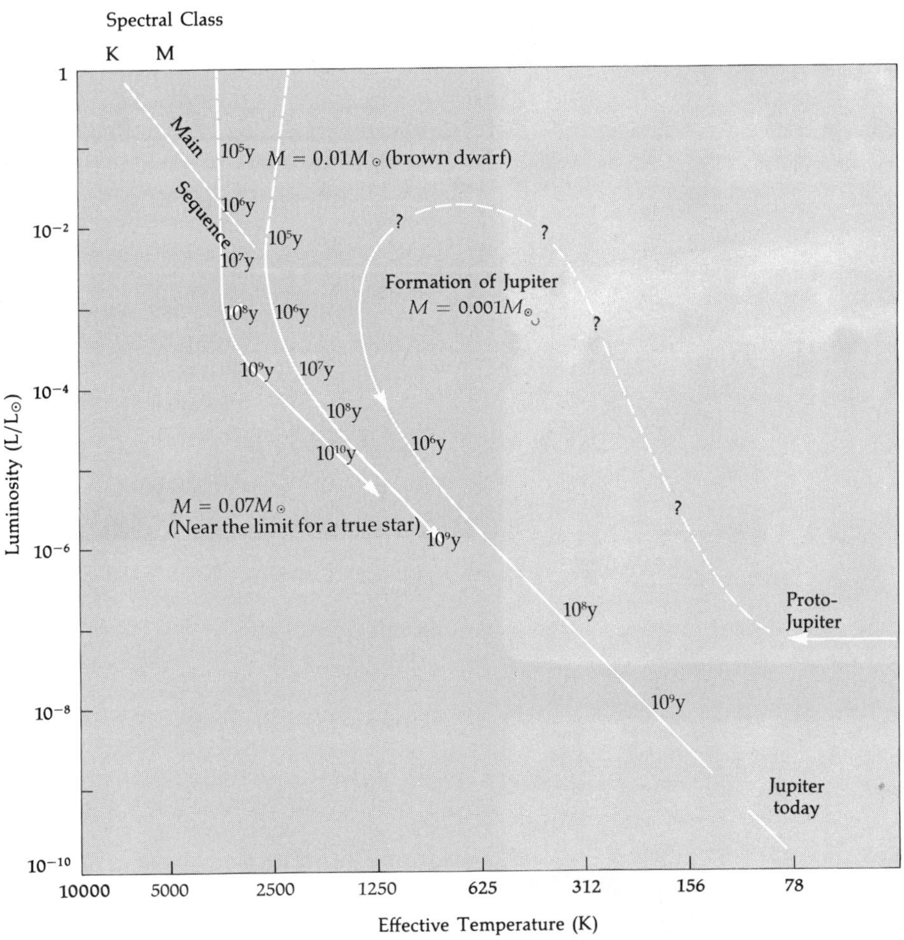

***Figure 16-6.** Extension of the H–R diagram into the low-luminosity low-temperature region to show evolutionary tracks of newly forming objects of very low mass. Bodies of 1, 10, and 70 Jupiter masses are shown. These are too small to initiate nuclear reactions and hence do not become true main-sequence stars. Jupiter may have followed such a track before reaching its present state (lower left). (Data from Bodenheimer, 1976; Staller and de Jong, 1981; and Van der Linden and Staller, 1983.)*

they are faint, small objects radiating their own heat of contraction in the infrared.

Still smaller objects would look like planets, which radiate only modest amounts of internal energy. Note how this ties into Chapter 10. Jupiter must have gone through a moderately high-luminosity phase that heated its inner satellites, and indeed it is still radiating more energy than it receives from the sun. Figure 16-6 shows that Jupiter (0.001 $M_\odot$) can be considered a black dwarf. Astrophysicist P. Bodenheimer (1976) has calculated the evolutionary track shown in Figure 16-6, showing how Jupiter, a "star that never made it," took 4.6 billion years to cool to its present state. This gives an interesting example of the link between stellar astronomy and the exploration of our own solar system.

WHAT HAPPENS IF THE MASS IS TOO BIG?

If the cloud is dense enough to contract but is more than about 100 $M_\odot$, the contraction is violent and produces an extremely high central temperature and pressure. Under these conditions, so much energy is generated inside the new star that the star is very luminous and may blow itself apart almost immediately without spending much time on the main sequence. This rapid destruction is more appropriate to the topic of the next chapter, and will be taken up again there. We will see that explosions of massive stars explain many features of our starry surroundings.

Examples of Pre-Main-Sequence Observed Objects

Several types of observed objects are believed to relate to late protostellar stages or pre-main-sequence stages of star formation. **Bok globules** are dense clouds of dust and gas that may be contracting protostars. They are made visible by being silhouetted against brighter nebulosity in the background; examples are the Coalsack nebula, silhouetted against the southern Milky Way (Figure 16-7), and small, dark blobs seen in Color Plate 25.

Still more interesting are infrared stars, mentioned earlier as predicted from theoretical work. Many are actually observed and may be examples of dusty cocoon nebulae around new stars—the same environments in which the earth and the other planets formed around the early sun, as described in Chapter 12.

The first detection of a cocoon nebula was announced in 1966, only a year after some of the first predictions of such objects. American astronomers Frank Low and Bruce J. Smith made this announcement based on their measures of R Monocerotis (star R in the constellation of Monoceros the Unicorn), a starlike object about 700 pc from the solar system (Figure 16-8). Low and Smith found that this object showed a large amount of infrared radiation and a faint amount of visible light. They hypothesized that the visible light came from a new star in the center of a cocoon nebula 200 AU in diameter, which absorbed most of the starlight. The dust

a

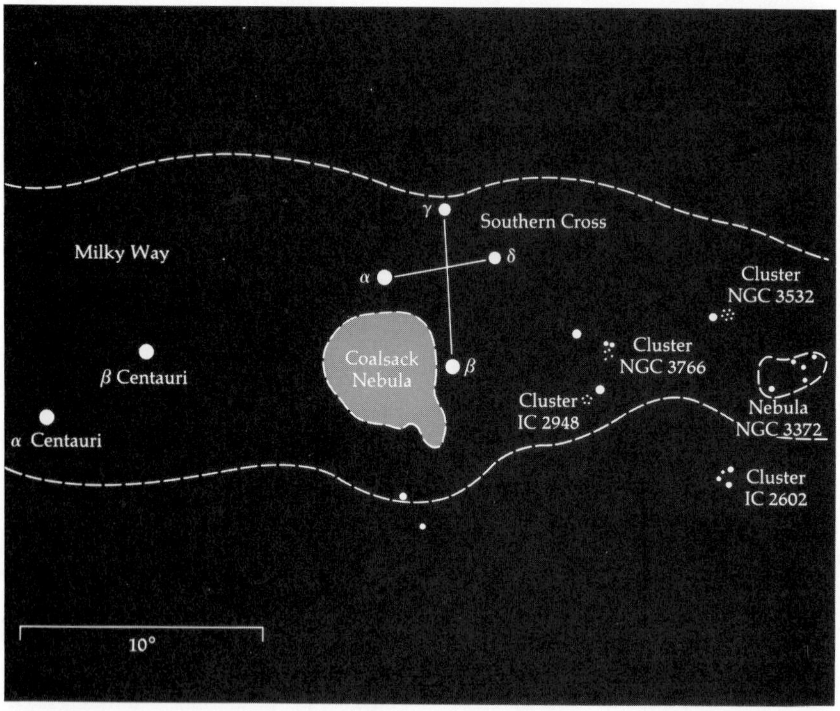

b

Figure 16-7. *The dark, round cloud in the center of this view is the Coalsack nebula, silhouetted against the Milky Way in the Southern Hemisphere. It is estimated to be about 8 pc across and 170 pc away. This view shows about 30° of the southern Milky Way. (Harvard Observatory.)*

grains in the nebula are heated to a temperature of about 850 K and radiate infrared light. Observations reported in 1984 suggest the nebula is disk-shaped and contains perhaps five earth-masses of dust. These data raise the possibility that planets are forming around R Monocerotis today, just as they did in our sun's dust-laden cocoon nebula 4½ billion years ago! What planets and landscapes will exist there 100 million years from now, when our civilization on Earth is only a dim memory?

R Monocerotis lies at the tip of a triangular nebula about 50 000 AU long, which reflects the light of the young star and its cocoon. As revealed in photos over more than six decades (Figure 16-8), the light, shape, and spectrum of the system change, again indicating that the system may be unstable and rapidly evolving, in accordance with the theory of star formation.

Many other infrared cocoon nebulae have been studied, ranging from examples that totally obscure any interior star to examples where most of the young star's light escapes.

The best examples of infrared stars and other probable newly formed stars are found in the **Orion star-forming region**, a large area of the sky around the constellation Orion. This region, which includes the R Monocerotis infrared star, is about 400 to 700 pc away and seems to be a hotbed of dense clouds and star-forming activity. When we look in this direction on a starry night, many of the prominent features of the sky are the results of recent star-forming activity, as seen in Figure 16-9.

These and other objects exhibit many features that tie in well not only with the theory of star formation but also with evidence about conditions in the nebula that surrounded our own sun as it formed. Among these features are:

1. Spectra indicating the presence of dust, including olivine-rich silicate and ices

2. Variable brightnesses, indicating instability

3. Disklike shapes for the surrounding nebulae, with dimensions from a few to 1000 AU

4. Detection of H_2O and OH molecules in some nebulae

5. Temperatures estimated at about 350 to 1800 K in the dust of the nebulae

The most important pre-main-sequence stars are the **T Tauri stars**, named after the variable star T in the constellation Taurus. Although a great many of them are found in the region of Taurus and the neighboring constellation Orion, they can be found in many other parts of the sky. They were first recognized as a group in 1943–44 in work almost unnoticed because the author, K. Himpel, was publishing inside Nazi Germany.

T Tauri stars may represent a transitional stage between infrared stars surrounded by opaque nebulae and stable stars that have lost their cocoons and settled on the main sequence. There are many signs of their youth and instability. The number of T Tauri stars per cubic parsec in a star-forming area may exceed the number of *all* stars per cubic parsec near the sun by more than a factor 10 (Cohen and Kuhi, 1979)! T Tauri stars vary irregularly in brightness. Penn State astronomer Joan Schmeltz (1984) finds these variations stem from changes both in the star's surface layers and in the enshrouding dust clouds. T Tauri stars lie to the right of the main sequence in the H–R diagram, where young stars are supposed to lie. Many have infrared radiation characteristic of cocoon nebulae, as pointed out by the Mexican astronomer E. Mendoza (1968).

T Tauri stars are typically 100 000 to a million years old, judging from all available evidence. Some may be even younger. A 1979 survey of star-forming regions counted T Tauri stars with ages ranging from only 10 000 years to 6 million years (Cohen and Kuhi, 1979). Stars that have evolved beyond the T Tauri stage would probably be indistinguishable from main-sequence stars.

FAINT INFRARED COMPANIONS OF INFRARED AND T TAURI STARS

As infrared detectors in ground-based and satellite telescopes improved, astronomers were able to map newly formed stars in ever-greater detail. Soon they found that many infrared stars originally cataloged as single objects were multiple infrared sources less than ¼ pc (50 000 AU) apart. These may be examples of double or multiple systems into which excessively massive protostars have fragmented.

Further studies reveal even more detailed structure. Many infrared stars, viewed at a resolution of one second of arc or less, are double. The separation distances are around 700 to 3000 AU in a number of cases—some tens of times the size of the planet orbits in our solar system. Most dramatic, however, is the discovery that the prototype T Tauri star, T Tauri itself, has a faint infrared companion orbiting some 80 to 150 AU away—a separation distance about two to three times the size of our solar system. The temperature was found to be 600–800 K, and the mass may be too small to be a full-fledged star. Some astronomers speculate that this companion may resemble a newly forming giant planet—a sort of proto-Jupiter (Hanson et al., 1983). Further work will be needed to confirm whether this can be taken as the long-sought evidence of a planetary system around another star, but the observations taken together indicate that small groupings of new stars, including low-mass companions, are common.

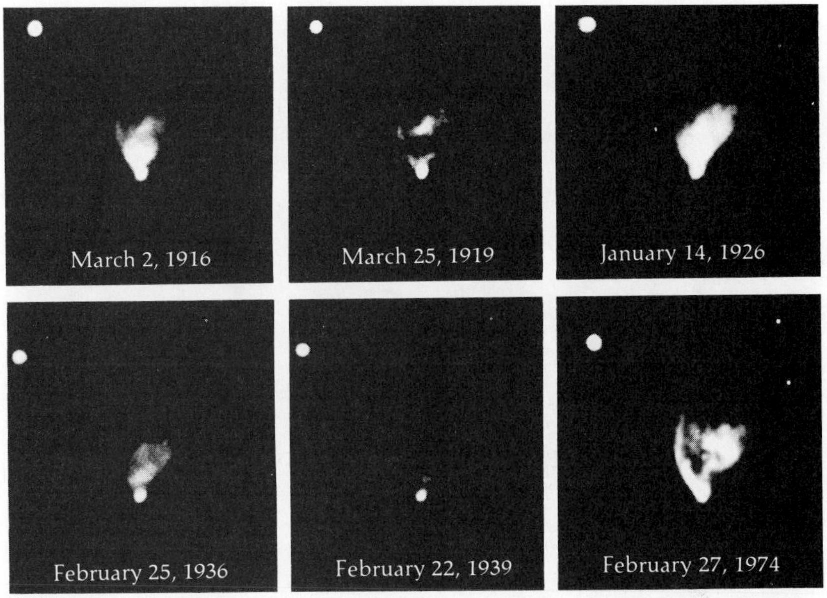

Figure 16-8. *Exposures of the newly forming star R Monocerotis over a 58-y period. The star is the bright object at the bottom tip of the irregular nebulosity. The nebula, called Hubble's variable nebula, changes shape from year to year, possibly due to changes in the illuminating radiation escaping from the cocoon nebula around the associated star. The changes indicate instability in the configuration. (Lowell Observatory; 1974 photo courtesy Alan Stockton, Mauna Kea Observatory, University of Hawaii.)*

THE MYSTERY OF MASS EJECTION FROM NEW STARS

During the 1970s and early 1980s, astronomers debated the fate of the cocoon nebula around new stars. Does it get blown outward, fall into the central star, or what? Studies of gas velocities (from Doppler shifts) around T Tauri stars gave conflicting answers. Observations in the 1970s indicated chaotic motions of gas around these new stars, sometimes infalling but more often blowing outward as fast as 50–200 km/s. The evidence suggested that T Tauri stars blow off as much as 0.4 $M_\odot$ of

a

Figure 16-9. (a) *A panorama of star formation in the midwinter evening sky, covering about 80° around the constellation Orion. Many of the features seen are unusually young. (22-min exposure with a 35-mm camera, wide-angle lens at f2.8, 2475 recording film; photo by author.)* (b) *Identification of features in* a. *Distances (parsecs) and ages (millions of years) of selected features are marked. Features in the constellation Monoceros, shown in Figure 16-8, are at top center. Dashed lines show major nebulae; radiating dashes are associations of T Tauri stars, not prominent to the naked eye. Compare also with Orion "close-up" in Figure 14-4, p. 260, and Color Plate 24.*

gas and dust before they reach the main sequence. Astronomers visualized very strong "solar winds" blowing out from T Tauti stars, but no one was sure how the mass was ejected (see nontechnical review by Lada, 1982).

Observations in the 1980s indicate that the outflow involves a mysterious phenomenon we will encounter again and again: *bipolar mass ejection.* This means that the gas flows out in two opposite directions, apparently along the polar axes of a disk, as seen in Figure 16-10. Gas clouds moving in opposite directions have actually been observed near new infrared stars (Bailly and Lada, 1983). The theoretical picture of these objects is that a cocoon nebula forms a disk around the central star, and its gas spirals turbulently inward (as if along the grooves of a phonograph record). Somewhere near the central star, it gets accelerated by an unknown means in diffuse jets "upward" and "downward," away from the disk, as seen in the figure.

This theoretical picture explains why Doppler shifts reveal inward motions of gas around some T Tauri stars (the ones viewed from a direction along the disk), while revealing an outward motion among

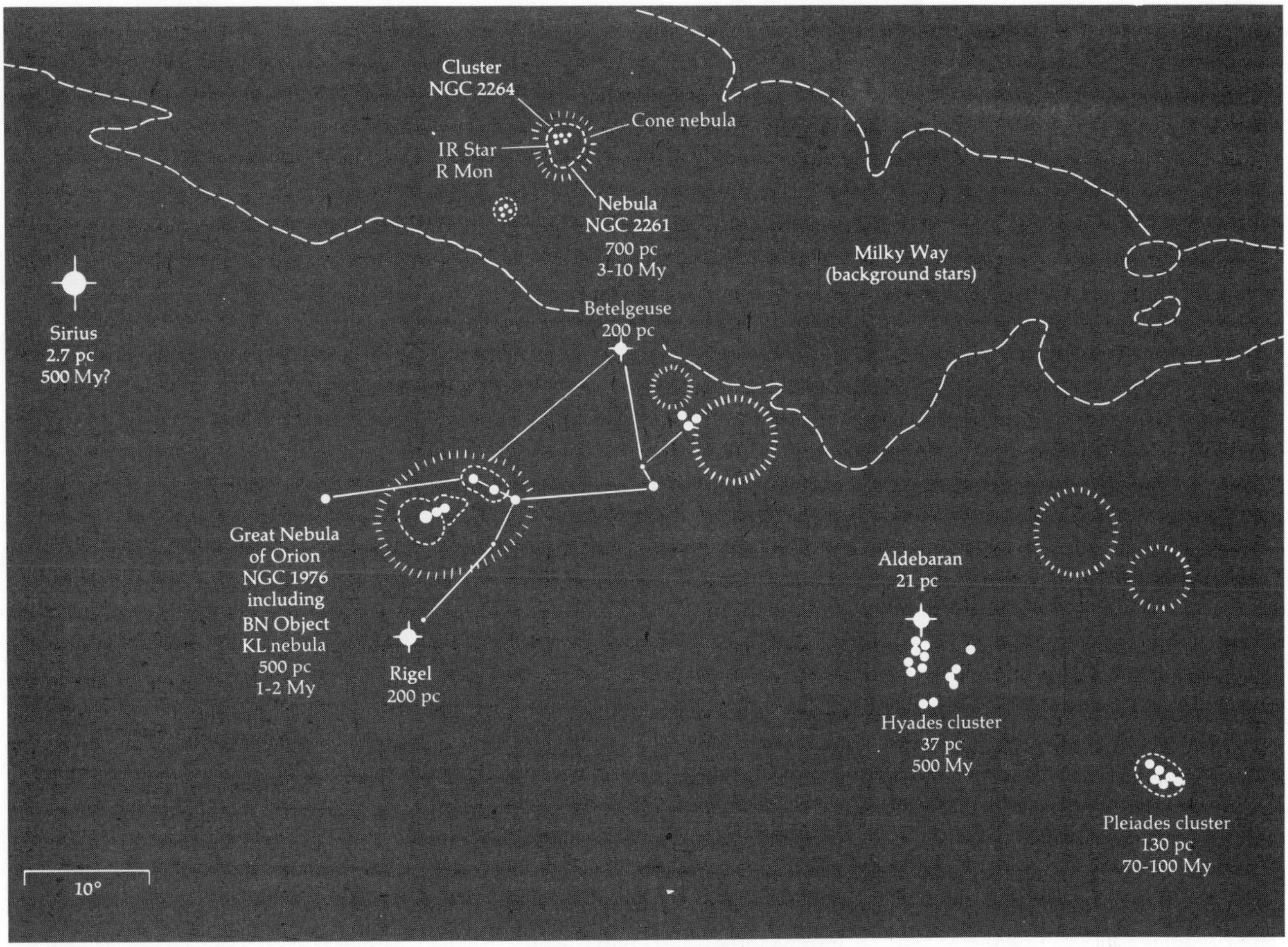

b

the majority (viewed from a direction along the diffuse, outward moving jets). The model also fits beautifully with evidence about the dusty, disk-shaped solar nebula in which the earth and other planets were born near the newly formed sun (review Chapter 12). Further work may clarify whether planets are forming from cocoon dust near other new stars, and how the jets of gas are expelled, eventually dissipating the cocoon.

The model also clarifies certain puzzling nebulae called *Herbig–Haro objects.* These are small, variable nebulae, sometimes appearing or disappearing during a period of a few years, as shown in Figure 16-11. Astronomers once thought they might be new stars, but recent mapping of bipolar flows of gas near infrared stars indicate that Herbig–Haro objects are clots of material associated with the outflowing gas (Bailly and Lada, 1983). Their luminosity derives from shock waves that excite the ejected gas as it collides at high speed (100^+ km/s) with nearby interstellar gas. (A shock wave is a violent heating and compression of gas as it collides with neighboring gas at faster than the local speed of sound; the sonic boom around a supersonic jet derives from a shock wave in gas compressed in front of the plane.) Additional luminosity may come from varying illumination from the nearby cocoon-shrouded star as the cocoon nebula rotates and breaks apart. In summary, the gas shed by infrared stars and T Tauri stars, and the associated creation of bipolar mass ejection and Herbig–Haro objects, may shed light on exciting new physical phenomena unknown until a few years ago (Schwartz, 1983; Hartmann and Raymond, 1984).

YOUNG CLUSTERS AND ASSOCIATED YOUNG STARS

Because stars form in groups, T Tauri and infrared stars are often found in groupings associated with star-forming regions. In the beginning of this chapter, we described clusters of stars proven to be very young by their dynamical properties and H–R diagram position. Because low-mass stars take the longest to evolve to

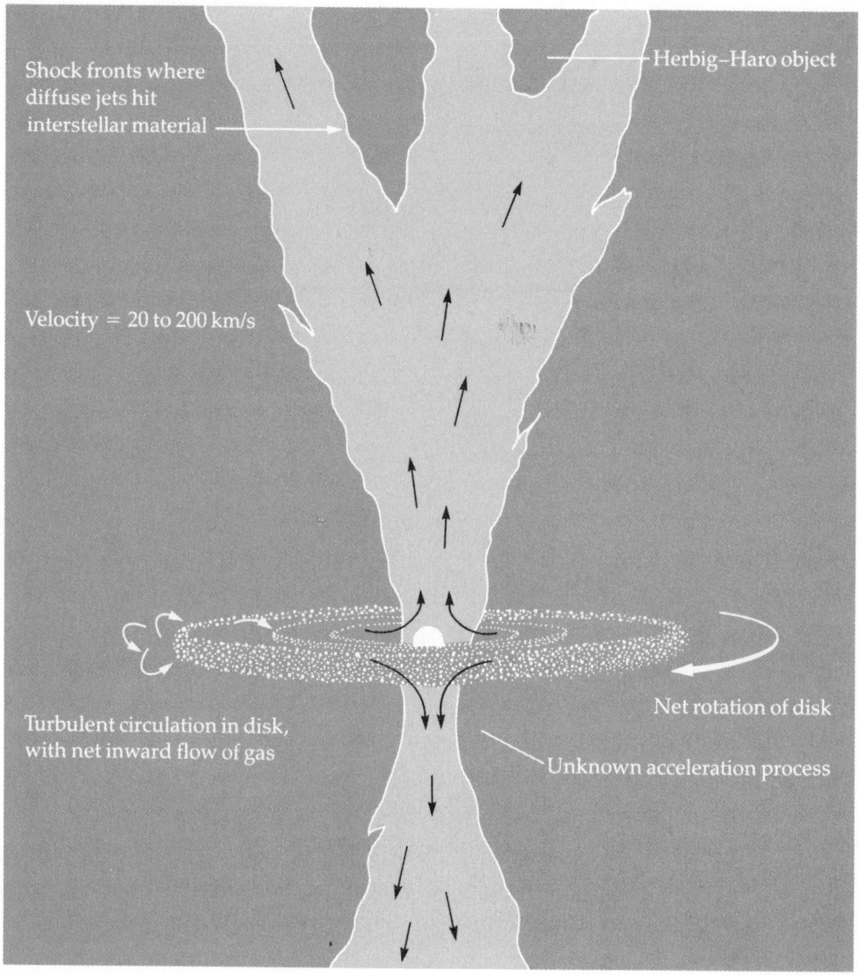

Figure 16-10. *Recent evidence suggests that as a new star turns on in the center of its turbulent disk-shaped cocoon nebulae, gas is blown outward along the disk's axes by an unknown mechanism. Outflowing clouds move "upward" and "downward," at speeds up to 200 km/s, ramming into surrounding interstellar gas.*

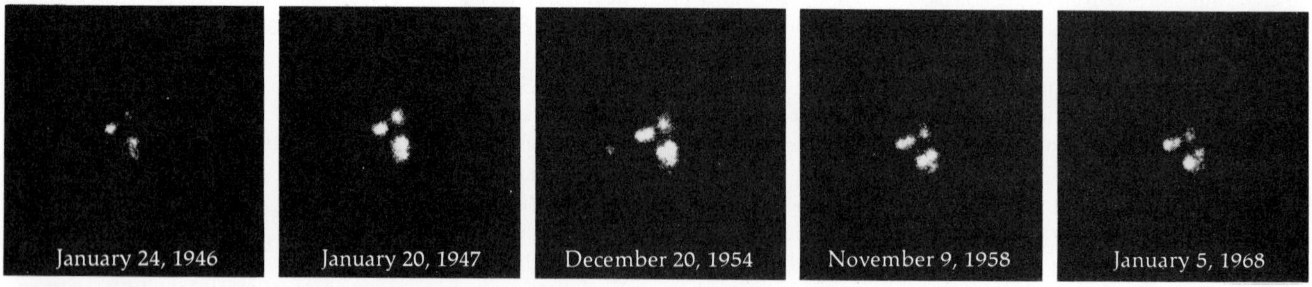

Figure 16-11. *This cluster of Herbig–Haro objects occurs near a site of star-forming activity. The bottom grouping appears to have changed from two to three objects between 1947 and 1954, while the left-hand grouping changed from one to two objects. Such changes probably involve variable glows from supersonic gas masses near young stars, together with variable illumination from young stars partly dimmed by cocoon nebulae. (G. Herbig, Lick Observatory.)*

the main sequence, low-mass stars in these clusters ought to be found in the T Tauri stage. This has been abundantly confirmed by observations.

One of the best-known examples is the young star cluster NGC 2264 (meaning object 2264 in the New General Catalog of astronomical objects), shown in Figure 16-12. Most of the low-mass stars (spectral classes A to K) in this cluster lie distinctly to the right of the main sequence, as shown in Figure 16-13, and many of these are identified as T Tauri stars. The calculated lines of constant age (dashed lines in Figure 16-13) show that the T Tauri stars match the positions predicted for an age of 3 to 30 million years. *This demonstrates how the H–R diagram can be used, together with theoretical data, to determine the* **ages of star clusters.** A 1983 study of 300 pre-main-sequence stars in this cluster indicated that many of them have dusty cocoons and that the low-mass stars formed first, at least 20 million years ago (Adams, Strom, and Strom, 1983).

Another very young group of stars is found in the center of the Orion star-forming region, the great Orion nebula shown in Color Plate 24. Signs of star formation in this region include T Tauri stars, infrared nebulae, and outrushing gas. The H–R diagram of stars within a ½° by ½° section of sky in the Orion nebula is shown in Figure 16-14. Here again we see the characteristic departure of the lower-mass stars (spectral classes B to K in this case) to the right of the main sequence. For this group, the theoretical lines of constant age suggest that the stars average only about a million years in age.

Recall our rule of thumb from Chapter 15 that massive stars evolve fastest. Brilliant, massive, bluish, O-type supergiants last only a few million years. For this reason they are found only in star-forming regions a few million years old. By the time stars have dispersed from such regions, the O supergiants have already burnt out. Thus they are rare among representative field stars but dominate pictures of star-forming areas such as Figure 16-12 or Color Plates 24, 25, 27, and 28.

Debris of Star Formation

Many star-forming regions include clouds of gas and dust that may be remnants of the star-forming process or clouds that did not get quite dense enough to fully collapse. Often, spectacular foreground clouds are seen in black silhouette against bright star groups or clouds illuminated by bright stars. Long, tubular, dark nebulae called *elephant-trunk nebulae* are sometimes found pointing toward the star-forming region as seen in Figure 16-1, a front endpaper, and Color Plate 25. These may be formed as gas rushing out from the star-forming center sweeps past dense clots of dusty gas.

The Main-Sequence Stage

The **main-sequence stage** is reached when the star begins to generate its energy by consuming hydrogen in nuclear reactions deep in the star's central regions. Prior to that time, the star generates energy primarily by its Helmholtz contraction (see p. 228), which raises the temperature and pressure in the central regions. As it enters the T Tauri pre-main-sequence stage, it still radiates by this means and still contracts slowly, as seen by comparing the evolutionary tracks and lines of constant radius in Figure 16-5. As the pressure near the star's center increases, the atoms jam together more tightly, and as the temperature increases, the atoms collide at greater speeds. Eventually individual nuclei of atoms begin hitting each other so hard that they begin to fuse together. These nuclear fusion reactions generate new heat and pressure that stop the contraction. Toward the end of the T Tauri stage, as the star is just settling onto the main sequence, there may be some minor nuclear reactions involving elements that react at lower temperatures than hydrogen, but hydrogen fusion soon becomes the dominant energy source as the main sequence is reached.

In the H–R diagram, an imaginary array of stars with different masses, which have all just reached the main sequence and just begun to consume hydrogen, is called the **zero-age main sequence.** Calculations show that it would be a very narrow band of stars on the H–R diagram. In reality, the main sequence on the H–R diagram is a bit broader because it contains stars of different ages, which have converted different fractions of their hydrogen to helium and thus have slightly different structures.

Once the star reaches the zero-age main sequence, it begins its life as a true star and commences a long sequence of various nuclear reactions. During each such reaction, tiny amounts of matter are converted to energy, providing the heat and light of the star. We can't overemphasize the basic cause of further stellar evolution: The nuclear reactions convert certain elements into other elements, changing the star's composition and thus the energy generation rates; this in turn causes the star to change to new forms. This is the essence of the Russell–Vogt theorem, which you should recall from Chapter 15. To clarify these changes, we will now review the nuclear reactions in stars in more detail.

NUCLEAR REACTIONS IN THE SMALLER MAIN-SEQUENCE STARS

Studies of our own main-sequence star, the sun, have revealed that its energy comes from a series of nuclear reactions called the **proton–proton cycle.** The main steps in this cycle can be listed as a series of reactions:

$$^1H + {}^1H \rightarrow {}^2H + e^+ + \text{neutrino}$$
$$^2H + {}^1H \rightarrow {}^3He + \text{photon}$$
$$^3He + {}^3He \rightarrow {}^4He + {}^1H + {}^1H + \text{photon}$$

In the first reaction, two hydrogen nuclei fuse to form a nucleus of "heavy" hydrogen (deuterium), which contains one proton and one neutron. This reaction liberates a *positron* (positively charged, electron-sized particle) and a chargeless

Figure 16-12. *Swirling mass of glowing gas, young stars, and dark dust clouds is seen in this star-forming region in the constellation Monoceros. Silhouetted dust cloud called the "Cone nebula," bottom, is mapped in Figure 16-9b. The dark "cone" is a dust-grain-and-gas concentration estimated to be about 50 000 AU in width, silhouetted against more distant glowing clouds. Molecules of formaldehyde (H_2CO) and other gases have been found in the region. (Copyright, Anglo-Australian Telescope Board.)*

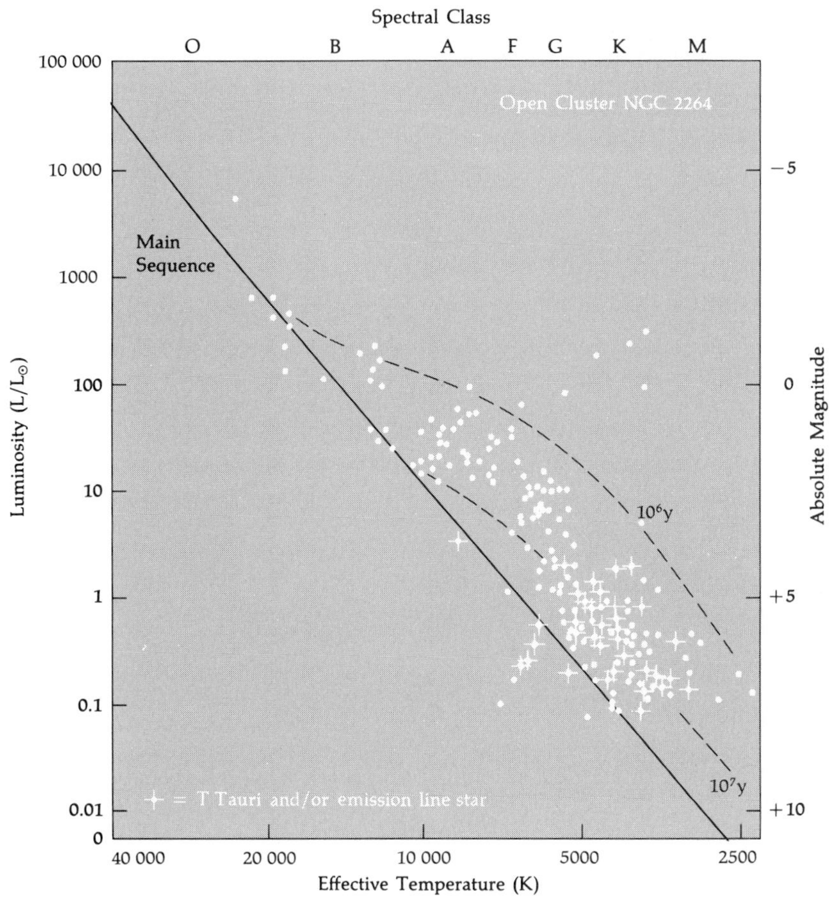

Figure 16-13. *This H–R diagram of stars in the region around Figure 16-12 reveals stars lying to the right of the main sequence along age lines (dashed) suggesting ages of only a few million years. (Data from M. Walker, 1972, with isochrons from I. Iben, 1965.)*

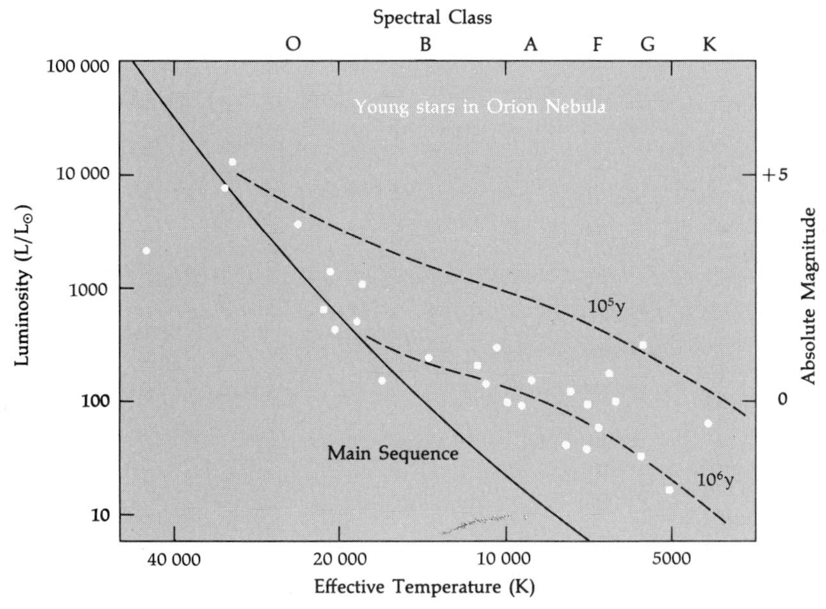

Figure 16-14. *This H–R diagram for 28 stars within ½° of the central Orion nebula (see Color Plate 24) shows positions to the right of the main sequence, along time lines (dashed) indicating an age of about 1 million years. (Data from M. Penston, 1973.)*

particle called a *neutrino*. Since a hydrogen nucleus is just a proton, the first reaction uses two protons and gives the name to the cycle. In the second reaction, the deuterium nucleus fuses with another hydrogen nucleus to form a helium 3 nucleus, releasing a *photon,* or unit of radiation. Finally, two of these helium 3 nuclei fuse to form a helium 4 nucleus, the normal form of helium, as well as two hydrogen nuclei and another photon of radiation. The net result is that four protons are used up to create one helium 4 nucleus. The helium 4 nucleus has slightly less mass than the original four protons, and the difference in mass is converted into photons, or light.

The proton–proton chain is the primary energy-producing process not only inside the sun, but also inside all main-sequence stars smaller than F stars of about $1\frac{1}{2}$ $M_\odot$. It dominates if the central temperatures are less than about 15 million Kelvin.

NUCLEAR REACTIONS IN THE LARGER MAIN-SEQUENCE STARS

In main-sequence stars larger than about $1\frac{1}{2}$ $M_\odot$ where interior temperatures are higher than about 15 million Kelvin, another reaction series dominates in producing energy. This is the **carbon cycle,** sometimes called the *CNO cycle* to reflect the involvement of carbon, nitrogen, and oxygen. The reactions are as follows:

$$
\begin{aligned}
{}^{12}C + {}^{1}H &\rightarrow {}^{13}N + \text{photon}\\
{}^{13}N &\rightarrow {}^{13}C + e^{+} + \text{neutrino}\\
{}^{13}C + {}^{1}H &\rightarrow {}^{14}N + \text{photon}\\
{}^{14}N + {}^{1}H &\rightarrow {}^{15}O + \text{photon}\\
{}^{15}O &\rightarrow {}^{15}N + e^{+} + \text{neutrino}\\
{}^{15}N + {}^{1}H &\rightarrow {}^{12}C + {}^{4}He
\end{aligned}
$$

Again, the net result is that hydrogen atoms are used up to produce helium 4 atoms, with an associated release of energy. In a sense, carbon acts as a catalyst (a stimulant of change), because carbon 12 reappears at the end of the cycle, to be used again in the first reaction of a subsequent cycle.

Summary

Stars are forming even today within a few hundred parsecs of the sun and in more distant regions of space. The starry sky is not a static scene, but the site of continual births of new objects. Many stars and star systems are less than a few million years old—much less than 1% as old as our galaxy. Some have become visible since humanity evolved, though prominent newly formed stars have probably not appeared in our sky during recorded history.

Star formation begins with protostars, which are clouds of dust and gas that begin to contract due to their own gravity. They collapse fairly rapidly to stellar dimensions and become pre-main-sequence starlike objects. Theoretical calculations in the last two decades have revealed what features these objects probably have. Many of these features have been confirmed by observation, especially by infrared equipment, which detects the radiation from relatively low-temperature dust in nebulae around the newly formed stars.

Among the objects revealed in this way are groups of stars evolving toward the main sequence. Many groups of stars and individual stars are surrounded and obscured by cocoon nebulae consisting of dust particles (probably silicates and ices similar to those that formed the first planetary material in our own solar system). More evolved objects, such as the T Tauri stars, appear to be shedding their cocoon nebulae and have almost reached the main sequence.

Once the main sequence is reached, energy is generated in the star as hydrogen converts into helium by means of the proton–proton cycle for smaller stars and the carbon cycle for larger stars.

Concepts

evidence for ongoing star formation
theory of star formation
gravitational contraction
subfragmentation
protostar
collapse
pre-main-sequence star
high-luminosity phase
cocoon nebula
infrared star
black dwarf (brown dwarf)
Bok globule
Herbig–Haro object
Orion star-forming region
T Tauri star
ages of star clusters
main-sequence stage
zero-age main sequence
proton–proton cycle
carbon cycle

Problems

1. Did the sun and solar system form in the first half or last half of our galaxy's history?

2. Suppose that you magically smoothed out all inhomogeneities in the interstellar gas so that it was all uniform.

a. Would this help or hinder star formation?
b. What processes would keep the gas from staying uniform indefinitely?
c. Would star-forming conditions tend to return to normal?

3. Why do stars form in groups instead of alone?

4. How do theories of solar system formation and theories of star formation support each other? Contrast the sources of information on these two subjects.

5. Compare the time scale for significant evolution of massive pre-main-sequence stars with the time during which astronomers have recorded observations of such stars (see Figures 16-4 and 16-5). Is it reasonable that some young, pre-main-sequence stars might show evolution-related fluctuations in their properties within the time that they have been observed?

6. Suppose a cluster of stars formed 3 million years ago. Why would you expect the H–R diagram of the cluster to show no stars on either the very high mass end of the main sequence or its very low mass end?

7. Suppose a new 1-$M_\odot$ star began forming about 10 pc from the sun. What would earth-based observers see during the next few million years? Consider infrared observers as well as naked-eye observers.

8. Why doesn't gravity immediately cause the collapse of all interstellar clouds?

9. Why does the structure of a star stabilize when it reaches the main sequence?

ADVANCED PROBLEM

10. Suppose you observe an infrared nebula whose strongest radiation comes at wavelength 10 μm and which has absorption lines of solid silicates in its spectrum, as well as faint lines indicating a G-type star. Suppose the silicate lines are blue-shifted by about 10^{-3} of their wavelength compared with the star's wavelength. What conclusions can you draw about this system?

Project

1. On a clear night (an early evening in February or a late evening in December is ideal) scan the region of Orion with your naked eyes and compare it to other regions of the sky. Note the concentration of bright blue O-, B-, and A-type stars (such as Sirius and Rigel) and star clusters (such as the Pleiades, Hyades, and the Orion belt region) in this broad area. How do these features indicate that star formation has been going on in this general direction from the sun in the last few percent of cosmic time?

CHAPTER 17

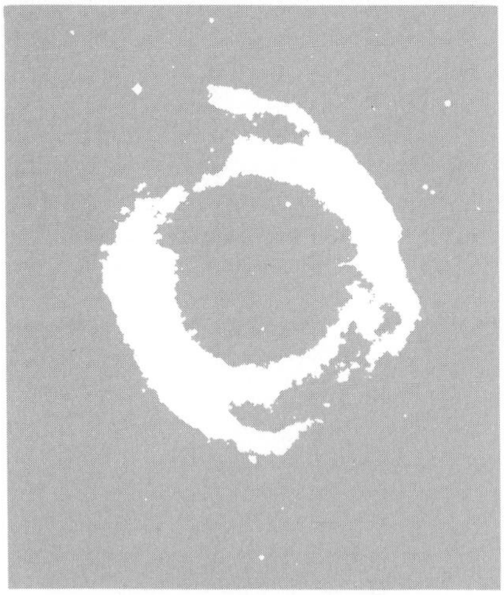

STELLAR EVOLUTION II: DEATH AND TRANSFIGURATION

Stars are born and stars die. By this we mean that the nuclear reactions converting mass to energy in stars have a beginning and an end. The sun, for example, spent some 10^7 y in its formative stages; it will spend about 10^{10} y on the main sequence, roughly 10^9 y in the giant state, and a shorter time in later unstable states consuming heavier elements until energy generation stops (see Figure 15-10). This evolutionary sequence might be likened to the periods of human life: 9 months in the womb, 65 years of normal life, 6 years of rapid aging, and perhaps a year of pathological terminal illness. As we will soon see, stars do evolve to pathological terminal states, featuring incredible forms of dense matter and processes unimagined until a few years ago.

Writer Ben Bova (1973) summed up star deaths by quoting one of Ernest Hemingway's characters, who is asked how he went bankrupt. "Two ways," he says. "Gradually and then suddenly." The gradual expenditure of a star's hydrogen is scarcely noticeable to an observer, but then the star suddenly goes into fits of activity, searching for new sources of fuel until, as Bova says, "gravity forecloses all the loans."

Gravity forecloses the loans by causing a continual tendency toward contraction. When a main-sequence star runs out of hydrogen, the energy-producing hydrogen reactions wind down and contraction of the central core begins, raising the temperature until helium nuclei react. After helium, certain heavier elements serve as fuels, but contraction is always required to raise the temperature to the point where these fuels react. Eventually, no more fuels react, and the star contracts to a very dense state. The chapter will deal with four main conditions of stellar old age: the giant stage, variable stages, explosive stages of several types, and the terminal high-density stages. Individual stars representing examples of these stages are listed in Table 17-1. We will see that some of the dense terminal stages include some of the most exotic objects yet discovered by astronomers, such as pulsars and black holes.

The Giant Stage

As described briefly in Chapter 15, a star evolves off the main sequence as hydrogen fuels run out. Exactly what happens as the hydrogen is exhausted in the central core? At this time, a thin shell in which hydrogen is still burning surrounds the core. Inside this shell, there is virtually no hydrogen left—nearly all nuclei are helium. Outside the shell, temperatures have never risen high enough to "ignite" the hydrogen nuclei (that is, to fuse them into helium) so there is still plenty of hydrogen fuel. Since nuclear reactions in the core are now inadequate to keep the core expanded, the core starts to contract. As always, this contraction raises the temperature (see the discussion of Helmholtz contraction, p. 228). This allows the hydrogen-burning shell to migrate further toward the surface. The increased temperatures cause great expansion of the outer layers, making the star evolve rapidly toward the giant state. The outermost atmosphere becomes huge, thin, and cool, even though the inner core is smaller, hotter, and denser than ever. From the outside, the outer atmospheric layers are seen to glow with a dull red color, and the star is perceived as an enormous **red giant**, such as shown in Figure 15-6.

As charted on the H–R diagram (Figure 17-1), stars from all points along the main sequence display a **funneling effect**: Their evolutionary tracks funnel into the red giant region. They resemble patients from all walks of life crowding into the same hospital because they are victims of the same malady—hydrogen exhaustion.

Meanwhile, the star cores contract until they reach temperatures near 200 million Kelvin. This is hot enough to begin to fuse helium nuclei in the central regions of

Table 17-1. ***Examples of Stars in Middle and Late Evolution***

Stage of Evolution[a]	Distance from Earth (pc)	Mass ($M/M_\odot$)	Radius ($R/R_\odot$)	Luminosity ($L/L_\odot$)	Spectral Type	Density[b] (kg/m^3)
Main sequence						
Sun	<1	1	1	1	G2	1420
α Centauri B	1.3	0.85	1.2	0.36	K4	700
Procyon A	3.5	1.7	2.3	6	F5	200
Algol A	31	4	3.0	100	B8	210
Giant						
Arcturus	11	~4	25	115	K1	0.36
Aldebaran A	21	~4	45	125	K5	0.07
β Pegasi	64	~5	140	400	M2	0.003
Supergiant						
Antares A	160	~12	700	9000	M1	0.000 05
Betelgeuse	170	~18	~700	11 000	M2	0.000 07
VV Cephei A	200–1200?	20–80?	400–1600?	200–10 000?	M2	0.000 03?
Variable and Explosive Stars						
Mira A (long period variable)	77	2	230	Up to 1100	M6	~0.000 2
Polaris A (Cepheid variable)	120?	~6	~25	830	F8	~0.5
HD 193576B (Wolf–Rayet)	?	12	~7		"WR"	~50
DQ Herculis B (nova)	?	0.2	~0.1	<0.1	M?	300 000
White Dwarf						
Sirius B	2.7	1.0	0.022	0.002	A5	130 000 000
Procyon B	3.5	0.65	0.02	0.000 5	F	120 000 000
Pulsar (neutron star)						
Crab nebula pulsar	1100	~2??	<0.000 02	High in UV, X ray	—	$\sim 10^{17}$?
Possible Black Hole						
Cygnus X-1	⩾2500	6–15?	<0.000 004	0	—	$\sim 10^{22}$?

Source: Data from Burnham (1978), Liebert (1980), Baize (1980), and Bonneau et al. (1982; Mira).

[a]Listed in order of increasing age.

[b]For comparison, the following are densities of familiar materials: atmosphere at sea level, 1.2 kg/m^3; water, 1000 kg/m^3; lead, 11 300 kg/m^3.

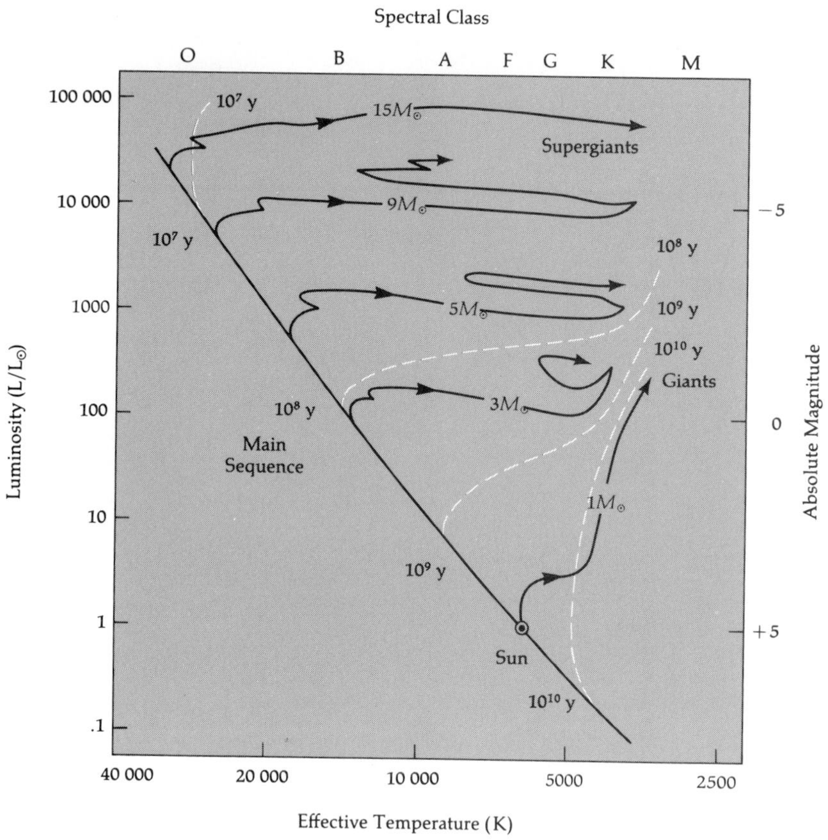

***Figure 17-1.** Evolutionary tracks off the main sequence toward the giant region, plotted on the H–R diagram for stars of different mass. Dashed lines give the length of time since star formation; massive stars evolve fastest. All the stars evolve toward the upper right corner of the H–R diagram, the region of giant stars. (After calculations by Iben.)*

most stars, primarily by the **triple-alpha process** (named after the alpha particle, another name for the helium 4 nucleus):

$$^4\text{He} + {^4\text{He}} \leftrightarrows {^8\text{Be}} + \text{photon}$$
$$^8\text{Be} + {^4\text{He}} \rightarrow {^{12}\text{C}} + \text{photon}$$

In this process, three helium 4 nuclei combine to produce a carbon 12 nucleus. (Because beryllium 8 is unstable, some beryllium atoms may break up before completing the process—but this merely reduces the efficiency of the process.) Notice that we are now burning helium to make an even heavier element, carbon.

The triple-alpha process produces prodigious energy, which at once heats the rest of the core, creating a sudden burst of helium burning often called the **helium flash**. In some stars, the helium flash may consume the central core's helium in only a few seconds. Since the energy from the flash diffuses through the star slowly, the heating effects seen at the surface may last thousands of years. The flash occurs as the star enters the giant region. In a sun-sized star, it may occur 300 million years after evolution off the main sequence.

After the helium flash, "squiggles" may develop in the evolutionary track. Reduction of pressure in the outer atmosphere may allow the star to contract, temporarily moving it back toward the main sequence. Further contraction of the interior core leads to still higher temperatures and reactions involving even heavier elements. Carbon-burning may occur. Some of the reactions create floods of neutrons that strike nearby atoms. These initiate the so-called **s-process reactions**, in which neutrons are *slowly* added to nuclei to build up still heavier elements, and the **r-process reactions**, in which *rapid* neutron addition builds up very heavy elements. The s-process and r-process reactions are very important because of the production of the heavier elements, such as metals. Without such reactions, the universe would still consist of almost pure hydrogen and helium and would lack sufficient heavy elements to make planets such as the earth. Table 15-3, in Chapter 15, reviewed this sequence of element evolution in stars.

STARS THAT BLOW OFF MASS

Some of the more massive stars (approaching 6 $M_\odot$) get so hot that they not only expand, they blow off their outer layers. Gas may be ejected either in explosions or in strong stellar winds, similar to (but stronger than) the solar wind (see Chapter 13). These stars leave the main sequence at a luminosity already brighter than giants and are thus called **supergiants**. Stars brighter than absolute magnitude −6 typically have too much mass to evolve smoothly or slowly and are among those that explosively blow off an appreciable fraction of their initial mass even as they become giants. Smaller stars may also lose mass; they have steadier stellar winds before, during, and after their giant stages. Examples of various types of mass-ejecting stars are described below.

O and B Stars. These are among the most massive stars. Ultraviolet observations from rockets and satellites, beginning in 1967, have revealed that massive O stars (greater than 20 $M_\odot$) have strong stellar winds from the time they leave the main sequence, blowing away as much as 10^{-5} $M_\odot$ of gas per year at speeds of 1500 to 1900 km/s (Conti and McCray, 1980). They could blow off several solar masses in less than a million years as they evolve toward the giant state. Plione, a B star in the Pleiades (see Color Plate 28), blew off glowing clouds revealed by spectral emission lines in 1888, 1938, and 1972.

FG Sagittae. This star, a supergiant 2500 to 4000 pc away, has the fastest known evolutionary rate. It has been moving toward the giant region in the H–R diagram for decades, brightening by 4 to 6 magnitudes since 1894. From about 1960 to 1978, it has cooled from spectral class B4 to G and expanded from 24 to 70 solar radii ($R_{\odot}$). A glowing nebula around it is expanding at 34 km/s, indicating that gas was blown out about 6000 to 9600 y ago. From 1969 to 1972, certain heavy elements in the star's atmosphere, such as zirconium, have increased 3 to 4 times in abundance, indicating that gas from the interior (where the heavy elements are created) has erupted. This star is interpreted as a star that originally had 3 to 5 $M_{\odot}$ but is losing mass as it evolves from the main sequence to the giant region.

M Giants. **M giants** often have blue-shifted absorption lines, arising in dusty gas blown off the star and expanding in stellar winds toward outside observers.

Wolf–Rayet Stars. The **Wolf–Rayet stars**, named after their discoverers, are very hot stars resembling O stars except for their expanding gaseous shells and compositions. Their spectra show that their surface gases are much richer in helium, carbon, and nitrogen than most stars. As indicated in Figure 17-2, Doppler shifts show that they have thrown off surrounding clouds of gas at velocities as fast as 1000 km/s! Wolf–Rayet stars are believed to be stars that began with very high mass and eventually blew away most of their outer layers of gas, exposing their cores in which helium and heavier elements had accumulated due to hydrogen burning (Humphreys and Davidson, 1984). More than 200 have been found since their discovery in 1867, and many are members of binary systems (pairs of coorbiting stars).

In most of these examples, only a fraction of the total mass of the star is being blown away, but as shown in Figure 17-3 and Color Plate 26 (top), the expelled debris may be spectacular. The helium-rich cores left behind by some of these stars are so hot that they radiate much of their light in the ultraviolet (UV) part of the spectrum. Recent UV astronomical satellites have observed these objects and measured their temperatures. The photons of UV light, which are more energetic than visual photons, help excite the atoms in expanding clouds of debris such as that in Figure 17-3, making them glow.

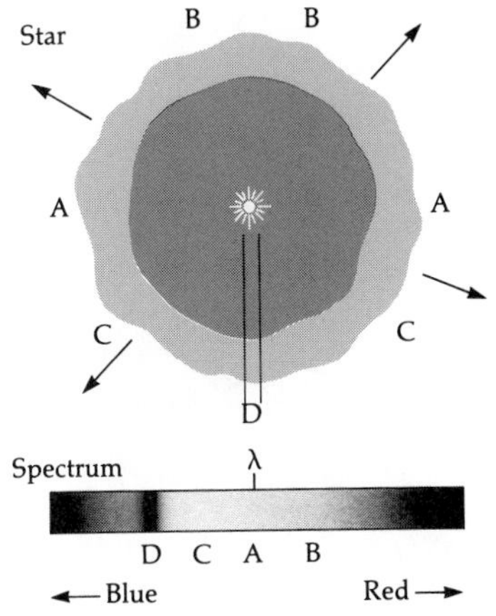

Figure 17-2. *Expanding shells of thin gas, such as those around Wolf–Rayet stars or novae, can be detected spectroscopically, even when unresolved photographically. Glowing gas produces spectral emission lines of the shape shown at the bottom. In region A, light with no Doppler shift (wavelength λ) is added to the continuum spectrum of the star. In region B, red-shifted light is added, and in region C, blue-shifted light. In region D, the maximum blue shift occurs (especially in Wolf–Rayet stars), and starlight is absorbed by the nebula, producing a dark absorption band at the blue end of the emission line.*

The Variable Stage

A **variable star** is a star that varies in brightness. Most variable stars apparently represent *postgiant stages* of evolution (although, as we saw in the last chapter, some variables are pre-main-sequence stars). Some pulse with a constant period; others flare up sporadically, often brightening by many magnitudes and then fading again. Such stars were recorded as long ago as 134 B.C. when Hipparchus recorded the flare-up of a "new star," proving that the heavens were not eternally constant. The star Omicron Ceti was known to the ancients as Mira (Wonderful) for its brightening from invisibility (magnitude 8–10) to the second magnitude every 331 d.

Shakespeare used the simile "constant as the northern star." He might have been interested to know that Polaris is constant neither in brightness nor in marking the north celestial pole. Polaris is a variable star that changes from magnitude 2.5 to magnitude 2.6 in just less than 4d. Because of precession (Chapters 1 and 2), it is the North Star for only a few centuries every 26 000 y.

Some 22 650 variable stars have been cataloged and divided into as many as 28 types. The type most important to astronomers is the **Cepheid variable**, which has regular variations in brightness, usually with periods from 5 to 30 d. In 1784, 19-year-old, handicapped English astronomer John Goodricke[1] discovered the first Cepheid variable, Delta Cephei. It varies from magnitude 4.4 to 3.7 with a period of 5.4 d; Cepheids were named after that star.

Cepheids are important for two reasons. First, because their variations are

[1]Goodricke offers an inspirational example of accomplishment during a short, difficult life. Deaf from infancy, he lived during the first European generation to recognize that deafness was not the same as idiocy. He attended the first English school for deaf children, took up astronomy, and made numerous interesting observations, but died at age 21 from pneumonia, possibly contracted from nighttime exposure during prolonged observing efforts.

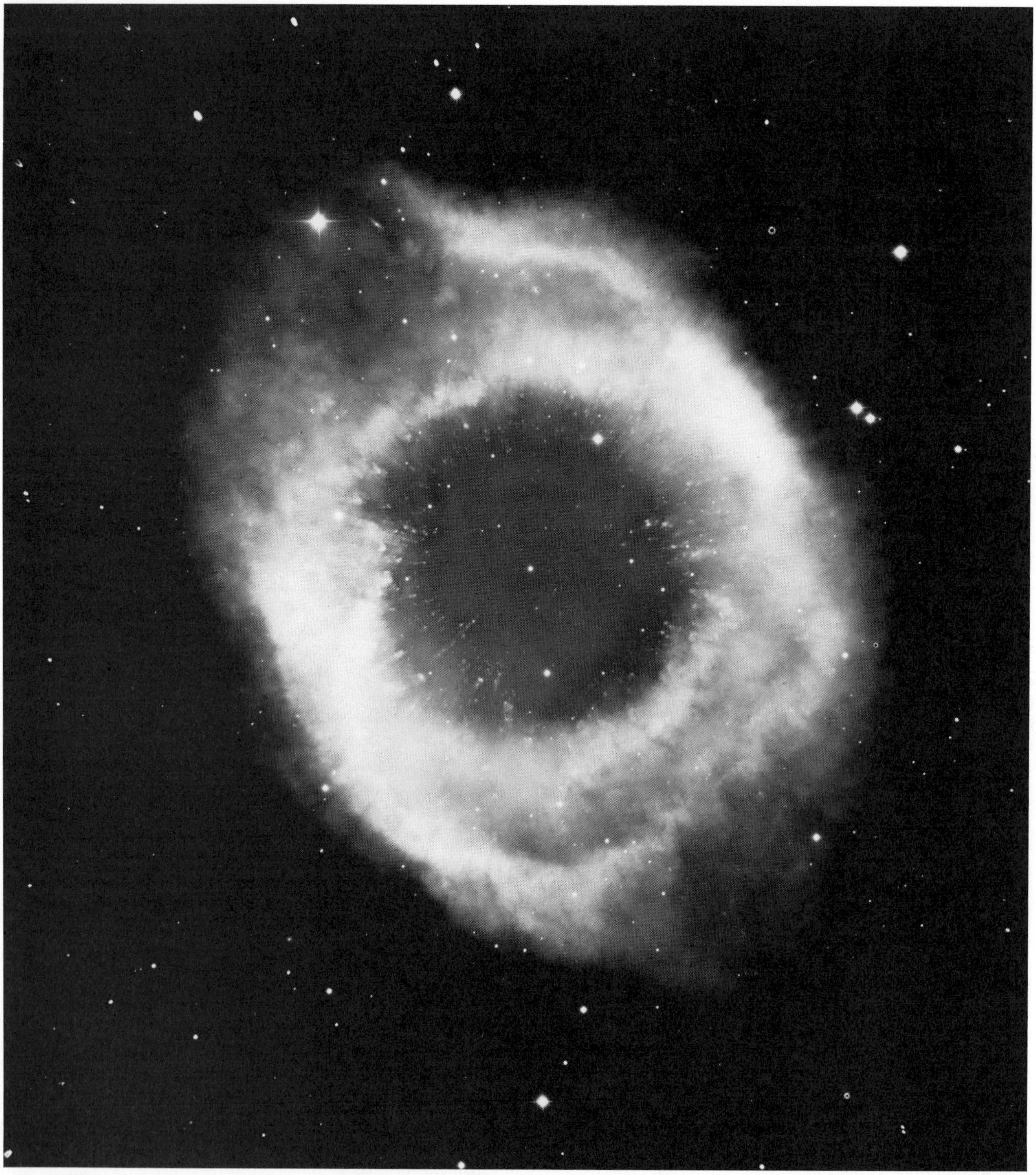

Figure 17-3. *Nebula NGC 7293 in Aquarius is a glowing spheroidal shell of gas probably blown out of the hot, moderate-mass central star by very strong stellar winds. Delicate filaments radiate from the center. (Hale Observatories.)*

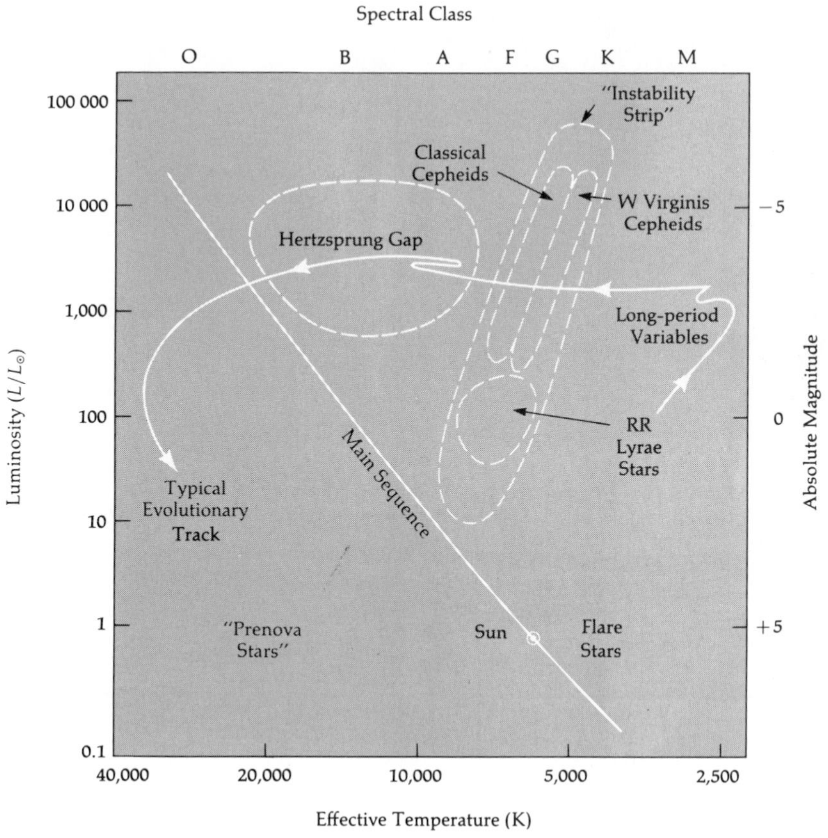

Figure 17-4. *An H–R diagram showing the later stages of evolution, indicated by a typical evolutionary track (in white). The star evolves across an "instability strip" that contains Cepheid variable stars and RR Lyrae stars and then rapidly across the Hertzsprung gap. The diagram also shows the probable position of prenova stars.*

regular, they are somewhat better understood than stars whose brightness changes are unpredictable (called **irregular variables**). Second, and more important, the period of variation of each Cepheid directly correlates with its average luminosity. This relation was discovered in 1912 by one of the most famous women astronomers, Henrietta S. Leavitt, who measured hundreds of photos of Cepheid variables in the first years of this century at Harvard Observatory. Two types of Cepheids and a related type of variable called an RR Lyrae star were eventually found to have distinct period–luminosity relationships. These relationships allow astronomers to determine the luminosity of any Cepheid at any distance simply by measuring its period. This, in turn, leads to a new way to measure the distances of stars[2]: find a Cepheid; measure its period and hence its luminosity, or absolute magnitude; then measure its apparent magnitude and derive its distance from the relations shown in Figure 14-12.

CAUSE OF BRIGHTNESS VARIATIONS

Most variables change brightness because of changes in their interior nuclear reactions or in the flow of radiation from their centers. Variables vary in physical size and other properties as well as in brightness. Delta Cephei itself varies in diameter (from roughly 38 million to 41 million kilometers) and surface temperature (from 5600 to 6500 K) during its cycle.

Astrophysicists have long sought the precise cause of the internal physical disturbances that make Cepheid variable stars pulsate. Various researchers, especially the Soviet astronomer S. Zhevakin, have established that the H–R diagram contains a nearly-vertical *instability strip,* shown in Figure 17-4. Stars become unstable where their evolutionary tracks cross this strip. Inside such stars ionized helium absorbs outgoing radiation and thus becomes doubly ionized. In the instability strip, this absorption occurs not far below the star's surface, where it temporarily "dams up" the radiation.

Just like a bell (or any other object), a star has a certain frequency or time period in which it tends to vibrate in response to a disturbance. If the time required to dam up the radiation is close to the star's natural oscillation period, the star will begin to oscillate repeatedly. This is what happens in a Cepheid. As the radiation is dammed, pressure builds up and the outer layers of the star expand. But once the expansion gets going, momentum and the star's natural tendency to oscillate at this speed carry the expansion too far, allowing radiation to escape easily. Then the star subsides, and the radiation once again begins to be dammed, restarting the cycle. In other types of stars, where the time scales of radiation damming and oscillation are not synchronized, the variation occurs irregularly.

THE HERTZSPRUNG GAP

Just to the left of the instability strip in the H–R diagram (Figure 17-4) is the *Hertzsprung gap,* a vertical band nearly empty of stars. Why do no stars have the combination of luminosity and temperature that would put them in this gap? Apparently the gap represents an evolutionary stage so unstable that stars evolve very quickly across it.

[2]For discussion of this very important technique in mapping distances of star clusters and similar galactic features, see Chapter 20, p. 369.

Novae: Exploding Members of Binary Pairs

A **nova** ("new star"; plural, *novae)* is a *brilliant exploding star* that blows off about 0.001 of its mass in a single eruption lasting a few days. Figure 17-5 shows the brilliant flare-up and subsequent fading of Nova Herculis, a nova that appeared in the constellation Hercules in 1934. Novae appear to mark a penultimate stage of evolution for certain stars. Their observable history is usually the following:

Prenova state: *faint blue star below main sequence in H–R diagram.*

↘

Sudden brightening: *often 9 magnitudes (factor 4000) in first day.*

↘

Peak brightness: *often 12 magnitudes (factor 60 000) brighter than originally. Emission lines indicate ejected gas.*

↘

Decline in brightness: *about 3–7 magnitudes over hundreds of days.*

↘

Continuing slow decline for years: *often irregular variations in brightness.*

↘

Postnova state: *expanding gaseous nebula may be visible in large telescope.*

Of the 100 billion stars in our galaxy, probably 30 to 50 explode as novae each year. Table 17-2 lists some of these "new stars" that have reached second magnitude or greater since A.D. 1000. Some of these stars were important links in humanity's discovery that the heavens are not immutable, as once thought, but are actually evolving.

The best-observed nova in history was Nova Cygni, which provided Cygnus the Swan with a prominent extra star for a few days in 1975.[3] Most novae have not been

Figure 17-5. *Nova Herculis, showing the decline in brightness from March 10, 1935 to May 6, 1935. Later studies showed that this and other novae are binary stars. (Lick Observatory.)*

Table 17-2. *Selected "Guest Stars" (Novae, Supernovae, and Variables That Have Become Brighter Than Apparent Magnitude 2)*

Star	Date Observed	Maximum Brightness (apparent magnitude)	Type
Lupus supernova	1006	−5?	Supernova
Crab nebula explosion	1054	−2 to −6?	Supernova
Tycho's star	1572	−4?	Supernova
Kepler's star	1604	−2?	Supernova
Cassiopeia A[a]	1680?	+2 to +5?	Supernova
Eta Carinae	1843	−0.8	Nova?
T Corona Borealis	1866	+1.9	Recurrent nova[b]
GK Persei	1901	+0.2	Nova
DQ Herculis	1934	+1.3	Nova
Nova Cygni	1975	+1.9	Nova
o Ceti (Mira)	—	+1.0	Long-period variable (period 331 d)

Source: Data from Glasby (1968); Lupus data from *Sky and Telescope* (July 1976).

Note: "Guest Star" is the ancient Chinese term for temporarily visible stars. Fourteen supernovae ranging back 2000 y have been tentatively identified from ancient records. Certain expanding nebulae must also mark prehistoric supernovae.

[a]Cassiopeia A is the brightest radio source, and its expanding cloud has been imaged by X-ray telescopes in space. It may be the transient "star" observed by Flamsteed in 1680 [*Nature 285* (1980): 132]. See Figure 17-10.

[b]Second recorded flare-up in 1946 reached only +3 magnitude.

[3]Note that it would be incorrect to say "exploded in 1975." Since the nova was estimated to be 1300 pc (4200 ly) away, the explosion really occurred around 2225 B.C.! The light just didn't get here until 1975.

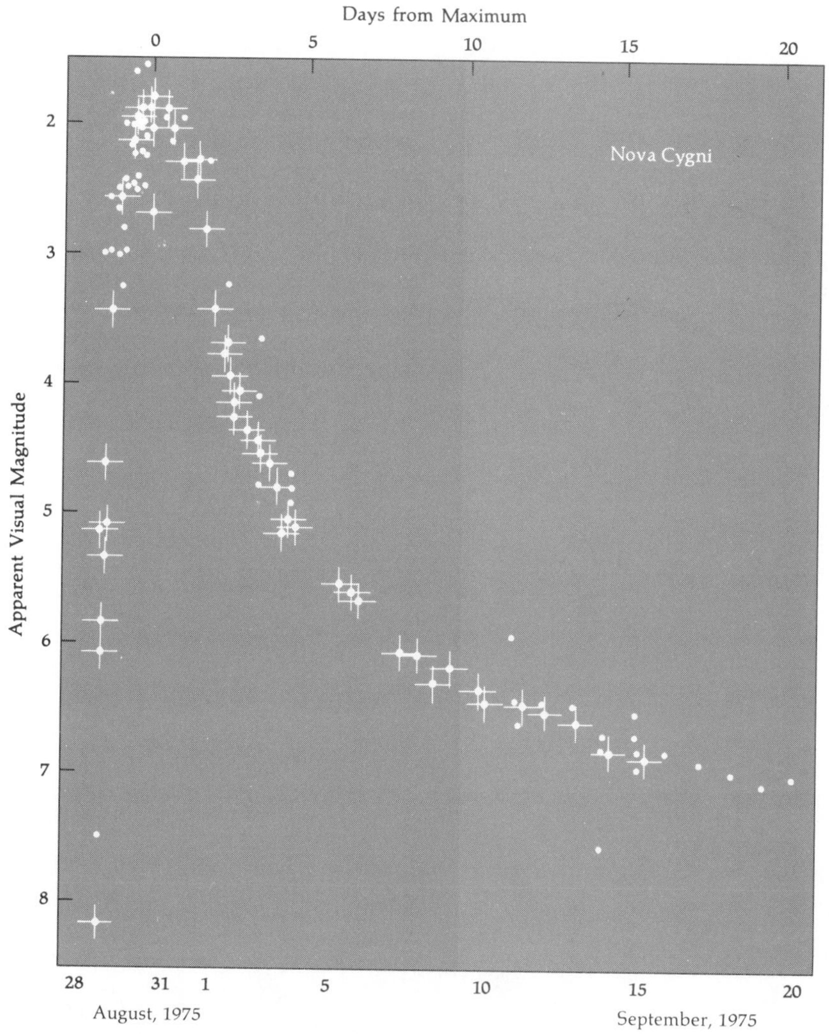

Figure 17-6. *The brightness fluctuations of well-observed Nova Cygni in 1975 include observations made both before and after its discovery on August 29. Small dots are visual estimates of brightness by amateur and professional astronomers; larger symbols are more accurate photoelectric or photographic data. Nova Cygni is estimated to be 1300 pc away. (From International Astronomical Union data.)*

observed during the early stages of the explosion, but during the Cygnus event, amateur astronomers in California and Maryland were photographing the sky and actually recorded the nova as it brightened. As shown in Figure 17-6, the star was doubling in luminosity about every 2 h during the explosion! Measures of Doppler shifts showed that the ejected gas cloud expanded at about 1000–2000 km/s.

What causes novae? Prenova stars, though poorly observed, are believed to be small, hot stars lying in the dwarf region below the main sequence, as shown in Figure 17-4. How can such seemingly insignificant stars produce, in a few days, outbursts that equal the sun's energy output in 10 000 y? A clue came when the American astronomer Merle Walker showed that a 1934 nova was one star of a very close binary pair that coorbit in only 4½ h. In 1955, Russian–American astronomer Otto Struve theorized that all novae might be members of binaries that are too close together to be seen separately from the earth. In 1964, Robert Kraft showed by spectroscopy that at least seven of ten other novae are binaries. Astronomers now think all novae are members of close binary systems.

What is it about membership in a close binary that promotes nova explosions? The more massive star of the pair evolves faster and goes through the giant stage, eventually burning all its fuel and contracting into a white dwarf. Later the less massive star expands to the giant stage and begins to blow off gas. Some of the gas may fall onto the surface of the dormant dwarf, adding new hydrogen and causing intermittent explosions. The expanding giant may even engulf the dwarf in some cases. There is some evidence that novae may occur in cycles. Hydrogen may accumulate until nova explosions occur, perhaps 10 000 y apart. The individual explosions blow off only a small fraction of the star's mass, leaving the rest intact so that the process can start over again. Precise details remain unclear.

Among the closer novae, the cloud of expanding debris can sometimes be seen in telescopes a few years after the explosion, as seen in Figure 17-7.

Terminal Stages: The March Toward High Density

As a star (or the remnant core of an exploded star) collapses, its density increases, atoms are fused into heavier atoms, and eventually a dense state is reached where no more nuclear reactions occur among the heavy elements in the star's interior. At this point no new heat is produced to resist the star's inexorable tendency to shrink. Only the forces between nuclear particles counteract gravity to determine what final, dense state the star will reach. These states produce some of the most remarkable objects known in the universe: white dwarfs, neutron stars (pulsars), and black holes.

WHITE DWARFS

In 1844 the German astronomer Friedrich Bessell studied the motions of the brightest star in the sky, Sirius, and found that it is being perturbed back and forth by a faint, unseen star orbiting around it. This star was not glimpsed until 1862, when American telescope maker Alvin Clark detected it. It is almost lost in the glare of Sirius, as shown in Figure 17-8. In 1915, Mt. Wilson observer W. S. Adams discovered that it was a strange, hitherto-unknown type. It is hot, bluish-white, and lies below the main sequence on the H–R diagram. It has about the mass of the sun, but is so faint that its total radiating surface cannot be much more than that of the earth.

A star the size of the earth? If a solar mass were packed into an earth-sized ball, it would be astonishingly dense. A cubic inch would weigh a ton! A thimbleful would collapse a table!

Dutch–American astronomer W. J. Luyten, who discovered a number of these stars, described what happened as astronomers first recognized Adams's amazing results (quoted by Shapley, 1960):

The figures were too staggering to be believed without further evidence. Could there be an error somewhere? What had happened? But while we pondered these things, two more stars of the same kind were found: small, extremely faint, and white, and all three were named white dwarfs.

A **white dwarf**, then, is a planet-sized, superdense star. About 200 are known. They apparently represent a terminal stage of evolution after a star has run out of nuclear fuel and gravitationally contracted to a very small size.

Theoretical astrophysicists such as Sir Arthur Eddington and the Indian–American theoretician S. Chandrasekhar explained the evolution of white dwarfs. Once no more energy is available to generate outward pressure, a star collapses until all its atoms are jammed together to make a very dense material. But what does "jammed together" really mean? At these densities, some electrons move at almost the speed of light and matter loses its familiar properties. Stellar matter stops behaving like a perfect gas. In fact, it is no longer either gas, liquid, or solid, but a new form known as **degenerate matter.** By a physical law called the **Pauli exclusion principle**, only a certain number of electrons can be jammed into a given small volume of space. Thus, the atomic nuclei are held apart by a sea of electrons—a so-called degenerate electron gas—at densities of about 10^8 to 10^{11} kg/m^3. White dwarfs are sometimes called *degenerate stars.*

Astronomical research on white dwarfs spurred more general physical research in the 1930s on the problem of high-density forms of matter. Luyten has gone so

Figure 17-7. ***Gaseous debris, blasted out of Nova Persei during an explosion observed in 1901, formed this expanding nebula around the star. This photo was made 58 y after the observed outburst. (Lick Observatory.)***

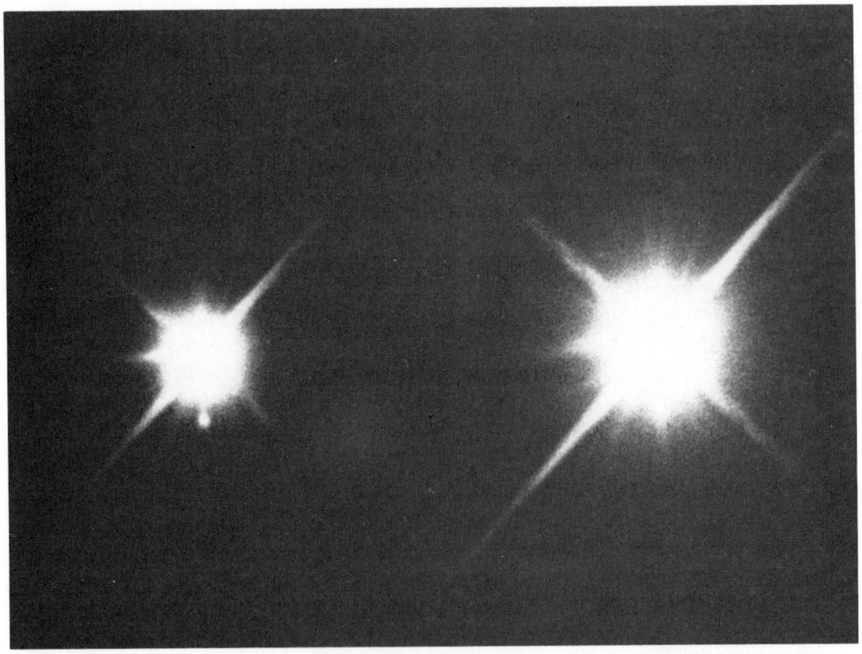

Figure 17-8. *Two exposures of Sirius with the Lick 3-m telescope. Short exposure (left) reveals the faint white dwarf (below Sirius) that is the binary companion roughly 25 AU from Sirius; in the long exposure (right) the white dwarf is lost in the overexposed image. "Rays of light" are artifacts of the telescope optics. (Lick Observatory.)*

far as to say that analyses of white dwarfs "forged another important link in the chain of events which eventually led to the atomic bomb."

In the 1970s astronomical satellites above the atmosphere began to record UV and other shortwave radiation from sources such as white dwarfs, allowing better determinations of their temperatures, which are often high. Sirius B, for example, has a temperature of 27 000 to 32 000 K (Liebert, 1980). The U.S.–U.S.S.R. Apollo–Soyuz joint astronaut flight in 1975 yielded the highest reported white dwarf temperature, 150 000 K, for white dwarf HZ 43. This star is a possible "missing link" between prenova explosive dwarfs and ordinary white dwarfs. BE Ursae Majoris is close behind at 100 000 K.

The coolest white dwarfs have temperatures around 5000 K, and theory indicates these may have cooled for about 7 billion years to reach these temperatures (Helfand, 1983).

Although the interiors of white dwarfs may have densities of billions of kilograms per cubic meter, the outer layers of some may consist of ordinary matter—possibly hot gas at the surface and crystalline rocklike or glasslike solids in a crust 20 to 75 km deep. At the base of these crusts, densities may be as high as 3 million kilograms per cubic meter! Theoreticians predicted pulsations with periods of a few minutes in white dwarfs of certain temperatures, and these pulsations have been confirmed observationally (Helfand, 1983).

White dwarfs cannot have more mass than 1.4 $M_{\odot}$, because the white dwarf structure becomes unstable at this point. This critical mass, 1.4 $M_{\odot}$, is called the **Chandrasekhar limit**, after its Indian–American astrophysicist discoverer. Stars with masses from about 0.08 to 1.4 $M_{\odot}$ apparently evolve smoothly to the white dwarf state. Current data suggest that stars from 1.4 to about 6 $M_{\odot}$ also evolve into white dwarfs by developing strong stellar winds (like the solar wind) or eruptive explosions that blow off mass until they are below the Chandrasekhar limit (Liebert, 1980).

Supernovae: Exploding Unstable Stars

What happens to stars more massive than the 6 $M_{\odot}$ limit? They explode catastrophically in some of the most titanic outbursts known in astronomy. Although such an event in our night sky looks roughly like a nova, detailed studies showed that such events are much more energetic than novae, and permanently disrupt the exploding star. A nova explosion blows off only a small fraction of the star's mass and may recur as more mass accumulates on the unstable member of the binary pair. The superexplosions that disrupt evolved stars are called **supernovae.** A supernova explosion takes the star to the amazing luminosity of 10 billion times that of the sun! For a few days, the star blazes with the light of an entire galaxy!

Many supernovae (and some novae) have been close enough to the solar system to produce temporarily prominent "new stars" that were recorded by ancient people. The ancient Chinese called them "guest stars," as indicated in Table 17-2. Some astronomers have suggested such an explanation for the Star of Bethlehem.

The most famous supernova was the explosion that produced the Crab nebula (Color Plate 26). It was visible in broad daylight for 23 days in July, 1054, and at night for the subsequent 6 months. It was recorded in Chinese, Japanese, and Islamic documents, and perhaps in American Indian rock art, as seen in Color Plate 26 and Figure 17-9. The nebula is the expanding, colorful gas shot out of this supernova. Other supernova remnants are scattered throughout our galaxy (see Figure 17-10). Records indicate about 14 supernovae in our galaxy during the last 2000 y, an average of 1 every 140 y. Any of us might live to see a bright one.

Supernovae occur in other galaxies, too.

Figure 17-9. *Sketch of the appearance of the crescent moon and the Crab nebula supernova rising in the predawn sky 2 to 3 h before sunrise in the American West on the first morning of its visibility, July 5, 1054. For a possible eyewitness painting of this scene by American Indians, see Color Plate 26. The sketch is based on computer calculations of the moon's position. According to Chinese records, the supernova was visible in daylight for 23 d. Its initial apparent magnitude was about −5, about 10 times as bright as Venus. See also caption for Color Plate 26.*

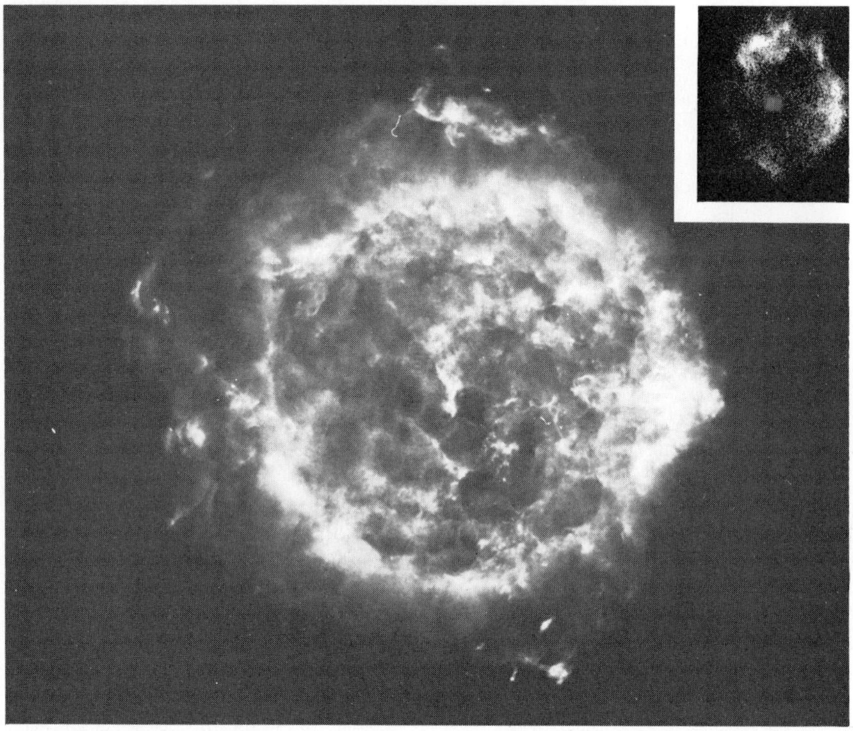

Figure 17-10. *Cassiopeia A, a spherical, expanding shell of hot gas blown out of a supernova visible around* A.D. *1680 (see Table 17-2). This nebula is not prominent visually, but has been imaged by its radio and X-ray radiation. Observations indicate that the gas has a temperature of a few million Kelvin and is rich in silicon, sulfur, and other heavy elements blown out of the inside of the supernova star, which had an original mass around 10–30* $M_\odot$. *The cloud diameter is probably a few parsecs. Radio image at wavelength 6 cm. (National Radio Astronomy Observatory, R. Tuffs, R. Perley, M. Brown, and S. Gull.) (Inset) X-ray image from unmanned orbiting Einstein Observatory. (Courtesy S. S. Murray, Harvard/Smithsonian Center for Astrophysics.)*

In 1885, a supernovae in the nearby Andromeda galaxy briefly doubled that galaxy's brightness. More than 300 have been recorded in other galaxies.

What happens inside a star to cause a supernova explosion? This is still a problem of current research, but one clue is that there are two types of supernovae, distinguished by spectra and rate of fading. Type I is probably a special case where mass from one evolved star gets dumped on a closeby, carbon-rich white dwarf binary companion, near the Chandrasekhar limit of 1.4 $M_\odot$. This is similar to the nova-producing situation, but instead of causing intermittent outbursts, it detonates a sequence of reactions producing enough energy to blow the white dwarf apart entirely. These reactions create debris consisting of the isotope nickel-56, whose decay over a period of weeks gives off the slowly declining light common to Type I supernovae (Wheeler and Nomoto, 1985).

Type II supernovae are of more interest because they mark the common fate of all stars too massive to be stable—more than about 6 to 8 $M_\odot$. As described earlier, each round of nuclear fusion reactions in a star produces a new heavier element. The interior of the star actually acquires an onionlike structure with shells composed of different elements. The early hydrogen burning at a few million Kelvin produces a core of helium. As hydrogen fuel is exhausted in this core, it contracts, heats up, and ignites at a few hundred million Kelvin in new reactions that produce a carbon core, surrounded by a helium shell, overlaid by hydrogen. Neon and oxygen burning follows at still higher temperature until an inner core of iron is built up, surrounded by layers of silicon-rich material, oxygen–neon, carbon, helium, and hydrogen. Nuclear structure is such that no more energy can be produced by "burning" iron. Energy is consumed, not produced, when iron atoms fuse into heavier atoms. Thus, iron doesn't ignite. As astrophysicist Frank Shu (1982) comments, "iron . . . is the ultimate slag heap of the universe." The iron core of a massive star thus contracts and heats. It just collapses, reaching ever higher temperatures, higher densities, and smaller sizes. When the incredible temperature of several billion Kelvin is reached, the iron nuclei hit each other so hard that they break apart into protons, neutrons, and other subatomic particles such as neutrinos. Calculations indicate that the core overcollapses and then rebounds outward, blowing away the outer layers in a titanic supernova explosion (Shu, 1982; Reddy, 1983; Wheeler and Nomoto, 1985). Astronomers are continuing such calculations to clarify the process, but it is clear that only the most massive stars, above 6 to 8 $M_\odot$, create supernovae.

NEUTRON STARS (PULSARS): NEW LIGHT ON OLD STARS

What is left after a supernova explosion? As early as 1934, American astronomers W. Baade and F. Zwicky made a correct guess:

> *With all reserve we advance the view that a supernova represents the transition of an ordinary star into a neutron star, consisting mainly of neutrons. Such a star may possess a very small radius and an extremely high density.*

Nuclear physicists were already working on the problem of very dense forms of matter, and they quickly took up the challenge of dense star forms. By 1938, J. Robert Oppenheimer[4] and R. Serber showed how stars, after exhausting their nuclear fuel, could collapse to states much denser than white dwarfs.

The concept of a **neutron star** was merely an extension of the concept of a white dwarf. In an ordinary gas, atoms are held apart by their motions, and when they collide, their nuclei are held apart by their surrounding electron swarms. In an ordinary star's core, the electrons are stripped off and nuclei begin to hit each other, reacting and releasing nuclear energy. In a white dwarf, the nuclei that are left can no longer react and are crowded ramdomly among a dense sea of electrons. A white dwarf, therefore, might be called an electron star since electrons control the spacings of particles.

But if a burnt-out star is still more massive than a white dwarf, the gravity is so strong that even the repulsion between electrons is overcome. Theory indicates that this could happen in objects between 1.4 $M_\odot$ and about 3 to 5 $M_\odot$. In such objects, the nuclei themselves begin to be jammed into contact. Since whatever reactions might occur have already run their course, there are no new energy-generating reactions. Instead, the nuclei may be broken into their constituent neutrons and protons. The positively charged protons may coalesce with negatively charged electrons to make more of the neutrally charged neutrons. The important particles are now the neutrons, and the star is a neutron star, smaller than the white dwarf. Its density approaches the incredible densities of atomic nuclei themselves: roughly 10^{17} kg/m^3. A thimbleful would weigh 100 million tons! A skyscraper-full could contain all the mass of the moon! A whole neutron star could contain the mass of the sun but be no larger than a small asteroid—a few kilometers across!

For decades, astrophysicists talked about neutron stars, but, like the weather, nobody did anything about them, because nobody *could* do anything. No known observational technique could detect them and no one could prove they existed.

But in November 1967, a 4.5-acre array of radio telescopes in England detected a strange new type of radio source in the sky. Analyzing the surveys (each equaling a 120-m roll of a paper chart), a sharp-eyed graduate student, Jocelyn Bell, was astonished to find that one celestial radio source (about a centimeter of data on the chart) emitted "beeps" every 1.33733 s!

At first, project scientists speculated that they might have actually discovered an artificial radio beacon placed in space by some alien civilization! But further work soon led away from this speculation. By January, another source was found, pulsing at a different frequency, arguing against the beacon hypothesis. Analysis showed that the first source was less than 4800 km across, much smaller than ordinary stars.

These pulsing radio sources came to be called **pulsars**. In Feburary, Anthony Hewish and his colleagues published an analysis suggesting that the pulsars might be superdense vibrating stars that could "throw valuable light on the behavior of compact stars and also on the properties of matter at high density."

The mysterious pulsars turned out to be the long-sought neutron stars! In an exciting burst of research, the number of scientific papers on pulsars jumped from zero in 1967 to 140 in 1968. By 1973, about 100 pulsars had been discovered. Codirectors of the original discovery project, Anthony Hewish and Martin Ryle, shared the 1974 Nobel Prize in physics.

Why do neutron stars pulse? After collapsing to a small size, supernova remnant stars have very strong magnetic fields and very fast spins, rotating once every second or so. Recall from Chapter 12 and Figure 12-2 that any spinning object—even a figure skater—spins faster as it contracts. Ions trapped in the magnetic fields spin around with velocities near the speed of light. In 1968, Cornell researcher Thomas Gold showed how ions trapped in magnetic fields of spinning neutron stars produce strongly beamed radio radiation, so that the pulsar acts like a lighthouse with a beam sweeping around every sec-

Table 17-3. Selected Theoretical Calculations of Star Deaths

Initial Mass of Star ($M_\odot$)	Death Processes	Calculated Mass of Final Remnant Star ($M_\odot$)	Probable Form of Final Remnant Star
<0.085	Gradual cooling	Same	Black dwarf
0.085–0.4	Gradual cooling and collapse	Same	White dwarf
0.4–6	Gradual mass loss	<1.4	White dwarf
6–8	Violent explosions	0.5–1.4	White dwarf
8–30	Violent explosions	1.4–3	Neutron star pulsar
30–100	Very violent explosions; nuclear "burning" of oxygen	3–100	Black hole?
⩾100	Catastrophic explosion?	~1?	?

Source: Data from various sources, including specific calculations on supernovae by six authors, summarized by Novikov and Thorne (1973); also Wheeler and Nomoto (1985).

---

[4]Physicist Oppenheimer later became famous for additional nuclear research and his work as leader of the Los Alamos team that developed the atomic bomb in 1945. His physics research was halted when he lost his security clearance in the anticommunist political wrangles of the 1950s.

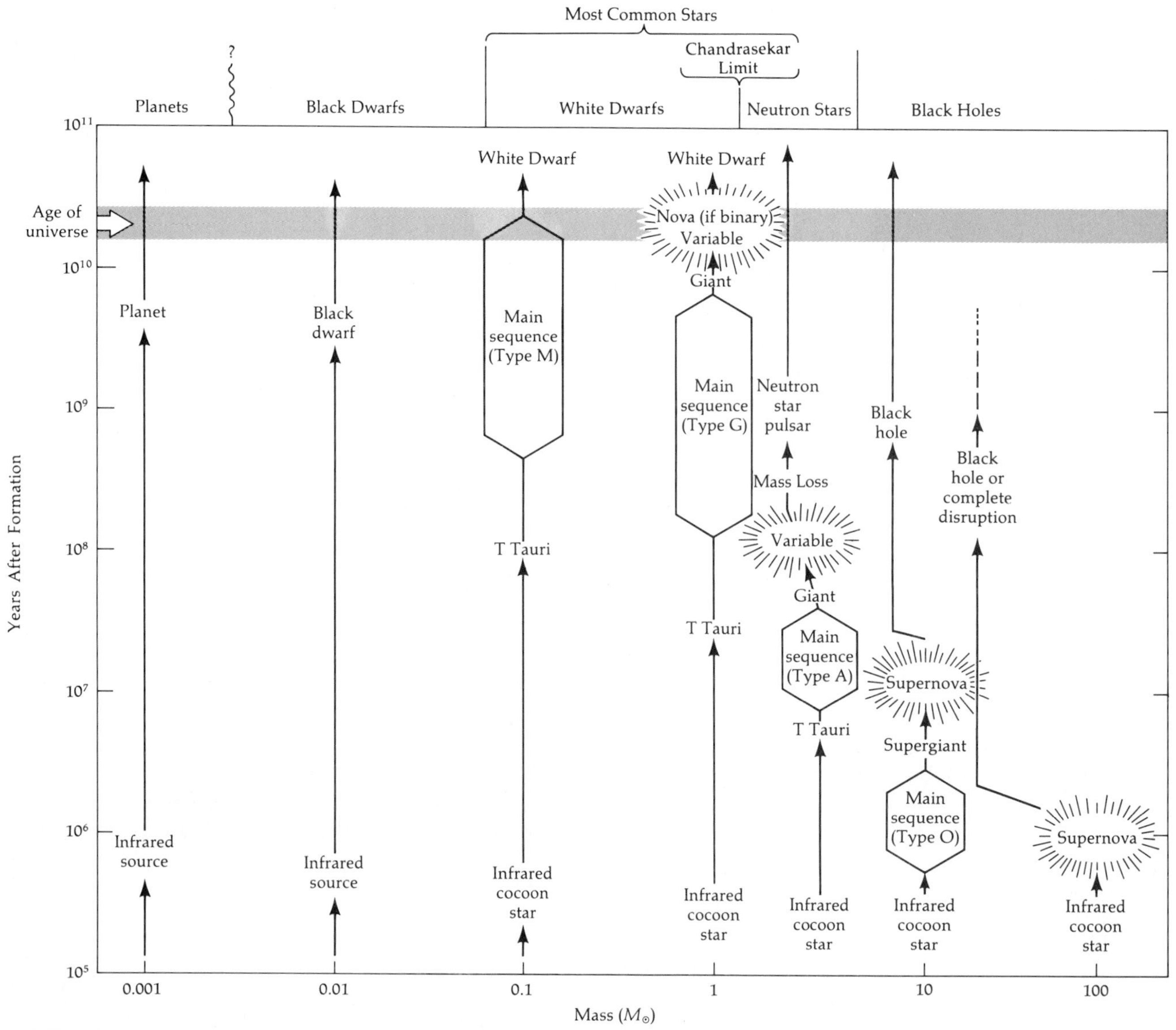

Figure 17-11. ***A schematic summary of stellar evolution, showing approximate evolutionary histories for objects with different initial masses and the associated terms described in this and preceding chapters. Some stages are hypothetical and are still being intensively researched by theoreticians and observers.***

ond. Most pulsars have periods of ¼ to 13. The fastest-spinning pulsar found thus far was discovered in 1982 and flashes with a 0.0016-s period. It is a mountain-sized ball of nuclear matter that spins 642 times every second (Waldrop, 1983)!

The discovery of a pulsar in the center of the Crab nebula (Color Plate 26, bottom) and in other supernova remnants proves that pulsars are related to supernovae. Careful studies of such pulsars show that they pulse not only in radio waves, but also in X rays and visible light, as shown in Figure 17-12.

A curious type of celestial object related to neutron star pulsars has been called the **burster**. In the 1970s, satellites designed to monitor the nuclear test ban treaty by detecting gamma-ray flashes from nuclear bombs discovered gamma-ray flashes coming from the sky! In at least one case,

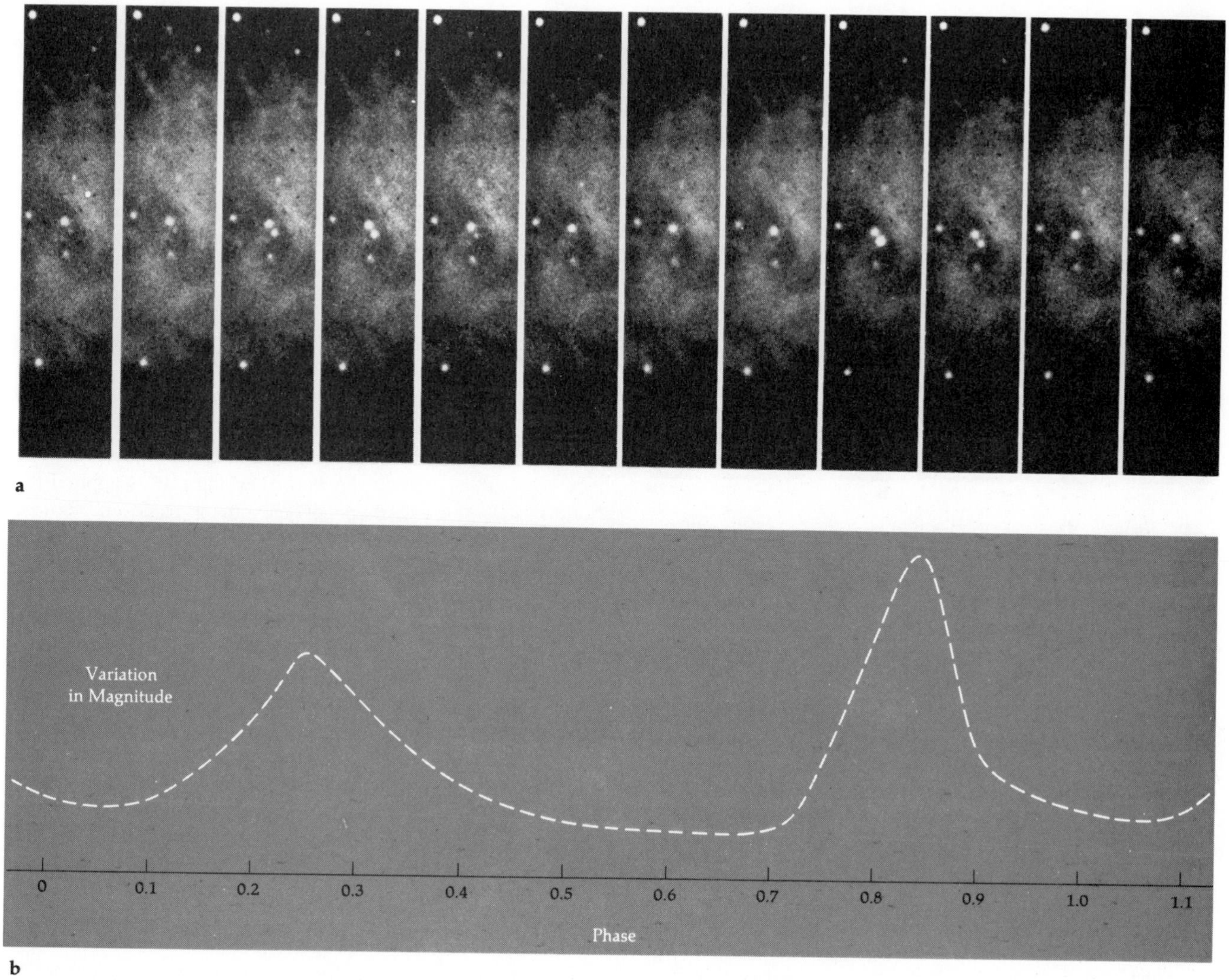

Figure 17-12. *A complete sequence of flashes from the pulsar NP 0532 in the center of the Crab nebula. Inner portions of the surrounding nebula can be seen. The entire cycle, including two flashes, lasts about 1/30 s, equaling one rotation of the pulsar. The pulsar and nebula are about 1100 to 2200 pc away. Color Plate 26 shows the full nebula. (Kitt Peak National Observatory.)*

the gamma-ray burst came from the position of a known neutron star. Researchers believe that bursts of gamma rays from the bursters may also be accompanied by flashes of visible light, X rays, and other radiation. One such visible flash may have been photographed, as shown in Figure 17-13. Such stellar flashes (lasting perhaps as much as a second) may be visible to the naked eye in the night sky, but would be very rare and difficult to confirm. Theorists have suggested that bursters may be caused by buildup of matter from a binary companion onto a neutron star until an explosion occurs (X-ray burster 4U1519-05 in Aquila seems to be in a binary system), or perhaps by sudden infall of clouds of debris or individual planetesimals onto the surface of a neutron star (Schorn, 1982).

Neutron stars may have solid crusts (1

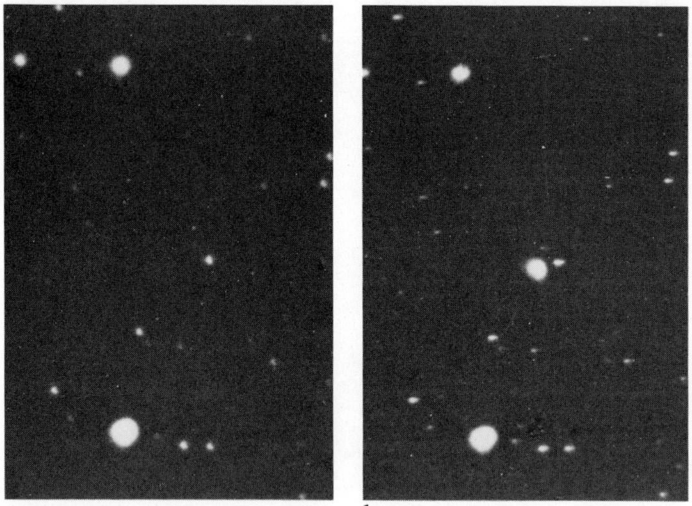

a b

Figure 17-13. *An accidental photo of a gamma-ray burster. During a squence of 45-min exposures made in South Africa in 1928, one of the images* (b) *apparently caught the optical flash of a gamma-ray burster in action. After a gamma-ray burster was discovered in this area of the sky in 1978, astronomers made a check of old plates in search of possible evidence of optical flashes. Photo* (a)*, along with other photos in the series, shows no star in the position of the gamma-ray burster, but photo* (b) *shows a bright image there (center). The image is not trailed left-to-right like the other images (due to imperfect telescope guiding during the 45-min exposure), proving that it must have been a brief flash of light. (Courtesy Bradley E. Schaefer, Massachusetts Institute of Technology.)*

km thick?) overlying a fluid "neutron soup." Tiny abrupt changes in the spin rates of pulsars have been detected and attributed to "starquakes," or fractures in the crust, which change the mass distribution and hence the rotation rate. It is strange to think of earthquake-like events happening in the solid crusts of star corpses!

BLACK HOLES

As soon as pulsars were discovered, the search for still denser objects began. From this search came evidence for the strangest of all astrophysical concepts—**black holes.** Black holes are bodies so dense that their gravitational fields can keep most light (or other forms of energy and matter) from escaping.

Black holes can be visualized through an analogy with Newtonian physics made as early as 1798 by the French astronomer–mathematician Pierre Laplace. He reasoned that some bodies might be dense enough to have an escape velocity (at their surfaces) faster than the speed of light. Laplace thus assumed light could never escape from such bodies and that they would be permanently black and opaque. Although this basic idea is nearly correct, we now know that the situation is more complex because of Einstein's theory of relativity. It more correctly predicts phenomena involving speeds near that of light. But theorists using Einstein's theory have concluded that black holes probably do exist.

Theorists believe that a black hole would be a very dense mass surrounded by a so-called **event horizon**, or imaginary surface from which no radiation or matter could escape. The black hole and any other matter inside the event horizon thus could not emit light or gas into the rest of the universe. Calculations indicate that black holes would form from the collapse of only the most massive stars (25 $M_\odot$ or more?) or the most massive exhausted post-supernova remnants (5 $M_\odot$ or more). This ultimate type of collapse could yield sizes of only a few kilometers or less!

Two theoretical results on black holes came in the mid-1970s, especially from the English physicist Stephen W. Hawking. First, he pointed out that since matter was more densely packed in the earliest days of the universe, objects much smaller than stars could have gravitationally contracted to form black holes as small as subatomic particles, like protons and neutrons. Some of these tiny, primitive black holes might still exist in space.

Second, Hawking (1977) applied the theory of quantum mechanics and showed that *black holes need not be entirely black,* as was once thought. Quantum mechanics shows that subatomic particles often act in ways unpredicted by older theories; thus, proton-sized black holes (containing up to 10^{12} kg—microscopic motes with the mass of mountains) could radiate energy, often as an explosion emitting gamma rays. Large, kilometer-scale black holes (containing as much mass as a star, or about 10^{30} kg) would radiate virtually nothing, just as in the original black hole theory.

If black holes exist, they must have truly unfamiliar properties. If an instrumented probe were dropped toward it, we would stop receiving signals after the probe fell through the event horizon, beyond which virtually no radiation or matter can escape. The probe, then, would seem to disappear from the observable universe. Where would it go?

Pondering this strange question, some scientists speculate that black holes amount to separate universes. Others feel that when mass "pops out of existence" by collapsing to superdensity, new mass or energy emerges somewhere else in the universe. But perhaps the probe would merely "squish" to very high density and collide with the black hole, adding a bit to its mass. Could our own universe have begun as a black hole in some other "universe"? Do black holes make fact out of the "space warps" invented decades ago by science fiction writers to allow their spaceships to wink out of sight in one place and re-emerge instantaneously at some distant point? No one knows. As the British geneticist John B. S. Haldane put it, "My suspicion is that the universe is not only queerer than we suppose, but queerer than we *can* suppose."

If a black hole itself cannot emit radiation or mass, can we ever hope to detect one? Yes. Outside their event horizons, black holes have gravity fields indistinguishable from those of ordinary stars of the same mass. Thus, they can orbit around stars just like planets or binary star companions. If we observed such a star from a distance, we would not see the black hole, but we could see the star's orbital motion and calculate the mass of the unseen companion, just as astronomers routinely do in the case of ordinary faint companions. The result would indicate an unusually high mass companion for an unseen star—maybe 5 or 10 $M_{\odot}$—which should tip us off that we are dealing with a black hole candidate. Furthermore, such strong gravity exists close to black holes that any matter falling into them undergoes terrific acceleration.

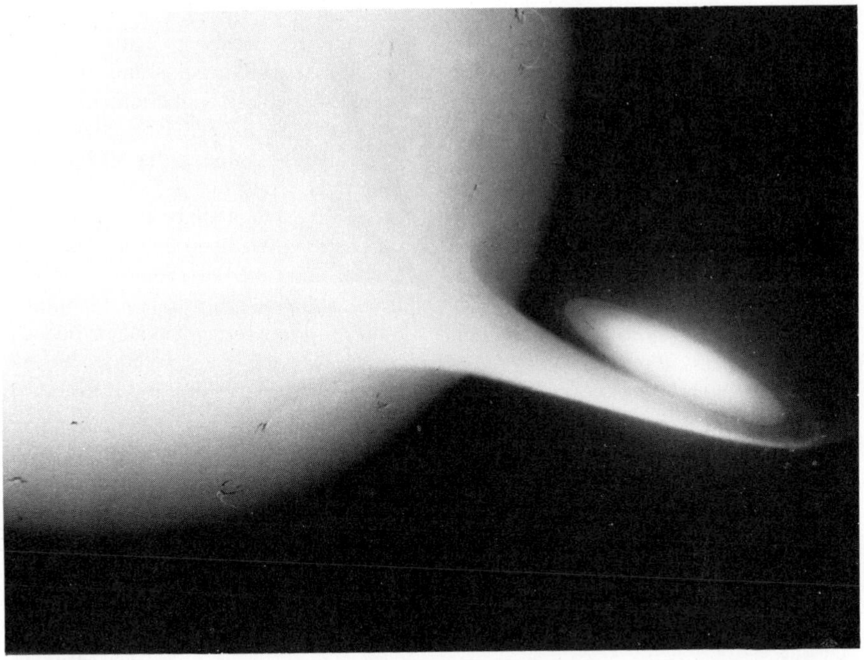

Figure 17-14. ***An imaginary close-up view of a star system, hypothesized to explain the X rays and gamma rays from systems such as Cygnus X-1. Gas flows from the surface of a supergiant star (left) toward a very dense star, possibly a black hole, where it accumulates into a disk-shaped nebula at very high speed, causing extremely high temperatures and the emission of X rays. (Painting by Adolf Schaller.)***

Suppose the black hole is orbiting around an evolved star that has expanded into the giant state and is shedding mass, as discussed earlier in this chapter. Some of the expanding gas would fall toward the black hole at terrific speed. Because this gas would, on the average, have some angular momentum around the star, rather than falling directly toward it, it would form a disk of gas spiraling inward toward the black hole (Figure 17-14). This disk is called an **accretion disk**. Its gas would be extremely hot, because it would be constantly hit by new gas streaming in from the other star. Because of the high temperature, the disk would radiate very short-wave radiation, such as UV, X-ray, and gamma-ray radiation. Therefore, X-ray and gamma-ray telescopes launched into orbit around the earth have played an important role in searching for black holes.

The best evidence for a black hole, then, would be a massive, high-temperature X-ray or gamma-ray source orbiting around another normal star. Two such candidates have been discovered. Cygnus X-1 (so named because it is the first X-ray source discovered in the constellation Cygnus) is a binary system about 2500 parsecs away. The visible star is a supergiant O or B star with a mass of 15 to 30 $M_{\odot}$. Orbiting around it is a small, hot, X-ray source with a mass of 10 to 15 $M_{\odot}$. The X-ray emissions suggest that the gas temperatures reach a billion Kelvin in the X-ray source (Nolan and Matteson, 1983). The second candidate is much more distant—55 000 parsecs away in a nearby galaxy called the Large Magellanic Cloud. It is called LMC X-3, and was first detected by the Uhuru X-ray satellite in the early 1970s. The visible star is a type B main sequence star, which orbits in 1.7 days around a small, hot, X-ray source with a mass of 6 to 14 $M_{\odot}$.

Many astronomers believe that the hot X-ray source in each of these systems is the accretion disk surrounding a black hole. Continuing observations will document these and other black hole candidates in more detail.

Black holes and stellar explosions may also account for such unexplained phenomena as strong infrared radiation from the center of certain galaxies, unexpectedly large red shifts of some galaxies, and reported gravitational disturbances in our galaxy. An explosion powerful enough to blow outer layers off massive stars, converting the star to a black hole, could produce enormous amounts of energy. As astrophysicist R. Penrose (1972) notes, "There is no shortage of unexplained phenomena in astronomy today that might conceivably be relevant" to black holes.

Practical Applications?

People often ask, "What good is astronomy? How can it be of use to know about places too far away to see, to touch, or almost to imagine?" We've argued that we attain firmer philosophical footing if we know about our cosmic surroundings and that exploration of the planets helps us understand the climate and resources of the earth. This chapter provides another reason. Astronomers' understanding of dense star forms like white dwarfs, neutron stars, and black holes has developed hand in hand with physicists' understanding of dense forms of matter in the lab. But there are limits to the experiments that can be conducted with such weird forms of matter. The discovery of actual examples of such matter in space, together with the measurement of masses, temperatures, magnetic fields, and so on, has provided physicists with examples impossible to create in their labs. Physicist D. Pines (1980) comments that such discoveries "may constitute an almost unique probe of... high-temperature plasma in super-strong magnetic fields,... neutron solids," and other superdense materials. If research on such material still sounds esoteric to you, recall that one of the most promising long-range solutions to the energy shortage on the earth is to control nuclear fusion in the lab by jamming nuclei together in dense plasmas constrained by strong magnetic fields. Fuel for hydrogen fusion could be the cheapest and most accessible material on the earth: seawater! Experimental work in this direction is already under way in the Soviet Union, the United States, and Europe. Thus, a direct line stretches from studies of neutron stars, to our understanding of matter on the earth, and on to our manipulation of matter to create new forms of usable energy.

Summary

The search for the forms of aging stars has yielded some of the most fascinating objects now being studied both by physicists and astronomers. Stellar old age leads to two basic phenomena: high-energy nuclear reactions, as ever-more-massive elements interact, and inexorable contraction, as energy sources are eventually exhausted. Some evolutionary histories of stars of different mass are summarized in Figure 17-11.

During the first stages of old age, as stars evolve off the main sequence, the high energy production can cause expansion of stars' outer atmospheres, producing giants and supergiants. As heavier elements go through quick reaction sequences, various kinds of instabilities may produce variable stars or explosive novae. After energy generation declines to a rate too low to resist contraction, low-mass stars contract to a dense state known as a white dwarf, with final mass less than $1.4\,M_\odot$. Rarer massive stars, which start out with as much as $30\,M_\odot$ or more, undergo supernova explosions and blow off much of their initial material.

If the remnant cores of stars end up between $1.4\,M_\odot$ and about $5\,M_\odot$, they form dense, rapidly rotating neutron stars known as pulsars. If the remnant cores have more than about $5\,M_\odot$, they may form black holes. Although virtually no radiation escapes from these strange objects, they may be detectable by orbital motions of their companion stars and by high-energy radiation from material falling into them. The detailed physics of these dense, small star forms is an area of intense current research.

Complications may arise if the evolving star has a nearby coorbiting companion. Mass may be blown off the one and fall onto accretion disks around the other, changing the second star's mass and causing sudden instabilities. This probably accounts for nova-type explosions and some sources emitting X rays and gamma rays. Ultraviolet, X-ray, and gamma-ray astronomy conducted on board orbiting satellites is an exciting area of new research on accretion disks, neutron stars, and black holes.

Concepts

red giant
funneling effect
triple-alpha process
helium flash
s-process reaction
r-process reaction
supergiant
Wolf–Rayet star
variable star
Cepheid variable
irregular variable
nova
white dwarf
degenerate matter
Pauli exclusion principle
Chandrasekhar limit
supernova
neutron star
pulsar
burster
black hole
event horizon
accretion disk

Problems

1. Why do stars just moving off the main sequence expand to become giants instead of starting to contract at once? Why does contraction ultimately win out?

2. Many red giants are visible in the sky, even though the red giant phase of stellar evolution is relatively short-lived. Why are so many red giants visible?

3. List examples of evidence that certain stars can lose mass.

4. Which stars will eventually become
  a. white dwarfs?
  b. neutron stars?
  c. black holes?

What will be the ultimate fate of the sun?

5. A main-sequence B3 star has about 10 times the mass of the sun and therefore has about 10 times as much potential nuclear fuel. Why then does it have a main-sequence lifetime only 1/200 as long as that of the sun?

6. According to the law of conservation of angular momentum, a figure skater spins faster as she pulls in her arms. How does this principle help explain why neutron stars spin much faster than main-sequence stars?

7. Comment on the roles and relations of theorists and observers in the three decades of work on white dwarfs, pulsars, and black holes. Are black holes fully understood today?

ADVANCED PROBLEMS

8. Suppose a 1-$M_\odot$ star reached a terminal evolutionary state where it had the same diameter as the earth.
  a. What would be the velocity of any possible material (such as captured meteoritic debris) in a circular orbit just above the star's surface?
  b. What velocity would be needed to blow material off its surface?
  c. Compare these values with values for the earth.
  d. What type of star would this object be?

9. Use the Stefan–Boltzmann law (p. 264) to prove that the surface of a star such as FG Sagittae, evolving to the right on the H–R diagram (keeping constant total luminosity but decreasing in temperature), must be expanding.

10. Many white dwarfs have spectral types and surface temperatures similar to A or F main-sequence stars, but are much smaller. Use the Stefan–Boltzmann law to prove that this statement requires white dwarfs to lie below the main sequence on the H–R diagram.

11. Suppose a nova occurred in the nearby star-forming region of Orion, 500 pc away. If the cloud expanded at an average velocity of 1000 km/s, how long would terrestrial observers have to wait before they could see details of the cloud's shape with telescopes resolving ½ second of arc? (*Hint:* Use the small-angle equation; 1 y is about $\pi \times 10^7$ s.)

12. Prove that a gamma-ray telescope (sensitive to wavelengths less than about 0.1 nm) would be best suited to observing radiation from a billion Kelvin accretion disk surrounding a black hole. Could such observations be made from a ground-based observatory?

13. The discussion of supernovae noted that the nuclei in stellar cores begin to break apart into protons and neutrons and other subatomic particles at temperatures of a few billion Kelvin. Laboratory experiments show that simple nuclei will fragment in this way when struck by particles moving fast enough to carry about 3.5×10^{-13} joules of energy. Prove that a gas would have to have a temperature of several billion degrees before its particles collided hard enough to begin to fragment nuclei of atoms. (*Hint:* In deriving Optional Basic Equation VI, we saw that the mean kinetic energy of a particle in an ordinary gas is $\frac{3}{2}$ kT. Gases in supernovae, accretion disks, and so on, at billions of degrees are unlikely to behave as perfect gases, but assume that they do for the purpose of this estimate. Assume also that the faster particles in the gas have four times the energy of the average particle mentioned above. Then you can estimate that the energies mentioned would be reached by the faster atoms in a gas at some 4 billion Kelvin.)

Projects

1. Locate the star Mira (R.A. = 2^h14^m; Dec. = $-3°.4$) with a small telescope and determine whether it is in its faint or bright stage. If it is bright enough to see with the naked eye, record its brightness from night to night by comparing it with other nearby stars of similar brightness. By checking brightnesses of these stars with star maps showing magnitudes, plot a curve of Mira's brightness over time.

2. With binoculars or a small telescope locate the star Delta Cephei (R.A. = 22^h26^m; Dec. = $+58°.1$) and compare it from night to night with other nearby stars of similar brightness. Can you detect its variations from about 4.4 to 3.7 magnitude in a period of 5.4 d?

P A R T E

Environment and Groupings of Stars

Every cubic inch of space is a miracle.

Walt Whitman

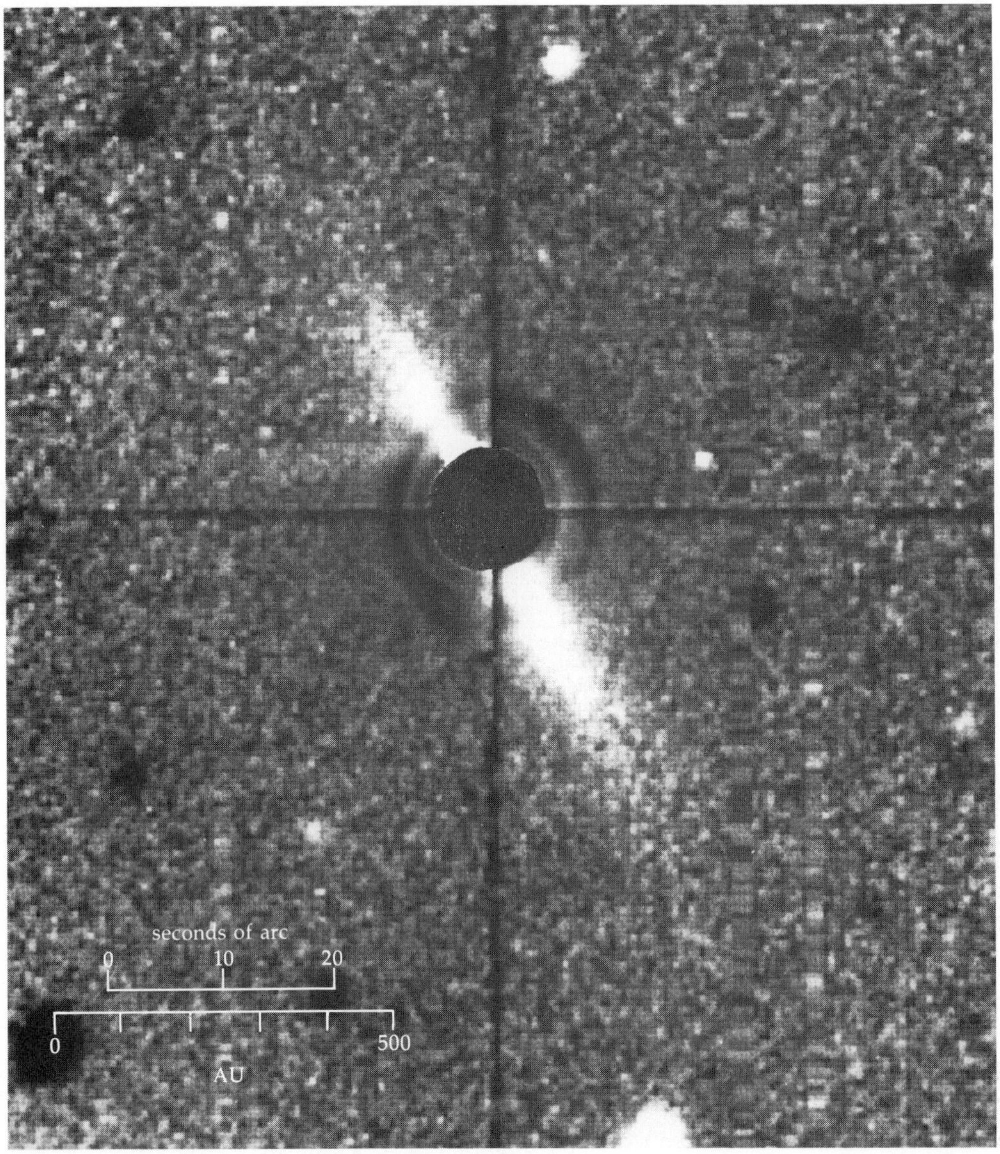

Here is the best pictorial evidence yet available of planetary dust around a nearby star, Beta Pictoris, 16 pc away. Thermal infrared radiation from the dust had been detected in 1983 by the IRAS infrared satellite telescope in space, leading astronomers to make this 1984 visual-light image from a ground-based telescope in Chile. The black central disk and crosshairs represent a system used to block the light of the star itself. With the star glare blocked, sensitive imaging equipment recorded bright material extending on either side of the star. This is believed to be an edge-on view of dust extending out to about 400 AU from the star—the distance of the inner Oort swarm of comets in our own planetary system. The Beta Pictoris system's dust is much thicker than the zodiacal dust of our own system. Researchers suggest that planets may have recently formed in the Beta Pictoris system, though none have yet been detected. (Photo by Bradford A. Smith and Richard J. Terrile, courtesy B. A. Smith)

INTERSTELLAR ATOMS, DUST, AND NEBULAE

The preceding chapters discussed stars as if they were isolated individual objects. But they are not isolated. They are engulfed in a thin but chaotic medium of gas, dust, and radiation. They form from this thin material, interact with it, recycle it, and expel it to form new interstellar material.

We sometimes casually say "Space is a vacuum," but this is not quite true. While space is a better vacuum than can be achieved in labs (Table 18-1), its material cannot be neglected. What significance can this thin material have for us? For one thing, it dots our sky with **nebulae**, or vast clouds of dust and gas, some twisted into beautiful wispy forms, some dark, and some glowing with different colors. For another thing, as Joni Mitchell said in her song "Woodstock," "We are stardust." As Chapter 12 pointed out, our solar system, our earth, and we ourselves are formed from atoms that were once part of the interstellar gas and dust. More provocatively, recent discoveries have demonstrated that interstellar material does contain complex organic molecules, and some scientists have speculated that primitive biochemical processes, perhaps related to the origin of life, may have occurred in nebulae.

The Effects of Interstellar Material on Starlight

Dispersed particles, whether floating in space or in the atmosphere, interact with radiation. When light from a distant star passes through clouds of interstellar atoms, molecules, and dust grains, the interaction changes the light's properties, such as intensity and color. Several complex

Table 18-1. *Gas Densities in Different Environments*

Locale	Density[a] (kg/m^3)	Particles per m^3	Typical Distance Between Particles
Air at sea level	1.2	10^{25}	1 nm
Circumstellar cocoon nebula	10^{-5}	10^{22}	50 nm
"Hard vacuum" in terrestrial lab	10^{-9}	10^{18}	1 μm
Orion nebula	10^{-18}	10^{9}	0.1 cm
Typical interplanetary space	10^{-20}	10^{7}	0.5 cm
Typical interstellar space	10^{-21}	10^{6}	1 cm
Interstellar space near edge of galaxy	10^{-25}	10^{2}	20 cm
Typical intergalactic space	10^{-28}	10^{-1}	2 m

[a]Average density of observable matter in the whole universe is estimated to be about 3×10^{-28} kg/m^3 (Shu, 1982), but many astronomers believe the average density may be somewhat higher (3×10^{-27} kg/m^3?) due to nonluminous unseen matter.

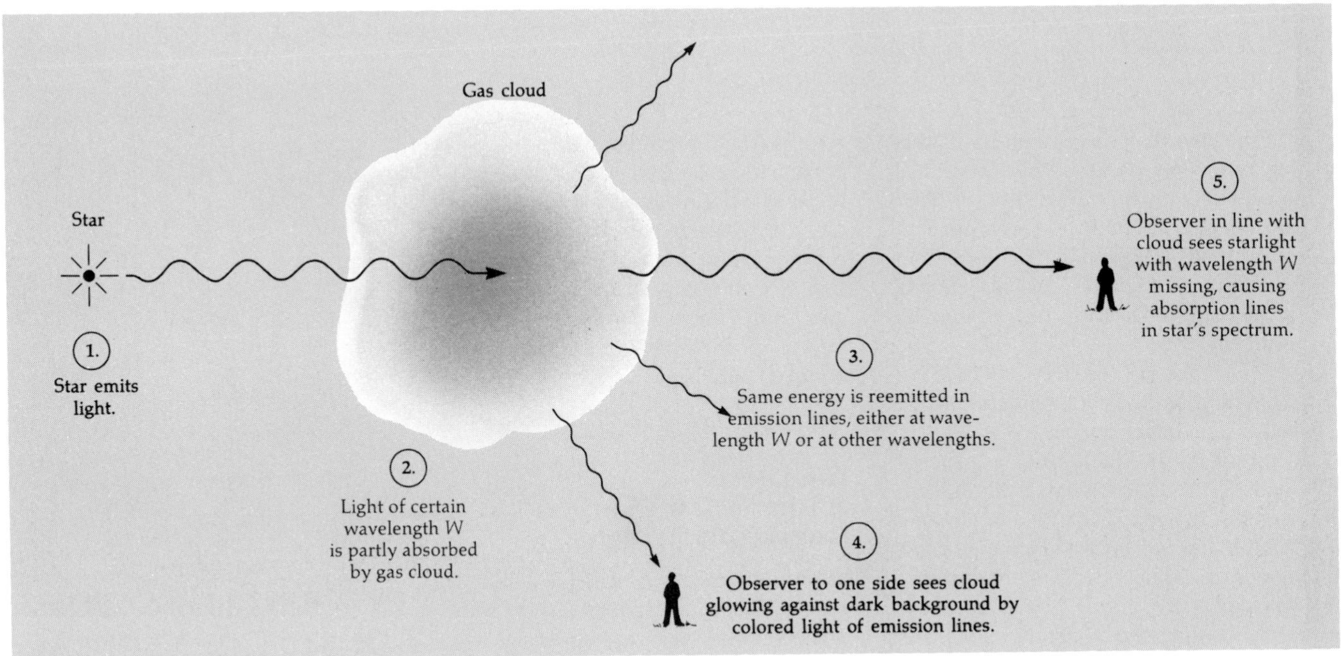

Figure 18-1. *A cloud of gas (atoms and molelcules) illuminated by a star and seen by an observer to one side (item 4) and an observer in line with the cloud and the star (item 5).*

physical laws describe these changes in some detail, but the changes can be grouped under two main principles:

1. When radiation (ultraviolet light, visible light, infrared, radiowaves, or any other type) interacts with particles, the type of interaction depends on the types of particles and their sizes relative to the wavelength of the light.

2. The appearance of the light and the particles may depend on the direction from which the observer looks.

Three important types of interaction between radiation and matter involve atoms, molecules, and dust grains. In reality, the interstellar material is always a mixture of gas and dust, but it is easier to understand the effects if we imagine separate interactions of light with atoms, molecules, and dust grains.

INTERACTION OF LIGHT WITH INTERSTELLAR ATOMS

As shown in Figure 18-1, several things can happen if starlight passes through a cloud of interstellar gas atoms. Suppose a star radiates light of all wavelengths, and the photons of light enter a cloud of gas. Photons corresponding to certain wavelengths will have just enough energy to **excite** this gas, or knock electrons from lower to higher energy levels. They may even **ionize** the gas, or knock electrons clear out of the atoms. Each time a photon excites or ionizes an atom, that photon is consumed and disappears from the light beam. Thus, an observer looking at the star through the cloud would see absorption lines created by the interstellar material, as shown in Figure 18-1.

But the energy absorbed by the cloud must be reradiated, assuming that the cloud is in equilibrium. This reradiation occurs as the electrons cascade back down through the energy levels of the atoms, creating emission lines. The photons in these emission lines leave the cloud in all directions, as shown in Figure 18-1, so that an observer off to one side would see the cloud glowing in the various colors corresponding to the emissions. If a red emission line is especially strong, the cloud might look red. In another cloud, struck by photons of different wavelengths or containing atoms at different levels of excitation, a green emission line might be strongest, and the cloud would glow with green light. Photos in different colors reveal different patterns (see Figure 18-2). Since hydrogen is the most abundant gas, and since the red Hα emission line is one of its strongest emissions, many clouds of excited gas glow with a beautiful deep red color, as seen in Color Plate 25. Other colors also appear in nebulae (see Color Plates 24 to 27).

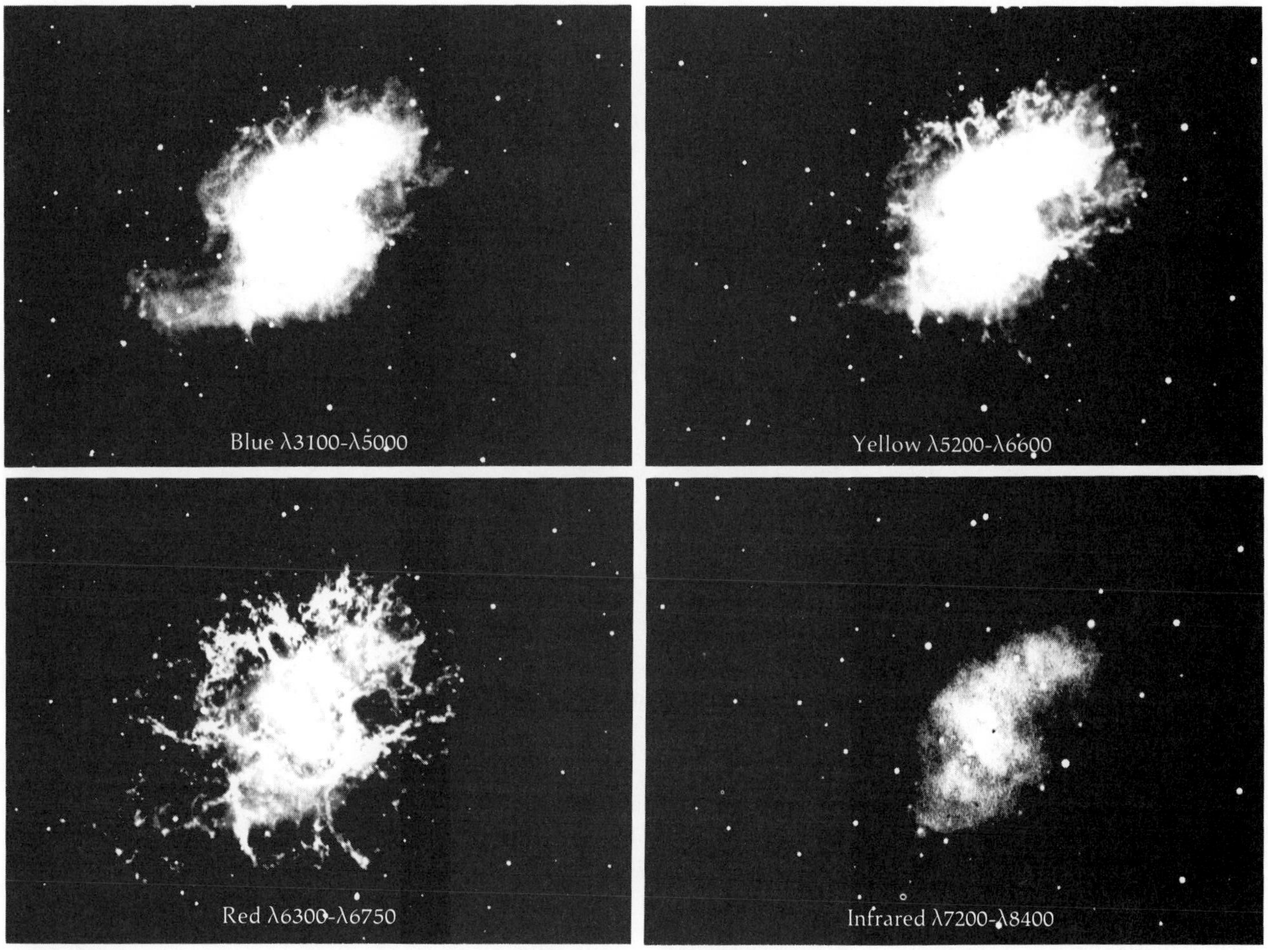

Figure 18-2. *Four views of the Crab nebula (Messier catalog number M1), photographed in light of different colors. Red and yellow views show the outer filamentary structure corresponding to clouds of excited gas giving off spectral emission lines (especially red lines from hydrogen). Infrared view emphasizes inner amorphous clouds. (Hale Observatories.)*

INTERACTION OF LIGHT WITH INTERSTELLAR MOLECULES

Molecules have similar interactions with starlight to those of atoms except that each absorption or emission affects a range of wavelengths, creating an absorption or emission *band* instead of a narrow line. Thus a real gas cloud, containing atoms and molecules involving different elements, produces a variety of absorption lines and bands and might glow with emission lines and bands of different colors (see Color Plate 26, top, for example).

INTERACTION OF LIGHT WITH INTERSTELLAR DUST GRAINS

Whereas atoms are as small as 0.001 times the wavelength of visible light, and molecules are a few percent of this wavelength, many interstellar dust grains are comparable in size to light waves, and their interactions are quite different. They absorb some starlight, dimming distant stars. They also affect colors over a much broader range of wavelengths than individual spectral lines or bands. The most important effect is that redder light (longer wavelengths) passes through clouds of dust, whereas bluer light (shorter wavelengths) is scattered out to the side of the beam (see Figure 18-3.)[1]

[1] Scattering of light is different from absorption, mentioned a few lines earlier, but we will not emphasize the difference. Think of absorption as a disappearance of a photon into an atom (or ion or molecule), making the atom more energetic. Think of scattering as a bouncing of a photon off an atom out of the light beam into a new direction. Either process reduces the amount of light in the beam.

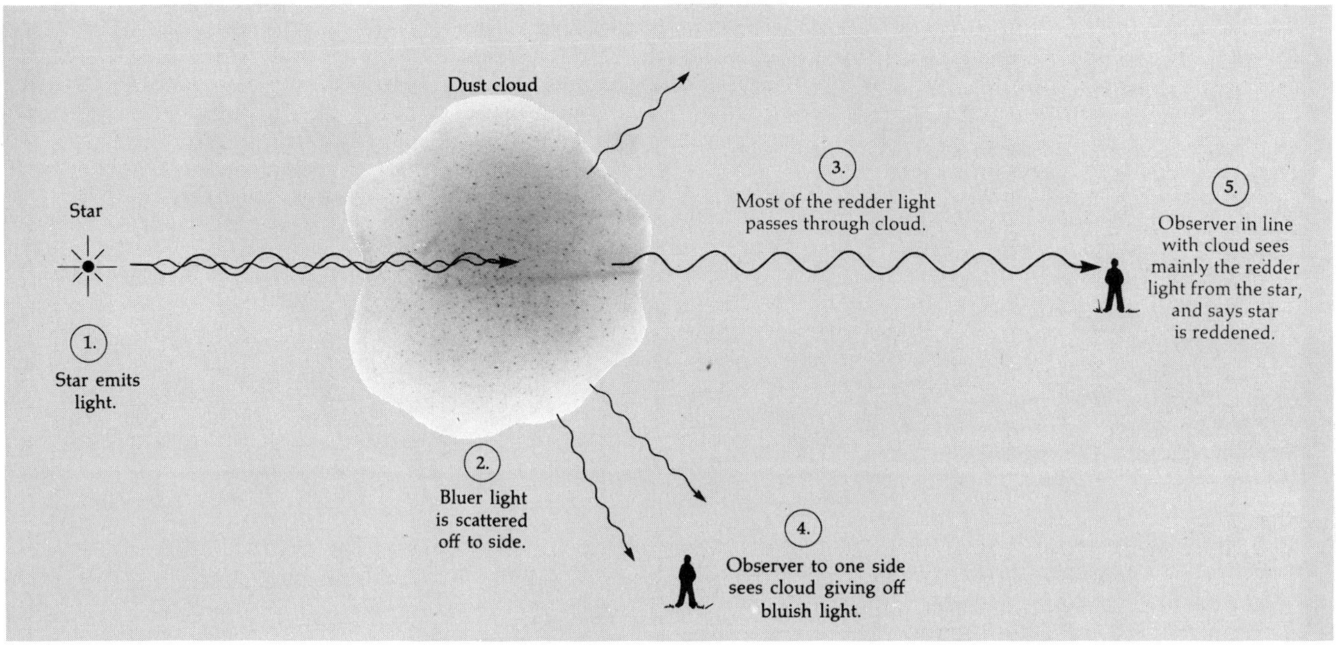

Figure 18-3. *A cloud of dust grains illuminated by a star and seen by an observer to one side (item 4) and an observer in line with the cloud and the star (item 5).*

The scattering of blue light results from a property of all particles smaller than the wavelength of light: they scatter more blue light than red light out of the beam. This preferential scattering of blue light is called **Rayleigh scattering**, after its discoverer. It occurs in interstellar dust and gas because many grains and all atoms are smaller than the wavelengths of visible light.

Thus, the observer who looks through the dust cloud at a distant star sees most of its red light, but not much of its blue light. In this way, interstellar dust makes distant stars look redder than they really are— an effect called **interstellar reddening**. However, an observer who looks at the dust cloud from the side will see the blue light scattered out of the beam, so that a nebula illuminated in this way will have a bluish color. Prominent examples of these blue nebulae are seen in Color Plates 27 and 28.

WHY IS THE SUNSET RED AND THE SKY BLUE?

These same rules also apply to material in our own atmosphere. The lower few kilometers of the atmosphere are full of dust and large molecules, including many that are slightly smaller than the wavelength of light. When we look at the nearest star, our sun, through these particles, the same effects can be seen. If the sun is high in the sky, we look through the minimum amount of dust, as shown in Figure 18-4a; thus the reddening is minimal, and the sun is perceived as white.

At sunset, as shown in Figure 18-4b, the sunlight passes through much more gas and dust, and much of the blue light is lost from the beam by Rayleigh scattering. This strongly reddens the sun and adjacent parts of the sky. At any time of day, if we look at some other part of the sky, as in Figure 18-4c, the light we see is the blue light scattered out of the beam of sunlight and then scattered by air molecules and dust back toward our eyes. In contrast, as seen in Color Plate 11, the sky on Mars is reddish because many of the reddish Martian dust particles are bigger than the wavelength of light; hence there is no Rayleigh scattering of blue light, but simply the reflection of their own red color.

Observed Types of Interstellar Material

Interstellar material includes gas (that is, atoms and molecules), microscopic dust grains, and possibly larger objects.

INTERSTELLAR ATOMS

Atoms of interstellar gas were discovered in 1904, when German astronomer Johannes Hartmann detected their absorption lines, such as those shown in Figure 18-5. While studying the spectra of a binary star, he accidentally discovered absorption lines caused by interstellar calcium atoms. Certain other **interstellar atoms**, such as sodium, were soon found to produce additional prominent interstellar absorption lines. These lines are identified as interstellar by the fact that they have different Doppler shifts than the stars in whose spectra they appear.

Further studies have convinced astronomers that, although atoms such as calcium and sodium have prominent absorptions, the most common interstel-

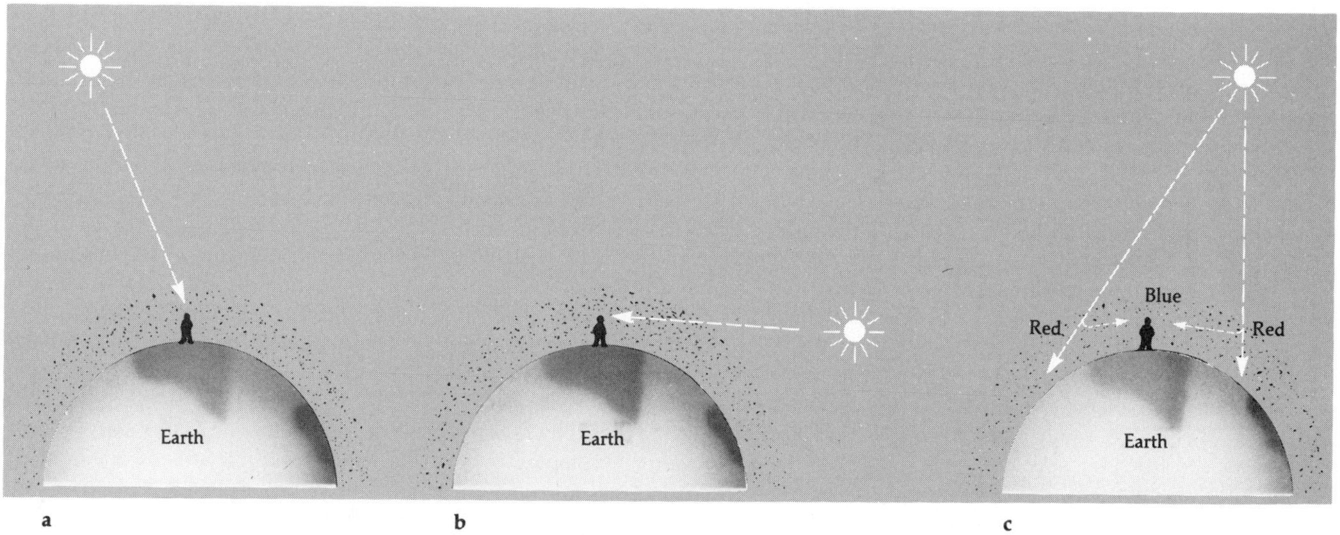

Figure 18-4. *Explaining colors in the sky.* (a) *Light from the sun high in the sky passes through minimal dust and is minimally reddened.* (b) *Light from the sun at sunset passes through the maximum amount of dust and is strongly reddened.* (c) *Light from the sky at any time of day is the blue light scattered out of the sunlight beam.*

lar gas is the ubiquitous hydrogen. Like the sun and stars, interstellar gas is about three-fourths hydrogen and nearly one-fourth helium.

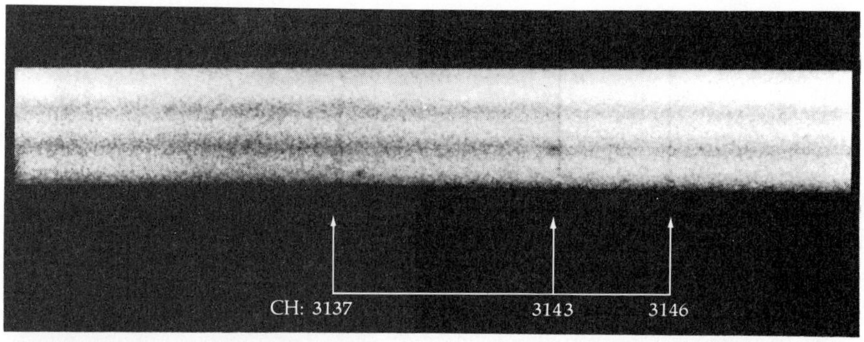

Figure 18-5. *Spectral absorption lines caused by interstellar molecules of CH gas. Background light is part of the blue portion of the spectrum of the star Zeta Ophiuchi. (G. Herbig, Lick Observatory.)*

Forbidden Lines from Interstellar Atoms. A scientific mystery soon arose from interstellar gas. All experience prior to the 1920s had indicated that the distant universe was composed of the same elements as found in the solar system. However, as early as 1864, English observer William Huggins had found three greenish and bluish emission lines from the light of certain nebulae. Further work showed that these did not match the spectra of any elements known on the earth, and they were assigned to a hypothetical new element named "nebulium." By the 1920s, it was clear that there was no room in the periodic table of elements for a new element like nebulium. What was creating the strange greenish light of nebulae?

In 1927, California astronomer Ira S. Bowen showed that in deep interstellar space, atoms could exist in so-called **metastable states**, which are states with unusual electron distributions that last for several minutes. These states are unknown on the earth, because in our dense atmosphere the atoms are struck by air molecules too fast to remain in these states. The spectral lines arising from these unusual states are called **forbidden lines**, because they cannot occur in the earth's atmospheric gas. The nebular light turned out to be coming from strong forbidden lines of very rarefied, ionized forms of oxygen—a gas familiar in our own air. (The astronomer Henry Norris Russell thereupon quipped that the mysterious nebulium had vanished into thin air.)

Radio Radiation from Interstellar Atoms. In 1944, the Dutch astronomer

H. C. van de Hulst predicted that the most important type of interstellar radiation would be an emission line with wavelength 21-cm, caused by a change in the spin of hydrogen atoms' electrons. These electrons can have only certain spin rates and directions, and emit the 21-cm radiation when they change from one state to another. Such a long wavelength is not visible light, but radio radiation. The predicted emission was confirmed in 1951 when Harvard astronomers detected this **21-cm emission line** of atomic hydrogen, using radio equipment.

This discovery not only confirms the importance of hydrogen as a main constituent of interstellar gas, but gives radio astronomers a tool to detect where clouds of interstellar gas are concentrated. Because of the long wavelength of this radiation, it can penetrate much greater distances through the interstellar gas and dust than ordinary light. The 21-cm line of interstellar hydrogen is thus the most important emission line in radio astronomy.

INTERSTELLAR MOLECULES

By 1940, astronomers at Mt. Wilson observatory in California had built spectrographs that detected absorptions due not only to interstellar atoms but also to **interstellar molecules**, such as CH and CN. Figure 18-5 shows spectral absorption lines caused by interstellar CH.

Because of molecular structure, a great many of the molecular absorptions lie in the infrared or radio parts of the spectrum. Not until after World War II did astronomers have available the technology of infrared and radio detectors to search for interstellar molecules. After being developed in the 1960s, some of the new detectors were put in satellites and large telescopes, sparking an explosion of interstellar discovery in the late 1960s and 1970s. For a time each new issue of astrophysics journals seemed to carry news of another molecule identified in interstellar space by spectral studies.

For example, the hydroxyl (OH) molecule was the only one found from 1963 to 1967, but water (H_2O), ammonia (NH_3), and formaldehyde (HCHO) were found in 1968–69, while 18 more molecules were found in 1970–71. Some 53 varieties were cataloged by 1982. These include molecular hydrogen (H_2) and carbon monoxide (CO). There is a trend toward discovering more and more complex forms, such as the nine-atom molecule ethyl alcohol (C_2H_5OH), found in 1974.

The atoms recurring again and again in these large molecules comprise the quartet carbon, hydrogen, oxygen, and nitrogen—the "building blocks of life"! Repeatedly, these building blocks have been found in space in complex large molecules. These elements are common, and hydrogen is the most common of all, so the surprise is not that they exist in space, but that some process brought these elements together in their tenuous environment to make these complex forms.

Still more provocative is the fact that two of the molecules, methylamine (CH_3NH_2) and formic acid (HCOOH), can react to form glycine (NH_2CH_2COOH), one of the **amino acids**. These large molecules can join to form the huge protein molecules that occur in living cells.

Thus, two exciting questions have come from research on interstellar molecules. First, does the existence of complex, carbon-rich molecules in space suggest that life could have originated elsewhere in space? The answer may be yes, and we will discuss this possibility in more detail in Chapter 26. Second, how do so many atoms come together to form these molecules? They cannot form in ordinary interstellar gas, because collisions there between atoms are extremely rare. Instead, the molecules are believed to form in the denser clouds, such as the cocoon nebulae around young stars.

INTERSTELLAR GRAINS

Still bigger than interstellar molecules are **interstellar grains**. They cause two observational effects: the reddening already explained, and a general dimming of starlight at all wavelengths, called **interstellar obscuration**. Some grains are clumped in distinct clouds, such as those in Figure 18-6, but others are widely distributed throughout the interstellar gas, producing a general haze or "interstellar smog." All distant stars are harder to observe because of the haze caused by interstellar grains.

Interstellar obscuration played a curious role in humanity's recognition that we do not necessarily live at the center of the universe. After the Copernican revolution, astronomers assumed that the solar system did not occupy a central position. Then, studies from the 1700s to the early 1900s of fainter and fainter stars seemed to indicate that the farther from the solar system we probe, the fewer stars we find. By 1922 astronomers began to wonder: Was the solar system in the *center* of a swarm of stars surrounded by empty space?

In 1931, American astronomer R. J. Trumpler published a very important paper proving that vast quantities of interstellar dust were merely obscuring the more distant stars. Trumpler proved this by studying distant star clusters; their apparent sizes and other properties enabled him to show that the more distant a cluster, the more it is dimmed by dust in the line of sight.

Further research indicates that if we look along the Milky Way plane, dust grains dim stars by an average of about 1.9 magnitudes for every 1000 pc traversed by the beam of light. Of this amount, about 1.6 magnitudes of dimming are caused by grains concentrated in clouds, and about 0.3 magnitude by grains dispersed between clouds. Over a distance of 10 000 pc, stars would be dimmed by 19 magnitudes! No wonder early observers found fewer and fewer stars the farther out they looked!

The Nature and Origin of the Grains. Astronomical studies reveal various properties of the interstellar grains: (1) They range in size from one-sixth to twice the wavelength of visible light; (2) they are concentrated in clouds; (3) compositions may vary somewhat from cloud to cloud; (4) many grains are elongated; and (5) many elongated grains are aligned parallel to other nearby grains, possibly

a

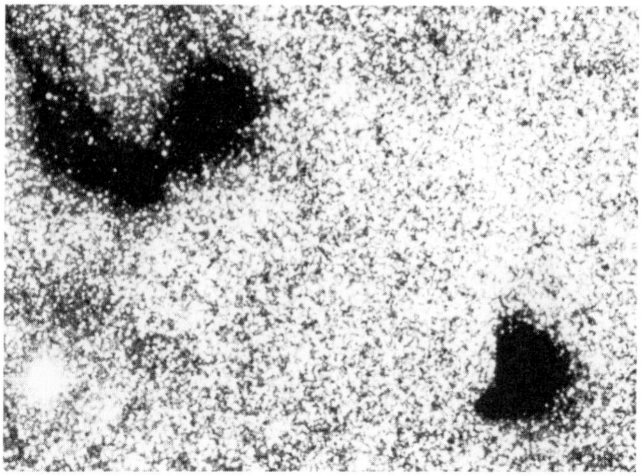

b

Figure 18-6. (a) *Unilluminated clouds of interstellar material, named Barnard 68 and 72, obscure distant stars of the Milky Way in the constellation Sagittarius. Similar clouds are often sites of concentrations of molecules and dust grains. (Steward Observatory, University of Arizona.)* (b) *A long exposure with higher contrast, made with a small telescope, enhances cloud silhouettes. (Hale Observatories.)*

because they are iron rich and aligned with magnetic fields in space.

Grain composition has long been debated. In 1967, Indian astrophysicist N. Wickramasinghe proposed that grains are a form of carbon condensed in cooling gas blown off carbon-rich giant stars. Others suggested silicates and ices, likening grains to the dust in our primordial solar system. All these materials probably exist in interstellar grains. Carbon compounds, silicates, and ices have been identified spectroscopically, especially in star-forming regions. Reactions initiated by ultraviolet light in these materials apparently create complex organic molecules on many grains' surfaces (Greenberg, 1984).

In the early 1980s, studies of interstellar grains came much closer to home! Geochemists found that primitive carbonaceous meteorites contain grains of carbon (in various forms) and spinel (Al_2MgO_4), about 0.4 to 100 μm in size, whose isotopes indicate that they are not solar system material (Whittaker et al., 1980; Lewis et al., 1980). Further studies of the isotope compositions show that many of these grains condensed in gas blown out of red giants, novae, and supernovae, confirming carbon condensates as one grain source (Lewis and Anders, 1983; Swart et al., 1983). Magnetite grains similar to those in meteorites have been thought to contain 16% of all the iron atoms in the galaxy, and to account for the magnetically controlled alignment. Only a few years ago, astronomers would have judged it implausible to study interstellar grains at firsthand. Now we seem to have them in our labs!

INTERSTELLAR SNOWBALLS?

Astronomers are now looking for **interstellar snowballs**—hypothetical bodies much larger than grains. These could be BB-sized, baseball-sized, or even kilometer-scale bodies in interstellar space. If light interacted with such particles, it would be blocked equally at all wavelengths much less than the particle size. While microscopic grains are revealed by reddening, large particles would be difficult to detect because virtually no color effects would occur. Nonetheless, if single atoms join into molecules, and molecules into dust grains, why not expect still larger particles?

There are some indications that interstellar snowballs exist (Greenberg, 1974; Herbig, 1974). For example, the interstellar gas is strangely lacking in certain elements. Aluminum has not been found at all. Calcium is 400 times rarer than we would predict from its abundances elsewhere in the universe. Titanium, iron, and magnesium are also rarer than expected. Because the interstellar grains apparently do not contain enough of these materials to make up for the atoms missing in the gas, another class of interstellar material—snowballs—may contain the missing atoms. They might look like our present conception of comet nuclei—icy bodies with silicates and other "dirt" mixed in—justifying the term "snowball" coined by Greenberg.

Four Types of Interstellar Regions

Interstellar gas and dust are far from uniform. There are thick clouds of material

Table 18-2. *Characteristics of Selected Nebulae*

Name	Constellation	Approx. Distance from Earth (pc)	Approx. Diameter (pc)	Estimated Atoms per m^3	Mass ($M_\odot$)	Spectral Type of Associated Star
Nebulae Probably Associated with Young Objects						
NGC 2261 Hubble's (R Mon)	Monoceros	700	10^{-5}	10^{18}	10^{-1}	F
Kleinmann–Low IR	Orion	500	0.1	10^{12}	100	Protostar?
Dark Nebulae						
Coalsack	Crux	170	8	2×10^6	15	None
IC 434 Horsehead	Orion	350	3	2×10^7	0.6	B
Emission Nebulae						
M42 Orion (central)	Orion	460	5	6×10^8	300	O
Eta Carinae	Carina	2 400	80	2×10^8	1 000	Peculiar
M8 Lagoon	Sagittarius	1 200	9	8×10^7	1 000	O
M20 Trifid	Sagittarius	1 000	4	10^8	1 000	O
Reflection Nebulae						
M45 Pleiades	Taurus	126	1.5	?	?	B
Cocoon	Cygnus	1 600	2	7×10^7	7	B
Planetary Nebulae						
M57 Ring	Lyra	700	0.2	10^9	0.2	White dwarf?
M27 Dumbbell	Vulpecula	220	0.3	2×10^8	0.2	White dwarf?
NGC 7293 Helix	Aquarius	140	0.5	4×10^9	0.2	White dwarf?
Supernovae Remnants						
M1 Crab	Taurus	2 200	3	10^9	0.1	Pulsar
NGC 6960/2 Veil (Loop)	Cygnus	500	22	?	?	?
Gum	Puppis–Vela	460	360	10^5	100 000	Pulsar?

Source: Data from Allen (1973); Maran, Brandt, and Stecher (1973); and other sources.

(nebulae), and regions of different temperature. Broadly, astronomers recognize four different types of regions, defined by the condition of the gas: (1) Cold regions of dark clouds of dust and gas at about 10 K are called **molecular clouds.** Here much gas exists as molecules, including not only molecular hydrogen (H_2) but many of the more complicated molecules discussed earlier. (2) Broad regions at around 100 K contain neutral hydrogen atoms and are called **HI regions** after the symbol for neutral hydrogen (HI). (3) Hot regions at around 10 000 K surround hot O and B stars, whose ultraviolet photons knock electrons out of the H atoms, ionizing them. These are called **HII regions**, after the symbol for ionized hydrogen (HII). (4) Very hot regions, at about 10^6 K, exist where gas has been superheated by expanding blasts from supernovae. These are called **superbubbles.** We will discuss these four types of regions in the following sections.

Radio telescopes and satellite X-ray telescopes helped reveal molecular clouds and superbubbles, respectively, only in recent years. Hence, it is not yet clear exactly what fractions of space are occupied by the four types of regions, although 10–50% of our galaxy's gas may be molecular (Blitz, 1982), and perhaps 10% in superbubbles.

Among the individual clouds of gas and dusts, particularly vivid nebulae were cataloged as early as 1781 by the French astronomer Charles Messier. They are thus known by their **Messier numbers,** or M numbers. The Orion nebula (Color Plate 24), for example, is M 42. Others are known by **NGC numbers** or **IC numbers,** based on the more recent New General Catalog and Index Catalog, respectively. Tradition also names most bright nebulae according to their appearance in small telescopes; examples include the Crab, Dumbbell, and Ring nebulae (Color Plate 26). Some examples are listed in Table 18-2.

Why has interstellar material not reached some uniform distribution? Suppose we magically smoothed the interstellar material to uniform density. Supernova explosions, gas ejected from giants, radiation pressure, and differential rotation of the galaxy would quickly create local dense regions. Some of these would be dense enough to contract gravitationally, making still denser clouds. Within a brief 100 million years or so the material would have reformed into a new distribution of nebulae.

MOLECULAR CLOUDS

In localized regions where gas gets compressed to higher-than-average densities (perhaps by winds from neighboring expanding clouds), atoms collide more often and molecules may tend to grow. Dust grains get concentrated along with the gas. They often make the cloud opaque and dark, shielding its inner parts from nearby hot stars. More importantly, the dust grains can radiate heat more easily than the gas, and so a region rich in dust grains gets cooler than surrounding clouds.

As we saw in Chapter 16 (especially Figure 16-2), the denser and colder a region of gas, the easier it is for stars to form in it by gravitational collapse. Thus, molecular clouds are extremely important as regions of star formation. They may persist for 30 million years or so, with new stars arising here and there within them. Eventually, however, enough bright hot O and B stars arise that they heat portions of the molecular cloud, creating glowing HII regions. This explains the apparent paradox that cold molecular clouds are found in close association with hot HII regions. The most massive, short-lived stars may soon explode, blowing out million-degree gas and creating expanding superbubbles that blow the molecular cloud apart. Molecular clouds and their histories are thus intimately bound up with star births and star deaths (Blitz, 1982; Scoville and Young, 1984).

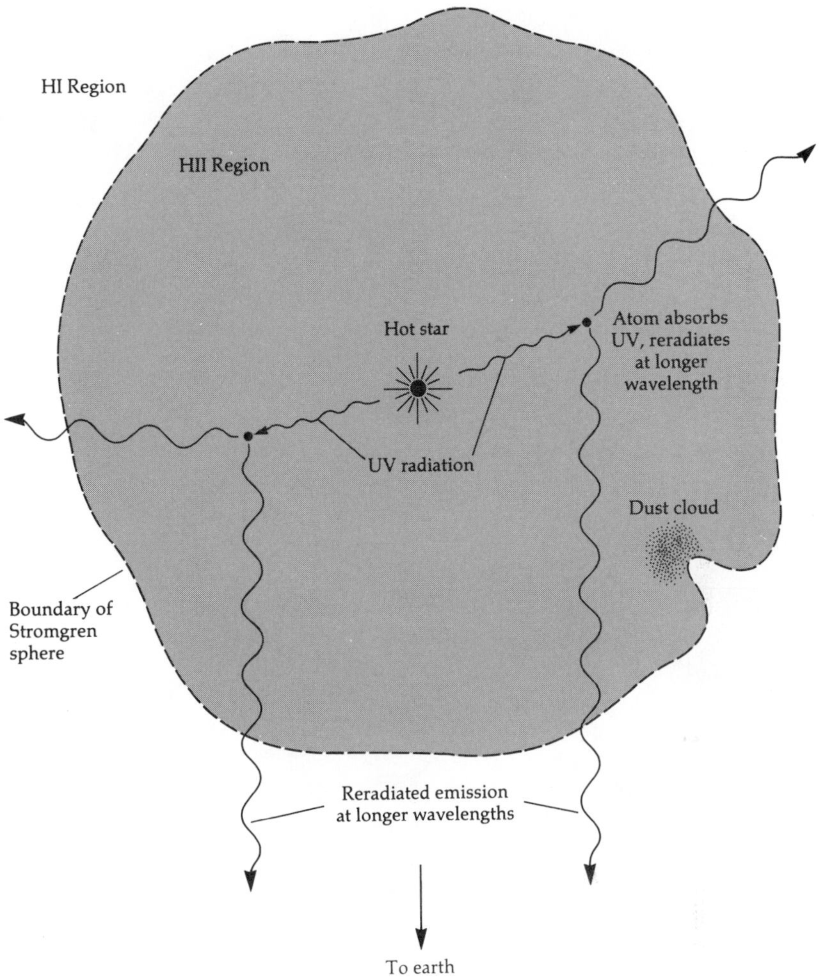

Figure 18-7. *Creation of an HII region. UV radiation from the central hot star is absorbed by hydrogen atoms, ionizing all of them out to a certain distance (dashed line). Recombination of electrons with atoms causes reradiation of various spectral emission lines in all directions, including earthward, causing the HII region to be visible as a glowing nebula.*

HI AND HII REGIONS: EFFECTS OF STARLIGHT ON INTERSTELLAR GAS

One of the most important processes in creating different kinds of regions is the action of light from hot stars on the neighboring gas. The appearance and spectrum of an interstellar gas cloud depend on the states of its atoms. These states depend on the source of illumination and the range of excitations and ionizations it causes in the cloud. Two important physical laws, **Planck's law** and **Wien's law**, control the effect.

Planck's Law
The bluer the light (that is, the shorter the wavelength) the more energy each photon contains. Thus, photons of ultraviolet or blue light excite or ionize atoms to a greater degree than photons of red light.

Wien's Law
Hotter sources radiate more blue light than cooler sources. (See also p. 123.)

These two laws predict that the most energetic photons will be encountered near the hottest stars. Thus, excitation and ionization in interstellar gas will generally be greatest near the hot, young stars of type O, and least in interstellar space, far from any stars.

Photons of ultraviolet light with wavelength shorter than 91.2 nm are energetic enough to ionize hydrogen. By Wien's law, stars hotter than about 30 000 K produce most of their radiation at these wavelengths, and even stars hotter than about 20 000 K produce many such photons. Therefore, hot, blue stars of spectral classes O and B are surrounded by large regions in which the interstellar hydrogen is ionized, as shown in Figure 18-7. If the gas were uniform, these regions would be spherical. Since the gas is nonuniform, these regions have ragged shapes.

In 1939, the Danish–American astronomer Bengt Stromgren calculated the average radii of the ionized regions around various types of stars, showing that only the hottest stars create large ionized regions. According to Stromgren's calculations, the diameters of the ionized regions in typical interstellar gas are:

Figure 18-8. *The Lagoon nebula (M8) is an emission nebula associated with a cluster of hot, young stars (lower center). Note the complexity of dark wisps and knots, some with glowing rims, silhouetted in front of the main nebula. The complex is about 25 pc across and 1200 pc away. (Kitt Peak National Observatory.)*

Figure 18-9. *The middle portion of this nebula, NGC 2359, is a spheroidal stellar bubble, inflated by strong stellar winds blowing out of a Wolf–Rayet star in its center. This photo was made using light radiated by doubly ionized oxygen atoms, which concentrate in filaments on the surface of the bubble. (M. H. Schneps and E. L. Wright, Harvard/Smithsonian Center for Astrophysics.)*

Figure 18-10. *Long exposure of the Dumbbell nebula shows a spheroidal bubble blown off a central star. Fainter outer portion is probably a shell of gas created by an earlier mass ejection from the central star. (Kitt Peak National Observatory.)*

Class	Diameter (pc)
O5	280 pc
B0	52 pc
B5	7.4 pc
A0	1.0 pc

Since hydrogen is the most common gas in all parts of interstellar space, regions where hydrogen is ionized around hot stars are often called HII regions; all other regions of interstellar space are called HI regions. The gas in HII regions is heated by stars to temperatures around 8000 to 10 000 K. In HI regions, farther from hot stars, temperatures are around 100 K. The most common HII regions are clouds with masses about 10 to 1000 $M_{\odot}$.

In and near HII regions, electrons in various elements lead an up-and-down life. On the one hand, they are likely to be hit by an energetic photon from the nearby hot star and knocked out of the atom. But on the other hand they tend to recombine with atoms and cascade down through the energy levels to reach the ground state. Each time they cascade down from a higher to a lower energy level, they lose energy by radiating light of a certain color. For this reason, gas in the HII regions emits light of varying colors. Especially strong is the bright red Hα light emitted when electrons cascade down from the $n = 3$ to $n = 2$ levels in hydrogen atoms (see Figure 14-8). Also strong is invisible UV light emitted when electrons go from the $n = 2$ to $n = 1$ levels. *Therefore, HII regions are also glowing nebulae called emission nebulae.* Many HII regions are among the most impressive features of the sky, as illustrated by the great swirls of gas in the Lagoon nebula (Figure 18-8) or the colorful clouds seen in Color Plate 27.

Stars with strong winds can push matter outward, making spheroidal expanding nebulae (usually HII gas); astronomers in 1975 coined the term **bubbles** for such objects, which have stronger expansion and symmetry than ordinary HII regions. Examples of various forms of bubbles are seen in Figures 18-9, 18-10, and 18-11. Wispy debris of a blown-apart bubble are seen in Figure 18-12.

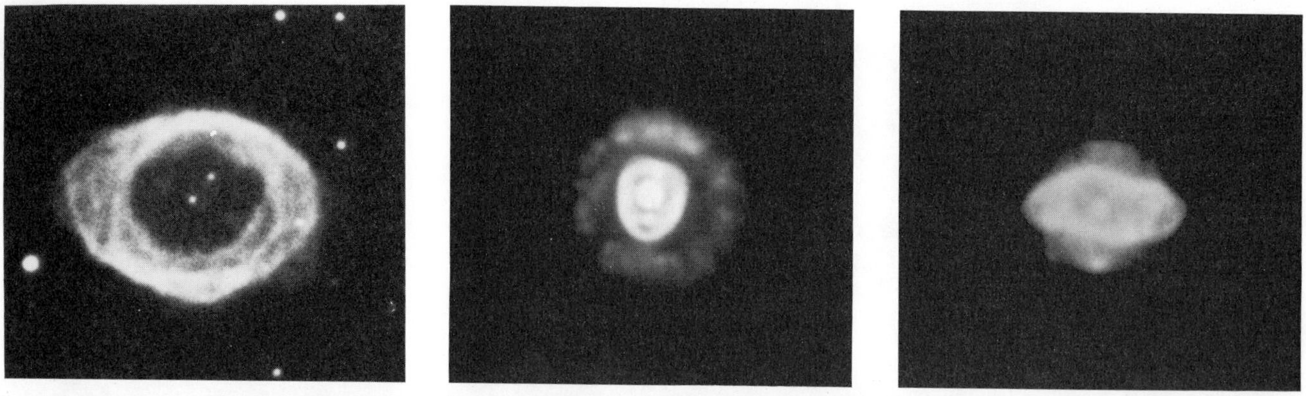

a b c

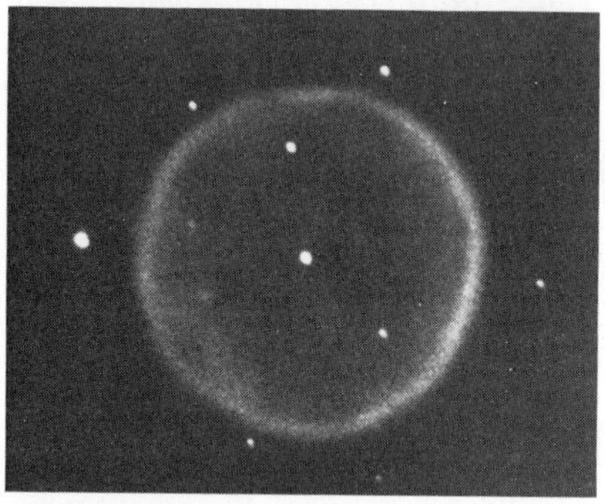

d

Figure 18-11. *Three examples of planetary nebulae, showing various shapes.* (a) *Ring nebula about 0.2 pc in diameter and 700 pc distant. Photographed in light emitted by neutral oxygen atoms. See also Color Plate 26. (Steward Observatory, courtesy R. Williams.)* (b) *NGC 2392, showing two rings, possibly indicating two distinct eruptions. About 0.2 pc in diameter and 1000 pc distant. (Mauna Kea Observatory, courtesy A. Stockton.)*
(c) *Saturn nebula, showing unusual lateral extensions. About 0.1 pc in diameter and 700 pc distant. (Kitt Peak National Observatory.)* (d) *In the absence of turbulent explosive activity, a central star may smoothly blow off a nearly perfect spheroidal shell, or "bubble," of gas. This example is Abell 39. (Kitt Peak National Observatory, courtesy George Jacoby.)*

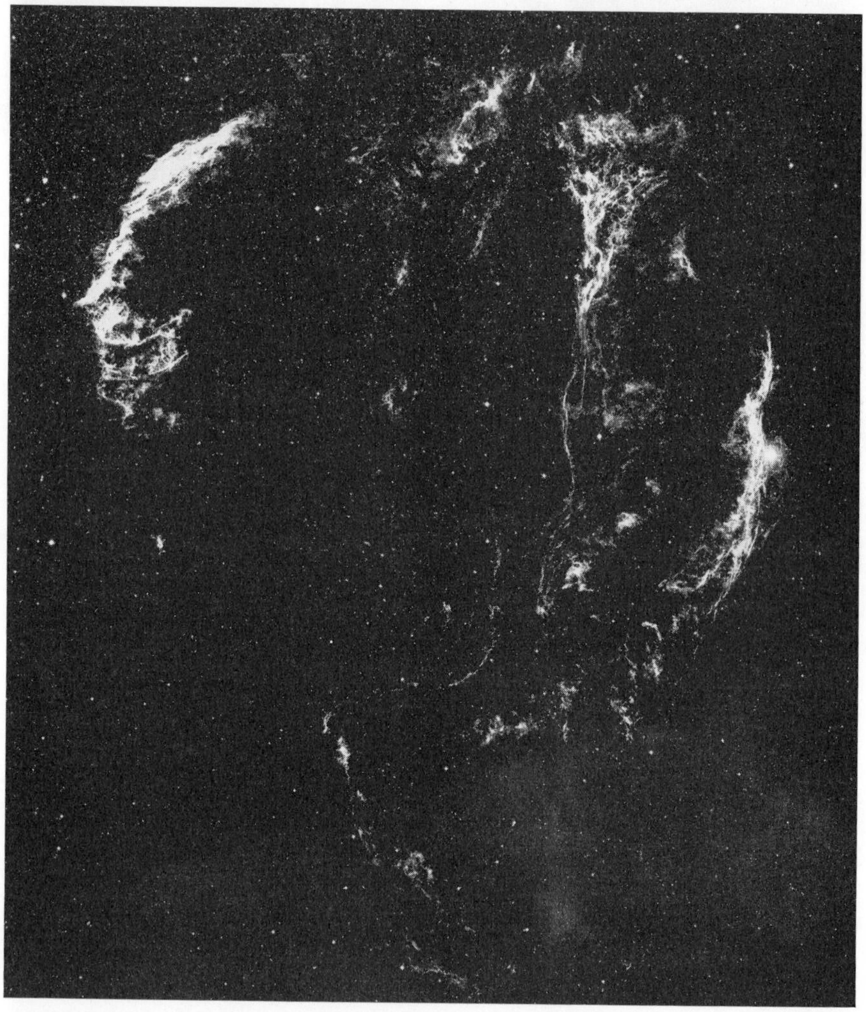

Figure 18-12. *The Veil nebula in Cygnus is made from wispy remnants of a spheroidal cloud blown into interstellar space by a supernova explosion. Distance is about 500 pc; diameter, about 22 pc; apparent diameter, $2\frac{1}{2}°$. (Hale Observatories.)*

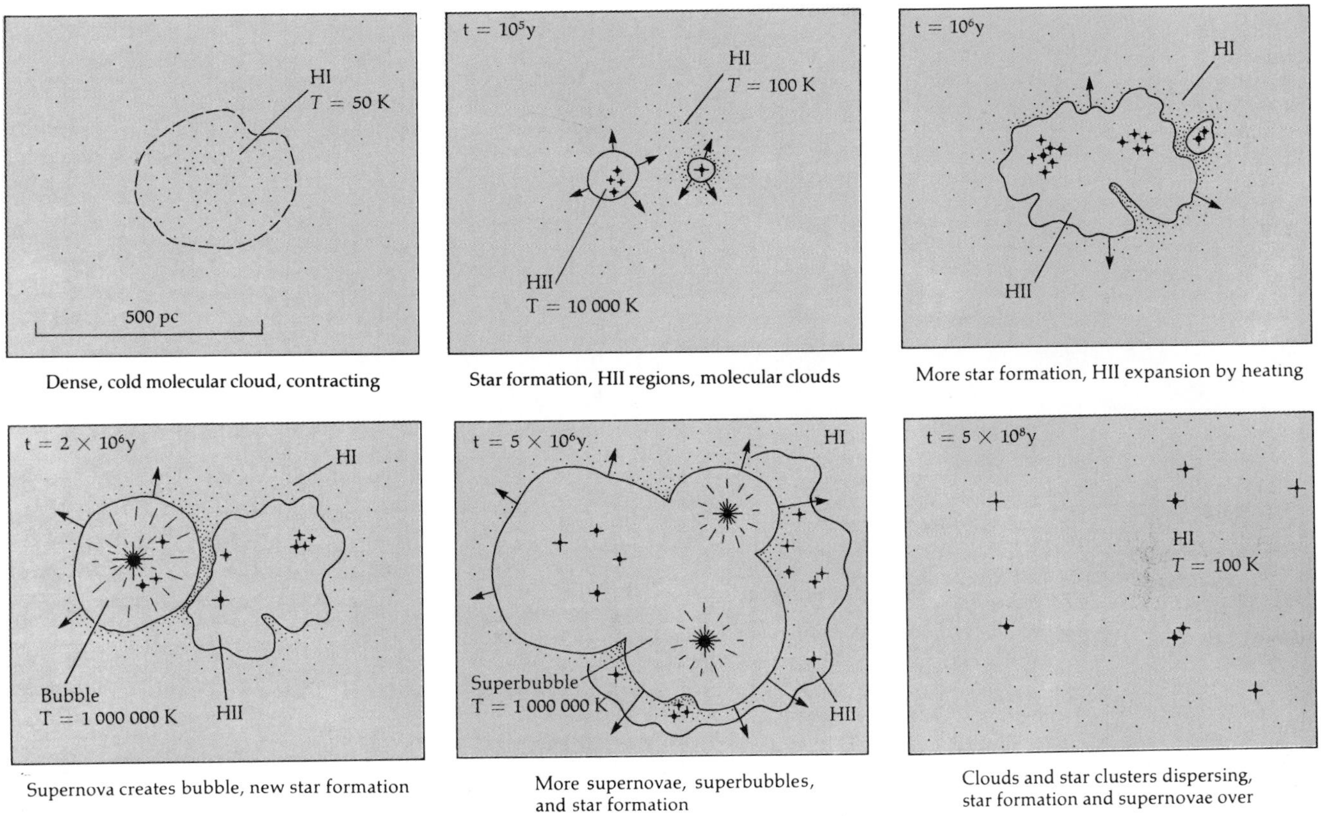

Figure 18-13. *Schematic drawings showing the evolution of interstellar material in a certain region. The first box shows a random concentration of gas and dust that initiates star formation. The largest of the new stars then explode as supernovae. This expands the gas and eventually creates superbubbles of very hot expanding gas. The surrounding gas is finally blown away and the star formation ends.*

EFFECTS OF SUPERNOVAE ON INTERSTELLAR MATERIAL: SUPERBUBBLES

Astronomers were surprised when X-ray telescopes in space in the 1970s revealed large clouds of much hotter gas radiating X rays at wavelengths such as 1.0 nm. Wien's law shows that X-ray emission from such hot gas must require temperatures over 1 million Kelvin! What could cause such high temperatures? These clouds are heated by violent expansion of gas blasted out of supernovae into surrounding space. Such events can create hot superbubbles. Figure 17-10 showed an X-ray image of such a cloud. A superbubble in Cygnus is about 500 pc across, has a temperature of about 2 million Kelvin, contains 10^{45} J of energy, would have required about 100 supernovae to be produced, and (based on its rate of expansion) is about 3 million years old, in contrast to the 1-million-year old group of young O and B stars inside it. From these characterictics, astronomers infer that a sequence of many supernovae expanded this bubble (Cash and Charles, 1980). Direct evidence of supernova explosions include several supernova remnants and the black hole candidate Cygnus X-1, all located inside the Cygnus superbubble.

As shown in Figure 18-13, the endless cycle of star birth and death is bound to produce superbubbles that play a key role in redistributing the interstellar material. Star formation produces a few very massive stars that rapidly explode; this heats and expands local gas (see Figure 17-10), which pushes against adjacent gas. The resulting compression starts new star formation on the outskirts of the expanding cloud. This leads to new explosions and expansions until the whole star-forming complex blows itself apart.

TYPES OF NEBULAE

By tradition, nebulae are often categorized by their superficial appearance in telescopes. This in turn depends on whether hot stars are near them and

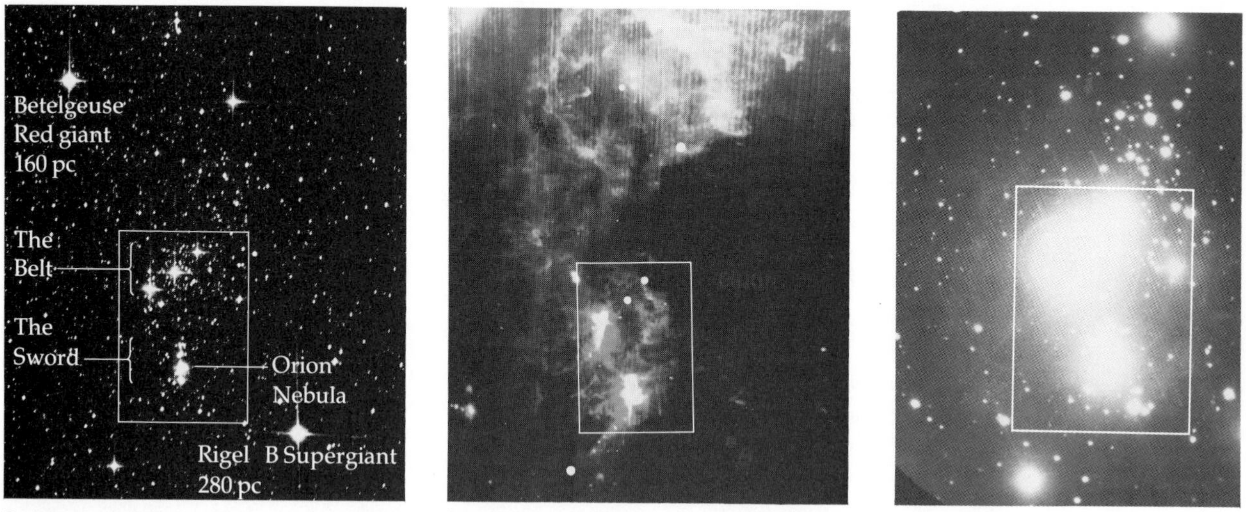

a b c

Figure 18-14. *Three faces of the constellation Orion: visible light, far infrared light, and ultraviolet light. Box shows star-forming region of belt and sword, seen in more detail in Figure 18-15.* (a) *Visible light view with "star effect" filter that emphasizes brightest stars. (Photo by author, 21-min exposure on 2475 Recording film, 55-mm lens,* f2.8.*)* (b) *Infrared view at wavelengths of 12 to 100 μm, emphasizing clouds of dust at temperatures about 30 K. At top, circling Orion's head, is a ring of warmer gas probably blown out from mass-shedding stars. Note that most individual stars are too hot to radiate much light at these wavelengths; they are shown by white dots. (NASA photo from IRAS Infrared Astronomical Satellite.)* (c) *Ultraviolet view at wavelengths 125–160 mm, emphasizing the hottest O stars at temperatures around 20 000 K. Note that the red giant Betelgeuse (upper left) is too cool to show up at these wavelengths. (Naval Research Laboratory photo from high-altitude rocket, 1975, courtesy George Carruthers.)*

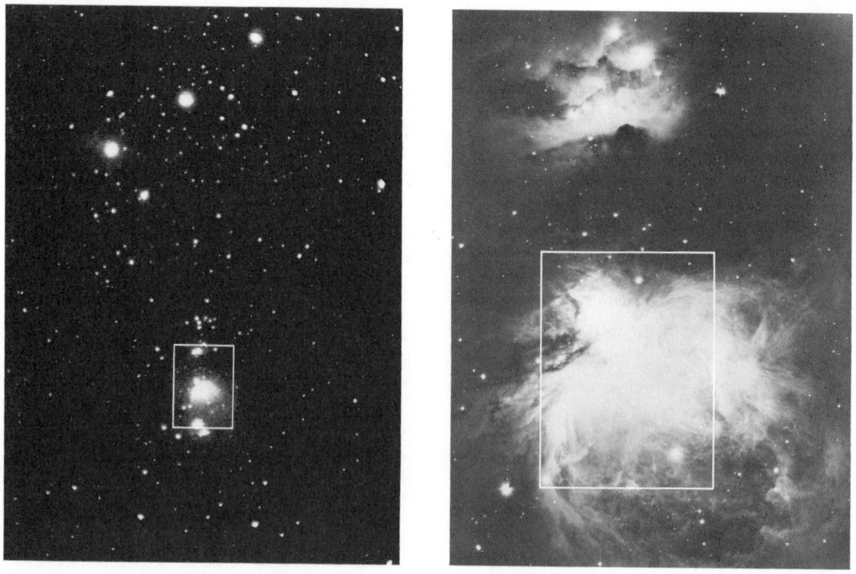

a b

Figure 18-15. *Zeroing in on the inner Orion nebula* (a) *Orion's belt (top) and sword (bottom). Vertical length is about 9° and box shows region of* b, *containing nebula that forms middle "star" in the sword. (Photo by author, 8-min exposure on 35-mm 2475 Recording film, 135-mm telephoto,* f2.8.*)* (b) *The Orion nebula and adjacent parts of sword. Nebula is about 5 pc wide. (Lick Observatory.)* (c) *Central nebula region. Compare color view in Color Plate 24. (Copyright Anglo-Australian Telescope Board.)* (d) *Region of the Trapezium (arrows) four very young stars about 0.04 pc (8300 AU) apart in the heart of the nebula. Width of picture is about 0.7 pc. (Allegheny Observatory, courtesy W. A. Feibelman.)*

whether we happen to see them against a background of dark space or distant bright nebulae. Table 18-2 lists characteristics of some specific nebulae.

Two main categories are dark nebulae (clouds silhouetted against brighter backgrounds) and bright nebulae. Dense parts of molecular dust clouds, for example, often appear as dark nebulae (see Figure 18-6). Bright nebulae include glowing HII regions (see Figure 18-8, Color Plate 25; often reddish due to the Hα emission line in the spectrum) and reflection nebulae (Color Plate 28 top; often bluish, with dust and gas reflecting and scattering light of a nearby star). Expanding bubbles often appear as disks in telescopes, and hence were given the misleading name of "planetary nebulae," though they have nothing to do with planets. Some well-known examples are shown in Figure 18-11 and Color Plate 26, top.

Example: The Orion Nebula, a Nearby Star-Forming Region

In the direction of the constellation Orion lies a region about 400 pc away dominated by nebulae and star-forming activity. Orion itself is the constellation of the hunter—a great figure raising a club over his head. As can be seen in the sketch map of Figure 16-9, three bright stars mark his belt and three below it mark his sword. Orion's shoulder on the left side is the bright red giant star Betelgeuse, 180 pc away. His opposite knee is the blue B-type star Rigel, 270 pc away.

As shown in Figure 16-9, if you look toward Orion on a February evening, swinging your head back and forth past this part of the sky to compare it with other regions, you will see a great concentration of bright stars. Orion contains 7 of the 100 brightest stars in the sky, and except for Betelgeuse, all of them are massive, hot O- and B-type stars 140 to 500 pc away. Since massive stars are short-lived, their mere existence shows that they and other stars must have formed recently in the nebulosity around Orion.

Since O and B stars are extremely hot, they radiate prodigious amounts of ultraviolet radiation. For this reason, UV photos of Orion, such as Figure 18-14c, reveal it to be one of the most dazzling regions in the "UV sky." The massive stars have ionized much hydrogen, and the hot gas is being blown outward from the bright stars in central Orion. Some filaments of it show up in the outer parts of the overexposed region of central stars in Figure 18-14. Note that the red giant Betelgeuse, though brilliant to our eyes, is absent in Figure 18-14 because it is not hot enough to radiate much UV light.

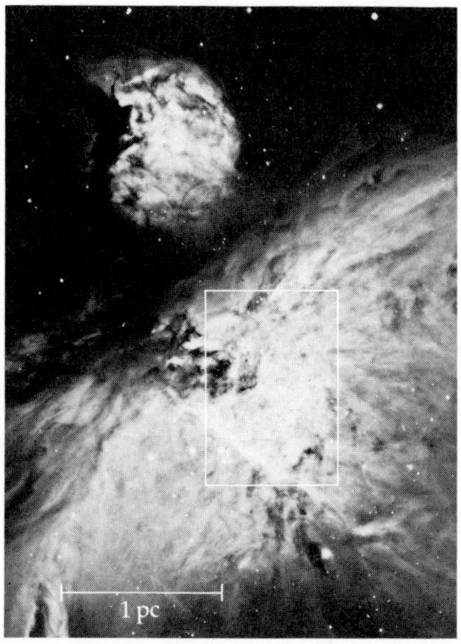

c

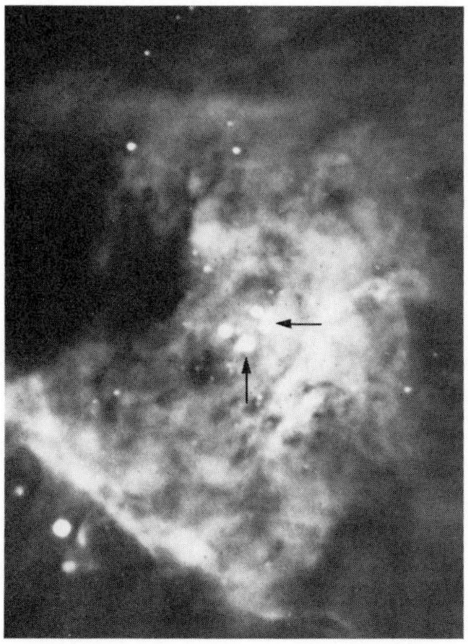

d

In Figures 18-15 a–d we "zero in" on the central star-forming area of Orion. Figure 18-15a, taken with an ordinary camera, shows Orion's belt and sword, and reveals that the central "star" in the sword is not a star at all but the great **Orion nebula**, an emission nebula about 460 pc away. Figures 18-15b and 18-15c show it in more detail, and Color Plate 24 shows the main nebula in color, revealing the red of Hα emission and other emission colors as well.

By carefully studying this region, astronomers have actually begun to piece together the arrangement and motions of material in the seething space around Orion. A backyard telescope reveals a quartet of faint stars, called the Trapezium, in the nebula's brightest core (shown in Figure 18-15d). The glowing region visible in a small telescope is about 8 pc across. Star formation is probably occurring in this region, where the gas and dust are unusually dense.

Figure 18-16 shows a cross section of the inner Orion region. On the far side from the solar system is an especially massive (about 1000 $M_\odot$) HI cloud of dust and gas, rich in molecules. According to one interpretation, expanding HII gas from the main nebula runs up against the HI cloud, limiting the size of the ionized region (Zuckerman, 1973). Especially dense clots of dust and gas were discovered inside the HI cloud by infrared observers in the 1960s. One of these, the Kleinmann–Low nebula, may mark a collapsing cloud about to form a whole cluster of stars. Various clouds move at about 8 to 10 km/s relative to each other, and backward tracing of their motions suggests some may have formed within the last 100 000 y.

A few degrees north of the Orion nebula is another spectacular, though less bright, nebulosity, shown in Figure 18-17. The Horsehead nebula is one of Orion's

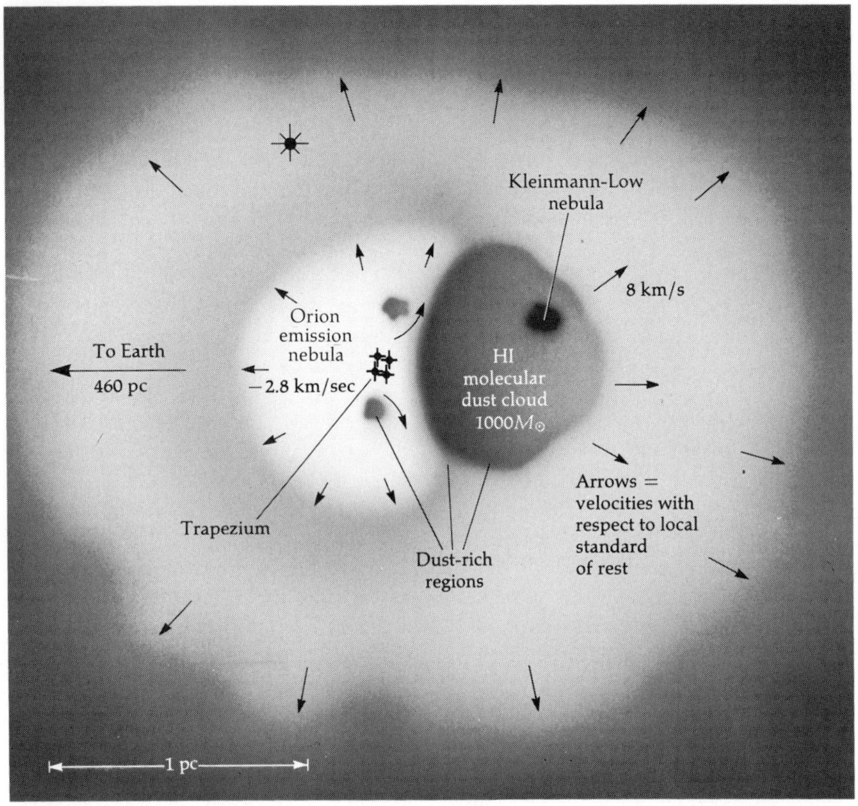

***Figure 18-16.** A hypothetical side view of the inner Orion nebula showing the Trapezium, dust clouds, and expansion.*

dense dust clouds, silhouetted against a background of glowing gas in such a way as to resemble an ominous cosmic chess piece. It is the most striking of all dark nebulae.

Radio astronomers have estimated that some 110 000 $M_\odot$ of HI and HII are involved in the expanding gas shell around the extended Orion star-forming region. Backward tracing of the motions indicates that the shell started to expand about 6 million years ago, when very massive, hot stars must have formed, heated nearby gas, and created a giant HII region that has been expanding ever since.

A time-lapse movie of Orion beginning about 6 to 10 million years ago, with frames every 10 000 years, would reveal star formation followed by a cosmic explosion of gas in the region of Orion's sword. The movie would also show certain stars racing out from central Orion like sparks from a blast. These high-speed stars are called **runaway stars.** Three bright O and B stars, for example, are racing out from the nebula at 70 to 130 km/s, and Betelgeuse itself is moving away from a region near Orion's belt. Their paths all point like accusing fingers back to the region of dense gas around the nebula. Something unknown happened in that region to create high-velocity stars. They may have been accelerated during near encounters between new stars in clusters or during disruptions of co-orbiting pairs of stars.

Orion has changed dramatically in the last few million years, since humanlike creatures emerged on the plains of Africa. New stars have blazed up, clouds of hydrogen have been expelled, and star formation is apparently continuing there today. As illustrated by Figure 18-18, a planet located in the midst of the Orion nebulosity would have a dramatic skyscape.

Example: Eta Carinae, a Nebula Around a Strange Star

Not many degrees from the Coalsack nebula (see Figure 16-7) lies an object that was recorded in the 1600s as an ordinary 4th magnitude star, which came to be called Eta Carinae.[2] It lies in a distant star-forming region about 2800 parsecs away. But in the 1800s its behavior was extraordinary. In 1827 it brightened to 1st magnitude and then dimmed. During a few weeks in 1837, English astronomer John Herschel, observing in South Africa, saw it become brighter than magnitude zero, finally reaching −0.7 in 1843! After observing with a large telescope, Herschel wrote in 1847 (quoted by Lovi, 1972):

It would ... be impossible by verbal description to give any just idea of the capricious forms and irregular gradations of light affected by the different branches and appendages of this nebula.... Nor is it easy... to convey a full impression of the beauty and sublimity of the spectacle it offers.

By 1857 the object had faded to +1; by 1900 it had dropped to +8. It has remained variable, rising to +6 in 1967.

Spectra and photos from 1892 to the present show that the nebula within about 0.05 pc of the central star is blasting outward at 630 km/s, apparently from the eruptions observed from 1827 to 1843. Several parsecs from the star, ionized gas is still expanding at about 30 km/s (Walborn and others, 1984). Neutral gas expands less rapidly.

[2]Eta Carinae lies at declination −59° and hence rises above the horizon only for observers south of 31° north latitude.

Figure 18-17. *The Horsehead nebula is a dark dust cloud silhouetted against glowing gas about 350 pc away. The whole complex is part of the Orion star-forming region, and is located just south of Orion's belt. Note that while background stars are numerous in the top half of the picture, they are mostly obscured by dust in the bottom half. (Kitt Peak National Observatory.)*

Modern photographs show the beautiful nebula described by Herschel (Figure 18-19). It exceeds 1° width and is roughly 60 pc across. Its red color comes from the strong light of the Hα emission line ar 656.3 nm. The nebula is surrounded by an enormous hydrogen complex, rich in hot, blue stars, with luminosities around 5 million $L_\odot$. Only a few hundred parsecs from the nebula is a cluster of at least 130 O and B stars, known as NGC 3293, and estimated to be only 8 million years old. The nebula is also bright in infrared light radiating from abundant warm (200 K) dust, probably grains condensing in material blasted out of the central star. The grains dim the star's light, but the total infrared and visual luminosity is 5 million $L_\odot$, like that of the nearby O stars (Humphreys and Davidson, 1984).

Like the Orion region, the Eta Carinae region seems to involve star formation. But much of the **Eta Carinae nebula** seems to have been thrown off a recently formed massive star that has already reached an unstable, old-age state. What kind of star? In the center of the nebula is a tiny reddish cloud about 0.02 pc across that may contain an unstable star nearing the nova stage. Its surface temperature has been estimated at 30 000 K (Humphreys and Davidson, 1984). However, the variable star expert D.J.K. O'Connell notes (quoted by Lovi, 1972):

Eta Carinae is certainly not an ordinary nova. Its behavior is indeed not paralleled by any other known star. Further observations are badly needed. . . . It may once again become one of the brightest stars in the sky.

Figure 18-18. *A planet associated with a newly formed star would be located in a star-forming region similar to that around the Orion nebula. The system would probably be embedded in major nebulosity and a cluster of young stars. This imaginary view from a crater rim on such a planet shows the sky dominated by bright nebulae and hot, young stars. A similar view might have been seen in the earth's night sky soon after the sun formed 4.6 billion years ago. (Painting by author.)*

Example: The Gum Nebula, an Ancient Spectacular

In the 1950s, a graduate student named Colin Gum surveyed the Southern Hemisphere sky in a search for HII nebulae. He used photographs sensitive to hydrogen alpha radiation and discovered a ring of nebulosity with the amazingly large angular diameter of 60°, which has come to be called the **Gum nebula**, shown in Figure

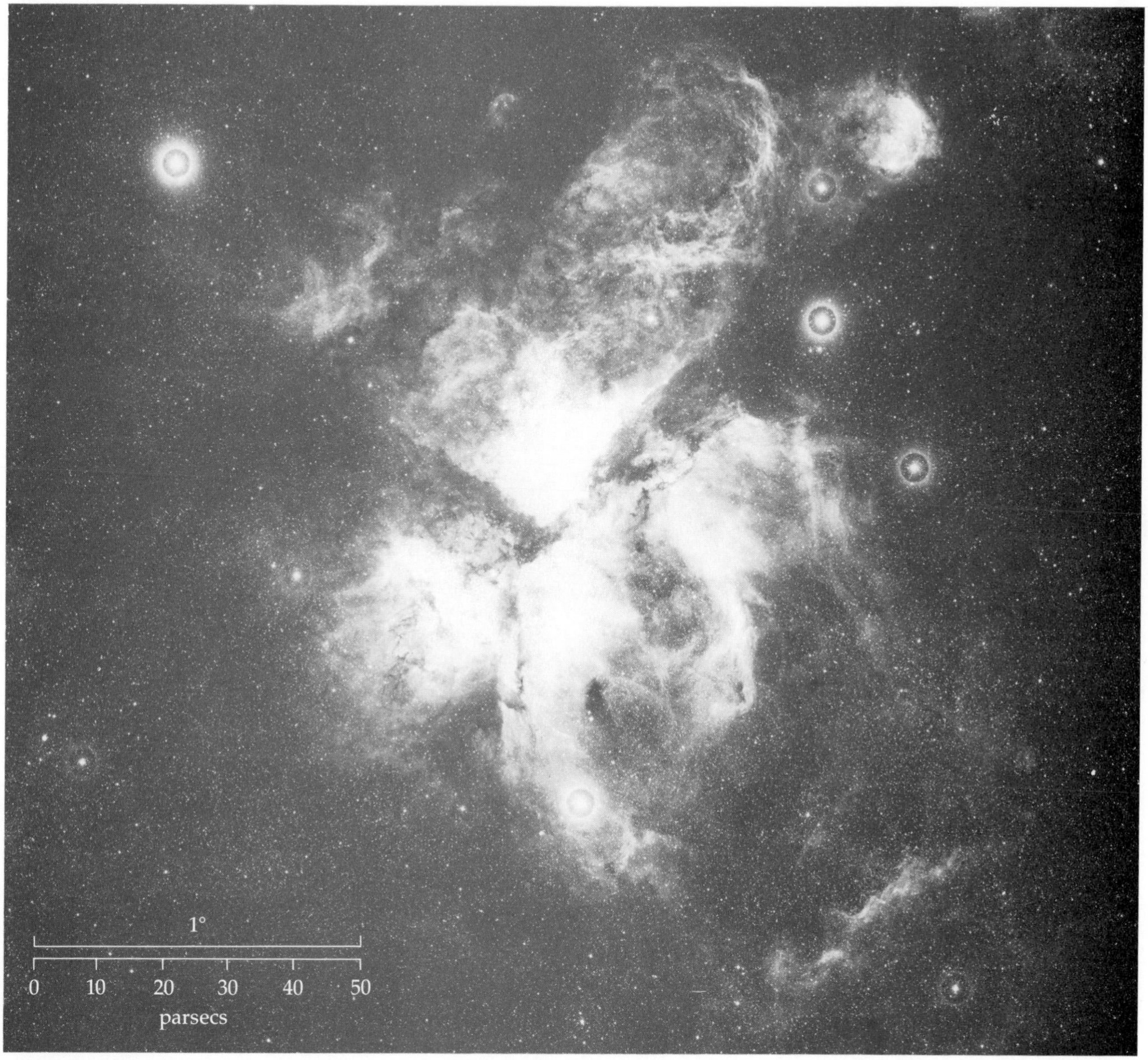

Figure 18-19. *The Eta Carinae nebula (NGC 3372) is the site of peculiar brightness changes. Swirling gas, individual bright clouds and dust lanes are prominent. The nebulosity extends across about 60 pc and is about 2800 pc away. Halation rings around the bright star images are produced by the photographic system; the star in the upper left is about 3rd magnitude. (Cerro Tololo Observatory, Chile; Kitt Peak National Observatory.)*

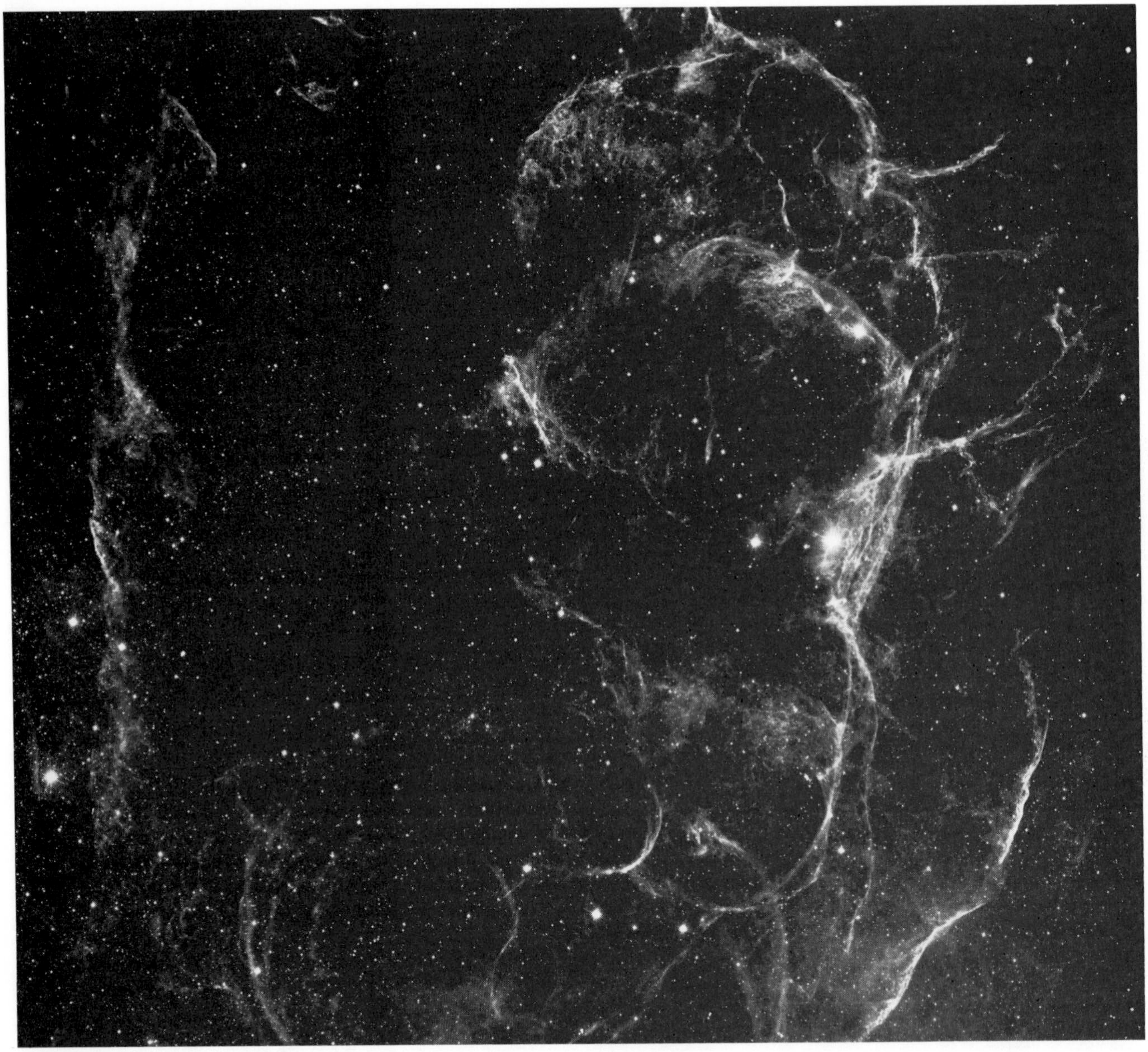

***Figure 18-20.** The inner part of the Gum nebula, here photographed in UV light, believed to mark remnants of an expanded supernova shell. A pulsar—the remnant of the supernova—lies near the center of the roughly circular ring of nebulosity (near the bright filament above the center). Nebular diameter is about 5°. Figure 18-21 shows the nebula's geometry. Compare also with Figure 18-12, which shows a similar filamentary nebula. (Kitt Peak National Observatory; Cerro Tololo Observatory.)*

18-20. It has been identified as an expanding bubblelike shell of gas from an ancient supernova explosion.

Figure 18-21 shows it in cross section. Its front side lies only about 100 pc from the solar system, and its center only about 460 pc away, accounting for its large angular size. The supernova remnant, a pulsar, has been found in the center of the expanding bubble.

Studies of the expansion rate suggest the explosion might have been witnessed around 9000 B.C. Because the explosion was so close (about one-third the distance of the Crab supernova) and had such high energy, it was probably extremely brilliant, reaching an estimated apparent magnitude of −10 (about as bright as the first quarter moon). This raises the intriguing prospect that late Stone Age humans (south of latitude +47°) may have seen a brilliant celestial beacon blaze forth for a year or two, rivaling the moon. What effect might this have had on their concepts of the heavens and gods in the sky?[3]

[3]The effect of the Vela supernova may have been more than a dramatic skyshow. Theorists have concluded that the gamma-ray and X-ray radiation from such a close supernova would deplete the ozone layer, increase the nitrogen oxide production, and cool the atmosphere. The fallout of the NO in rain would have a fertilizing effect, increasing photosynthesis and deposition of organic sediments. G. R. Brackenridge (1981, *Icarus, 46,* 81) describes eleven dramatic cases of black, organic sediments deposited around 9000 B.C.. and suggests that they were caused when the Vela supernova produced a world-wide but brief climatic change. The earth may be affected more than we know by "winds" that blow from interstellar space.

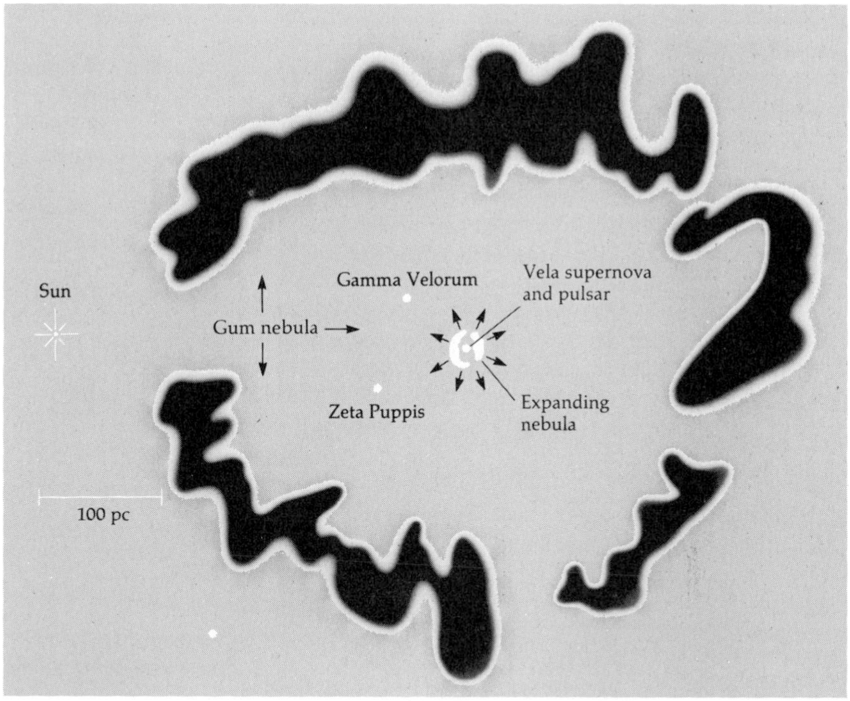

Figure 18-21. *A side view of Gum nebula geometry, showing relative positions of the sun, Vela pulsar, and stars Zeta Puppis and Gamma Velorum (compare with Figure 18-20).*

Summary

Space is not empty, but thinly filled with atoms and molecules of gas, grains of dust, and possibly bigger debris, which partly obscure distant stars and redden their light. Concentrations of these materials are nebulae. Some are excited to fluorescence by nearby stars, forming emission nebulae; some merely reflect the light of nearby stars; and some are dark silhouettes against distant backgrounds. Various processes of emission, Rayleigh scattering, and absorption give nebulae different colors. Various processes of exansion and turbulence give nebulae different shapes.

Nebulae are the media from which stars are born and into which the larger stars blow some of their material when their fuel runs out. Their existence shows that matter in our galaxy has not dispersed uniformly and stably, but rather is continually stirred, formed into clouds, dispersed, and disturbed by influences such as the formation of new stars, the explosions of old stars, and movements of all local material around our galaxy's center.

From nebulae and the general interstellar complex we know that all local material visible from the earth has participated in a vast cosmic recycling. Interstellar matter forms clouds; clouds may contract to form stars; young and old stars blow out their material, creating a new interstellar medium. Nebulae also reveal clear evidence of cosmic events that have markedly changed our celestial, and perhaps even terrestial, environment within the last million years—a period less than 0.01% of cosmic time.

Concepts

nebula	molecular cloud
excitation	HI region
ionization	HII region
Rayleigh scattering	superbubble
interstellar reddening	Messier number
interstellar atom	NGC number
metastable state	IC number
forbidden line	Planck's law
21-cm emission line	Wien's law
interstellar molecule	bubble
amino acid	Orion nebula
interstellar grains	runaway star
interstellar obscuration	Eta Carinae nebula
interstellar snowball	Gum nebula

Problems

1. Since sunlight is white (a mixture of all colors), why does the sun look red at sunset? What happens to the blue light? Why does the part of the sky away from the sun look blue?

2. Why was the sky red as photographed on Mars by Viking cameras?

3. How do massive stars help keep interstellar gas stirred up?

4. How do interstellar molecules help illustrate the fact that complex organic chemistry is likely elsewhere in the universe?

5. How do masses of prominent nebulae compare with masses of single stars—are stars more likely to form singly or in groups?

6. Why are O-type supergiant stars likely to be associated with large emission nebulae, whereas solar-type stars are not?

7. Why do planetary nebulae often have simple, nearly spherical forms, whereas typical larger emission nebulae, such as the Orion nebula, are ragged, irregular masses?

ADVANCED PROBLEMS

8. If the sun's material were redispersed into space at a density of 10 atoms/cm^3, typical of some clouds, how big a cloud would it make? (*Hint:* The sun contains about 10^{57} atoms. 1 pc = 3×10^{16} m = 2×10^5 AU.)

a. Compare with the size of the solar system.

b. Compare with prominent nebulae.

9. Suppose spectral lines from gas on the near side of the Crab nebula have a Doppler blue shift of 0.5%.

a. At what velocity is it expanding?

b. If the Crab nebula is 1500 pc away (or 4.5×10^{19} m), use the small-angle equation to derive the expansion velocity of the nebula if it is observed to expand at an angular rate of 0.2 seconds of arc per year. (Hint: 1 y $\approx \pi \times 10^7$ s.)

c. Confirm that the above two methods of estimating expansion velocity in the Crab nebula give consistent results.

10. Suppose a tight cluster of newly formed O stars totals 100 solar masses within a region a parsec wide (like the heart of the Orion nebula). Several of them explode as supernovae, creating a superbubble. The bubble expands to a point 10 pc from the center of the cluster, and has a temperature of one million Kelvin. Compare the thermal velocity of a hydrogen atom at the edge of the bubble with the velocity it would need to escape from the gravitational pull of the star cluster into interstellar space. Would you expect the bubble to continue to expand and dissipate, assuming it did not run into dense clouds of adjacent material?

Projects

1. Observe a cloud of cigarette or match smoke illuminated by a single light source, preferably a shaft of sunlight in a darkened room, or a strong reading lamp. Compare the color of light transmitted through the smoke (by looking into the beam) with the color of light scattered out of the smoke (by looking at right angles across the light beam). Are there any differences? What can you conclude about the size of particles in the smoke cloud, assuming that the light wavelength is mostly 400 to 800 nm? Compare forms in the drifting smoke cloud with forms of nebulae illustrated in this book.

2. In a dark area away from city lights, by naked eye, observe and sketch the Milky Way in the region of Cygnus. Can you observe the dark "rift" that divides the Milky Way into two bright lanes in this region? The rift is caused by clouds of obscuring interstellar dust close to the galactic plane that are between us and the more distant parts of the Milky Way galaxy.

3. Observe the Orion nebula with a telescope. Sketch its appearance. Locate the Trapezium (four stars near the center). The dark wedge radiating from the Trapezium is a dense mass of opaque dust. If a large telescope (50–100 cm) is available, look carefully for color characteristics. Generally, the eye is unresponsive to colors of very faint light, but large telescopes gather enough light so that colors can sometimes be perceived, especially with fairly low magnifications, giving a compact, bright image.

4. Observe the Ring nebula in Lyra or other nebulae such as the Crab (in Taurus), the Trifid (in Sagittarius), or others as available. Comment on differences in form and origin.

CHAPTER 19

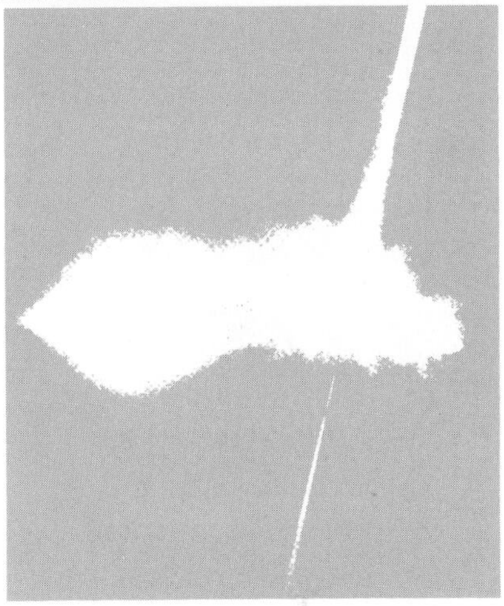

BINARY AND MULTIPLE STAR-SYSTEMS

We have been soft-pedaling a fundamental fact: Most stars are not solitary wanderers in space. As seen in Table 15-1, five of the first six stars we encounter beyond the solar system have known companion stars. Surveys of this kind suggest that more than half of all "stars" are really coorbiting star systems.

Each pair of coorbiting stars is called a **binary star system**, or sometimes just a binary star. Each system with more than two stars is called a **multiple star system.** How do these systems affect our understanding of the universe? For one thing, we would like to know if our own multiple system, the sun and its planets, is related to other multiple star systems, or whether it is a different kind of phenomenon. Second, we must be sure that our theories of star formation and evolution account for binaries.

A review of the preceding chapters shows that the theories are consistent with the existence of binary and mutiple systems. Chapter 16 showed that stars must form in groups, and Chapter 18 gave examples of groups of young stars (such as the Orion nebula's Trapezium) embedded in star-forming nebulae. Chapter 17 required companion stars to explain certain facts of late evolution, including transfer of mass in binary systems to explain ordinary novae (not supernovae) and X-radiation caused by dumping of mass from giants onto coorbiting dense stars or black holes.

Optical Doubles Versus Physical Binaries

To understand binaries and multiples in more detail, we face an observational problem. Among the star pairs that appear to be close together in the sky, some are at different distances and merely aligned by chance (as seen from the earth)—they are not actually coorbiting. Such star pairs are called **optical double stars.** They can be identified by the absence of any orbital motion around each other (revealed by photos or Doppler shifts) and are of little consequence in astronomy.

Pairs that are close enough to orbit around each other are called **physical binary stars.** They are the ones that clarify our knowledge of star properties and are the ones we will discuss here.

Early observers thought that all close star pairs were merely optical doubles, but in 1767 John Michell pointed out that so many chance alignments were unlikely. He proposed a physical association. This was confirmed in 1804 when William Herschel discovered that Castor (brightest star in the constellation Gemini) has a companion orbiting around it. In 1827 F. Savary showed that the orbit of the physical binary Xi Ursae Majoris is an ellipse fitting Kepler's laws. These results marked the *first discovery of gravitational orbital motion beyond the solar system,* an important confirmation that gravitational relations are universal.[1]

Among physical binaries, the brighter star is usually designated A, and the fainter B—for instance, Castor A and Castor B. Analysis of orbits reveals which is the more massive star, and it is usually called the *primary.* The less massive one is called the *secondary.* Normally, the primary is also the brighter, or star A.

DIFFERENT TYPES OF PHYSICAL BINARIES

Astronomers classify physical binaries according to how thay are detected. To

---

[1]Herschel realized that this discovery was much more important than his original goal of simply measuring distances, because the orbital motions reveal many properties of the stars. This illustrates how important, unexpected scientific results often derive from more mundane research on another topic. Herschel reportedly likened himself to the biblical character Saul, who went out to find his father's mules and discovered a new kingdom.

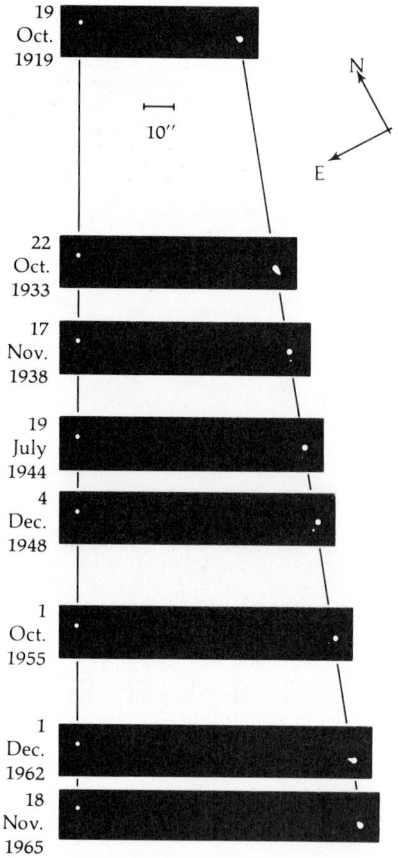

Figure 19-1. *Forty-six years' photography of the visual binary Krüger 60. Binary pair Krüger 60, on the right, is moving past a background star, on the left. The photos show the orbital motion of one component of the binary around the other with a period of 45 y and a separation of 1.4 to 3.4 seconds of arc. The two stars have magnitudes about 9 and 11 and masses of 0.3 and 0.2* $M_\odot$. *(Leander McCormick Observatory and Sproul Observatory, after J. F. Wanner,* Sky and Telescope, *1967.)*

understand what we can learn from binaries, it helps to understand these different methods of detection.

A **visual binary** is a physical binary in which both members can be resolved (seen separately) with the eye, the telescope, or the camera. Some 65 000 have been studied, and a good example is shown in Figure 19-1.

In a **spectroscopic binary**, orbital motion is revealed by periodic Doppler shifts in the spectral lines, but the individual stars cannot be resolved. There are two subtypes: those in which only *one* spectrum can be detected (Figure 19-2), and those in which *two* sets of spectral lines are seen (Figure 19-3). The latter, displaying lines of both stars, yield more information. About 1000 pairs have been measured.

An **eclipsing binary** is a binary pair (generally unresolved) whose orbit is seen nearly edgewise and is revealed by eclipses. Because our line of sight lies in, or nearly in, the orbital plane, the stars alternately eclipse each other. (Such events are more properly called *occultations* and *transits,* but the term *eclipse* is widely used.) The eclipses are detected by plotting **light curves**, or plots of brightness versus time, as shown in Figure 19-4. Depending on the relative brightness and size of the stars, the eclipse of the primary may produce a marked, short-term decrease in brightness. The star then returns to its normal brightness. The most famous example is Algol (β Persei), discovered in 1699. Algol dims to about one-third its normal brightness every 2.9 d.

Most eclipsing binaries are really **eclipsing–spectroscopic binaries**, in which both Doppler shifts and eclipses can be detected. This is the most informative type of binary, permitting very detailed analysis of the motion, mass, and size of the stars.

An **astrometric binary** is one detected not necessarily by sight, Doppler shifts, or eclipses, but by its orbital motions with respect to those of the background stars. According to Kepler's laws, stars in a coorbiting system revolve around their center of gravity, as described in Chapter 14. Each star therefore describes an oscillating path in three-dimensional space, moving back and forth around the system's center of gravity. Because *astrometry* is the study of stellar positions and motions, these systems are called astrometric binaries. Figure 19-5 shows a visual binary that is also a suspected astrometric binary: 61 Cygni. For some years, 61 Cygni was reported to have an unseen companion only 8 times as massive as Jupiter. More recent studies (Heintz, 1978) question the mass and even reality of the companion. Until the case is settled, 61 Cygni remains a beautiful system with an aura of mystery.

Spectrum binaries make up a relatively unimportant class, where the presence of an unresolved binary is revealed by a unique spectrum consisting of lines from stars of two different temperatures, but with no detectable Doppler shifts.

Table 19-1 lists examples of the most important types. The famous star Mizar, in the middle of the Big Dipper's handle, illustrates several types of doubles. Mizar forms an optical double with a fainter star, Alcor, less than a degree away. The Arabs called this pair the "horse and rider" and regarded it as a test of good eyesight. In 1650, the Italian observer Jean Riccioli discovered that Mizar itself is a physical binary with a visible 4th-magnitude companion, Mizar B, just 14 seconds of arc away. In 1889 Mizar A was found to have double spectral lines with periodic Doppler shifts, revealing that it is a spectroscopic binary. In 1908, the same was found true of Mizar B. Thus, Mizar is really a quadruple star system, with one close coorbiting pair revolving around another close coorbiting pair.

What Can We Learn from Binary Stars?

The various types of physical binaries yield much information. *The most important application of binary star studies is in determining star masses.* As explained in Chapter 14, the Kepler–Newton laws of orbital motion allow us to calculate the masses of astronomical bodies from measures of their revolution periods and orbit sizes (pp. 115–117).

As an example of a different application, consider an eclipsing–spectroscopic system. Suppose one star moves in a circular orbit at 100 km/s, and the time it takes to pass behind the other star (eclipse duration) is 3 h, or about 10 000 s. Then the diameter of the larger, obscuring star is 1 million kilometers (10^2 km/s × 10^4 s = 10^6 km), or about the size of the sun. This type of observation is the most accurate measure of stellar size, and hence

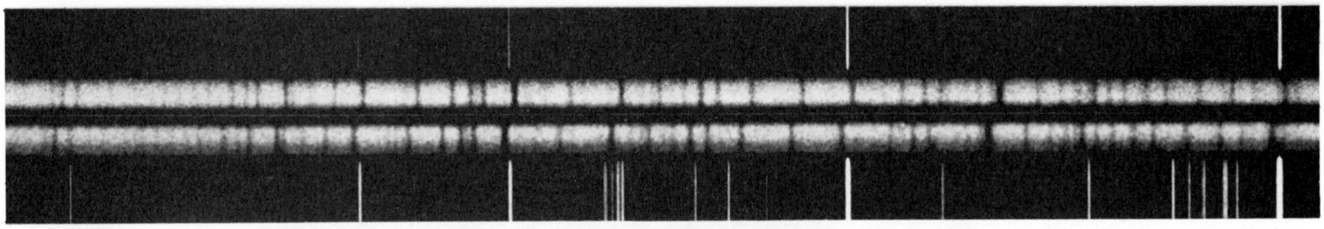

Figure 19-2. *A portion of the spectrum of a single-line spectroscopic binary, Alpha Geminorum. The bright vertical lines at top and bottom are reference emission lines produced in the spectrograph; the middle two bright spectra crossed by dark, vertical absorption lines are spectra of the star on two dates. Offsets of these lines to the red and then the blue show that the star is receding and then approaching because it is orbiting around another star. (Lick Observatory.)*

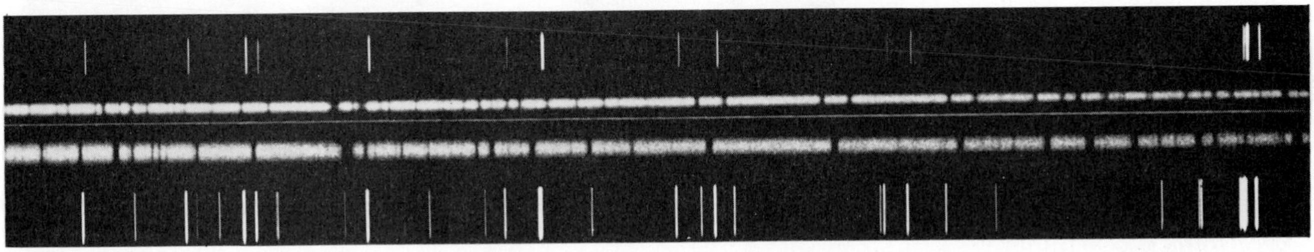

Figure 19-3. *A portion of the spectrum of a double-line spectroscopic binary, Kappa Arietis. Arrangement as in Figure 19-2. Two sets of absorption lines in the top stellar spectrum reveal two stars, one receding (red-shifted) and one approaching (blue-shifted). The lower stellar spectrum, when orbital motions are perpendicular to the line of sight, shows lines merged, with no Doppler shift. (Lick Observatory.)*

one of the most basic checks on theories of stellar structure.

The same eclipse provides another example of how binaries contribute to knowledge of stellar structure. As the smaller star is eclipsed, the larger star successively obscures different parts of the smaller star's surface. It first covers part of the limb, then sweeps across the center parts, and finally covers the other limb. Later, as the eclipse ends, the process is reversed. The larger star acts as a screen, successively covering and uncovering different parts of the other star. Thus, by studying how the total brightness of the system changes during the eclipse, we can study the brightness *distribution* across the face of the hidden star, thereby learning more about its atmospheric structure.

How Many Stars Are Binary or Multiple?

This question is a challenge to observers. The nearest stars are easiest to observe but give too small a statistical sample to be reliable. At greater distances there are more stars, but faint companions might not be detected. Spectroscopic binary statistics are biased toward pairs with small separation distances, because according to Kepler's laws these have the fastest velocities and greatest Doppler shifts, thus being the most likely to be discovered. Visual binary statistics are biased toward wide separation distances, which make the two stars easier to resolve. All these biases, which tend to make the data nonrepresentative of the whole population, are called **selection effects**.

Table 19-2 lists three estimates of the **incidence of multiplicity** for systems ranging from single to sextuple. (Of 1200 cataloged binaries and multiples, only two sextuple systems are known.) By the time we reach six-member systems, definitions of multiple systems become hazy. It is unclear whether close groupings like the Trapezium in the Orion Nebula should be counted as multiple systems. We don't know if there is a physical relation between large multiple systems and small clusters or if they are different phenomena. While Table 19-2 illustrates our uncertainty about the frequency of binaries, it does show that single stars are a minority.

Kitt Peak astronomers Helmut Abt and Saul Levy (1976) surveyed stars to determine the incidence not only of multiplic-

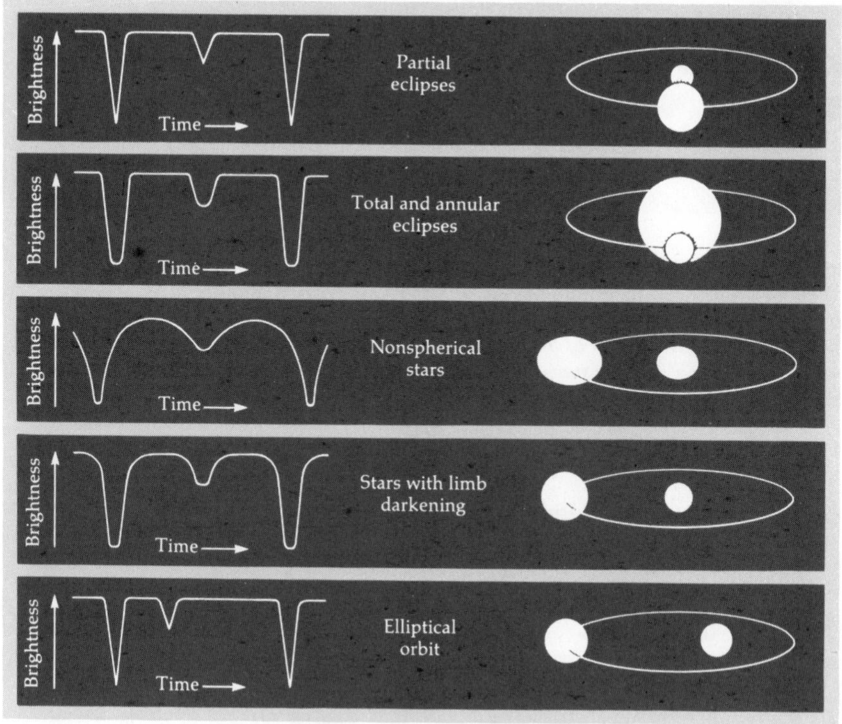

***Figure 19-4.** Different types of light curves (left) from eclipsing binaries reveal different geometric properties of the eclipsing stars and their orbits (right), even though the stars themselves cannot be resolved with the telescope.*

ity but also of different masses. They found that about two-thirds of all stars have detectable companions, consistent with Table 19-2. But from statistics of companions' masses, they estimated that the other seemingly single stars probably all have companions too small to detect! Some companions might be black dwarfs with only a few percent of a solar mass; still smaller ones may be planets. According to this estimate, virtually all stars have at least one companion.

Thus, although many people mistakenly assume that most stars in the night sky are single, *multiple systems are more common than single stars.* Obviously, then, we must understand the origin of systems of two, three, four, and more stars in order to claim any understanding of stars in general. We will return to the problem of origin after reviewing some related observational data.

Orbits and Relationships to Planetary Systems

How do multiple star systems relate to our own multiple systems of planets? As shown in Figure 19-6, orbits of binary stars are often highly elliptical. In contrast, the nine planets of the solar system have nearly circular orbits. Thus, the formation processes of typical multiple star systems and planetary systems probably differ.

Another test of the relation lies in the inclinations of orbits in multiple star and planetary systems. Among the nine planets of the solar system, Pluto has the highest inclination to the common plane, about 17°. None of the other eight is inclined more than 7°. In other words, as Figure 19-7a shows, planets' orbits are highly *coplanar*. Of 10 triple and quadruple star

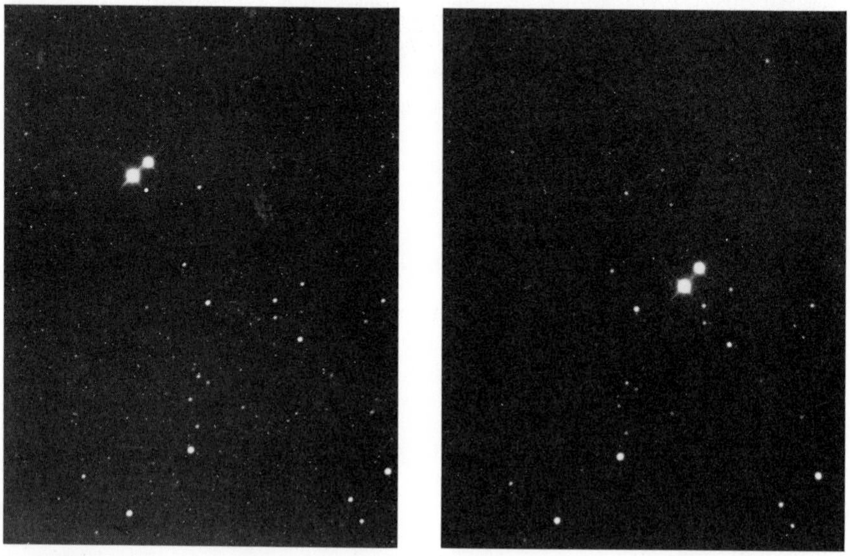

a b

***Figure 19-5.** Two views of the visual binary star, 61 Cygni. Taken 35 years apart with the same telescope, these two photos show the motion of this relatively nearby star pair past our solar system, relative to the background stars. One of the closest star systems, 61 Cygni in 1838 became the first star to have its distance measured. An unseen astrometric companion has been reported but questioned. View (a) dates from 1916; view (b) from 1951. (Lowell Observatory photographs.)*

Table 19-1. *Selected Binary and Multiple Stars*

	Component A		Component B			Component C		
System Name	Distance (pc)	Mass ($M_\odot$)	Separation from A (AU)	Mass ($M_\odot$)	Eccentricity	Separation from A (AU)	Mass ($M_\odot$)	Eccentricity
Visual Binaries								
α Centauri	1.3	1.1	24	0.9	0.52	10 000	0.1	
Sirius	2.6	2.2	20	0.9	0.59			
Procyon	3.5	1.8	16	0.6	0.31			
Eclipsing Binaries								
α Coronae Borealis	22	2.5	0.19	0.9?				
Algol (β Persei)	27	3.7	0.73	0.8	0.04	2?	1.7	0.13
β Aurigae	27	2.4	0.08	2.3				
ϵ Aurigae	1350	15?	35	15?				
Eclipsing–Spectroscopic Binaries								
β Scorpii	118	13	0.19	8.3	0.27			
η Orionis	175	11	0.6	11	0.02			
Astrometric Binaries								
Krüger 60	4.0	0.28	9.5	0.16	0.41			
Barnard's star	1.8	0.14	2.7?	0.001?				
Visual and Astrometric Binaries								
L 726-8	2.6	0.11	5.3	0.11				
61 Cygni	3.5	0.6	83	0.06			0.008?	
BD 66°34	10	0.4	41	0.13	0.05	1.2	0.12	0.00
Solar System								
Sun, Jupiter, Saturn	0.0	1.0	5.2	0.001	0.05	9.5	0.0003	0.06

Source: Heintz (1978); Popper (1980).

systems studied, all orbit pairs have inclinations of more than 18°, and half more than 40° (Batten, 1973). Another system, BD +66°34, may have three low-mass stars in coplanar orbits (Hershey, 1973). Nevertheless, while most multiple star orbits are corevolving, they are usually not coplanar (Heintz, 1978), again suggesting that typical multiple star systems are unlike planetary systems.

According to Kepler's laws, the motions of coorbiting bodies are related to the masses of the bodies around which orbital motion occurs. In many multiple systems the stars are of comparable mass, thus disturbing each other's motions. Multiple orbits of similar sizes would result in unstable systems, raising the possibility that perturbations or collisions would change the system. The most commonly observed arrangements in multiple systems are thus not concentric, but hierarchical. For example, in triple systems, such as the Alpha Centauri system, two stars commonly revolve in a *close* orbit, and the third star revolves around the pair at a *great distance,* almost always more than eight times the separation distance of the close pair (Heintz, 1978), as seen in Figure 19-7b. In quadruple systems, such as Mizar, two close pairs are likely to revolve around each other, with the pairs widely separated (Figure 19-7c).

Thus, in orbital shape, inclination, and arrangement, widely separated binaries appear to be fundamentally different from planetary systems. Abt (1979) found that while masses of widely separated pairs have the same statistical distribution as masses of ordinary representative stars, masses of close companions (within a few tens of astronomical units, or solar system dimensions) have a different mass distribution. Abt concluded that the wide pairs probably formed by separate collapses of two protostars; they were either near each other initially or they were captured into orbits around each other during a near approach in the crowded environment of the parent star cluster. The close pairs formed by a separate mechanism. Probably they resulted when a single protostar had too rapid a spin (too much angular momentum) to collapse into one star and split instead into two.

The transition between the wide types and the close types occurs around separations of about 10 to 50 AU. Possibly our solar system is an unusual example of the close type, where the mass ratio of primary to secondary (sun : Jupiter) is about 1:1000. Was our sun a proto-binary that didn't make it? We don't yet know. Perhaps further studies of binary and multiple star systems will help us understand how planetary systems form.

Table 19-2. Incidence of Multiplicity Among Stars (Estimated Fraction of Systems Containing n Members)

n	25 Systems Within 4 pc	Average of 7 Estimates by Various Authors[a]	Estimate by Batten (1973)[b]
1	0.48	0.41	0.30
2	0.36	0.41	0.53
3	0.12	0.14	0.13
4	0.4[c]	0.03	0.03
5	—	0.01	0.008
6	—	—	0.002

[a]Data from Batten (1973); Abt, 1983, p. 345.

[b]Batten's estimates attempt to average over all stars, using data from various sources. Differences between the estimates are measures of our uncertainty about multiple systems.

[c]Solar system.

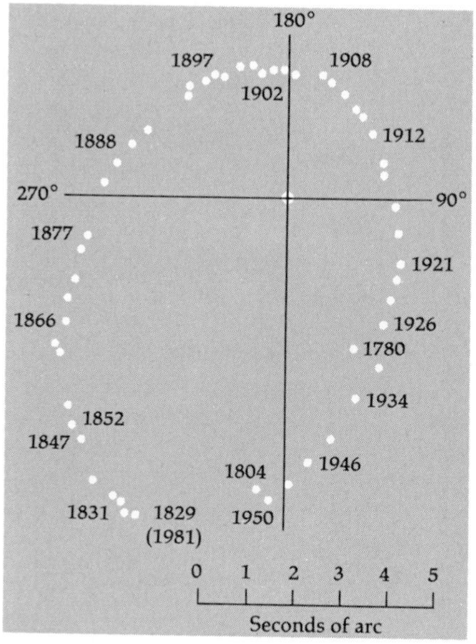

Figure 19-6. *Observed positions of the fainter component of the visual binary Xi Boötes from 1780 to 1950, plotted relative to the brighter component, reveal Keplerian elliptical motion. (The apparent ellipse is the projection of the true elliptical orbit in the plane of the sky.) (After a diagram courtesy* Sky and Telescope.*)*

Evolution of Binary Systems

The evolution of individual systems provides a natural way to classify them. The theory of dynamics of binary stars indicates that a system of two coorbiting stars is an imaginary **Lagrangian surface** of special importance (often called the Roche surface). Its cross section is a figure 8, with one lobe around each star, as shown in Figure 19-8a. Slow-moving material inside either lobe orbits around the star in that lobe. Material that moves out of either lobe is not gravitationally bound to either star individually. It may transfer from one to the other or, with enough velocity, eventually leave the system entirely. When either star evolves toward a red giant state, it expands until it fills its lobe, as in Figure 19-8b. Its outer layers then assume the peculiar teardrop shape of the lobe, and the Lagrangian surface becomes the real surface of the star. Therefore three **classes of binaries** exist, as shown in Figure 19-8.

1. Systems in which neither star fills its Lagrangian lobe

2. Systems in which one star fills its Lagrangian lobe

3. Systems in which both stars fill their Lagrangian lobes; that is, stars in contact with each other (**contact binaries**)

Evolution can take a single binary through all three types. The pair starts as class 1. The more massive star swells into a giant as it evolves. This giant may fill its Lagrangian lobe, creating class 2. Any further tendency to expand causes matter to be shed, mostly through the point common to the two lobes, as shown in Figure 19-8b. Like sand in an hourglass, matter lost through this point enters the lobe of the smaller star. Now the first (larger) star loses mass as the second (smaller) one gains mass. Obviously, the evolution of both stars is altered from what it would have been if they had been isolated. At some later time, the second star evolves toward its own giant stage and expands, either because of mass gain, or because of its own normal evolution. Both stars may now simultaneously fill their lobes, producing class 3, a contact binary. The two stars actually touch as they orbit around each other as a single unit! If tidal forces cause the two stars to become closer, the contact binary may merge into a new, rapidly spinning single star with a dual core and a flattened, egglike, or disklike shape. If two stars start very far apart, the giant expansion may not fill either Lagrangian lobe, and the binary may never evolve out of class 1.

NOVAE AND OTHER PECULIAR STARS AS BINARIES

Now we are better able to explain why Chapter 17 said that **novae** are probably results of binary star evolution. A prenova star is probably a class 2 binary, in which one star has already evolved to the white dwarf state, while the second is expanding into a red giant, filling its Lagrangian lobe.[2] Hydrogen-rich matter leaks off from the outer layers of the red giant and spirals into a disk surrounding the dwarf, as was shown in Figure 17-14. The dwarf, having exhausted its own nuclear fuel, now

[2]The colorful nomenclature of star types has encouraged shameless wags to concoct such descriptions as "degenerate dwarf in the company of young starlet."

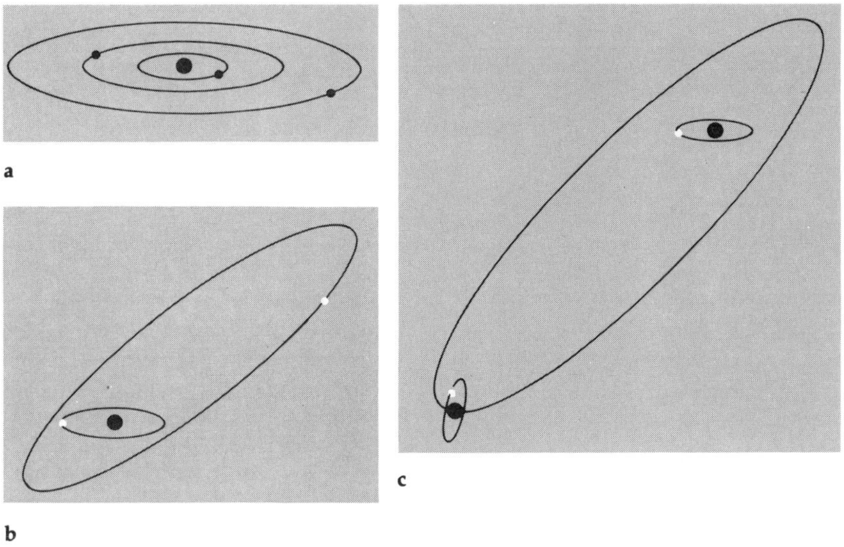

Figure 19-7. Oblique views of different types of orbital arrangements. (a) In the solar system, orbits are circular and coplanar, each about twice as far from the sun as its inner neighbor. (b) In triple systems, a distant companion often accompanies a close pair in an elliptical, inclined orbit. (c) Quadruple systems often have two remote pairs in elliptical, inclined orbits. Differences between the planetary orbits and the multiple star orbits suggest different origins.

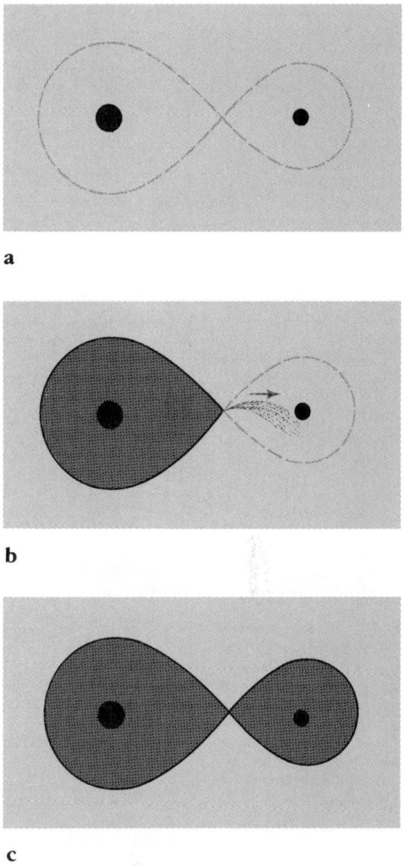

Figure 19-8. Evolution of a binary pair is related to the configuration of Lagrangian lobes. (a) In the first class of systems, neither star fills its lobe (dashed line). (b) After the larger star expands to become a giant, it may fill its lobe, taking on a teardrop shape and perhaps ejecting some mass through the tip, which interacts with the second star. (c) In the third class of systems, both stars fill their lobes.

finds itself receiving a new layer of "burnable" hydrogen. This causes catastrophic instabilities in the dwarf, leading to nova explosions.

Observations suggest that binary evolution accounts for not only novae but also other unusual stars. Gamma-ray bursts, X-ray bursts, and irregular brightness fluctuations are probably associated with mass transfer from an evolving member of a binary pair to a companion, for example (Ventura et al., 1983). Many irregular variables are binary systems in which matter has streamed from a giant into a disk surrounding a smaller companion. Several stars once thought to be short-period variables have been recognized as probable short-period binaries in which matter is streaming from one star to another, causing irregular variations in brightness. In some systems, matter streaming onto one side of a secondary star makes a hot, bright spot on its surface, causing unusual effects. Certain peculiar stars of spectral type A having unusually strong spectral lines of metals—and therefore known as Am stars—are mostly binaries. Tidal effects in these pairs apparently slow the rotation and cause unusual rates of diffusion among metal atoms in the stars' atmospheres. As we learn more, we may find that many peculiar types of stars are explained by strange processes of evolution in binary or multiple systems.

CONTACT BINARIES—W URSAE MAJORIS STARS

Among the most unusual stars are the contact binaries (Figure 19-8c). The most famous of these are the **W Ursae Majoris stars**, named after the prototype in the constellation Ursa Major (better known as the Big Dipper). These consist of stars of rather similar mass, both filling their Lagrangian lobes. They have total masses ranging from 0.8 to 5 $M_{\odot}$, and because they are so close together, their common revolution periods are very short, less than 1.5 d.

How did such a pair form? Some theorists believe they formed from protostars rotating so rapidly that they split, or *fissioned*, into two components. In contrast to the widely separated types of binaries, which have mass ratios expected from random pairings of field stars, close binaries show a tendency toward mass ratios of 1:1 (Abt, 1979), consistent with

Figure 19-9. *An imaginary view of a contact binary system as seen from a nearby planet. The planet is imagined to circle a third star ("offstage" right). The planet and star orbit the contact pair of red giants at a large distance. The planet's surface is imagined to be deeply fissured by tidal forces acting among the three stars. (Painting by author.)*

the fission theory of origin. Theorists have also proposed mechanisms that might cause widely spaced pairs to evolve into contact binaries.

Possibly related are stars with very close orbits and extremely short orbital periods. The shortest known period is that of the peculiar helium-rich star AM Canum Venaticorum. In this irregular variable, hydrogen-rich outer layers have apparently been transferred from one star to a dwarf companion and "blown away" in nova explosions. As mass was blown out of the system, the semimajor axis and period decreased (by Kepler's laws). As both stars lost mass, their helium-rich cores were revealed. AM Canum Venaticorum's revolution period is only 17^m31^s. Theoretically, mass loss could produce contact binaries with periods as short as 2 min! Somewhere in our galaxy there might be a planet in whose sky is a giant glowing figure 8 doing cartwheels like some bizarre advertising gimmick, as illustrated in Figure 19-9.

Examples of Binary and Multiple Systems

ALGOL AND SIMILAR SYSTEMS: ALTERED MASSES

Algol, a variable star with a regular period of 2.9 d, is an eclipsing–spectroscopic binary. The light curve, with its dramatic dimming during eclipse, is seen in Figure 19-10. The primary is a hot, blue B8 star of about 5 $M_\odot$. The secondary is a K0 giant of about 1.0 $M_\odot$. Here lies a paradox: The more massive B or A star should be the one to expand first; yet the less massive star is the more evolved giant. Why? Is there a fundamental mistake in our ideas of stellar evolution?

Theoretical studies have resolved the paradox: What is now the smaller star was *originally* the more massive star. Reaching the giant stage first, it transferred so much mass to the other star that it became the less massive of the pair. In one theoretical study of a similar system, the more massive star starts out with 9 $M_\odot$ and is reduced to a mass of about 2 or 3 $M_\odot$ in only about 50 000 y! During this very rapid mass exchange, the stars reached more nearly similar masses.

SS-433: A BIZARRE X-RAY BINARY

As emphasized in Chapter 17, observations from space have revealed binary systems that strongly emit X rays and gamma rays, probably from gas heated to extremely high temperatures as it transfers from one evolved star onto the surface of a dense companion, such as a neutron star. One of the most exciting is SS-433, a binary in which mass transfer not only heats the gas but also somehow accelerates the atoms into two jets that shoot out of the system in opposing directions at 26% of the speed of light.

The story of SS-433 illustrates the sometimes laborious process of astronomical discovery. SS-433 was first cataloged in the 1950s and 1960s as one of many stars with visible emission lines as well as radio emissions; but its special nature was not recognized. Satellites in the 1970s showed X-ray radiation from this area, and in 1978 astronomers realized that the radio waves and X-radiation were all coming from a single star with a peculiar spectrum. Now the race was on to figure out its nature. Detailed spectra showed strange emission lines eventually identified as lines shifted far out of their normal positions by Doppler shifts much larger than ever found in any other star. Analysis revealed that the emission lines originate in two jets of 20 000-K hydrogen and helium gas, one approaching us (blue-shifted) and one receding (red-shifted). (SS-433 has 100 times the blue shift of any other known object in the universe!) Further study showed that the line positions oscillate in a 13-d cycle, as in a spectroscopic binary, proving that the star producing the jets is orbiting around another star every 13 d.

Figure 19-11 shows an imaginary view of the system, as interpreted by some astronomers. Gas is being transferred from one normal, evolved star into an accretion disk surrounding a dense stellar core, probably a neutron star (Margon, 1980, 1982).

The most important thing about SS-433 is that it somehow produces two 100-AU-long, brilliant, narrow, high-speed jets shooting out of the accretion disk. As mentioned in Chapter 16 (Figure 16-10), diffuse bipolar jets have been found emanating from various accretion disks, but nothing like the tightly beamed, relativistic jets of SS-433. What mechanism creates the jets? In 1984, using gamma-ray observations from a satellite telescope, Iowa astronomer Richard Lamb announced the discovery that thermonuclear reactions of the carbon cycle (see p. 306) may occur in the jets. Could the jet source regions contain dense matter in which these reactions occur, as in stellar interiors? How does nature accelerate material to a sizable fraction of the speed of light without producing a diffuse explosion of million-degree gas? The answers aren't known.

Astronomers continue to study SS-433 with considerable excitement. Some models call for central objects as small as a 1-$M_\odot$ neutron star. However, Harvard astronomer J. E. Grindlay and colleagues (1984) studied X rays from SS-433 with the Einstein Orbiting Observatory and concluded the central object is a black hole of 10 $M_\odot$, surrounded by a gaseous accretion disk some 10 to 100 AU across (see Figure 19-12). Time will tell who is right!

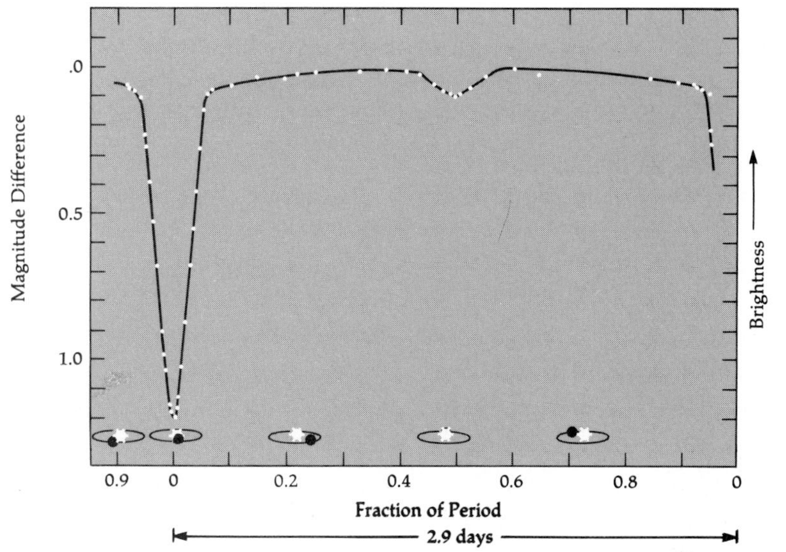

Figure 19-10. *Light curve of the eclipsing–spectroscopic binary Algol. The curve plots brightness versus time, and shows that Algol's brightness drops about 1.4 magnitude when the small, bright B star passes behind the fainter, cooler K giant.*

EPSILON AURIGAE: A FORMING SYSTEM?

Epsilon Aurigae is a strange pair whose brighter component, an F-type supergiant, undergoes an eclipse every 27.1 y by an invisible companion. This eclipse takes 714 d and absorbs only about half the light of the F star. It is "so peculiar as to defy interpretation in terms of any conventional model" (Kopal, 1971). The two stars are estimated to be 35 AU apart, and the primary supergiant is about as big as Mars' orbit. The secondary is believed to be an evolving disk in which a central star is only beginning to form. As early as the 1950s, Kopal and others suggested that the dimming of the F star was caused by meteoroids in the disk—planetary material—because there were no color changes that would be caused by smaller, microscopic grains. During the eclipse, the F star's light shines dimly *through* the disk. The disk has a diameter of about 40 AU and thus is a virtual duplicate of the solar nebula—the disk in which our planets formed!

Epsilon Aurigae eclipses come but twice in an astronomer's lifetime. The probability that we can study light shining through a newly forming planetary system thus led to excited anticipation of the 1982–84 eclipse. Infrared, ultraviolet, and visible-light observations from both orbiting and ground-based telescopes supported the idea that small solid bodies are involved in obscuring the primary star, but they also showed that the whole double star system is surrounded by thin, hot gas. The mysterious star hidden in the disk seems to be a relatively hot, 10,000-degree, newly forming star with an estimated 4 to 28 $M_{\odot}$ whose light is hidden by the disk's dust. The disk itself is heated to about 1100 K on the side facing the F star but is only about 500 K on the side facing away (Backman et al., 1984; Hack, 1984; Thomsen, 1985).

Astronomers continue to analyze the new data from this 1350-pc distant system, which resembles an ancient solar system with an intensely bright extra sun at the distance of Neptune.

BARNARD'S STAR, T TAURI, AND RELATED SYSTEMS: PLANETS?

Barnard's star, the fourth-closest star system, is an unresolved astrometric binary only 1.8 pc from the earth. It has caused a flurry of debate because it may contain at least one planet-sized companion, thus possibly being another planetary system near our own. In 1963 and 1968 American astronomer Peter van de Kamp announced that Barnard's star has an unseen companion only 50% more massive than Jupiter with an orbital period of 25 y. Van de Kamp regarded it as a planet, but its orbit appeared highly elliptical, unlike a planet's orbit. In 1969 van de Kamp re-interpreted the data in terms of two planets in circular orbits with masses about 1.1 and 0.8 times Jupiter's mass. This interpretation was questioned, because orbital motion attributed to one companion can always be re-interpreted in terms of many smaller companions in circular orbits.

Other workers got inconsistent results. Gatewood and Eichhorn (1973) analyzed different photos and found no evidence of any planetary system! Jensen and Ulrych (1973) reported that van de Kamp's data

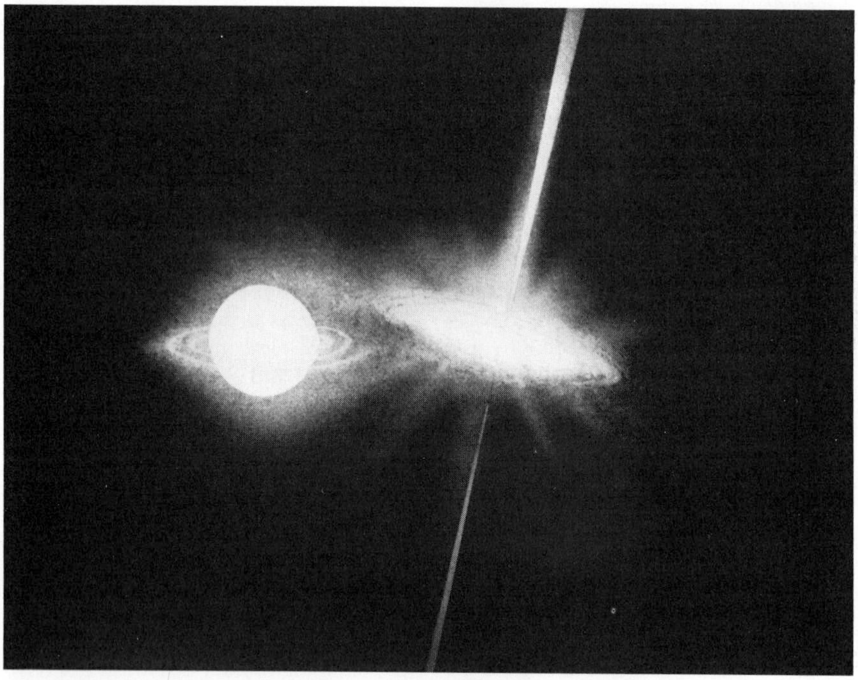

Figure 19-11. *Imaginary view of the strange binary SS-433. Interpretation of spectra suggests that gas is flowing out of an evolved star (left) into an accretion disk of hot gas surrounding a neutron star (right). Gas spiraling into the center of the disk is intensely heated and somehow ejected into 78 000-km/s jets moving outward in two opposing directions. Hydrogen alpha emission from the glowing jets is Doppler red-shifted to a deep red color in the receding jet (bottom) and blue-shifted to a more yellowish color in the approaching jet (top). (Painting by author.)*

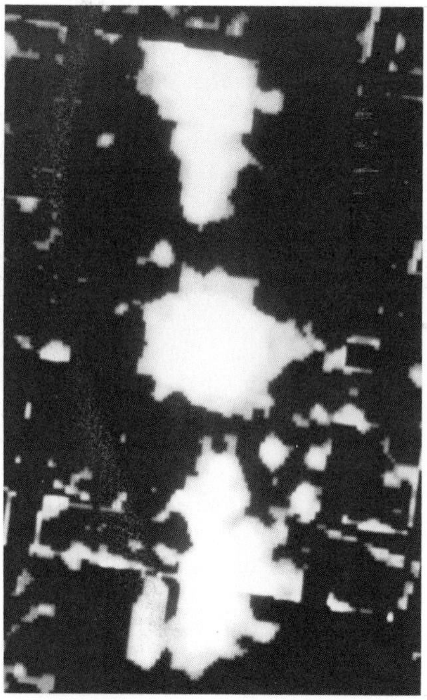

Figure 19-12. *An X-ray image from an orbiting telescope, called the Einstein Observatory, shows the two jets (top and bottom) emitted from SS-433 (center). X rays emitted by hot gas in the SS-433 system were used to form this image. (Harvard/Smithsonian Center for Astrophysics, courtesy Fred Seward.)*

could just as well be interpreted in terms of four planets, ranging from 1.6 to 0.7 Jupiter masses. Heintz (1976, 1978) concluded there were no planets, only errors in interpretation! As recently as 1981 van de Kamp still believed that two nearly coplanar planets of about 0.7 and 0.6 Jupiter masses, 3.0 and 3.7 AU from the 0.14-$M_{\odot}$ star, respectively, best explain the observations. Most astronomers consider these proposed planets as unconfirmed at best, however, and the story illustrates the difficulties of detecting worlds near other stars.

More provocative results have come from recent observations of faint companion stars radiating only infrared light. In 1982, Hawaiian observers announced discovery of a faint infrared companion to the prototype T Tauri star, T Tauri itself (Dyck et al., 1982). Recently formed, T Tauri was already known to have a surrounding cocoon nebula, with evidence of silicate dust grains. The companion has been interpreted as an accreting protoplanet with temperature 800 K and mass of 5–80 Jupiter masses (Hanson et al., 1983). Similar infrared companions have been found for at least two other stars (see Chapter 26).

Tables 15-1 and 19-1 include more stars with small companions around 10 to 30 times as massive as Jupiter. The star BD +66°34, which is 10 pc from the earth, consists of two main-sequence M-type stars of mass 0.4 $M_{\odot}$ and 0.12 $M_{\odot}$, and a third unseen star of mass 0.13 $M_{\odot}$ (about 130 Jupiter masses). The orbits may be circular and in the same plane, more akin to the solar system's configuration than most other binaries (Hershey, 1973).

Although no star is now regarded as proved to have another planet, various techniques are being developed that should increase the sensitivity of our search dramatically within a decade. These include state-of-the-art advances in ground-based astrometry as well as space telescope applications.

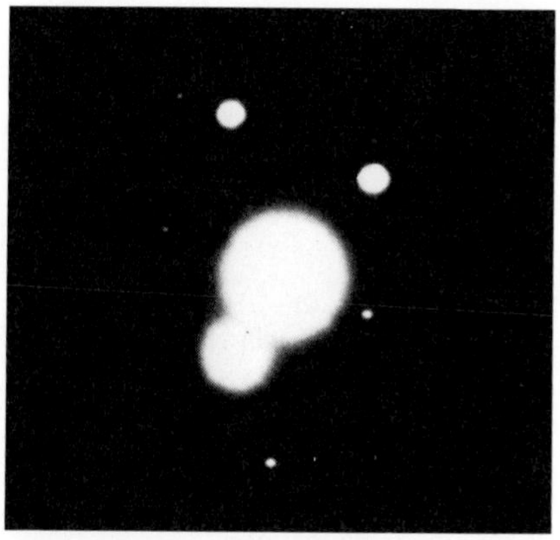

Figure 19-13. *The multiple star system Beta Lyrae. The four brightest images are known members of the system, including the 3rd-magnitude primary (overexposed, center), a 7th-magnitude companion 46" (seconds of arc) away (lower left), and 9th- and 10th-magnitude companions 67" and 86" away (top). The bright primary is an eclipsing–spectroscopic binary. Other faint stars (magnitude 13–15) might be associated with the system or be background stars. The primary is a spectroscopic binary consisting of a 12-$M_{\odot}$ star transferring mass into a dusty disk-shaped nebula surrounding a star of about 2 or 3 $M_{\odot}$. The system may be only about 10 million years old. (Allegheny Observatory, courtesy W. A. Feibelman.)*

DUST AROUND MAIN-SEQUENCE STARS

Meanwhile an intriguing result has emerged: Systems of dust grains have been found not only around newly formed stars but also around main-sequence stars. In the former case, dust is often regarded as left-over debris and is often seen blowing away from the star as the cocoon nebula is shed; but near main-sequence stars, like the sun, it may be a permanent part of the system, hinting at the presence of still larger unseen "particles," like asteroids, comets, or even planets.

One example is Beta Pictoris, whose edge-on dust disk, reported in 1984 (Smith and Terrile, 1984), is shown on page 328. It has about twice the sun's mass and its dust disk is about ten times the dimensions of our planetary system.

Other examples have been found by an infrared space telescope. Near several well-known bright main-sequence stars, such as Vega and the southern star Fomalhaut, this telescope in 1983 detected infrared thermal radiation from unseen dust orbiting around the star. Analyses have suggested grains from sand-sized to one-tenth that size. Some analysts believe the small grains would be quickly blown away from the star and hence must be replenished from larger, permanent bodies, like asteroids or comets. This new research is continuing.

The Origin of Binary and Multiple Stars

Our hopes of understanding all stars would brighten if we could explain exactly how binary and multiple stars form. Of course, we must explain not only how they form, but how the proportions of each type—binaries, triplets, quartets, and so on—arise. Unfortunately, we cannot.

Astrophysicists are confident that most individual stars arise from contracting gas clouds, as explained in Chapter 16. Coorbiting stars may be the product of *breakup of the forming, protostellar clouds,* or of *interactions between stars after they form.* Theories can be divided into three groups reminiscent of the three classes of theories on the origin of the "binary" earth–moon system.

Fission Theories. Theories that picture a fast-spinning protostar as splitting in two are called **fission theories.** Many researchers believe that very close pairs such as the W Ursae Majoris contact binaries are formed in this way.

Capture Theories. According to **capture theories,** systems of widely separated stars that share weak gravitational attraction may be chance configurations arising when one star approaches another. The chance of encounters among random field stars is far too slight to explain the observed numbers of binaries. Furthermore, randomly paired stars would be of widely different ages, but this is not observed among binaries. Where could stars of similar ages interact in a closely packed group? In a newly formed star cluster. Massachusetts astronomers T. Arny and P. Weissman (1973) showed that fully half the protostars in a cluster probably undergo collisions or close encounters. Therefore, astronomers believe at least some binaries and multiples formed inside newly formed open clusters as protostars approached, and, by various possible gravitational interactions, became bound in orbit around one another. Many wide pairs probably formed in this way.

Common Condensation Theories. Depending on rotational properties, subfragmentation of a prestellar cloud may produce multiple systems with either two high-mass stars or several low-mass stars in coplanar orbits. Some moderately close systems may have formed in this way. The description in Chapter 12 of our planetary system's origin can be classed as one of these **common condensation theories.** Small groups of several dozen stars, perhaps resembling the Trapezium in Orion, may form near each other and then finally break apart into binary and multiple star systems.

Each of these three theoretical processes may produce binaries of a certain type. A complication in sorting out binaries of these different types is that orbits of binary and multiple stars evolve through gravitational influences. Mass transfer in close pairs can alter orbits. Widely spaced pairs formed inside larger star clusters can evolve into closely spaced pairs as the clusters break apart. In one theoretical study of some 800 imaginary triple star systems, about 97% were found to be gravitationally unstable, eventually kicking out one star and becoming binary systems. Sometimes the ejection velocities are quite high; 58 km/s is a typical example. This in turn helps explain the runaway stars observed speeding out of some young star-forming areas, as described in Chapter 18. Thus, the observed statistics of binaries and multiples may not reflect their original characteristics. Many systems are quite complex, as illustrated by Figure 19-13.

Nonetheless, astronomers who have attempted to correct for selection effects have found real statistical differences between the orbits of closely spaced systems and widely separated systems; they suggest that such systems form by more than one process.

Summary

At least half of all the seemingly single stars in the sky are binaries or multiple systems. Many of these may have formed by interactions of stars in crowded new clusters, but some may have formed by other means such as fission or common condensation.

Binary and multiple systems can be detected in different ways. Binaries were once classified by these different methods of detection, which yield different types of knowledge. Examples include spectroscopic binaries, eclipsing binaries, and astrometric binaries. Certain types of binaries give the best available data on certain properties of stars, such as mass and diameter (see pp. 268–270).

A classification based on evolution includes binaries that have filled neither Lagrangian lobe, one Lagrangian lobe, or both Lagrangian lobes. Once one lobe has been filled by expansion of a star to the red giant stage, gas may flow from one star to the other, causing flare-ups, X-ray emission, and nova explosions.

Studies of astrometric binaries have revealed unseen companions with masses only a few times greater than Jupiter, and many astronomers believe that companions of some such stars may be true planets. Whether companions as small as planets form in a contracting protostar may depend on the protostar's rotational properties. If this idea is correct, there may be other earthlike planets in the universe.

Concepts

binary star system
multiple star system
optical double star
physical binary star
visual binary
spectroscopic binary
eclipsing binary
light curve
eclipsing–spectroscopic binary
astrometric binary
spectrum binary
selection effect
incidence of multiplicity
Lagrangian surface
classes of binaries
contact binary
nova
W Ursae Majoris stars
fission theory
capture theory
common condensation theory

Problems

1. Describe verifications of Kepler's laws other than the planets' motions around the sun. What was the first verification outside the solar system?

2. How do binary and multiple star systems generally differ from planetary systems?

3. Give evidence that at least a subclass of binary and multiple star systems might be generically related to planetary systems.

4. How are novae related to binaries? Are supernovae related to binaries?

5. How will the evolution of a 1-$M_\odot$ star in orbit close to a 3-$M_\odot$ star differ from the evolution of a 1-$M_\odot$ star by itself?

6. Why are binaries more likely to have formed in star clusters than as isolated field stars?

ADVANCED PROBLEMS

7. If a star of low mass (about 0.05 $M_\odot$, for example) were orbiting in the earth's orbit around the sun, what would be its period of revolution?

8. Use the circular velocity equation to derive the orbital velocity of a Jupiter-sized body around a 1-$M_\odot$ star if the separation distance is 5.2 AU (7.8×10^{11} m). (*Hint:* 1-$M_\odot = 2 \times 10^{30}$ kg.)

a. How does this compare with the actual orbital velocity of Jupiter?

b. What would be the orbital velocity if the central star had two solar masses?

9. If you observed a 0.5-$M_\odot$ star in circular orbit around a 5-$M_\odot$ star and could tell from Doppler shifts that the orbital velocity is 47 km/s, what would you conclude is the separation distance between the stars in astronomical units?

10. A star of mass m is in a circular orbit around a star of mass $M = 10\ M_\odot$. Star M explodes and rapidly blows away 0.6 of its mass.

a. What is the orbital velocity of m before the explosion?

b. How does the total force of attraction on m from M change during the explosion?

c. Describe the future history of m. Will it remain in orbit around the remnant of M? (*Hint*: This sort of event has been suggested as a source of runaway stars; see p. 344.)

Projects

1. Observe Mizar and Alcor with the naked eye. They are the middle "star" (actually a close pair) in the handle of the Big Dipper. Can you see the faint star Alcor? Sketch its position. (Inability to see Alcor may be due to insufficiently keen eyesight, a hazy sky, or a sky illuminated by city light.)

2. Observe the eclipsing binary Algol with a telescope or binoculars each evening for 10 to 20 days in a row. (This can be done as a class project with rotating observers.) Using neighboring stars as brightness reference standards, estimate the brightness of Algol. Can you detect the eclipses, which occur at 2.9-d intervals?

3. The star Epsilon Lyrae is famous as the "double double." It consists of a binary pair 208 seconds of arc apart, easily seen in a small telescope. But each of these is a binary only 2 to 3 seconds apart. These pairs are a test of good optics and good atmospheric observing conditions. Does your telescope reveal the two close pairs? Sketch them.

CHAPTER 20

STAR CLUSTERS AND ASSOCIATIONS

If you could roam through space to ever-greater distances, you would eventually lose track of individual stars and see the galactic disk defined primarily by clusters of stars. Writing about clusters in 1930, Harvard astronomer Harlow Shapley pointed out that "their problems are intimately interwoven with the most significant questions of stellar organization and galactic evolution."

Yet at the same time Shapley noted that scientific study of them had hardly begun. Nobody knew how to measure their distances or plot their distribution in space until the 1920s. Although some clusters, such as the Pleiades (or Seven Sisters), are easy to recognize with the unaided eye, others are so far away that they require large telescopes to detect. Still others are so close that they cover much of our sky

Table 20-1. *Selected Star Clusters and Associations*

Name	Distance (pc)	Z^a (pc)	Diameter (pc)	Estimated Mass ($M_\odot$)	Estimated Age (y)
Open Clusters					
Ursa Major	21	18	7	300	2×10^8
Hyades	42	18	5	300	5×10^8
Pleiades	127	54	4	350	1×10^8
Praesepe	159	84	4	300	4×10^8
M 67	830	450	4	150	4×10^9
M 11	1 900	99	6	250	2×10^8
h Persei	2 250	156	16	1 000	1×10^7
χ Persei	2 400	167	14	900	1×10^7
O Associations					
I Orionis	470	150	?	3 000	?
I Persei	1 900	164	?	180	?
T Associations					
Ori T2	400	132	28	800	?
Tau T1	180	52	?	50?	?
Globular Clusters					
M 4	2 800	800	9	60 000	$\sim 1 \times 10^{10}$
M 22	3 100	430	9	7 000 000	$\sim 1 \times 10^{10}$
47 Tuc	4 600	3 300	5		$\sim 1 \times 10^{10}$
M 13	8 200	5 400	11	300 000	$\sim 1 \times 10^{10}$
M 5	9 200	6 700	12	60 000	$\sim 1 \times 10^{10}$
M3	13 000	12 000	13	210 000	$\sim 1 \times 10^{10}$

[a] Z = perpendicular distance north or south of Milky Way plane.

and were not even recognized until recent years.

In other words, our cosmic journey has brought us to clusters so far-flung that they were recognized as a class only in this century. Clusters have been of major importance because they have revealed to us the shape and age of our own galaxy, as will be clarified in this chapter.

Three Types of Star Groupings

The three basic types of clusters are *open clusters, associations,* and *globular clusters.* Examples are listed in Table 20-1.

OPEN STAR CLUSTERS

Open star clusters are moderately close-knit, irregularly shaped groupings of stars. They usually contain 100 to 1000 members and are usually about 4 to 20 pc in diameter. Our sun is possibly inside or on the edge of a loose open cluster centered only about 22 pc away toward the constellation Ursa Major, many of whose stars belong to this cluster, as shown by Figure 20-1. The best-known clusters, the Hyades and the Pleiades (see Color Plate 28), lie 12° apart in our winter evening sky, about 42 and 127 pc away, respectively. Figure 20-2 shows the relation of the Hyades and Pleiades in the sky, while Figures 20-3 and 20-4 show the Pleiades in more detail. About 900 open clusters are concentrated along the Milky Way band, indicating that they lie in the plane of our galaxy. (Open star clusters are sometimes called *galactic clusters* for this reason.)

As described in Chapter 19, stars form in open clusters, and most open clusters have prominent young stars or associated clouds of star-spawning gas. Then why aren't all stars in clusters? The reason is that most open clusters break apart into individual stars within only a few hundred million years because of dynamical forces acting on them. In comparison with most cosmic lifetimes, open clusters are short-lived.

ASSOCIATIONS

Associations are cousins of open clusters. They often have fewer stars, but are larger in size and have a looser structure. Some large associations include an open star cluster within them. They may have 10 to a few hundred members and diameters of about 10 to 100 pc. They are rich in very young stars, such as O and B stars (which burn their fuel too fast to last long), or T Tauri stars (which evolve toward the main sequence too fast to last long). Associations are classified as **O associations** or **T associations**, depending on whether the prominent stars are O and B blue stars or T Tauri variables. The general region of Figure 20-2 includes a T association.

The smallest associations grade into small, multiple-star-like groups, such as the Trapezium in the Orion nebula. Sometimes called *trapezium systems,* these might be a link between multiple stars and small clusters. Like open clusters, associations are involved with regions of recent star formation and are short-lived. More than 60 associations were tabulated in a 1963 catalog.

GLOBULAR STAR CLUSTERS

Globular star clusters are quite different from the other two types. They are much more massive, more tightly packed, more symmetrical, and very old. Figure 20-5 emphasizes the remarkable symmetry and compactness of globulars. They typically contain 20 000 to several million stars, although many of these stars crowd too close to be resolved by earth-based telescopes, especially in the central regions.

Typical diameters of the central concentrations range from only 5 to 25 pc. To imagine conditions inside a globular cluster, picture 10 000 stars placed around the sun at distances no farther than Alpha Centauri, our nearest star!

Even the nearest globular clusters are thousands of parsecs away from us. It is only because they have so many and such very bright stars that we see them at all. Yet a modest backyard telescope can reveal many prominent examples. The total of

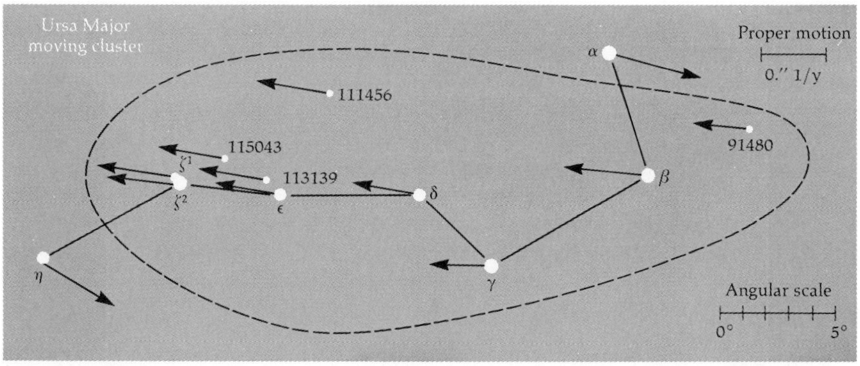

Figure 20-1. ***Some members of the closest open cluster, the Ursa Major cluster, form part of the familiar figure of the Big Dipper (solid lines). Arrows show proper motions of the stars, or the rates of angular motion relative to distant background stars. All but two of the stars in the figure are moving together at almost exactly the same rate, thus defining the cluster (dashed line). Stars at each end of the dipper are moving in different directions and are not true cluster members. Compare with Figure 1-3.***

Figure 20-2. *Two open clusters are prominent in this portion of the January evening sky. The Pleiades is the tight group in the lower center, about 127 pc away. Closer to us is the Hyades, a V-shaped group 12° to the upper left, about 42 pc away. The brightest star, at the upper left tip of the Hyades, is the 1st magnitude red giant Aldebaran. Box shows the region of Figure 20-3. (Ten-min exposure with 35-mm camera,* f1.4, *50-mm lens, 2475 Recording film.)*

Figure 20-3. *The Pleiades. Field of view width is 5°. Box shows the region of Figure 20-4. (5-min exposure with 35-mm camera,* f2.8, *135-mm lens, 2475 Recording film.)*

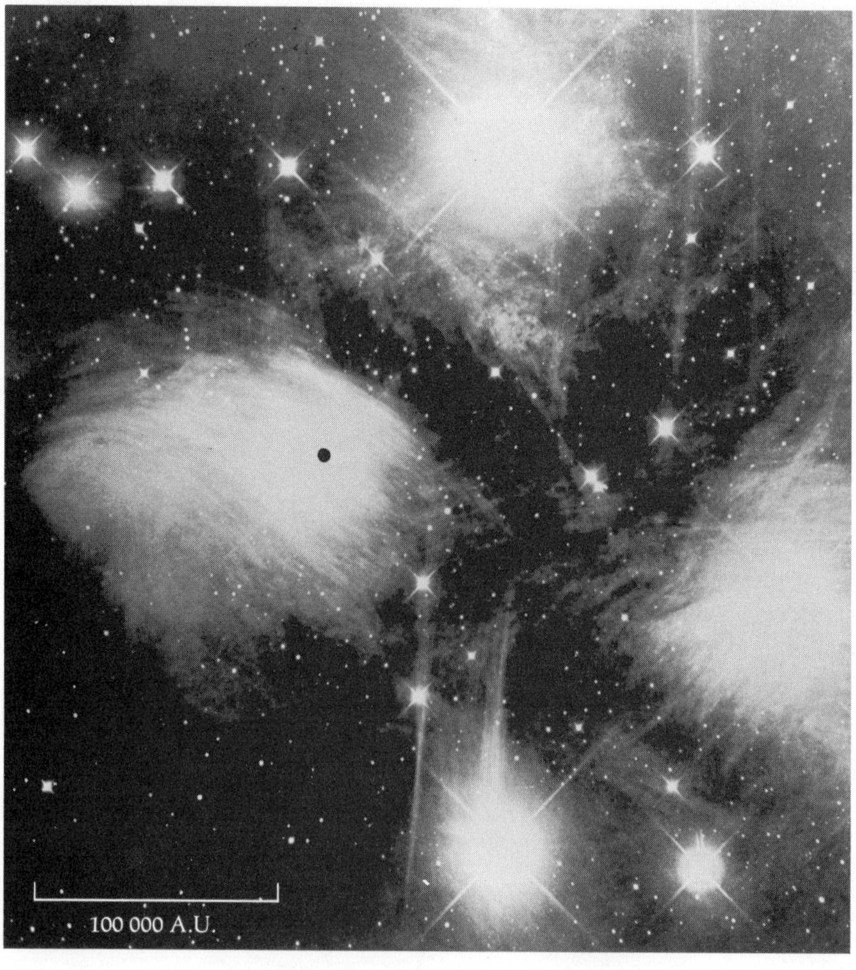

Figure 20-4. *A portion of the Pleiades photographed with a large telescope, revealing clouds of wispy nebulosity concentrated around each star. See also Color Plate 28. Solar system would fit well within black dot. (Lowell Observatory.)*

Figure 20-5. *Globular star cluster M 5 (NGC 5904). The cluster is estimated to be about 12 pc across and 9200 pc away. It contains roughly 60 000 stars. (Kitt Peak National Observatory.)*

known globulars around our galaxy is about 120, and the total around our sister galaxy, M31 in Andromeda, is around 200. In three-dimensional space they are not confined to the galactic plane, but are distributed in a spherical *halo* surrounding our galaxy. Similar distributions are found around other galaxies. They are among the oldest objects in our Milky Way galaxy, with ages estimated around 12 billion years.

Discoveries and Catalogs of Clusters

Recognition of clusters is an enterprise spread out over many centuries, as shown in Table 20-2.

When the French astronomer Charles Messier tabulated nebulae in 1781, he included many star clusters that are still known by their "M numbers." Between 1864 and 1908 many clusters were cataloged in the New General Catalog (NGC) and the Index Catalog (IC), and became known by their NGC and IC numbers. Not until the 1920s was the difference between various types of clusters and the still more remote galaxies understood, and only then could reasonable research begin. Shapley's 1930 book was the first substantial work on star clusters.

Measuring Distances of Clusters

Because of clusters' enormous range of distances, astronomers have devised varied and sometimes ingenious methods to determine their distances. Details of all the methods are beyond the scope of this book, but a few examples can be given.

Parallax. The most basic method of measuring star distances, the measurement of parallax (described in Chapter 14), is almost useless on clusters because it is not very accurate for distances beyond about 20 pc. Parallaxes give checks on distances of some stars in the Ursa Major cluster, but they mostly provide a foundation on which other distance-measuring methods are built.

Star Luminosity. If observers can determine the luminosity of any star in a cluster, the distance from the earth to the cluster can be estimated if there are no complications from obscuring interstellar matter. (This method is similar to the example in Chapter 14 of estimating the distance of a certain light if we know that it is a candle and that there is no intervening fog.) One way to determine luminosity is simply to use spectra to identify spectral types and to estimate luminosities from H–R diagrams that plot luminosity versus spectral type, as derived from nearby stars.

Cluster Diameter. Once the linear diameters (in parsecs) of a few nearby clusters were known and found to be roughly comparable, remote clusters could be assumed to have the same linear diameter. Then the distances of remote clusters were estimated from their angular diameters (seconds of arc) by applying the small-angle equation and assuming that all clusters of similar appearance have similar linear size. For example, if a globular cluster 10 pc across subtends $\frac{1}{10}°$, it must be about 6000 pc away.

THE COMPLICATION OF INTERSTELLAR OBSCURATION

Before 1930, these methods were carried out under the assumption that all intervening interstellar space is transparent. However, early estimates of distances, brightnesses, and sizes of clusters gave inconsistent results. In 1930 Robert Trumpler used the cluster results to show that the problem lay in the erroneous assumption that space was clear. He showed that diffuse interstellar dust dims stars and clusters that are more than a few dozen parsecs away. This throws off the estimates of stars' luminosities and distances. Fortunately, the total amount of dimming can be estimated by measuring the amount of interstellar reddening, or color change caused by the dust. The more reddening is observed, the greater the degree of dimming. Once the obscuration is measured and taken into account, distances can be accurately measured if the luminosity of any star or class of stars in the cluster is known.

Table 20-2. *First Recognition of Selected Clusters*

Cluster	Type	First Recognition
Pleiades	Open	Prehistoric[a]
Hyades	Open	Prehistoric[a]
ω Centauri	Globular	Ptolemy, c. A.D. 140[b]
Praesepe	Open	Galileo, c. 1611
M 22	Globular	Ihle, 1665
M 11	Open	Kirch, 1681
M 5	Globular	Kirch, 1702
M 13	Globular	Halley, 1714
27 globular clusters and 29 open clusters, cataloged		Messier, 1781
I Persei	Association	Ambartsumian, 1949

Source: Data from Shapley (1930).

[a]First recognition as a star cluster similar to telescopic examples was in Messier's 1781 list.
[b]Not resolved into stars, but listed as a bright, fuzzy patch.

for closure (see p. 457), the estimated age would be two-thirds as much, or about 7 to 10 billion years, which seems embarassingly small, since this is less than the ages estimated for globular clusters.[1]

Ages of the Elements. In Chapters 3, 5, and 12 we discussed how the radioactive elements can be analyzed to reveal the 4.6-billion-year age of the solar system. Chapters 12, 17, and 21 added evidence that the heavier elements were mostly synthesized by fusion reactions inside stars and distributed by supernova explosions during a period stretching back further than 4.6 billion years ago. The big bang theory allows physicists to model conditions inside the primeval fireball and calculate how many elements of various types synthesized there. By comparing today's abundances of radioactive elements with their decay rates and with their calculated production rates inside stars and the primeval fireball, it is possible to calculate the total **duration of element formation** in the universe.

For example, if all uranium formed in the big bang, the universe should be about 7 billion years old to account for uranium abundances today. But we know that such a heavy element did not form efficiently in the big bang. If it all formed slowly inside successive generations of stars, the universe should be about 18 billion years old (Peebles, 1971). By this calculation, the age of the universe must be between 7 and 18 billion years.

From the rough agreement among these three lines of evidence, most astronomers conclude that the big bang occurred 9 to 18 billion years ago; perhaps the best estimate is 14 billion years ago. Some unique explosive event must have occurred at this time, apparently creating matter and sending it flying out on its expanding journey. Many astronomers refer to this time as the **age of the universe**, marking the creation of the universe as we know it.

The term *explosive* may be misleading because the big bang was not an ordinary explosion. According to many models, no one could have floated in nearby space and watched the fireball expand, because, paradoxically, the fireball always filled all space. *Space itself is viewed as contracted* at the beginning. An analogue to these models would be the expanding balloon, whose surface represents space. An ant living on a spot on its surface could detect the expansion and other spots moving away, but it could not get off the surface to look at the whole balloon expanding. At the beginning, the surface of the balloon would have been "small," but it still would have constituted the ant's whole universe. In this sense, the "small" balloon—the hot, high-density fireball—would fill all space perceived by the ant.

Instead of picturing an explosion, it may be more helpful to think of the big bang just as a singular moment of high-temperature compression of the universe's matter, whose earlier history, if any, we can't trace.

The First Hour of the Universe

One of the astonishing things about the big bang theory is that it lets us describe **initial conditions** during the first moments in the history of the universe. The assumptions are simple, because if we say that all the observable mass was once as concentrated as possible, then at the initial moment the density should have been nearly infinite; it has been declining ever since. Packing all mass at high density would lead to extraordinary temperatures. A simple theoretical model developed around 1950 by George Gamow assumed infinite temperature and density at the zero instant! Since we know the behavior of expanding hot material, Gamow was able to calculate subsequent conditions. After 1 s, the temperature was 15 billion Kelvin and the density about that of the air we are breathing. This model also predicts a present temperature of about 20 K and a density of about 10^{-27} kg/m^3, which are close to the observed values.

PREDICTING ABUNDANCES OF ELEMENTS IN POPULATION II

The big bang theory also succeeds in describing how the primordial particles joined together to form atoms of the various elements, and it also predicts the abundances of those elements. In the high-temperature gas of the first seconds, matter was broken down into its simplest components—not grains or molecules or even atoms, but subatomic particles such as neutrons, protons, and electrons. Calculations based on known atomic physics show how these particles would interact under conditions after 1 s, 2 s, and so on.

The first step in making atoms was to make their nuclei, or inner cores. The outer electrons could be added only later, under cooler conditions. The nucleus of a hydrogen atom is simply a proton, so it could be said that the universe was already full of hydrogen nuclei, since protons were an abundant basic particle. The next-heaviest atomic nucleus would be that of deuterium, or heavy hydrogen, which consists of a joined proton and neutron.

[1]Research around 1980 nearly doubled the best estimate of the Hubble constant from the 50s to around 90 km/s per megaparsec, reducing the estimated age to a value around 10 billion years. This figure is difficult to reconcile with globular cluster ages of 13 billion years. The prestigious journal *Science* paraphrased Shakespeare: "Double Hubble, age in trouble." Two papers in 1983 gave H values of 74 and 82 km/s per Mpc, more consistent with this age.

Gamow and his associates[2] in 1948 found that protons and neutrons could collide and form deuterium:

$$\text{proton} + \text{neutron} = \text{deuterium} + \text{gamma radiation}$$

The higher the temperature, the faster particles collided. During the first 3 min or so, collisions would have been so energetic that deuterium nuclei would have broken apart faster than they would have formed. After about 3 min, according to this model, deuterium began to accumulate.

Gamow at first thought that successive fusing of neutrons would build up heavier and heavier elements, forming nearly all the elements by the end of the first hour. However, later calculations (especially by Margaret Burbidge, Geoffrey Burbidge, William Fowler, and Fred Hoyle in 1957) showed that nuclei of atomic mass 5 (heavy helium) would not form in this way and that most elements heavier than helium built up inside stars as outlined in Chapter 17.

Nonetheless, Gamow's basic idea seems to account for the formation of most of the material in the universe—the hydrogen, its heavy form deuterium, and helium—during the universe's first hour. A recent calculation based on these principles (Reeves et al., 1972) gives these abundances at the end of the first hour:

Element	Percent Mass
Hydrogen (^{1}H)	75
Helium 4 (^{4}He)	25
Deuterium (^{2}H)	0.1?
Helium 3 (^{3}He)	0.001?
Lithium 6, lithium 7 (^{6}Li, ^{7}Li)	0.000001?
Heavier nuclei	Negligible

[2]Gamow deliberately collaborated with Ralph Alpher and Hans Bethe (pronounced like *beta*) so that their paper could come under the byline Alpher Bethe Gamow, a pun on "alpha beta gamma," a phrase suitable to a milestone in cosmology.

Perhaps it seems presumptuous to talk of events during the first hour some 14 billion years ago, but the results can be confirmed by observation. Remember that the oldest stars we can find, the extreme Population II stars, appear to have formed from exactly such material. This finding shows that when stars began to form, the heavy elements did not yet exist. Thus we can be pretty certain that shortly after the big bang, the universe consisted almost entirely of very hot hydrogen and helium.

RADIATION IN THE EARLY UNIVERSE

According to the big bang model, radiation was more important than matter in the first few thousand years. According to Einstein's famous equation $E = mc^2$, energy, E, is equivalent to mass, m (c is the velocity of light). This means that each photon of radiation, carrying a certain energy E, will have an equivalent mass $m = E/c^2$. Thus any given amount of radiation corresponds to an equivalent mass of material. At the billion-Kelvin temperatures in the first moments of the **primeval fireball**, there must have been intense radiation, hence many photons per unit volume, and hence a large equivalent density. Under these conditions, the equivalent mass density of the radiation was greater than the density of the mass itself! This is indicated in the upper left part of Figure 25-1, which shows the schematic history of the universe.

Today, on the other hand, radiation density is negligible. As Gamow pointed out, the mass of radiation in a cubic kilometer of air today is only about 10^{-17} kg. Even in the center of the sun or an exploding atomic bomb it is only about 1 kg/m^3. But in the first tenth of a second it may have exceeded 1000 kg/m^3!

The importance of this result is twofold. First, the intense radiation agitated the particles of matter so violently that neither galaxies nor other clumps of matter could form. As time went on and everything rushed apart, the radiation density dropped rapidly. After some thousands of years, radiation density dropped to below the density of matter. Within the first billion years, recognizable masses of material began to form.

The second consequence of the high radiation density is that in the far distant past (seen by looking toward distant, high-red-shifted regions), a different kind of radiation from that found near our own galaxy should be seen. It would be a trace of the dense radiation in the primeval fireball. The early universe was permeated by the kind of radiation that would come from high-temperature gas. As matter flew outward, the radiation was diluted and red-shifted. The red-shifting would lengthen the apparent wavelength of the radiation, and, according to Wien's law, the peak energy would thus come at a wavelength corresponding to quite a low temperature. Starting as early as 1948, this temperature has been variously predicted to be about 1 to 5 K. In other words, if the big bang theory is right, the sky should be uniformly filled with faint radio radiation resembling radiation from an object at about 1 to 5 K.

This line of thought was foreseen by the father of the big bang theory, Georges Lemaître, who wrote in 1931:

The evolution of the universe can be compared to a display of fireworks that has just ended: some few red wisps, ashes, and smoke. Standing on a cooled cinder, we see the slow fading of the suns, and we try to recall the vanished brilliance of the origin of the worlds.

But why just try to recall it? Why not point detectors at the sky to see if there is long-wavelength radiation left over from the primeval fireball?

DISCOVERING THE PRIMEVAL RADIATION

Indeed, direct observational evidence confirms this radiation. In the 1960s a research team at Princeton built a sensitive radio detector to search for it. The predicted radiation would be microwave radio noise. However, even before the Princeton device could be applied, Bell

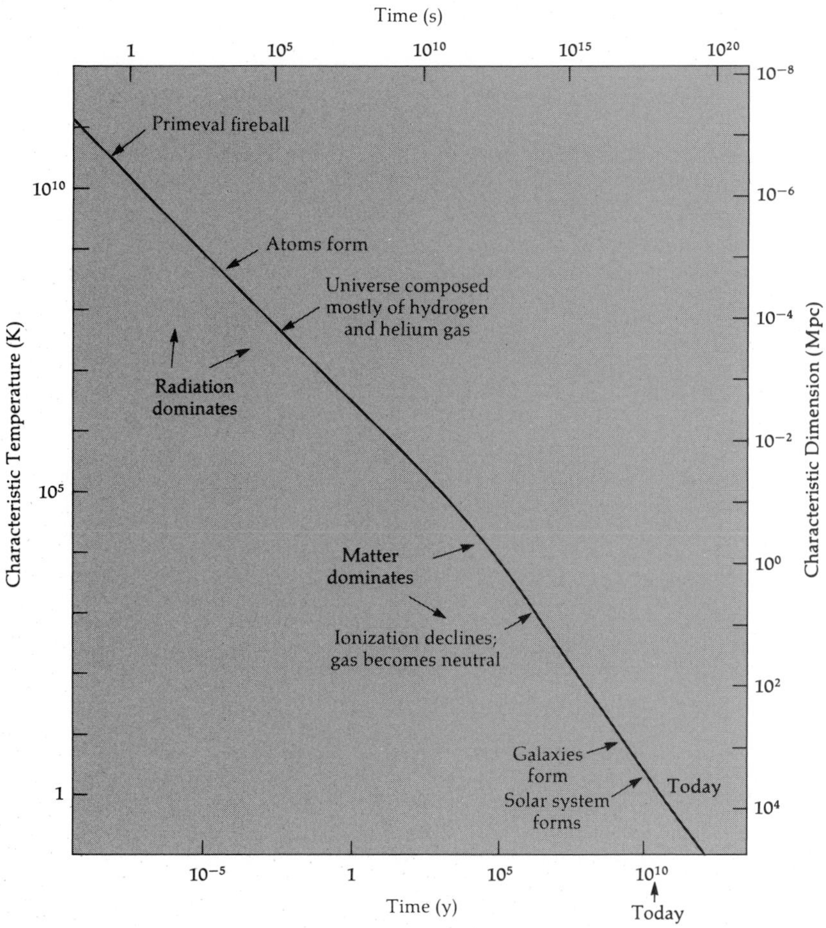

Figure 25-1. *A brief history of the universe, based on current research. The logarithmic time scale, bottom, expands the early history of the universe to fill most of the graph, allowing discrimination of early events. The curve shows declining temperature (left scale) and increasing distances to which internal observers could see (right scale). Radiation dominated the early stages. Atoms formed by combinations of subatomic particles. Gravitational contraction of gas clouds led to the formation of galaxies. Characteristic temperatures corresponding to radiation have declined to 3 K at present.*

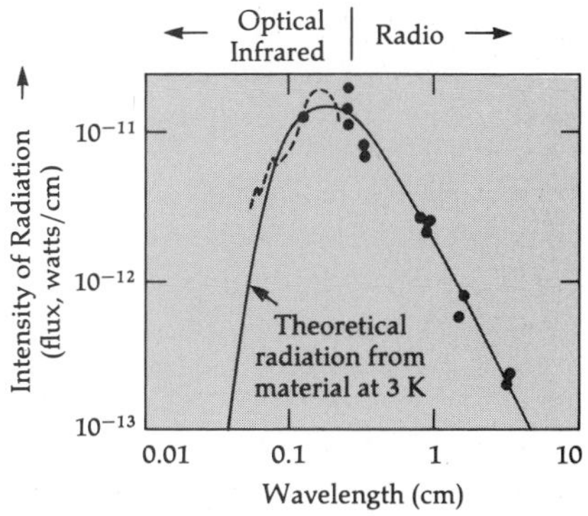

Figure 25-2. *Observational detection of radiation attributed to the primeval fireball of the big bang. The dots and the dotted line are observed amounts of radiation at different wavelengths; the solid curve is the amount predicted for radiation emitted from material at a temperature of 3 K.*

Laboratory researchers Arno Penzias and Robert Wilson, using radio telescopes in experiments on the first Telstar communications satellite in 1965, found puzzling microwave noise. At first they thought it was a problem in the instruments, but finally they established that it was coming from all over the sky. Analyzing these results, the Princeton physicists realized that this was the predicted remnant radiation permeating the universe from the primeval fireball. The radiation corresponds to 3-K blackbody emission and has been confirmed at many wavelengths, as shown in Figure 25-2. It is sometimes called "3-degree radiation." Penzias and Wilson later received the Nobel Prize for this important discovery about the universe's beginnings.

This discovery is the strongest confirmation of the big bang theory, especially because the **3-K radiation** seems to be uniform in all directions. Just as predicted, observations at many radio wavelengths indicate no variations in intensity within an accuracy of a few percent.

WHY DO WE SEE THE 3-K RADIATION?

Radio radiation from all over the sky, emanating from material that appears to be at the specific but extremely low temperature of 3 K (three degrees above absolute zero) may seem like a strange concept at first. But it is not hard to understand, given the principles we have discussed. Consider the early history of the universe, shown in Figure 25-1. For the first few hundred thousand years, the temperature was so high that the hydrogen gas was ionized and opaque, like the gas in the sun. When the temperature dropped to around 3000 K, the electrons could join with the protons to form neutral hydrogen atoms; by a million years after the big bang, the gas was no longer ionized. We know from studying the sharply defined surface of the sun that ionized hydrogen can be quite opaque, even though interstellar cool hydrogen between the stars is quite transparent. So, for the first million years, the universe was filled with a brilliant glowing fog of plasma, and after a million years, "the fog lifted" (Silk, 1980).

Now consider astronomers looking out through space. As they look out past nearby galaxies, they look through transparent space, very thinly populated by neutral hydrogen. But as they look farther away, they look back in time. This is shown schematically in Figure 25-3. At several thousand megaparsecs, their line of sight encounters the region where light left the brilliant fog about a million years after the big bang. This region is like a wall. It is opaque, and the astronomers can't see beyond it. But why doesn't it look like a 3000-K brilliant fog, resembling the surface of the sun? The wavelength of the light they see from it is Doppler-shifted by about a factor of 1000 (note the universe has expanded in dimension by a factor of about 1000 since this era, as seen in the right-hand scale of Figure 25-1). So the temperature they perceive (following Wien's law) is about 1000 times lower, or about 3 K. The 3-K radiation is really the light from the million-year flash that accompanied the big bang.

A SUPER-RAPID EXPANSION WITHIN THE FIRST SECOND?

While observing the 3-K radiation from all over the sky, astronomers identified an interesting challenge to the big bang theory. Recall from the last chapter that clusters of galaxies show a filamentary structure when mapped across the sky; this means that by the time galaxies began to form, some 10^8 or 10^9 y after the big bang, the gaseous matter had aggregated into filaments or clouds. But the 3-K glow is virtually uniform, all over the sky. This means that at the time the glow was being emitted, only 10^5 or 10^6 after the big bang, the glowing material was virtually uniform. As we noted earlier, this would have been at a time when the material had cooled to about 3000 K. However, the traditional big bang theory predicted that by the time the universe reached these conditions, different regions would have had different temperatures and densities, consistent with the filamentary pattern of galaxy clusters. The theory would thus predict a spotty 3-K glow in the sky, and this is not observed! What is wrong?

During the late 1970s and early 1980s, physicists came to the rescue with interesting new insights about the events of the first second after the big bang. These physicists were working on a variety of problems, such as the behavior of subatomic particles (electrons, neutrinos, and so on) under extraordinary temperatures, like 10^{27} K! These conditions are not accessible in the lab, but did prevail in the first instants after the big bang. Also, these physicists were working on so-called "Grand Unified Theories" about the relationships among basic forces in the universe. Their theories made predictions about what happened during the big bang.

According to these theories, the universe did not expand smoothly outward from a point, as in the traditional big bang theory. Instead, it expanded at one rate for only the first 10^{-35} s (a brief moment indeed!). At that point the temperature had cooled to 10^{27} K (not very cool!), and the universe underwent a "change of state," something like the change of state from liquid water to gaseous water vapor when water starts to boil. In the example of the water, the temperature stays at 100°C (212°F) during the change of state, (because the heat input from the stove is used by the water as it changes state), but the water expands drastically as it changes to a gas. Similarly, the universe stayed very hot at the moment of inflation but expanded drastically. Various versions of this theory, by MIT physicist Alan Guth and others, have come to be called **inflationary big bang theories**, because of the rapid inflation near the beginning. Different versions of the theories differ somewhat on the duration of the inflation; but they agree that this sudden rapid expansion produced a sudden jump in volume by as much as 10^{25} times.

The inflationary big bang theories explain many properties of the universe that are otherwise puzzling (Barrow and Turner, 1982). First, they explain why the 3-K radiation is so uniform. The continued slower expansion of the universe, after the rapid inflation stopped, follows that modelled in the traditional big bang theory, but the portion of the universe we observe today arose from a *preinflation* volume 10^{25} times smaller and much more uniform than the volume pictured at the corresponding instant by traditional big bang theorists. According to some versions of this still-developing theory, our universe may even have been a single uniform "bubble" of ordinary space that arose in the preinflation state of space as the change of state proceeded.

Second, the inflationary theory explains why space is in the special condition of being nearly uncurved and having the special density between an open universe and a closed universe. The theory predicts that the universe would have emerged from the inflation in this condition. The special, flat condition was hard to understand in the traditional big bang theory—just as it would be hard to understand how a pencil could fall off a table and land in the special state of being balanced on its point. But this special state seems to arise naturally in the inflationary theories.

Formation of Galaxies

According to the big bang theory, including the inflationary versions, the densities of both radiation and matter decreased as material rushed apart after the big bang and the inflation. Probably within a few thousand years the density of matter became greater than that of radiation, allowing stronger gravitational forces between particles of matter. Clouds of gas formed in the bubble-like pattern detected in galaxy cluster maps. After many millions of years, temperature and density were probably such that gravitational contraction began in the denser clouds of gas, as indicated in Figure 25-1. Certain clouds or turbulent eddies had enough density and self-gravity to contract, perhaps to masses as large as entire superclusters of galaxies. Denser subregions eventually

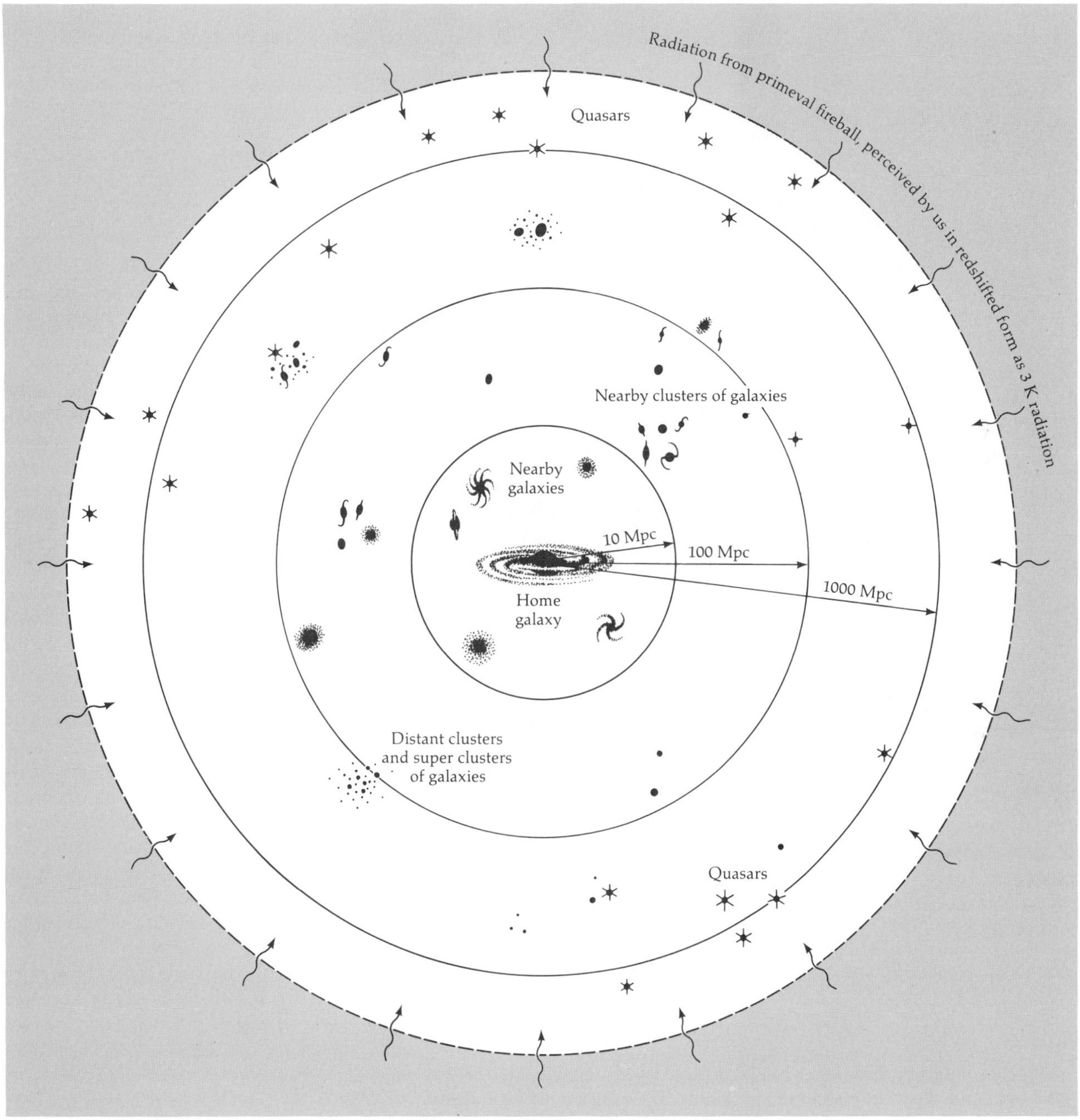

Figure 25-3. *A schematic representation of the universe as seen from any galaxy, including ours. Note change of scale with increasing distance from the center. Clusters and superclusters populate space out to great distances. At very great distances, corresponding to a large red shift, we see light that was emitted billions of years ago, and the only visible objects are quasars. From an extremely red-shifted region beyond them comes radio radiation emitted during the era of the opaque fireball, about a million years after the big bang.*

Figure 25-4. *Galaxies without end. A view of part of the Centaurus cluster of galaxies symbolizes the aggregation of matter from the expanding big bang fireball into galaxies. The view shows elliptical galaxies of various sizes and a few spirals, reminding us that space is filled with galaxies in many forms. (Cerro Tololo Inter-American Observatory.)*

contracted to form individual galaxies, as seen in Figure 25-4. Thus, **formation of galaxies** and galaxy clusters is a predictable result of the big bang.

Galaxies in large clusters show alignment of their rotation axes—they may simply be fragments of single, rotating, primitive clouds (Ozernoy, 1974). This supports the idea that the galaxies themselves are just one more case of the fragmentation of contracting clouds (as discussed in Chapter 16).

Surprisingly enough, the era of galaxy formation can be dated by the statistics of quasars. A 1982 survey indicated a rapid drop-off in numbers of quasars as the red shift exceeds 3.53. This corresponds to an era about 1 billion to 4 billion years after the big bang, depending on the cosmological model. We see few earlier quasars. Therefore, astronomers infer a rapid turn-on in new galaxies (some now seen as quasars) began about 1 to 4 billion years after the big bang, or perhaps 10 to 13 billion years ago (Schmidt, 1982). This is shown in Figure 25-1 and seems consistent with a formation of our own galaxy roughly 13 ± 4 billion years ago.

Formation of Heavy Elements

The first galaxies and stars presumably formed from the gas produced in the big bang, about 75% hydrogen and 25% helium by mass. This statement agrees with direct observations of the ancient Population II stars. A review of the theory of stellar evolution reveals the **origin of heavy elements.** They were created later inside stars by nuclear reactions such as the s-process and the r-process, mentioned in Chapter 15.

It is the massive stars especially that synthesize heavy elements in their high-pressure interiors and then blast them into space when they explode as supernovae. When the next generation of stars forms from the resulting interstellar clouds, the new stars incorporate some of these ready-made heavy elements. Then, as they evolve, they cook up still more heavy elements in their "stellar pressure cookers."

As shown in Table 25-1, most heavy elements have probably been created in this way. The great stability of certain elements, such as iron, explains why they are especially abundant: They rarely broke apart as other atoms joined and refragmented during atomic collisions inside stars.

Thus, the big bang theory, combined with our knowledge of stellar processes, also succeeds in explaining the observed abundances of elements. As Figure 25-5 shows, the heavier elements have gradually increased in relative abundance from virtually none before about 12 billion years ago to a few percent today.

A Cosmic Reference Frame

We have commented that both observations and the big bang theory imply that all galaxies are moving smoothly away from

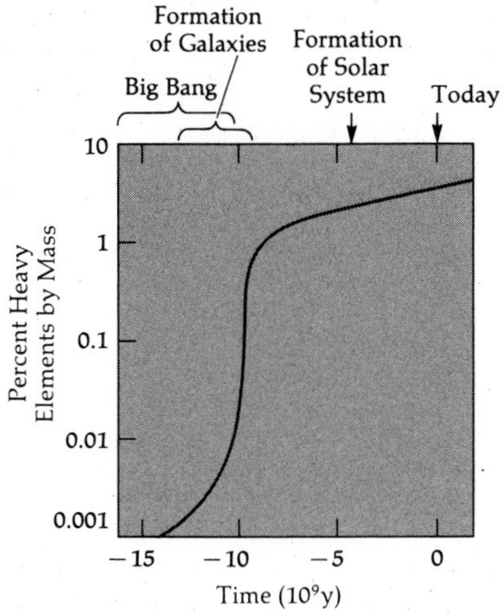

Figure 25-5. ***The abundance of heavy elements as a function of time. Shortly after the big bang, heavy elements constituted virtually none of the total mass. After the formation of galaxies, they increased to about 1% of total mass because they were produced inside individual stars and dispersed by supernova explosions. They have since risen to several percent due to repeated generations of star formation and disruption. (After Reeves et al., 1972.)***

Table 25-1. *Probable Origins of Selected Elements*

Element	Big Bang	Supermassive Exploding "Pregalactic" Objects	Massive Unstable Stars	Ordinary Stars	Cosmic Ray Interactions
Hydrogen (^{1}H)	Yes	—	—	—	—
Deuterium (^{2}H)	Yes	?	?	—	—
Helium 3 (^{3}He)	Yes	—	?	?	—
Helium 4 (^{4}He)	Yes	?	?	?	—
Lithium 6 (^{6}Li)	—	—	—	—	Yes
Lithium 7 (^{7}Li)	?	?	?	?	Some
Beryllium 9 (^{9}Be)	—	?	—	—	Yes
Boron 10 (^{10}B)	—	—	—	—	Yes
Boron 11 (^{11}B)	—	?	?	—	Yes
Heavier elements	—	?	Yes	Yes	—

Source: Data adapted from Reeves et al. (1972).
Note: "Yes" indicates strong mathematical evidence that the observed atoms formed in the given environment; "?" indicates the possibility or probability that some observed atoms formed; a blank indicates the unlikelihood that significant numbers of atoms formed.

each other and that the view from any one galaxy is similar to that from any other. By measuring Doppler shifts of galaxies, we can measure our galaxy's motion relative to those of the others. For example, the "Whirlpool" galaxy, M51, is moving away from us at 550 km/s.

The discovery of the 3-K radiation gives us a larger frame of reference: We can measure Doppler shifts in the 3-K radiation in different directions. In this way, astronomers have found that our galaxy is moving at about 600 km/s relative to the average defined by the universal 3-K radiation. This is interpreted partly as a small random motion of our galaxy among the swarm of flying galaxies, just as the sun has its own small random motion among the other stars as they all orbit around the center of the Milky Way. It may also be partly a result of the motion or rotation of our supercluster of galaxies. For example, some data suggest that we belong to a supercluster system extending over about 15 Mpc, our rotational speed being around 400 km/s (Silk, 1980).

Before the Big Bang

What existed before the big bang? No one is sure that this question has meaning or that any observations could reveal an answer.

According to classic big bang theory, the mysterious explosive event at the beginning is *assumed* to have mixed matter and radiation in a primordial soup of nearly-infinite density, erasing any possible evidence of earlier environments. If so, science would have no way of answering the question.

Conceivably, the big bang was not unique, but only the latest in a series of big bangs, as in curve D of Figure 24-4. This variant of big bang cosmogony is called the hypothesis of the **oscillating universe**. In this view, after the primeval fireball explodes, gravity eventually pulls back the distant material, and the universe collapses again. Another high-density period would follow, then another expansion, and so on. The possibilities of future recollapse or everlasting expansion were discussed further in the last chapter.

In a variant of the oscillating universe hypothesis, the high-density era was not the infinite-density epoch of the classical big bang, but only a period of maximum, finite density, during which some structures from a preceding era might have survived. At least one rotating neutron star has been interpreted as slowing its spin at a rate that may make it older than 16 billion years, and some astronomers have speculated that it might be a survivor from an earlier era.

In the classic big bang theory, the big bang is the true beginning of the universe as we know it, a creation like those in the creation myths of many cultures. The oscillating universe model—that the big bang was not a true beginning, but only the beginning of the current cycle—resembles some Eastern mythologies.

The new inflationary big bang theories add an interesting twist to these ideas. The physicists who developed these theories note that the properties of all the particles in the universe tend to cancel each other out. For example, there seem to be as many positively-charged particles as negatively-charged ones, so that the net charge in the universe is zero. Similarly, the net angular momentum, and other special properties of subatomic particles, seem to add up to zero. This has led some physicists to postulate that the big bang began from a special vacuumlike energy field in which all properties added up to zero, and there were no material properties. It may have been, in cosmologist Alan Guth's words, a situation that "you can't distinguish from nothing." In other words, big bang theorists can now postulate, in a sense, that the universe started from "nothing"—a state that exploded in a shower of positive and negative particles leading to the state we now see.

Are There "Other Places"?

We have defined the universe as all matter and energy that exist anywhere. Until recent decades, most scientists had assumed that all parts of the universe had to be continuously connected, in the sense that light or a spaceship could go from one region to any other region and back again (even if the trip took a long time). However, if there are black holes, there *might* be some regions permanently detached or unobservable from anywhere else. Some scientists speculate that there might be "other regions," such as regions "inside" black holes, in which events can occur but never be seen by us.

A still more intriguing speculation is that we might be in a black hole in someone else's universe. Recent work suggests that—contrary to earlier theories that black holes could never emit anything—a black hole might be able to explode in a burst of gamma-ray radiation and subatomic particles. Thus, the big bang itself might have been some kind of black hole explosion within some sort of larger universe.

Since black holes are thought to be permanently out of touch with each other and (so far) subject only to theoretical investigation, speculations on these problems are at the borders of what the scientific method can now deal with.

Why?

Why did the universe come into being? If the preceding questions of "how?" seem difficult—if they seem at the borders of the scientific method—questions involving "why?" are clearly beyond the scientific method. Why was there a big bang? What started the bang? What was there before? What is the purpose behind it all? When thinking about these questions, we may find it helpful to recall a whimsical but deep comment by Mark Twain: "Why shouldn't truth be stranger than fiction? Fiction, after all, has to make sense . . . " By this he meant that our cultural tradition tends to make us want to look for "sense" in a story—even the story of the universe. We like our stories to have a beginning, a middle, and an end—and perhaps even a nice moral that is consistent with what we believed before we heard the story. When we seek the story of the universe, we have no guarantee of this kind of sense. But we do have a virtual guarantee, based on experience, that if we keep making observations—for example, with improved telescopes in space—they will draw us on to new, fascinating discoveries about the universe. Perhaps they will clarify still further what the big bang was all about.

Remember that the scientific method does not answer "why?" questions, except to show how events in nature follow from laws we have already discovered. The scientific method is basically a procedure for analyzing observations and predicting phenomena slightly beyond those we know. Thus, people practicing the trade of science can only ask questions that can be answered by specific observations.

To the questions of "why?", astronomers' answers may be little better than anyone else's.[3] One can always make up answers, but it is more interesting to admit that there are many things we do not understand—and things we do not even understand *how* to understand.

[3]In *The Hitchhiker's Guide to the Galaxy,* a satire on intergalactic epics, the most giant of computers succeeded in deducing the "ultimate answer" to this ultimate philosophical question—and the answer turned out to be "42." However, the computer noted that the ultimate question itself was so vague that the computer was unable to explain the method by which it deduced this ultimate answer.

Summary

Three lines of astronomical evidence—ages of globular clusters, the expansion age calculated from Hubble's constant, and estimated ages of elements—indicate that the formative conditions existed in the universe roughly 9 to 18 billion years ago—probably close to 14 billion years ago. Most astronomers believe that an explosionlike event called the big bang occurred then. It takes some audacity to claim that we humans know something about this mysterious event, but there are several compelling observations:

1. Because galaxies are receding from each other, they must have once been concentrated in a much smaller volume; this is consistent with the big bang model of initial high density.

2. The big bang model explains the abundances of elements that we see both in Population II stars and, with the addition of later element-forming processes inside stars, in Population I stars as well.

3. The big bang model predicted a pervasive weak radio radiation, known as the 3-K radiation, *later* found by radio astronomers.

4. Theories of galaxy formation, counts of distant galaxies, and observations of the most ancient galaxies and quasars, while not uniquely predicted by the big bang theory, are consistent with it.

The big bang theory takes us close to the limits of the ability of science to explain the properties of matter and energy in the universe. Questions such as why the universe came to exist are beyond the scope of science.

Concepts

cosmogony	primeval fireball
ages of globular clusters	3-K radiation
Hubble's constant	inflationary big bang theories
duration of element formation	formation of galaxies
age of the universe	origin of heavy elements
initial conditions	oscillating universe

Problems

1. Which phrase describes the big bang theory best?
   a. An assumption with logical consequences that turn out to be verified by observation
   b. A fact proven by repeated observation
   c. A revelation

2. Why was the discovery of 3-K radiation from all over the sky heralded as strong evidence in favor of the big bang theory? Does comparing the radiation from different parts of the sky give any evidence of asymmetry or inhomogeneity in the universe?

3. Why would planets such as the earth be unlikely to exist in globular clusters? (*Hint:* Consider the earth's composition.)

4. How would spectroscopic observations help reveal that a group of distant galaxies, such as Abell 18 (illustrated in Figure 24-8), is not a group of stars inside our own galaxy?

5. Why does the radiation from the primeval fireball as observed today have such long wavelength (radio waves) instead of the very short wavelengths (ultraviolet) that might be expected from the high temperatures theorized for the fireball?

6. Do you find any fundamental disagreement between the description of the universe's origin and structure, as described in this and the last chapter, and any philosophical or religious beliefs you may hold? If so, do you believe such a disagreement might be clarified by further observations, or do you believe further observations are superfluous?

ADVANCED PROBLEM

7. Hubble's constant H can be thought of as specifying the speed at which galaxies have traveled to reach any specified distance from the (arbitrarily chosen) central site of the big bang.
   a. Show by logical deduction that $1/H$ ought to equal the age of the universe (the time since the big bang).
   b. Confirm that $1/H$ has the dimensions of time.
   c. Confirm numerically that $1/H$ equals 13 billion years, if $H = 78$ km/s per kpc. (*Hint:* The problem really deals with conversion of units. Note that $1\ y \approx \pi \times 10^7$ s.)

CHAPTER 26

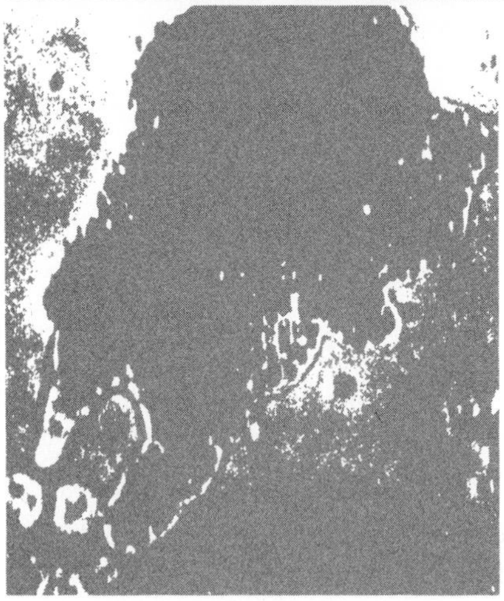

LIFE IN THE UNIVERSE

One of the most intriguing questions in astronomy is whether planets elsewhere in the universe harbor what we are pleased to call "intelligent life." Extraterrestrial life must either exist or not exist. Either case has striking consequences. The nineteenth-century Scottish writer Thomas Carlyle sardonically said that other worlds offer "a sad spectacle. If they be inhabited, what a scope for misery and folly. If they be not inhabited, what a waste of space."

If *intelligent* aliens exist, our society (if it survives) is someday likely to be influenced by them, for better or worse. Anthropologist D. K. Stern notes that discovering such creatures "would irreversibly destroy man's self-image as the pinnacle of creation." If alien life does not exist, we would be the only living creatures in the universe—a remarkable situation that implies a universe wide open for us to utilize and colonize.

What can astronomy say about these two possibilities? Some scientists argue that astronomy, lacking any definitive detections of extrasolar planets or life, can say nothing. However, we certainly have data related to life's origins: **Organic molecules** (complex, carbon-based molecules) have been found in interstellar clouds; meteorites containing amino acids show that organic chemistry has been active in our own solar system outside the earth; and the paucity of organic molecules on Mars places intriguing limits on the biochemical possibilities of planets. Admittedly, discussions of the existence, intelligence, psychology, or appearance of higher alien life forms are almost entirely speculative. Nonetheless, the search for alien life seems to be a new scientific adventure in the making, and a growing number of astronomers, biologists, chemists, anthropologists, and theologians are taking an interest in this quest.

In this chapter, we consider the subject in several steps. First, we discuss what we mean by *life* and what conditions are necessary for life to exist on planets. Second, we review the long process that led to the evolution of life on the earth, as best we understand it. Third, we ask whether there are planets that could support a similar evolution of life elsewhere in the universe. Finally, we try to estimate the probability of intelligent life actually existing on such planets elsewhere and whether we might contact it, or it, us.

The Nature of Life

What do we mean by *life?* **Life** is not a status but a *process*—a series of chemical reactions using carbon-based molecules, by which matter is taken into a system and used to assist the system's growth and reproduction, with waste products being expelled. The system in which the processes occur is the cell. All known living things are composed of one or more cells.[1] A **cell** is, in essence, a container filled with an intricate array of organic and inorganic molecules (protoplasm). Codes for cellular processes are contained in very complex molecules (such as the famous DNA) located in a central body called the nucleus. The elements most prominent in the organic molecules are carbon, hydrogen, oxygen, and nitrogen—all very common in regions populated by evolved (Population I) stars. Phosphorus, also important to life (in small amounts), is widely available. Carbon is especially critical because it can combine to make long chains of atoms—large molecules that encourage the complicated chemistry of genetics, reproduction, and

[1] Viruses might be an exception. They are simpler than cells, yet can reproduce themselves using materials from host cells. Biologists disagree on whether viruses should be considered a form of life.

so on. That is why the term **organic chemistry** (as well as *organic molecules*) refers not specifically to life forms but more generally to all chemistry (and molecules) involving carbon.

We often make the mistake of thinking of ourselves as static beings instead of dynamic systems. We casually assume that we are constant entities, as if our identity depended solely on the form of our bodies. But our bodies today are not the same ones we had seven years ago. Hardly a cell is still alive that was part of that body. Even our seemingly inert skeletons are living and changing, always replacing their cells. We *must* keep changing—the cells must keep processing new materials to stay alive. When the processing stops, we call it death.

The nature of living beings is illustrated in an analogy from the Russian biochemist A. I. Oparin (1962). Consider a bucket that has water pouring in at the top from a tap and flowing out at the same rate through a tap in the bottom. The water level in the bucket stays constant, and a casual observer would call it a "bucket of water." But it is not like an ordinary bucket standing full of water. The water at any instant is not the same water as at any other instant, yet the outward appearance is constant. We are like buckets with water, nutrients, and air flowing through us, but with other, much more complex attributes, such as the ability to reproduce and be affected by genetic changes that let us evolve from generation to generation.

The flow and change of material inside living organisms give us a clue to the kinds of processes involved in the origin of life. We are looking for a process in which complex, carbon-based molecules can enter cell systems and enable them to create new molecules, incorporate these molecules into new structures, eject unused material, and reproduce themselves.

Scientists therefore usually choose to define *life* by these specific carbon-based processes. Often at this point people ask, "What about some unknown form of consciousness?[2] Or what about some unknown chemistry based on other elements, such as silicon, that can form big molecules?" The answer is that we have never observed or experimented with such life forms, so we can say nothing substantive about them. If we admit they are plausible, we simply increase the probability of life or consciousness in the universe. But researchers usually restrict their discussion to carbon-based life. As physicist Gary Feinberg remarked on this question, "Some of my best friends are made of carbon." We have no noncarbon friends to contemplate.

Whatever other conceptions we invent—civilization, religion, technology, art, war, love, communication—it is the chemical processes of life that define us, just as they define the spiders, sea urchins, elephants, moths, amoebas, redwoods, and all the other incredibly varied living creatures around us. To judge whether life may exist on other planets—whether other planets are already "taken"—we must find out how those processes got started on the earth and what sort of environments fostered them (Figure 26-1).

***Figure 26-1.** Sun and sea symbolize the earth's unusual environment, which favors the origin of life. Our planet is far enough from its central star, the star is stable enough, and the orbit is circular enough, so that the temperature remains fairly uniform, and some water remains liquid. Energy, water, and the right chemical environment are believed necessary to initiate life. (Photo by author.)*

The Origin of Life on the Earth

According to the preceding discussion, understanding the origin of life requires that we understand the early history of interactions among the elements carbon, hydrogen, oxygen, and nitrogen—the most important elements in living organisms. How did these elements get together into complex molecules such as **amino acids**, the molecules that join to form **proteins**, the huge molecules in cells? And how did proteins aggregate into cells that could absorb surrounding materials, grow, and reproduce?

[2]For example, in his novel, *The Black Cloud*, astronomer Fred Hoyle imagined an interstellar nebular cloud with matter and electromagnetic fields organized in such a way that it developed a consciousness or will of its own.

PRODUCTION OF COMPLEX MOLECULES

Research in the last few decades has proven that the first steps toward life were surprisingly easy. Chapter 18 revealed that

Figure 26-2. *Fragments of a carbonaceous chondrite meteorite in which chemists have discovered extraterrestrial amino acids. Similar carbon-based compounds are important in living cells on the earth. Although no fossils or life forms have been found in the meteorite, its dark color is due to the abundance of carbon and carbon compounds, showing that building blocks of life can form outside the earth. This meteorite fell in France in 1964. (NASA.)*

complex molecules such as ethyl alcohol (C_2H_5OH) and even the amino acid glycine ($C_2H_5O_2N$) have been discovered in molecular clouds in interstellar space. Extraterrestrial amino acids have also been found inside carbonaceous chondrites (Figure 26-2), the primitive type of meteorite described in Chapter 11. That these acids originated outside the earth is proven because their distribution of symmetries differs from that of the same types of molecules on the earth. Carbonaceous chondrites once contained liquid water in their carbon-rich soil; the amino acids must have formed in this moist environment in the cometlike or asteroidlike parent bodies of these meteorites.

Amino acids would have quickly formed in the earth's early environment. This was proven by an experiment designed to simulate chemical conditions in the earth's primitive environment. This experiment, called the **Miller experiment**, was first carried out by chemist S. L. Miller (1955), who bottled materials such as methane (CH_4) and ammonia (NH_3), representing the earth's hydrogen-rich primitive atmosphere, together with water, representing the ocean. He passed sparks through this material to represent lightning. After a short time, Miller found that a sludge containing amino acids, such as glycine, had formed in the flask. This experiment has been successfully repeated many times with different "primitive atmospheres" and energy sources. Even on a planet with fewer hydrogen compounds and more volcanic gases, such as carbon dioxide (CO_2), nitrogen (N_2), and carbon monoxide (CO, see Figure 3-11, page 60), sunlight tends to initiate photochemical reactions that produce water, hydrogen cyanide (HCN), and eventually cyanamide (CN_2H_2) and the amino acid glycine (Goldsmith and Owen, 1979).

In 1983 researcher C. Ponnamperuma announced that all five of the critical organic compounds—called "bases"—that are responsible for coding genetic information in the DNA and RNA of living cells were found in a single carbonaceous chondrite. At the same time, he and his colleagues were able to synthesize them in a Miller-type experiment in a primitive atmosphere. This work seems to clinch the likelihood that many of life's building blocks arose naturally in extraterrestrial bodies.

In Chapter 10 we saw that similar processes have produced a photochemical smog on the interesting satellite Titan. With its reddish clouds of organic compounds and its possible rains of liquid methane, Titan may give us an interesting natural laboratory for further study of these processes.

This is not to say that all planets are oozing with organic slime. Sunlight acts to encourage reactions that not only *form* organic molecules but also *break down* organic molecules. When energetic UV photons strike molecules, the molecules can break apart. On worlds like the moon and Mars, where there is no ozone layer to block UV light, the rate of destruction is high, and gases are insufficient to provide raw materials to form new organics. As noted in Chapter 9, sunlight has apparently destroyed any and all organic molecules that may have formed on Mars in the past. Indeed, fossil and geochemical evidence suggests life did not emerge on the earth's land surface until ocean plants, which were shielded from sunlight by sea water, emitted enough oxygen to build up an ozone (O_3) layer in the earth's atmosphere. This ozone layer then shielded land-based life from solar UV rays.

Although the Viking landers found no organic molecules on Mars, they did reveal interesting clues about the development of such molecules on planets. Unlike the earth, Martian soil is exposed to strong solar UV rays, which apparently produce unfamiliar chemical reactive states in minerals. The soil contains material that can synthesize organic molecules from atmospheric carbon dioxide and release gas once nutrients are added. Because the Martian soil in its natural state contains virtually no organic molecules, most scientists believe these processes are not caused by Martian life forms. In fact, laboratory experiments in 1977 duplicated most Viking results using simulated Martian soil (iron oxide minerals exposed to UV light in carbon dioxide) without any organisms. Nonetheless, the processes may indicate how complicated chemical reactions create material from which life could form on other planets.

THE PRODUCTION OF CELLS

The next step toward life—aggregating complex organic molecules into cell-like structures—is less certain. Although some of the processes can occur in dry environments, liquid water was probably critical to extensive biochemical evolution, because it provided a fluid medium in which materials could move and aggregate. After all, we evolved from the sea, most of our body weight is water, and the earliest fossils are of sea creatures. One botanist recently commented that "all cells, of all living organisms, are strictly aquatic creatures." He described any land-based organism as "merely a protective shell" filled up with millions of aquatic cells.

Florida biologist Sidney W. Fox has shown that simple heating of dry amino acids (as might happen on a dry planet) can create protein molecules. Once water is added, these proteins assume the shape of round, cell-like objects called **proteinoids**, which take in material from the surrounding liquid, grow by attaching to each other, and divide. Though they are not considered living, they resemble bacteria so much that experts have trouble distinguishing them by appearance.

Possibly related to proteinoids are objects discovered in the 1930s by Dutch chemist H. G. Bungenberg de Jong. When proteins are mixed in water solutions with other complex molecules, both sets of substances spontaneously accumulate into cell-sized clusters called **coacervates.** The remaining fluid is almost entirely free of complex organic molecules.

The next step toward recognizable life forms is still more uncertain. If organic molecules or coacervates are present in a pool of water, they will be left in the pool as the water evaporates. In this way, evaporation of the water in tidewater pools (Figure 26-3) probably produced high, localized concentrations of amino acids, proteins, and other molecules, allowing cell-like structures to form. The cell-like structures in the primeval pools of "organic broth" could have begun reacting with fluids in the pools and with each other, accumulating more molecules and growing more complex, as suggested by Figure 26-4. Eventually these could have evolved into biochemical systems capable of reproducing.

Figure 26-3. *Broths rich in amino acids and complex organic molecules formed in long-lived tidewater pools on the ancient earth. Evaporation of water could have concentrated the remaining organic materials, allowing complex reactions. The resulting products could "fertilize" oceans with living organisms or protoliving materials. (Photo by author.)*

THE EARTH'S EARLIEST LIFE FORMS

Whatever the processes, microscopic cellular life must have arisen between 4.5 and 3.5 billion years ago, because fossils have been discovered in rocks after that period. A 1981 conference on the subject estimated 4.0 ± 0.1 billion years as the age of Earth's first life (Ponnaperuma, 1983). For example, fossils of methane-producing bacteria were found in 3.4-billion-year-old rocks from South Africa in 1977 (Gould, 1978). The precise date of the earliest life on the earth remains controversial. One review of the data (Nisbet, 1980) lists the earliest "compelling" evidence as 2.7-billion-year-old fossil **stromatolites** (colonies of blue-green algae with a cabbagelike structure) from Canada and Zimbabwe. The earliest "probable" evidence is stromatolites from Western Australia between 3.4 and 3.5 billion years old; the oldest "possible" evidence is a 3.7-billion-year-old rock containing carbon isotopes of possible biological origin from western Greenland.

Life probably couldn't have evolved much before 4.1 billion years ago because of the intense early meteoritic bombardment and the possible magma ocean covering much of the earth's crust. Yet biochemical evolution must have begun in that period. Primitive life evidently arose "as soon as it could; perhaps it was as inevitable as quartz or feldspar" (Gould, 1978).

For the earth's first 2 billion years, most life remained in the oceans, where liquid water provided a supporting and protective environment. Life forms were mostly soft bodied and rarely produced fossils, so their development is hard to trace. Stromatolites became more abundant 2.5 to 2.0 billion years ago, living at the boundary of water and rock along seacoasts, as shown in Figure 26-5. Today they are found only in rare, oxygen-poor salt marshes on some seacoasts, suggesting that the early atmosphere in which they flourished was also oxygen poor. Stromatolites produce oxygen by photosynthesis.

It is strange to realize that during most of the earth's history, life would have been hard to find in most *land*scapes, as Figure 26-5 suggests. Even during the earth's middle age, the land was barren. Some areas must have looked like today's deserts or like Mars. Some areas were moist and washed by rains, but instead of luxurious forests, there were only bare acres,

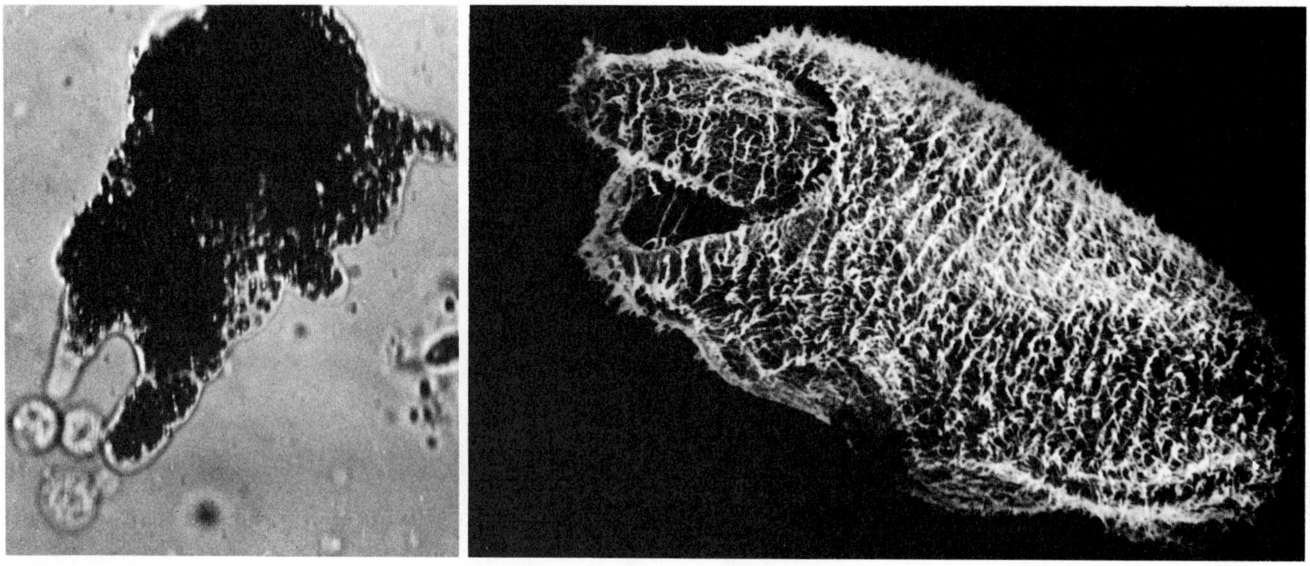

a b

***Figure 26-4.** The importance of a fluid medium for primitive evolution is suggested by these photos of single-celled organisms engulfing food from surrounding fluid. (a) An amoeba flows to surround a nearby food particle. (Optical microscope photo; S. L. Wolfe.) (b) The protozoan* Woodruffia *ingests a* Paramecium. *(Electron microscope photo; T. K. Golder.)*

eroded gullies, and grand canyons. Brown vistas stretched to the sea.

The flourishing of stromatolites and early plant forms 2.5 to 2.0 billion years ago helped boost oxygen (O_2) production (Walker, 1977; Goldsmith and Owen, 1979). The atmosphere therefore evolved from more *reducing conditions* (dominated by hydrogen compounds) toward *oxidizing conditions* (dominated by oxygen compounds). Oxygen content in the initial atmosphere was probably only a few percent, but began to rise rapidly toward the modern value of 21%. Evidence for this is that sediments that formed and were buried more than about 2.5 billion years ago are less oxidized than sediments formed today (Goldsmith and Owen, 1979). Oxidized red sediments are rare before 2 billion years ago and common afterward (Walker, 1977). Oxygen production by plants helped modify the whole environment. Solar UV light broke some O_2 molecules, and the free oxygen atoms joined with other O_2 molecules to make *ozone* (O_3) molecules. This led to the formation of an **ozone layer** high in the earth's atmosphere, which absorbs solar UV light and thus protects organisms on the earth's surface as mentioned before.

Living things could scarcely be unaffected by these major changes in the environment, as fossils testify. Until as recently as 1.5 billion years ago, most organisms lived best in oxygen-poor environments and were relatively resistant to genetic changes by means of UV radiation. Cells through this time period were of a type lacking a distinct nucleus (called prokaryotes); about 1.4 billion years ago, cells with nuclei (called eukaryotes) first appeared (Vidal, 1984). After 1.5 to 1.0 billion years ago, we also find new types of organisms in the fossil record, capable of oxygen metabolism, less resistant to UV damage, but able to flourish because the new ozone layer protected the surface from UV (Windley, 1977, pp. 44, 302). This was the first big step in the expansion of life from the sea onto the land (Figure 26-6)—a step as momentous as the contemplated colonization of other planets by us! The emergence of us oxygen-consuming landlubbers was spurred by the planetary evolution of the earth itself.

Biologists agree that the mechanism by which new species of life forms evolved to fit the changing environment and the new environmental niches, such as land and air, was **natural selection**: Random changes in the genetic codes (as might be caused by UV irradiation or other processes) occasionally produced offspring with improved survival traits. In turn, these offspring lived longer and had more offspring of their own, promoting retention of the new trait.

A population of organisms on one arid continent, for example, might eventually evolve in a different direction from a population of the same organisms on another, wetter continent (witness the different animal populations on Australia and Asia). A rapid change in global climate toward drier conditions might lead the first group to flourish while the second group died out. The first group would then spread rapidly from their home continent to other regions. The fossil record of a region thus might show the sudden emergence of the first group and extinction of the second. All these steps are probably part of natural selection, though the details of the process are still debated by biologists. Some details are uncertain because we often sample only discontinuous layers of fossil-bearing strata. However, many fossil sequences show clear, step-by-step evolution, as with the 50-million-year sequence showing the evolution of a ½-m-tall, three-toed mammal into the modern horse.

As Darwin pointed out long ago, the overall process is like the artificial selection practiced by breeders who develop new strains by producing more offspring between individuals with traits the breeders favor.

Evidence of the adaptability of life forms is the rapid proliferation of advanced species once they evolved, as indicated in the geologic time scale shown in Table 3-1. While it took about half the available time to go from complex molecules to algae and bacteria, it took only the last 12% of earth history to go from the first hard-bodied sea creatures (such as trilobites, Figure 26-7) to humans.

Whatever the specific mechanisms, we can say from the fossil record that the earth experienced a few-billion-year evolution from nonliving organic chemicals to small organisms, and then a much more rapid evolution to species with self-conscious intelligence.

Planets Outside the Solar System?

From all we have just said, we conclude that if planetary surfaces with the necessary conditions—liquid water and the "CHON" chemicals (carbon, hydrogen, oxygen, and nitrogen)—exist long enough anywhere, life is likely to evolve. Advanced species are likely to appear eventually. But are there such planets?

In view of the hundred billion stars in our galaxy, not to mention the innumerable other galaxies, we should not limit our search for life to the solar system alone.

Figure 26-5. *The first signs of life encroaching on land did not occur until the earth reached middle age. A patch of stromatolites is visible (left) at low tide, and early lichens can be seen on the rocks (right). Most of the land was still barren. (Painting by Ron Miller.)*

Figure 26-6. *Lichens, symbiotic combinations of algae and fungi, were probably some of the first organisms on land. As shown here, they grow on rock surfaces, weathering the rock to produce soil. They are widely adaptable, existing in both arctic tundra and hot deserts. Penny shows scale. (Photo by author.)*

Figure 26-7. *Fossil trilobite, a hard-bodied sea animal that appeared about 600 million years ago, providing diagnostic fossils for recognizing and dividing Cambrian time. (Gayle Hartmann and Joyce Rehm.)*

There are four kinds of **evidence for planets near other stars**.

First, most newly formed stars are immersed in cocoon nebulae in which mineral grains have condensed. These grains may aggregate into planetary bodies as a natural by-product of star formation.

Second, even some main-sequence stars have recently been found to have systems of dust particles orbiting around them. Vega, for example, has a Jupiter-mass or more of dust particles at least a millimeter across, revealed by infrared satellite telescope. Page 328 shows a photo of a similar dust system around the star Beta Pictoris.

Third, statistics of masses among binary and multiple stars suggest the likelihood of many stars having low-mass companions that are planets.

Fourth, unseen low-mass companions have actually been detected around a few nearby stars (Table 26-1). Some are between the mass of Jupiter and that of the smallest stars; they may be planetlike.

Table 26-1. *Components of the Nearest Systems and Other Selected Systems*

Name of Largest Star in System	Distance: Parsecs	Distance: Light-years	Mass of Components ($M_\odot$): 10^{-4}	10^{-3} (Jupiter)	10^{-2}	10^{-1}	1 (Sun)	10
Nearest Stars								
Sun	0	0		x x			x	
Alpha Centauri	1.3	4.3				x	xx	
Barnard's star	1.8	5.9		?		x		
Wolf 359	2.3	7.6				x		
+36°2147	2.5	8.1			x	x		
Sirius	2.6	8.6					x	x
Luyten 726-8	2.7	8.9				xx		
Ross 154	2.9	9.4				x		
Ross 248	3.2	10.3				x		
Other Selected Systems								
61 Cygni	3.5	11.2			x		xx	
Krüger 60	3.9	12.9				xx		
Ross 614	4.0	13.0				x x		
B.C. +68°946	4.8	15.6			x		x	
B.C. +20°2465	4.9	16.0			?		x	
Stein 2051(G175-34)	5.5	17.8			x	x	x	
VB10 (Van Biesbroeck's Star)	5.7	18.7			x	x		
Eta Cassiopeiae	5.9	19.2			x		xx	
VB8	6.5	21.1			x	x		
T Tauri	140	450			x			x

Source: Data from Allen (1973), Harrington and Kalla/akal (1983), Hanson and others (1983), Strand (1977), and other sources.

At the distances of most stars, planets smaller than Jupiter are undetectable by current methods, but only marginally so. Astronomers have proposed new devices and techniques to detect even earth-sized planets near nearby stars. Among these are space telescopes and sensitive spectrometers that would detect radial-velocity oscillations of stars, caused by planets going around them (O'Leary, 1980). The search for planets near other stars is an exciting enterprise that will be pursued in coming years; it may finally clarify whether our own planetary system is commonplace or freakish.

For the present, an informed guess might be that between 1% and 30% of all stars have at least one planet nearby.

HABITABLE PLANETS?

Even if planets exist near some other stars, there is no guarantee that they are habitable. Astronomers have proposed several conditions needed to make a planet habitable:

1. The central star should not be more than about 1.5 $M_\odot$, so that it will last long enough for substantiated life to evolve (at least 2 billion years), and so that it will not kill evolving life with too much UV radiation, which breaks down organic molecules.

2. The central star should be at least 0.3 $M_\odot$ to be warm enough to create a reasonably large orbital zone in which a planet could retain liquid water.

3. The central star should not flare violently or emit strong X rays. It should be on the main sequence in order to have been stable long enough to give its planets long-term climatic stability. Conditions 1, 2, and 3 restrict us to main-sequence stars of spectral types F, G, and K.

4. The planet must orbit at the right distance from the star so that liquid water will neither evaporate nor permanently freeze (Figure 26-8).

5. The planet's orbit must be circular, stable enough to keep it at a proper distance and prevent drastic seasonal changes. Such stable, habitable orbits are less probable, but still possible in common types of binary systems (Harrington, 1977).

6. The planet's gravity must be strong enough to hold a substantial atmosphere.

The American physicist Stephen Dole (1964) reviewed these criteria and speculated that only a few percent of all stars, mostly from 0.9 to 1.0 $M_{\odot}$, have habitable planets. However, the figure remains highly uncertain.

Has Life Evolved Elsewhere?

If habitable planets exist, and if life evolves readily under habitable conditions, shall we immediately conclude that life must be abundant throughout the universe? There are several additional factors to consider. For one thing, planetary environments change with time, so that today's habitable planet may not be habitable tomorrow. This raises the question of the ability of life to adapt to change. The big problem is that we don't know how unusual the earth's history has been. Has life narrowly escaped complete extinction by climatic change in the past? Or was there never any question that natural selection would allow a few species to survive each step of the earth's evolution?

EFFECTS OF ASTRONOMICAL PROCESSES ON BIOLOGICAL EVOLUTION

Basic planetary or stellar processes may be involved in encouraging or hindering biological evolution. For example, the earth's fossil records indicate an episode called the "great dying," when as many as 96% of all existing land and sea species became extinct in less than a few million years (Raup, 1979). This happened about 230 million years ago (see Table 3-1, page 64); many paleontologists believe widespread climate changes were involved.

A number of geological and astronomical processes have been identified that

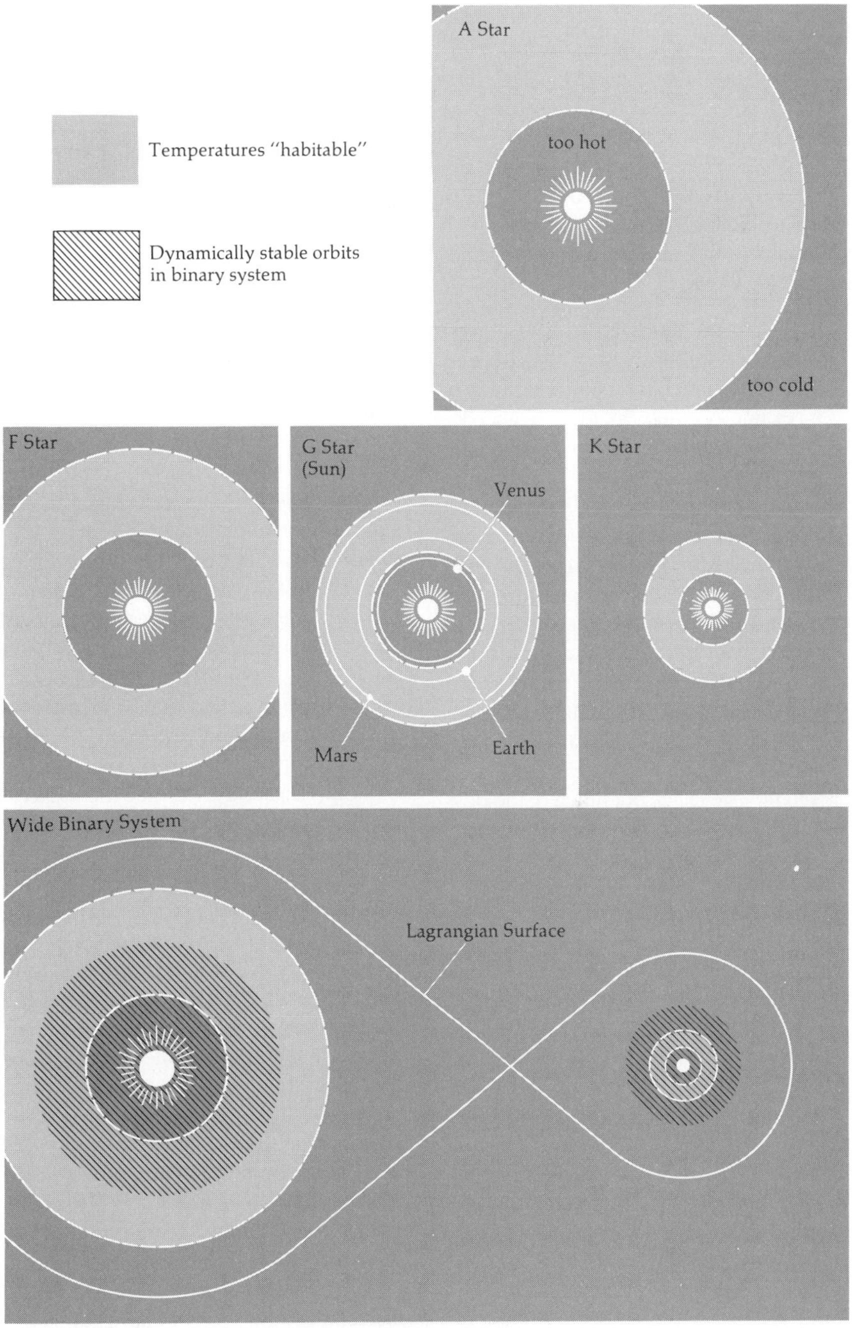

Figure 26-8. *Doughnut-shaped light areas show the relative sizes where water would remain liquid in planetary systems around stars of different spectral types. In a binary system (bottom), the situation is complicated because orbits outside hatched regions suffer large perturbations, causing evolution away from circular orbits, in turn causing temperature extremes. (After Huang, 1960.)*

could cause massive climate changes that could affect the course of biological evolution. Among these are:

1. Planetary convection could cause plate tectonic crustal splitting; change sea levels, ocean currents, wind patterns, and seasonal extremes. This process has been responsible for isolating land masses such as Australia and allowing different species to evolve there.

2. Volcanic eruptions may have spewed enough dust into the high atmosphere to reduce the sunlight reaching the surface. For example, widespread sunlight decreases of up to 25% occurred after the 1883 Krakatoa (Sumatra), 1912 Katmai (Alaska), and 1982 El Chichon (Mexico) eruptions. Rarer larger eruptions could cut sunlight for some years, lowering summer and/or winter temperatures and causing the decline of some species and ascendancy of new ones. Current discoveries of dust layers in active ice packs are beginning to allow dating and reconstruction of prehistoric volcanic cataclysms. The Mount St. Helens eruption of 1980 dropped ash in nearby states, but much of the dust also reached the stratosphere and scattered around the world.

3. Slight changes in Earth's orbit and the tip of the planetary axis to the plane of the ecliptic, caused by gravitational forces, are believed to have caused major climatic changes, such as the ice ages (for example, Imbrie and Imbrie, 1980; Covey, 1984).

4. Slight changes in the sun's radiation may have changed climates.

5. Irradiation from a nearby supernova could have affected the climate or directly damaged organisms.

6. An asteroidal or cometary impact or atmospheric explosion could have damaged the ozone (O_3) layer, exposing organisms to enhanced radiation. Turco and others (1981) found that the Tunguska asteroid or comet explosion of 1908 (Sekanina and Yeomans, 1984) generated as much as 30 million T of nitrogen oxide (NO) in the stratosphere and mesophere, reacting to deplete up to 45% of the O_3 in the northern hemisphere, consistent with Smithsonian measures made in 1909–1911.

7. As illustrated in Chapter 3 (p. 65), a catastrophic asteroidal impact probably occurred about 65 million years ago, causing the dramatic break in the fossil record between the Cretaceous and the Tertiary periods. The idea of a large impact causing widespread extinction of species was suggested as early as 1969–70 (McLaren, 1970) but ignored for lack of evidence. Alvarez et al. (1980) presented the first evidence: They found the iridium (Ir) content of 65-million-year-old limestones equalling 20 to 160 times that of limestones of other ages. Because Ir is enriched in meteorites relative to Earth, they suggested that a 6- to 14-km-diameter asteroid struck Earth and ejected massive amounts of meteoroid-enriched dust into the atmosphere. This hypothesis has had a tremendous impact on the paleontological community and caused a burst of interesting research. Within months, other investigators confirmed the Ir excess in 63-million-year-old sediments from many parts of the world. The effects of the impact are hard to contemplate. Some 75% of species (notably dinosaurs) became extinct within a few million years, and new species (notably mammals) became dominant (Russell, 1982). However, detailed studies of the fossil record show that this did not happen all at once, and various species were affected differently (Alvarez et al., 1984). Discovery of a worldwide soot layer in the boundary layer suggests that much of the vegetation of the world was consumed in fire (Wolbach et al., 1985)! No land animals larger than 25 kg survived, but freshwater animals and plants were scarcely affected (Russell, 1982). Ejected dust and soot may have blanketed the stratosphere and caused 3 to 6 months of daytime darkness consistent with a reported extinction of 44% to 49% of all genera of (light-dependent) floating marine organisms, but only 20% of (less light-dependent) land genera (Ahrens and O'Keefe, 1982). Consistent with this is a drop in calcium carbonate in the 65-million-year-old stratum, due to the disappearance of plankton that would normally drop their calcium-rich skeletons onto the seafloor. Further isotopic studies indicate a warming of surface ocean waters by as much as 10K at this time. If marine plankton were immediately decimated by the impact, they would have stopped consuming CO_2 and the consequent increase in the CO_2 of the air could have caused a greenhouse warming that led to massive land organism extinctions over the next 50 000 years or so. Because an impact into the ocean is most probable, Ahrens and O'Keefe (1982) modeled the tsunami produced; they envision a wave hundreds of meters high inundating all low-lying continental areas within 27 h of impact.

The impact that produced certain North American *tektites* may have caused the profound climatic event that ended the Eocene 34 million years ago, causing widespread extinction of radiolarians (minute marine protozoans). An iridium excess has been found at this transition stratum (Alvarez et al., 1984).

Would these kinds of changes necessarily hinder biological evolution? Biological disasters may happen only when changes are too extreme (for instance, the freezing of all water as on Mars) or too sudden (massive infusion of dust into the stratosphere by a huge impact explosion). Many biologists are coming to believe that changes of the magnitude that have occurred on the earth, while causing the decline of some species, have promoted the emergence of advanced species. For example, if some aspects of intelligence had first appeared in a benign, constant environment where food was plentiful, these traits would have had little value. But if ice ages destroyed the mild environment, then the more clever groups might have emerged from their previous obscurity.

Perhaps it is no coincidence, then, that the time scale of geologic change is comparable to the time scale of biological change. In response to changing environments, life changes itself, on time scales as short as a million years.

ADAPTABILITY AND DIVERSITY OF LIFE

The great variety of ancient and modern species on the earth and the variety of environments in which they have thrived suggest that, given time, life could also have evolved to fit a wide range of conditions on other planets. Even humans have a remarkable adaptability. We thrive from equatorial wet jungles to dry deserts to arctic plains to Andean summits where air pressure is barely half that at sea level. In the past, during ice ages that glaciated New York, humans survived by migrating. We can survive a 3% variation in body temperature, from about 303 to 313 K. Simpler organisms can withstand much wider variations. Certain bacteria that appeared about 3 billion years ago can withstand 18% temperature variations, over a range of 70 K (126°F).

The environmental range covered by *all* terrestrial species is even greater. For example, some microorganisms live in Antarctic ponds that remain liquid at 228 K (−49°F) because of dissolved calcium salts, and others live in Yellowstone Hot Springs at temperatures of 363 K (194°F). Habitable pressures range over a factor of 1000. Bacteria exist at altitudes where the atmospheric pressure is only about 0.2 atm, while more advanced organisms live at ocean depths with pressures of hundreds of atmospheres.

Although sunlight is the ultimate energy source for all familiar organisms (since the ecological chain leads back to vegetable foods that flourish by photosynthesis), recent discoveries show that not even sunlight is necessary to terrestrial life. Oceanographers in 1980 announced the discovery of colonies of strange deep sea animals clustered in the darkness around seafloor volcanic vents. Their metabolism is based on volcanic heat and hydrogen sulfide (H_2S) gas, utilized by bacteria that form the bottom rung of the local food chain (Edmond and Von Damm, 1983). The local pressure is 250 atm, nearly three times that on Venus. Some bacteria recovered from these environments thrive and reproduce at 523 K (482°F), extending the temperature range for known life to a value more than two-thirds that on Venus. But the 523 K figure has been suggested as a rough upper limit for life (Brock, 1985), because known amino acids break down at higher temperatures. The seafloor vent life forms show that a pleasant, sunny planet with a clear atmosphere is not the only possible locale for life!

Could any terrestrial organisms withstand the environments of other planets? For simple organisms, probably yes. Grass seed can germinate in a variety of atmospheres composed of simple compounds of the common elements carbon, hydrogen, oxygen, and nitrogen. Eight species of insects (relatively complicated creatures) studied at different pressures behaved normally all the way down to 10 to 16% of normal atmospheric pressure on the earth (Siegel, 1972).

For these reasons, spacecraft have been at least partially sterilized to avoid contaminating other planets. Such contamination might not only alter planetary environments but also ruin any chance for us to discover whether simple organisms or complex organic molecules ever formed independently on those planets.

It works the other way, too. Simple organisms from other planets might have devastating effects on the earth. There are historical examples of similar events. The plague caused by bacteria introduced into Europe in the 1300s killed about a quarter of all Europeans and as many as three-quarters of the inhabitants in some areas. Diseases introduced into Hawaii after the first European contact in 1778 killed about half of all Hawaiians within 50 years. Some 95% of the natives of Guam were wiped out by disease within a century of continued European contact. For these reasons, early Apollo astronauts were quarantined until it was clear they carried no lunar organisms.

With these facts as well as cultural competition in mind, anthropologist D. K. Stern (1975) has remarked, "It is likely that the meeting of two alien civilizations will lead to the subordination of one by the other." Thus, change and evolution in life populations are likely not only from life's adaptability to new environments but also by the invasion and destruction of some populations by others.

WOULD THEY LOOK LIKE US?

Opinion is divided on this point. MIT physicist Philip Morrison (1973) emphasized that species with similar capabilities in similar habitats may evolve to look alike. For example, three different species of large sea creatures "designed" for fast ocean swimming look alike: an extinct reptile, the ichthyosaur; a fish, the shark; and a mammal that returned to the sea, the dolphin. The idea is that aliens living on planetary surfaces in gaseous atmospheres and using "intelligence" to manipulate their surroundings with tools might well have bilateral symmetry, appendages used as hands, a pair of eyes designed (like ours but unlike most animals') for stereo vision, and so on.

On the other hand, the famous paleontologist George Gaylord Simpson (1964, 1973) argued that although life is likely to start, the long chain of environmental changes and evolutionary steps required to produce humans is unlikely to be approximated elsewhere, so that there is likely to be a "nonprevalence of humanoids." Simpson criticized the whole attempt to estimate the probabilities of such life as nearly meaningless because of our lack of experiments or examples.

Physicist W. G. Pollard (1979) counters that we do have examples of independent evolution that show divergence toward different forms. He notes that about 180 million years ago Australia broke off Gondwanaland and can be thought of as an earthlike planet, "A," where evolution continued independently from a primarily reptilian stock. Similarly, South America broke off Africa 130 million years ago and can be viewed as an independent planet, "S." Independent evolution also continued on planet "E" (the rest of the earth, especially Africa and the adjoining land). During the last 130 to 180 million years, independent evolution *diverged* toward different kinds of animals on these three "planets," rather than converging. Humans appeared only on planet "E," primates on "S," and marsupials on "A." The humans on "E" appeared only about 4 million years ago and have existed for only 0.1% of the earth's history.

In summary, natural selection seems to produce species capable of occupying any habitable environment. Thus, we should not be surprised if life has evolved on another planet. But this life may look very strange to us, even if it displays recognizable intelligence. After all, if mushrooms and corals and woolly mammoths and Venus flytraps all evolved on one planet, how much greater may be the differences between life forms on two different planets? Feathers and fur and sex and seeds and symphonies may be the products of the earth only.

EFFECTS OF TECHNOLOGICAL EVOLUTION ON BIOLOGICAL EVOLUTION

If life will evolve when the right conditions exist, and if the conditions probably do exist elsewhere, then what can we predict about that life? Should we predict intelligence and civilizations? What do these terms mean? Should we assume that other civilizations will achieve spaceflight or might visit us some day? This raises the question of technology and its role in the evolution of life. Just as cosmologists have only one universe as an example, **exobiology** (the study of possible life on other worlds) has only one inhabited world as an example, so the answers to these questions are uncertain. Many exobiologists have assumed that intelligence involves, among other things, use of tools to modify the environment and hence technology.

But as we have seen, while limited environmental change can be helpful, environmental change that is too much or too fast can be fatal! As Pulitzer Prize-winning naturalist René Dubos points out, we are umbilically connected to the earth, and if we alter our planet too much before acquiring an ability to leave it, we are finished. For this reason, we conclude that the development of technology could actually end civilization on some worlds. This is hardly wild speculation, since we see a few nominally moral, intelligent technologists on our own world spending entire careers devising weapons solely to deal death to our own species.

Past wars did not threaten our whole species because conflicts involved only a small percentage of the world. But today, nuclear, biological, and other types of weapons could involve the whole world. For example, the radioactive strontium 90 produced by nuclear explosions has a half-life of 28 y. It was blasted freely into the atmosphere before the Nuclear Test Ban Treaty of 1963. A year after the first H-bomb tests by the United States in the Pacific, strontium 90 deposits in American soil increased soil radioactivity by ½%. A few weeks after a 1976 Chinese nuclear test, airborne radioactive debris fell onto the United States, increasing radiation levels. Obviously, a sufficiently massive nuclear exchange could devastate not only civilization but also future forms of life, whose genetic pool would be exposed to high radiation levels for decades. Humanity has thus proved that a planetary culture could wipe itself out by conscious design of weapons, as irrational as that may seem.

We have also proved that this disaster *could* happen by mistake. As our technology reaches planetary scale, our accidents can also involve the whole planet. Problems as diverse as nuclear power and aerosol spray cans illustrate the issue. Some currently planned nuclear power plants will produce radioactive plutonium wastes, among the most toxic of known materials. Although safe when sealed, kilograms of plutonium dust accidentally spilled into the air could devastate whole states. Yet many kilograms are already being processed, and several governments promote greater dependence on nuclear power plants until other energy forms become available. Accidents are not the only danger: We have seen our society spawn terrorists and madmen. With some grams of stolen plutonium 238, such characters could threaten whole cities. Our creation of highly radioactive substances presents a technological danger that our society has allowed to develop in spite of knowing about it ahead of time.

The seemingly harmless aerosol can is a different story. In 1974, several planetary scientists realized from theoretical calculations that when Freon (the propellant gas used in such cans) is released into the air, it reaches the earth's protective ozone layer, high in the stratosphere. Here its breakdown products act to destroy the ozone layer. Depletion of the ozone would let more solar UV radiation reach the surface, increasing the risks of skin cancer. Even as more Freon was being sold, planetary astronomers and chemists published refined calculations showing that to deplete the Freon already in the atmosphere could take as long as a century. After a 1975 report by the National Academy of Sciences confirmed this danger, moves were made in 1976 and 1977 by American regulatory agencies to phase out Freon aerosol propellants—an example of a technological danger that almost slipped by.[3] Thus, our generation's consumer gimmicks could be creating future health hazards.

Our purpose here is not just to toll fashionable bells of doom and gloom. We all hope that by exercising a bit of intelligence and caution, humanity can recognize these dangers and avoid them. Nonetheless, if humanity is any example, long-term survival of a planetary culture is not assured. Although we have been around for less than 1% of the age of our planet, we are already beginning to have brushes with global disaster. Thus, we can speculate that if evolution produces intelligent societies that remain tied to one planet, some may last less than a percent of the age of the universe—reducing the chance that a nearby culture will be around at the same time that we are.

[3]But we should be haunted by the words of Harvard planetary physicist Michael McElroy, who performed some of the earliest Freon calculations: "What the hell else has slipped by?"

More optimistically, we have succeeded in identifying the **cultural hurdle** that we (and perhaps intelligent species on any planet) must surmount: the transition from scattered, competing nation-states with the capability to damage the planet to stable global or interplanetary societies of intelligence and imagination. The situation is not unlike George Lucas's conflict between the good and dark sides of "the force": We face a race between the good side of our technological abilities, which make civilization possible (plumbing, electricity, stereos, and interplanetary colonies that could bring new resources back to Earth), and the dark side of technology, which threatens civilization (carcinogenic by-products, pollution, and hydrogen bombs). If the good side can win the race, we can perhaps break out of our muddle of competitive strife over resources and ideological space on a finite planet. Perhaps some alien cultures have crossed this cultural hurdle and spread across many planets, thus insuring their survival against ecological disaster on any one planet. Such cultures might last and be detectable for millions of years instead of thousands.

Our conclusion so far is that while some fraction of the stars have planets, most of these planets may be inclement for life, as symbolized by Figure 26-9a and b. But a fraction of the planets may have produced life, as symbolized in Figure 26-9c. Of these, a fraction may have produced either civilizations or relics of destroyed civilizations. The next question is whether any actual evidence for extraterrestrial carbon-based life as we know it exists today.

ALIEN LIFE IN THE SOLAR SYSTEM?

Spacecraft exploration of the planets has virtually eliminated any chance of advanced extraterrestrial **life in the solar system** and has probably ruled out any extraterrestrial life whatsoever in our system. The terrestrial planets except for the earth seem devoid of life, according to our landers, and the outer planets seem too cold.

The most interesting environments seem to be those of Mars, Titan, the middle atmospheres of giant planets, and the sub-ice ocean of Europa. Although UV light has probably destroyed any organic material on the Martian surface, the Martian soil has the ability to synthesize organic molecules from artificial nutrients, and some Viking scientists still leave open the possibility of Martian microbes below the surface. Florida State University biologist E. I. Friedmann in 1978 reported the discovery of microorganisms thriving in pore spaces inside rocks in seemingly lifeless, Mars-like dry valleys in Antarctica, giving some support to the possibility of hidden organisms on Mars. Titan's thick atmosphere and methane chemistry remain interesting, especially if "hot springs" of any sort exist on the surface. Giant planets' atmospheres may have hydrogen-dominated layers with temperatures and pressures not too different from the atmosphere of the earth; Carl Sagan has speculated about the evolution of organisms that could float in such atmospheres.

The discovery of the sun-independent organisms near deep-sea vents on Earth, coupled with the discovery of tidal heating of the interiors of Jupiter satellites, has raised the possibility of organisms that might have evolved in liquid water zones under the ice of satellites such as Jupiter's moon, Europa (a theme developed in Arthur C. Clarke's novel, *2010*). It might be interesting in the future if we can drill through the icy crust of this moon and study the "buried ocean" believed to exist there.

Although we don't expect to meet creatures of any great intellect in our solar system, there are clearly places left to explore that may shed light on biochemical and biological evolution.

ALIEN LIFE AMONG THE STARS?

There is no direct evidence on whether there is life among the stars, but American, Soviet, and other scientists in international meetings have put together a method for considering the possibilities (Sagan, 1973). The logic is to try to estimate the various fractions—of stars having planets, of those planets that are habitable, of those where conditions remain favorable long enough for life to evolve, of those where life does evolve, of those where intelligence evolves, and of the planet's life during which intelligence lasts (see Figure 26-9). The product of all these fractions is the estimated fraction of all stars harboring intelligence. This statement is sometimes called the **Drake equation**, after radio astronomer Frank Drake, who first formalized this scheme for discussing alien intelligence.

Table 26-2 shows "optimistic" and "pessimistic" estimates of the various fractions, and consequent estimates of the upper and lower limits on the fraction of stars that might have civilizations today. In the optimistic case, the answer comes out a few percent; in the pessimistic case, only one star in 10^{14} (one in 100 million million). We stress that these are little more than educated guesses, because we lack the data to make better scientific estimates. But this need not stop us from considering the consequences of our conjectures.

For example, how far away would the civilizations be? Figure 26-10 shows the answer by plotting the radial distance required to include a given number of stars.

In the first case, the nearest civilization might be only 15 ly away—amazingly close. At the speed of light it might take only one generation to reach it. In the pessimistic case, the closest civilization would probably not be in our own galaxy, but roughly 10 million light-years away (a few megaparsecs) in a distant galaxy. Nonetheless, there are so many galaxies, that it is hard to avoid the conclusion—even with the pessimistic view—that **life outside the solar system** is likely, with millions or billions of technological civilizations. If so, then at this moment intelligent creatures may be pursuing their own ends in unknown places under unknown suns.

WHERE ARE THEY?

Radio astronomers have listened for radio messages from alien civilizations and heard none. Nor do our skies seem to be overrun with alien visitors trying to contact us. Nonetheless, theorists have recently pointed out that if *any one* civilization began sending out multigeneration slower-than-light starships that set up colonies, which in turn sent out new starships in a few generations to other stars, that civilization could populate all the habitable planets in the galaxy on a timescale short compared to the age of the galaxy (but long compared to the timescale of evolution, so that the original civilization might produce many new species as it went). So why haven't we heard from aliens?

One answer could be that there are no other aliens—intelligent life on Earth is some sort of unique accident. Another answer might be that we *are* being visited—witness the "flying saucer" reports. Some UFO reports are intriguing, but the verifiable evidence for actual alien spaceships is abysmally poor (for further discussion of UFOs, see Enrichment Essay A). If any alien visits have occurred, they are so rare that we have no proof of them.

A third possible answer is that the pessimistic figures are right and that the nearest civilizations are in distant galaxies. Even their radio messages, if any, could be 10 million years old by the time we receive them, and their spaceships would be unlikely to reach the earth if limited to speeds less than that of light, as current physics requires. However, at a 1971 Soviet–American conference on this problem, the favored estimate came out to be one civilization per 10^5 stars. This would put a million civilizations in our galaxy and the nearest civilizations only a few hundred light years away. Why, then, are there no frequent visits to Earth?

A fourth possible answer is that biological evolution need not produce creatures who have a desire to build "civilizations" or travel through space. Not even all human societies necessarily evolve toward technology. Are humans fated to

a

b

Figure 26-9. *Scenes on three imaginary planets hint at the range of possible alien environments. From the point of view of life, these three planets are too hot, too cold, and just right.* (a) *A planet close to a luminous binary star is too hot for liquid water, and is an arid wasteland. (Painting by Don Dixon.)* (b) *A planet on the outskirts of a galaxy is so far from its star that all water is frozen, producing a desolate landscape of rock and ice. (Painting by Ron Miller.)* (c) *On this planet with two moons (one is eclipsing the sun), liquid water is abundant and primitive life has evolved. If conditions remain stable, higher life forms could evolve within a few hundred million years—a brief moment in cosmic history. (Painting by Rick Sternbach.)*

c

Table 26-2. *Estimated Fraction of Stars with Planets Having Intelligent Life*

	Plausible Lower Limit	Plausible Upper Limit
Stars having planets	10^{-2}	0.3
Those stars ever having habitable conditions on at least one planet	10^{-1}	0.7
Those planets on which such conditions last long enough for life to evolve	10^{-1}	1
Those planets on which life does evolve	10^{-1}	1
Those planets on which habitable conditions last long enough for intelligence to evolve	10^{-1}	0.9
Those planets on which intelligence does evolve	10^{-1}	1
Those planets on which intelligent life endures	10^{-7}	10^{-1}
Product: Fraction of all stars with planets having intelligent life	10^{-14}	2×10^{-2}
Implication: Distance to nearest civilization	10^{7}ly	15 ly

be explorers, bridgebuilders, and businesspeople rather than artists, athletes, or daydreamers? Is the stereotyped aggressive Westerner more representative of the essence of humanity than the stereotyped contemplative Easterner? May not our aggressive technocracy be just one type of *cultural* activity rather than a universally achieved stage of *biological* evolution? Historically, patterns we once assumed to be biological have turned out to be merely cultural. (Confusion of these two has led to the racist and sexist biases we are still trying to overcome.)

If humanity is not predestined to develop a technological civilization, how much less certain is the course of social development on other worlds. It is absurdly anthropocentric to suppose that beings on other planets would resemble us physically, psychologically, or socially. Consider again the variety of highly evolved life forms on our planet alone. Ants live in ordered societies that do not appear to regard individual survival as important. Dolphins communicate and have brains that seem almost comparable to ours, but they have no manipulative organs and hence no technology. Perhaps we cannot expect aliens to be motivated by emotions that mean much to us. We certainly cannot expect, as always happens in grade C science fiction movies, that humanlike aliens will walk out of saucers and invite us to join their democratically constituted United Planets, a galactic organization structured by documents that are curiously reminiscent of the U.S. Constitution. So why assume that other civilizations would even try to visit us?

This brings us to a fifth and perhaps most significant answer: We may be farther from aliens in evolutionary time than in physical space. Even if another planet started evolving at exactly the same time as ours, and even if its biochemistry produced creatures like us, those creatures are not likely to be in a phase of evolution similar to ours. If the evolutionary "clocks" on the two planets got only 1% out of synchronization, they would be 40 million years ahead of us or behind us—as far from us evolutionarily as we are from early mammals. Thus, even in the unlikely event that other planets produce civilizations we can recognize, contact might have to come in a very narrow time interval in order for us to see any recognizable common interests.

Evolution may pass through a brief **explorative interval** during which societies on one planet try to reach other planets; beyond that stage communication or space exploration might be no more attractive than a national program on our part to communicate with chimpanzees, ants, or dolphins. To be sure, a few of our scholars try this, but they "contact" an infinitesimal fraction of these lower creatures. By the same token, our solar system might be ignored by advanced aliens. Aliens a million years ahead of us might be no more interested in us than we are in ants.

How long might an explorative interval last? We have used tools for about two million years, and it appears certain that we will progress far beyond current technology in another million years, if we survive. Our explorative interval might be a few million years, then, or less than 0.1% of the history of the planet. Thus, if the last factor in Table 26-2 is interpreted as "time during which intelligence is recognizable to us and interested in communicating," the optimistic, upper limit would have to be revised downward to three stars in 100 000, about the figure cited in the Soviet–American conference of 1971. If civilizations are at least 300 ly apart, then interstellar voyages and messages would take around half a millennium. There might be little incentive for the effort. Any spaceships that did arrive on the earth could well arrive many thousands or millions of years apart.

This brings us to a sixth possible answer to the question of alien visits: They may have happened in the remote past, as symbolized in Figure 26-11. This is the "ancient astronaut" hypothesis, popularized in several pseudoscientific books (see Enrichment Essay A). There is no good evidence for this possibility. The Soviet and American collaborators I. S. Shklovskii and Carl Sagan (1966) surveyed archaeological and mythological literature even before the hypothesis was a popular fad and found no compelling evidence for ancient astronauts. Neither the earth, the moon, Mars, nor Venus is littered with ancient alien artifacts, and not a single mysterious artifact has been advanced as physical evidence of ancient astronauts. Recent popular books about ancient astronauts misrepresented some ancient artifacts belonging to known terrestrial cultures; such pseudoscientific books have been resoundingly debunked (Story, 1976).

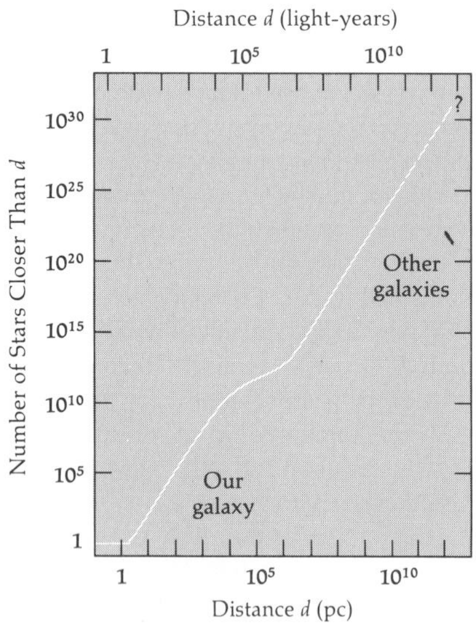

Figure 26-10. *Distance required to encounter the number of stars indicated at left. If one star in 10^9 has life-bearing planets, the nearest one might be within a few thousand parsecs.*

Figure 26-11. *Even if 10 000 alien expeditions have visited the earth at random times during the earth's history, visits would have averaged hundreds of thousands of years apart. Even the most recent expedition in such a scenario would therefore be unlikely to have left any readily recoverable historic or archaeological traces. (Painting by Jim Nichols.)*

RADIO COMMUNICATION

Possibly our period of isolation may be nearing an end. For half a century we have been broadcasting radio communications among ourselves. Already our unintentional, but weak alert is more than 60 ly out from Earth. Aliens may one day pick up our signals and send radio messages (or an expedition?) in return. Radio astronomers in both the United States and the Soviet Union are therefore still conducting modest programs to listen for such messages with large radio telescopes, such as shown in Figure 26-12. In one such search, for example, radio astronomers listened for broadcasts near 1420 MHz (the 21-cm wavelength that marks an astronomically important radiation from neutral hydrogen atoms in space), targeting on all solar-type stars (185 in all) within 25 pc and known to be members of multiple star systems. No artificial broadcasts were identified (Horowitz, 1978). Goldsmith and Owen (1979) summarize other searches involving more than 600 stars. All were negative. But in terms of total listening time and the total number of possible wavelengths to tune in, the search has hardly begun. Larger listening instruments have been proposed in both the United States and the Soviet Union.

A message from 1000 ly away would come from a civilization 1000 years in the past, and no answers to our questions could come back for 2000 years. Such communication would be unlike dialogue but, as physicist Philip Morrison has pointed out, more like our receipt of "messages" (books, letters, plays, and art) from ancient civilizations such as Greece.

Meanwhile, we have sent a few messages of our own. Several space vehicles, such as Pioneer 10 (which flew by Jupiter and left the solar system in 1973), carry plaques designed to convey our appearance and location to possible alien discoverers of the derelict spacecraft sometime in the future. The first radio message was a test message sent from the large radio telescope at Arecibo, Puerto Rico, in 1974, beamed toward globular star cluster M 13, which is 27 000 ly away.[4]

[4]Astronomers in these projects have been deluged with letters ranging from support to complaints about the nudity of the human figures on the Pioneer 10 plaque. The radio astronomers promptly received a telegram: "Message received. Help is on the way—M 13." Its authenticity might be questioned, since the round-trip message time to M 13 is 54 000 y!

Figure 26-12. *The Very Large Array, or VLA, radio telescope went into operation in 1977 in central New Mexico and is the world's largest. The system consists of 27 movable radio telescopes like that in the foreground, arranged along three 21-km-long arms. In addition to studying problems of distant stars and galaxies, this system is expected to listen for radio signals hypothetically broadcast by aliens. Note figure for scale, and note background line of telescopes. (Photo by author.)*

a

b

Figure 26-13. *Discovery of positive evidence for extraterrestrial civilizations could be a pivotal and dramatic event. It could occur by* (a) *us going out and finding it (discovery of an alien artifact on the moon, as dramatized in* 2001—A Space Odyssey, *MGM Studios) or* (b) *it arriving on the earth (arrival of mother ship, as dramatized in* Close Encounters of the Third Kind, *©1977 Columbia Pictures Industries, Inc.).*

Summary

Discovery of firm evidence for alien civilizations or even alien life forms could be a pivotal development in our view of ourselves, as suggested in Figure 26-13. Yet questions of exobiology, especially of intelligent life on other worlds, leave us with a mystery. Experimental evidence indicates rather strongly that life should start on other planets if liquid water, energy, and the right chemicals are present. Astronomical evidence suggests, but does not prove, that habitable planets ought to exist elsewhere in the universe. Biological evidence shows that life is adaptable and species can evolve to fit different environments, from ocean depths to low-pressure atmospheres of different composition. Yet advanced life has not evolved to survive on Mars.

While the limited evidence indicates that other life forms should exist, there is no evidence that they do or that they have tried to communicate with us. We can only speculate about the reasons. Perhaps they are too far away. Perhaps most civilizations destroy themselves before successfully expanding into the universe. Perhaps evolution carries them beyond a stage where they would care to communicate with us. Perhaps they are unrecognizable.

Arthur C. Clarke has remarked that any technology much advanced beyond your own looks like magic. Perhaps we are too limited by our own concept of civilization. After all, one creature's civilization may be another's chaos, as shown by Mohandas Gandhi's remark when asked what he thought of Western civilization: He said it wouldn't be such a bad idea. It seems likely that our first contact with aliens (if they exist) might be as incomprehensible as the dramatized contact in the closing segment of Clarke's novel and the Kubrick–Clarke film, *2001*.

Clearly we have been reduced to speculation by a lack of facts. Indeed, the whole field of exobiology has been criticized as a science without any subject matter. Exobiology recalls Mark Twain's comment "There is something fascinating about science. One gets such wholesale returns of conjecture from such a trifling investment of fact." The only way to reduce the conjecture and increase the proportion of fact is to pursue research in many related fields—physics, chemistry, geology, meteorology, and biology—and listen to the skies with radio telescopes. There may be surprises waiting out there.

Concepts

organic molecule
life
cell
organic chemistry
amino acid
protein
Miller experiment
proteinoid
coacervate
stromatolite
ozone layer
natural selection
evidence for planets near other stars
exobiology
cultural hurdle
life in the solar system
Drake equation
life outside the solar system
explorative interval

Problems

1. Compare the probability of detecting radio broadcasts from intelligent creatures on Alpha Centauri, Barnard's star, and Sirius.

2. Describe several ways in which the earth's internal evolution has affected the evolution of life.

3. Describe several ways in which extraterrestrial events, such as solar or stellar evolution, might have affected the evolution of life on the earth or other planets.

4. Describe ways technology could affect the survival of intelligent life on the earth or other planets. Construct scenarios of (a) the possible destruction of life and (b) the assured survival of life. (*Hint:* In case [b], consider the impact of space travel.)

5. In your own opinion, what would be the long-range consequences of the following:

a. Arrival of an alien spacecraft and visitors at a prominent place, such as the UN Building

b. Discovery of radio signals arriving from a planet about 10 ly distant and asking for two-way communication

c. Proof (by some unspecified means) that life existed *nowhere else* in the observable universe

6. Given the assumption that our technology has the potential for creating planetwide changes in environment, defend the proposition that *if* life on other worlds produces technologies like ours, then that life is likely either to have become extinct or to be widely dispersed among many planets by means of space travel.

7. In view of the devastation wrought on many terrestrial cultures by contact with more technologically advanced cultures, which would you say is the safest course: (1) aggressive broadcasting of radio signals to show where we are in hopes of attracting friendly contacts; (2) careful listening with large radio receivers to see if there are any signs of intelligent life in space; or (3) neither broadcasting nor listening, but just waiting to see what happens?

a. How would the results of our listening program be affected if other intelligent species had reached the first, second, or third conclusion?

b. If we listen at many frequencies and pick up no artificial signals, does that prove that life has not evolved elsewhere in the universe?

8. Why would native life be unlikely on a planet associated with

a. an O or B star?
b. a red giant star?
c. a white dwarf?
d. a pulsar?

9. For each known extraterrestrial planet in the solar system, list several environmental factors believed to be adverse to the existence of advanced life forms.

10. According to present astronomical theory, the sun should eventually use all its hydrogen and turn into a red giant.

a. When will this happen (see Chapter 13)?
b. How does this compare with the time scale of biological evolution?
c. Would you expect the human species to be recognizable by the time this happens?
d. In what way might life that originated on the earth conceivably escape such an event?

ADVANCED PROBLEMS

11. If an earth-sized planet (diameter 12 000 km) was circling a star 1 pc away (3×10^{16} m),

a. what would be its angular diameter when seen from the earth?
b. Could this be resolved by existing telescopes?
c. If the planet orbited 1 AU (1.5×10^{11} m) from its star, what would be its maximum angular separation from the star as seen from the earth?
d. Could this angle be resolved?

12. Aside from its distance, why is the globular cluster M 13 a poor choice of target if we are really trying to send a message to inform carbon-based, intelligent organisms, like ourselves, of our existence?

a. Using the distance listed for M 13 in Table 20-1, confirm the round-trip travel time for radio waves given in the footnote on page 485.

13. If an alien spacecraft happened to pass through the solar system at 99% the speed of light, how long would it take to traverse the system (assumed to have the diameter of Pluto's orbit)?

a. If it was broadcasting on the 21-cm radio wavelength of hydrogen, describe how its radio transmissions would be received by us, taking into account Doppler shifts.
b. Why would detection be difficult?

THE COSMIC PERSPECTIVE

Just as a vacation in an interesting place can help us view our own workaday cares from a clearer perspective, our astronomical explorations to the ends of the known universe help us view our earthly cares from a clearer, cosmic perspective. In this epilogue I will give some personal thoughts about how our cosmic journey reflects on our daily lives.

The cosmic perspective has had a deep impact on humanity during the last few centuries. Astronomical experience has outmoded once popular ideas, such as:

1. The idea that there are physical heavens and hells directly above and below the earth.

2. The idea that the earth is the universe's center, around which all other bodies revolve.

3. The idea that humans are the inevitable lords of creation.

4. The idea that the heavens are unchanging, or that they are constant except for occasional sudden creation of new features.

5. The idea that the earth is the sole and inexhaustible reservoir of raw materials.

What has the cosmic perspective given us in exchange for these cherished ideas? Space extends without known limit around us—a new frontier. Planets circle suns; suns circle mysterious galactic cores. Humans are a special part of a great chain of life, in which many lifeforms have been linked. We don't know if this chain of life is strong and stretches among many planets in our galaxy, or if it is weak and exists only in very rare, scattered segments. The heavens evolve constantly. Humans have watched new starlike objects appear and old stars explode in month-long flashes that blast out clouds that last for millennia. Even the earth itself wobbles so that Polaris will be the north star for only a few centuries. While we are exhausting the earth's finite resource-rich surface layer, we may be able to utilize cosmic sources of energy and material with less damage to the earth than is caused by our present resource-seeking and resource-consuming activities.

Some critics have said that science has only demeaned humans by putting us on a satellite of an average star on one side of an average galaxy. Certainly these ideas shocked people during the Renaissance. But as astronomer Harlow Shapley asked:

Are we debased by the greater speed of the sparrow, the larger size of the hippopotamus, the keener hearing of the dog, the finer detectors of odor possessed by insects? We can easily get adjusted to all of these. . . . We should also take the stars in stride. We should adjust ourselves to the cosmic facts.

Science has also been criticized for replacing outmoded ideas only with machines, which have brought us the evils of pollution, nuclear weapons, and a dehumanized existence. This criticism seems to me to confuse science (knowing) with technology (using knowledge and material). There is a difference between knowing and using. To me there is something appealing about knowing as much as possible and using as little as necessary. The popular criticism of "science and technology" (lumped together as if one) is really a criticism of how we have *used* knowledge and materials.

And it is justified. There is great irony in the fact that in spite of 10 000 years of effort to produce labor-saving devices, many people have trouble getting by on 40-hour-per-week jobs where they labor to make products that do not even interest them and that they themselves may label as "consumer gimmicks." This is a failure of our overall culture, rather than a failure of science or technology. The cosmic perspective is making this clearer, as more people begin to think in terms of our total planetary budget of materials, energy, and creative talent. The cosmic perspective helps us see that the political and social difficulties of maintaining a stable, workable civilization are at least as challenging as the problems of scientific exploration.

Besides giving us a new view of our place in the universe and our planetary society, the cosmic perspective gives us facts on which to base our actions. Facts, of course, are merely conclusions drawn from observations, usually verified by generations of observers. They provide a foundation for practical life in the real world. They grade indistinguishably into hypotheses about how the universe acts and how it is put together.

If a philosopher–critic argues that we can never really know facts, a scientist can agree and argue that acting on hypotheses is like betting. If you are forced to put your money on one thing or another, you try to put it on what is most likely to be successful. The world forces us to act, and we want to act on the basis of the ideas that are most likely to be true. So we *act* on our understanding of physical laws when we design bridges, and we *act* on our understanding of the solar system when we choose not to waste time or money on astrological advice. The scientific method of learning about the universe is simply a scheme for accumulating hypotheses on which to act until better information comes along. Hypotheses are useful because they suggest possibilities of new discoveries, new principles, and new places to go. They suggest goals, tests, and experiments.

This brings up the attempts, mentioned in Chapter 12, to legislate a retreat from scientific debate about our cosmic environment. According to some critics of science, the last century of scientific exploration has demeaned human and religious values, and society would be better served if the theories resulting from scientific research could be suppressed or force-fitted into one ideological mold or another. Key among the theories criticized is the theory of biological evolution. For some decades in the Soviet Union, for example, party leaders insisted on a Marxist version of the evolutionary process, and this ideological intrusion stunted Soviet biological research for years.

In the United States, lawsuits and legislation in several states have called for bans on public school teaching of evolution, or "equal time" presentations of some alternative hypothesis, usually a "creationist" hypothesis involving the sudden, one-time creation of the universe, the earth, geologic features, and/or species. The proposed creationist hypothesis is generally derived from fundamentalist (literal) interpretations of the Judeo–Christian Bible, raising the question of why a science teacher (if he or she *must* teach a tradition outside the scientific method) should choose this particular creation tradition over the creation traditions of other cultures, such as Hindu, Buddhist, or native American Indian. In 1981 the legal attack on the theory of evolution shifted to the grounds that the theory of evolution itself is more like a religious faith than a scientific theory (for example, see *Science 211:* 1331; 20 March 1981). Nonetheless, "equal-time" creationist laws have been struck down by U.S. courts on the constitutional grounds that the state is not to be involved in teaching religious traditions.

These trends leave unclear whether only the theory of biological evolution of species will come under legal attack, or whether the attack will one day spread to concepts of stellar evolution, planet growth, and the evolution of the earth's atmospheric and geologic environment, since some creationist traditions propose instantaneous or seven-day formation of all present features in the physical universe. Many backers of the current "creationist" legal cases reportedly believe the world was created only 10 000 years ago.

One can sense that the intellectual thrust of this movement is to abandon our two-way dialogue with nature in favor of a one-way dialogue in which some set of ancient writings is chosen not only as a guide to ethical conduct but also the literal guide to the nature of the physical universe. To a scientist, this seems a dangerous trend. The popular press has presented it as an argument of "science versus religion," but I fail to see it in quite such an adversary way. It seems to me that real religious awe and scientific curiosity are very similar; both involve the recognition of something "bigger than we are," about which we don't have final answers. Perhaps the "something" can best be expressed as what the American naturalist Thoreau called "this vast, savage, howling mother of ours, Nature." Our philosophical ideas about life ought to mature in a way consistent with what we know about nature. When we ask the right questions, nature suggests answers that we can act on. To turn away from this dialogue with the universe and insist that science courses include someone's interpretation of old writings seems both an antiintellectual and antireligious abandonment of the talents we've been given.

Earlier chapters showed examples of civilization getting into trouble in this way before. For example, Chapter 6 described how medieval scholars accepted the views of Aristotle and Ptolemy on the earth's place among the planets, arresting and even executing dissenters who later turned out to be correct. The Renaissance came as scholars began to debate these questions not with old manuscripts but with *new observational evidence.*

We must remember that science and science courses involve the discussion of *evidence*—that is, nature. Any hypothesis—about evolution of species, evolution of stars, the sudden appearance of humanity in the Garden of Eden, the sudden creation of everything in modern form 10 000 years ago, the big bang, relativity—can be suggested and discussed scientifically as long as the discussion centers around observation, experiment, calculation, and prediction. It is reasonable that the theories most discussed should be the ones that have emerged as consensus views from decades of scientific debate. The answer to the current attack on science is not, one hopes, to legislate that favorite theories must be taught, but rather to continue the most open, rich, vigorous discussions of hypotheses as tested against evidence. Out of such a free discussion of evidence, one is free to shape one's philosophical beliefs—a practical cosmic perspective. (See Enrichment Essay A for further thoughts on separating science and non-science.)

Does the cosmic perspective have further bearing on economic, political, social, or philosophical problems of our civilization? Each person brings his or her own experience into answering such a question, but here are some loosely connected thoughts that might be considered:

1. A classic science fiction plot deals with a vast interstellar spaceship, kilometers long, in which generations of men and women live and die, maintaining a stable environment and culture while en route from one part of the galaxy to another. This plot turns out to be the truth. We have only recently realized that we are on board just such a vehicle—Spaceship Earth—with its finite area and finite resources.

2. Places in the universe must be very rare where we can walk naked away from our machines, breathe in the atmosphere, let the light of the nearest star fall on us, and find water to drink on the surface. In other words, the earth may be a Hawaii in a universe of Siberias. While we can take technology with us to adapt to new places (as we have adapted to the extremes of the earth), it would be a good idea to take care of the one known place in the universe where we can do these things. Astronaut William Anders summed this up while describing his view during a trip around the moon: The earth "was the only color in the universe ... very delicate indeed ... It reminded me of a Christmas-tree ornament—colorful and fragile—something that we needed to learn to handle with care."

3. Twentieth-century global society is not living within its planetary means. Certain resources, such as metals, coal, and natural gas, are being consumed at rates hundreds of thousands of times faster than their replacement rates. In the past, frontiers have swept west from China, east and west from India, out from the Mediterranean, and around the world from Europe. Humanity has lived by eating its frontier. But the earth is spherical and the frontier has closed in on itself. We are now groping our way through the required period of adjustment, which may or may not be pleasant depending on our ability to sense the proper direction and make the necessary changes. Throughout the last decade, projections of mineral resource consumption, even assuming 5 times the present known supply on Earth, predict exhaustion of most metals and other resources in the next century (2000–2100), if present consumption growth patterns continue. Therefore, global social changes are in the wind.

4. It seems technologically feasible to engineer a stable society, but it also seems possible to destroy civilization and all life on the earth either by design or by accident.

5. Pressures on humanity and the earth's environment will increase the longer we draw on the earth's finite resources. Even an era of conservation only stretches the time available unless new materials, energy, or ideas to solve the underlying problem of consumption in a finite world are developed. Pursuit of human capabilities in space therefore may offer a way out of these problems. A gradual shift of heavy industrial activity into space, where solar energy could be used instead of fossil fuels, could help reverse the polluting trends on the earth today by reducing the mining of fossil fuels and some metals, reducing the emission of pollutants into the atmosphere, and fostering an ideal of earth stewardship. Resulting pollution of the space environment would be minimal, both because of its vastness (the volume of the solar system is about 100 billion billion times that of our atmosphere) and because gases and dust are flushed out of the solar system by the solar wind. Space exploration may be a rare area of human activity that is challenging, exciting, and compatible both with environmental concerns and with human economic growth.

6. We can describe with some accuracy the arrangement and history of the planetary and galactic systems in which we find ourselves, but we do not really understand the arrangement or history of the largest-scale clusters of galaxies. The most fundamental structure of the universe is thus still a mystery.

7. We may be prohibited, by the Hubble recession and the finite speed of light, from ever obtaining information about places beyond about 10 000 Mpc.

8. To some philosophers, the only reality is the reality directly witnessed by human observers. Yet three centuries of repeated, reliable observations give us the right to talk about real events that happened over a billion years ago and real places where no person has ever walked. Time and distance can be objectively probed, giving results that remain consistent (within our limits of measurement), regardless of who does the measuring. Certain places on Mars have looked the same whether photographed in 1971 by an American spacecraft or in 1973 by a Russian spacecraft. Thus, it is reasonable to suppose that there are real places where things happen, even though there are no humans to see them. In a way, it is astonishing to think that rocks roll down hills on the far side of the moon when no one is there; that winds churn dust across empty Martian landscapes just as black winds of water churn the bottoms of our seas; that gas heaves upward in stars; that starquakes crack crusts of unseen neutron stars; that unseen cataclysms play themselves out silently in space in front of no audiences; and that waves crashed on the earth's beaches for billions of years before there were land animals to see them.

9. No place in the universe is permanent. The seemingly changeless panorama of stars in the sky will alter in a few thousand years as the stars move with respect to the sun. The patterns of nebulae will change as the gas clouds are torn and stirred during their 250-million-year circuits around the galactic center, just as the mountains and continents of the earth are torn and stirred by plate tectonic processes in similar periods of time. Probably the sun will run out of hydrogen within the next 6 to 10 billion years, expanding to the red giant state, engulfing the earth, and finally collapsing to a dwarf. New stars will form near other stars, and perhaps other earth-like planets will form around some of these stars.

10. As the physical universe changes, the

biological universe also changes. In relatively brief periods of millions of years, new creatures may evolve, just as humanity itself appeared only in the last few million years. A philosophy that sees mankind as a permanent fixture in an unchanging environment, while it may be practical for short periods such as human lifetimes or centuries, cannot be defended on a cosmic scale.

11. In spite of our seeming sophistication and ability to modify our local surroundings, the universe remains implacable. We humans live on a small planet in the midst of forces and processes, some far greater than any we can create or modify. Known life and intelligence are adaptable and ingenious, but their present ability to affect current events in the universe is limited. Beyond our mortal processes, there are greater processes of death and birth that we have to accept as part of nature. Whole worlds are demolished and whole worlds are created.

The naturalist Thoreau commented with regret that most modern people have forgotten that the Greek word for the universe, **κοσμος** (cosmos), meant beauty or order. We need a sense of beauty and majesty that not only responds to a pleasant sunset, but also can encompass that more permanent cosmic sunset: a supernova. Such chaotic explosions are also part of the *order of things* in the cosmos.

The question of creators is a mystery—exciting, puzzling, beyond our abilities to explain. Here astronomy establishes a point of contact with both old and new religious perceptions. Beyond this point, astronomers are no better equipped than anyone else to make definite statements.

12. Many mysteries remain, and much space is left still to explore. Such explorations would seem to be an endeavor worthy of human energy—worthier, for example, than warfare. It is good to admit that there is still mystery in the universe. There is more to life than what we see on this particular planet at this particular time.

PART H

Enrichment Essays

We have loved the stars too fondly to be fearful of the night.

Inscription on the crypt of John and Phoebe Brashear, Allegheny Observatory

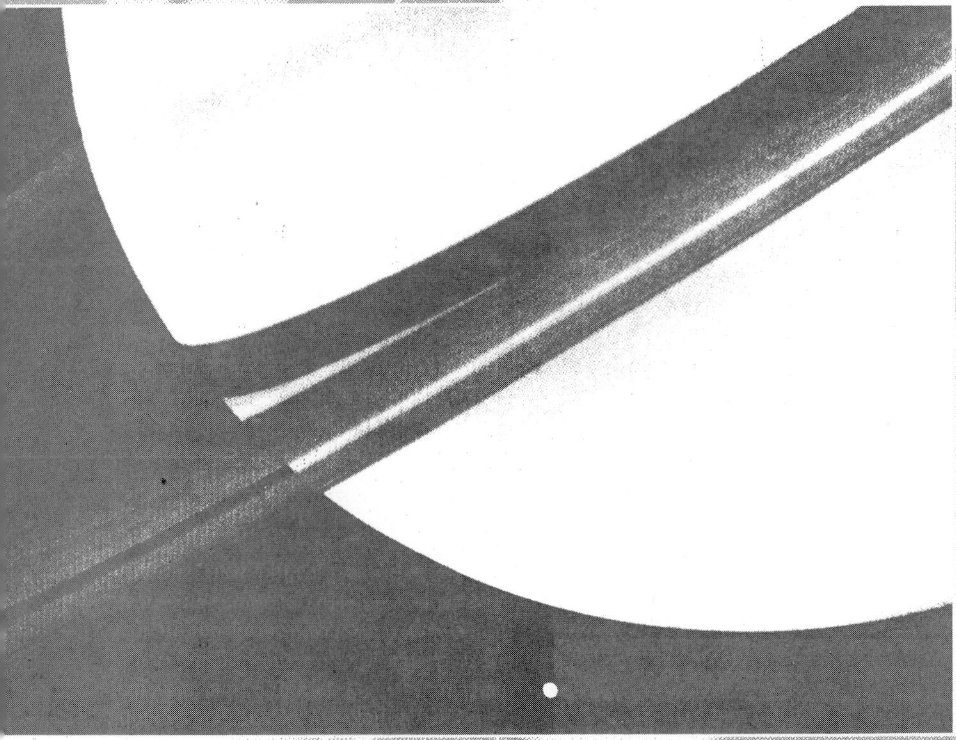

E S S A Y A

PSEUDOSCIENCE AND NONSCIENCE

When confronted with a hypothesis that sounds farfetched, some scientists may immediately label it pseudoscience or even nonsense. This is often unfair. Any hypothesis has as much right to vie for verification as any other. The true scientific method is to test hypotheses. A good hypothesis must predict some things about nature, and if the predictions are wrong, the hypothesis must be rejected or modified. Though a hypothesis can never be proven "ultimately true," if experiments keep turning up consistent results, it is considered more and more reliable and comes to be an accepted theory or law.

Science and Pseudoscience

A **pseudoscience** is a body of hypotheses treated as if true (usually with commercial intent) but without any consistent body of supporting experimental or observational evidence. Pseudoscience may appear to be backed by the trappings of real science, such as quotations of evidence, but the evidence is often hearsay, the references are often to other poorly researched commercial books, and the work is rarely reviewed by other professional researchers before publication. Note that pseudoscience is not determined by subject matter but by its method of dealing with *evidence.*

The dangers of pseudoscience are twofold. First, its practitioners often bilk consumers of their money by falsely promising new discoveries or mystic knowledge. Second, it misrepresents real scientific discovery and often contributes to antiintellectual attitudes that exchange mysticism and magic for exploration and discovery. The exchange is a poor one because, as I hope I have shown in this book, real discoveries about the universe are as exciting as the erroneous claims of pseudoscience.

Astrology

Astrology is a classic example of pseudoscience. We can trace its roots to ancient superstitions about the sky that developed at about the same time as ancient astronomical observing. One of a class of ancient systems for attempting to predict events from patterns in nature (such as patterns in tea leaves or flocks of migrating geese), astrology tries to predict events by using the patterns made by the moving planets against the background of starry constellations. Chapter 1 traces the roots and shortcomings of astrology in more detail.

During the tortuous evolution of history, many old superstitions have been abandoned while others have survived; it is ironic and sad to realize that while most citizens of the modern Western world reject using tea leaves to plan their day, many of them pore over an asrology column in the morning paper. Old traditions (whether sensible or not) die hard. New knowledge comes slowly, but faster in our century than ever before.

Pseudoscience Based on Extraordinary Reports

Another form of pseudoscience has grown up around alleged eyewitness reports of strange events. Once again, it is perfectly permissible to hypothesize that strange events, such as an alien monster's visit to the earth, will sometimes occur. We cannot dismiss the hypothesis, but we can ask if research programs have turned up verifiable evidence. To answer this question, we need to understand the processes that lead to eyewitness reports.

PERCEPTION, CONCEPTION, AND REPORTING

One-time events in the sky are hard to confirm. It is easier to learn from repeatable experiments in the laboratory than from sporadic views of celestial phenomena, which is one reason why less occult lore accumulates around a science like hydraulics than around astronomy.

The first step in generating an eyewitness observation is *perception*—the observer's intake of external sensory stimuli. The problem is that this perception must be converted in the observer's brain into a *conception,* a step that involves subjective factors, such as the association the person may make between the object and concepts prevailing in the culture. For example, witnesses almost invariably conceive of meteors in terms of the distances of aircraft. They say, "It landed just behind the barn," when the meteor may in fact have been hundreds of miles away.

Reporting, the third step, transmits conceptions to other people. Throughout most of history, this process has been by word of mouth, with secondhand reports and hearsay blending into long-lived oral traditions, which may incorporate extraneous incidents or myth. Even today, it is hard to transmit conceptions accurately by words. There is a classic UFO anecdote about a man seeing a large orange object on the ground, with flashing red lights, rows of windows, and small people inside. Does this account make you visualize a flying saucer? The object was a school bus. This joke shows how, in the appropriate context, simple, accurate words can lead many people to the wrong conception.

The quality of reporting cannot be overemphasized because it affects the beliefs of millions of people. We must retain a healthy skepticism, because many news or entertainment media thrive on sensational stories with minimal documentation. Astronomer Carl Sagan (Sagan and Agel, 1975) recounts an example of public overacceptance. He saw a woman reading a pseudoastronomy bestseller that contained many errors and flagrantly misquoted Sagan himself. Sagan asked the woman if she knew the book contained inaccuracies. "It couldn't," she said, "because they wouldn't let him publish it if it weren't true." Not so. Many books see print simply because they *will* sell, and many bookstands carry as much fantasy as fact in their so-called nonfiction selections.

In summary, the processes of perception, conception, and reporting of rare events could in principle be a source of scientific information, but in practice these processes often produce misleading data. Reports of weird, rare events are likely to circulate and be published whether these events really occurred or not.

UFOs

UFOs are any form of unidentified flying object reported in the sky. Undeniably, there have been thousands of reports of different kinds of UFOs, ranging from alleged metal disks to amorphous glows to unusual objects detected by radar. The previous discussion shows why they are hard to study scientifically.

Most of the literature on UFOs is pseudoscience, not because of the subject, but because of the way the evidence has been treated. Most popular UFO books abound in exaggerated claims and distortions. A 1968 University of Colorado study (in which I personally participated in hopes of identifying exciting evidence of unusual phenomena) established that many of the classic UFO photos used on covers of UFO publications are fakes or photos of known natural phenomena. Nevertheless, these photos reappear, year after year, on new UFO publications.

A sociological, fadlike element in the UFO problem is shown by the waves of UFO reports following space events such as the first Sputniks and the first photos from Mars (see Figure A-1), and by UFO hoaxes that followed within weeks of the first "flying saucer" report in 1947. Waves of UFO reports have even occurred in earlier eras, when the UFOs were reported to look not like the saucers popularized in our movies but like images popular in those times. A UFO wave in the 1890s reported "airships" looking like Victorian dirigibles, with fanlike propellers; this wave came a few years after the 1886 publication of Jules Verne's *Robur the Conquerer,* in which a mad scientist terrifies the world from . . . you guessed it . . . his newfangled dirigible. One such UFO reportedly crashed into a windmill in Texas in 1897, killing its Martian occupants. A prosaucer research organization in 1973 labeled this case a hoax.

When a satellite reentered the atmosphere and broke into a group of burning pieces in the night sky over the eastern United States in 1968, many people witnessed the phenomenon. Some correctly identified it, but many others misconceived it. Of a group of 30 extensive reports collected by the Air Force, 57% said that the objects were flying in formation, implying intelligent control. About 17% conceived that the glowing objects must be attached to an unseen black object, and they reported a "cigar-shaped" or "rocket-shaped" object that did not exist. Others conceived the glowing objects as windows and reported a dark object with glowing windows. One totally erroneous report said, "It was shaped like a fat cigar. . . . It appeared to have rather square shaped windows along the side . . . the fuselage was constructed of many pieced or flat sheets . . . with a 'riveted together look'" (Condon et al., 1969).

The abundance of such misconceptions and misreporting does not, of course, prove that "flying saucers" or alien spaceships do not exist. But it does show that proof of their existence is nearly impossible to establish from the kinds of reports available. While a few unsolved UFO cases may suggest unusual atmospheric phenomena, none give good evidence for spaceships. If there really are any spaceships around, one might predict, in view of the increasing numbers of surveillance devices and tourist cameras, that good

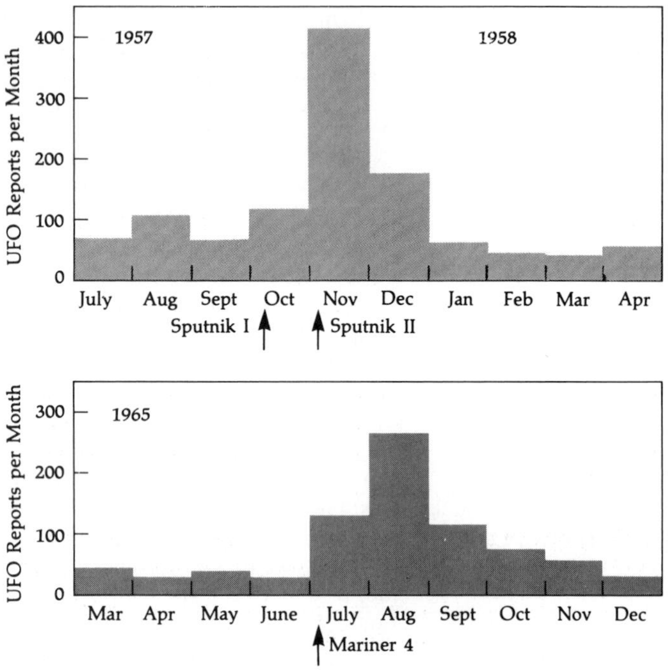

Figure A-1. *Examples of sudden increases in the number of UFO reports correlated with social factors such as the first satellite launches (top) and the first close-up photos of Mars (bottom). This correlation suggests that "waves" of sightings, discussed in many popular books, have social rather than physical causes.*

evidence should appear. So far, none has.

ANCIENT ASTRONAUTS

A similar area of pseudoscience is the literature on possible ancient astronauts. The basic technique in many of these books is to misrepresent astronomical or archaeological evidence. One best-seller, for example, claims that studies of the 1908 Siberian meteorite impact (Chapter 11) "confirmed a nuclear explosion," which is not true. The same book presents ancient lines laid out in the soil in Peru as a mystery requiring supertechnology to construct. I have visited them, and they were made by simply pushing aside an inch-thick cover of dark stones. Several scientists examined the evidence for ancient alien visitors years before these books became popular and found no convincing evidence (Shklovskii and Sagan, 1966). The ancient astronaut books have now faded from popularity after being discredited, but not until author and publisher gained considerable income from gullible readers.

ASTROCATASTROPHES

A more interesting case was a study of ancient myths by naturalist and psychiatrist Immanuel Velikovsky. He used a pseudoscientific method of interpreting all ancient documents literally. For example, where the Book of Joshua says the sun stood still (10:13), Velikovsky assumed that the earth indeed suddenly stopped turning and then started again, ignoring modern evidence that this is unlikely. Compiling his results, Velikovsky (1950) concluded that two astronomical supercatastrophes occurred around 1500 B.C. and 750 B.C., during which Venus first appeared, passed near Mars, and then passed near the earth, reaching its present orbit only after 750 B.C.

Velikovsky's work was interesting in its compilation of strange ancient myths, but when astronomers criticized his work, he was celebrated as an antiestablishment underdog and his books became best-sellers. However, his conclusions have been solidly refuted by scientists. For example, records of Venus' motions that are usually dated around 1600 B.C. and records of old eclipses show no signs of these disturbances; computer calculations that can trace planetary motions back millions of years reveal no effects suggesting the hypothesized disturbances; the lunar lava flows that Velikovsky attributed to the catastrophes date from 3 billion years ago, not 3000.

Although Velikovsky was far too literal in interpreting old records, astronomical events such as meteorite falls may indeed have influenced early cultures and myths. Scientific methods can be applied to this problem. For example, Florida geologist Cesare Emiliani and his coworkers (1975) described physical evidence of worldwide coastal flooding around 9600 B.C., which they suggest may be the basis of the worldwide flood myth.

"Scientific Creationism"

Renewed attacks on the theory of evolution raise the issue of a new kind of pseudoscience, often called *scientific creationism*. The history of this creationist movement since the 1920s is traced in some detail by Numbers (1982). It apparently grew out of a feeling of some parents that scientific evidence (usually being taught in biology classes) conflicts with the family's religious faith. This faith may be limited to the idea that humans did not evolve from other animals, but were placed on the earth in modern form. It may also involve ideas such as the creation of the earth only a few thousand years ago or the lack of evolution among *any* animal species.

In its most extreme form, "scientific creationism" strikes at the heart of science not by citing faulty evidence, but by proposing a "scientific" study in which evidence means nothing at all! Humanity (or the solar system, or whatever) was created all at once along with fossils in the ground, radioisotopes in their present distribution, and so on. In this view, the fossils that suggest evolution have no meaning—they were put there to fool us. While this is a logically consistent hypothesis in which one might choose to put one's faith, it cannot be dealt with scientifically. Does it predict the existence of any hitherto unobserved types of fossils for which we can search? Can it be tested by new observations? Does it encourage new exploration and theorizing? Evidently not.

This point can be understood by considering the theory of "fairies in my rose garden." Suppose I hypothesize that shy fairies live in my rose garden and help tend the roses, but are so clever that they always know if I put a camera or recorder out to document them; thus they avoid leaving any record. This explains why no evidence of fairies exists. This is a fine hypothesis, but it is not a part of rational science, because it has the built-in proviso that no meaningful evidence can be gathered. Once the assumption is made that fairies exist, many books could be published about where they might live, how they might dress, and so on. These books could be fine literature and brilliant exercises in logic, but they would be speculation, not science. (Some of the more enthusiastic books about life on other worlds have been criticized on these same grounds; but at least we have evidence of amino acids in meteorites, biochemical evidence about laboratory reactions, and astronomical evidence about the environments of other star systems!)

What is the main danger of these ideas? Why not simply integrate some of the creationist ideas into science courses, as the creationists have advocated in their lawsuits? The main problem is not with the hypotheses, whose merits should always be tested against natural evidence. The main problem is with the underlying premise that the answers are already known from ancient sources and that science merely studies nature to verify those answers. Historically, this has always been a terribly ineffective, if not dangerous, way to approach science and nature. Many scientists have fallen into the trap of becoming sure that their favorite theory is right and ended up (in the worst cases) trying to force data to fit incorrect conclusions. The business of science is an open, two-way dialogue with nature. If we ask the right questions, nature not only shows us the answers but may lead us on to the next question. The things that we discover may not only be interesting in themselves but also lead to practical applications. These results are less likely in a science that is conducted only to verify conclusions that are already "known" from ancient authorities. We commit a sin of pride if we presume to tell nature what the answer is, before we even attempt our observations!

Ideally, a science course should be a synthesis of evidence derived from nature all around us—from observation, experimentation, and calculation. How this evidence about the nature of the universe is then organized into a system of philosophical or religious thought is the subject of philosophy and religion courses—which are exciting enterprises in themselves, but not science. Many of the world's great thinkers have concluded that systems of ethical thought based not only on natural evidence but also on assumptions (for example, that human activities are all right as long as they don't damage other people or their property) are valuable in helping us define ethical conduct. But these systems of thought have always been grouped in the field of philosophy and religion.

Having such systems is not inconsistent with the scientific study of nature, as realized by philosophers during the Renaissance, when application of the scientific method began to change the world. Yet, in the 1980s lawsuits have attempted to force the teaching of scientific creationism under the guise of science in science courses. In the opinion of most educators, scientific creationism does not belong in a science classroom because it does not deal in a testable way with evidence. It is not part of a dialogue with nature; it is not science. It might be judged on its merits in a course on comparative religion.

How to React to Pseudoscience

When considering generations of low-income people, both in America and around the world, wasting meager resources on dubious cures, astrological advice, and superstitious cults, we may grow impatient and demand that such activities be abolished. Yet, as pointed out in Chapter 1, one of the beauties of Western democracy is that we attempt, at least, to allow freedom of speech—freedom of competing ideas—in a hope that the right ideas will emerge as the incorrect ideas are found to be flawed. We try to encourage healthy debate, and hope to find what seems closest to the truth.

Perhaps in this approach we do not pay enough attention to the fact that the human mind has an amazing ability to conceive intricate systems of thought that seem self-consistent, but in the final analysis have little to do with reality. Sometimes it is labeled fiction: Tolkien's novels of Hobbits, for example, create an intricate world of gnomes with their own language, mythology, and history—but no one claims they are reality. Sometimes we think our intricate thought-systems *may* reflect reality and label it a scientific hypothesis—but then we repeatedly test it against observation and experiment to see if we can disprove it. If so, we replace it with something *better*. Sometimes we take an intricate thought system and simply claim it *is* reality, and urge people to give their hearts to it. Sometimes these ideas are untestable; uncritical acceptance may be claimed as a virtue. It is these thought systems—pseudoscience and superstition—that threaten to drain away energy, resources, and imaginations.

What we can do with any thought system—fiction, scientific hypothesis, superstition, or some unclear mixture—is to use *evidence* to test it against experience and reality.

In the meantime, scientists and science teachers for their part must avoid being dogmatic about interpreting natural evidence, and people everywhere should cheerfully anticipate continuing refinement of scientific knowledge with new discoveries that deepen our understanding of the universe around us and our relation to it.

ASTRONOMICAL COORDINATES AND TIMEKEEPING SYSTEMS

Two kinds of astronomical systems are related to the rotation of the earth. The first is the system of astronomical coordinates, whereby astronomers map positions of objects in the sky. The second is the timekeeping system employed around the world—an outgrowth of the early timekeeping and calendar systems described in Chapter 1.

Right Ascension and Declination Coordinates

We use the earth's spin to define coordinate systems for locating objects both on the ground and in the sky. The ground system, the system of latitude and longitude, is based on the earth's spin: latitude lines are perpendicular to the axis of rotation, whereas longitude lines intersect at the poles of rotation.

A nearly identical system describes locations of stars and other celestial objects. It involves coordinates called *right ascension* and *declination* (often abbreviated "R.A." and "dec."). *Right ascension and declination are merely longitude and latitude, respectively, projected from the center of the earth through the earth's surface and onto the sky and fixed with respect to the stars.* The celestial equator (see p. 8) is the line of 0° declination, and the north celestial pole (see p. 8) is at declination 90°N, usually written +90°; the south celestial pole is at −90°.

To visualize this system, imagine the earth as a giant nonrotating glass sphere with only the longitude and latitude lines engraved on it, and imagine yourself at the center of the sphere. You look out and see the longitude and latitude lines, by which you can locate features on the surface of the earth. But these lines are also the lines of right ascension and declination, and you see them projected onto the stars. The earth's equator overlaps the celestial equator. The earth's North Pole overlaps the north celestial pole and almost obscures the star Polaris.

Figures B-1 and B-2 help in visualizing the right ascension–declination system of sky coordinates. Figure B-1 shows two views of the northern sky an hour apart as the stars wheel around the north celestial pole. Figure B-2 shows two similar views of the eastern sky as Orion rises there.

As zero right ascension, astronomers have chosen the right ascension line running through the vernal equinox: the point where the sun appears to cross the celestial equator going north. Instead of measuring the right ascension in degrees, astronomers measure it in hours, because they are interested in the time required for apparent motions of stars across the sky. For example, if a star's right ascension places it 1^h below the eastern horizon, astronomers know that it will rise in an hour. (The earth turns 15° in an hour, so 1 h = 15°.)

Stars are cataloged by their right ascension and declination. Sirius, for example, lies at R.A. 6^h41^m, dec. −17°; Polaris, which is virtually on the north celestial pole, is at R.A. 1^h23^m, dec. +89°.

The Effect of Precession on Right Ascension and Declination

Now we can more clearly describe the effects of the precession Hipparchus discovered around 130 B.C. (p. 40 ff). Because the right ascension–declination system is anchored on the earth, and because the earth wobbles with respect to the stars every 26 000 y, the system shifts among the stars. The coordinates of stars shift slightly every year, and astronomers must continually update the right ascension and declination positions listed in catalogs. Catalogs generally specify "epoch 1950" or "epoch 2000" as the year for which the positions are listed. The same shift of the celestial equator and vernal equinox invalidated the original system of astrology by shifting the right ascension–

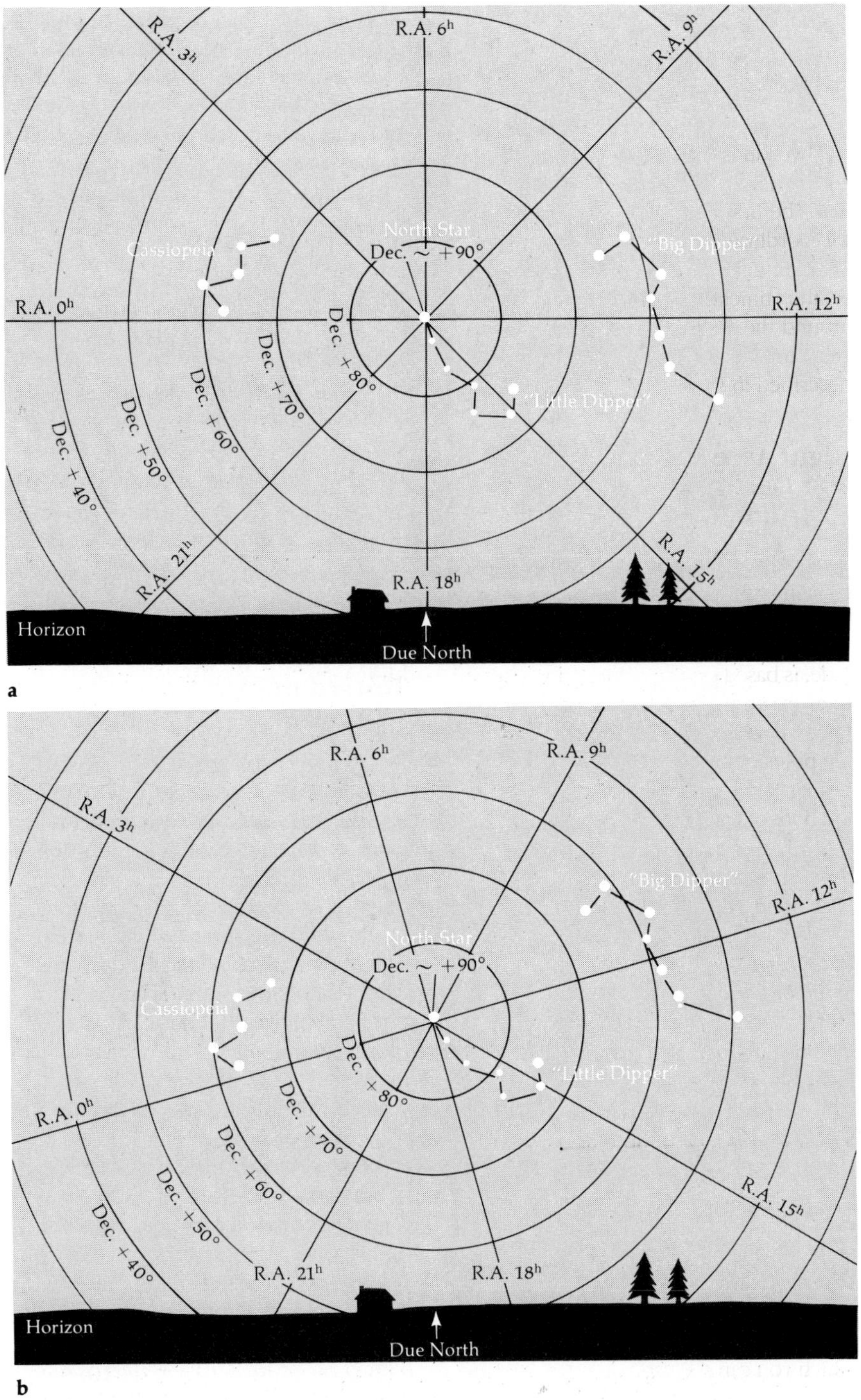

***Figure B-1. (a)** Schematic sky view looking north from the mid–United States on a February evening. Prominent constellations and the right ascension–declination system of imaginary coordinates are shown. Note that the leading star in Cassiopeia is nearly on right ascension 0^h and declination 60°. **(b)** The same view about an hour later. Note how the right ascension–declination system moves with the stars as they rise in the east and set in the west.*

anchored astrological signs away from their original constellations, as described in Chapter 1. For an example of the changing position of the north celestial pole among the stars, see Figure 2-10, pp. 40–41.

Another Celestial Coordinate System

A system useful for describing objects in the sky during both day or night is the *altitude–azimuth system. Altitude* is the number of degrees above the horizon. *Azimuth* is the number of degrees around the horizon, starting from the north point (0°) and moving through the east (90°). Since azimuth is counted from the north to the east, an azimuth of, say, 60° is often written N60°E for maximum clarity. Thus, an object on the horizon due east would have altitude 0°, azimuth N90°E. An object halfway up the sky in the west would have altitude 45°, azimuth N270°E. This system is clarified in Figure B-3, which shows two views of Orion rising in the east with altitude and azimuth lines superimposed.

An object at the *zenith* has altitude 90°. The most important azimuth line is the *meridian* (see p. 8).

Systems of Timekeeping

Before clocks were invented, time was kept by watching the apparent motion of the sun around the earth, caused by the earth's rotation. This is called *apparent solar time.* Noon was the moment when the sun crossed the meridian. A.M. *(antemeridiem)* is the period before the sun crosses the meridian; P.M. *(postmeridiem),* the period after. Once clocks were invented, it became clear that the sun's apparent motions were not exactly uniform, because the combination of the earth's rotation and orbital revolution produces days of varying length. Later, more careful measurements revealed slight, sudden speedups and slowdowns in the earth's rotation due to earthquakes, tides, and winds.

For these reasons, today's clocks run on *mean solar time,* a time system based on the uniform average rate of the sun's

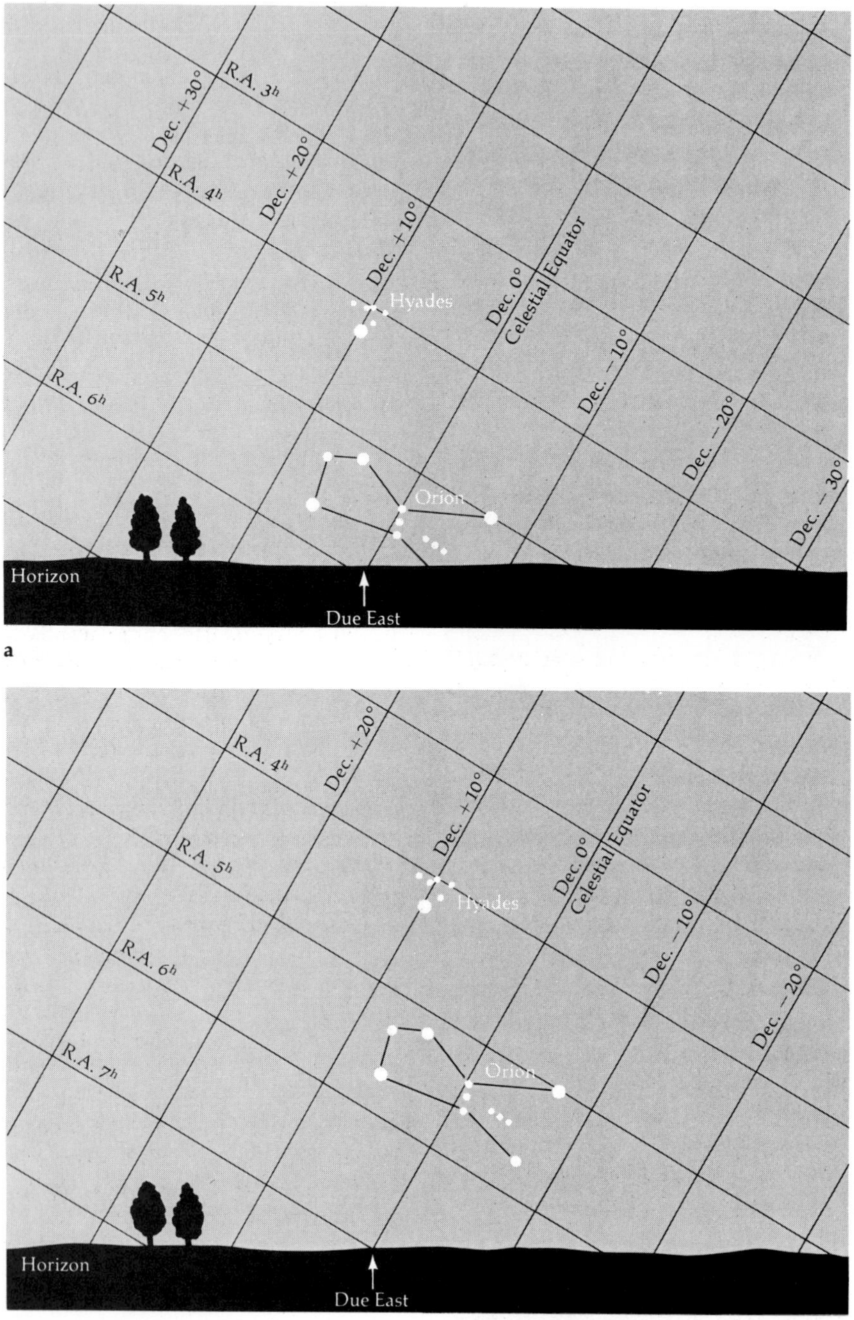

***Figure B-2. (a)** Schematic sky view looking east on a November evening. Prominent constellations and the right ascension–declination system of imaginary coordinates. Note that Orion lies on the celestial equator. **(b)** The same view about an hour later. Note how the right ascension–declination system moves with the stars as they rise in the east.*

motion rather than its actual position in the sky. The most accurate calibration of clocks for scientific work is now based on an internationally determined definition of the second based on constant atomic processes: 1s is 9 192 631 770 cycles in certain radiation from the cesium 133 atom. This standard is called *atomic time.*

For long-term prediction of planetary positions or eclipses, astronomers have defined an additional time system based on the motions of the planets, called *ephemeris time.*

Astronomers, who sight their telescopes on the stars every night, need a star-based time system rather than the sun-based system of mean solar time by which most clocks run. This system, called *sidereal time,* is simply a clock reading equal to the right ascension crossing the local meridian at any moment.

Toward the Modern Calendar

Just as timekeeping based on the earth's rotation has been continually refined, so has the calendar system based on the earth's orbital revolution. Chapter 1 described how early people developed calendar systems and estimated the length of the year. The year was often officially started on an astronomically determined date, such as the summer solstice. One problem with early calendars was accurately determining the number of days in the year. We now know that the sidereal year (the earth's motion around the sun with respect to the stars) is 365.256 ephemeris days, and the tropical year (cycle of seasons) is 365.242 ephemeris days.

If early timekeepers adopted a calendar with, say, 365 days in a year, they discovered that the astronomically determined New Year's Day was one day in error after four years. This problem was dealt with by a process called *intercalation:* Astronomers, government officials, or priests announced extra days, often celebrated as holidays, every few years to keep the calendar in step with astronomical events.

As long as society was stable, this process could be kept up for many years.

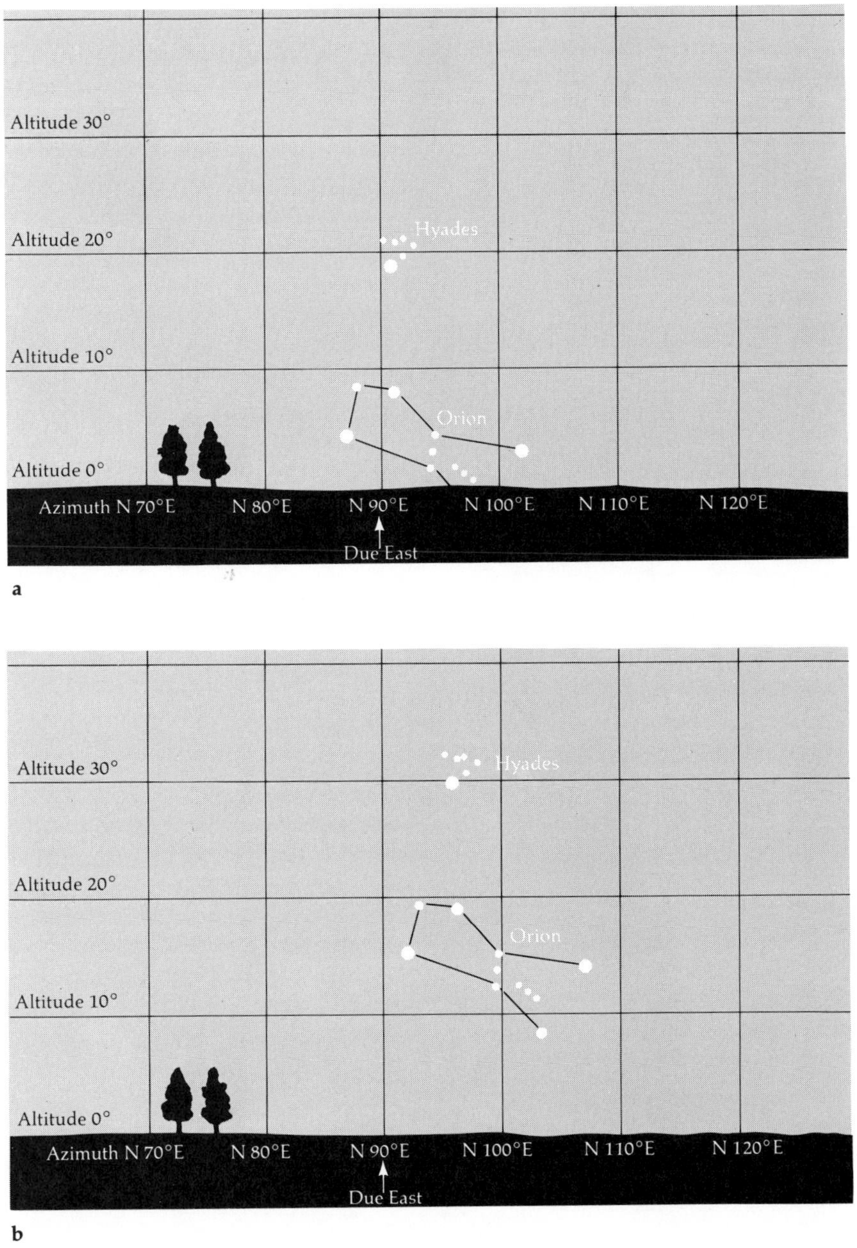

***Figure B-3. (a)** This figure is the same as Figure B-2a, but uses the altitude–azimuth system of coordinates instead of the right ascension–declination system. In this view, the trees reach an altitude of about 5°, and the Hyades are at an altitude of about 20°. **(b)** The same view about an hour later. Note that the altitude–azimuth system of coordinates is tied to the horizon and does not move with the stars. The trees still reach an altitude of about 5°, while the Hyades have risen to an altitude of 30°.*

The Romans, for example, developed the ancestor of our calendar, beginning the year at the spring equinox. (The Latin-derived names *September* through *December* represent month positions [7 through 10] in that calendar.) The extra days were intercalated at the end of the last month, February, where we still insert our extra day every leap year.

However, by 46 B.C., local officials in different parts of the empire were inserting extra days as they pleased, and the calendar became confused. Julius Caesar had the calendar revised in 44 B.C. under the Alexandrian astronomer Sosigenes, and the modern leap year scheme was formalized. The fifth month of the new *Julian* calendar (Quintilis) was renamed July in honor of Julius Caesar. August was named after Augustus Caesar, who made some additional improvements in 7 B.C.

The Julian calendar worked well for more than a thousand years, but since the year is not exactly 365¼ days long, new errors crept in. In the 1200s, European scientists noted that there was a full week discrepancy in the date of the vernal equinox. After several earlier attempts, Pope Gregory XIII commissioned a new revision of the calendar during the period from 1576 to 1603.

This Gregorian calendar was adopted quickly in Roman Catholic countries, more slowly in Protestant countries, and still more slowly in Eastern Orthodox and other countries. England and the colonies converted in 1752, dropping 11 days. (Riots and legislation against rent abuse followed because of the resulting 19-day month.) Russia converted during the 1917 revolution and had to drop 13 days. Because of the different dates of conversion to the Gregorian calendar in different countries, scholars must beware of historical records of dates, and not compare different dates in different countries without considering the calendar in effect at the time that records were made.

To clear up historical confusion about dates, astronomers have devised a system called the *Julian day count,* which designates any date in history, anywhere in the world, by the number of days since January 1, 4713 B.C. On July 4, 1976, to pick a nonrandom date, the Julian day count was 2 443 964. Computer-calculated dates of ancient eclipses and other astronomical events are expressed in this system.

Appendixes

APPENDIX 1

POWERS OF TEN

It is no accident that the word *astronomical* has become a synonym for "enormous, almost beyond conception." Astronomy is full of extraordinary numbers designating great ages and distances. In astronomy (and other sciences), therefore, a convenient shorthand system of writing numbers is favored. In this system, an exponent, or superscript, designates the number of factors of 10 that have to be multiplied together to give the desired quantity—that is, the number of zeros in the quantity. For example:

$$1 = 10^0$$
$$10 = 10^1$$
$$100 = 10^2$$
$$1\,000 = 10^3$$
$$10\,000 = 10^4, \text{ and so on.}$$

The most often used of these large numbers are

one thousand = 1 000 = 10^3
one million = 1 000 000 = 10^6
one billion[1] = 1 000 000 000 = 10^9
one trillion = 1 000 000 000 000 = 10^{12}

[1]Beware—the British use the term *billion* to refer *not* to 10^9 but to 10^{12}. This book uses *billion* to mean 10^9.

The usefulness of this system is illustrated by expressing the age of the earth

$$4\,600\,000\,000 \text{ y} = 4.6 \times 10^9 \text{ y}$$

or the distance that light travels in a year

$$6\,000\,000\,000\,000 \text{ mi} = 6 \times 10^{12} \text{ mi}$$

A similar system is used for very small numbers. Here the exponent is negative and refers to the number of decimal places. Thus:

$$0.0001 = 10^{-4}$$
$$0.001 = 10^{-3}$$
$$0.01 = 10^{-2}$$
$$0.1 = 10^{-1}$$
$$1.0 = 10^0$$

The density of gas in interstellar space, which is:

$$0.000\,000\,000\,000\,000\,000\,000\,002 \text{ g/cm}^3$$

can be more conveniently written as

$$2 \times 10^{-24} \text{ g/cm}^3$$

One of the beauties of the metric system (discussed further in Appendix 2) is that the units are related by powers of 10. For instance, a kilometer is 10^3 m (as opposed to our own confusing English system, where 1 mile is 5280 ft). To make the system easier to use, the International System of Units (known as the SI metric system) has a standardized set of prefixes, such as *kilo-,* that represents multiples of 10^{-3}, 10^3, 10^6, etc. These prefixes are shown in Table A1-1.

Table A1-1. ***Prefixes Used in the SI Metric System of Units***

Prefix	Symbol	Multiple
Tera- (TER-a)	T	10^{12}
Giga- (JIG-a)	G	10^9
Mega- (MEG-a)*	M	10^6
Kilo- (KILL-low)*	k	10^3
Hecto- (HECK-toe)	h	10^2
Deka- (DECK-a)	da	10^1
Deci- (DESS-ee)	d	10^{-1}
Centi- (SEN-tee)*	c	10^{-2}
Milli- (MILL-ee)*	m	10^{-3}
Micro- (MY-crow)*	μ	10^{-6}
Nano- (NAN-oh)	n	10^{-9}
Pico- (PEE-koh)	p	10^{-12}
Femto- (FEM-toe)	f	10^{-15}
Atto- (AT-oh)	a	10^{-18}

**Note:* Starred items are the most important in ordinary applications.

UNITS OF MEASUREMENT

Most scientists express distances and other measurements in metric units, and the world is in the process of converting to this system. This conversion will save time and money, since publications, quantities of goods, and machine parts will be measured in one uniform system. The metric system is also easier to learn and use, since it expresses units in multiples of 10 (like our money system), instead of in unpredictable multiples such as 12 inches per foot, 5280 feet per mile, and 16 ounces per pound.

Table A2-1 gives some common units in both systems. This book has emphasized the International System of metric units (abbreviated "SI" for *Système Internationale*), which uses the meter, kilogram, and second as the basic measures of length, mass, and time, respectively. When using the SI system to solve problems involving complex quantities or constants, such as energy or the gravitational constant G, one must make sure that these quantities are expressed in terms of the basic units kg, m, and s. Equivalents in the centimeter–gram–second (cgs) system and in larger units are also given in Table A2-1.

While solving certain problems, such as the advanced problems in this book, we need to use certain physical constants, such as Newton's universal constant of gravitation *G*, described in Optional Math Equation II on page 70. It is important to remember that in solving any such numerical problems, all measurement units must be converted into a single system of units, and any physical constants must also be expressed in that system. The SI metric system is recommended for this purpose, since it is the international standard. Table A2-2 gives some of the more useful physical constants expressed in the SI system.

***Table A2-1.** Units of Measurement*

English System	SI (m-kg-sec)	cgs System	Astronomical Measurements
1 inch	0.025 m	2.54 cm	—
1 foot	0.305 m	30.5 cm	—
1 yard	0.914 m	91.4 cm	—
1 mile	1609 m	1.609×10^5 cm	—
1 pound	0.454 kg	454 g	—
	1.50×10^{11} m	1.50×10^{13} cm	1 astronomical unit (AU)
	9.46×10^{15} m	9.46×10^{17} cm	1 light-year (ly)
	3.08×10^{16} m	3.08×10^{18} cm	1 parsec (pc) = 3.26 ly
	3.08×10^{19} m	3.08×10^{21} cm	1 kiloparsec (kpc)
	3.08×10^{22} m	3.08×10^{24} cm	1 megaparsec (Mpc)
	1.99×10^{30} kg	1.99×10^{33} g	1 solar mass (1 $M_\odot$)

***Table A2-2.** Some Useful Physical Constants (SI system of metric units)*

Constant	Numerical Value	Page Reference (Optional Basic Equation)
Seconds of arc in one radian	206 265	38–39 (I)
G (Newton's universal gravitational constant)	6.67×10^{-11} N · m²/kg²	70, 73 (II, III)
Wien's constant	2.90×10^{-3} m · K	125 (V)
k (Boltzmann constant)	1.38×10^{-23} J/K	200 (VI)
c (velocity of light)	3.00×10^{8} m/s	265, 437 (VII, IX)
σ (Stefan–Boltzmann constant)	5.67×10^{-8} W/m² · K⁴	266 (VIII)
Mass of a proton	1.67×10^{-27} kg	
Mass of an electron	9.11×10^{-31} kg	
Solar constant	1390 W/m²	
$L_\odot$ Solar luminosity	8×10^{26} watts = 8×10^{26} Joules/s	

SUPPLEMENTAL AIDS IN STUDYING ASTRONOMY

Astronomical Calendar. Published annually by Guy Ottewell; sponsored by Department of Physics, Furman University, Greenville, SC 29613, (803) 294-2207. $8.00 each; 5–24 copies, subtract 20%; 25–49 copies, subtract 30%; 50–499 copies, subtract 40%; 500 copies or more, subtract 50% (1981 prices including tax and postage). Star maps for each month; lists of astronomical events for each month; explanation of phenomena and various terms; astronomical glossary.

Astronomy Magazine. Published monthly; 411 East Mason St., 6th Floor, Milwaukee, WI 53202. $21/year (12 issues; 1984 price). Popular articles on astronomical phenomena and observing techniques; astronomical paintings and photographs; monthly star charts.

Comet News Service. Published quarterly with irregularly issued bulletins; McDonnell Planetarium, 5100 Clayton Rd., St. Louis, MO 63110. $4/year (4 issues, 1981 price). News of recent discoveries about comets; predicted visibilities. Mostly useful for telescope observers who want news about comet science and observations.

Griffith Observer. Published monthly; Griffith Planetarium, 2800 E. Observatory Rd., Los Angeles, CA 90027. $5/year (12 issues; 1981 price). Popular articles about astronomy.

Journal of the Association of Lunar and Planetary Observers. Published bimonthly; Box 3 AZ, University Park, NM 88003. $8.00/year membership (6 issues; 1981 price). Journal of an association of amateur astronomers devoted to reports of amateur research and observing projects concerning the moon, planets, and comets. Publishes drawings and reports by members and coordinates research projects.

Mercury Magazine (Journal of the Astronomical Society of the Pacific). Published bimonthly; Astronomical Society of the Pacific, 1290 24th Avenue, San Francisco, CA 94122. $18.00/year membership (6 issues; 1981 price). Popular articles about astronomy and the astronomical scene; news of the society, which encourages astronomy and holds occasional public functions.

Minor Planet Bulletin. Contact Prof. R. G. Hodgson, Dort College, Sioux Center, IA 51250, for information. Bulletin of the Minor Planet Section of the Association of Lunar and Planetary Observers with news of recent discoveries, predicted visibilities, and observing projects in which amateur observers can make scientific contributions. Mostly useful for advanced telescope observers.

The Planetary Report. Published bimonthly; 116 S. Euclid Ave., Pasadena, CA 91101. Journal of a society started in 1980 by Carl Sagan and others to promote popular and congressional support for space exploration programs; news of NASA mission plans, recent discoveries about the solar system; recent spacecraft photos; paintings. Membership ($20, 1984 price) includes journal subscription.

Sky and Telescope Magazine. Published monthly; 49-50-51 Bay State Rd., Cambridge, MA 02138. $18/year (12 issues; 1984 price). Popular and semitechnical articles on astronomical phenomena, history, and observing techniques; photographs; monthly star charts.

Space World. Published monthly. Box 7535, Ben Franklin Station, Washington, D.C. 20044. $30/year (12 issues, 1984 price). Magazine of a space advocacy group reporting national and international space progress.

AU: (See *astronomical unit*)

absolute: Intrinsic; not dependent on the position or distance of the observer.

absolute brightness: Any measure of the intrinsic brightness or luminosity of a celestial object.

absolute magnitude: The absolute brightness (luminosity) of a star expressed in the magnitude system. The sun's absolute magnitude is +5.

absorption: The loss of photons as light passes through a medium. A photon is lost when it strikes an electron, and the photon's energy is consumed in knocking the electron to a higher energy level.

absorption bands: A dark or dim region of the spectrum, caused by absorption of light over a moderate range of wavelengths, typically about 0.1 nm, usually by molecules or crystals.

absorption line: In a spectrum, a reduction in intensity in a narrow interval of wavelength, caused by absorption of the light by atoms between the source and the observer.

accretion disk: A disk of gas and dust, usually hot, surrounding a star. Usually used in the context of material which has been thrown off a nearby star and is falling into a companion star.

achondrite: A type of stony meteorite in which chondrules have been destroyed, probably by heating or melting.

age of the earth: The period since the earth's formation from planetesimals, estimated to be 4.5 billion years.

age of globular clusters: About 13 ± 3 billion years.

age of the (Milky Way) galaxy: Estimated to be roughly 13 ± 3 billion years.

age of open clusters: Time since formation of open clusters, judged by their H–R diagrams. Most are less than 100 million years. Ages from 1 million to a few billion years have been reported.

age of stars: Time since star formation, typically billions of years for smaller stars, but less than a million years for some massive stars in recently formed clusters. Age is difficult to measure for individual stars, but possible to measure for clusters of stars.

age of the universe: The time since the big bang, 9 to 18 billion years ago, probably close to 14 billion years ago.

airglow: Visible and infrared glow from the atmosphere produced when air molecules are excited by solar radiation.

Airy disk: In the telescopic image of a star, a small disk caused by optical effects.

Alexandrian library: The research institution and collection of ancient works preserved after the fall of Rome at Alexandria, Egypt. Alexandrian knowledge passed to the Arabs with the Arab conquest of Alexandria, and eventually back into Europe around A.D. 100–1500.

Alpha Centauri: (1) The nearest star system, composed of three members; (2) the brightest of these three.

amino acid: A complex organic molecule important in composing protein and called a "building block of life."

Andromeda galaxy: The nearest spiral galaxy comparable to our own, about 660 kpc away.

angular measure: Any measure of the size or separation of two objects as seen from a specified point, expressed in angular units (degrees, minutes of arc, or seconds of arc), but not linear units (such as kilometers, miles, or parsecs).

angular size: The angle subtended by an object at a given distance.

annular solar eclipse: An eclipse in which the light source is almost, but not quite, covered, leaving a thin ring of light at mid-eclipse.

aperture: The diameter of the light-gathering objective in a telescope.

apogee: The point in an orbit around the earth that is farthest from the earth.

Apollo asteroids: Asteroids that cross the earth's orbit.

Apollo program: The U.S. program to land humans on the moon, 1961–1972; first landing July 20, 1969.

apparent: Not intrinsic, but dependent on the position or distance of an observer.

apparent brightness: The brightness of an object as perceived by an observer at a specified location (but not measuring the object's intrinsic, or absolute, brightness).

apparent magnitude: Apparent brightness of one star relative to another as expressed in the magnitude system (see examples in Table 14-2).

apparent solar time: Time of day determined by the sun's actual position in the sky. Apparent solar noon occurs as the sun crosses the meridian. Apparent solar time is different at each different longitude.

apparition: The period of a few weeks during which a planet is most prominent or best placed for observation from earth.

association: A loosely connected grouping of young stars.

asteroid: A rocky or metallic interplanetary body (usually larger than 100 m in diameter).

asteroid belt: The grouping of asteroids orbiting between Mars and Jupiter.

asthenosphere: In a planetary body, a subsurface layer which is more plastic than adjacent layers because the combination of pressure and temperature places it near (or slightly above) the melting point. Asthenospheric movements may disrupt the planet's surface.

astrology: The superstitious belief that human lives are influenced or controlled by the positions of planets and stars; this belief is rejected by modern astronomers and other scientists.

astrometric binary: A binary star system detectable from the orbital motion of a single visible component.

astrometry: The study of positions and motions of the stars.

astronomical unit (AU): The mean distance from the earth to the sun, about 150 million kilometers.

astronomy: The study of all matter and energy in the universe.

atom: A particle of matter composed of a nucleus surrounded by orbiting electrons.

aurora: Glowing, often moving colored light forms seen near the north and south magnetic poles of the earth; caused by radiation from high-altitude air molecules excited by particles from the sun and Van Allen belts.

B-type shell star: A star of spectral type B that occasionally blows off a cloud of gas, forming a gaseous shell around the star.

Balmer alpha line: A brilliant red spectral line at wavelength 656.3 nm, caused by transition of the electron in the hydrogen atom from the third-level orbit to the second-level orbit.

barred spiral galaxy: A spiral galaxy whose spiral atoms attach to a barlike feature containing the nucleus.

basalt: A type of igneous rock, often formed in lava flows, common on the moon and terrestrial planets.

basaltic rock: Igneous rock (including basalt) with a composition resembling basalt and a relatively low content of quartz (SiO_2).

basin: Large impact crater on a planet, usually several hundred kilometers across, flooded with basaltic lava and surrounded by concentric rings of faulted cliffs.

belts: Dark cloud bands on giant planets.

big bang: The hypothetical explosion-like event that initiated the universe as we know it, probably between 10 and 18 billion years ago.

binary star system: A pair of coorbiting stars.

black dwarf (brown dwarf): A starlike object too small to achieve nuclear reactions in its center; any stellar object smaller than about 0.08 $M_{\odot}$.

black hole: An object whose surface gravity is so great that no radiation or matter can escape from it. Some black holes discussed in astronomy are collapsed stars, but much smaller ones are theoretically possible.

blue shift: A Doppler shift of spectral features toward shorter wavelengths, indicating approach of the source.

Bode's rule: A convenient memory aid for listing the planets' distances from the sun.

body tide: A tidal bulge raised in the solid body of a planet.

Bok globule: A relatively small, dense, dark cloud of interstellar gas and dust, usually silhouetted against bright clouds.

bolometric luminosity: The total energy radiated by an object at all wavelengths, usually given in joules per second (identical to watts).

Tycho Brahe (1546–1601): Danish astronomer who recorded planetary positions, ultimately enabling Kepler to deduce laws of planetary orbits.

breccia: A rock made from angular fragments cemented together.

brecciated meteorite: A meteorite formed from cemented fragments of one or more meteorite types.

bubble: A roughly spheroidal shell of interstellar gas blown outward from a star by a stellar explosion or strong stellar wind.

burster: A stellar object (neutron star?) that emits flashes of light lasting only some seconds. Many emit gamma rays, but visible light bursters are also suspected.

cgs system: Metric system of measurement using centimeters, grams, and seconds as the fundamental units.

Cambrian period: A period from 570 to 500 million years ago during which fossil-producing species of plants and animals first proliferated.

canals: Alleged straight-line markings on Mars found not to exist by spacecraft visits to Mars.

capture theory: A theory of origin of a planet–satellite or binary star system in which one body captures another body by gravity.

carbonaceous chondrite: A type of carbon-rich and volatile-rich meteorite, believed to be a nearly unaltered example of some of the earliest-formed matter in the solar system.

carbon cycle: A series of nuclear reactions in which hydrogen is converted to helium, releasing energy in stars more massive than about 1.5 $M_{\odot}$ Carbon is used as a catalyst.

Cassini's division: The most prominent gap in Saturn's rings.

catastrophic theories: Theories invoking sudden or very short (cosmically or geologically speaking) energetic events to explain observed phenomena.

catastrophism: An early scientific school which held that most features of nature formed in sudden events, or catastrophes, instead of by slow processes.

cause of eclipses: The falling of a shadow of one body onto another body.

cause of the seasons: The tilt (obliquity) of the earth's axis to its orbit plane causes first the North Pole and later the South Pole to be tipped toward the sun during the course of a year.

celestial equator: The projection of the earth's equator onto the sky.

celestial poles: The projections of the two poles of the earth's rotation onto the sky.

cell: The unit of structure in living matter.

center of mass: The imaginary point of any system or body at which all the mass could be concentrated without affecting the motion of the system as a whole; the balance point.

Cepheid variable: Any of a group of luminous variable stars with periods of 5 to 30 d (depending on their population). The periods are correlated with luminosity, allowing distance estimates out to about 3 mpc.

chain reaction theory: A theory of the cause of spiral arms in spiral galaxies. Star formation in a region leads to a chain reaction of adjacent star formation, and the region of young stars gets sheared into a spiral arm.

Chandrasekhar limit: A mass of about 1.4 $M_\odot$, the maximum for white dwarfs; stars of greater mass have too great a central pressure, causing formation of a star type denser than a white dwarf.

channel: One of the riverbedlike valleys on Mars, which are possible sites of ancient Martian rivers.

chemical reaction: Reaction between elements or compounds in which electron structures are altered; atoms may be moved from one molecule to another, but nuclei are not changed and thus no element is changed to another.

chondrite: Stony meteorite containing chondrules, believed to be little altered since their formation 4.6 billion years ago.

chondrule: BB-sized spherule in certain stony meteorites, believed among the earliest-formed solid materials in the solar system.

chromosphere: A reddish-colored layer in the solar atmosphere, just above the photosphere.

circular velocity: Velocity of an object in circular orbit.

$$V = \sqrt{\frac{GM}{R}}$$

G = Newton's gravitational constant. In SI units it is $6.67 \times 10^{-11}\ \mathrm{N \cdot m^2/kg^2}$
M = mass of central body
R = distance of orbiter from center of central body

circumpolar zone: The zone of stars, centered on a celestial pole, that never sets, as seen from a given latitude.

circumstellar nebula: Gas and dust surrounding a star.

closed space: Space that is curved in such a way that the total volume is finite.

closed universe: A theoretical model of the universe with finite volume and curved space. In most cosmological models, the universe, if closed, will eventually recollapse.

closure density: A critical density of matter in the universe, above which the universe would be closed.

cluster of galaxies: A relatively close grouping of galaxies, often with some members coorbiting or interacting with each other.

coacervate: Cell-sized, nonliving globule of proteins and complex organic molecules that forms spontaneously in a water solution.

Coalsack nebula: A prominent dark nebula about 170 kpc away silhouetted against the Milky Way.

cocoon nebula: A dust-rich nebula enclosing and obscuring a star during its formation, but later shed.

collapse: Rapid contraction, esp. of a cloud of gas and dust during star formation.

colliding galaxies: Galaxies undergoing interpenetration or close enough to cause major gravitational distortion of each other.

coma: (1) The diffuse part of the head of a comet surrounding the nucleus; (2) a type of distortion in some telescopes and optical systems.

comet: An ice-rich interplanetary body that, when heated by the sun in the inner solar system, releases gases that form a bright head and diffuse tail. (See also *coma*)

comet head: The coma and nucleus regions of a comet.

comet nucleus: The brightest starlike object near the center of a comet's head; the physical body (believed to be icy and a few kilometers across) within a comet.

comet tail: Diffuse streamers of gas and dust released from a comet and blown in the direction away from the sun by the solar wind.

common condensation theory: A theory of origin of a planet–satellite or binary star system by simultaneous growth of two coorbiting bodies.

comparative planetology: An interdisciplinary field of astronomy and geology attempting to discover and explain differences between planets in properties such as climate and interior structure.

composition of Population II stars: About ¾ hydrogen and ¼ helium, by mass, with virtually no heavier elements.

compound telescope: A telescope that combines lenses and mirrors in the light-gathering system.

condensation sequence: The sequence in which chemical compounds condense to form solid grains in a cooling, dense nebula.

conditions in the galactic center: A dense concentration of stars (mostly Population II), dust, and gas, possibly with one or more large black holes marking the nucleus.

conduction: One of three processes that transfer heat from hot to cold regions; occurs as fast-moving molecules in the hot region agitate adjacent molecules.

conjunction: The period when a planet lies at zero or minimum angular distance from the sun, as seen from the earth.

conservation of angular momentum: A useful physical rule which states that the total angular momentum in an isolated system remains constant

constellations: Imaginary patterns found among the stars, resembling animals, mythical heroes, etc.; different cultures map different constellations.

contact binary: A coorbiting pair of stars whose inner atmospheres or surfaces touch.

continental drift: The motion of continents due to (convective?) motion of underlying material in the earth's mantle.

continental shield: A stable, ancient region, usually flat and oval shaped, in a continent.

continuous creation theory: The hypothesis that matter is being created in interstellar or intergalactic space during the current era (and always).

continuum: In a spectrum with absorption or emission lines, the background continuous spectrum.

convection: One of three modes of transmission of heat (energy) from hot regions to cold regions; involves motions of masses of material.

Copernican revolution: The intellectual revolution associated with adopting Copernicus' model of the solar system, which displaced the earth from the center of the universe.

core: The densest inner region of the earth, probably of nickel–iron composition; in other planets, similar high-density central regions; in the sun or stars, a dense central region where nuclear reactions occur; in galaxies, the densest, brightest central regions.

Coriolis drift: A departure from a straight-line trajectory, perceived by an observer in a rotating system; Coriolis effects in clouds were early evidence of the earth's rotation.

corona: The outermost atmosphere of the sun, having a temperature of about 1 to 2 million Kelvin.

coronograph: An instrument permitting direct observation of the solar corona without an eclipse.

cosmic fuels: Nonfossil energy sources provided by cosmic processes; for example, solar and geothermal energy.

cosmic rays: High-energy atomic particles (85% protons) that enter the earth's atmosphere from space. Many may originate in supernovae and pulsars.

cosmogony: Any theory of the origin of the universe or one of its component systems, such as galaxies or the solar system.

cosmological principle: The assumption (unproved) that the universe is homogeneous and isotropic.

cosmological red shift: Any redward Doppler shift attributed to the mutual recession of galaxies or the expanding universe.

cosmology: The study of the structure of the universe. The term is often broadened to include the origin of the universe as well.

crust: The outermost, solid layer of a planet, with composition distinct from the mantle and differentiated by a seismic discontinuity.

cultural hurdle: The hypothetical survival requirement for a planetary culture between the time it achieves technology capable of quickly altering its planetary environment and the time it can establish viable bases off its planet; the uncertainty of the probability of survival affects our estimates of the probability of intelligent life elsewhere in space.

curved space: Space in which Euclidean solid geometry is not valid.

daughter isotope: An isotope resulting from radioactive disintegration of a parent isotope.

declination: Angular distance north or south of the celestial equator. (Abbreviation: Dec.)

degenerate matter: Matter in a very high density state in which electrons are freed from atoms and pressure is a function of density but not temperature.

degree: An angle equalling $\frac{1}{360}$ of a circle.

density–wave theory: The leading theory for explaining the formation of spiral arms in galaxies, which posits periodicities in star, dust, and gas motions.

deposition: The accumulation of eroded materials in one place.

desert: Any of the brighter regions of Mars.

differential rotation: The differences in speed for stars at different distances from the center of the galaxy. Orbital velocities are actually slower at 5000 pc from the center than at the sun's distance, which is 8000 to 10 000 pc from the center.

differentiation: Any process that tends to separate different chemicals from their original mixed state and concentrate them in different regions.

diffraction: The slight bending of light rays as they pass edges, producing spurious rays and rings in telescopic images of stars.

dimensions of the galaxy: As early as 1935 astronomers agreed that the sun is about 8000 to 10 000 pc from the center, and the overall diameter of our galaxy is about 30 000 pc.

dirty iceberg model: A theoretical description of a comet nucleus as a large icy body with bits of silicate "dirt" embedded in it.

disruption of open clusters: Gradual dispersion of stars from an open cluster as the cluster is sheared by differential galactic rotation, and as high speed stars escape. Disruption time is usually a few hundred million years.

distance limit for reliable parallaxes: 20 parsecs.

Doppler effect: The shift in wavelength of light or sound as perceived by the observer of an approaching or receding body. For speeds well below that of light, the shift is given by the equation

$$\text{Original wavelength} \times \frac{\text{radial velocity}}{\text{velocity of light}}$$

Drake equation: The statement that the fraction of stars harboring intelligent life equals the number of all stars times a sequence of fractions, such as the fraction of all stars having planets, the fraction of planets that are habitable, etc. Named after radio astronomer Frank Drake.

duration of element formation: For each element, the time interval during which it formed. Most atoms of light elements, like H, formed in the first minutes of the big bang, but heavy elements have been synthesized inside stars for billions of years.

dwarf elliptical galaxy: An ellipsodial galaxy resembling a globular cluster but usually at least a few times larger.

early intense bombardment: The very intensive bombardment of planetary bodies by meteorites, from 4.6 to about 4 billion years ago, following planet formation.

earthquake: Vibration or rolling motion of the earth's surface accompanying the fracture of underground rock.

eclipse: An event in which the shadow of one body falls on another body.

eclipsing binary: A binary star system seen virtually edge-on so that the stars eclipse each other during each revolution.

eclipsing–spectroscopic binary: An eclipsing binary whose motions are measurable from spectral Doppler shifts. The most informative type of binary star.

ecliptic: (1) The plane of the earth's orbit and its projection in the sky as seen from the earth; (2) approximately, the plane of the solar system.

effective temperature: Temperature of an object as calculated from the properties of the radiation it emits.

ejecta blanket: A layer of debris thrown out of a crater onto a planet's surface.

electromagnetic radiation: Light, radio waves, x rays, and other forms of radiation that propagate as disturbances in electric and magnetic fields, travel at the speed of light, and combine to make up the electromagnetic spectrum.

electron: One of the negatively charged particles orbiting around the atomic nucleus, with mass 9.1×10^{-31} kg.

element: A chemical material with a specified number of protons in the nucleus of each atom. Atoms with one proton are hydrogen; with two protons, helium; and so on.

ellipse: A closed, oval-shaped curve (generated by passing a plane through a cone) describing the shape of the orbit of one body around another.

emission: Release of electromagnetic radiation from matter.

emission bands: Narrow wavelength intervals in which molecules emit light.

emission lines: Very narrow wavelength intervals in which atoms emit light.

energy: In physics, a specific quality equal to work or the ability to do work. Energy may appear in many forms, including electromagnetic radiation, heat, motion, and even mass (according to the theory of relativity).

energy level: The orbit of an electron around the nucleus of an atom.

English system: A nondecimal system of units using pounds, inches, and seconds, now being replaced by the more convenient metric system.

ephemeris: A table of predicted positions of a planet, asteroid, or other celestial body.

ephemeris time: A timekeeping system based on the motions of planets, more regular than conventional systems based on the earth's rotation.

epicycle: A small circular motion superimposed on a larger circular motion.

epicycle model: An early theory by Ptolemy that the planets moved around the earth in epicycles.

equatorial zone: On Jupiter, Saturn, and possibly other great planets, a bright cloud zone near the equator.

equinox: The date when the sun passes through the earth's equatorial plane (occurs twice annually).

erg: The unit of energy in the cgs metric system.

erosion: Removal of rock and soil by any natural process.

escape velocity: The minimum speed needed to allow a projectile to move away from a planet and never return to its point of launch. It equals $\sqrt{2} \times$ circular velocity (q.v.).

Eta Carinae nebula: A nebula around a peculiar novalike variable star about 2 kpc away.

event horizon: The theoretical surface around a black hole from which matter and energy do not escape.

evidence for ongoing star formation: The fact that the solar system is much younger than the galaxy; existence of young open clusters; existence of short-lived massive stars.

evidence for planets near other stars: Although planets the size of Jupiter or less near other stars are beyond our current detection capability, new techniques of imaging and astrometry will allow their detection within a few years, if they exist. Already some objects larger than Jupiter but smaller than stars have been found.

evidence of spiral structure (in Milky Way): Spiral distribution of hydrogen gas mapped by 21-cm radio line; spiral distribution of open clusters; spiral patterns observed in other galaxies.

evolution of galaxies: Changes in form and stellar populations of galaxies as a result of consuming gas and dust during star formation and production of heavy elements during star evolution.

evolutionary theory: A theory in which changes occur by relatively slow processes or processes commonly growing out of the initial conditions, rather than by sudden or unusual processes.

evolutionary track: The sequence of points on the H–R diagram occupied by a star as it evolves.

excited atoms and molecules: Atoms or molecules in which electrons are not all in the lowest possible energy levels.

excited state: The state of an atom or molecule when not all electrons are in the lowest possible energy levels.

exobiology: Study of life beyond the earth.

expanding universe: A term popularized by Eddington to describe the mutual recession of galaxies.

explorative interval: The hypothetical interval of time during which a species actively engages in the exploration of other planets.

extragalactic standard of rest: An assumed stationary frame of reference defined by using the nearby galaxies as reference objects.

fault: A fracture along which displacement has occurred on the solid surface of a planet or other celestial body.

fields: Entities dispersed in space, but having a measurable value or magnitude that can be measured at any point in space. Examples are gravitational, electric, and magnetic fields.

fireball: An unusually bright meteor, which may yield meteorites.

fission theory: A theory of origin of a planet–satellite or binary star system by breakup of a single original body.

flare: (1) On the sun, a sudden, short-lived, localized outburst of energy, often ejecting gas at speeds exceeding 1000 km/s; (2) an outburst from certain types of variable stars, sometimes called *flare stars*.

focal length: The distance from a lens or mirror to the point where it focuses the image of a very distant object, such as the moon.

focus: One of the two interior points around which planets or stars move in an elliptical orbit.

forbidden line: Spectral line arising from a metastable state in atoms.

force: In physics, a specific phenomenon producing acceleration of mass. Forces can be generated in many ways, such as by gravity, pressure, and radiation.

formation of galaxies: Processes that led to subdivision of the universe's gas after the big bang and its collapse into individual galaxies.

free-fall: Motion under the influence of gravity only, without any other force or acceleration, such as rocket firing.

free-fall contraction: Contraction of a cloud or system of particles by gravity only, unresisted by any other force.

frequency: Number of electromagnetic oscillations per second corresponding to electromagnetic radiation of any given wavelength.

funneling effect: Concentration of all evolutionary tracks into the giant region of the H–R diagram.

galactic equator: The plane of the Milky Way galaxy projected on the sky.

galactic halo: A spherical swarm of globular clusters above and below the galactic disk, centered on a point in the direction of the constellation Sagittarius.

galactic jets: Narrow beams of extremely hot, ionized gas shooting out of some active galaxies in two opposite directions, above and below the plane of the disk. They radiate enormous amounts of energy and their origin is uncertain.

galactic latitude: Angular distance around the galactic equator from the galaxy's center.

galactic longitude: Angular distance from the galactic equator.

galactic nucleus: The center of the galaxy.

galaxies with active nuclei: Galaxies whose centers emit more energy than other, normal galaxies. They are generally strong radio sources.

galaxy: Any of the largest groupings of stars, usually of mass 10^8 to 10^{13} $M_\odot$.

Galilean satellites: The four large satellites of Jupiter, discovered by Galileo.

Galileo Galilei (1564–1642): Italian scientist who first applied the telescope to observe other planets, discovering lunar craters, Jupiter's moons, and other celestial phenomena.

geologic time scale: The sequence of events in the history of the earth.

giant elliptical galaxy: Galaxies with diffuse elliptical form, somewhat resembling globular clusters but much larger.

giant planet: (1) Jupiter, Saturn, Uranus, or Neptune; (2) any planet much more massive than the earth.

giant star: Highly luminous star larger than the sun. O and B main-sequence stars are sometimes called *blue giants*; evolved stars of extremely large radius are called *red giants*.

globular star cluster: A dense spheroidal cluster of stars, usually old, with mass of 10^4 to 10^6 $M_\odot$.

grains: Small (usually microscopic) solid particles in space.

granite: A rock type of modest density and high silica content, formed in association with differentiation processes, and therefore, found primarily on the earth.

granitic rock: A silica-rich, light-colored rock type common in the earth's continents. Granites, being low in density, tend to float to the surfaces of planets that have had extensive melting in the outer layers.

granules: Convection cells 1000 to 2000 km across, rising from the subphotospheric layers of the sun. Each granule rises at a speed of 2 to 3 km/s and lasts for a few minutes.

gravitational contraction: Slow contraction of a cloud, star, or planet due to gravity, causing heat and radiation.

gravitational red shift: A red shift caused by light emitted from a region of very high gravity, such as a neutron star or the environs of a black hole.

gravity: the force by which all masses attract all other masses.

great circle: Any circle on the surface of a sphere (especially the earth or sky) generated by a plane passing through the center of the sphere; the shortest distance between two points on a sphere.

Great Red Spot: A large, reddish, oval, semipermanent cloud formation on Jupiter.

greenhouse effect: Heating of an atmosphere by absorption of outgoing infrared radiation.

Gregorian calendar: Essentially the modern calendar system, introduced around A.D. 1600 under Pope Gregory XIII, and containing the modern system of reckoning leap years.

ground state: The lowest energy state of an atom, in which all electrons are in the lowest possible energy levels.

Gum nebula: A large, relatively nearby nebula in the southern hemisphere sky, detected in hydrogen alpha light and formed by a supernova explosion estimated to have occurred around 9000 B.C.

HI region: Interstellar region in which hydrogen is predominantly neutral.

HII region: Interstellar region in which hydrogen is predominantly ionized.

half-life: In any phenomenon, the time during which the main variable changes by half its original value: often used loosely to indicate the characteristic time scale of a phenomenon. In radioactive decay, the time for half the atoms in any system to disintegrate.

Hayashi track: A sharply descending evolutionary track in the H–R diagram covering the early period of stellar evolution from the high-luminosity phase to the main sequence.

heliacal rising: A star's first visible rising during the yearly cycle.

heliacal setting: A star's last visible setting during the yearly cycle.

helium flash: Runaway helium "burning" inside a star as it evolves off the main sequence and into the giant phase of evolution. It occurs when degenerate gas at the star's center reaches a temperature of about 10^8 K.

Helmholtz contraction: Slow contraction of a cloud or system of particles by the force of gravity, which is retarded by outward gas pressure and the limited rate at which radiation can escape.

hierarchical universe: A theoretical model of the universe organized in ever-larger clusters of galaxies, plausibly with a density that approaches zero as larger volumes are considered.

high-luminosity phase: A star's short-lived stage of maximum brightness during pre-main-sequence evolution.

high-velocity star: A star with a high velocity relative to the sun; generally associated with the galactic halo.

homogeneous: Uniform in composition throughout the volume considered.

hour angle: The number of hours since a star (or other body) last crossed the local meridian.

H–R diagram: A technique for representing stellar data by plotting spectral type (or color or temperature) against luminosity (or absolute magnitude), named after its early proponents, Hertzsprung and Russell.

Hubble's constant: The ratio between a galaxy's recession speed and its distance, measured to be about 50 to 100 km/s/mpc.

Hubble's relation: Expression indicating how the recession velocity of a galaxy increases with distance.

hydrogen alpha line: The designation of hydrogen's red spectral line at 656.3 nm, more properly called *hydrogen Balmer alpha.*

hydrogen Balmer series: The series of all hydrogen spectral lines from 364.6 to 656.3 nm, caused by electron transitions between the second and higher energy levels.

hyperbola: The orbital curve followed by any free-falling body moving faster than escape velocity.

hypothesis: A proposed explanation of an observed phenomenon or a proposal that a certain observable phenomenon occurs.

IC number: The catalog number of a cluster, nebula, or galaxy in the *Index Catalog.*

igneous rock: Rock crystallized from molten material.

impact crater: A roughly circular depression of any size (known examples range from microscopic size to diameters greater than 1000 km) caused by a meteorite impact.

incidence of multiplicity: Among stars, the fraction of systems that contain more than one star. Probably 50 to 70% of systems have companion stars, many of these having more than one companion.

inferior planet: Mercury or Venus.

inflationary big-bang theories: Models of the big bang, calling for a super-rapid expansion of space in the first fraction of a second after the big bang, due to a "change of state" of space.

infrared light: Radiation of wavelength too long to see, usually about 1 to 100 μm.

infrared star: A star detected primarily by infrared light.

initial conditions: (1) In any scientific problem, the conditions that define the beginning of the process being studied; (2) in cosmogony, the conditions at the moment of, or immediately after, the big bang.

intergalactic globular cluster: A globular star cluster or a small galaxy resembling same, found in intergalactic space.

interstellar atom: Atom of gas in interstellar space.

interstellar grains: Microscopic solid grains in interstellar space; interstellar dust.

interstellar molecule: Molecule of gas in interstellar space.

interstellar obscuration: Absorption of starlight by interstellar dust, causing distant objects to appear fainter.

interstellar reddening: Loss of blue starlight due to interstellar dust, causing distant objects to appear redder and fainter.

interstellar snowball: Hypothetical interstellar particle larger than an interstellar grain.

inverse square law: The relation describing any entity, such as radiation or gravity, that varies as $1/r^2$, where r is the distance of the entity from the source.

ions: Charged atoms or molecules.

iron meteorite: Meteorite composed of a nearly pure nickel–iron alloy.

irregular galaxy: A galaxy of amorphous shape. Most have relatively low mass (10^8–10^{10} $M_\odot$).

irregular variable: A star that fluctuates in brightness irregularly.

isotope: A form of an element with a specified number of neutrons in the nucleus. Each element may have many possible isotopic forms, but only a few are stable.

isotropic: Appearing uniform no matter what the direction of view.

joule: The unit of energy in the SI metric system of units.

Julian date: The date based on a running tabulation of days, starting January 1, 4713 B.C.

Jupiter's atmospheric composition: Mostly hydrogen, with additional helium, hydrogen-based compounds, and other gases.

Jupiter's infrared thermal radiation: Infrared radiation from Jupiter due to slow contraction of its interior and exceeding the incoming solar radiation.

Jupiter's interior: Beneath Jupiter's atmosphere, a high-pressure region of liquid hydrogen, liquid metallic hydrogen, and a small central core of rocky material.

Jupiter's temperature: Temperatures around −200°F are measured at and above the cloud tops in Jupiter's atmosphere, but the air temperature increases below the clouds to values around room temperature and even warmer.

Kelvin scale: The absolute temperature scale, with 0K = absolute zero. A Kelvin degree is the same size (some temperature difference) as a Centigrade degree.

Johannes Kepler (1571–1630): Astronomer who first deduced the shapes and relations of planets' elliptical orbits.

Kepler's laws: The three laws of planetary motion that describe how the planets move, show that the sun is the central body, and allow accurate prediction of planetary positions.

Kirchhoff's laws: Laws describing conditions that produce emission, absorption, and continuous spectra.

$L_\odot$: The luminosity of the sun, 4×10^{26} watts.

LSR: (See *local standard of rest*)

Lagrangian points: In an orbiting system with one large and one small body, an array of five points where a still smaller body would retain a fixed position with respect to the other two.

Lagrangian surface: An imaginary surface with a figure 8 cross section surrounding two coorbiting bodies in circular orbits and constraining motions of particles within the system.

lava: Molten rock on the surface of a planet.

life: A process in which complex, carbon-based materials organized in cells take in additional material from their environment, replicate molecules, reproduce, and do other weird things like writing books.

life on Mars: Although long sought and believed possible by many scientists, biological processes on Mars were apparently ruled out in 1976 when Viking landers found no organic material there.

life in the solar system: Hypothetical biological activity on other planetary bodies, such as on Mars, under Europa's ice, or in Jupiter's atmosphere. Now regarded as unlikely.

life outside the solar system: Hypothetical biological activity beyond the outskirts of our solar system, such as on hypothetical planets around other stars. Such life is regarded as plausible, even if sparsely scattered, by most astronomers.

light curve: A plot of brightness of a star (or other object) versus time.

light-gathering power: The ability of a telescope or binoculars to gather light. It is proportional to the area of the objective, that is, the square of the aperture.

light-year: The distance light travels in one year, 9.46×10^{12} km.

limb: The apparent edge of a celestial object.

linear measure: Measurement involving linear distances, as opposed to angles or angular distances.

line of nodes: A line formed by the intersection of an orbit and some other reference plane, such as the plane of the solar system.

lithosphere: The solid rocky layer in a partially molten planet.

local group: The cluster of galaxies to which the Milky Way and 26 other nearby galaxies belong.

local standard of rest (LSR): A frame of reference moving with the average velocity of the nearby stars (out to about 50 pc from the sun).

luminosity: The total energy radiated by a source per second. The luminosity of the sun ($L_\odot$) is 4×20^{26} watts.

lunar eclipse: Dimming of the moon as it passes into the earth's shadow.

$M_\odot$: The mass of the sun, 2×10^{30} kg.

M giant: A giant star of spectral class M.

Magellanic clouds: The two galaxies nearest the Milky Way, irregular in form and visible to the naked eye in the Southern Hemisphere.

Magellanic stream: Gas filaments connecting the Magellanic clouds to the Milky Way.

magma: Underground molten rock.

magma ocean: Primordial layer of molten lava on the initial surface of the moon and (by inference) planets.

magnetic braking: The slowing of rotation of a star or planet by interaction of its magnetic field with surrounding ionized material.

magnetic field: Region of space in which a compass (or other detector) would respond to magnetism of some body.

magnifying power: Apparent angular size of a telescope image divided by the angular size of the object seen by naked eye.

main-sequence star: One of the group of stars defined on the H–R diagram that have a relatively stable interior configuration and are consuming hydrogen in nuclear reactions; a star on the main sequence.

mantle: A region of intermediate density surrounding the core of planets.

mare (pl. *maria*)**:** A dark-colored region on a planet or satellite; a region of basaltic lava flow on the moon.

Mars-crossing asteroids: Asteroids whose orbits cross that of Mars.

Mars' rotation period: The day on Mars—24h 37m, only slightly longer than the earth's.

Martian air pressure: The pressure exerted by the very thin Martian air. On Mars' surface, the air pressure is only about 0.7% of that at the earth's surface.

Martian air temperature: Soil temperatures approach or exceed freezing in the day; the air is colder. Night temperatures around -123°F are recorded.

Martian atmosphere: The thin gases around Mars, composed almost entirely of carbon dioxide (CO_2).

maser (*m*icrowave *a*mplification by *s*imulated *e*mission of *r*adiation): (1) A device that amplifies microwave radio waves through special electronic transitions in atoms; (2) an interstellar cloud that acts in this way.

mass: (1) Material; (2) the amount of material.

mass/luminosity ratio: The mass per unit of light or total radiation emitted from an object such as a galaxy.

mass–luminosity relation: The relation between the mass of a main-sequence star and its total radiation rate; the more massive, the greater the luminosity.

mass of the galaxy: About 4×10^{41} kg, or $2 \times 10^{11}\ M_\odot$, as estimated from the circular velocity equation.

Maunder minimum: The interval from 1645 to 1715, when solar activity was minimal.

mean density: Mass of an object divided by its volume.

mean solar time: Time shown by conventional clocks, determined by the sun's mean rate averaged over the year.

megaparsec: One million parsecs.

meridian: (1) A north–south line on a planet, moon, or star; (2) a great circle through the celestial pole and the zenith.

Messier number: The catalog number of a nebula, star cluster, or galaxy in *Messier's Catalog.*

metallic hydrogen: A high-pressure form of hydrogen with free electrons.

metastable state: In an atom, a configuration of electrons that is relatively long-lived, but is rarely found on the earth because it is disrupted by collisions with other atoms; it may be found in interstellar atoms, creating forbidden spectral lines.

meteor: A rapidly moving luminous object visible for a few seconds in the night sky (a "shooting star").

meteorite: An interplanetary rock or metal object that strikes the ground.

meteorite impact craters: Circular depressions in planetary surfaces, caused by explosions as meteorites crash into the surfaces at high speeds.

meteoroid: A particle in space, generally smaller than a few meters across.

meteor shower: A concentrated group of meteors, seen when the earth's orbit intersects debris from a comet.

meter: 39.4 in.

Milky Way galaxy: The spiral galaxy in which we live.

Miller experiment: An experiment in which amino acids are created in laboratory conditions simulating the conditions of the early earth.

minerals: Chemical compounds, usually in the form of crystals, that constitute rocks.

minute of arc: An angle equaling 1/60 of a degree.

missing mass: Mass suspected to exist distributed through the universe (in intergalactic space?) that, when added to the currently observed mass, would bring the universe closer to closure density.

mks system: A metric system of units expressing length in meters, mass in kilograms, and time in seconds. (See also *cgs system*)

molecular cloud: An interstellar cloud of gas and dust, with greater than average density, dust content, and high concentration of molecules.

moon: (1) The earth's natural satellite; (2) any satellite.

multiple scattering: Redirection of electromagnetic radiation (such as light waves) by repeated interaction with atoms, molecules, or dust grains in space or in an atmosphere.

multiple star system: A system of three or more stars orbiting around each other.

mutual recession of galaxies: The phenomenon that all distant galaxies are moving away from us, and the farther away they are, the faster they are receding.

natural selection: The theory which states that those individuals best adapted to the ever-changing environment produce a greater number of offspring.

nebula: A cloud of dense gas and/or dust in interstellar space or surrounding a star.

negative hydrogen ions: Hydrogen atoms that have temporarily captured an extra electron, responsible for opacity in the photosphere of the sun and many stars.

neutron: One of the two major particles constituting the atomic nucleus; it has zero charge and mass 1.6749×10^{-27} kg.

neutron star: A star with a core composed mostly of neutrons, with density 10^{16} to 10^{18} kg/m^3. Many or most neutron stars are pulsars.

Isaac Newton (1642–1727): English physicist who discovered the spectrum and laws of gravitation and motion; he also developed calculus and made other discoveries. Possibly the greatest physicist of history.

Newtonian–Euclidean static cosmology: A hypothetical model of the universe with infinite volume, no expansion, and Euclidean geometry.

Newton's laws of motion: Three rules describing motion and forces. Briefly, (1) a body remains in its state of motion unless a force acts on it; (2) force equals mass times acceleration; (3) for every action there is an equal and opposite reaction.

NGC number: The catalog number of a nebula, cluster, or galaxy in the *New General Catalog*.

node: One of two points where an orbit crosses a reference plane.

noncosmological red shift: A hypothetical red shift of distant galaxies *not* caused by the Doppler effect.

non-Euclidean geometry: A hypothetical geometry in which Euclid's relations are not true; a geometry of curved space.

nonthermal radiation: Radiation *not* due to the heat of the source; e.g., synchrotron radiation.

North Star: (1) Polaris; (2) any bright star that happens to be within a few degrees of the north celestial pole during a given era.

nuclear reaction: Reaction involving the nuclei of atoms in which a nucleus changes mass.

nucleus: (1) The matter at the center of an atom, composed of protons and neutrons; (2) the bright central core (or solid body) of a comet; (3) the bright central core of a galaxy.

O association: An association of O-type stars.

objective: The major light-gathering element of a telescope; the mirror in a reflector and the lens in a refractor.

oblate spheroid: The shape assumed by a sphere deformed by rotation.

obliquity: The angle by which a planet's rotation axis is tipped to its orbit.

Occam's razor: The principle that the simplest hypothesis, with the fewest assumptions, is most likely to be correct. Named after its use to cut away false hypotheses.

ocean tide: The moon's tidal stretching of the earth, as observed in the ocean surface (as opposed to body tide).

Olbers' paradox: The problem of why the sky is dark at night if the universe is filled with stars.

Oort cloud: The swarm of comets surrounding the solar system.

opacity: The extent to which gaseous (or other) material absorbs light.

open star cluster: A grouping of relatively young Population I stars (usually 10^2–10^3 $M_\odot$), sometimes called a *galactic cluster*.

open space: Space that is uncurved or curved in such a way as to have infinite volume and no boundaries.

open universe: A universe with infinite volume and no boundaries. In most cosmological models, the universe, if open, will continue expanding forever.

opposition: The period when a superior planet lies in an opposite direction from the sun as seen from the earth. At opposition, a planet appears in the midnight sky, well placed for observation.

optical double star: A pair of stars that have small angular separation but are not coorbiting.

orbital precession: A slow, cyclic change in the orientation of the plane of an orbit.

organic chemistry: Chemistry involving organic molecules.

organic molecule: Molecule based on the carbon atom, usually large and complex, but not necessarily part of living organisms.

origin of the heavy elements: The process by which light elements fused to form nuclei of heavy elements, primarily inside massive stars.

Orion nebula: Several hundred solar masses of gas and dust composing the core of a star-forming region about 460 parsecs away. One of the most prominent nebulae in our sky, it forms the central "star" in Orion's sword.

oscillating universe: A hypothetical model of the universe with cycles of contraction and expansion.

ozone layer: An atmospheric layer rich in ozone (O_3), created by the interaction between oxygen molecules (O_2) and solar radiation. On earth, its altitude is about 20–60 km.

parabola: (1) The curved trajectory followed by a particle moving at escape velocity; (2) the curve of a Newtonian telescope's primary mirror.

parallax: An angular shift in apparent position due to an observer's motion; more specifically, a small angular shift in a star's apparent position due to the earth's motion around the sun. *Stellar parallax,* used to measure stellar distance, is defined as the angle subtended by the radius of the earth's orbit as seen from the star.

parent body: A body in which a meteorite formed and later broke off as a fragment.

parent isotope: A radioactive isotope that disintegrates and forms a daughter isotope.

parsec: A distance of 206 265 AU, 3.26 ly or 3.09×10^{13} km; defined as the distance corresponding to a parallax of 1 second of arc.

partial solar eclipse: An eclipse in which the light source is not totally obscured from an observer.

particlelike properties of light: Characteristics of light, such as concentration of energy and momentum in discrete microscopic packets (photons) that mimic the properties of particles.

Pauli exclusion principle: A principle of subatomic physics specifying that no two electrons in a very small volume have exactly the same properties of energy, motion, etc.

peculiar velocity: A star's velocity with respect to the local standard of rest.

penumbra: (1) The outer, brighter part of a shadow, from which the light source is not totally obscured; (2) the outer, lighter part of a sunspot.

perigee: The point in an orbit around the earth that is closest to the earth.

permafrost: Semipermanent underground ice.

photometry: The measurement of the amount of light, either total or in different specified colors, coming from an object.

photon: The quantum unit of light, having some properties of a wave. For each wavelength of radiation, the photon has a different energy.

photosphere: The light-emitting surface layer of the sun.

physical binary stars: Two stars orbiting around a common center of mass.

Planck's law: A formula that describes the energy associated with each wavelength (color) in the spectrum. It shows that photons of blue light are more energetic than photons of red light.

planet: A solid (or partially liquid) body orbiting around a star but too small to generate energy by nuclear reactions.

planetesimals: One of the small bodies from which planets formed, usually ranging from micrometers to kilometers in diameter.

planetology: The study of the planets' origins, evolution, and conditions.

Planet X: A term used sometimes for a hypothetical tenth planet in the solar system, beyond Pluto.

plasma: A high-temperature gas consisting entirely of ions, instead of neutral atoms or molecules. Because of the high temperature, the atoms strike each other hard enough to keep at least the outer electrons knocked off.

plates: Moving units of the earth's lithosphere, typically of continental scale.

plate tectonics: Motions of a planet's lithosphere, causing fracturing of the surface into plates. Primary example occurs on the earth.

Polaris: The North Star.

Population I: Stars with a few percent heavy elements (heavier than helium), found in the disks of spiral galaxies and in irregular galaxies.

Population II: Stars composed of nearly pure hydrogen and helium, found in the halo and center of spiral galaxies, in elliptical galaxies, and to a limited extent in irregulars.

powers of 10: The number of times 10s must be multiplied together to give a specific number; the exponent of 10. (Example: $10^2 = 100$; the power, or exponent, is 2; see also Appendix 1.)

precession: The wobble in the position of a planet's rotation axis caused by external forces. Also, the change in a coordinate system (tied to any planet) caused by such a wobble.

pre-main-sequence: Evolutionary state of stars prior to arrival on the main sequence, esp. just before the main sequence is reached.

primary atmosphere: The atmosphere of a planet (if any) just after planet formation. (See also *secondary atmosphere*)

primeval fireball: The hypothetical, expanding, initial cloud of high temperature plasma during the big bang.

principle of relativity: The principle that observers can measure only relative motions, since there is no absolute frame of reference in the universe by which to specify absolute motions.

problem of the missing mass: Motions of stars around galaxies imply that these galaxies have more mass than accounted for by the visible stars. Thus, there must be unseen mass in some uncertain form. Astronomers are still searching for this "missing" mass.

prograde rotation: Spinning on an axis from west to east, or counterclockwise as seen from the North Pole (as in the case of the earth).

prominence: A radiating gas cloud extending from the solar surface into the thinner corona.

prominent star: One of the brightest stars in our sky, but not necessarily one of the nearest.

proper motion: The angular rate of motion of a star or other object across the sky (most stars have proper motions less than a few seconds of arc per year).

protein: Any of several types of complex organic molecules made from amino acids inside plants and animals, which are essential in living organisms.

proteinoid: Cell-like, nonliving spheroid of protein molecules created in the laboratory by heating amino acids and adding water; a possible step in the evolution of life.

protogalaxy: A gravitationally stable cloud of galactic mass contracting toward galactic dimension.

proton: One of the two basic particles in an atomic nucleus, with positive charge and mass $1.6726 = 10^{-27}$ kg.

proton–proton cycle: A series of thermonuclear reactions that convert hydrogen nuclei to helium nuclei, converting a tiny amount of mass into energy.

protoplanet: A planet shortly before its final formation. Sometimes hypothesized to have a massive atmosphere and greater mass than in its present state.

protostar: A gravitationally stable cloud of stellar mass contracting in an early pre-main-sequence evolutionary state.

pseudoscience: Research that has the trappings of science but does not follow the scientific method, usually lacking review and repetition of observations by independent researchers.

Ptolemaic model: The ancient earth-centered model of the solar system, with the sun, moon, and other planets moving in epicycles.

pulsar: A rapidly rotating neutron star with a strong magnetic field, observed to emit pulses of radiation.

QSO: Quasi-stellar object; faint bluish star with a spectrum similar to that of a quasar.

quantum: A small indivisible unit of some quantity such as energy or mass.

quasar: Any of a group of starlike, faint celestial objects with very large red shifts. Many astronomers believe they are extremely distant galaxies of unusually energetic form.

r-process reactions: Rapid reactions, probably occurring inside supernovae, in which heavy elements are formed as atomic nuclei capture neutrons. (See also *s-process reactions*)

radial velocity: The velocity component along the line of sight toward or away from an observer. Recession is positive; approach is negative.

radiation: (1) Any electromagnetic waves or atomic particles that transmit energy across space; (2) one of three modes of heat (energy) transmission through stars or planets from warm regions to cool regions.

radiation pressure: An outward pressure on small particles exerted by electromagnetic radiation in a direction away from the light source.

radioactive atom: Any atom whose nucleus spontaneously disintegrates.

radio galaxy: A galaxy that emits unusually large amounts of radio radiation.

radioisotopic dating: Dating of rock or other material by measuring amounts of parent and daughter isotopes.

ray: A bright streak of material ejected from a crater on the moon or other planet.

Rayleigh scattering: Scattering of light by particles smaller than the light's wavelength. This process favors scattering of blue light.

red giant: A post-main-sequence star whose surface layers have expanded to many solar radii and have relatively low temperatures.

red shift: A Doppler shift of spectral features toward longer wavelengths, indicating recession of the source.

red shift of galaxies: The shift toward longer wavelengths in light of distant galaxies, due to their recession from the solar system. It increases with galaxies' distances.

reflected radiation: Radiation that has arrived from outside a body and bounced off, as opposed to thermal radiation.

reflector: A type of telescope using a mirror as the light collector.

refractor: A type of telescope using a lens as the light collector.

refractory element: An element least likely to be driven out of a material by heating. These elements are usually concentrated in the last components to melt when a material such as rock is heated.

regolith: A powdery soil layer on the moon and some other bodies caused by meteorite bombardment.

regression of nodes: A shifting of the R.A. = Dec. coordinate system, relative to the stars, as a result of the 26 000 year wobble of the earth's rotation axis.

relativistic: Moving at speeds near that of light.

relativity: (See *principle of relativity*)

representative stars: A sample of stars randomly drawn from the total population of stars in space.

resolution: The smallest angle that can be discerned with an optical system; for example, the eye can resolve about 2 minutes of arc.

retrograde motion: Revolution or rotation from east to west contrary to the usual motion in the solar system.

retrograde rotation: Spinning on an axis from east to west, or clockwise as seen from the North Pole (opposite to the spin direction of the earth).

rift: A major split in a planet's lithosphere due to active or incipient plate tectonic stresses.

right ascension: Longitude lines projected onto the celestial sphere. (Abbreviation: R.A.)

rille: A type of lunar valley.

Roche's limit: The distance from a large body within which tidal forces would disrupt a satellite.

rock: Solid aggregation of minerals.

rotational line broadening: Broadening of spectral lines due to rotation of the source.

rotation curve: Orbital velocity as a function of distance from the center of a galaxy.

RR Lyrae star: A type of variable star similar to the Cepheids that has been found associated with Population II and not Population I.

runaway star: A star rapidly moving away from a region of recent star formation.

Russell–Vogt theorem: The theorem stating that the equilibrium structure of a star is determined by its mass and chemical composition.

s-process reactions: Slow reactions in giant stars in which heavy elements are built up as atomic nuclei capture neutrons. (See also *r-process reactions*)

Saha equation: An equation derived in 1920 by the Indian physicist, Saha, that tells the percent of atoms in each different excited state, given the conditions in a gas. This in turn controls what spectral lines are emitted or absorbed by that gas.

saros cycle: An interval of 6585 d (about 18 y) separating cycles of similar eclipses, used by ancient people to predict eclipses.

satellite: Any small body orbiting a larger body.

Saturn's atmosphere: The thick, cloudy gases around Saturn, composed mostly of hydrogen.

Saturn's ring system: A system of innumerable icy particles orbiting Saturn.

Saturn's satellite system: A family of at least 17 moons orbiting Saturn, ranging from 20 km diameter up to a size slightly exceeding that of the planet Mercury.

science: Study of nature using the scientific method (q.v).

scientific method: The method of learning about nature from making observations, formulating hypotheses, and constructing observational or experimental tests to see if the hypotheses are accurate.

seasonal changes in Mars' features: Changes in shape and darkness of the dusky patches on Mars from summer to winter and year to year. Once thought to indicate Martian vegetation, the changes are now known to result from blowing dust deposits.

secondary atmosphere: A planet's atmosphere after modification by outgassing and other processes. (See also *primary atmosphere*)

second of arc: An angle equaling 1/3600 of a degree.

sedimentary rock: Rock formed from sediments.

seeing: The quality of stillness or lack of shimmer in a telescopic image, associated with atmospheric conditions. If the atmosphere is very still and the image is sharp, the seeing is said to be good.

seismic waves: Waves passing through the interior or surface layers of a planet due to s disturbance, such as an earthquake or large meteorite impact.

seismology: Study of vibrational waves passing through planets, revealing internal structure.

selection effect: Any effect that systematically biases observations or statistics away from a correct understanding.

Seyfert galaxy: A type of galaxy with a bright, bluish nucleus, possibly marking a transition between ordinary galaxies and quasars.

short-period comet: A comet with revolution period less than about 100 y.

SI metric system: The internationally standardized scientific system of units, in which length is given as meters, mass as kilograms, and time as seconds.

sidereal: Referring to stars.

sidereal period: A period of rotation or revolution where the movement is measured relative to the stars.

sidereal time: Time measured by the apparent motion of the stars (instead of the sun), used by astronomers to point telescopes toward celestial targets; it is the right ascension that is on the meridian at any given location.

significant figures: The number of digits known for certain in a quantity.

small-angle equation: The equation giving the relation between the distance D of an object, its diameter d, and its angular size α (expressed in seconds of arc):

$$\frac{\alpha}{206\,265} = \frac{d}{D}$$

solar apex: The direction toward which the sun is moving relative to nearby stars.

solar constant: The amount of energy reaching a sun-facing square meter at a given planet's (usually the earth's) orbit per unit time; for the earth, it is $1390 W/m^2$.

solar core: The sun's central region of high-pressure gases where nuclear energy is produced.

solar cycle: 22-y cycle of solar activity.

solar eclipse: Partial or total blocking of the sun's light by an astronomical body (in most usages, by the moon).

solar nebula: The cloud of gas around the sun during the formation of the solar system.

solar rotation: Turning of the sun on its axis in 25.4 days.

solar system: The sun and all bodies orbiting around it.

solar wind: An outrush of gas past the earth and beyond the outer planets. Near the earth, the solar wind travels at velocities near 600 km/s, sometimes reaching 1000 km/s.

solstice: The date when the sun reaches maximum distance from the celestial equator (occurs twice annually).

solstice principle: According to this principle, the sunrise and sunset positions of the sun on the eastern and western horizons (respectively) shift positions according to season and the observer's latitude. The principle can be used to optimize passive solar energy input into a home or other building.

space velocity: A star's velocity with respect to the sun.

spectral class: A class to which a star belongs because of its spectrum, which in turn is determined by its temperature. The spectral classes are O, B, A, F, G, K, and M, from hottest to coolest.

spectral line strength: Measure of the total energy absorbed or emitted in a spectral line.

spectrograph: An instrument for recording a photographic image of a spectrum.

spectroheliograph: An instrument for observing the sun in certain specified wavelengths.

spectrometer: An instrument for tracing the intensity of a spectrum at different wavelengths; the result is a graph.

spectrophotometry: The study of the amount of radiation at each wavelength in the spectrum.

spectroscopic binary: A binary star revealed by varying Doppler shifts in spectral lines.

spectroscopy: Study of spectra, especially as revealing the properties of the light source.

spectrum binary: A binary revealed by mixture of two spectral classes in the spectrum.

spectrum–luminosity diagram: The H–R diagram.

speed of light: Designated as *c*, the speed of light is about 300 000 km/s and is constant as perceived by all observers.

spicule: Narrow jet of gas extending out of the solar chromosphere, with a lifetime of about 5 min.

spiral arm: In spiral galaxies, one of the arms lying at an angle to the sun–center line. The arms contain open clusters, O and B stars, and nebulae.

spiral galaxy: A disk-shaped galaxy with a spiral pattern, typically containing 10^{10}–$10^{12}\ M_{\odot}$ of stars, dust, and gas.

standard time: Solar time appropriate to the given local time zone.

star: A mass of material, usually wholly gaseous, massive enough to initiate (or to have once initiated) nuclear reactions in its central region.

Stefan–Boltzmann law: A law giving the total energy E radiated from a surface of area A and temperature T per second: $E = \sigma T^4 A$. Sigma (σ), the Stefan–Boltzmann constant, equals 5.67×10^{-8} W/m^2 · K^4 in SI units.

stellar evolution: Evolution of every star from one form to another forced by changes in composition as nuclear reactions proceed.

stellar populations in galaxies: Groupings of stars with different composition and age. Population I consists of young stars with a few percent of heavy elements; stars near the sun are Population I. Population II includes older stars with virtually no heavy elements.

Stonehenge: A prehistoric English ruin with built-in astronomical alignments.

stony–iron meteorite: Stony meteorite that probably comes from deep within the parent body, where melted stony and metallic material coexisted.

stromatolite: A primitive life form, colonies of blue-green algae that were among the first to appear along shorelines. They are among the most abundant fossils from 3.5 to 2 billion years ago.

subfragmentation: Breakup of a contracting cloud into smaller condensations.

subtend: To have an angular size equal to a specified angle. For example, the moon subtends to ½°.

sun: The star orbited by the earth.

sun's composition: 78% hydrogen, 20% helium, 2% other gases.

sunspot: A magnetic disturbance on the sun's surface that is cooler than the surrounding area.

superbubble: A large volume of hot gas in interstellar space, formed by coalescence of bubbles blown around supernovae.

supercluster of galaxies: Cluster of clusters of galaxies.

supergiant star: An extremely luminous star in the uppermost part of the H–R diagram.

supergranulation: Large-scale (15 000–30 000 km in diameter) convective cell patterns in the solar photosphere.

superior planet: Any planet with an orbit outside the earth's orbit.

supernova: A very energetic stellar explosion expending about 10^{42} to 10^{44} joules and blowing off most of the star's mass, leaving a dense core.

symbiotic stars: A pair of stars whose evolutions are affecting each other, especially by mass transfer.

synchronous rotation: Any rotation such that a body keeps the same face toward a coorbiting body.

synchrotron radiation: Radiation emitted when electrons move at nearly the speed of light in a magnetic field.

synodical month: One complete cycle of lunar phases, 29.53 d.

tangential velocity: The velocity component perpendicular to the line of sight.

Tarantula nebula (30 Doradus): A huge HII emission nebula in the Large Magellanic Cloud.

T association: An association of T Tauri stars.

tectonics: Disruption of planetary or satellite surfaces by large-scale mass movements, such as faulting.

telescope: An instrument for collecting electromagnetic radiation and producing magnified images of distant objects.

temperature: A measure of the average energy of a molecule of a material.

terminator: The dawn or dusk line separating night from day on a planet or satellite.

terrestrial planet: (1) Mercury, Venus, Earth, or Mars; (2) a planet composed primarily of rocky material.

theory: A body of hypotheses, often with mathematical backing and having passed some observational tests; often implying more validity than the term *hypothesis*.

theory of cosmological red shifts: The theory that galaxies' red shifts are all due to recessional motion, increase with distance, and thus give an indicator of distance.

theory of noncosmological red shifts: The theory that at least some galaxies' red shifts are not Doppler shifts due to recession, but are due to some other cause.

thermal escape: Escape of the fastest-moving gas atoms or molecules from the top of a planet's atmosphere by means of their thermal motion.

thermal motions: Movements of atoms and molecules associated with the temperature of the material; they grow faster as the temperature increases.

thermal radiation: Electromagnetic radiation emitted by a body and associated with an object's temperature; it grows greater and bluer in color as the temperature increases.

3-kpc arm: An inner arm of our galaxy expanding from the center at about 53 km/s.

3-K radiation: Radio radiation coming uniformly from all over the sky, believed to be a red-shifted remnant of the big bang radiation.

thrust: The force generated by a high-speed discharge, as from a rocket or airplane.

tidal recession: Recession of the moon (or other satellite) from the earth (or other planet) caused by tidal forces.

tide: A bulge raised in a body by the gravitational force of a nearby body.

total solar eclipse: (1) An eclipse in which the light source is totally obscured from a specified observer; (2) an eclipse in which a body is entirely immersed in another's shadow. (See also *eclipse*)

transit: (1) Passage of a planet across the sun's disk; (2) any passage of a body with a small angular size across the face of a body with a large angular size.

triple-alpha process: A nuclear reaction in which helium is transformed into carbon in red giant stars.

Trojan asteroids: Asteroids caught near the Lagrangian points in Jupiter's orbit, 60° ahead of and 60° behind the planet.

tsunami: A large ocean wave generated by earthquake or volcanic activity (the correct name for a tidal wave).

T Tauri star: A type of variable star, often shedding mass, believed to be still forming and contracting onto the main sequence.

"tuning fork" diagram: A classification scheme for galaxies.

21-cm emission line: The important radio radiation at 21-cm wavelength from interstellar neutral atomic hydrogen.

21-cm radio waves: Produced by neutral hydrogen, these waves are especially useful for galactic mapping because they allow us to detect HI clouds, which are concentrated in the spiral arms.

ultrabasic rock: A rock of high density, low silica content, and high iron content, often derived from the upper mantle of a planet or satellite.

ultraviolet light: Radiation of wavelength too short to see, but longer than that of X rays.

umbra: (1) The dark inner part of a shadow, in which the light source is totally obscured; (2) the dark inner part of a sunspot.

universe: Everything that exists.

Uranus' rotation axis: Notable for its almost right-angle tilt (obliquity) of 97° to Uranus' orbital plane.

Urey reaction: Reaction by which the earth's carbon dioxide was concentrated in carbonate rocks after dissolving in seawater.

Van Allen belt: A doughnut-shaped belt around the earth (or another planet with a strong magnetic field) that traps energetic ions from the sun.

variable star: A star that varies in brightness.

Venera 7: First spacecraft to land successfully on another planet; it transmitted data from the surface of Venus in 1970.

Viking 1: The first successful probe to land on Mars (July 20, 1976). It made the first surface photos and measures of soil composition.

Viking 2: The second successful probe to land on Mars (September 3, 1976).

visible light: Electromagnetic radiation at wavelengths that can be perceived by the eye.

visual apparent magnitude: An apparent magnitude estimate based only on visual radiation from an object (excluding infrared, ultraviolet, X-rays, etc.).

visual binary: A binary in which both components can be seen.

volatile element: Element easily driven out of a material by heating.

volcanic crater: A circular depression caused by volcanic processes such as explosion or collapse.

volcanism: Eruption of molten materials at the surface of a planet or satellite.

wavelength: (1) The length of the wavelike characteristic of electromagnetic radiation; (2) in any wave, the distance from one maximum to the next.

wavelike properties of light: Characteristics of light, such as frequency and diffraction, that mimic properties of waves.

white dwarf star: A planet-sized star of roughly solar mass and very high density (10^8 to 10^{11} kg/m^3) produced as a terminal state after nuclear fuels have been consumed.

white light: A mixture of light of all colors in proportions as found in the solar spectrum.

Wien's law: A formula giving the wavelength W at which the maximum amount of radiation comes from a body of temperature T. The formula is $W = 0.00290/T$.

Wolf–Rayet star: A type of very hot star ejecting mass.

W Ursae Majoris stars: Contact binary stars.

X ray: Electromagnetic radiation of wavelength about 0.01 to 10 nm.

X-ray source: Celestial objects emitting X rays; many are probably binary systems where mass is transferred.

zenith: The point directly overhead.

zero-age main sequence: The main sequence defined by a population of stars all of which have just evolved onto the main sequence. (Further evolution modifies the main sequence shape on the H–R diagram slightly.)

zodiac: A band around the sky about 18° wide, centered on the ecliptic, in which the planets move.

zodiacal light: A glow, barely visible to the eye, caused by dust particles spread along the ecliptic plane.

zone: Light cloud band on a giant planet.

zone of avoidance: A band around the sky, centered on the Milky Way, in which galaxies are obscured by the Milky Way's dust.

Asterisked references are less technical, more readily available, and recommended for general reading. They are good general references for preparation of term papers.

Chapter 1

*__Ashbrook,__ J. 1973. "Astronomical Scrapbook." *Sky and Telescope 46:* 300.

__Aveni,__ A. F. 1975. "Possible Astronomical Orientations in Ancient Mesoamerica." In *Archaeoastronomy in Pre-Columbian America,* ed. A. F. Aveni. Austin: University of Texas Press.

__Aveni,__ A. F., S. L. Gibbs, and H. Hartung. 1975. "The Astronomical Significance of the Caracol of Chichén Itzá." *Science 188:* 977.

__Baity,__ E. C. 1973. "Archaeoastronomy and Ethnoastronomy So Far." *Current Anthropology 14:* 389.

__Bjorkman,__ J. K. 1973. "Meteors and Meteorites in the Ancient Near East." *Meteoritics 8:* 91.

*__Bok,__ B. 1975. "A Critical Look at Astrology." *The Humanist,* September/October, p. 5.

__Bok,__ B., and M. Mayall. 1941. "Scientists Look at Astrology." *Scientific Monthly 52:* 233.

__Castetter,__ E. F., and W. H. Bell. 1942. *Pima and Papago Indian Agriculture.* Albuquerque: University of New Mexico Press.

__Crommelin,__ A. 1925. "The Ancient Constellation Figures." In *Splendour of the Heavens,* ed. T. Phillips and W. Steavenson. New York: McBride.

*__Doig,__ P. 1950. *A Concise History of Astronomy.* London: Chapman and Hall.

__Eddy,__ J. A. 1974. "Astronomical Alignment of the Big Horn Medicine Wheel." *Science 184:* 1035.

__Farrington,__ B. 1973. "Astrology." In *Encyclopaedia Britannica.* Chicago: Encyclopaedia Britannica, Inc.

*__Gingerich,__ O. 1967. "Musings on Antique Astronomy." *American Scientist 55:*88.

*__Gingerich,__ O. 1984. "The Origin of the Zodiac." *Sky and Telescope 67:* 218.

__Gleadow,__ R. 1963. *The Origin of the Zodiac.* New York: Atheneum.

*__Harber,__ H. E. 1969. "Five Mayan Eclipses in Thirteen Years." *Sky and Telescope 37:* 72.

__Hartner,__ W. 1965. "The Earliest History of the Constellations in the Near East, and the Motif of the Lion–Bull Contest." *Journal of Near Eastern Studies 24:* 1.

______. 1969. "Eclipse Periods and Thales' Prediction of a Solar Eclipse: Historic Truth and Modern Myth." *Centaurus 14:* 60.

*__Hawkins,__ G., and J. B. White. 1965. *Stonehenge Decoded.* New York: Doubleday.

*__Hoyle,__ F. 1972. *From Stonehenge to Modern Cosmology.* San Francisco: W. H. Freeman.

__Humphreys,__ C. J., and W. G. Waddington. 1983. "Dating the Crucifixion." *Nature 306:*743.

*__Jerome,__ L. E. 1975. "Astrology: Magic or Science?" *The Humanist,* September/October, p. 10.

__Lockyer,__ J. N. 1894. *The Dawn of Astronomy.* Cambridge, Mass.: M.I.T. Press. (Reprint of edition copyrighted 1893, published 1894.)

__Lowell,__ P. 1906. *Mars and Its Canals.* 2nd ed. New York: Macmillan.

__Luce,__ G. G. 1975. "Trust Your Body Rhythms." *Psychology Today,* April, p. 52.

*__Marshack,__ A. 1972. *The Roots of Civilization.* New York: McGraw-Hill.

__Neugebauer,__ O. 1945. "History of Ancient Astronomy." *Journal of Near Eastern Studies 4:* 1.

*______. 1957. *The Exact Sciences in Antiquity.* Providence, R.I.: Brown University Press.

__Ovenden,__ M. 1966. "The Origin of Constellations." *Philosophical Journal 3:* 1.

__Owen,__ N. K. 1975. "The Use of Eclipse Data to Determine the Maya Correlation Number." In *Archaeoastronomy in Pre-Columbian America,* ed. A. F. Aveni. Austin: University of Texas Press.

*__Pannekoek,__ A. 1961. *A History of Astronomy.* London: Allen and Unwin.

__Smiley,__ C. H. 1975. "The Solar Eclipse Warning Table in the Dresden Codex." In *Archaeoastronomy in Pre-Columbian America,* ed. A. F. Aveni. Austin: University of Texas Press.

*__Stephenson,__ F. R. 1982. "Historical Eclipses." *Scientific American,* October, p. 170.

*__Stephenson,__ F. R., and D. H. Clark. 1977. "Ancient Astronomical Records from the Orient." *Sky and Telescope 53:* 84.

__Thom,__ A., and A. S. Thom. 1971. *Megalithic Lunar Observatories.* Oxford: Oxford University Press.

______. 1974. "Stonehenge." *Journal of the History of Astronomy 5:* 71.

*__Thomsen,__ D. E. 1984. "Calendric Reform in Yucatan." *Science News 126:* 282.

Chapter 2

*__Durant,__ W. 1926. *The Story of Philosophy.* New York: Simon & Schuster.

*__Gingerich,__ O. 1967. "Musings on Antique Astronomy." *American Scientist 55:* 88.

*______. 1986. "Islamic Astronomy." *Scientific American,* April, p. 74.

*__Gribbon,__ J. 1975. "Did Chinese Cosmology Anticipate Relativity?" *Nature 256:* 619.

*__Lewis,__ D. 1973. *We, the Navigators.* Honolulu: University of Hawaii Press.

__Mozans,__ H. J. 1913. *Woman in Science.* New York: Appleton.

__Needham,__ J. 1959. *Science and Civilization in China.* Vol. 3. London: Cambridge University Press.

*__North,__ J. D. 1974. "The Astrolabe." *Scientific American,* January, p. 96.

*__Pannekoek,__ A. 1961. *A History of Astronomy.* London: Allen and Unwin.

__Pritchard,__ J. B., ed. 1955. *Ancient Eastern Texts Relating to the Old Testament.* 2nd ed. Princeton, N.J.: Princeton University Press.

__Wilson,__ J. A. 1951. *The Culture of Ancient Egypt.* Chicago: University of Chicago Press.

__Zeilik,__ M. 1985. "The Ethnoastronomy of the Historic Pueblos, I: Calendrical Sun Watching." *Archaeoastronomy* 8: S1.

Chapter 3

Alvarez, L. W., et al. 1980. "Extraterrestrial Cause for the Cretaceous–Tertiary Extinction." *Science 208:* 1095.

***Davies,** G. L. 1969. *The Earth in Decay.* New York: American Elsevier.

***Dewey,** J. F. 1972. "Plate Tectonics." *Scientific American,* May, p. 56.

Dunbar, C. O. 1963. *Historical Geology.* New York: John Wiley.

Einstein, A. 1961. *Relativity.* New York: Crown.

Faggart, B. E., A. Basu, and M. Tatsumoto. 1985. "Origin of the Sudbury Complex by Meteoritic Impact: Neodymium Isotopic Evidence." *Science 230:* 436.

Ganapathy, R. 1980. "A Major Meteorite Impact on the Earth 65 Million Years Ago: Evidence from the Cretaceous-Tertiary Boundary Clay." *Science 209:* 921.

***Hurley,** P. M. 1968. "The Confirmation of Continental Drift." *Scientific American,* April, p. 52.

Jeffreys, H. 1970. *The Earth.* 5th ed. New York: Cambridge University Press.

Jordan, T. H. 1975. "The Continental Tectosphere." *Review of Geophysics and Space Physics 13:* 1.

Kozlovsky, Y. A. 1984. "The World's Deepest Well." *Scientific American,* p. 436.

***Nelkin,** D. 1976. "The Science-Textbook Controversies." *Scientific American,* April, p. 33.

Pepin, R. O. 1976. "The Formation Interval of the Earth." *Abstracts Seventh Lunar Science Conference,* Houston, Lunar and Planetary Science Institute.

***Renfrew,** C. 1971. "Carbon-14 and the Prehistory of Europe." *Scientific American,* October, p. 63.

Schneider, S. H., and R. Dickinson. 1974. "Climate Modeling." *Review of Geophysics and Space Physics 12:* 447.

Schopf, J. W. 1975. "The Age of Microscopic Life." *Endeavour 34:* 51.

***Tazieff,** H. 1964. *When the Earth Trembles.* London: Hart–Davis.

Toon, O. B., et al. 1982. "Evolution of an Impact-generated Dust Cloud and Its Effects on the Atmosphere." In *Geological Society of America Special Paper 190,* ed. L. Silver and P. Schultz. Boulder, Colo.: Geological Society of America.

Wyllie, P. J. 1971. *The Dynamic Earth.* New York: John Wiley.

***______.** 1975. "The Earth's Mantle." *Scientific American,* March, p. 50.

Chapter 4

***Arnold,** J. R. 1980. "The Frontier in Space." *American Scientist 68:* 299.

Ball, W.W.R. 1972. *An Essay on Newton's* Principia. New York: Johnson Reprint Collection. (Originally published 1893.)

***Burbidge,** E. Margaret. 1983. "Adventure into Space." *Science 221:* 421.

***Clarke,** A. C. 1951. *The Exploration of Space.* New York: Harper & Bros.

***Dyson,** F. 1969. "Human Consequences of the Exploration of Space." *Bulletin of Atomic Scientists,* September, p. 8.

***Heinlein,** R. 1950. *The Man Who Sold the Moon.* New York: New American Library.

***Lewis,** R. S. 1969. *Appointment on the Moon.* New York: Viking Press.

***Logsdon,** J. 1970. *The Decision to Go to the Moon.* Cambridge, Mass.: M.I.T. Press.

National Commission on Space. 1986. *Pioneering the Space Frontier.* New York: Bantam Books.

Newton, I. 1962. *Principia.* A. Motte, trans. Berkeley: University of California Press. (Originally published 1687.)

***Nicholson,** M. 1949. *Voyages to the Moon.* New York: Macmillan.

***O'Neill,** G. K. 1977. *The High Frontier.* New York: William Morrow.

***Sagan,** C., and J. Agel. 1973. *The Cosmic Connection.* New York: Doubleday.

***Verne,** J. 1949. *From the Earth to the Moon.* New York: Didear. (Originally published 1865.)

Chapter 5

Abetti, G. 1952. *The History of Astronomy.* New York: Henry Schuman.

Darwin, G. H. 1962. *The Tides and Kindred Phenomena in the Solar System.* San Francisco: W. H. Freeman. (Originally published 1898.)

El-Baz, F. 1975. "The Moon After Apollo." *Icarus 25:* 495.

***Goldreich,** P. 1972. "Tides and the Earth–Moon System." *Scientific American,* April, p. 42.

Hartmann, W. K., and D. Davis. 1975. "Satellite-Sized Planetesimals and Lunar Origin." *Icarus 24:* 504.

Hartmann, W. K., R. Phillips, and J. G. Taylor. 1986. *Origin of the Moon.* Houston: Lunar and Planetary Institute.

Kaula, W. K., and A. Harris. 1975. "Dynamics of Lunar Origin and Orbital Evolution." *Review of Geophysics and Space Physics 13:* 363.

Lammlein, D., et al. 1974. "Lunar Seismicity, Structure, and Tectonics." *Review of Geophysics and Space Physics 12:* 1.

***Lewis,** R. S. 1977. "Space Prospect: Factories and Electric Power." *Smithsonian 8* (9): 94.

***Mason,** B., and W. G. Melson. 1970. *The Lunar Rocks.* New York: Wiley-Interscience.

***Mutch,** T. 1973. *Geology of the Moon.* 2nd ed. Princeton, N.J.: Princeton University Press.

Ringwood, A. E. 1972. "Some Comparative Aspects of Lunar Origin." *Physics of Earth and Planetary Interiors 6:* 366.

Shoemaker, E., et al. 1970. "Lunar Regolith at Tranquillity Base." *Science 167:* 452.

Taylor, S. R. 1973. "Geochemistry of the Lunar Highlands." *Moon 7:* 181.

______. 1975. *Lunar Science: A Post-Apollo View.* New York: Pergamon Press.

Toksoz, M. N., et al. 1974. "Structure of the Moon." *Review of Geophysics and Space Physics 12:* 539.

Wood, J. 1972. "Thermal History and Early Magmatism in the Moon." *Icarus 16:* 229.

Chapter 6

Allen, C. W. 1973. *Astrophysical Quantities.* Cambridge, Mass.: Athlone Press.

***Chapman,** Clark R. 1982. *Planets of Rock and Ice.* New York: Charles Scribner's.

***de Santillana,** Georgio. 1962. *The Crime of Galileo.* New York: Time.

***Dreyer,** J.L.E. 1953. *A History of Astronomy from Thales to Kepler.* New York: Dover. (Reprint of 1906 edition.)

***Gingerich,** O. 1973a. "Copernicus and Tycho." *Scientific American,* December, p. 86.

*______. 1973b. *Crisis Versus Aesthetic in the Copernican Revolution.* Cambridge, Mass.: Smithsonian Astrophysical Observatory.

***Hartmann,** W. K. 1983. *Moons and Planets.* Belmont, Calif.: Wadsworth.

***Lerner,** L. S., and E. A. Gosselin. 1973. "Giordano Bruno." *Scientific American,* April, p. 86.

***Pannekoek,** A. 1961. *A History of Astronomy.* New York: Interscience.

***Rosen,** E. 1973. "Copernicus' Place in the History of Astronomy." *Sky and Telescope 45:* 72.

***Taylor,** S. R. 1982. *Planetary Science: A Lunar Perspective.* Houston, Tex.: Lunar and Planetary Institute.

Chapter 7

Bahcall, J. N., and L. Spitzer. 1982. "The Space Telescope." *Scientific American,* July, p. 40.

Kristian, J., and M. Blouke. 1982. "Charge-Coupled Devices in Astronomy." *Scientific American,* October, p. 66.

Chapter 8

Colin, L., et al. 1979. Special issue on Pioneer Venus results. *Science 203:* 743 ff.

Connes, P., et al. 1967. "Traces of HCl and HF in the Atmosphere of Venus." *Astrophysical Journal 147:* 1230.

***Cruikshank,** D. P., and C. R. Chapman. 1967. "Mercury's Rotation and Visual Observations." *Sky and Telescope 34:* 24.

Donahue, T. M. 1979. Special issue on Pioneer Venus results. *Science 205:* 41 ff.

Duncombe, R. 1969. "Report on Numerical Experiment on the Possible Existence of an 'Anti-Earth.'" In *Scientific Study of Unidentified Flying Objects,* ed. E. U. Condon. New York: Bantam.

Hunten, D. M., L. Colin, T. Donahue, and V. Moroz (editors). 1983. *Venus.* Tucson: University of Arizona Press.

Marov, M. Y. 1972. "Venus: A Perspective at the Beginning of Planetary Exploration." *Icarus 16:* 415.

***Moore,** P. 1954. *A Guide to the Planets.* New York: W. W. Norton.

Morrison, D. 1970. "Thermophysics of the Planet Mercury." *Space Science Review 11:* 271.

***Murray,** B. C. 1975. "Mercury." *Scientific American,* September, p. 58.

Oyama, V. I., et al. 1979. "Venus Lower Atmospheric Composition: Analysis by Gas Chromatography." *Science 203:* 802.

***Petengill,** G. H., D. B. Campbell, and H. Masursky. 1980. "The Surface of Venus." *Scientific American,* April, p. 54.

Sagan, C. 1962. "Structure of the Lower Atmosphere of Venus." *Icarus 1:* 151.

Sagdeev, R., et al. 1986. "Overview of VEGA Venus Balloon in Situ Meteorological Measurements." *Science 231:* 1411.

***Schubert,** G., and C. Covey. 1981. "The Atmosphere of Venus." *Scientific American,* June, p. 66.

Sill, G. T. 1973. "Sulfuric Acid in the Venus Clouds." *Communications of the Lunar and Planetary Laboratory, University of Arizona 9:* 191.

Walker, J. C. C. 1977. *Evolution of the Atmosphere.* New York: Macmillan.

Young, A. T. 1973. "Are the Clouds of Venus Sulfuric Acid?" *Icarus 18:* 564.

Chapter 9†

Arvidson, R., et al. 1983. "Three Mars Years: Viking Lander 1 Imaging Observations." *Science 222:* 463.

***Bradbury,** R. 1950. *The Martian Chronicles.* New York: Doubleday.

Burns, J., and M. Harwit. 1973. "Towards a More Habitable Mars—or—The Coming Martian Spring." *Icarus 19:* 126.

***Burroughs,** E. R. 1972. *The Chessmen of Mars.* New York: Doubleday. (Originally published 1922.)

Haberle, R. M. 1986. "The Climate of Mars." *Scientific American,* May, p. 54.

***Hartmann,** W. K., and O. Raper. 1974. *The New Mars.* Washington, D.C.: National Aeronautics and Space Administration.

Levin, G. V., and Patricia Stroat. 1981. "A Search for a Nonbiological Explanation of the Viking Labeled Release Life Detection Experiment." *Icarus 45:* 494.

***Lowell,** P. 1906. *Mars and Its Canals.* 2nd ed. New York: Macmillan.

***Murray,** B. C. 1973. "Mars from Mariner 9." *Scientific American,* January, p. 48.

National Commission on Space. 1986. *Pioneering the Space Frontier.* New York: Bantam Books.

Owen, T., et al. 1977. "The Composition of the Atmosphere at the Surface of Mars." *Journal of Geophysical Research 82:* 4635.

Sagan, C., O. Toon, and P. Gierasch. 1973. "Climatic Change on Mars." *Science 181:* 1045.

Soffen, G. A., et al. 1977. Special issue on Viking results. *Journal of Geophysical Research 82:* 3959 ff.

***Wells,** H. G. 1898. *The War of the Worlds.* London: W. Heinemann.

†Note: Early results from Viking are reported in special issues of *Science,* 27 August 1976 and 1 October 1976.

Chapter 10

Bodenheimer, P. 1976. "Contraction Models for the Evolution of Jupiter." *Icarus 29:* 165.

***Chapman,** C. R. 1968. "The Discovery of Jupiter's Red Spot." *Sky and Telescope 35:* 276.

Cruikshank, D. P., and R. E. Murphy. 1973. "The Post-Eclipse Brightening of Io." *Icarus 20:* 7.

***Cruikshank,** D. P., and D. Morrison. 1976. "The Galilean Satellites of Jupiter." *Scientific American,* May, p. 108.

Cruikshank, D. P., C. Pilcher, and D. Morrison. 1976. "Pluto: Evidence for Methane Frost." *Science 194:* 835.

Cruikshank, D. P., and P. Silvaggio. 1979. "Triton: A Satellite with an Atmosphere." *Astrophysical Journal 233:* 1016.

Cruikshank, D. P., R. Brown, and R. Clark. 1983. "Nitrogen on Triton." *Bulletin of the American Astronomical Society 15:* 857.

***Grosser,** M. 1962. *The Discovery of Neptune.* Cambridge, Mass.: Harvard University Press.

Hanel, R., et al. 1979. "Infrared Observations of the Jovian System from Voyager 1." *Science 204:* 972.

***Hartmann,** W. K. 1983. *Moons and Planets.* Belmont, Calif.: Wadsworth.

***Hunten,** D. M. 1975. "The Outer Planets." *Scientific American,* September, p. 131.

Khare, B. N., and C. Sagan. 1973. "Red Clouds in Reducing Atmospheres." *Icarus 20:* 311.

Lebofsky, L., T. Johnson, and T. McCord. 1970. "Saturn's Rings: Spectral Reflectivity and Compositional Implications." *Icarus 13:* 226.

Lewis, J., and R. G. Prin. 1970. "Jupiter's Clouds: Structure and Composition." *Science 169:* 472.

Minton, R. B. 1973. "The Red Polar Caps of Io." *Communications of the Lunar and Planetary Laboratory, University of Arizona 10:* 35.

Morrison, David, ed. 1982. *Satellites of Jupiter.* Tucson: University of Arizona Press.

Peale, S. J., P. Cassen, and R. Reynolds. 1979. "Melting of Io by Tidal Dissipation." *Science 203:* 892.

***Peek,** B. M. 1958. *The Planet Jupiter.* New York: Macmillan.

Pollack, J. 1973. "Greenhouse Models of the Atmosphere of Titan." *Icarus 19:* 43.

***Stone,** E., et al. 1979a. Special issue on Voyager 1 Jupiter results. *Science 204:* 945 ff.

***Stone,** E., et al. 1979b. Special issue on Voyager 2 Jupiter results. *Science 206:* 925 ff.

***Stone,** E., et al. 1981. Special issue on Voyager 1 Saturn results. *Science 212:* 159 ff.

***Stone,** E. et al. 1982. Special issue on Voyager 2 Saturn results. *Science 212:* 499 ff.

Tombaugh, C. W. 1961. "The Trans-Neptunian Planet Search." In *Planets and Satellites,* ed. G. Kuiper and B. Middlehurst. Chicago: University of Chicago Press.

Van Flandern, T., et al. 1981. "The Renewal of the Trans-Neptunian Planet Search." Paper presented at the meeting of the American Astronomical Society, Albuquerque. Abstract in *Bulletin of the American Astronomical Society 12:* 830 (1980).

Veverka, J. 1973. "Titan: Polarimetric Evidence for an Optically Thick Atmosphere?" *Icarus 18:* 657.

***Wolfe,** J. H. 1975. "Jupiter." *Scientific American,* September, p. 118.

Zellner, B. 1972. "On the Nature of Iapetus." *Astrophysical Journal 174:* L107.

Chapter 11

***Brandt,** J. C., and R. Chapman. 1981. *Introduction to Comets.* Cambridge: Cambridge University Press.

***Chapman,** C. R. 1975. "The Nature of Asteroids." *Scientific American,* January, p. 24.

Gaffey, M. J., and T. McCord. 1977. "Mining the Asteroids." *Mercury 6:* 1.

Gehrels, T. 1972. "Physical Parameters of Asteroids and Interrelations with Comets." In *From Plasma to Planet,* ed. A. Elvius. New York: John Wiley.

***Gehrels,** Tom, ed. 1979. *Asteroids.* Tucson: University of Arizona Press.

***Hartmann,** W. K. 1975. "The Smaller Bodies of the Solar System." *Scientific American,* September, p. 143.

*———. 1982. "Mines in the Sky Are Not So Wild A Dream." *Smithsonian,* September, p. 70.

*———. 1983. *Moons and Planets.* Belmont, Calif.: Wadsworth.

Kamoun, P. G., et al. 1982. "Comet Encke: Radar Detection of Nucleus." *Science 216:* 293.

Krinov, E. L. 1966. *Giant Meteorites.* New York: Pergamon Press.

Lebofsky, L. A., et al. 1981. "The 1.7 to 4.2-m Spectrum of Asteroid 1 Ceres: Evidence for Structural Water in Clay Minerals." *Icarus 48:* 453.

Nininger, H. H. 1938. "Meteorite Collecting Among Ancient Americans." *American Antiquity 4:* 39.

Oort, H. H. 1963. "Empirical Data on the Origin of Comets." In *The Moon, Meteorites, and Comets,* ed. B. Middlehurst and G. P. Kuiper. Chicago: University of Chicago Press.

***Sagan,** C. 1975. "Kalliope and the Kaa'ba: The Origin of Meteorites." *Natural History 84:* 8.

Sagdeev, R. Z., et al. 1986. "Encounters with Comet Halley—The First Results." *Nature 321:* 259.

Sekanina, Z. 1983. "The Tunguska Event: No Cometary Signature in Evidence." *Astronomy Journal 88:* 1382.

Shapley, H., and H. Howarth, eds. 1929. *A Source Book in Astronomy.* New York: McGraw-Hill.

Wasson, J. T. 1974. *Meteorites.* New York: Springer-Verlag.

***Wilkening,** Laurel, ed. 1982. *Comets.* Tucson: University of Arizona Press.

Chapter 12

***Cameron,** A. G. W. 1975. "The Origin and Evolution of the Solar System." *Scientific American,* September, p. 32.

Clark, B. C., et al. 1977. "The Viking X-ray Fluorescence Experiment: Analytical Methods and Early Results." *Journal of Geophysical Research 82:* 4577.

Clayton, R., L. Grossman, and T. Mayeda. 1973. "A Component of Primitive Nuclear Composition in Carbonaceous Chondrites." *Science 182:* 485.

Fairbridge, R. W. 1972. *The Encyclopedia of Geochemistry and Environmental Sciences.* New York: Van Nostrand Reinhold.

Gehrels, T., ed. 1978. *Protostars and Planets.* Tucson: University of Arizona Press.

Greenberg, R., et al. 1978. "Planetesimals to Planets: Numerical Simulation of Collisional Evolution." *Icarus 35:* 1.

***Grossman,** L. 1975. "The Most Primitive Objects in the Solar System: Carbonaceous Chondrites." *Scientific American,* February, p. 30.

Hartmann, W. K. 1978. "Planet Formation: Mechanism of Early Growth." *Icarus 33:* 50.

***Herbst,** W., and G. Assousa. 1979. "Supernovas and Star Formation." *Scientific American,* August, p. 138.

***Lewis,** J. S. 1974. "The Chemistry of the Solar System." *Scientific American,* March, p. 50.

***Mason,** B., and W. G. Melson. 1970. *The Lunar Rocks.* New York: John Wiley.

Page, T. L. 1973. "Notes on the 4th Lunar Science Conference." *Sky and Telescope 46:* 88.

Papanastassiou, D., and G. Wasserburg. 1968. "Initial Strontium Isotopic Abundance and the Resolution of Small Time Differences in the Formation of Planetary Objects." *Earth and Planetary Science Letters 5:* 361.

Pollack, J. B., and D. C. Black. 1979. "Implications of the Gas Compositional Measurements of the Pioneer Venus for the Origin of Planetary Atmospheres." *Science 205:* 56.

Safronov, V. S. 1972. *Evolution of the Protoplanetary Cloud and Formation of the Earth and Planets.* Springfield, Va.: National Technical Information Service.

Shapley, H., and H. Howarth, eds. 1929. *A Source Book in Astronomy.* New York: McGraw-Hill.

Taylor, S. R. 1973. "Chemical Evidence for Lunar Melting and Differentiation." *Nature 245:* 203.

Chapter 13

Anders, E., and M. Ebihara. 1982. "Solar-system Abundances of the Elements." *Geochimica et Cosmochimica Acta 46:* 2365.

***Arnold,** J. R. 1980. "The Frontier in Space." *American Scientist 68:* 299.

Babcock, H. W. 1961. "Topology of the Sun's Magnetic Field and the 22-Year Cycle." *Astrophysical Journal 133:* 572.

***Bahcall,** J. N. 1969. "Neutrinos from the Sun." *Scientific American,* July, p. 28.

Cameron, A. G. W. 1973. "Major Variations in Solar Luminosity?" *Review of Geophysics and Space Physics 11:* 505.

Dicke, R. H. 1979. "Solar Luminosity and the Sunspot Cycle." *Nature 280:* 24.

Eddy, J. A. 1976. "The Maunder Minimum." *Science 192:* 1189.

Gibson, E. G. 1973. *The Quiet Sun.* Publication no. SP-303. Washington, D.C.: National Aeronautics and Space Administration.

Gilliland, R. 1982. "Modeling Solar Variability." *Astrophysical Journal 253:* 399.

***Herbst,** W., and G. Assousa. 1979. "Supernovas and Star Formation." *Scientific American,* August, p. 138.

***Hubbert,** M. 1971. "The Energy Resources of the Earth." *Scientific American,* September, p. 61.

Kerr, R. A. 1986. "The Sun Is Fading." *Science 231:* 339.

***Kreith,** F., and R. T. Meyer. 1983. "Large-Scale Use of Solar Energy with Central Receivers." *American Scientist 71:* 598.

Leighton, R. B. 1969. "A Magneto-Kinematic Model of the Solar Cycle." *Astrophysical Journal 156:* 1.

***Lewis,** R. S. 1977. "The Space Prospect: Factories and Electric Power." *Smithsonian,* December, p. 94.

Newkirk, G., and K. Frazier. 1982. "The Solar Cycle." *Physics Today,* April, p. 25.

***Parker,** E. N. 1983. "Magnetic Fields in the Cosmos." *Scientific American,* August, p. 44.

***Pasachoff,** J. M. 1973. "The Solar Corona." *Scientific American,* October, p. 68.

***Pasachoff,** J. M. 1980. "Our Sun." In *Astronomy Selected Readings,* ed. M. A. Seeds. Menlo Park, Calif.: Benjamin/Cummings.

Pollack, J. B., and D. C. Black. 1979. "Implications of the Gas Compositional Measurements of the Pioneer Venus for the Origin of Planetary Atmospheres." *Science 205:* 56.

Schneider, S. H., and C. Mass. 1975. "Volcanic Dust, Sunspots, and Temperature Trends." *Science 190:* 741.

Shapley, H., and H. Howarth, eds. 1929. *A Source Book in Astronomy.* New York: McGraw-Hill.

***Snell,** J. E., P. Achenbach, and S. Peterson. 1976. "Energy Conservation in New Housing Design." *Science 192:* 1305.

***Wilcox,** J. M. 1976. "Solar Structure and Terrestrial Weather." *Science 192:* 745.

***Wolfson,** R. 1983. "The Active Solar Corona." *Scientific American,* February, p. 104.

Wood, C. A., and R. R. Lovett. 1974. "Rainfall, Drought, and the Solar Cycle." *Nature 251:* 594.

Chapter 14

Allen, C. W. 1973. *Astrophysical Quantities.* 3rd ed. London: Athlone Press.

***Aller,** L. H. 1971. *Atoms, Stars, and Nebulae.* Cambridge, Mass.: Harvard University Press.

***Bahcall,** J., and L. Spitzer. 1982. "The Space Telescope." *Scientific American,* July, p. 40.

Neugebauer, G., and 11 others. 1984. "Early Results from the Infrared Astronomical Satellite." *Science 224:* 13.

***Shapley,** H., ed. 1960. *Source Book in Astronomy, 1900–1950.* Cambridge, Mass.: Harvard University Press.

***Shapley,** H., and H. Howarth, eds. 1929. *A Source Book in Astronomy.* New York: McGraw-Hill.

Weissman, Paul. 1984. "The Vega Particulate Shell: Comets or Asteroids." *Science 224:* 987.

Chapter 15

Allen, C. W. 1973. *Astrophysical Quantities.* 3rd ed. London: Athlone Press.

***Aller,** L. H. 1971. *Atoms, Stars, and Nebulae.* Cambridge, Mass.: Harvard University Press.

Boesgaard, A., and W. Hagen. 1974. "The Age of Alpha Centauri." *Astrophysical Journal 189:* 85.

***Burnham,** R. 1978. *Burnham's Celestial Handbook.* New York: Dover Publications.

Eddington, A. S. 1926. *The Internal Constitution of the Stars.* Cambridge, Mass.: Harvard University Press.

***Humphreys,** R., and K. Davidson. 1984. "The Most Luminous Stars." *Science 223:* 243.

Iben, I. 1967. "Stellar Evolution: Within and Off the Main Sequence." *Annual Review of Astronomy and Astrophysics 4:* 171.

Kamper, K., and A. J. Wesselink. 1978. "Alpha and Proxima Centauri." *Astronomical Journal 83:* 1653.

Larson, R. B. 1969. "Numerical Calculations of the Dynamics of a Collapsing Proto-Star." *Monthly Notices of the Royal Astronomical Society 145:* 271.

Liebert, J. 1980. "White Dwarf Stars." *Annual Review of Astronomy and Astrophysics 18:* 363.

Popper, D. M. 1980. "Stellar Masses." *Annual Review of Astronomy and Astrophysics 18:* 115.

Schwarzchild, M. 1958. *Structure and Evolution of the Stars.* Princeton, N.J.: Princeton University Press.

Shapley, H. ed. 1960. *Source Book in Astronomy, 1900–1950.* Cambridge, Mass.: Harvard University Press.

van de Kamp, P. 1981. *Stellar Paths.* Dordrecht, Netherlands: Reidel.

Westbrook, C., and C. B. Tarter. 1975. "On Protostellar Evolution." *Astrophysical Journal 200:* 48.

Chapter 16

Adams, M., K. Strom, and S. Strom. 1983. "The Star-forming History of the Young Cluster NGC 2264." *Astrophysical Journal,* Supplement Series *53,* No. 4.

Arny, T., and W. Weissman. 1973. "Interaction of Proto-Stars in a Collapsing Cluster." *Astronomical Journal 78:* 309.

Bailly, J., and C. Lada. 1983. "The High-Velocity Molecular Flows Near Young Stellar Objects." *Astrophysical Journal 265:* 824.

Becklin, E., G. Neugebauer, and C. Wynn-Williams. 1973. "On the Nature of the Infrared Point Source in the Orion Nebula." *Astrophysical Journal 182:* L7.

Bodenheimer, P. 1976. "Contraction Models for the Evolution of Jupiter." *Icarus 29:* 165.

Bok, B. J., and E. Reilly. 1947. "Small Dark Nebulae." *Astrophysical Journal 105:* 255.

Bok, B. J., C. Cordwell, and R. Cromwell. 1970. "Globules." In *Dark Nebulae, Globules, and Protostars,* ed. B. Lynds. Tucson: University of Arizona Press.

Cohen, M., and L. Kuhi. 1979. *Astrophysical Journal,* Supplement Series, December.

Hanson, R. B., B. Jones, and D. Ling. 1983. "The Astrometric Position of T Tauri and the Nature of Its Companion." *Astrophysical Journal 270:* L27.

Hartmann, L., and J. Raymond. 1984. "A High-Resolution Study of Herbig-Haro Objects 1 and 2." *Astrophysical Journal 276:* 560.

Hartmann, W. K. 1967. "On the Nature of the Infrared Nebula in Orion." *Astrophysical Journal 149:* L87.

Hayashi, C. 1961. "Stellar Evolution in the Early Phases of Gravitational Contraction." *Publications of the Astronomical Society of Japan 13:* 450.

Henyey, L., R. LeLevier, and R. Levée. 1955. "The Early Phases of Stellar Evolution." *Publications of the Astronomical Society of the Pacific 67:* 154.

Herbig, G. 1968. "The Structure and Spectrum of R Monocerotis." *Astrophysical Journal 152:* 439.

———. 1975. "The Spectrum and Structure of 'Minkowski's Footprint': M 1-92." *Astrophysical Journal 200:* 1.

Iben, I. 1965. "Stellar Evolution I. The Approach to the Main Sequence." *Astrophysical Journal 141:* 993.

Jeans, J. H. 1902. "The Nebular Theory of the Origin of the Solar System." *Philosophical Transactions of the Royal Society of London 199:* 1. Reprinted in K. Lang and O. Gingerich, eds., *A Source Book in Astronomy and Astrophysics, 1900–1975* (Cambridge, Mass.: Harvard University Press, 1979).

***Lada,** C. J. 1982. "Energetic Outflow from Young Stars." *Scientific American,* June, p. 82.

Larson, R. B. 1969. "Numerical Calculations of the Dynamics of a Collapsing Protostar." *Monthly Notices of the Royal Astronomical Society 145:* 271.

———. 1972. "The Evolution of Spherical Protostars with Masses 0.25 to 10 $M_{\odot}$." *Monthly Notices of the Royal Astronomical Society 157:* 121.

Low, F. J., and B. Smith. 1966. "Infrared Observations of a Preplanetary System." *Nature 212:* 675.

***Lynds,** B. T., ed. 1970. *Dark Nebulae, Globules, and Protostars.* Tucson: University of Arizona Press.

Mendoza, V. E. 1968. "Infrared Excesses in T Tauri Stars and Related Objects." *Astrophysical Journal 143:* 1010.

***Neugebauer,** G., and E. Becklin. 1973. "The Brightest Infrared Sources." *Scientific American,* April, p. 28.

Penston, M. 1973. "Photoelectric UBV Photometry of Stars in Ten Selected Areas." *Monthly Notices of the Royal Astronomical Society 164:* 121.

Poveda, A. 1965. "The H–R Diagram of Young Clusters and the Formation of Planetary Systems." *Boletin de los Observatorio Tonantzintla y Tacubaya* (Mexico City) *4:* 15.

Schmelz, J. 1984. "An Investigation of T Tauri Variability." *Astronomical Journal 89:* 108.

Schwartz, R. D. 1983. "Herbig–Haro Objects." *Annual Reviews in Astronomy and Astrophysics 21:* 209.

***Scoville,** N., and Judith Young. 1984. "Molecular Clouds, Star Formation and Galactic Structure." *Scientific American,* April, p. 42.

Spitzer, L., Jr. 1941. "The Dynamics of the Interstellar Medium. II. Radiation Pressure." *Astrophysical Journal 94:* 232.

Staller, R. F. A., and T. de Jong. 1981. "Theoretical Luminosity Functions of Red and Black Dwarf." *Astronomy and Astrophysics 98:* 140.

Stockton, A., D. Chesley, and S. Chesley. 1975. "Spectroscopy of R Monocerotis and NGC 2261." *Astrophysical Journal 199:* 406.

***Strom,** S., and K. Strom. 1973. "The Early Evolution of Stars." *Sky and Telescope 45:* 279, 359.

Van der Linden, T., and R. Staller. 1983. "Evolution of Very Low-Mass Stars." *Astronomy and Astrophysics 118:* 285.

Walker, M. 1972. "Studies of Extremely Young Clusters. VI." *Astrophysical Journal 175:* 89.

Weaver, W. 1974. "The Coalsack III: A Search for T Tauri Stars." *Astrophysical Journal 189:* 263.

Wolstencroft, R., and T. Simon. 1975. "The Variable Circular Polarization of V1057 Cygni." *Astrophysical Journal 199:* L169.

Wright, A. 1970. "Results of a Computer Program for Gravito-Gas Dynamic Collapse." *Proceedings of the Liège Colloquium* Mémoires de la Société Royale des Sciences de Liège *19:* 75.

Chapter 17

Baade, W., and F. Zwicky. 1934. "Cosmic Rays from Super-Novae." *Proceedings of the National Academy of Science 20:* 259.

Baize, P. 1980. "Les Masses des Étoiles Variables a Longue Période." *L'Astronomie 94:* 71.

Bonneau, D., and others. 1982. "The Diameter of Mira." *Astronomy and Astrophysics 106:* 235.

***Bova,** B. 1973. "Obituary of Stars: A Tale of Red Giants, White Dwarfs, and Black Holes." *Smithsonian 4:* 54.

***Burnham,** Robert. 1978. *Burnham's Celestial Handbook.* New York: Dover Publications.

***Christy-Sackman,** J., and K. Despan. 1974. "An Interpretation of the Puzzling Observations of FG Sagittae." *Astrophysical Journal 189:* 523.

Conti, P. S., and R. McCray. 1980. "Strong Stellar Winds." *Science 208:* 9.

Glasby, J. S. 1968. *Variable Stars.* London: Constable.

Gold, T. 1968. "Rotating Neutron Stars as the Origin of Pulsating Radio Sources." *Nature 218:* 731.

***Hawking,** S. W. 1977. "The Quantum Mechanics of Black Holes." *Scientific American,* January, p. 34.

***Helfand,** D. 1983. "Theory Points to Pulsating White Dwarfs." *Physics Today,* January, p. 21.

Hewish, A., et al. 1968. "Observation of a Rapidly Pulsating Radio Source." *Nature 217:* 709.

Hoxie, D. 1969. "The Structure and Evolution of Stars of Very Low Mass." Ph.D. dissertation, University of Arizona.

***Humphreys,** R., and K. Davidson. 1984. "The Most Luminous Stars." *Science 223:* 243.

Iben, I. 1967. "Stellar Evolution: Within and Off the Main Sequence." *Annual Review of Astronomy and Astrophysics 5:* 571.

Kraft, R. 1964. "Binary Stars Among Cataclysmic Variables. III." *Astrophysical Journal 139:* 457.

Kumar, S. 1964. "On the Nature of Planetary Companions of Stars." *Zeitschrift für Astrophysic 58:* 248.

Langer, G., R. Kraft, and K. Anderson. 1974. "FG Sagittae: The *s*-Process Episode." *Astrophysical Journal 189:* 509.

Leavitt, H. S. 1912. *Periods of 25 Variable Stars in the Small Magellanic Cloud.* Harvard College Observatory Circular no. 173, p. 1.

***Liebert,** J. 1980. "White Dwarf Stars." *Annual Review of Astronomy and Astrophysics 18:* 363.

Nolan, P., and J. Matteson. 1983. "A Feature in the X-ray Spectrum of Cygnus X-1: A Possible Positron Annihilation Line." *Astrophysical Journal 265:* 389.

Norman, C. A., and D. ter Haar. 1973. "On the Black Hole Model of Galactic Nuclei." *Astronomy and Astrophysics 24:* 121.

Novikov, I., and K. Thorne. 1973. "Astrophysics of Black Holes." In *Black Holes,* ed. C. DeWitt and B. S. DeWitt. London: Gordon and Breach.

Oppenheimer, J. R., and R. Serber. 1938. "On the Stability of Stellar Neutron Cores." *Physics Review 54:* 540.

***Penrose,** R. 1972. "Black Holes." *Scientific American,* May, p. 38.

Pines, D. 1980. "Accreting Neutron Stars, Black Holes, and Degenerate Dwarf Stars." *Science 207:* 597.

Popper, D. M. 1980. "Stellar Masses." *Annual Review of Astronomy and Astrophysics 18:* 115.

Reddy, F. 1983. "Supernovae: Still a Challenge." *Sky and Telescope 66:* 485.

***Ruderman,** M. 1971. "Solid Stars." *Scientific American,* February, p. 24.

*______. 1972. "Pulsars: Structure and Dynamics." *Annual Review of Astronomy and Astrophysics 10:* 427.

***Schorn,** R. A. 1982. "The Gamma-Ray Burster Puzzle." *Sky and Telescope 63:* 560.

Schwarzschild, M. 1958. *Structure and Evolution of the Stars.* Princeton, N.J.: Princeton University Press.

Shapley, H., ed. 1960. *Source Book in Astronomy, 1900–1950.* Cambridge, Mass.: Harvard University Press.

Shu, Frank H. 1982. *The Physical Universe.* Mill Valley, Calif.: University Science Books.

Strohmeier, W. 1972. *Variable Stars.* New York: Pergamon Press.

***Wade,** N. 1975. "Discovery of Pulsars: A Graduate Student's Story." *Science 189:* 359.

Waldrop, M. M. 1983. "The 0.001557806449023-Second Pulsar." *Science 219:* 831.

Walker, M. 1954. "Nova D. Q. Herculis (1934): An Eclipsing Binary with Very Short Period." *Publications of the Astronomical Society of the Pacific 66:* 230.

Wheeler, J., and K. Nomoto. 1985. "How Stars Explode." *American Scientist 73:* 240.

Witten, T. A. 1974. "Compounds in Neutron-Star Crusts." *Astrophysical Journal 188:* 615.

Zwicky, F. 1939. "On the Theory and Observation of Highly Collapsed Stars." *Physics Review 55:* 726.

Chapter 18

Allen, C. W. 1973. *Astrophysical Quantities.* 3rd ed. London: Athlone Press.

***Aller,** L. 1971. *Atoms, Stars and Nebulae.* Cambridge, Mass.: Harvard University Press.

***Blitz,** L. 1982. "Giant Molecular-Cloud Complexes in the Galaxy." *Scientific American,* April, p. 84.

Bowen, I. S. 1927. "The Origin of the Chief Nebular Lines." *Publications of the Astronomical Society of the Pacific 39:* 295.

Brackenridge, G. R. 1981. "Terrestrial Paleoenvironmental Effects of a Late Quaternary-Age Supernova." *Icarus 46:* 81.

***Cash,** W., and P. Charles. 1980. "Stalking the Cygnus Superbubble." *Sky and Telescope 59:* 455.

Deharveng, L., and M. Maucherat. 1975. "Optical Study of the Carina Nebula." *Astronomy and Astrophysics 41:* 27.

Dombrovsky, W. A. 1954. "On the Nature of the Crab Nebula Radiation." *Doklady Academy of Sciences USSR 94:* 1021.

Gaustad, J. E. 1970. "The Composition of Interstellar Dust." In *Dark Nebulae, Globules, and Protostars,* ed. B. T. Lynds. Tucson: University of Arizona Press.

***Gehrz,** R., and E. Ney. 1972. "The Core of Eta Carinae." *Sky and Telescope 44:* 4.

Greenberg, J. M. 1974. "The Interstellar Depletion Mystery, or Where Have All These Atoms Gone?" *Astrophysical Journal 189:* L81.

______. 1984. "The Structure and Evolution of Interstellar Grains." *Scientific American,* June, p. 124.

Hartmann, J. F. 1904. "Investigations on the Spectrum and Orbit of σ Orionis." *Astrophysical Journal 19:* 268.

Hartmann, W. K. 1970. "Growth of Planetesimals in Nebulae Surrounding Young Stars." In *Evolution stellaire avant la séquence principale.* Mémoires de la Société Royale des Sciences de Liège. Collection in −8°, 5th sér., *19,* 215.

Herbig, G. H. 1970. "Introductory Remarks." In *Evolution stellaire avant la séquence principale.* Mémoires de la Société Royale des Sciences de Liège. Collection in −8°, 5th sér., *19,* 13.

*______. 1974. "Interstellar Smog." *American Scientist 62:* 200.

Johnson, H. L. 1968. "Interstellar Extinction." In *Nebulae and Interstellar Material,* ed. B. Middlehurst and L. Aller. Chicago: University of Chicago Press.

***Lewis,** R., and E. Anders. 1983. "Interstellar Matter in Meteorites." *Scientific American,* August, p. 66.

Lewis, R. S., et al. 1980. "Stellar Condensates in Meteorites: Isotopic Evidence from Noble Gases." *Astrophysical Journal 234:* L165.

***Lovi,** G. 1972. "Rambling Through May Skies." *Sky and Telescope 43:* 306.

***Lynds,** B. T., ed. 1971. *Dark Nebulae, Globules and Protostars.* Tucson: University of Arizona Press.

***Maran,** S. 1971. "The Gum Nebula." *Scientific American,* December, p. 20.

Maran, S., J. Brandt, and T. Stecher, eds. 1973. *The Gum Nebula and Related Problems.* Publication no. SP-332. Washington, D.C.: National Aeronautics and Space Administration.

***Miller,** J. S. 1974. "The Structure of Emission Nebulae." *Scientific American,* October, p. 34.

Ney, E. P. 1972. "Infrared Excesses in Supergiant Stars: Evidence for Silicates." *Publications of the Astronomical Society of the Pacific 84:* 613.

Oort, J., and H. C. van de Hulst. 1946. "Gas and Smoke in Interstellar Space." *Bulletin of the Astronomical Society of the Netherlands 10:* 187.

Poveda, A. 1965. "G. W. Orionis, a 20,000 Year Old T Tauri Star?" *Boletin de los Observatorio Tonantzintla y Tacubaya 4:* 77.

Scoville, N. Z. 1975. "Molecular Clouds in the Galaxy." *Astrophysical Journal 199:* L105.

***Scoville,** N., and J. Young. 1984. "Molecular Clouds, Star Formation and Galactic Structure." *Scientific American,* April, p. 42.

Shapiro, P. R. 1975. "Interstellar Polarization: Magnetite Dust." *Astrophysical Journal 201:* 151.

Shlovsky, I. S. 1954. "On the Nature of Radio Radiation of Nebular Masses." *Astronomical Journal of the USSR 31:* 533.

Shu, F. H. 1982. *The Physical Universe: An Introduction to Astronomy.* Mill Valley, Calif.: University Science Books.

Simon, T., and H. M. Dyck. 1975. "Silicate Absorption at 18 μm in Two Peculiar Infrared Sources." *Nature 253:* 101.

Stromgren, B. 1939. "The Physical State of Interstellar Hydrogen." *Astrophysical Journal 89:* 526.

Swart, P. K., et al. 1983. "Interstellar Carbon in Meteorites." *Science 220:* 406.

Trumpler, R. J. 1930. "Preliminary Results on the Distances, Dimensions, and Space Distributions of Open Star Clusters." *Lick Observatory Bulletin 14:* 154.

***Turner,** B. E. 1973. "Interstellar Molecules." *Scientific American,* March, p. 50.

***Vandervoort,** P. O. 1965. "The Age of the Orion Nebula." *Scientific American,* February, p. 90.

Walborn, N., J. Heckathorn, and J. Hesser. 1984. "The High-ionization and Excited-state Interstellar Lines in the Carina Nebula: A Giant HII Region in Absorption." *Astrophysical Journal 276:* 524.

Whittaker, A. G., et al. 1980. "Carbynes: Carriers of Primordial Noble Gases in Meteorites." *Science 209:* 1512.

***Wickramasinghe,** R. M. 1967. "Interstellar Molecules." *Scientific American,* March, p. 50.

Zuckerman, B. 1973. "A Model of the Orion Nebula." *Astrophysical Journal 183:* 863.

Chapter 19

Abt, H. A. 1979. "The Frequencies of Binaries on the Main Sequence." *Astronomical Journal 84:* 1591.

______. 1983. "Normal and Abnormal Binary Frequencies." *Annual Review of Astrophysics 21:* 343.

Abt, H., C. Bolton, and S. Levy. 1972. "1C 4665, A Cluster of Binaries." *Astrophysical Journal 171:* 259.

Abt, H., and S. Levy. 1976. "Multiplicity Among Solar-Type Stars." *Astrophysical Journal,* Supplement series *30:* 273.

Arny, T., and P. Weissman. 1973. "Interaction of Proto-Stars in a Collapsing Cluster." *Astronomical Journal 78:* 310.

Backman, D., et al. 1984. "Infrared Observations of the Eclipse of Epsilon Aurigae: Direct Measurement of the 500 K Secondary at 5, 10, 20 Microns." Submitted to *Astrophysical Journal.*

***Batten,** A. H. 1973. *Binary and Multiple Systems of Stars.* Oxford: Pergamon Press.

Dyck, H., T. Simon, and B. Zuckerman. 1982. "Discovery of an Infrared Companion to T Tauri." *Astrophysics Journal Letters 225:* L103.

Feibelman, W. A. 1963. "Stars near Beta Lyrae." *Astrophysical Journal 137:* 701.

Gatewood, G., and H. Eichhorn. 1973. "An Unsuccessful Search for a Planetary Companion of Barnard's Star." *Astronomical Journal 78:* 769.

Grindlay, J. E., et al. 1984. "The Central X-ray Source in SS 433." *Astrophysical Journal 277:* 286.

Hack, 1984. "Epsilon Aurigae." *Scientific American,* October, p. 98.

Hanson, R., B. Jones, and D. Lin. 1983. "The Astrometric Position of T Tauri and the Nature of Its Companion." *Astrophysics Journal Letters 270:* L27.

Heintz, W. D. 1976. "Systematic Trends in the Motions of Suspected Stellar Companions." *Monthly Notices of the Royal Astronomical Society 175:* 533.

***______.** 1978. *Double Stars.* Dordrecht, Netherlands: Reidel.

Hershey, J. L. 1973. "Astrometric Analysis of the Triple Star BD +66°34." *Astronomical Journal 78:* 935.

Jensen, O. G., and T. Ulrych. 1973. "An Analysis of the Perturbations on Barnard's Star." *Astronomical Journal 78:* 1104.

Kopal, Z. 1955. "'The Classification of Close Binary Systems." *Annals of Astrophysics 18:* 379.

______. 1971. "The Eclipsing System of Epsilon Aurigae and Its Possible Relevance to the Formation of a Planetary System." *Astrophysics and Space Science 10:* 332.

Kraft, R. P. 1964. "Cataclysmic Variables as Binary Stars." *Advances in Astronomy and Astrophysics 2:* 43.

***Margon,** B. 1980. "The Bizarre Spectrum of SS 433." *Scientific American,* October, p. 54.

***______.** 1982. "Relativistic Jets in SS 433." *Science 215:* 247.

Popper, D. M. 1980. "Stellar Masses." *Annual Review of Astronomy and Astrophysics 18:* 115.

Radzievskii, V., and E. Radzievskaya. 1973. "Coplanar System of Binary Stars in Aquila." *Soviet Astronomy 17:* 239.

Smith, B. A., and Terrile, R. 1984. "A Circumstellar Disk Around Beta Pictoris." *Science 226:* 1421.

Standish, E. M., Jr. 1972. "The Dynamical Evolution of Triple Star Systems: A Numerical Study." *Astronomy and Astrophysics 21:* 185.

***Struve,** O. 1955. "The Duplicity of Nova Herculis." *Sky and Telescope 14:* 275.

Thomsen, D. 1985. "A 'Brickbat' in the Sky." *Science News 127:* 154.

Van de Kamp, P. 1963. "Astrometric Study of Barnard's Star with Plates Taken with the Sproul 24-Inch Refractor." *Astronomical Journal 68:* 515.

______. 1969. "Alternate Dynamical Analysis of Barnard's Star." *Astronomical Journal 74:* 757.

______. 1975. "Astrometric Analysis of Barnard's Star from Plates Taken with the Sproul Observatory 61-cm Refractor." *Astronomical Journal 80:* 658.

______. 1979. Quoted in *Sky and Telescope 57:* 247.

______. 1981. *Stellar Paths.* Dordrecht, Netherlands: Reidel.

Ventura, J. et al. 1983. "Can X-ray Bursts Originate from Low-Mass Binaries?" *Nature 301:* 491.

Wanner, J. F. 1967. "The Visual Binary Krüger 60." *Sky and Telescope 33:* 16.

***Warner,** B. 1972. "Six Ultra-Short Period Binary Stars." *Sky and Telescope 44:* 358.

Chapter 20

Ambartsumian, V. A. 1947. *Stellar Evolution and Astrophysics.* U.S.S.R.: Erevan.

______. 1960. "Expanding Stellar Associations." In *Source Book in Astronomy, 1900–1950,* ed. H. Shapley. Cambridge, Mass.: Harvard University Press.

Anthony-Twarog, B. J., B. Twarog, and R. D. McClure. 1979. "The Old(est) Open Cluster: Melotte 66." *Astrophysical Journal 233:* 188.

***Iben,** I. 1970. "Globular-Cluster Stars." *Scientific American,* July, p. 26.

Jones, K., and P. Demarque. 1983. "The Ages and Composition of Old Clusters." *Astrophysical Journal 264:* 206.

***Jones,** K. G. 1969. *Messier's Nebulae and Star Clusters.* New York: American Elsevier.

King, I. R. 1985. "Globular Clusters." *Scientific American,* July, p. 79.

***Metz,** W. D. 1975. "Astronomy from Space: New Class of X-Ray Sources Found." *Science 189:* 1073.

Ostriker, J., L. Spitzer, and R. Chevalier. 1972. "On the Evolution of Globular Clusters." *Astrophysical Journal 176:* L51.

Peebles, P., and R. Dicke. 1968. "Origin of the Globular Star Clusters." *Astrophysical Journal 154:* 891.

Ruzmaikina, T. 1972. "On the Cosmological Origin of Globular Clusters." *Astronomical zhurnal nauk SSSR 49:* 1229.

***Shapley,** H. 1930. *Star Clusters.* Cambridge, Mass.: Harvard University Press.

Spitzer, L., Jr. 1984. "Dynamics of Globular Clusters." *Science 225:* 465.

Trumpler, R. 1930. "Absorption of Light in the Galactic System." *Publications of the Astronomical Society of the Pacific 42:* 214.

Chapter 21

Allen, C. W. 1973. *Astrophysical Quantities.* London: Athlone Press.

Blaauw, A., and M. Schmidt, eds. 1965. *Galactic Structure.* Chicago: University of Chicago Press.

***Bok,** B. J., and P. Bok. 1981. *The Milky Way.* 5th ed. Cambridge, Mass.: Harvard University Press.

Dicke, R. H. 1969. "The Age of the Galaxy from the Decay of Uranium." *Astrophysical Journal 155:* 123.

***Geballe,** T. R. 1979. "The Central Parsec of the Galaxy." *Scientific American,* July, p. 60.

Genzel, R. et al. 1984. "Far-infrared Spectroscopy of the Galactic Center: Neutral and Ionized Gas in the Central 10 Parsecs of the Galaxy." *Astrophysical Journal 276:* 551.

Hawkins, M.R.S. 1983. "Direct Evidence for a Massive Galactic Halo." *Nature 303:* 406.

***Huang,** S.-S. 1972. "Disk and Ring Structure in the Universe." *Sky and Telescope 43:* 225.

Jansky, K. 1935. "A Note on the Source of Interstellar Interference." *Proceedings of the Institute of Radio Engineers 23:* 1158.

Kapteyn, O. C. 1918. "On the Parallaxes and Motions of the Brighter Galactic Helium Stars." *Astrophysical Journal 47:* 107, 146, 255.

Klare, G., and T. Neckel. 1970. "Polarization of Southern OB Stars." In *The Spiral Structure of Our Galaxy,* ed. W. Becker and G. Contopolos. Dordrecht, Netherlands: Reidel.

***Kraft,** R. P. 1959. "Pulsating Stars and Cosmic Distances." *Scientific American,* July, p. 48.

Lacy, J. H., et al. 1980. "Observations of the Motion and Distribution of the Ionized Gas in the Central Parsec of the Galaxy. II." *Astrophysical Journal 241:* 132.

Leventhal, M., et al. 1982. "Time-variable Positron Annihilation Radiation from the Galactic Center Direction." *Astrophysical Journal Letters 260:* L1.

Lo, K., and M. Claussen. 1983. "High-resolution Observations of Ionized Gas in Central 3 Parsecs of the Galaxy: Possible Evidence for Infall." *Nature 306:* 647.

Maraschi, L., A. Treves, and M. Tarenghi. 1973. "Accretion by Neutron Stars at the Galactic Center." *Astronomy and Astrophysics 25:* 153.

Moffat, F. J., and N. Vogt. 1973. "An Up-to-Date Picture of Galactic Spiral Features Based on Young Open Star Clusters." *Astronomy and Astrophysics 23:* 317.

O'Connell, D.J.K., ed. 1958. *Stellar Populations.* New York: Wiley-Interscience.

Oort, J., and G. Rougoor. 1960. "The Position of the Galactic Center. V." *Monthly Notices of the Royal Astronomical Society 121:* 171.

Reber, G. 1944. "Cosmic Static." *Astrophysical Journal 100:* 279.

Sanders, R., and K. Prendergast. 1974. "The Possible Relation of the 3-kpc Arm to Explosions in the Galactic Nucleus." *Astrophysical Journal 188:* 489.

***Saunders,** J. 1963. "The Globular Cluster Omega Centauri." *Sky and Telescope 26:* 133.

***Shapley,** H. 1930. *Star Clusters.* Cambridge, Mass.: Harvard University Press.

Shapley, H., and H. Howarth, eds. 1929. *A Source Book in Astronomy.* New York: McGraw-Hill.

***Struve,** O. 1960. "A Historic Debate About the Universe." *Sky and Telescope 19:* 398.

Walborn, N. R. 1973. "The Space Distribution of the O Stars in the Solar Neighborhood." *Astronomical Journal 78:* 1067.

Waldrop, M. M. 1985. "The Core of the Milky Way." *Science 230:* 158.

***Whitney,** C. A. 1971. *The Discovery of Our Galaxy.* New York: Alfred A. Knopf.

Chapter 22

Allen, C. W. 1973. *Astrophysical Quantities.* London: Athlone Press.

***Bok,** B. J. 1966. "Magellanic Clouds." *Annual Review of Astronomy and Astrophysics 4:* 95.

Freedman, W., and B. Madore. 1983. "Time Evolution of Disk Galaxies Undergoing Stochastic Self-propagating Star Formation." *Astrophysical Journal 265:* 140.

***Hirshfeld,** A. 1980. "Inside Dwarf Galaxies." *Sky and Telescope 59:* 287.

Hubble, E. 1925. "N. G. C. 6822, a Remote Stellar System." *Astrophysical Journal 62:* 409.

***Humphreys,** R., and K. Davidson. 1984. "The Most Luminous Stars." *Science 223:* 243.

Light, E., R. Danielson, and M. Schwarzschild. 1974. "The Nucleus of M 31." *Astrophysical Journal 194:* 257.

Mathewson, D. 1985. "The Clouds of Magellan." *Scientific American,* April, p. 107.

Mathewson, D., M. Cleary, and J. Murray. 1974. "The Magellanic Stream." *Astrophysical Journal 190:* 291.

Mathis, J., B. Savage, and J. Cassinelli. 1984. "A Superluminous Object in the Large Cloud of Magellan." *Scientific American,* August, p. 52.

Ostriker, J. P. 1981. "Some Thoughts on Galaxy Formation." *Bulletin of the American Astronomy Society 12:* 770.

***Rubin,** V. C. 1973. "The Dynamics of the Andromeda Nebula." *Scientific American,* June, p. 30.

***Rubin,** V. 1983. "Dark Matter in Spiral Galaxies." *Scientific American,* June, p. 96.

***Sandage,** A. 1961. *The Hubble Atlas of Galaxies.* Washington, D.C.: The Carnegie Institution.

Sandage, A., and G. A. Tammann. 1974. "Steps Toward the Hubble Constant. III." *Astrophysical Journal 194:* 223.

***Shapley,** H. 1957. *The Inner Metagalaxy.* New Haven, Conn.: Yale University Press.

***Strom,** K., and S. Strom. 1982. "Galactic Evolution: A Survey of Recent Progress." *Science 216:* 571.

Thackeray, A. D. 1971. "Survey of Principal Characteristics of the Magellanic Clouds." In *The Magellanic Clouds,* ed. A. Muller. Boston: Reidel.

van den Bergh, S. 1972. "Search for Faint Companions to M31." *Astrophysical Journal 171:* L35.

***Whitney,** C. A. 1971. *The Discovery of Our Galaxy.* New York: Alfred A. Knopf.

Chapter 23

Aaronson, M., and J. Mould. 1983. "A Distance Scale from the Infrared Magnitude/H Velocity–width Relation. IV. The Morphological Type Dependence and Scatter in the Relation; the Distances to Nearby Groups." *Astrophysical Journal 265:* 1.

Allen, C. W. 1973. *Astrophysical Quantities.* London: Athlone Press.

Arp, H. 1974. "Evidence for Non-Velocity Red-shifts—New Evidence and Review." In *The Formation and Dynamics of Galaxies,* ed. J. Shakeshaft. Dordrecht, Netherlands: Reidel.

Balick, B., and T. Heckman. 1983. "Spectroscopy of the Fuzz Associated with Four Quasars." *Astrophysical Journal 265:* L1.

Balkowski, C., et al. 1974. "Observational Evidence for Non-Velocity Red-shift in Stephan's Quintet." In *The Formation and Dynamics of Galaxies,* ed. J. Shakeshaft. Dordrecht, Netherlands: Reidel.

Bartel, N., et al. 1985. "Hubble's Constant Determined Using Very-long Baseline Interferometry." *Nature 318:* 25.

Blandford, R., et al. 1982. "Cosmic Jets." *Scientific American,* May, p. 124.

Burns, J., and R. Price. 1983. "Centaurus A: The Nearest Active Galaxy." *Scientific American,* November, p. 56.

Chincarini, G., and H. Rood. 1980. "The Cosmic Tapestry." *Sky and Telescope 59:* 364.

de Vaucouleurs, G. 1974. "Structure, Dynamics, and Statistical Properties of Galaxies." In *The Formation and Dynamics of Galaxies,* ed. J. Shakeshaft. Dordrecht, Netherlands: Reidel.

Djorgovski, S., et al. 1985. "Discovery of a Probable Galaxy with a Redshift of 3.218." *Astrophysical Journal 299:* L1.

Eddington, A. 1933. *The Expanding Universe.* Cambridge: Cambridge University Press.

***Hodge,** P. W. 1972. "Some Current Studies of Galaxies." *Sky and Telescope 44:* 23.

———. 1984. "The Cosmic Distance Scale." *American Scientist 72:* 474.

***Hoyle,** F. 1972. *From Stonehenge to Modern Cosmology.* San Francisco: W. H. Freeman.

Hubble, E. P. 1931. "The Velocity Distance Relation Among Extra-Galactic Nebulae." *Astrophysical Journal 74:* 43.

———. 1936. "A Relation Between Distance and Radial Velocity Among Extra-Galactic Nebulae." Reprinted in *Source Book in Astronomy, 1900–1950,* ed. H. Shapley. Cambridge, Mass.: Harvard University Press, 1960.

———. 1958. *The Realm of the Nebulae.* New York: Dover.

Khachikian, E. Y., and D. Weedman. 1974. "An Atlas of Seyfert Galaxies." *Astrophysical Journal 192:* 581.

Kronberg, P., and R. Sramek. 1985. "Discovery of New Variable Radio Sources in the Nucleus of the Nearby Galaxy M82." *Science 227:* 28.

Kuhr, H., et al. 1973. "The Most Luminous Quasar: S5 0014+81." *Astrophysical Journal 275:* L33.

Oort, J. H. 1974. "Recent Radio Studies of Bright Galaxies." In *The Formation and Dynamics of Galaxies,* ed. J. Shakeshaft. Dordrecht, Netherlands: Reidel.

Osmer, P. 1982. "Quasars as Probes of the Distant and Early Universe." *Scientific American,* February, p. 126.

Peterson, B. 1982. "PKS 2000-300: A quasi-stellar Radio Source with a Red Shift of 3.78." *Astrophysical Journal 260:* L27.

Petrosian, V. 1974. "The Hubble Relation for Non-Standard Candles and the Origin of the Redshift of Galaxies." *Astrophysical Journal 188:* 443.

Sandage, A., and G. A. Tammann. 1974. "Steps Toward the Hubble Constant. IV." *Astrophysical Journal 194:* 559.

Saslaw, W. C. 1974. "Theory of Galactic Nuclei." In *The Formation and Dynamics of Galaxies,* ed. J. Shakeshaft. Dordrecht, Netherlands: Reidel.

Schweizer, F. 1986. "Colliding and Merging Galaxies." *Science 231:* 227.

Shapley, H., ed. 1960. *Source Book in Astronomy, 1900–1950.* Cambridge, Mass.: Harvard University Press.

Shu, F. H. 1982. *The Physical Universe: An Introduction to Astronomy.* Mill Valley, Calif.: University Science Books.

Silk, J. 1980. *The Big Bang.* San Francisco: W. H. Freeman.

Smith, M., et al. 1983. "Bubbles, Jets, and Clouds in Active Galactic Nuclei." *Astrophysical Journal 264:* 432.

***Toomre,** A., and J. Toomre. 1973. "Violent Tides Between Galaxies." *Scientific American,* December, p. 38.

Turner, E. 1984. "Quasars and Gravitational Lenses." *Science 223:* 1255.

van der Hulst, J., W. Golisch, and A. Huschick. 1983. "The H1 Absorption on NGC 5128 (Centaurus A)." *Astrophysical Journal 264:* L37.

Waldrop, M. 1983. "NGC 4151: The Monster in the Middle." *Science 222:* 1003.

———. 1985. "Why Do Galaxies Exist?" *Science 228:* 978.

***Weymann,** R. J. 1969. "Seyfert Galaxies." *Scientific American,* January, p. 28.

***Whitney,** C. A. 1971. *The Discovery of Our Galaxy.* New York: Alfred A. Knopf.

Wolfe, A. M. 1974. "On the Steady Flow of Gas from the Nuclei of Seyfert Galaxies." *Astrophysical Journal 188:* 243.

Chapter 24

Brown, G. S., and B. Tinsley. 1974. "Galaxy Counts as a Cosmological Test." *Astrophysical Journal 194:* 555.

***deBoer,** K., and B. Savage. 1982. "The Coronas of Galaxies." *Scientific American,* August, p. 54.

De Lapparent, V., M. Geller, and J. Huchra. 1986. "A Slice of the Universe." *Astrophysical Journal 302:* L1.

***de Vaucouleurs,** G. 1970. "The Case for a Hierarchical Cosmology." *Science 167:* 1203.

***Dicus,** D., et al. 1983. "The Future of the Universe." *Scientific American,* March, p. 90.

Einstein, A. 1952. "Cosmological Considerations on the General Theory of Relativity." In *The Principle of Relativity.* New York: Dover. (Originally published 1917.)

***Gaster,** T. H. 1952. *The Oldest Stories in the World.* Boston: Beacon Press.

Gott, J. R., et al. 1974. "An Unbound Universe?" *Astrophysical Journal 194:* 543.

***Gregory,** S., and L. Thompson. 1982. "Superclusters and Voids in the Distribution of Galaxies." *Scientific American,* March, p. 106.

***Harrison,** E. R. 1974. "Why the Sky Is Dark at Night." *Physics Today 27:* 5.

***Hoyle,** F. 1972. *From Stonehenge to Modern Cosmology.* San Francisco: W. H. Freeman.

Kant, I. 1968. *Kant's Cosmology.* Translated by W. Hastie. New York: Greenwood.

Kraus, J. 1981. "Universe Beyond the Galaxies." Letter in *Science 212:* 610.

North, J. D. 1965. *The Measure of the Universe.* Oxford: Oxford University Press.

Overbye, D. 1980. "The Ghost Universe of Neutrinos." *Sky and Telescope 60:* 115.

Peebles, P. J. E. 1971. *Physical Cosmology.* Princeton, N.J.: Princeton University Press.

Raychaudhuri, A. K. 1979. *Theoretical Cosmology.* Oxford: Clarendon Press.

***Rubin,** V. 1983. "Dark Matter in Spiral Galaxies." *Scientific American,* June, p. 96.

Ryle, M. 1968. "The Counts of Radio Sources." *Annual Review of Astronomy and Astrophysics 6:* 249.

***Schatzman,** E. S. 1965. *The Origin and Evolution of the Universe.* New York: Basic Books.

Shu, F. 1982. *The Physical Universe.* Mill Valley, Calif.: University Science Press.

Silk, J., 1980. *The Big Bang.* San Francisco: W. H. Freeman.

***Silk,** J., A. Szalay, and Y. Zel'dovich. 1983. "The Large-Scale Structure of the Universe." *Scientific American,* October, p. 72.

***Waldrop,** M. M. 1983. "The New Inflationary Universe." *Science 219:* 375.

***Wilson,** J. A. 1949. "Egypt." In *Before Philosophy,* ed. H. Frankfort and H. A. Frankfort. Baltimore: Penguin Books.

Chapter 25

Alpher, R. A., and R. C. Herman. 1948. "Evolution of the Universe." *Nature 162:* 774.

***Barrow,** J., and M. Turner. 1982. "The Inflationary Universe—Birth, Death, and Transfiguration." *Nature 298:* 801.

Dicke, R. H., et al. 1965. "Cosmic Black Body Radiation." *Astrophysical Journal 142:* 414.

Gamow, G. 1952. *The Creation of the Universe.* New York: Viking Press.

***Hawking,** S. W. 1977. "The Quantum Mechanics of Black Holes." *Scientific American,* January, p. 34.

***Morrison,** D., and N. Morrison. 1975. "An Infinitely Expanding Universe." *Mercury,* September, p. 27.

Ozernoy, L. M. 1974. "Dynamics of Superclusters as the Most Powerful Test for Theories of Galaxy Formation." In *The Formation and Dynamics of Galaxies,* ed. J. Shakeshaft. Dordrecht, Netherlands: Reidel.

Peebles, P. J. E. 1971. *Physical Cosmology.* Princeton, N.J.: Princeton University Press.

Penzias, A. A., and R. W. Wilson. 1965. "A Measurement of Excess Antenna Temperature at 4080 Mc/s." *Astrophysical Journal 142:* 419.

Sandage, A., and G. A. Tammann. 1974. "Steps Toward the Hubble Constant. IV." *Astrophysical Journal 194:* 559.

Schmidt, M. 1982. "Quasar Boundary." *Science 216:* 6.

***Silk,** J. 1980. *The Big Bang.* San Francisco: W. H. Freeman.

Thorstensen, J. R., and R. B. Partridge. 1975. "Can Collapsed Stars Close the Universe?" *Astrophysical Journal 200:* 527.

***Webster,** A. 1974. "The Cosmic Background Radiation." *Scientific American,* August, p. 26.

Chapter 26

Ahrens, T. J., and J. D. O'Keefe. 1982. "Impact of an Asteroid or Comet in the Ocean and Extinction of Terrestrial Life." *Lunar Planet Science Abstracts 13:* 3.

Allen, C. W. 1973. *Astrophysical Quantities.* London: Athlone Press.

***Alvarez,** W., et al. 1984. "Impact Theory of Mass Extinctions and the Invertebrate Fossil Record." *Science 223:* 1135.

***Berendzen,** R., ed. 1973. *Life Beyond Earth and the Mind of Man.* Publication no. SP-328. Washington, D.C.: National Aeronautics and Space Administration.

Berkner, L. V., and L. C. Marshall. 1965. "On the Origin and Rise of Oxygen Concentration in the Earth's Atmosphere." *Journal of Atmospheric Science 22:* 225.

Brock, T. D. 1985. "Life at High Temperatures." *Science 230:* 132.

***Covey,** C. 1984. "The Earth's Orbit and the Ice Ages." *Scientific American,* February, p. 58.

***Dole,** S. H. 1964. *Habitable Planets for Man.* Waltham, Mass.: Blaisdell.

***Dubos,** R. 1968. *So Human an Animal.* New York: Charles Scribner's.

***Edmund,** J. M., and K. Von Damm. 1983. "Hot Springs on the Ocean Floor." *Scientific American,* April, p. 78.

Goldsmith, D., and T. Owen. 1979. *The Search for Life in the Universe.* Menlo Park, Calif.: Benjamin/Cummings.

***Gould,** S. J. 1974a. "The Great Dying." *Natural History,* October, p. 22.

*______. 1974b. "An Unsung Single-Celled Hero." *Natural History,* November, p. 33.

______. 1978. "An Early Start." *Natural History,* February, p. 10.

Hanson, R., B. Jones, and D. Lin. 1983. "The Astrometric Position of T Tauri and the Nature of Its Companion." *Astrophysical Journal 270:* L27.

Harrington, R. S. 1977. "Planetary Orbits in Binary Stars." *Astronomical Journal 82:* 753.

Harrington, R. S., V. V. Kallarkakal, and C. Dahn. 1983. "Astrometry of the Low-Luminosity Stars VB8 and VB10." *Astronomical Journal 88:* 1038.

Holland, H. D. 1962. "Model for Evolution of the Earth's Atmosphere." In *Petrologic Studies.* New York: Geological Society of America.

Horowitz, N. H., G. Hobby, and J. Hubbard. 1976. "The Viking Carbon Assimilation Experiments: Interim Report." *Science 194:* 1321.

Horowitz, P. 1978. "A Search for Ultra-Narrowband Signals of Extraterrestrial Origin." *Science 201:* 733.

Huang, S.-S. 1965. "Rotational Behavior of the Main-Sequence Stars and Its Plausible Consequences Concerning Formation of Planetary Systems." *Astrophysical Journal 141:* 985.

Imbrie, J., and J. Z. Imbrie. 1980. "Modeling the Climatic Response to Orbital Variations." *Science 207:* 943.

Karl, D. M., et al. 1980. "Deep-Sea Primary Production at the Galápagos Hydrothermal Vents." *Science 207:* 1345.

***Kurten,** B. 1969. "Continental Drift and Evolution." *Scientific American,* March, p. 54.

Kvenvolden, K., et al. 1970. "Evidence for Extraterrestrial Amino Acids and Hydrocarbons in the Murchison Meteorite." *Nature 228:* 923.

Levin, G. V., and P. A. Straat. 1976. "Viking Labeled Release Biology Experiment: Interim Results." *Science 194:* 1322.

McLaren, D. J. 1970. "Presidential Address." *Journal of Paleontology 44:* 801.

***Maruyama,** M., and A. Harkins. 1975. *Cultures Beyond the Earth.* New York: Vintage.

Miller, S. L. 1955. "Production of Some Organic Compounds Under Possible Primitive Earth Conditions." *Journal of the American Chemical Society 77:* 2351.

Nisbet, E. G. 1980. "Archaean Stromatolites and the Search for the Earliest Life." *Nature 284:* 394.

O'Leary, B. 1980. "Searching for Other Planetary Systems." *Sky and Telescope 60:* 111.

***Oparin,** A. I. 1962. *Life: Its Nature, Origin, and Development.* New York: Academic Press.

***Pollard,** W. B. 1979. "The Prevalence of Earth-like Planets." *American Scientist 67:* 653.

Ponnamperuma, C., ed. 1983. *Cosmochemistry and the Origin of Life.* Dordrecht, Netherlands: D. Reidel.

Raup, D. M. 1979. "Size of the Permo-Triassic Bottleneck and Its Evolutionary Implications." *Science 206:* 217.

***Russell,** D. A. 1982. "The Mass Extinctions of the Late Mesozoic." *Scientific American,* January, p. 82.

Sagan, C. ed. 1973. *Communication with Extraterrestrial Intelligence.* Cambridge, Mass.: M.I.T. Press.

Sekanina, Z., and D. Yeomans. 1984. "Close Encounters and Collisions of Comets with the Earth." *Astronomical Journal 89:* 154.

Shklovskii, I. S., and C. Sagan. 1966. *Intelligent Life in the Universe.* San Francisco: Holden-Day.

***Siegel,** S. M. 1972. "Experimental Biology of Extreme Environments and Its Significance for Space Bioscience—2." *Spaceflight 12:* 256.

Simpson, G. G. 1964. "The Nonprevalence of Humanoids." *Science 143:* 769.

———. 1973. Added comments on "The Nonprevalence of Humanoids." In *Communication with Extraterrestrial Intelligence,* ed. C. Sagan. Cambridge, Mass.: M.I.T. Press.

Strand, K. A. 1977. "Triple System Stein 2051 (G175-34)." *Astronomical Journal 82:* 745.

Turco, R., et al. 1981. "Tunguska Meteor Fall of 1908: Effects on Stratospheric Ozone." *Science 212:* 19.

***Underwood,** J. 1975. *Biocultural Interactions and Human Variation.* Dubuque, Iowa: W. C. Brown.

Van Valen, L., and R. Sloan. 1972. "Ecology and the Extinction of Dinosaurs." *24th International Geological Congress Abstracts.*

Vidal, G. 1984. "The Oldest Eukaryotic Cells." *Scientific American,* February, p. 48.

Walker, J. C. G. 1977. *Evolution of the Atmosphere.* New York: Hafner.

Windley, B. F. 1977. *The Evolving Continents.* New York: Wiley and Sons.

***Woese,** Carl R. 1981. "Archaebacteria." *Scientific American,* June, p. 98.

Wolbach, W., R. Lewis, and E. Anders. 1985. "Cretaceous Extinctions: Evidence for Wildfires and Search for Meteoritic Material." *Science 230:* 167.

Enrichment Essay A

***Abetti,** G. 1951. *The History of Astronomy.* New York: Henry Schuman.

***Ashbrook,** J. 1973. "Astronomical Scrapbook." *Sky and Telescope 46:* 300.

***Condon,** E. U., et al. 1969. *Scientific Study of Unidentified Flying Objects.* New York: Dutton.

Emiliani, C., et al. 1975. "Paleoclimatological Analysis of Late Quaternary Cores from the Northeastern Gulf of Mexico." *Science 189:* 1083.

***Gould,** S. J. 1975. "Velikovsky in Collision." *Natural History,* February, p. 20.

Mathur, K. D. 1974. "Indian Astronomy in the Era of Copernicus." *Nature 251:* 283.

***Minnaert,** M. 1954. *Light and Colour in the Open Air.* New York: Dover.

***Numbers,** R. L. 1982. "Creationism in 20th-Century America." *Science 218:* 538.

Pannekoek, A. 1961. *A History of Astronomy.* London: George Allen and Unwin.

***Shklovskii,** I. S., and C. Sagan. 1966. *Intelligent Life in the Universe.* San Francisco: Holden-Day.

***Stephenson,** F. R., and D. H. Clark. 1977. "Ancient Astronomical Records from the Orient." *Sky and Telescope 53:* 84.

***Story,** R. 1976. *The Space Gods Revealed.* New York: Harper & Row.

Velikovsky, I. 1950. *Worlds in Collision.* New York: Doubleday.

von Däniken, E. 1971. *Chariots of the Gods?* New York: Bantam.

Historically important astronomers, literary figures, contributing artists, and selected others.

Key **Concepts** appear in **boldface** type.

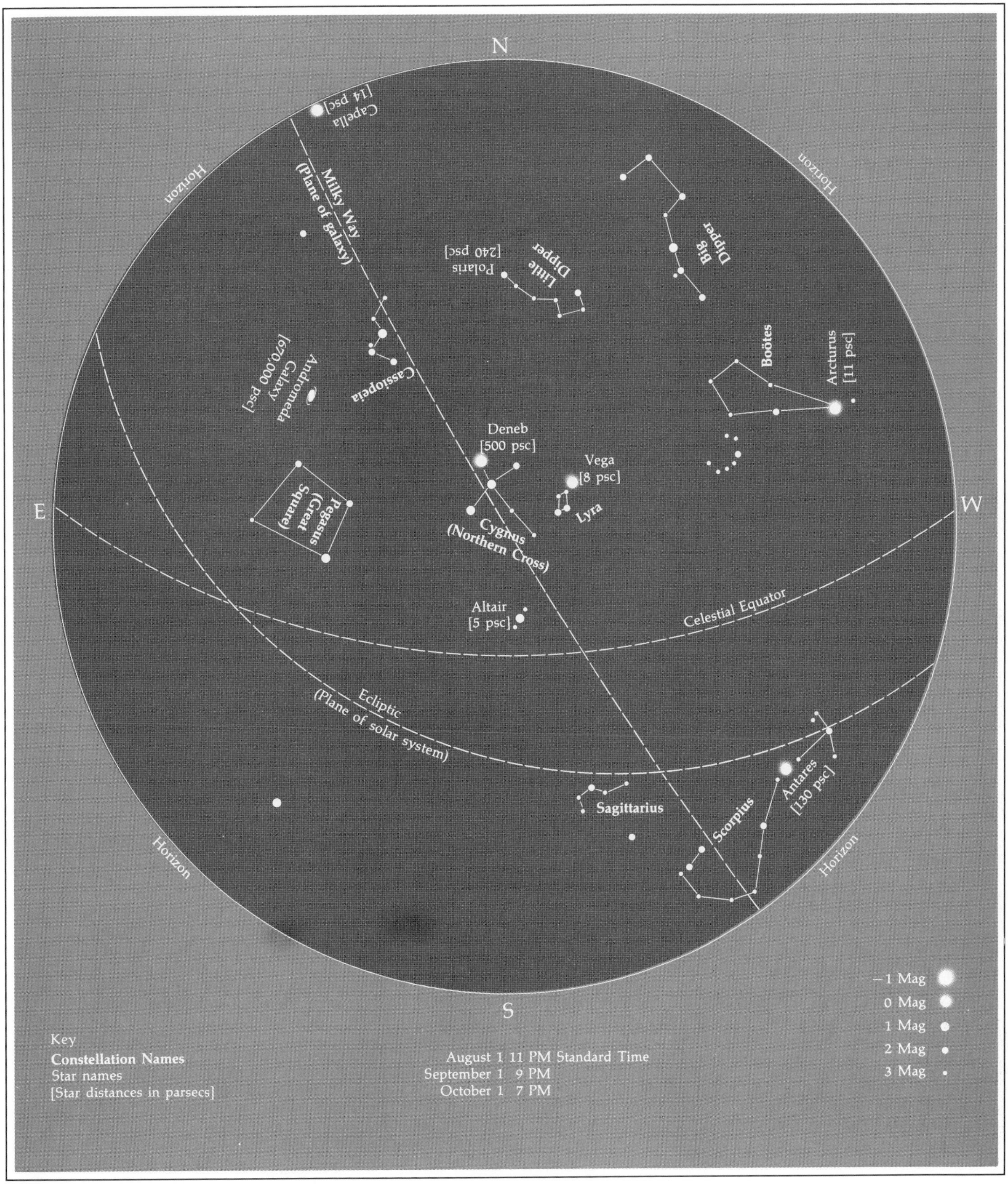
N
S
E
W
Horizon
Horizon
Horizon
Horizon
Capella [14 psc]
Milky Way (Plane of galaxy)
Little Dipper
Polaris [240 psc]
Big Dipper
Cassiopeia
Andromeda Galaxy [670,000 psc]
Boötes
Arcturus [11 psc]
Deneb [500 psc]
Vega [8 psc]
Lyra
Pegasus (Great Square)
Cygnus (Northern Cross)
Altair [5 psc]
Celestial Equator
Ecliptic (Plane of solar system)
Sagittarius
Scorpius
Antares [130 psc]
−1 Mag
0 Mag
1 Mag
2 Mag
3 Mag
Key
Constellation Names
Star names
[Star distances in parsecs]
August 1 11 PM Standard Time
September 1 9 PM
October 1 7 PM

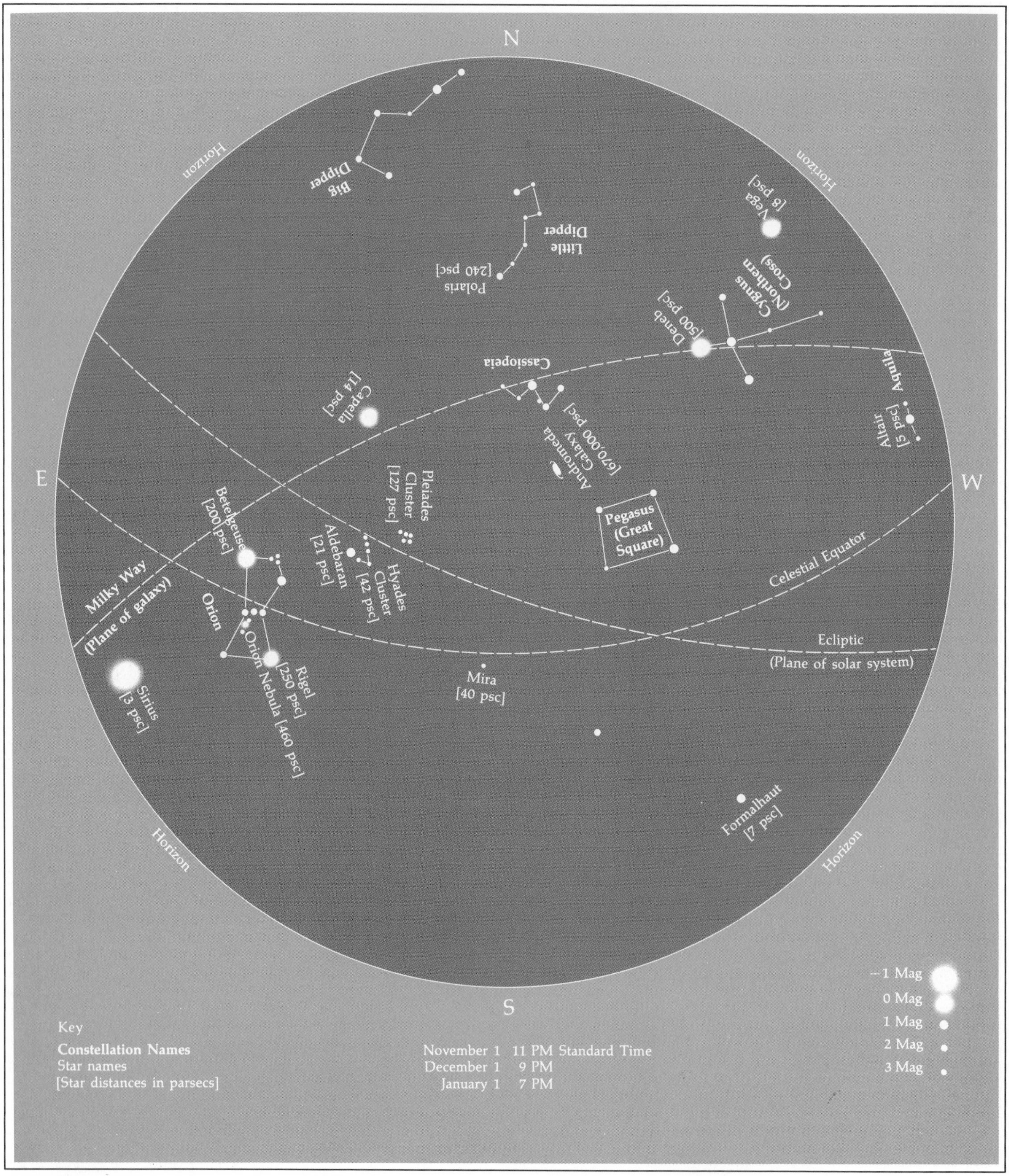
N
S
E
W
Horizon
Big Dipper
Little Dipper
Polaris [240 psc]
Vega [8 psc]
Cygnus (Northern Cross)
Deneb [500 psc]
Cassiopeia
Capella [14 psc]
Aquila
Altair [5 psc]
Andromeda Galaxy [670,000 psc]
Pleiades Cluster [127 psc]
Pegasus (Great Square)
Betelgeuse [200 psc]
Aldebaran [21 psc]
Hyades Cluster [42 psc]
Celestial Equator
Milky Way (Plane of galaxy)
Orion
Orion Nebula [460 psc]
Rigel [250 psc]
Ecliptic (Plane of solar system)
Mira [40 psc]
Sirius [3 psc]
Formalhaut [7 psc]
−1 Mag
0 Mag
1 Mag
2 Mag
3 Mag
Key
Constellation Names
Star names
[Star distances in parsecs]
November 1 11 PM Standard Time
December 1 9 PM
January 1 7 PM

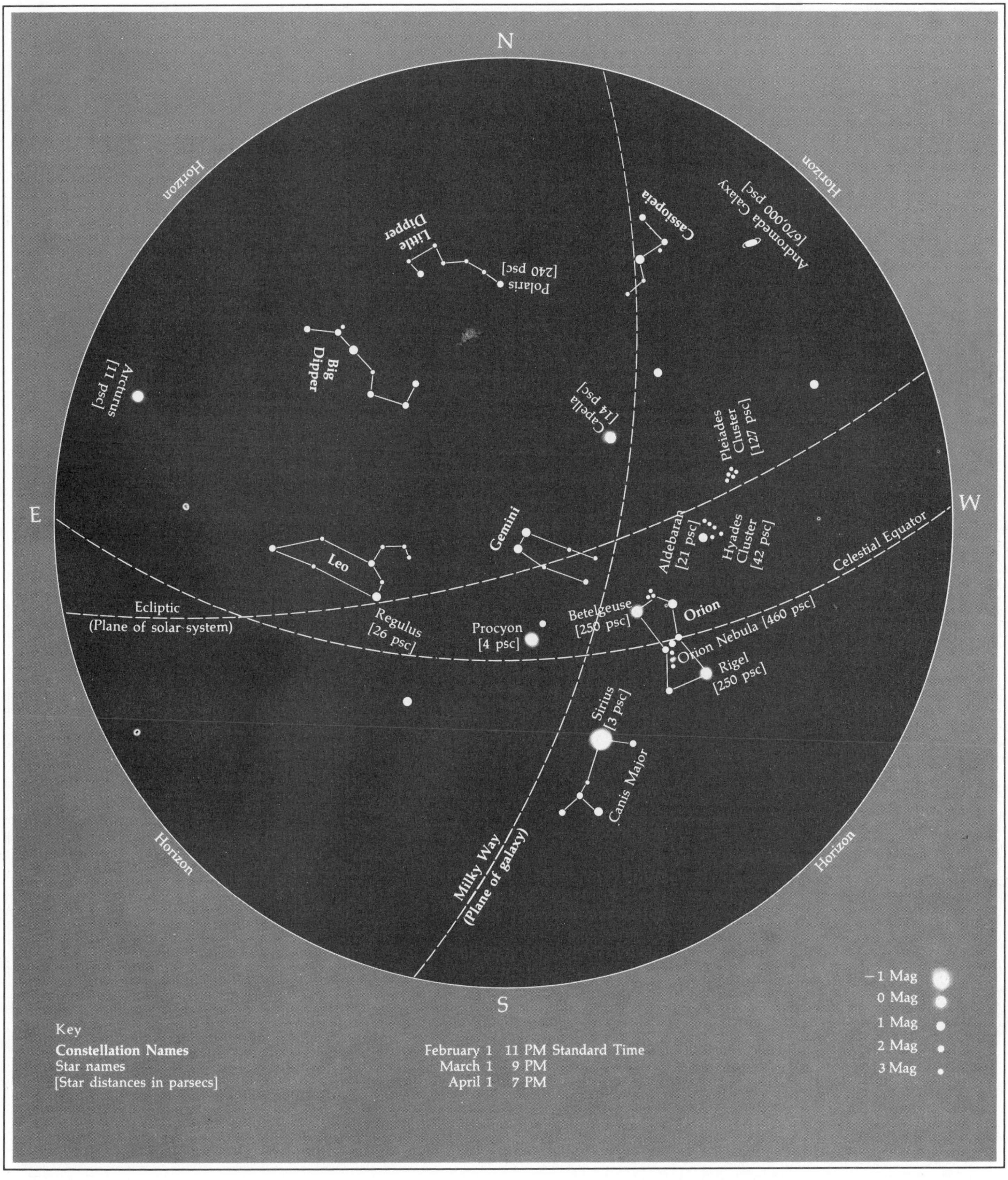

Key
Constellation Names
Star names
[Star distances in parsecs]

February 1 11 PM Standard Time
March 1 9 PM
April 1 7 PM

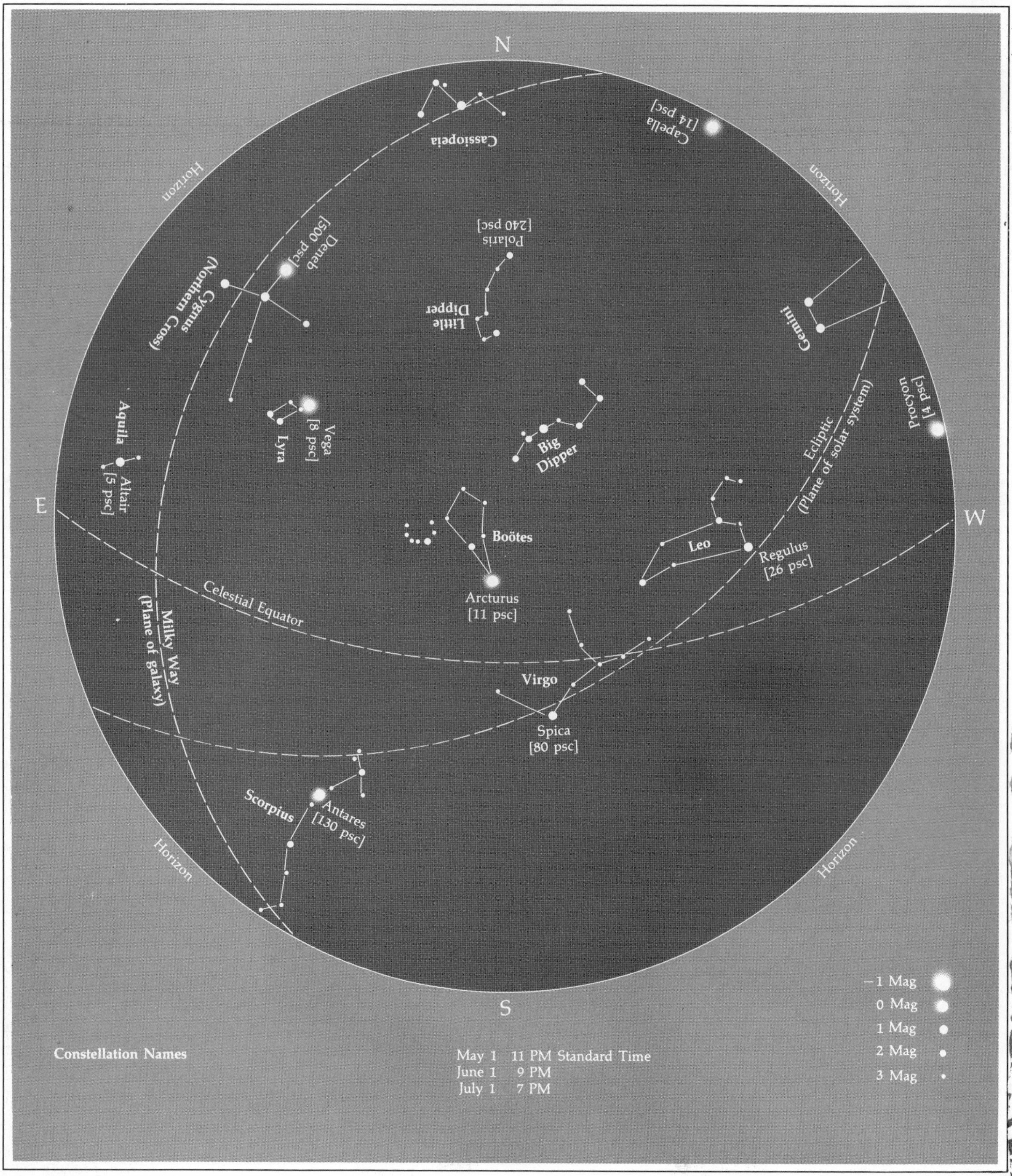
N
Cassiopeia
Capella
[14 psc]
Horizon
Horizon
Polaris
[240 psc]
Deneb
[500 psc]
Cygnus
(Northern Cross)
Little
Dipper
Gemini
Procyon
[4 psc]
Aquila
Vega
[8 psc]
Lyra
Big
Dipper
Ecliptic
(Plane of solar system)
Altair
[5 psc]
E
W
Boötes
Leo
Regulus
[26 psc]
Arcturus
[11 psc]
Celestial Equator
Milky Way
(Plane of galaxy)
Virgo
Spica
[80 psc]
Scorpius
Antares
[130 psc]
Horizon
Horizon
S
−1 Mag
0 Mag
1 Mag
2 Mag
3 Mag
Constellation Names
May 1 11 PM Standard Time
June 1 9 PM
July 1 7 PM

▲ Open cluster NGC 6520 and dust cloud Barnard 86

▼ Zodiacal light and Milky Way at dawn